Climate Change: Science, Impacts and Policy

PROCEEDINGS OF THE SECOND WORLD CLIMATE CONFERENCE

WMO

UNEP

UNESCO

Climate Change:
Science, Impacts and Policy

Proceedings of the Second World Climate Conference

Edited by:

J. JÄGER *(Stockholm Environment Institute)* and
H. L. FERGUSON *(Co-ordinator, SWCC)*

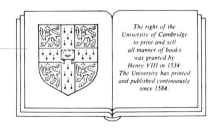

The right of the
University of Cambridge
to print and sell
all manner of books
was granted by
Henry VIII in 1534.
The University has printed
and published continuously
since 1584.

CAMBRIDGE UNIVERSITY PRESS

Cambridge
New York   Port Chester
Melbourne   Sydney

Published by the Press Syndicate of the University of Cambridge
The Pitt Building, Trumpington Street, Cambridge CB2 1RP
40 West 20th Street, New York, NY 10011, USA
10 Stamford Road, Oakleigh, Melbourne 3166, Australia

First Published 1991

Printed in Great Britain at the University Press, Cambridge

British Library cataloguing in publication data

Library of Congress cataloguing in publication data available

ISBN   0 521 41631 0   hardback
ISBN   0 521 42630 8   paperback

The Second World Climate Conference

Sponsors

World Meteorological Organization (WMO)
United Nations Environment Programme (UNEP)
United Nations Educational, Scientific and Cultural Organization (UNESCO)
and its Intergovernmental Oceanographic Commission (IOC)
Food and Agriculture Organization (FAO)
International Council of Scientific Unions (ICSU)

Financial Supporters

The Second World Climate Conference has benefited from the encouragement and support of many countries and organizations. The sponsors are pleased to acknowledge in particular the substantial financial support of: Canada, France, Germany, Italy, Japan, The Netherlands, Norway, Switzerland, the United Kingdom, the United States of America, the European Community, the Stockholm Environment Institute and the Environmental Defense Fund (USA).

Contents

CLIMATE CHANGE AND SOCIO-ECONOMIC ACTIVITIES

Forests

Integrated Studies

SPECIAL PRESENTATION

TASK GROUPS AND CONSULTATION GROUP REPORTS

SUMMARIES OF PANEL DISCUSSIONS

CONFERENCE STATEMENT

MINISTERIAL SESSIONS

OPENING CEREMONY

CLOSING CEREMONY

Foreword

The World Climate Programme (WCP) was established in 1979 following a World Climate Conference convened in Geneva by the World Meteorological Organization (WMO) and the United Nations Environment Programme (UNEP). Since its inception the WCP has been the major international programme for co-ordinating global climate monitoring and research, and studies of climate change and its impacts.

In 1986 WMO initiated planning for a Second World Climate Conference (SWCC), with an end-of-decade review of the WCP as its main theme.

By 1988 global climate change was becoming a major public and political issue. During that year the Intergovernmental Panel on Climate Change (IPCC) was established by WMO and UNEP and its was decided that the IPCC would publish an Assessment Report in August of 1990. Plans for the SWCC were adjusted accordingly. The Conference was rescheduled to follow the IPCC Report and to take it fully into account in reviewing the World Climate Programme and related international programmes. It was also decided that the Conference would include a ministerial meeting as well as scientific and technical sessions.

The organization of the Conference was guided by a distinguished international committee chaired by Prof. J.C.I. Dooge of Ireland. The overall Conference Chairman was Mr Zou Jingmeng of China, President of WMO. Details of planning and preparations were carried out by the SWCC Co-ordinator's Office established in the WMO Secretariat in August, 1989. Plans were continually reviewed by the Heads of the six sponsoring bodies.

The Second World Climate Conference was held at the International Conference Centre in Geneva, Switzerland, from 29 October to 7 November 1990. It was co-sponsored by WMO, UNEP, the United Nations Educational, Scientific and Cultural Organization (UNESCO) and its Intergovernmental Oceanographic Commission (IOC), the Food and Agriculture Organization (FAO) and the International Council of Scientific Unions (ICSU). Substantial financial support was also provided by Canada, France, Germany, Italy, Japan, the Netherlands, Norway, Switzerland, the United Kingdom, the United States of America, the European Community, the Stockholm Environment Institute and the Environmental Defense Fund (USA).

The primary goals of the Conference were:

- To formulate recommendations for the continuing World Climate Programme, taking into account the report of the IPCC, with a primary goal of insuring that authoritative scientific information is continually obtained and provided to governments for evaluating impacts, implementing responses, and developing additional international policies to address the issues of global climate change, environmentally sound and sustainable development, the survival of species, and the quality of human life.
- To provide an opportunity for Ministers:
  1) To consider specific follow-up actions pertaining to the IPCC Report and declarations of various international conferences relevant to climate change in an effort to identify elements that could give impetus to negotiations of the framework climate convention
  2) To consider the special needs of developing countries including improvement of their access to climate data and information and to technology and additional financial resources
  3) To consider specific goals for enhancing intergovernmental co-operation in monitoring, detecting and predicting global climate change.

The first six days consisted of scientific and technical sessions of a highly interdisciplinary nature. World-class experts in meteorology, oceanography, agriculture, energy planning, water resource management, land use, forestry, law, health and environmental protection were drawn from all regions of the world. There were 747 participants from 116 countries.

Participants took full advantage of the opportunity to review the findings of the IPCC and other bodies, to raise awareness and to gain a higher priority for action on climate change related issues across the broad spectrum of relevant international programmes.

The final two days consisted of ministerial sessions. The Opening Ceremony at the Palais des Nations on 6 November was addressed by the King of Jordan, the President of Switzerland, and the Prime Ministers of France, Malta, Tuvalu and the United Kingdom. There were 908 participants from 137 countries at the ministerial sessions, representing more than 80% of the United Nations.

About 40% of the participants in the scientific and technical sessions were from the developing world while over 55% of the delegates to the ministerial sessions were from developing countries.

Global media coverage was assured by the presence of 466 accredited media representatives. This represented by far the largest-ever media coverage of a United Nations-sponsored environmental conference in Geneva.

Both the Conference Statement and the Ministerial Declaration acknowledged and endorsed the work of the World Climate Programme and related global programmes and the Intergovernmental Panel on Climate Change.

The Statement and Declaration called for the urgent negotiation of a Framework Convention on Climate Change, with a view to signing such an agreement in time for the UN Conference on Environment and Development in Rio de Janeiro in 1992.

The special needs of developing countries were clearly recognized, with various recommendations stressing the need for technical and financial support to encourage sustainable economic development along environmentally-beneficial pathways.

Both documents advocate actions leading to the stabilization of atmospheric concentrations of greenhouse gases and specifically note that scientific uncertainties, though significant, should not be used as a reason for delaying action to minimize adverse impacts of climate warming.

Both documents strongly urge increased support for research and enhanced global monitoring of climate. In particular, the Ministerial Declaration states:

> "We recognize the importance of supporting the needs of the World Climate Programme, including contributions to the WMO Special Fund for Climate and Atmospheric Environmental Studies;
> We invite the 11th Congress of the World Meteorological Organization, in the formulation of plans for the future development of the World Climate Programme, to ensure that the necessary arrangements are established in consultation with UNEP, UNESCO, (and its IOC), FAO, ICSU and other relevant international organizations for

effective co-ordination of climate and climate change related research and monitoring programmes."

In their statements during the ministerial sessions of the SWCC the governments of Canada and the United States committed, respectively, $100,000 and $500,000 to the WMO Special Fund for Climate and Atmospheric Environmental Studies.

This book focuses on the scientific/technical sessions of the Conference, including the papers prepared by well-known experts from a broad range of disciplines. In many cases co-authors were invited, one from an industrialized country and one from a developing country, to promote discussion of differing perceptions of the issues. Given the recent publication of the IPCC's authoritative scientific assessment, many SWCC authors provided overviews of selected topics in a form readily understood by individuals in other disciplines rather than previously-unpublished, highly-specialized research results.

A total of twelve Task Groups and a Consultation Group on the Special Needs of Developing Countries were formed several months in advance of the Conference. Eight of the Task Groups dealt with specific socio-economic sectors, two focused specifically on the World Climate Programme and the remaining two examined broader strategic questions including legal and institutional issues. Most Task Groups had between 10 and 15 invited members selected for their knowledge of the subject and relevant international programmes. Efforts were made to ensure that each Task Group had as broad a regional representation as possible. At the Conference, each group provided a short summary of its recommendations. These summaries were a major input to the Conference Statement which was prepared for discussion at the final scientific/technical plenary session. Each group has also provided a longer report for inclusion in this volume.

The report of the Consultation Group on the Special Needs of Developing Countries provided in these Proceedings includes the findings presented at a plenary Panel Discussion. Four other expert Panels were convened during the Conference and summaries of their lively discussions are included in this book.

These Proceedings are organized with the special interests of readers in mind. Following the initial overviews of the IPCC and the WCP, the invited papers are presented under the same major socio-economic topic headings as the Task Group reports.

The speeches by Heads of State and Heads of Government at the opening ceremony of the Ministerial Sessions on 6 November are included here, together with the Ministerial Declaration. A list of countries and organizations represented at the Ministerial Sessions is provided in Appendix 5. A report on the ministerial sessions, with a list of participants, has been published separately and is available on request from the World Meteorological Organization, Geneva.

Non-governmental organizations were well represented at the Second World Climate Conference and included industrial as well as environmental interests. Presentations were made by NGO's during both the technical and ministerial parts of the Conference.

An event such as the Second World Climate Conference derives its success from the skills, diligence, co-operation and dedication of many people. Those who contributed through participation in the International Organizing Committee and other inter-agency Committees of SWCC are listed in Appendix 1.

Special thanks are due to Dr G. Goodman and the Stockholm Environment Institute for making available the services of Dr W. J. Maunder and Dr J. Jäger. Dr Maunder was seconded to the SWCC Co-ordinator's Office for a six-month period. His responsibilities included arrangements for travel grants and for Poster Sessions. Dr Maunder also prepared the Climate Change Lexicon which was provided to participants. Dr Jill Jäger took on the task of co-editing the Proceedings.

Mr Alex Alusa was seconded from UNEP to the Co-ordinator's Office for several months to assist with conference preparations. Miss Shelagh Varney was made available by the United Kingdom and provided invaluable support on protocol and VIP arrangements and in the preparation of conference publications.

As lead agency for the SWCC, the World Meteorological Organization provided major infrastructure support. To facilitate this support, an internal committee chaired by the Deputy Secretary General of WMO, Dr D. N. Axford, met two or three times a month beginning in July, 1990.

Special thanks are due to the Co-ordinator, Mr. H. L. Ferguson, and members of his staff whose dedication and diligence were crucial to the success of the Conference. Ms

Joelee Joyce, the Assistant Co-ordinator, contributed to all aspects of planning and operations. Ms Francoise Martin and Ms Irma Nichols provided valuable support.

In preparations for the involvement of Heads of State, Heads of Government, Ministers and other Heads of Delegations, excellent support was provided by the Missions of countries to the United Nations in Geneva.

As the host country, Switzerland provided substantial assistance including security arrangements for the many national leaders and other VIP's who attended. From May until December, 1990, Mr Alain Clerc, a senior official of the Swiss government, was seconded to the SWCC Co-ordinator's Office, where he had a major responsibility for the development of the Ministerial Declaration.

Ministerial sessions were chaired by Mr Flavio Cotti, now President of the Swiss Federation. We are grateful to Switzerland, the host country, and to the Canton and City of Geneva for their generous support.

Intergovernmental meetings to prepare for the Ministerial Sessions were chaired by Mr Zou Jingmeng, President of WMO and Ambassador I. N. Topkov, President of the Governing Council of UNEP.

The Ministerial Declaration of the Conference, endorsed by 137 countries, will find its place in history as a landmark of man's efforts to reverse the serious damage being caused to the earth's atmosphere through human activity.

The Second World Climate Conference was a remarkable example of co-operation among the six sponsoring international agencies. We would like to take this opportunity to extend our sincere thanks to those countries that provided substantial encouragement and financial support and to the national leaders and the scientific, technical and policy experts whose participation made the Conference such an outstanding event.

Finally, we wish to thank Dr J. Jäger and Mr H. L. Ferguson for their work in editing this book.

G. O. P. Obasi
Secretary General
World Meteorological Organization

M. K. Tolba
Executive Director
United Nations Environment Programme

F. Mayor
Director General
United Nations Educational, Scientific and Cultural Organization

E. Saouma
Director General
Food and Agriculture Organization

M. G. K. Menon
President
International Council of Scientific Unions

Second World Climate Conference
29 October, 1990

Address by Professor G. O. P. Obasi
Secretary General, World Meteorological Organization (WMO)

It is a great honour to address you today on the occasion of the opening of the scientific and technical sessions of the Second World Climate Conference. I take pleasure in welcoming you to this Conference, and to Geneva, the city which hosts the headquarters of the World Meteorological Organization.

Today is a banner day, particularly for scientific and technical community concerned with the global climate. As the world watches, we begin our discussions of perhaps the most important aspects of the global environment, the changing atmosphere of our planet, and the climate it produces. We are well aware how climate has shaped our lives, our national resources and our economies. Through the sub-Sahelian droughts in Africa, desertification and widespread forest fires in many other regions, recurrent flooding in Bangladesh and other countries, increasing damage due to the slowly rising sea-level over the past century, we have a foretaste of the serious problems that can result from shifts in climatic conditions.

So, for the people of the world, the gathering here in Geneva of the scientific and technical leaders in the climate sciences and related fields offers promise of new understanding of the Earth's climate system and the forces that shape it. The Conference statement which will be produced at the end of this week will reflect this new understanding. Next week, the Ministerial session of the Conference will, we hope, give new indications to the world as to what measures governments are prepared to take in the combat against environmental degradation, in particular global warming.

As we begin this Second World Climate Conference, we recall that it is a little more than a decade since the First World Climate Conference was held in 1979. That Conference heralded the beginning of the World Climate Programme. Many of you have worked with enthusiasm and brilliance within the framework of the four components of World Climate Programme (WCP)—data, applications, research and impact studies. This work, along with monitoring and studies of the atmospheric composition, has brought us to our present understanding of climate and its forcing factors, including those due to human activities. Many of you have also laboured vigorously over the last two years on the assessment by the Intergovernmental Panel on Climate Change (IPCC) of our current understanding of the science, impacts and policy responses.

The IPCC Working Group and Special Committee reports and the Overview represent an unprecedented achievement, accomplished with the active participation of more than 1 000 specialists from some 70 countries. The IPCC's first assessment report will be the indispensable starting point for our discussions at this Conference on increasing greenhouse gases and their effects in changing climate.

For these achievements, I congratulate you all, and I salute all those who have contributed so much to the World Climate Programme and the IPCC process.

We are here today, and indeed the whole of this week, to critically review and evaluate our collective efforts under the World Climate Programme. Let us consolidate our scientific achievements of the past decade so that they form a firm foundation for our future workplan.

I also wish to acknowledge the spirit of co-operation which has prevailed, and without which our accomplishment would not have been possible.

In this connection, I should like to make some remarks on the collaboration extended by a number of international agencies, governmental and non-governmental, on various undertakings, including the organization of this present Conference. While WMO has provided overall international co-ordination for the WCP, the International Council of Scientific Unions (ICSU) has shared with us the responsibility for the research component. This has been a most successful and productive arrangement, bringing governmental and academic scientists together effectively. The United Nations Environment Programme (UNEP) has also been a partner from the outset, taking responsibility for the impact studies aspects of the WCP. This co-operation has developed and matured through the formation in November 1988 of the IPCC by WMO and UNEP. We are understandably proud of the achievements of this jointly supported body. Moreover, WMO and UNEP have been requested to prepare for the process leading to a global framework convention on climate change. This will serve as a further basis of strengthened co-operation.

For this Conference, we also enjoy the great advantage of support from our partners in hydrological and oceanographic matters, the United Nations Educational, Scientific and Cultural Organization (UNESCO) and its Intergovernmental Oceanographic Commission (IOC). The United Nations Food and Agriculture Organization (FAO), with which WMO has for many years had joint projects in agriculture and forest meteorology, has also co-sponsored this landmark Conference. In addition, we greatly appreciate the financial support and other forms of assistance from a number of governments, international organizations and non-governmental organizations which have made this Conference possible.

May I also especially thank the SWCC Organizing Committee for the valuable preparatory work it has done under the able chairmanship of Professor Jim Dooge of Ireland. The Co-ordination Office for the Conference and the WMO Special Committee have worked long and hard to make this event a true milestone. I am sure you will all agree with me that all the people associated with them deserve our deepest gratitude.

As we can note, the co-operative efforts in the WCP, the IPCC, and this Conference demonstrate the positive work among agencies and countries on climate. This has been truly remarkable; indeed, it could be considered a model for other global activities.

There will be a need to call on further collaboration as we now move into a new phase regarding the issue of climate change as influenced by man's activities. Many Heads of State or Governments and Ministers will, early next week, be advising us of their countries' views and their commitments to concerted international action concerning climate change. As mentioned earlier, WMO and UNEP, under instructions from the UN General Assembly and our respective governing bodies, have made preparations for the opening of more formal negotiations of a global framework convention on climate change. Indeed, the preparatory meeting for the development of a framework convention on climate change was held here last month. Negotiations will begin in Washington on 4 February 1991, at the invitation of the Government of the United States of America. They will be based on the best scientific knowledge we now have, produced by the WCP and summarized by the IPCC. The Second World Climate Conference can also significantly influence the coming negotiations. The UNEP Executive Director, Dr Tolba, and

I shall be reporting to the UN General Assembly by the middle of next month on the outcome of your discussions and of those of the Ministerial component.

Even as we forge ahead, we all recognize that scientific uncertainties remain, and that any agreement for international response in the face of the serious threat, based upon the 'precautionary principle', must be flexible and capable of adjustment as new scientific results become available. We are confident of the predictions of the general trend of climate but the reduction of uncertainties regarding timing, magnitude and regional variations of global warming and changes in hydrology remains even more urgent than before. Efforts to limit the effects of climate change, by means of greenhouse-gas emission control and biomass uptake of CO_2, and to prepare our adaptation to climate change, can gradually be made more effective with increasingly reliable predictions and detailed observations of climate change. Thus, the views of all of those present on scientific and technical priorities, especially as regards the future directions of the World Climate Programme, are of great importance to WMO and its partners. This week you will help set the course for future scientific activities concerning climate, which in turn can very well affect the very future of our planet Earth and its inhabitants.

In your considerations, I would also ask you to pay special attention to the needs of developing countries. If they are to become full partners in global agreements to address climate change, they must develop more vigorously their capability to measure and assess their climates and to use climatic information to improve their lives and economies. Your views will be greatly appreciated on how measurement, training, research, and related activities can best be promoted on a truly world-wide basis.

As UN General Assembly resolutions have noted, climate is 'the common concern of mankind.' You who have been invited to the Second World Climate Conference have a great responsibility. The world is counting on your knowledge and wisdom to provide the basis for an initial global action plan to address the issue of greenhouse gases and climate change. At the same time, you must consider how the scientific and technical work should best proceed to reduce the scientific, technical and economic uncertainties in our predictions and make maximum use of climatic information for various areas of application.

These are great challenges for you at this Conference; they also serve as rare opportunities. Looking over this room full of outstanding experts on the subject, from all over the world, I am confident that you will provide inspired and visionary guidance. This will be a long-lasting legacy you can bequeath, in the name of service to humanity.

Second World Climate Conference
29 October, 1990

Address by Dr Mostafa K. Tolba
Executive Director, United Nations Environment Programme (UNEP)

It is with great pleasure that I join with my dear friends and colleagues in welcoming you to this Second World Climate Conference.

We all know that the world faces a threat potentially more catastrophic than any other threat in human history: climate change and global warming. The scale of this threat may only recently have begun to filter into the public domain, but it has been at the forefront of concern for the international scientific community for more than a decade. It was articulated at the First World Climate Conference in 1979.

Despite limited resources, the World Climate Programme—adopted at that conference—has become one of two global authoritative programmes on climate and climate change, the other being the International Geosphere-Biosphere Programme (IGBP). In the past decade, these two programmes have underpinned the whole process of research into climate and its impacts.

During the next six days, you will review this research. And there will be discussion on reports from the Intergovernmental Panel on Climate Change (IPCC). In the light of all this you will chart the agenda for the World Climate Programme during the next ten years.

Much has been clarified since the First World Climate Conference in our understanding of climate and climate change. There is no longer any doubt that increasing emissions of greenhouse gases will spark a rise in global temperature of a greater scale and speed than any change experienced in the past tens of thousands of years. The available evidence suggests that global warming may already have begun. The resulting climate change may in many cases be catastrophic. Scientific studies indicate that a global temperature rise of up to three degrees Celsius and the ensuing climate change expected by the end of the next century may well be conservative.

Of course we all know that predictions of future temperature rise and likely climate change are based on current computer models, which cannot take account of many of the interactions within the complex network of systems that make up the dynamics of our planet. Nor can the models quantify the reactions of many planetary systems to increasingly high temperatures. But the IPCC Science Working Group indicates that these various feedback mechanisms are likely to magnify the overall warming and the severity of the impacts of climate change.

Working Group I further warns that, based on past records, it is at least possible that the coming climate change may occur in abrupt, drastic shifts rather than in a gradual and relatively comfortable manner.

But the IPCC correctly emphasizes the urgent need for major collaborative research efforts to fill in gaps, particularly in the areas of oceans, clouds and regional impacts, and the equally urgent need for extensively improving monitoring and assessments. This naturally includes the collection and analysis of data. All of this is the stronghold of WMO, UNESCO and ICSU. The IGBP is about to take off to fill a number of these gaps over the next ten years. And the programme you will agree here should be designed to fill as many of these gaps as is humanly possible within the shortest possible time.

Similar gaps need to be filled in the area of impacts of climate change. FAO and UNEP have to take a leading role in filling these gaps. They are not alone in this. Several other UN agencies should chip in as well. But let me make myself clear on all these points. International organizations within and outside the UN system do not normally run research centres. The research institutions are yours. They belong to your countries. We assist you in defining the global programme and in implementing it. We do not define programmes for you nor do we implement them for you.

Yes, there are gaps in our knowledge and some of them are large gaps. But what we know now is more than enough to act, and to act fast.

The sum of research into the science and impacts of climate change makes it clear that nothing less than dramatic reductions in emissions of greenhouse gases will stop the inexorable warming of the planet. Nothing short of action which affects every individual on this planet can forestall global catastrophe. Nothing less than a complete change in attitudes and lifestyles will succeed.

Preparations for intergovernmental negotiations of a global convention and related legal instruments to deal with the problem of climate change are now under way. The forty-fifth session of the United Nations General Assembly will consider, two weeks from now ways, means and modalities for further pursuing these negotiations.

The aim is to reach agreement on the convention as rapidly as possible. Our target date is 1992 in Brazil at the time of the United Nations Conference on Environment and Development. However, two main criteria should govern that process:

1) That the convention does not cause distress to developing countries. These nations will need time, additional financial resources and the transfer of the required technologies to prepare for the changes they must make if they are to play their part in reducing greenhouse gas emissions, and

2) That rapid action in developing the convention should not come at the expense of its content. We need a convention with clout, a convention with clear-cut commitments.

That is one side of the story. The other side is that we cannot wait for the international community to agree on a convention on climate change before we act. The need for immediate action is becoming increasingly pressing. The longer we delay, the worse the disaster will be for our children, and for our children's children.

Individual industrialized nations must act now to reduce their emissions of greenhouse gases, and to help the developing world begin to prepare itself to do the same in the foreseeable future. We should not wait until Rio de Janeiro to initiate action. Rather, at Rio de Janeiro there must be reporting of successful corrective action and, in the light of that, commitments to do more, much more.

Progress towards curbing emissions would be more rapid if nations shared expertise and knowledge. And it would be more rapid if there were fewer gaps in our knowledge of science and impacts of climate change.

This is where your work over the next few days really matters. You will be establishing the outline of a revised World Climate Programme. It is crucial that the priorities of this revised programme help the international community to make the hard socio-economic decisions that are urgently needed.

Curbing greenhouse gas emissions will cost vast sums, and will demand great changes in economic, legal and institutional systems. Your Second World Climate Programme, especially its applications and impacts components, should address at least some of these issues—the cost of action, and the cost of inaction with respect to these changes, the impacts of these changes on the various sectors of economic activity and their impacts on international economic relations and co-operation.

There is room for hope that such changes can be achieved. In agreeing on the Montreal Protocol, the international community showed it could reach consensus on difficult actions if the threat were great enough.

Negotiations for the Montreal Protocol began with fewer facts to back up the possibility of environmental disaster than we now know about global warming. The protocol was signed three years ago without full proof that ozone destruction was a reality, and with little evidence that substitute chemicals and technologies could become available for the broad range of applications in which chlorofluorocarbons seemed so essential.

I am fully aware that to reach consensus on action needed to halt climate change, minimize its impacts and adapt to it will be infinitely more difficult than dealing with ozone depletion. But I am also aware that you, the scientists, have indicated in an unambiguous way that the threat is great enough, that you have enough evidence on hand that cannot be ignored and that makes immediate action completely justifiable.

Of course there are still uncertainties about certain aspects of the science, and some impacts of climate change. Thus you, the scientists, are the lynch-pin in the continuing process of action in the face of this threat.

If policy-makers are to be able to react effectively to this expanding threat, they will need information that is clear, that is accurate, and that is supplied as rapidly as possible. Already, the near-impossible has been achieved by the IPCC, in collating the present knowledge about such a broad-ranging and complex subject within a very short time-frame.

The world now depends on you to continue this work—all of you, and your fellow scientists outside this hall, in every corner of the globe. Your work under a revised World Climate Programme, under the IGBP and by the IPCC, will feed into the forthcoming negotiations and give a solid basis to urgently needed government action. Your deliberations are crucial to saving our Earth, to saving humanity. It is a tough responsibility. But as a scientist, as one of you, I am confident you will carry it out with distinction.

Second World Climate Conference
29 October, 1990

Address by Dr. Federico Mayor
Director General, United Nations Educational, Scientific and Cultural Organization (UNESCO)

Today more than ever, decision-makers require—mainly because of the global nature, the complexity and the urgency of the issues to be addressed—the support of the scientific community, to provide them with the knowledge already available for application and, whenever time permits, to undertake a joint identification of the gaps in knowledge calling for more research. Certain threats to the environment demand the adoption of immediate measures, implying a long-sighted and courageous attitude. The postponement of such measures, a failure of vision and courage, could mean that the legacy we have a duty to pass on to future generations will be irreversibly damaged. Our descendants will not judge us on our hopes and recommendations, but on our actions.

I am therefore very pleased to have the opportunity of addressing the distinguished audience that has come together for the opening of this very important event, the Second World Climate Conference. This Conference is no longer—as it was originally planned to be—an exclusively scientific event. This is indeed a measure of its importance: climate change, its socio-economic impact and response strategies have today become major issues of national and international politics and economics. It is an honour for me to join this assembly of the world's experts on this crucial question, as well as the distinguished guests and my eminent colleagues from the other sponsoring organizations.

UNESCO is proud to be one of the co-sponsors of this Conference. The Organization's concern for climate and its interest in this Conference are demonstrated by the fact that it is sponsoring the Conference in two ways—first in its own right and, second, through its Intergovernmental Oceanographic Commission (IOC). I should like here to express my profound gratitude to my colleagues from the other sponsoring organizations of the Conference—and I offer a special word of thanks here to Mr Obasi and Mr Tolba—for the exemplary co-operation established between us. A special word of thanks must go to the Co-ordinator of the Second World Climate Conference, Mr Howard Ferguson, and his colleagues for their outstanding work in preparing for this Conference.

UNESCO and the IOC are deeply involved in many of the scientific aspects of the climate system, climate change and related impact assessments, and are concerned with three of the main topics to be discussed during this Conference, namely:

1) The major role of oceans in governing the world climate system and climate change
2) The role of the global water cycle in climate and the impact of climate change on freshwater resources
3) Training, education and public awareness related to those specific topics and to the climate in general.

The first two topics represent areas in which UNESCO has contributed significantly to the World Climate Programme. Several other issues to be discussed in this Conference are related to scientific programmes in UNESCO, e.g. the land-atmosphere interface and the impact of climate change on terrestrial ecosystems, the study of past climate changes in relation to geological processes, and the energy/climate link. In dealing with these matters, UNESCO and the IOC act in partnership with WMO, UNEP, FAO and ICSU, as well as other international bodies and programmes such as the Intergovernmental Panel on Climate Change (IPCC).

A subject of major concern to UNESCO, through the IOC, is the role of the ocean in the climate system. The ocean plays a role in the climate system which is complementary to that of the atmosphere and of comparable importance. The ocean stores heat and releases it later, usually in a different place, and it transports heat in amounts comparable with atmospheric transport. The ocean absorbs and releases carbon dioxide. It has a long memory;

the water now reaching the surface of the ocean from its deeps last felt the breath of the atmosphere about 800 years ago—at about the time the 'little ice-age' began! The water-vapour content of the atmosphere to a very large extent originates from the ocean—and water vapour of course is a most important 'greenhouse gas'. The climate system cannot be adequately understood, modelled, and predicted without taking the role of the ocean properly into account, as has also been pointed out by the IPCC. Many of the most severe potential impacts of climate change are also transmitted through the ocean, by the effects of sea-level and temperature rises on coastal zones and areas and small islands. I am pleased that UNESCO, mainly through the programme of the IOC, is helping to elucidate these questions.

Scientific assessments strongly point to the urgent need for a global ocean observing system providing adequate data which can be assimilated and fed into climate prediction models. We know this is needed—and the longer we delay action, the further we fall behind in our ability to make reliable climate change predictions. The IOC, together with WMO and UNEP, is working very hard on this, and I would urge that it be given due attention.

Let me now turn to another scientific area of interest to UNESCO: the hydrological cycle, which is a major vehicle of energy and mass transfers between land, ocean and atmosphere. What are the main consequences that might be expected of climate impacts on the hydrological cycle? In what areas do we need more knowledge?

To begin with, it must be understood that, while we may speak of 'global' changes, effects on the hydrological cycle occur in specific localities. To understand what would happen to any given river, lake, or aquifer, we need to understand how the rainfall would be affected on a specific river basin: this we are not yet capable of doing. Furthermore, our ability to connect a climatic model to a precipitation model is now barely in its infancy. Those models that have been used give a completely inadequate representation not only of local precipitation but, more importantly, of how that rainfall is converted into runoff, groundwater, and evaporation.

Other uncertainties concern the melting of glaciers and polar ice masses. If the sea-level were to rise, this could have extremely significant hydrological consequences, since many coastal aquifers are hydraulically connected to the seas, and thus would have their flow systems modified by such a rise. Similarly, in estuaries a rise in the sea-level would not only cause navigational and other social problems, but could drastically change the flow patterns of fresh-water and salt-water mixing. Within UNESCO, the International Hydrological Programme has undertaken to study many of these uncertainties in its present six-year programme, Hydrology and Water Resources for Sustainable Development in a Changing Environment.

The limited time available this morning does not permit me to present here in similar detail other scientific issues of great interest to UNESCO. Let me just mention that, while public and scientific attention has focused mainly on what global warming means for humanity, its effects on plant and animal life, on ecosystems and agro-ecosystems, have often been overlooked. Researchers are just beginning to understand how plants and animals, and consequently ecosystems, respond to the changes brought on by the greenhouse effect. This is a research area of the utmost interest to UNESCO's Man and the Biosphere (MAB) Programme.

Full participation in the World Climate Programme and in related scientific programmes remains at present restricted to a relatively small number of countries, those which have already a high scientific capacity in this field, whilst the vast majority of developing countries can participate only to a very limited extent in these efforts. The inadequate involvement of specialists from developing countries in the work of the IPCC is a further illustration of this regrettable situation. I trust that this Conference will address the serious problem of training and of building up the corresponding institutional capacities in developing countries and formulate recommendations on how best to remedy the situation. UNESCO and the IOC stand ready to help whenever it is within our institutional competence and capacity.

This Conference would not be doing what is required of it if it did not also address the problem of public information, of information for decision-makers, of awareness-raising and of education at all levels in relation to the global climate issue—a major task in industrialized and developing countries alike. By virtue of its Constitution, UNESCO ought certainly to be an active partner in these endeavours, and it is for this reason that it has started promoting the inclusion of issues related to climate change in education programmes at all levels.

Climate change is a world-wide problem with important regional and national implications, and its solution needs to be found through globally concerted action, for which an indispensable prerequisite is the active collaboration of all concerned at national, regional and international levels. In my view, global action must include particular intellectual efforts in three areas of UNESCO's competence: science, education and information, in particular information for decision-makers. Without science and technology no solution will be found that is socially and economically acceptable; without international co-operation no meaningful research and data collection can be undertaken.

In this regard, I very much hope that the Conference will come forward with concrete and specific proposals for follow-up and for the continuation of the World Climate Programme and related programmes, with appropriate reorientations. I would also wish and urge that the co-operation manifested through the preparations for, and co-sponsorship of, this Conference be strengthened as a result of the Conference, and continued in the follow-up. UNESCO and its IOC are committed to contributing to a revamped World Climate Programme, to the negotiations leading to a framework convention on climate change and, last but not least, to the continuation of the work of the Intergovernmental Panel on Climate Change.

Let us not forget that this Conference, important as it is in its own right, is also a precursor of the more comprehensive United Nations Conference on Environment and Development in 1992. We must endeavour—all together, synergically—to ensure that the 1992 Conference is a real turning-point, a moment of solemn commitment to redressing the present situation and trends. This is not only a physical and economic issue but an ethical one. We are used to war and its price. We must now learn to live in peace. And to know and pay its price.

Second World Climate Conference
29 October, 1990

Address by Professor M. G. K. Menon
President, International Council of Scientific Unions (ICSU)

I am honoured to be here, to address this important Conference on behalf of the International Council of Scientific Unions. This is a tribute to the role of the large international scientific community in this significant event. For the first time in history, people and governments are focusing a large fraction of their attention on the global environment and the impact of human actions on this environment. Humanity is clearly at a turning point. We need to redefine the direction of future development, which has to be sustainable and based on equity and social justice for all. It is our duty at this Conference to help to determine the new directions.

There have been many changes since the First World Climate Conference was held in February 1979. Confident hopes were expressed then that improvements in computers, observing systems and in scientific understanding would enable us to better predict climate fluctuations and climate change. These hopes were the basis for setting up the World Climate Programme (WCP) under WMO. Within this, the World Climate Research Programme (WCRP) was developed, as a specific area of co-operation between ICSU and ICSU bodies and WMO, UNESCO and IOC. The important large-scale experiments of TOGA and WOCE emerged through the WCRP.

A little over a decade later, it would be an understatement just to say that we have better computers; computing power and capabilities have grown in phenomenal fashion. Our understanding has improved. While improvements in observing systems have not fully come up to expectation, we do have some new systems in place. TOGA now has many of the characteristics of an operational programme, yielding data which are being regularly analysed in operational centres as well as by research groups. WOCE observations started this year. In sum: the science has advanced.

Our data have improved and expanded. Accurate measurements of carbon dioxide have been continued. An expanding series of measurements have been undertaken on other greenhouse gases: methane, CFCs and nitrous oxide. The ingenious use of tiny quantities of air trapped in glacial ice has enabled us to obtain data relating to earlier periods in history, and to assert with confidence that the recently observed increases in concentration are not 'natural fluctuations' but are increases of anthropogenic origin, built on a rather stable base of 'pre-industrial levels'.

Other observations have continued and their interpretation has been refined. There is now a wide consensus that global temperatures have increased by something like half a degree over the last century. There is also consensus that this increase is *consistent* with its having been caused by increased greenhouse gas concentrations; but is not conclusive of such an effect, since it remains within the range of the natural fluctuations which have also been revealed by these analyses.

Many careful studies of sea-level changes, and of the causes of sea-level changes, have been carried out. These have considerably narrowed the range of uncertainty. There is no longer talk, in responsible circles, of an imminent rise of many metres. A rise of about half a metre over the next century, with an uncertainty of a factor of two, is now considered to be a more reasonable projection. In this context, we must try to decouple the land and ocean changes in records of relative sea-level. Narrowing the range of uncertainty in this way has important economic and political consequences.

On the basis of these successes, we look with confidence to the future. We expect still further improvements in computer power and capability. We look with expectant hope at plans for new observing systems, both from satellites and *in situ*. We are aware of the bubbling of new ideas which will refine our understanding and reduce uncertainty. We trust that the heightened interest in the subject will result in the recruitment of many bright young minds, absolutely needed if the many tasks to be performed are to be carried out in reasonable time.

All this will, of course, require a continued and enhanced flow of resources. We are confident that the awakened interest of people and of governments will make these resources available.

The First World Climate Conference focused on the physics of the problem. In the interim, we have become increasingly aware that the biosphere plays an active and interactive role. Changes in the physical climate form only one aspect of a broader topic of global change, which ICSU is addressing. To deal with all the other related aspects on an integrated basis, we at ICSU have set up the International Geosphere-Biosphere Programme (IGBP), in the tradition of a long history of international co-operative programmes, including the International Geophysical Year (1957–1958) and the International Biological Programme (1964–1974).

The primary goal of the IGBP is to advance our ability to predict changes of the global environment so that society can be forewarned about the consequences of pathways being currently pursued, and about changes that need to be introduced if catastrophe is to be avoided and sustainable development assured.

It is clear that a programme of this nature has to be, by definition, global in character. One cannot understand the ecology of the globe, other than by measurements over a whole range of latitudes and longitudes on the Earth's surface, as well as at various altitudes in the atmosphere and depths in the oceans. Three-fifths of the land area of the Earth is covered by developing countries, which must therefore be participants in this programme. Scientists from different disciplines and areas of experience and expertise have to work on this programme; therefore a large number of our discipline-oriented Unions, scientists from our National Members, and Associates of ICSU with global and regional coverage, such as the Third World Academy of Sciences, will all have to play an important role in this activity.

A programme of this nature has become possible only now, with the availability of global observation systems based on satellites with advanced sensors, and of large-scale computational capabilities for analyses and development of models. We also know much more now on how to extract palaeo-information on environmental parameters of the distant past.

The programme will call for observations and data from many sources and locations. There will be need for a new global data and communication system. It will involve the development of complex two- and three-dimensional global models.

This programme will have to go on for several decades. Developing countries will be able to participate in this—in making observations, as well as in the analysis of one of the most complex problems ever tackled: an understanding of the global ecosystem. There are immense intellectual

challenges and rewards. IGBP can be a significant engine for scientific development in third world countries.

A programme of this nature must, of necessity, involve free availability and exchange of data and information. Therefore, it will demand a system of 'openness' which is at the very heart of scientific endeavour.

ICSU has moved forward considerably in regard to the planning of IGBP, which is now poised for implementation. The IGBP and the WCRP are significantly complementary and are the two major research programmes in the world identified by the Intergovernmental Panel on Climate Change (IPCC) as devoted to decreasing our uncertainties in relation to global climate change. The focus of the IGBP is on the biogeochemical aspects of Earth system modelling and recovery and interpretation of data dealing with global change of the past. The WCRP focuses on the physical aspects of climate change. Through these activities, ICSU is mobilizing a huge capability of scientific groups all over the world in a variety of disciplines, to bear on these questions. These capabilities relate to tools, techniques, analytical methods, observations and data, development and testing of models and so on. The enormous power that all this represents will be felt as the years roll by.

Awakening of concern relating to the environment led to the 1972 Stockholm Conference, which in turn resulted in the establishment of the United Nations Environmental Programme (UNEP). Many nations have created Departments of Environment. These actions were largely concerned with pollution 'hot spots', more local and obvious in nature. Now we have become much more aware of the nature and danger of more insidious, more generalized pollution problems. Science and technology have permitted us to detect and measure the pollution of the global atmosphere by increasing levels of greenhouse gases. Science tells us that this pollution can lead to important changes in climate.

The quality of life in any particular location is overwhelmingly influenced by the climate of that location. Climate affects life; and we are becoming increasingly aware of the manner in which life changes climate. The things that are new in this regard are:

1) A single species - ourselves - is importantly modifying the atmosphere and therefore climate
2) The pace of change is rapid
3) The dominant species - again ourselves - is capable of consciously changing the course of events. We, and only we, among the countless species which have inhabited the Earth, need not be helplessly carried along by the flow of events of our own making

That is why we are here.

The situation arises because our numbers are already large and increasing rapidly, as a result of the great advances in biology and medicine; and each of us is demanding more. To satisfy an apparently insatiable human desire for convenience and mobility, we have harnessed sources of energy, coal and oil, and gas, which were created over periods of hundreds of millions of years. The principal by-product, carbon dioxide, is changing the composition of the atmosphere. To deal with particular industrial problems, we have created CFCs and PCBs, which we now discover to be troublesome. To feed the burgeoning population, we have expanded agricultural areas and increased the use of fertilizers to a degree that these changes are also having global influence, some of which is negative.

It is important to remember that most of our present problems arise from our solutions to previous problems. We do not wish to go back. Indeed we cannot go back. The bridges behind us have been burned, if only because of the great increase of population already in place. Society has justifiable expectations of a better quality of life. There is the necessity to meet basic human needs. We must, therefore, go forward. However, 'forward' can be in many directions. We, as people, have the power to decide in which direction to move although we must remember that there are many less privileged people who have no such power. Those of us who have must decide with their interests in mind, not just our own. Decisions will have to be made, for even to choose not to decide is also a decision.

An important document for this Conference is the report by Working Group I of the IPCC. It is an impressive report, as I am sure you will agree. It is full of authoritative and clear statements that will help us to focus our discussions. It also expresses clearly the substantial range of uncertainty with which we are left, because of the complexity of the subject, the limitations of our understanding, and the limitations of our present ability to observe and to compute.

In developing a strategy for determining priorities and the allotment of resources, it is important to keep in mind the relative costs of: (1) research, (2) inaction, and (3) action.

Research is relatively inexpensive. The IPCC does not quantify the cost of implementing the research programmes identified in Chapter 11 of the WG I Report, but it is unlikely to exceed a few billion dollars per year. Much of this will have to be spent on satellite systems, and thus be a charge on the most highly developed nations.

The cost of inaction is likely to be orders of magnitude larger, even if expressed only in terms of misplaced investments and loss of property values. Of course, the real cost of inaction is not economic but human. The greater part of the damage will fall on those least able to sustain it: the poor and the underdeveloped. They have neither the resources nor the level of organization necessary to cope with large-scale changes not of their making.

When considering the cost of action, we must distinguish between prudent preventive measures that are valuable in their own right, such as saving energy and materials, and more dramatic measures based on incomplete scientific evidence. While the former are obviously recommendable and should replace 'business as usual', the latter can be very costly indeed, assuming that such action is both technically and politically possible. It will also put great strains on our political systems if such action is to be enforced internationally, for any changes are likely to affect different societies and nations differently. How will the world deal with the resentment of those whose development is hampered or slowed down, and who face greater financial difficulties in a so-called common interest?

The answers to this last question are crucial. If we act inappropriately, we might do no good and indeed might even aggravate the situation. If we take extreme action, on the basis of the upper ranges of possible effects, we may find that we have hugely wasted resources and disrupted lives unnecessarily. If we do little or nothing on the basis of the lower ranges of possible effects, we may find that we have neglected to take action with very large benefit-cost ratios.

We must, therefore, be willing to pay the relatively modest cost of research, in order to narrow the range of uncertainty to a more acceptable level. We must make the resources available to the scientists.

In return, for receiving the resources necessary for their work, scientists have many contributions to make:

1) To predict, as accurately as possible, how the climate will change

2) To identify the remaining uncertainties so as to help quantify risks and provide the basis for risk management

3) To explain what that means for humanity, in terms of habitability, food production, sea-level rise and highly adverse weather situations

4) To help identify policies which are both desirable in their own right and tend to reduce adverse future effects

5) To help design optimal procedures for reducing negative effects and coping with those which cannot be stopped

6) To develop benign substitute products.

In the case of global environmental science, there is a demand for resources of a nature which does not usually occur in other areas of science. This arises from the highly interdisciplinary nature of the work, and the need for data more extensive in space and more continuous in time than can be collected by scientists themselves. An outstanding example is that of meteorology. The need for global observations, taken very frequently for an indefinite period of time, and exchanged rapidly and globally for weather-forecasting purposes, has led to the international creation of the World Weather Watch and its Global Tele-communication System. In recent years, the proliferation of Earth-observing satellite systems has greatly expanded the range of data available repeatedly and globally. However, there is a need developing for other kinds of data, also globally and repeatedly, which cannot (at least yet) be obtained from space. For example, there is as yet no way of measuring sea-surface—let alone deep—salinity from space. There is no way to measure ocean temperature structure. There is no way to measure soil depth. There is no way to distinguish between mangroves and other types of tropical forest. These, and many others, require *in situ* determination, although the data, once determined, can be distributed by satellite communications.

To give more examples appropriate to this Conference, meteorology provides other outstanding illustrations. Surface pressure cannot be determined from space observations, although the data are needed globally and regularly. Hence a global surface measuring system, in some cases reporting by satellite, has been set up. To support this data-gathering effort there are meteorological services backed by WMO and the World Data Centres.

Some other sciences, notably oceanography, need this kind of service support but do not as yet have it in all countries and on the scale required.

The focus at this Conference is on long-term climate change, particularly that due to human activities. But there are also short-term significant variations that are natural. These have serious impacts on society. An example is the repeated devastation that has ravaged the Sahel. Developing countries are particularly susceptible to adverse impacts. The effects are especially serious in the tropics. Scientific efforts to understand and predict these variations are of great importance, so that action to mitigate their negative effects can be taken. While we need to look at the long-term future, we should not lose sight of the need for short- and medium-term prediction, which is within our grasp and is of such profound and immediate importance to such large numbers of humanity.

In closing, allow me to draw attention to the remarkable nature of this Conference. It is sponsored by the major intergovernmental bodies which have responsibility for various aspects of environmental science, plus the non-governmental organization for which I speak, and whose membership covers all the scientific disciplines which must be brought to bear. The Second World Climate Conference will be addressed by outstanding scientists, who have played an active role both in the advancement of their sciences and in the international organization of science. Behind them is the support of countless scientists not present, but whose weight is felt in the preparation of the IPCC report and in the studies upon which it is based. In its latter part, this Conference is to be attended by Ministers who will contribute to the making of the crucial decisions

about directions to be taken. In this sense, it is different from the First WCC. It demonstrates the changes that have taken place on the world scene. International science now needs to interact with its partners, the governments and intergovernmental organizations, to relate truly to society.

There is a potential here able to influence the course of events for the years to come.

We are entering into a new stage that calls for an extraordinary response. These environmental issues are linked to economic development from which they cannot be separated. The issue of technology required for environmentally sustainable economic growth and prosperity has not yet been tackled, and the issue of equity between people and between nations needs immediate attention. This calls for a spirit of co-operation between north and south. We have to bend the great powers of science and technology to bring about these new developments, rather than allow a continuation of acceleration in the direction set by the past. We have to aim for a direction which would lead to a better world for all. For this there is need as much for moral resources as for physical resources.

Second World Climate Conference
29 October, 1990

Address by Dr. H. De Haen

Assistant Director General, Agriculture Department
Food and Agriculture Organization of the United Nations (FAO)

First, on behalf of the Director-General of FAO, who unfortunately cannot be with us today, may I thank the Government of Switzerland and the other countries and organizations that have given financial support to this Conference. Second, may I express my appreciation for the tremendous efforts by Mr. Ferguson and the other members of the Secretariat to bring this Conference to life. Finally, may I extend my thanks to those of you who have invested time and intellectual efforts in the preparation of the scientific papers we are to consider this week.

Our primary task in the next five days is to provide the Ministerial Sessions next week with the best possible scientific understanding of climate change and its potential impacts. Good science and not conjecture is the prerequisite for good environment policy. We and succeeding generations cannot afford the risks implicit in the current 'experiment' being performed with global climate—an experiment which could conceivably put the lives of millions of people at risk and have irreversible environmental impacts. Hopefully our scientific advice will convince Ministers of the need to end this experiment.

You all should have received the FAO Position Paper when you registered. It is entitled 'Climate change and agriculture, forestry and fisheries', and gives a brief account of FAO's assessment of the situation and what we are doing to respond to the risks and uncertainties associated with climate change. I will not, therefore, repeat what is presented in that paper. Instead, I will focus on the overall scope of FAO's strategy, and particularly on a number of technical issues and open questions relating to the causes and to the potential impacts of climate change. I will also focus on response options to mitigate those impacts which FAO feels need greater attention by this Conference and by other bodies.

FAO's basic strategy is to concentrate on activities that clarify the issues and response options for developing countries, and alleviate or overcome present problems which contribute to or would be intensified by climate changes of the projected nature and magnitude. In our view the response options must be socially and economically justifiable on current grounds and not solely on the basis of any projected benefit following climate change. The objective, therefore, is to support countries to modify or give greater emphasis to those current policies which, in addressing today's problems, also help to limit or mitigate climate change and its impact on agriculture, forestry and fisheries.

Let me now turn to the question of the contribution by agriculture and forestry to greenhouse gas emissions and how they might be reduced. Clearly, agriculture and forestry make a significant contribution to total emissions, though a minor one compared with the 75 per cent contribution of the energy sector. There are, however, a number of important scientific uncertainties which need to be resolved before we can be emphatic about the magnitude of the contribution. I will give two examples.

Paddy rice is, without doubt, an important source of methane, but it is difficult at present to be sure about the size of its net contribution. This difficulty arises because a significant proportion of paddy rice production takes place on lands which, in their natural state, were also a source of methane. The lands were natural wetlands or consisted of soil types prone to be anaerobic and hence released methane. Their utilization for paddy rice merely added, to an unknown degree, to this natural release of methane.

The second example concerns the release of nitrous oxide from mineral fertilizers. Most of the measurements used to derive the emission values quoted by the Intergovernmental Panel on Climate Change (IPCC) were spot measures made at a limited point in time and space, and therefore difficult to generalize from. Moreover, scientific knowledge is lacking regarding the effect of temperature, cultivation practices, etc. on emissions from mineral fertilizers. So more research is needed in these areas before suggestions to restrict the use of fertilizers can be seriously considered, since they currently play an indispensable role in efforts to achieve food security for all.

Notwithstanding these uncertainties, there is a lot we can do to limit emissions from agriculture. Most of the emission limitation options are concerned with raising the productivity of input use in agriculture, and, as such, constitute a major proportion of FAO's work programme. I will highlight only two of the options which are economically valid in their own right, but may be given additional justification in the context of climate change.

The first option concerns improvements in fertilizer use efficiency, and here I refer to mineral and organic fertilizers, since they are both sources of greenhouse gases. Improvements in the determination of fertilizer requirements, in the formulation and coating of mineral fertilizers, and in application techniques, could lead to a 25 per cent or more reduction in emissions.

The second option is improvements in cattle feed to limit methane emissions from ruminants. In the short to medium term there are possibilities for ameliorating the current widespread problem of quantitatively and qualitatively inadequate feed supply in developing countries which cause unnecessary methane emissions. In the longer term there are possibilities for using recent progress in biotechnology to improve rumen performance.

Let me now come to the impact of climate change on agriculture. Much of the IPCC's analysis is focused on the potentially negative impacts of climate change, with little attention being given to the positive effects, which could pose a different set of policy problems, namely those of surplus disposal if appropriate adjustments are not made. We know more about the physiology of crop plants than the physics of the atmosphere, but there are still important gaps needing further research before clear policy advice can be given. Carbon dioxide enhancement of yields, for example, is a well-established process, but there are still important uncertainties regarding the concurrent impact on respiration and water use efficiency. Consequently, it is not yet possible to establish whether the combined effect of temperature, precipitation, and CO_2 changes on crop yields will be positive or negative. This clearly justifies a higher priority in research agendas for work on the net effect of carbon dioxide enhancement.

Whilst acknowledging the excellent analyses of the IPCC's Working Groups II and III, FAO is concerned that agriculture receives relatively limited attention in the IPCC's interim conclusions. Agriculture is more sensitive to climate change than any other sector of the economy. Over two billion people depend on agriculture for their livelihood. It is the driving force for economic and social development in many developing countries, providing in a number of countries more than 50 per cent of employment, 30–50 per cent of gross domestic product, and 30–90 per cent of foreign exchange earnings. FAO therefore considers that agricultural impact analysis should play a more central role in the follow-up to the IPCC's interim assessment.

This agricultural analysis, in our view, should be undertaken primarily through country case studies, since the nature of impacts and response options may vary considerably both within and between countries, and it will be national decision-makers who will carry the ultimate responsibility for policy action. FAO has already made a start on such country case studies, using an elaboration of the agro-ecological zone methodology that we first developed as part of the FAO/IIASA/UNFPA project Land Resources for Populations of the Future.

Our studies are on the western and central countries of the Sahel. The preliminary results bring out very clearly the serious threat posed by climate change to food security in this region. They indicate also that the tools and instruments of scientific assessment which underlie most of the analysis we will be considering this week must be complemented by those of the economists.

Let me now elaborate some of my reasons for recommending the wider use of economic analysis.

As yet, the Global Circulation Models cannot project regional changes in climate with a high degree of confidence. However, one of the few results which the various models have in common is the projected deterioration in rainfall or soil moisture availability in the Sahel. According to FAO's analysis, if this deterioration in rainfall follows a similar pattern to that actually experienced between 1965 and 1985, when rainfall was abnormally low, the region's food security will be seriously undermined. The area of land receiving adequate rainfall for crop production will be greatly reduced. In essence, the agro-ecological zones in some Sahelian countries will shift over 200 kilometres southwards.

What is of greater concern still is the impact on population and supporting capacity. In general, the projected climate deterioration would cause about a 30 per cent fall in the population that Sahelian countries could feed from their own resources and a decline in both food self-sufficiency and food self-reliance, given that the negative impacts would be on both food crops for the domestic market and cash crops for export.

Moreover, the impact will not be restricted to potential physical output. Costs of production will also rise appreciably in today's semi-arid countries, particularly as regards mineral fertilizers and irrigation. The profitability of fertilizer use is strongly dependent on rainfall and particularly on soil moisture. The projected decline in soil moisture will reduce fertilizer efficiency, thereby raising the costs of production and discouraging farmers from the use of mineral fertilizers. Similarly, increased evapotranspiration will lower the effective capacity of irrigation systems, thereby reducing the area that can be irrigated and raising the costs of irrigation. Thus the food security of considerable numbers of people in semi-arid areas might be endangered in two ways: reduced and more erratic food production, and declining incomes and food purchasing power, particularly amongst the rural poor.

FAO will continue to work on improving the analytical tools for impact assessment, both directly and in partnership with other organizations and institutions. We are already collaborating with the Geological Survey of the Netherlands on issues relating to sea-level rise, and with the Geographical Institute of the Swiss Federal Polytechnic on effects of climate change on irrigation. I am sure delegates are aware of the severe financial constraints facing FAO and other UN agencies, so our ability to expand work on impact assessment is severely constrained. Nonetheless, we will try to do so, and welcome suggestions for collaboration with other institutions, particularly where this raises the efficiency of our own investments in climate change analysis.

In this context, may I welcome on FAO's behalf the IPCC's recommendation for wider involvement of UN agencies in the follow-up to the interim assessment. FAO contributed to that assessment, notably by providing data, preparing long-term projections of agricultural production and input use, supplying its own estimates of deforestation rates, and drafting part of the report. In the future FAO is willing to play a stronger role in both the World Climate Programme and the IPCC to the extent that financial resources permit. I would see much of this support being in the fields of agrometeorology, impact analysis, food security, remote sensing and forest management, where FAO's mandate and long-standing experience place it in a unique position amongst the UN agencies.

This brings me to three other issues that FAO believes need closer attention by this Conference and in any follow-up actions to it.

The first issue concerns population growth. Analyses to date have tended to neglect the issues relating to the interplay of pressures arising from population growth and those from climate change. Current population pressures on the environment in many areas are already severe. They will become even more severe, at least in the medium term ahead, given the built-in momentum in population growth dynamics, even if population policies are successful in reducing fertility rates. Consequently the impact of climate change in some regions will be superimposed on an already unsustainable and deteriorating situation. More intense rainfall in the tropics could add to the serious soil erosion that is prevalent now as population growth leads to the cultivation of steep slopes, and will continue to be a major problem unless determined action is taken in the coming decades. Similarly, continued expansion in livestock numbers in semi-arid areas will add to the serious problem of over-grazing and desertification if rainfall or soil moisture availability declines.

The second issue is that of economic and environmental trade-offs implicit in some of the suggested response options. It has been suggested, for example, that there should be a change in production practices in order to limit methane emissions from paddy rice. The proposal is that paddy rice should be replaced by irrigated rice. This could have three serious economic or environmental trade-offs. First, costs of production would rise because paddy rice needs less mineral fertilizer and herbicides than irrigated rice. Secondly, to maintain the same output more mineral fertilizer may have to be used, thereby raising emissions of nitrous oxide, which is a more potent greenhouse gas than methane. Finally, the application of more herbicides could contaminate the environment.

Last, but not least, is the forestry issue where FAO feels that insufficient attention has been given to its multi-faceted role and its importance as a sink for carbon dioxide rather than a source. This multi-faceted role must be protected not simply because of forests' role as a sink for or a source of emissions, but also because of the benefits to human society, e.g. from their contributions to soil and water conservation, the availability of wood and non-wood forest products, genetic resources, amenity and wildlife habitat. This requires more than a narrow focus on afforestation or re-afforestation. Instead, there should be an integrated approach involving ecologically and silvi-culturally sound management to promote long-term sustained yield from existing or new forests and the development of reliable markets for the products of the forests. And any afforestation targets, such as the 12 million hectare per year target of the Noordwijk Declaration, should be related to realistic opportunities and local needs in the context of overall land-use planning.

In order to foster such an integrated approach, FAO is studying the possible scope and content of an international legal instrument on the conservation and development of forests. This instrument would apply to boreal, temperate and tropical forests, and could complement the proposed climate convention.

These issues need deeper consideration in the IPCC follow-up.

Finally, let me turn to two other items on the Conference agenda, namely international research and the special needs of the developing countries.

FAO, as co-sponsor of the Consultative Group on International Agricultural Research, has kept the secretariat of that body informed of the work of the IPCC. We are encouraging the International Agricultural Research Centres, which are the operational arm of that Group, to take account of climate change issues in their work programmes, particularly with regard to resource efficiency; to increased resilience to climate change and weather perturbations; and to the development of environmentally benign technologies.

Most of FAO's work programme is related in one way or another to the special needs of developing countries. I will therefore highlight only two of the ways FAO might contribute more to meeting these needs.

The first concerns monitoring and assessment, where FAO has a major collaborative programme to improve remote sensing techniques. The objective is to provide reliable estimates of evolving pressures on the environment, and to give advance warning of crop and pasture failures arising from drought and other climate perturbations. This programme is currently focused primarily on sub-Saharan Africa, but could be extended to other regions.

The second concerns the country case studies I mentioned earlier. Our work so far has allowed us to improve on the existing analytical tools for climate change impact analysis, and we would gladly share this experience with developing country institutions.

In conclusion, may I wish delegates a successful Conference, which FAO hopes will continue to have an important impact for a long time after you have all gone home. FAO, for its part, will seek to ensure that your conclusions and recommendations are duly reflected in our future work programme, and will endeavour to give full support to agricultural, forestry and fishery aspects which are followed up through the World Climate Programme and the IPCC.

Scientific and Technical Sessions

The Intergovernmental Panel on Climate Change (IPCC)

by Bert Bolin *(Sweden)*

Based on the Brundtland Commission Report in 1987 the UN adopted a resolution calling for major efforts to achieve sustainable development for all countries of the world. In that context it was also pointed out that environmental deterioration resulting from human activities represents a serious threat to the accomplishment of such development and that far-reaching protective measures may become necessary. The most challenging of the threats is man-induced climate change.

In 1988 the World Meteorological Organization and the United Nations Environment Programme jointly set up the Intergovernmental Panel on Climate Change (IPCC) to assess more closely the current state of knowledge on climate change. Since the first meeting of the IPCC in November 1988, such an assessment has been completed on our present knowledge of the science and impacts of climate change, and on the ways and means to limit and/or adapt to such change. The IPCC First Assessment Report is now available and will serve as a basis for the negotiations on a Framework Convention on Climate Change that will be launched in February 1991 in Washington D.C.

A more detailed account of the IPCC assessment will be presented in a series of talks that will follow my presentation. I wish for the moment only to bring out a few principal points.

i) There is a greenhouse effect, that is at present being enhanced by man due to emissions of a number of the so-called greenhouse gases.

ii) These emissions have so far increased the greenhouse effect by an amount that is equivalent to about 50% of that due to the pre-industrial concentrations of carbon dioxide in the atmosphere.

iii) Even if the man-induced emissions of greenhouse gases were stopped immediately, the increase that has occurred so far would be with us for a century or more. The man-induced greenhouse effect is eliminated by natural processes only very slowly.

iv) Although we cannot yet predict very accurately the change of climate that a given increase of the concentrations of greenhouse gases would cause, we can tell with confidence that it is going to be significant if present increases of the emissions continue without constraints.

v) The climate system responds rather slowly to the changes of the radiation balance that the increasing concentrations of greenhouse gases bring about. Therefore, the change of climate so far in the making due to the currently-observed increases in these concentrations is partly hidden; this undoubtedly is one reason for the fact that we as yet cannot tell for sure if man already has caused a change of the global climate. The observed increase of the global mean temperature of 0.3–0.6 degrees Celsius during the last 100 years or so is, however, largely consistent with the predictions made with the aid of climate models.

vi) The impact of a change of climate on the environment and global society cannot yet be assessed in quantitative terms. The IPCC has, however, presented sensitivity analyses that provide a qualitative picture of the kind of changes that can be expected.

vii) The IPCC has agreed that measures to limit the increases of greenhouse gases in the atmosphere may well have to be taken before we know with certainty either the magnitude or the timing of a man-induced change of climate and its geographical distribution. At the same time it is likely that global society will respond only rather slowly in implementing agreed measures to prevent even far-reaching consequences of a climate change because of the difficulties in assessing their socio-economic consequences, in overcoming the inertia of the industrialized society and in finding ways and means of achieving reductions of the greenhouse gas emissions in an optimum way, particularly in view of the major differences that exist between developed and developing countries.

As pointed out above, progress towards international agreements on preventive measures will be rather slow because of the complexity of the climate issue. A long-term commitment is therefore required. There is a need to develop a strategy for how to deal with this issue in parallel with a good number of other global issues that have been identified, e.g. the increasing world population and the imperative need for development in third world countries, or others that will emerge during the coming decades. There is a need for some steps to be taken in order not to delay action too far into the future, thereby making future action more difficult.

The conclusions summarized above and elaborated in the full IPCC Report give clear justification for the need to start the process of combating climate change now.

There is also agreement that significant steps can be taken to stabilize emissions of carbon dioxide and thereby prevent a further acceleration of the rate of climate change; such steps can be justified on grounds other than climate change and can be taken at modest cost.

In this context it is interesting to note the agreement that has recently been reached in the European Community to stablize the emissions of carbon dioxide at the present level by about the turn of the century. An agreement of this kind among a group of industrialized countries is important in several respects. The process of planning appropriate measures and implementing them will be of value in order to find out ways and means of doing so in the most efficient way. The process itself is likely to reveal a number of difficulties and also surprises that we need to learn about before taking steps to reduce carbon dioxide emissions further. It is also clear that agreements between developed and developing countries cannot be reached unless the former indeed first take the initiative to reduce their emissions, since it is generally recognized that they still produce about 75% of present emissions, while their populations are only about 25% of the world's population.

Although further assessments are needed with regard to particular aspects of the scientific problem of climate change and the impacts of climate change, it is clear that the IPCC efforts during the next year or two must focus on studies of the socio-economic implications of a change of climate as well as on implementing more far-reaching preventive measures. It is clear from the discussions in the Panel that very different views are held on these matters and also that the consequences are going to be very different for different nations; particularly large differences can be expected between developed and developing countries. Such studies will ultimately have to be based on much better knowledge about the expected impacts of climate change, but some general analyses have already

been made and can certainly be expanded considerably. Implicit in such studies is also the problem of the world's future energy supply, i.e. the likely role of nuclear energy, the requirements for accelerated development of renewable energy sources and again the particular concerns of the developing countries in this context.

The climate issue is a complex one and will be with us for decades to come. It is therefore essential that a clear structure for the future work to deal with it be established. I envisage three parallel streams of activities.

It is first of all most essential that the research and observational efforts regarding global change problems are substantially increased and that these are well organized internationally.

This concerns the efforts to understand how the global climate system operates in order to permit better projections of what might happen in the future due to increasing human activities and to explore optimum protective and/or adaptive measures. There are already major efforts under way. The World Climate Research Programme (WCRP) and the International Geosphere-Biosphere Programme (IGBP) are of particular importance for studies of the scientific problem of climate change; the World Climate Impact Studies Programme (WCIP) has the responsibility to foster impact studies. Much less has so far been done to try to co-ordinate research efforts in the socio-economic fields and increased activities of this kind are urgently needed. Such research programmes should be in the hands of scientists and the IPCC is anxious to establish the best possible working relationships with such research programmes, and not govern these. The present Second World Climate Conference is aimed at providing an overview of the present status of these research efforts and also at calling for increased resources for future activities.

On the other hand, negotiations for a climate convention are about to begin. This is obviously going to be a very difficult process, in which political considerations will and should be in the forefront of the debate. It is most essential that this process is not mixed up with the task given to the IPCC by its parent organizations, WMO and UNEP, i.e. assessments of the likelihood and the characteristics of an oncoming climate change, its possible impacts and the ways and means of limiting and/or adapting to such a change. It is obvious that a clear division of responsibilities in this way was not quite achieved in the final phase of the IPCC assessment that was completed in August 1990, primarily because there was no agreement on the organization of the negotiations.

The IPCC would also be available in the negotiating process for advice and should be able to respond rather quickly and effectively to requests from the negotiating parties. It is important, however, that in doing so, the IPCC proceed in an orderly manner. A distinction must be made between:

i) Requests that can be answered by more careful consideration of the material that has been assembled in the IPCC reports

ii) More complex requests that will require the examination of the question by one (or several) of the IPCC subgroups or special task groups created by the IPCC.

In the latter case some response may be possible within four to eight months and accordingly be of use in the negotiations that are to be completed before the 1992 UN Conference on Environment and Development. Other requests may well require more time for analysis and only be available for the negotiations in later years.

It is hoped that the future work of the IPCC will become simpler if this structure of the three parallel streams of activities is generally accepted.

The task that lies ahead of us is indeed a most difficult one and close co-operation between different sectors of society, within nations as well as between nations, is most essential in order to achieve what will be required. It should also be recognized that actions to prevent an unacceptable climate change must be developed with due regard to other major global issues, particularly that of a steadily increasing world population. The global issue of climate change is part of the much broader issue of management of the global resources for a sustainable development of the world as a whole.

Scientific Assessment of Climate Change: Summary of the IPCC Working Group I Report

by John T. Houghton *(United Kingdom)*

Executive Summary

We are certain of the following:

- there is a natural greenhouse effect which already keeps the Earth warmer than it would otherwise be

- emissions resulting from human activities are substantially increasing the atmospheric concentrations of the greenhouse gases: carbon dioxide, methane, chlorofluorocarbons (CFCs) and nitrous oxide. These increases will enhance the greenhouse effect, resulting on average in an additional warming of the Earth's surface. The main greenhouse gas, water vapour, will increase in response to global warming and further enhance it.

We calculate with confidence that:

- some gases are potentially more effective than others at changing climate, and their relative effectiveness can be estimated. Carbon dioxide has been responsible for over half the enhanced greenhouse effect in the past, and is likely to remain so in the future

- atmospheric concentrations of the long-lived gases (carbon dioxide, nitrous oxide and the CFCs) adjust only slowly to changes in emissions. Continued emissions of these gases at present rates would commit us to increased concentrations for centuries ahead. The longer emissions continue to increase at present day rates, the greater reductions would have to be for concentrations to stabilise at a given level

- the long-lived gases would require immediate reductions in emissions from human activities of over 60% to stabilise their concentrations at today's levels; methane would require a 15-20% reduction.

Based on current model results, we predict:

- under the IPCC Business-as-Usual (Scenario A) emissions of greenhouse gases, a rate of increase of global mean temperature during the next century of about 0.3°C per decade (with an uncertainty range of 0.2°C to 0.5°C per decade); this is greater than that seen over the past 10,000 years. This will result in a likely increase in global mean temperature of about 1°C above the present value by 2025 and 3°C before the end of the next century. The rise will not be steady because of the influence of other factors

- under the other IPCC emission scenarios which assume progressively increasing levels of controls, rates of increase in global mean temperature of about 0.2°C per decade (Scenario B), just above 0.1°C per decade (Scenario C) and about 0.1°C per decade (Scenario D)

- that land surfaces warm more rapidly than the ocean, and high northern latitudes warm more than the global mean in winter

- regional climate changes different from the global mean, although our confidence in the prediction of the detail of regional changes is low. For example, temperature increases in Southern Europe and central North America are predicted to be higher than the global mean, accompanied on average by reduced summer precipitation and soil moisture. There are less consistent predictions for the tropics and the Southern Hemisphere

- under the IPCC Business as Usual emissions scenario, an average rate of global mean sea level rise of about 6cm per decade over the next century (with an uncertainty range of 3 - 10cm per decade), mainly due to thermal expansion of the oceans and the melting of some land ice. The predicted rise is about 20cm in global mean sea level by 2030, and 65cm by the end of the next century. There will be significant regional variations.

There are many uncertainties in our predictions particularly with regard to the timing, magnitude and regional patterns of climate change, due to our incomplete understanding of:

- sources and sinks of greenhouse gases, which affect predictions of future concentrations

- clouds, which strongly influence the magnitude of climate change

- oceans, which influence the timing and patterns of climate change

- polar ice sheets which affect predictions of sea level rise.

These processes are already partially understood, and we are confident that the uncertainties can be reduced by further research. However, the complexity of the system means that we cannot rule out surprises.

Our judgement is that:

- Global - mean surface air temperature has increased by 0.3°C to 0.6°C over the last 100 years, with the five global-average warmest years being in the 1980s. Over the same period global sea level has increased by 10-20cm. These increases have not been smooth with time, nor uniform over the globe

- The size of this warming is broadly consistent with predictions of climate models, but it is also of the same magnitude as natural climate variability. Thus the observed increase could be largely due to this natural variability; alternatively this variability and other human factors could have offset a still larger human-induced greenhouse warming. The unequivocal detection of the enhanced greenhouse effect from observations is not likely for a decade or more

- There is no firm evidence that climate has become more variable over the last few decades. However, with an increase in the mean temperature, episodes of high temperatures will most likely become more frequent in the future, and cold episodes less frequent

- Ecosystems affect climate, and will be affected by a changing climate and by increasing carbon dioxide concentrations. Rapid changes in climate will change the composition of ecosystems; some species will benefit while others will be unable to migrate or adapt fast enough and may become extinct. Enhanced

levels of carbon dioxide may increase productivity and efficiency of water use of vegetation. The effect of warming on biological processes, although poorly understood, may increase the atmospheric concentrations of natural greenhouse gases.

To improve our predictive capability, we need:

- to **understand** better the various climate-related processes, particularly those associated with clouds, oceans and the carbon cycle

- to **improve** the systematic observation of climate-related variables on a global basis, and further investigate changes which took place in the past

- to **develop** improved models of the Earth's climate system

- to **increase** support for national and international climate research activities, especially in developing countries.

- to **facilitate** international exchange of climate data.

Introduction: what is the issue ?

There is concern that human activities may be inadvertently changing the climate of the globe through the enhanced greenhouse effect, by past and continuing emissions of carbon dioxide and other gases which will cause the temperature of the Earth's surface to increase - popularly termed the "global warming". If this occurs, consequent changes may have a significant impact on society.

The purpose of the Working Group I report, as determined by the first meeting of IPCC, is to provide a scientific assessment of:

1) the factors which may affect climate change during the next century, especially those which are due to human activity.
2) the responses of the atmosphere - ocean - land - ice system.
3) current capabilities of modelling global and regional climate changes and their predictability.
4) the past climate record and presently observed climate anomalies.

On the basis of this assessment, the report presents current knowledge regarding predictions of climate change (including sea level rise and the effects on ecosystems) over the next century, the timing of changes together with an assessment of the uncertainties associated with these predictions.

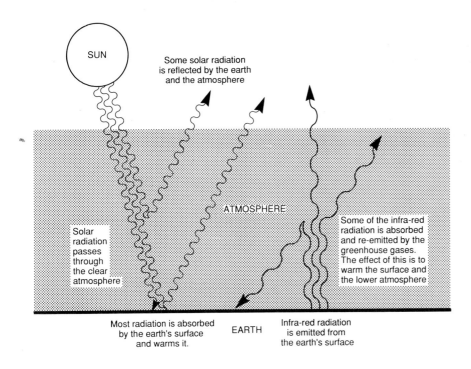

Figure 1: A simplified diagram illustrating the greenhouse effect.

This Policymakers Summary aims to bring out those elements of the main report which have the greatest relevance to policy formulation, in answering the following questions:

- What factors determine global climate?
- What are the greenhouse gases, and how and why are they increasing?
- Which gases are the most important?
- How much do we expect the climate to change?
- How much confidence do we have in our predictions?
- Will the climate of the future be very different ?
- Have human activities already begun to change global climate?
- How much will sea level rise?
- What will be the effects on ecosystems?
- What should be done to reduce uncertainties, and how long will this take?

This report is intended to respond to the practical needs of the policymaker. It is neither an academic review, nor a plan for a new research programme. Uncertainties attach to almost every aspect of the issue, yet policymakers are looking for clear guidance from scientists; **hence authors have been asked to provide their best-estimates wherever possible**, together with an assessment of the uncertainties.

This report is a summary of our understanding in 1990. Although continuing research will deepen this understanding and require the report to be updated at frequent intervals, basic conclusions concerning the reality of the enhanced greenhouse effect and its potential to alter global climate are unlikely to change significantly. Nevertheless, the complexity of the system may give rise to surprises.

What factors determine global climate ?

There are many factors, both natural and of human origin, that determine the climate of the earth. We look first at those which are natural, and then see how human activities might contribute.

What natural factors are important?

The driving energy for weather and climate comes from the Sun. The Earth intercepts solar radiation (including that in the short-wave, visible, part of the spectrum); about a third of it is reflected, the rest is absorbed by the different components (atmosphere, ocean, ice, land and biota) of the climate system. The energy absorbed from solar radiation is balanced (in the long term) by outgoing radiation from the Earth and atmosphere; this terrestrial radiation takes the form of long-wave invisible infrared energy, and its magnitude is determined by the temperature of the Earth-atmosphere system.

There are several natural factors which can change the balance between the energy absorbed by the Earth and that emitted by it in the form of longwave infrared radiation; these factors cause the **radiative forcing** on climate. The most obvious of these is a change in the output of energy from the Sun. There is direct evidence of such variability over the 11-year solar cycle, and longer period changes may also occur. Slow variations in the Earth's orbit affect the seasonal and latitudinal distribution of solar radiation; these were probably responsible for initiating the ice ages.

One of the most important factors is the **greenhouse effect**; a simplified explanation of which is as follows. Short-wave solar radiation can pass through the clear atmosphere relatively unimpeded. But long-wave terrestrial radiation emitted by the warm surface of the Earth is partially absorbed and then re-emitted by a number of trace gases in the cooler atmosphere above. Since, on average, the outgoing long-wave radiation balances the incoming solar radiation, both the atmosphere and the surface will be warmer than they would be without the greenhouse gases.

The main natural greenhouse gases are not the major constituents, nitrogen and oxygen, but water vapour (the biggest contributor), carbon dioxide, methane, nitrous oxide, and ozone in the troposphere (the lowest 10-15km of the atmosphere) and stratosphere.

Aerosols (small particles) in the atmosphere can also affect climate because they can reflect and absorb radiation. The most important natural perturbations result from explosive volcanic eruptions which affect con-centrations in the lower stratosphere. Lastly, the climate has its own **natural variability** on all timescales and changes occur without any external influence.

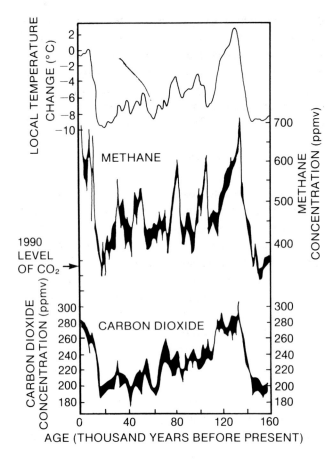

Figure 2: Analysis of air trapped in Antarctic ice cores shows that methane and carbon dioxide concentrations were closely correlated with the local temperature over the last 160,000 years. Present day concentrations of carbon dioxide are indicated

How do we know that the natural greenhouse effect is real?

The greenhouse effect is real; it is a well understood effect, based on established scientific principles. We know that the greenhouse effect works in practice, for several reasons.

Firstly, the mean temperature of the Earth's surface is already warmer by about 33°C (assuming the same reflectivity of the earth) than it would be if the natural greenhouse gases were not present. Satellite observations of the radiation emitted from the Earth's surface and through the atmosphere demonstrate the effect of the greenhouse gases.

Secondly, we know the composition of the atmospheres of Venus, Earth and Mars are very different, and their surface temperatures are in general agreement with greenhouse theory.

Thirdly, measurements from ice cores going back 160,000 years show that the Earth's temperature closely paralleled the amount of carbon dioxide and methane in the atmosphere (see Figure 2). Although we do not know the details of cause and effect, calculations indicate that

changes in these greenhouse gases were part, but not all, of the reason for the large (5-7°C) global temperature swings between ice ages and interglacial periods.

How might human activities change global climate?

Naturally occurring greenhouse gases keep the Earth warm enough to be habitable. By increasing their concentrations, and by adding new greenhouse gases like chloro-fluorocarbons (CFCs), humankind is capable of raising the global-average annual-mean surface-air temperature (which, for simplicity, is referred to as the "global temperature"), although we are uncertain about the rate at which this will occur. Strictly, this is an **enhanced** greenhouse effect - above that occurring due to natural greenhouse gas concentrations; the word "enhanced" is usually omitted, but it should not be forgotten. Other changes in climate are expected to result, for example changes in precipitation, and a global warming will cause sea levels to rise; these are discussed in more detail later.

There are other human activities which have the potential to affect climate. A change in the albedo (reflectivity) of the land, brought about by **desertification or deforestation** affects the amount of solar energy absorbed at the Earth's surface. Human-made **aerosols**, from sulphur emitted largely in fossil fuel combustion, can modify clouds and this may act to lower temperatures. Lastly, changes in **ozone in the stratosphere** due to CFCs may also influence climate.

What are the greenhouse gases and why are they increasing?

We are certain that the concentrations of greenhouse gases in the atmosphere have changed naturally on ice-age time-scales, and have been increasing since pre-industrial times due to human activities. Table 1 summarizes the present and pre-industrial abundances, current rates of change and present atmospheric lifetimes of greenhouse gases influenced by human activities. Carbon dioxide, methane, and nitrous oxide all have significant natural and human sources, while the chlorofluorocarbons are only produced industrially.

Two important greenhouse gases, water vapour and ozone, are not included in this table. Water vapour has the largest greenhouse effect, but its concentration in the troposphere is determined internally within the climate system, and, on a global scale, is not affected by human sources and sinks. Water vapour will increase in response to global warming and further enhance it; this process is included in climate models. The concentration of ozone is changing both in the stratosphere and the troposphere due

to human activities, but it is difficult to quantify the changes from present observations.

For a thousand years prior to the industrial revolution, abundances of the greenhouse gases were relatively constant. However, as the world's population increased, as the world became more industrialized and as agriculture developed, the abundances of the greenhouse gases increased markedly. Figure 3 illustrates this for carbon dioxide, methane, nitrous oxide and CFC-11.

Since the industrial revolution the combustion of fossil fuels and deforestation have led to an increase of 26% in carbon dioxide concentration in the atmosphere. We know the magnitude of the present day fossil-fuel source, but the input from deforestation cannot be estimated accurately. In addition, although about half of the emitted carbon dioxide stays in the atmosphere, we do not know well how much of the remainder is absorbed by the oceans and how much by terrestrial biota. Emissions of chlorofluorocarbons, used as aerosol propellants, solvents, refrigerants and foam blowing agents, are also well known; they were not present in the atmosphere before their invention in the 1930s.

The sources of methane and nitrous oxide are less well known. Methane concentrations have more than doubled because of rice production, cattle rearing, biomass burning, coal mining and ventilation of natural gas; also, fossil fuel combustion may have also contributed through chemical reactions in the atmosphere which reduce the rate of removal of methane. Nitrous oxide has increased by about 8% since pre-industrial times, presumably due to human

Table 1: *Summary of Key Greenhouse Gases Affected by Human Activities*

	Carbon Dioxide	Methane	CFC-11	CFC-12	Nitrous Oxide
Atmospheric concentration	ppmv	ppmv	pptv	pptv	ppbv
Pre-industrial (1750-1800)	280	0.8	0	0	288
Present day (1990)	353	1.72	280	484	310
Current rate of change per year	1.8 (0.5%)	0.015 (0.9%)	9.5 (4%)	17 (4%)	0.8 (0.25%)
Atmospheric lifetime (years)	(50-200)†	10	65	130	150

ppmv = parts per million by volume;
ppbv = parts per billion (thousand million) by volume;
pptv = parts per trillion (million million) by volume.
† The way in which CO_2 is absorbed by the oceans and biosphere is not simple and a single value cannot be given; refer to the main report for further discussion.

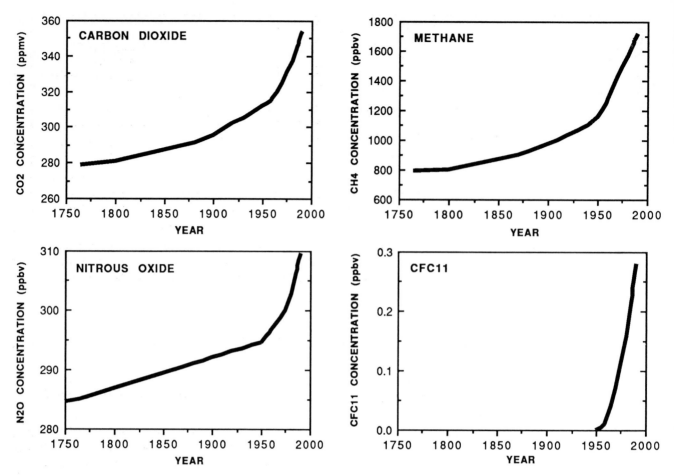

Figure 3: Concentrations of carbon dioxide and methane after remaining relatively constant up to the 18th century, have risen sharply since then due to man's activities. Concentrations of nitrous oxide have increased since the mid-18th century, especially in the last few decades. CFCs were not present in the atmosphere before the 1930s.

activities; we are unable to specify the sources, but it is likely that agriculture plays a part.

The effect of ozone on climate is strongest in the upper troposphere and lower stratosphere. Model calculations indicate that ozone in the upper troposphere should have increased due to human-made emissions of nitrogen oxides, hydrocarbons and carbon monoxide. While at ground level ozone has increased in the Northern Hemisphere in response to these emissions, observations are insufficient to confirm the expected increase in the upper troposphere. The lack of adequate observations prevents us from accurately quantifying the climatic effect of changes in tropospheric ozone.

In the lower stratosphere at high southern latitudes ozone has decreased considerably due to the effects of CFCs, and there are indications of a global-scale decrease which, while not understood, may also be due to CFCs. These observed decreases should act to cool the earth's surface, thus providing a small offset to the predicted warming produced by the other greenhouse gases. Further reductions in lower stratospheric ozone are possible during the next few decades as the atmospheric abundances of CFCs continue to increase.

Concentrations, lifetimes and stabilisation of the gases

In order to calculate the atmospheric concentrations of carbon dioxide which will result from human-made emissions we use computer models which incorporate details of the emissions and which include representations of the transfer of carbon dioxide between the atmosphere, oceans and terrestrial biosphere. For the other greenhouse gases, models which incorporate the effects of chemical reactions in the atmosphere are employed.

The atmospheric lifetimes of the gases are determined by their sources and sinks in the oceans, atmosphere and biosphere. Carbon dioxide, chlorofluorocarbons and nitrous oxide are removed only slowly from the atmosphere and hence, following a change in emissions, their atmospheric concentrations take decades to centuries to adjust fully. Even if all human-made emissions of carbon dioxide were halted in the year 1990, about half of the increase in carbon dioxide concentration caused by human activities would still be evident by the year 2100.

In contrast, some of the CFC substitutes and methane have relatively short atmospheric lifetimes so that their atmospheric concentrations respond fully to emission changes within a few decades.

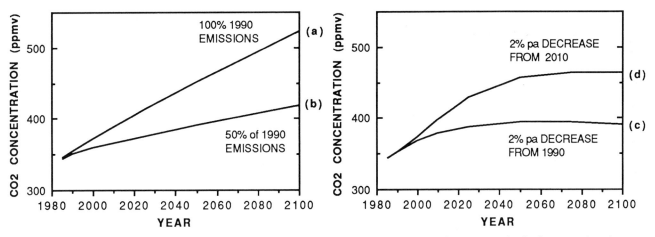

Figure 4: The relationship between hypothetical fossil fuel emissions of carbon dioxide and its concentration in the atmosphere is shown in the case where (a) emissions continue at 1990 levels, (b) emissions are reduced by 50% in 1990 and continue at that level, (c) emissions are reduced by 2% pa from 1990, and (d) emissions, after increasing by 2% pa until 2010, are then reduced by 2% pa thereafter.

To illustrate the emission-concentration relationship clearly, the effect of hypothetical changes in carbon dioxide fossil fuel emissions is shown in Figure 4: (a) continuing global emissions at 1990 levels; (b) halving of emissions in 1990; (c) reductions in emissions of 2% per year (pa) from 1990 and (d) a 2% pa increase from 1990-2010 followed by a 2% pa decrease from 2010.

Continuation of present day emissions are committing us to increased future concentrations, and the longer emissions continue to increase, the greater would reductions have to be to stabilise at a given level. If there are critical concentration levels that should not be exceeded, then the earlier emission reductions are made the more effective they are.

The term "**atmospheric stabilisation**" is often used to describe the limiting of the concentration of the greenhouse gases at a certain level. The amount by which human-made emissions of a greenhouse gas must be reduced in order to stabilise at present day concentrations, for example, is

shown in Table 2. For most gases the reductions would have to be substantial.

How will greenhouse gas abundances change in the future?

We need to know future greenhouse gas concentrations in order to estimate future climate change. As already mentioned, these concentrations depend upon the magnitude of human-made emissions and on how changes in climate and other environmental conditions may influence the biospheric processes that control the exchange of natural greenhouse gases, including carbon dioxide and methane, between the atmosphere, oceans and terrestrial biosphere - the greenhouse gas "feedbacks".

Four scenarios of future human-made emissions were developed by Working Group III. The first of these assumes that few or no steps are taken to limit greenhouse gas emissions, and this is therefore termed Business-as-Usual (BaU). (It should be noted that an aggregation of

Table 2: *Stabilisation of Atmospheric Concentrations. Reductions in the human-made emissions of greenhouse gases required to stabilise concentrations at present day levels:*

Greenhouse Gas	Reduction Required
Carbon Dioxide	>60%
Methane	15 - 20%
Nitrous Oxide	70 - 80%
CFC-11	70 - 75%
CFC-12	75 - 85%
HCFC-22	40 - 50%

Note that the stabilisation of each of these gases would have different effects on climate, as explained in the next section.

national forecasts of emissions of carbon dioxide and methane to the year 2025 undertaken by Working Group III resulted in global emissions 10-20% higher than in the BaU scenario). The other three scenarios assume that progressively increasing levels of controls reduce the growth of emissions; these are referred to as scenarios B, C, and D. They are briefly described in the Annex to this summary. Future concentrations of some of the greenhouse gases which would arise from these emissions are shown in Figure 5.

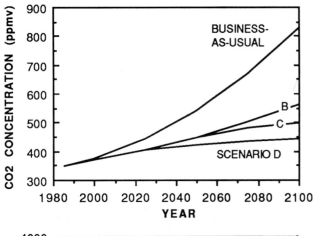

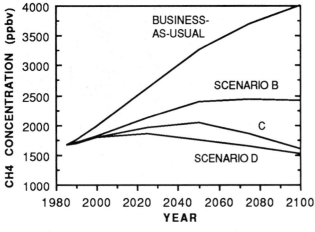

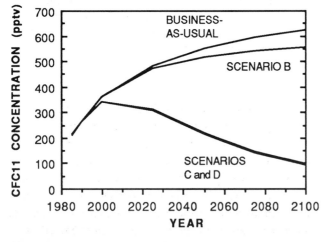

Figure 5: Atmospheric concentrations of carbon dioxide, methane and CFC-11 resulting from the four IPCC emissions scenarios

Greenhouse gas feedbacks

Some of the possible feedbacks which could significantly modify future greenhouse gas concentrations in a warmer world are discussed in the following paragraphs.

The net emissions of carbon dioxide from terrestrial ecosystems will be elevated if higher temperatures increase respiration at a faster rate than photosynthesis, or if plant populations, particularly large forests, cannot adjust rapidly enough to changes in climate.

A net flux of carbon dioxide to the atmosphere may be particularly evident in warmer conditions in tundra and boreal regions where there are large stores of carbon. The opposite is true if higher abundances of carbon dioxide in the atmosphere enhance the productivity of natural ecosystems, or if there is an increase in soil moisture which can be expected to stimulate plant growth in dry ecosystems and to increase the storage of carbon in tundra peat. The extent to which ecosystems can sequester increasing atmospheric carbon dioxide remains to be quantified.

If the oceans become warmer, their net uptake of carbon dioxide may decrease because of changes in (i) the chemistry of carbon dioxide in seawater, (ii) biological activity in surface waters, and (iii) the rate of exchange of carbon dioxide between the surface layers and the deep ocean. This last depends upon the rate of formation of deep water in the ocean which, in the North Atlantic for example, might decrease if the salinity decreases as a result of a change in climate.

Methane emissions from natural wetlands and rice paddies are particularly sensitive to temperature and soil moisture. Emissions are significantly larger at higher temperatures and with increased soil moisture; conversely, a decrease in soil moisture would result in smaller emissions. Higher temperatures could increase the emissions of methane at high northern latitudes from decomposable organic matter trapped in permafrost and methane hydrates.

As illustrated earlier, ice core records show that methane and carbon dioxide concentrations changed in a similar sense to temperature between ice ages and interglacials.

Although many of these feedback processes are poorly understood, it seems likely that, overall, they will act to increase, rather than decrease, greenhouse gas concentrations in a warmer world.

Which gases are the most important?

We are certain that increased greenhouse gas concentrations increase radiative forcing. We can calculate the forcing with much more confidence than the climate change that results because the former avoids the need to evaluate a number of poorly understood atmospheric responses. We then have a base from which to calculate the

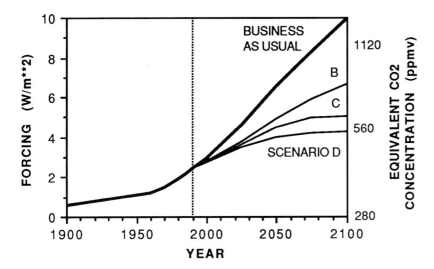

Figure 6: Increase in radiative forcing since the mid-18th century, and predicted to result from the four IPCC emissions scenarios, also expressed as equivalent carbon dioxide concentrations.

relative effect on climate of an increase in **concentration** of each gas in the present-day atmosphere, both in absolute terms and relative to carbon dioxide. These relative effects span a wide range; methane is about 21 times more effective, molecule-for-molecule, than carbon dioxide, and CFC-11 about 12,000 times more effective. On a kilogram-per-kilogram basis, the equivalent values are 58 for methane and about 4,000 for CFC-11, both relative to carbon dioxide. Values for other greenhouse gases are to be found in Section 2.

The total radiative forcing at any time is the sum of those from the individual greenhouse gases. We show in Figure 6 how this quantity has changed in the past (based on observations of greenhouse gases) and how it might change in the future (based on the four IPCC emissions scenarios). For simplicity, we can express total forcing in terms of the amount of carbon dioxide which would give that forcing; this is termed the **equivalent carbon dioxide concentration**. Greenhouse gases have increased since pre-industrial times (the mid-18th century) by an amount that is radiatively equivalent to about a 50% increase in carbon dioxide, although carbon dioxide itself has risen by only 26%; other gases have made up the rest.

The contributions of the various gases to the total increase in climate forcing during the 1980s is shown as a pie diagram in Figure 7; carbon dioxide is responsible for about half the decadal increase. (Ozone, the effects of which may be significant, is not included)

How can we evaluate the effect of different greenhouse gases?

To evaluate possible policy options, it is useful to know the relative radiative effect (and, hence, potential climate effect) of equal emissions of each of the greenhouse gases. The concept of relative **Global Warming Potentials**

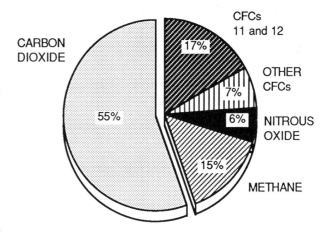

Figure 7: The contribution from each of the human-made greenhouse gases to the change in radiative forcing from 1980 to 1990. The contribution from ozone may also be significant, but cannot be quantified at present.

(GWP) has been developed to take into account the differing times that gases remain in the atmosphere.

This index defines the time-integrated warming effect due to an instantaneous release of unit mass (1 kg) of a given greenhouse gas in today's atmosphere, relative to that of carbon dioxide. The relative importances will change in the future as atmospheric composition changes because, although radiative forcing increases in direct proportion to the concentration of CFCs, changes in the other greenhouse gases (particularly carbon dioxide) have an effect on forcing which is much less than proportional.

The GWPs in Table 3 are shown for three time horizons, reflecting the need to consider the cumulative effects on climate over various time scales. The longer time horizon is appropriate for the cumulative effect; the shorter timescale will indicate the response to emission changes in the short term. There are a number of practical difficulties in devising and calculating the values of the GWPs, and the values given here should be considered as preliminary. In addition to these direct effects, there are indirect effects of human-made emissions arising from chemical reactions between the various constituents. The indirect effects on stratospheric water vapour, carbon dioxide and tropospheric ozone have been included in these estimates.

Table 3 indicates, for example, that the effectiveness of methane in influencing climate will be greater in the first few decades after release, whereas emission of the longer-lived nitrous oxide will affect climate for a much longer time. The lifetimes of the proposed CFC replacements range from 1 to 40 years; the longer lived replacements are still potentially effective as agents of climate change. One example of this, HCFC-22 (with a 15 year lifetime), has a similar effect (when released in the same amount) as CFC-11 on a 20 year time-scale; but less over a 500 year time-scale.

Table 3 shows carbon dioxide to be the least effective greenhouse gas per kilogramme emitted, but its contribution to global warming, which depends on the product of the GWP and the amount emitted, is largest. In the example in Table 4, the effect over 100 years of emissions of greenhouse gases in 1990 are shown relative to carbon dioxide. This is illustrative; to compare the effect of different emission projections we have to sum the effect of emissions made in future years.

There are other technical criteria which may help policymakers to decide, in the event of emissions reductions being deemed necessary, which gases should be considered. Does the gas contribute in a major way to current, and future, climate forcing? Does it have a long lifetime, so earlier reductions in emissions would be more effective than those made later? And are its sources and sinks well enough known to decide which could be controlled in practice? Table 5 illustrates these factors.

How much do we expect climate to change?

It is relatively easy to determine the direct effect of the increased radiative forcing due to increases in greenhouse gases. However, as climate begins to warm, various processes act to amplify (through positive feedbacks) or reduce (through negative feedbacks) the warming. The main feedbacks which have been identified are due to changes in water vapour, sea-ice, clouds and the oceans. The best tools we have which take the above feedbacks into account (but do not include greenhouse gas feedbacks) are

three-dimensional mathematical models of the climate system (atmosphere-ocean-ice-land), known as General Circulation Models (GCMs). They synthesise our knowledge of the physical and dynamical processes in the overall system and allow for the complex interactions between the various components. However, in their current state of development, the descriptions of many of the processes involved are comparatively crude. Because of this, considerable uncertainty is attached to these predictions of climate change, which is reflected in the range of values given; further details are given in a later section.

The estimates of climate change presented here are based on

i)   the "best-estimate" of equilibrium climate sensitivity (i.e the equilibrium temperature change due to a doubling of carbon dioxide in the atmosphere) obtained from model simulations, feedback analyses and observational considerations (see later box: "What tools do we use?")

ii)   a "box-diffusion-upwelling" ocean-atmosphere climate model which translates the greenhouse forcing into the evolution of the temperature response for the prescribed climate sensitivity. (This simple model has been calibrated against more complex atm-osphere-ocean coupled GCMs for situations where the more complex models have been run).

How quickly will global climate change?
a. If emissions follow a Business-as-Usual pattern
Under the IPCC Business-as-Usual (Scenario A) emissions of greenhouse gases, the average rate of increase of global mean temperature during the next century is estimated to be about 0.3°C per decade (with an uncertainty range of 0.2°C to 0.5°C). This will result in a likely increase in global mean temperature of about 1°C above the present value (about 2°C above that in the pre-industrial period) by 2025 and 3°C above today's (about 4°C above pre-industrial) before the end of the next century.

The projected temperature rise out to the year 2100, with high, low and best-estimate climate responses, is shown in Figure 8. Because of other factors which influence climate, we would not expect the rise to be a steady one.

The temperature rises shown above are **realised** temperatures; at any time we would also be **committed** to a further temperature rise toward the equilibrium temperature (see box: "Equilibrium and Realised Climate Change"). For the BaU "best-estimate" case in the year 2030, for example, a further 0.9°C rise would be expected, about 0.2°C of which would be realised by 2050 (in addition to changes due to further greenhouse gas increases); the rest would become apparent in decades or centuries.

Even if we were able to stabilise emissions of each of the greenhouse gases at present day levels from now on, the

Table 3 *Global Warming Potentials. The warming effect of an emission of 1kg of each gas relative to that of CO_2*
These figures are best estimates calculated on the basis of the present day atmospheric composition

| | TIME HORIZON | | |
	20 yr	100 yr	500 yr
Carbon dioxide	1	1	1
Methane (including indirect)	63	21	9
Nitrous oxide	270	290	190
CFC-11	4500	3500	1500
CFC-12	7100	7300	4500
HCFC-22	4100	1500	510

Global Warming Potentials for a range of CFCs and potential replacements are given in the full text.

Table 4 *The Relative Cumulative Climate Effect of 1990 Man-Made Emissions*

	GWP (100yr horizon)	1990 emissions (Tg)	Relative contribution over 100yr
Carbon dioxide	1	26000†	61%
Methane*	21	300	15%
Nitrous oxide	290	6	4%
CFCs	Various	0.9	11%
HCFC-22	1500	0.1	0.5%
Others*	Various		8.5%

* These values include the indirect effect of these emissions on other greenhouse gases via chemical reactions in the atmosphere. Such estimates are highly model dependent and should be considered preliminary and subject to change. The estimated effect of ozone is included under "others". The gases included under "others" are given in the full report.
† 26 000 Tg (teragrams) of carbon dioxide = 7 000 Tg (=7 Gt) of carbon

Table 5 *Characteristics of Greenhouse Gases*

GAS	MAJOR CONTRIBUTOR?	LONG LIFETIME?	SOURCES KNOWN?
Carbon dioxide	yes	yes	yes
Methane	yes	no	semi-quantitatively
Nitrous oxide	not at present	yes	qualitatively
CFCs	yes	yes	yes
HCFCs, etc	not at present	mainly no	yes
Ozone	possibly	no	qualitatively

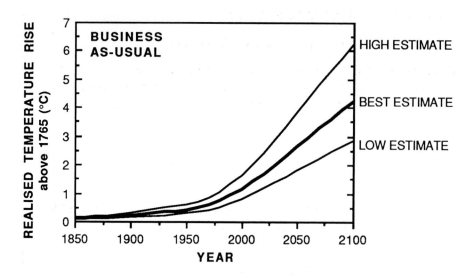

Figure 8: Simulation of the increase in global mean temperature from 1850-1990 due to observed increases in greenhouse gases, and predictions of the rise between 1990 and 2100 resulting from the Business-as-Usual emissions.

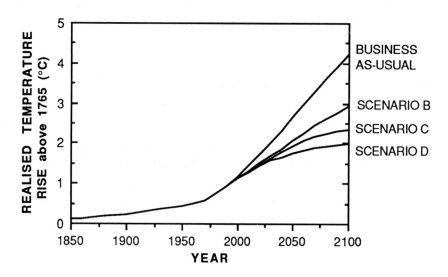

Figure 9: Simulations of the increase in global mean temperature from 1850-1990 due to observed increases in greenhouse gases, and predictions of the rise between 1990 and 2100 resulting from the IPCC Scenario B,C and D emissions, with the Business-as-Usual case for comparison.

temperature is predicted to rise by about 0.2°C per decade for the first few decades.

The global warming will also lead to increased global average precipitation and evaporation of a few percent by 2030. Areas of sea-ice and snow are expected to diminish.

b. If emissions are subject to controls

Under the other IPCC emission scenarios which assume progressively increasing levels of controls, average rates of increase in global mean temperature over the next century are estimated to be about 0.2°C per decade (Scenario B), just above 0.1°C per decade (Scenario C) and about 0.1°C per decade (Scenario D). The results are illustrated in Figure 9, with the Business-as-Usual case shown for

comparison. Only the best-estimate of the temperature rise is shown in each case.

The indicated range of uncertainty in global temperature rise given above reflects a subjective assessment of uncertainties in the calculation of climate response, but does not include those due to the transformation of emissions to concentrations, nor the effects of greenhouse gas feedbacks.

What will be the patterns of climate change by 2030?
Knowledge of the global mean warming and change in precipitation is of limited use in determining the impacts of climate change, for instance on agriculture. For this we need to know changes regionally and seasonally.

ESTIMATES FOR CHANGES BY 2030

(IPCC Business-as-Usual scenario; changes from **pre-industrial**)

The numbers given below are based on high resolution models, scaled to be consistent with our best estimate of global mean warming of 1.8°C by 2030. For values consistent with other estimates of global temperature rise, the numbers below should be reduced by 30% for the low estimate or increased by 50% for the high estimate. Precipitation estimates are also scaled in a similar way.

Confidence in these regional estimates is low

Central North America (35°-50°N 85°-105°W)
The warming varies from 2 to 4°C in winter and 2 to 3°C in summer. Precipitation increases range from 0 to 15% in winter whereas there are decreases of 5 to 10% in summer. Soil moisture decreases in summer by 15 to 20%.

Southern Asia (5°-30°N 70°-105°E)
The warming varies from 1 to 2°C throughout the year. Precipitation changes little in winter and generally increases throughout the region by 5 to 15% in summer. Summer soil moisture increases by 5 to 10%.

Sahel (10°-20°N 20°W-40°E)
The warming ranges from 1 to 3°C. Area mean precipitation increases and area mean soil moisture decreases marginally in summer. However, throughout the region, there are areas of both increase and decrease in both parameters throughout the region.

Southern Europe (35°-50°N 10°W- 45°E)
The warming is about 2°C in winter and varies from 2 to 3°C in summer. There is some indication of increased precipitation in winter, but summer precipitation decreases by 5 to 15%, and summer soil moisture by 15 to 25%.

Australia (12°-45°S 110°-115°E)
The warming ranges from 1 to 2°C in summer and is about 2°C in winter. Summer precipitation increases by around 10%, but the models do not produce consistent estimates of the changes in soil moisture. The area averages hide large variations at the sub-continental level.

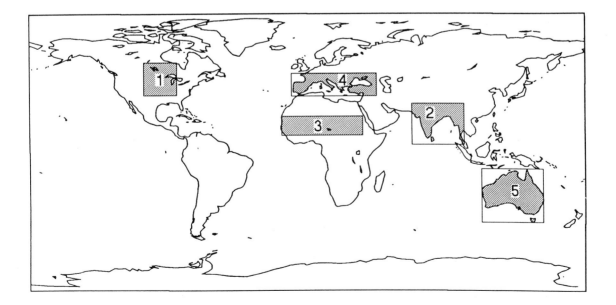

Figure 10: Map showing the locations and extents of the five areas selected by IPCC

Models predict that surface air will warm faster over land than over oceans, and a minimum of warming will occur around Antarctica and in the northern North Atlantic region.

There are some continental-scale changes which are consistently predicted by the highest resolution models and for which we understand the physical reasons. The warming is predicted to be 50-100% greater than the global mean in high northern latitudes in winter, and substantially smaller than the global mean in regions of sea-ice in summer. Precipitation is predicted to increase on average in middle and high latitude continents in winter (by some 5 - 10% over 35-55°N).

Five regions, each a few million square kilometres in area and representative of different climatological regimes, were selected by IPCC for particular study (see Figure 10). In the box (over page) are given the changes in temperature, precipitation and soil moisture, which are predicted to occur by 2030 on the Business-as-Usual scenario, as an average over each of the five regions. There may be considerable variations within the regions. In general, confidence in these regional estimates is low, especially for the changes in precipitation and soil moisture, but they are examples of our best estimates. We cannot yet give reliable regional predictions at the smaller scales demanded for impacts assessments.

WHAT TOOLS DO WE USE TO PREDICT FUTURE CLIMATE, AND HOW DO WE USE THEM?

The most highly developed tool which we have to predict future climate is known as a **general circulation model or GCM**. These models are based on the laws of physics and use descriptions in simplified physical terms (called parameterisations) of the smaller-scale processes such as those due to clouds and deep mixing in the ocean. In a climate model an atmospheric component, essentially the same as a weather prediction model, is coupled to a model of the ocean, which can be equally complex.

Climate forecasts are derived in a different way from weather forecasts. A weather prediction model gives a description of the atmosphere's state up to 10 days or so ahead, starting from a detailed description of an initial state of the atmosphere at a given time. Such forecasts describe the movement and development of large weather sys-tems, though they cannot represent very small scale phenomena; for example, individual shower clouds.

To make a climate forecast, the climate model is first run for a few (simulated) decades. The statistics of the model's output is a description of the model's simulated climate which, if the model is a good one, will bear a close resemblance to the climate of the real atmosphere and ocean. The above exercise is then repeated with increasing concentrations of the greenhouse gases in the model. The differences between the statistics of the two simulations (for example in mean temperature and interannual variability) provide an estimate of the accompanying climate change.

The long term change in **surface air temperature** following a doubling of carbon dioxide (referred to as the **climate sensitivity**) is generally used as a benchmark to compare models. The range of results from model studies is 1.9 to 5.2°C. Most results are close to 4.0°C but recent studies using a more detailed but not necessarily more accurate representation of cloud processes give results in the lower half of this range. Hence the models results do not justify altering the previously accepted range of 1.5 to 4.5°C.

Although scientists are reluctant to give a single best estimate in this range, it is necessary for the presentation of climate predictions for a choice of best estimate to be made. Taking into account the model results, together with observational evidence over the last century which is suggestive of the climate sensitivity being in the lower half of the range, (see section: "Has man already begun to change global climate?") a value of climate sensitivity of 2.5°C has been chosen as the best estimate. Further details are given in Section 5 of the report.

In this Assessment, we have also used much simpler models, which simulate the behaviour of GCMs, to make predictions of the evolution with time of global temperature from a number of emission scenarios. These so-called box-diffusion models contain highly simplified physics but give similar results to GCMs when globally averaged.

A completely different, and potentially useful, way of predicting patterns of future climate is to search for periods in the past when the global mean temperatures were similar to those we expect in future, and then use the past spatial patterns as **analogues** of those which will arise in the future. For a good analogue, it is also necessary for the forcing factors (for example, greenhouse gases, orbital variations) and other conditions (for example, ice cover, topography, etc.) to be similar; direct comparisons with climate situations for which these conditions do not apply cannot be easily interpreted. Analogues of future greenhouse-gas-changed climates have not been found.

We cannot therefore advocate the use of palaeo-climates as predictions of regional climate change due to future increases in greenhouse gases. However, palaeo-climatological information can provide useful insights into climate processes, and can assist in the validation of climate models.

How will climate extremes and extreme events change?

Changes in the variability of weather and the frequency of extremes will generally have more impact than changes in the mean climate at a particular location. With the possible exception of an increase in the number of intense showers, there is no clear evidence that weather variability will change in the future. In the case of temperatures, assuming no change in variability, but with a modest increase in the mean, the number of days with temperatures above a given value at the high end of the distribution will increase substantially. On the same assumptions, there will be a decrease in days with temperatures at the low end of the distribution. So the number of very hot days or frosty nights can be substantially changed without any change in the variability of the weather. The number of days with a minimum threshold amount of soil moisture (for viability of a certain crop, for example) would be even more sensitive to changes in average precipitation and evaporation.

If the large-scale weather regimes, for instance depression tracks or anticyclones, shift their position, this would effect the variability and extremes of weather at a particular location, and could have a major effect. However, we do not know if, or in what way, this will happen.

Will storms increase in a warmer world?

Storms can have a major impact on society. Will their frequency, intensity or location increase in a warmer world?

Tropical storms, such as typhoons and hurricanes, only develop at present over seas that are warmer than about 26°C. Although the area of sea having temperatures over this critical value will increase as the globe warms, the critical temperature itself may increase in a warmer world. Although the theoretical maximum intensity is expected to increase with temperature, climate models give no consistent indication whether tropical storms will increase or decrease in frequency or intensity as climate changes; neither is there any evidence that this has occurred over the past few decades.

Mid-latitude storms, such as those which track across the North Atlantic and North Pacific, are driven by the equator-to-pole temperature contrast. As this contrast will probably be weakened in a warmer world (at least in the Northern Hemisphere), it might be argued that mid-latitude storms will also weaken or change their tracks, and there is some indication of a general reduction in day-to-day variability in the mid-latitude storm tracks in winter in model simulations, though the pattern of changes vary from model to model. Present models do not resolve smaller-scale disturbances, so it will not be possible to assess changes in storminess until results from higher resolution models become available in the next few years.

Climate change in the longer term

The foregoing calculations have focussed on the period up to the year 2100; it is clearly more difficult to make calculations for years beyond 2100. However, while the timing of a predicted increase in global temperatures has substantial uncertainties, the prediction that an increase will eventually occur is more certain. Furthermore, some model calculations that have been extended beyond 100 years suggest that, with continued increases in greenhouse climate forcing, there could be significant changes in the ocean circulation, including a decrease in North Atlantic deep water formation.

Other factors which could influence future climate

Variations in the output of **solar energy** may also affect climate. On a decadal time-scale solar variability and changes in greenhouse gas concentration could give changes of similar magnitudes. However the variation in solar intensity changes sign so that over longer time-scales the increases in greenhouse gases are likely to be more important. **Aerosols** as a result of volcanic eruptions can lead to a cooling at the surface which may oppose the greenhouse warming for a few years following an eruption. Again, over longer periods the greenhouse warming is likely to dominate.

Human activity is leading to an increase in aerosols in the lower atmosphere, mainly from sulphur emissions. These have two effects, both of which are difficult to quantify but which may be significant particularly at the regional level. The first is the direct effect of the aerosols on the radiation scattered and absorbed by the atmosphere. The second is an indirect effect whereby the aerosols affect the microphysics of clouds leading to an increased cloud reflectivity. Both these effects might lead to a significant regional cooling; a decrease in emissions of sulphur might be expected to increase global temperatures.

Because of long-period couplings between different components of the climate system, for example between ocean and atmosphere, the Earth's climate would still vary without being perturbed by any external influences. This **natural variability** could act to add to, or subtract from, any human-made warming; on a century time-scale this would be less than changes expected from greenhouse gas increases.

How much confidence do we have in our predictions?

Uncertainties in the above climate predictions arise from our imperfect knowledge of:

- future rates of human-made emissions
- how these will change the atmospheric concentrations of greenhouse gases
- the response of climate to these changed concentrations

EQUILIBRIUM AND REALISED CLIMATE CHANGE

When the radiative forcing on the earth-atmosphere system is changed, for example by increasing greenhouse gas concentrations, the atmosphere will try to respond (by warming) immediately. But the atmosphere is closely coupled to the oceans, so in order for the air to be warmed by the greenhouse effect, the oceans also have to be warmed; because of their thermal capacity this takes decades or centuries. This exchange of heat between atmosphere and ocean will act to slow down the temperature rise forced by the greenhouse effect.

In a hypothetical example where the concentration of greenhouse gases in the atmosphere, following a period of constancy, rises suddenly to a new level and remains there, the radiative forcing would also rise rapidly to a new level. This increased radiative forcing would cause the atmosphere and oceans to warm, and eventually come to a new, stable, temperature. A commitment to this **equilibrium** temperature rise is incurred as soon as the greenhouse gas concentration changes. But at any time before equilibrium is reached, the actual temperature will have risen by only part of the equilibrium temperature change, known as the **realised** temperature change.

Models predict that, for the present day case of an increase in radiative forcing which is approximately steady, the realised temperature rise at any time is about 50% of the committed temperature rise if the climate sensitivity (the response to a doubling of carbon dioxide) is 4.5°C and about 80% if the climate sensitivity is 1.5°C. If the forcing were then held constant, temperatures would continue to rise slowly, but it is not certain whether it would take decades or centuries for most of the remaining rise to equilibrium to occur.

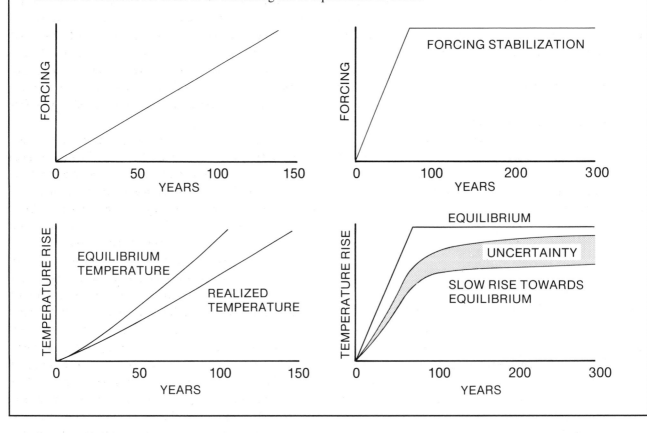

Firstly, it is obvious that the extent to which climate will change depends on the rate at which greenhouse gases (and other gases which affect their concentrations) are emitted. This in turn will be determined by various complex economic and sociological factors. Scenarios of future emissions were generated within IPCC WGIII and are described in the Annex to this Summary.

Secondly, because we do not fully understand the sources and sinks of the greenhouse gases, there are uncertainties in our calculations of future concentrations arising from a given emissions scenario. We have used a number of models to calculate concentrations and chosen a best estimate for each gas. In the case of carbon dioxide, for example, the concentration increase between 1990 and 2070 due to the Business-as-Usual emissions scenario spanned almost a factor of two between the highest and lowest model result (corresponding to a range in radiative forcing change of about 50%).

Furthermore, because natural sources and sinks of greenhouse gases are sensitive to a change in climate, they may substantially modify future concentrations (see earlier section: "Greenhouse gas feedbacks"). It appears that, as climate warms, these feedbacks will lead to an overall increase, rather than decrease, in natural greenhouse gas abundances. For this reason, climate change is likely to be greater than the estimates we have given.

Thirdly, climate models are only as good as our understanding of the processes which they describe, and this is far from perfect. The ranges in the climate predictions given above reflect the uncertainties due to model imperfections; the largest of these is cloud feedback (those factors affecting the cloud amount and distribution and the interaction of clouds with solar and terrestrial radiation), which leads to a factor of two uncertainty in the size of the warming. Others arise from the transfer of energy between the atmosphere and ocean, the atmosphere and land surfaces, and between the upper and deep layers of the ocean. The treatment of sea-ice and convection in the models is also crude. Nevertheless, for reasons given in the box overleaf, we have substantial confidence that models can predict at least the broad-scale features of climate change.

Furthermore, we must recognise that our imperfect understanding of climate processes (and corresponding ability to model them) could make us vulnerable to surprises; just as the human-made ozone hole over Antarctica was entirely unpredicted. In particular, the ocean circulation, changes in which are thought to have led to periods of comparatively rapid climate change at the end of the last ice age, is not well observed, understood or modelled.

Will the climate of the future be very different?

When considering future climate change, it is clearly essential to look at the record of climate variation in the past. From it we can learn about the range of natural climate variability, to see how it compares with what we expect in the future, and also look for evidence of recent climate change due to man's activities.

Climate varies naturally on all time-scales from hundreds of millions of years down to the year-to-year. Prominent in the Earth's history have been the 100,000 year glacial-interglacial cycles when climate was mostly cooler than at present. Global surface temperatures have typically varied by 5-7°C through these cycles, with large changes in ice volume and sea level, and temperature changes as great as 10-15°C in some middle and high latitude regions of the Northern Hemisphere. Since the end of the last ice age, about 10,000 years ago, global surface temperatures have probably fluctuated by little more than 1°C. Some

fluctuations have lasted several centuries, including the Little Ice Age which ended in the nineteenth century and which appears to have been global in extent.

The changes predicted to occur by about the middle of the next century due to increases in greenhouse gas concentrations from the Business-as-Usual emissions will make global mean temperatures higher than they have been in the last 150,000 years.

The **rate of change** of global temperatures predicted for Business-as-Usual emissions will be greater than those which have occured naturally on Earth over the last 10,000 years, and the rise in sea level will be about three to six times faster than that seen over the last 100 years or so.

Has man already begun to change the global climate?

The instrumental record of **surface temperature** is fragmentary until the mid-nineteenth century, after which it slowly improves. Because of different methods of measurement, historical records have to be harmonised with modern observations, introducing some uncertainty. Despite these problems we believe that a real warming of the globe of 0.3°C - 0.6°C has taken place over the last century; any bias due to urbanisation is likely to be less than 0.05°C.

Moreover since 1900 similar temperature increases are seen in three independent data sets: one collected over land and two over the oceans. Figure 11 shows current estimates of smoothed global-mean surface temperature over land and ocean since 1860. Confidence in the record has been increased by their similarity to recent satellite measurements of mid-tropospheric temperatures.

Although the overall temperature rise has been broadly similar in both hemispheres, it has not been steady, and differences in their rates of warming have sometimes persisted for decades. Much of the warming since 1900 has been concentrated in two periods, the first between about 1910 and 1940 and the other since 1975; the five warmest years on record have all been in the 1980s. The Northern Hemisphere cooled between the 1940s and the early 1970s when Southern Hemisphere temperatures stayed nearly constant. The pattern of global warming since 1975 has been uneven with some regions, mainly in the northern hemisphere, continuing to cool until recently. This regional diversity indicates that future regional temperature changes are likely to differ considerably from a global average.

The conclusion that global temperature has been rising is strongly supported by the retreat of most **mountain glaciers** of the world since the end of the nineteenth century and the fact that global **sea level** has risen over the same period by an average of 1 to 2mm per year. Estimates of thermal expansion of the oceans, and of increased melting of mountain glaciers and the ice margin in West

CONFIDENCE IN PREDICTIONS FROM CLIMATE MODELS

What confidence can we have that climate change due to increasing greenhouse gases will look anything like the model predictions? Weather forecasts can be compared with the actual weather the next day and their skill assessed; we cannot do that with climate predictions. However, there are several indicators that give us some confidence in the predictions from climate models.

When the latest atmospheric models are run with the present atmospheric concentrations of greenhouse gases and observed boundary conditions their simulation of present climate is generally realistic on large scales, capturing the major features such as the wet tropical convergence zones and mid-latitude depression belts, as well as the contrasts between summer and winter circulations. The models also simulate the observed variability; for example, the large day-to-day pressure variations in the middle latitude depression belts and the maxima in interannual variability responsible for the very different character of one winter from another both being represented. However, on regional scales (2,000km or less), there are significant errors in all models.

Overall confidence is increased by atmospheric models' generally satisfactory portrayal of aspects of variability of the atmosphere, for instance those associated with variations in sea surface temperature. There has been some success in simulating the general circulation of the ocean, including the patterns (though not always the intensities) of the principal currents, and the distributions of tracers added to the ocean.

Atmospheric models have been coupled with simple models of the ocean to predict the equilibrium response to greenhouse gases, under the assumption that the model errors are the same in a changed climate. The ability of such models to simulate important aspects of the climate of the last ice age generates confidence in their usefulness. Atmospheric models have also been coupled with multi-layer ocean models (to give coupled ocean-atmosphere GCMs) which predict the gradual response to increasing greenhouse gases. Although the models so far are of relatively coarse resolution, the large-scale structures of the ocean and the atmosphere can be simulated with some skill. However, the coupling of ocean and atmosphere models reveals a strong sensitivity to small-scale errors which leads to a drift away from the observed climate. As yet, these errors must be removed by adjustments to the exchange of heat between ocean and atmosphere. There are similarities between results from the coupled models using simple representations of the ocean and those using more sophisticated descriptions, and our understanding of such differences as do occur gives us some confidence in the results.

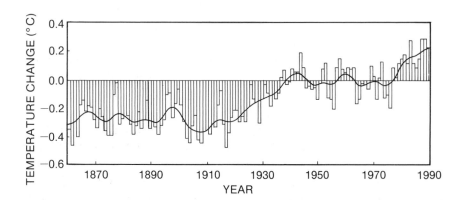

Figure 11: Global-mean combined land-air and sea-surface temperatures, 1861 - 1989, relative to the average for 1951-80.

Greenland over the last century, show that the major part of the sea level rise appears to be related to the observed global warming. This apparent connection between observed sea level rise and global warming provides grounds for believing that future warming will lead to an acceleration in sea level rise.

The size of the warming over the last century is broadly consistent with the predictions of climate models, but is also of the same magnitude as natural climate variability. If the sole cause of the observed warming were the human-made greenhouse effect, then the implied climate sensitivity would be near the lower end of the range

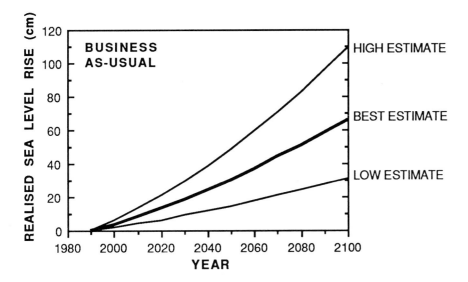

Figure 12: Sea level rise predicted to result from Business-as-Usual emissions, showing the best-estimate and range.

inferred from the models. The observed increase could be largely due to natural variability; alternatively this variability and other man-made factors could have offset a still larger man-made greenhouse warming. The unequivocal detection of the enhanced greenhouse effect from observations is not likely for a decade or more, when the committment to future climate change will then be considerably larger than it is today.

Global-mean temperature alone is an inadequate indicator of greenhouse-gas-induced climatic change. Identifying the causes of any global-mean temperature change requires examination of other aspects of the changing climate, particularly its spatial and temporal characteristics - the man-made climate change "signal". Patterns of climate change from models such as the Northern Hemisphere warming faster than the Southern Hemisphere, and surface air warming faster over land than over oceans, are not apparent in observations to date. However, we do not yet know what the detailed "signal" looks like because we have limited confidence in our predictions of climate change patterns. Furthermore, any changes to date could be masked by natural variability and other (possibly man-made) factors, and we do not have a clear picture of these.

How much will sea level rise ?

Simple models were used to calculate the rise in sea level to the year 2100; the results are illustrated below. The calculations necessarily ignore any long-term changes, unrelated to greenhouse forcing, that may be occurring but cannot be detected from the present data on land ice and the ocean. The sea level rise expected from 1990-2100 under

the IPCC Business-as-Usual emissions scenario is shown in Figure 12. An average rate of global mean sea level rise of about 6cm per decade over the next century (with an uncertainty range of 3 - 10cm per decade). The predicted rise is about 20cm in global mean sea level by 2030, and 65cm by the end of the next century. There will be significant regional variations.

The best estimate in each case is made up mainly of positive contributions from thermal expansion of the oceans and the melting of glaciers. Although, over the next 100 years, the effect of the Antarctic and Greenland ice sheets is expected to be small, they make a major contribution to the uncertainty in predictions.

Even if greenhouse forcing increased no further, there would still be a commitment to a continuing sea level rise for many decades and even centuries, due to delays in climate, ocean and ice mass responses. As an illustration, if the increases in greenhouse gas concentrations were to suddenly stop in 2030, sea level would go on rising from 2030 to 2100, by as much again as from 1990-2030, as shown in Figure 13.

Predicted sea level rises due to the other three emissions scenarios are shown in Figure 14, with the Business-as-Usual case for comparison; only best-estimate calculations are shown.

The West Antarctic Ice Sheet is of special concern. A large portion of it, containing an amount of ice equivalent to about 5m of global sea level, is grounded far below sea level. There have been suggestions that a sudden outflow of ice might result from global warming and raise sea level quickly and substantially. Recent studies have shown that individual ice streams are changing rapidly on a decade-to-century time-scale; however this is not necessarily related

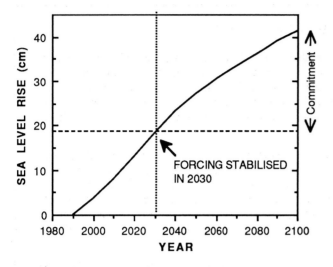

Figure 13: Commitment to sea level rise in the year 2030. The curve shows the sea level rise due to Business-as-Usual emissions to 2030, with the additional rise that would occur in the remainder of the century even if climate forcing was stabilised in 2030.

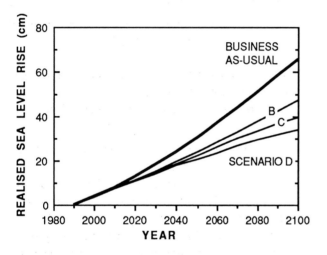

Figure 14: Model estimates of sea-level rise from 1990-2100 due to all four emissions scenarios.

to climate change. Within the next century, it is not likely that there will be a major outflow of ice from West Antarctica due directly to global warming.

Any rise in sea level is not expected to be uniform over the globe. Thermal expansion, changes in ocean circulation, and surface air pressure will vary from region to region as the world warms, but in an as yet unknown way. Such regional details await further development of more realistic coupled ocean-atmosphere models. In addition, vertical land movements can be as large or even larger than changes in global mean sea level; these movements have to

be taken into account when predicting local change in sea level relative to land.

The most severe effects of sea level rise are likely to result from extreme events (for example, storm surges) the incidence of which may be affected by climatic change.

What will be the effect of climate change on ecosystems?

Ecosystem processes such as photosynthesis and respiration are dependent on climatic factors and carbon dioxide concentration in the short term. In the longer term, climate and carbon dioxide are among the factors which control ecosystem structure, i.e., species composition, either directly by increasing mortality in poorly adapted species, or indirectly by mediating the competition between species. Ecosystems will respond to local changes in temperature (including its rate of change), precipitation, soil moisture and extreme events. Current models are unable to make reliable estimates of changes in these parameters on the required local scales.

Photosynthesis captures atmospheric carbon dioxide, water and solar energy and stores them in organic compounds which are then used for subsequent plant growth, the growth of animals or the growth of microbes in the soil. All of these organisms release carbon dioxide via respiration into the atmosphere. Most land plants have a system of photosynthesis which will respond positively to increased atmospheric carbon dioxide ("the carbon dioxide fertilization effect") but the response varies with species. The effect may decrease with time when restricted by other ecological limitations, for example, nutrient availability. It should be emphasized that the carbon content of the terrestrial biosphere will increase only if the forest ecosystems in a state of maturity will be able to store more carbon in a warmer climate and at higher concentrations of carbon dioxide. We do not yet know if this is the case.

The response to increased carbon dioxide results in greater efficiencies of water, light and nitrogen use. These increased efficiencies may be particularly important during drought and in arid/semi-arid and infertile areas.

Because species respond differently to climatic change, some will increase in abundance and/or range while others will decrease. Ecosystems will therefore change in structure and composition. Some species may be displaced to higher latitudes and altitudes, and may be more prone to local, and possibly even global, extinction; other species may thrive.

As stated above, ecosystem structure and species distribution are particularly sensitive to the rate of change of climate. We can deduce something about how quickly global temperature has changed in the past from palaeoclimatological records. As an example, at the end of the last glaciation, within about a century, temperature increased by up to 5°C in the North Atlantic region, mainly in Western

---

DEFORESTATION AND REFORESTATION

Man has been deforesting the Earth for millennia. Until the early part of the century, this was mainly in temperate regions, more recently it has been concentrated in the tropics. Deforestation has several potential impacts on climate: through the carbon and nitrogen cycles (where it can lead to changes in atmospheric carbon dioxide concentrations), through the change in reflectivity of terrain when forests are cleared, through its effect on the hydrological cycle (precipitation, evaporation and runoff) and surface roughness and thus atmospheric circulation which can produce remote effects on climate.

It is estimated that each year about 2 Gt of carbon (GtC) is released to the atmosphere due to tropical deforestation. The rate of forest clearing is difficult to estimate; probably until the mid-20th century, temperate deforestation and the loss of organic matter from soils was a more important contributor to atmospheric carbon dioxide than was the burning of fossil fuels. Since then, fossil fuels have become dominant; one estimate is that around 1980, 1.6 GtC was being released annually from the clearing of tropical forests, compared with about 5 GtC from the burning of fossil fuels. If all the tropical forests were removed, the input is variously estimated at from 150 to 240 GtC; this would increase atmospheric carbon dioxide by 35 to 60 ppmv.

To analyse the effect of reforestation we assume that 10 million hectares of forests are planted each year for a period of 40 years, i.e., 4 million km^2 would then have been planted by 2030, at which time 1 GtC would be absorbed annually until these forests reach maturity. This would happen in 40-100 years for most forests. The above scenario implies an accumulated uptake of about 20 GtC by the year 2030 and up to 80 GtC after 100 years. This accumulation of carbon in forests is equivalent to some 5-10% of the emission due to fossil fuel burning in the Business-as-Usual scenario.

Deforestation can also alter climate directly by increasing reflectivity and decreasing evapotranspiration. Experiments with climate models predict that replacing all the forests of the Amazon Basin by grassland would reduce the rainfall over the basin by about 20%, and increase mean temperature by several degrees.

Europe. Although during the increase from the glacial to the current interglacial temperature simple tundra ecosystems responded positively, a similar rapid temperature increase applied to more developed ecosystems could result in their instability.

What should be done to reduce uncertainties, and how long will this take?

Although we can say that some climate change is unavoidable, much uncertainty exists in the prediction of global climate properties such as the temperature and rainfall. Even greater uncertainty exists in predictions of regional climate change, and the subsequent consequences for sea level and ecosystems. The key areas of scientific uncertainty are:

- **clouds:** primarily cloud formation, dissipation, and radiative properties, which influence the response of the atmosphere to greenhouse forcing
- **oceans:** the exchange of energy between the ocean and the atmosphere, between the upper layers of the ocean and the deep ocean, and transport within the ocean, all of which control the rate of global climate change and the patterns of regional change
- **greenhouse gases:** quantification of the uptake and release of the greenhouse gases, their chemical

reactions in the atmosphere, and how these may be influenced by climate change
- **polar ice sheets:** which affect predictions of sea level rise.

Studies of land surface hydrology, and of impact on ecosystems, are also important.

To reduce the current scientific uncertainties in each of these areas will require internationally coordinated research, the goal of which is to improve our capability to observe, model and understand the global climate system. Such a program of research will reduce the scientific uncertainties and assist in the formulation of sound national and international response strategies.

Systematic long-term **observations** of the system are of vital importance for understanding the natural variability of the Earth's climate system, detecting whether man's activities are changing it, parameterising key processes for models, and verifying model simulations. Increased accuracy and coverage in many observations are required. Associated with expanded observations is the need to develop appropriate comprehensive global information bases for the rapid and efficient dissemination and utilization of data. The main observational requirements are:

i) the maintenance and improvement of observations (such as those from satellites) provided by the World Weather Watch Programme of WMO

ii) the maintenance and enhancement of a programme of monitoring, both from satellite-based and surface-based instruments, of key climate elements for which accurate observations on a continuous basis are required, such as the distribution of important atmospheric constituents, clouds, the Earth's radiation budget, precipitation, winds, sea surface temperatures and terrestrial ecosystem extent, type and productivity

iii) the establishment of a global ocean observing system to measure changes in such variables as ocean surface topography, circulation, transport of heat and chemicals, and sea-ice extent and thickness

iv) the development of major new systems to obtain data on the oceans, atmosphere and terrestrial ecosystems using both satellite-based instruments and instruments based on the surface, on automated instrumented vehicles in the ocean, on floating and deep sea buoys, and on aircraft and balloons

v) the use of palaeo-climatological and historical instrumental records to document natural variability and changes in the climate system, and subsequent environmental response.

The **modelling** of climate change requires the development of global models which couple together atmosphere, land, ocean and ice models and which incorporate more realistic formulations of the relevant processes and the interactions between the different components. Processes in the biosphere (both on land and in the ocean) also need to be included. Higher spatial resolution than is currently generally used is required if regional patterns are to be predicted. These models will require the largest computers which are planned to be available during the next decades.

Understanding of the climate system will be developed from analyses of observations and of the results from model simulations. In addition, detailed studies of particular processes will be required through targetted observational campaigns. Examples of such field campaigns include combined observational and small-scale modelling studies for different regions, of the formation, dissipation, radiative, dynamical and microphysical properties of clouds, and ground-based (ocean and land) and aircraft measurements of the fluxes of greenhouse gases from specific ecosystems. In particular, emphasis must be placed on field experiments that will assist in the development and improvement of sub grid-scale parametrizations for models.

The required program of research will require unprecedented international cooperation, with the World Climate Research Programme (WCRP) of the World Meteorological Organization and International Council of Scientific Unions (ICSU), and the International Geosphere-Biosphere Programme (IGBP) of ICSU both playing vital

roles. These are large and complex endeavours that will require the involvement of all nations, particularly the developing countries. Implementation of existing and planned projects will require increased financial and human resources; the latter requirement has immediate implications at all levels of education, and the international community of scientists needs to be widened to include more members from developing countries.

The WCRP and IGBP have a number of ongoing or planned research programmes, that address each of the three key areas of scientific uncertainty. Examples include:

- **clouds**:
  International Satellite Cloud Climatology Project (ISCCP)
  Global Energy and Water Cycle Experiment (GEWEX).
- **oceans**:
  World Ocean Circulation Experiment (WOCE)
  Tropical Oceans and Global Atmosphere (TOGA).
- **trace gases**:
  Joint Global Ocean Flux Study (JGOFS)
  International Global Atmospheric Chemistry (IGAC)
  Past Global Changes (PAGES).

As research advances, increased understanding and improved observations will lead to progressively more reliable climate predictions. However considering the complex nature of the problem and the scale of the scientific programmes to be undertaken we know that rapid results cannot be expected. Indeed further scientific advances may expose unforeseen problems and areas of ignorance.

Time-scales for narrowing the uncertainties will be dictated by progress over the next 10-15 years in two main areas:

- Use of the fastest possible computers, to take into account coupling of the atmosphere and the oceans in models, and to provide sufficient resolution for regional predictions.
- Development of improved representation of small-scale processes within climate models, as a result of the analysis of data from observational programmes to be conducted on a continuing basis well into the next century.

Annex

EMISSIONS SCENARIOS FROM WORKING GROUP III OF THE
INTERGOVERNMENTAL PANEL ON CLIMATE CHANGE

The Steering Group of the Response Strategies Working Group requested the USA and the Netherlands to develop emissions scenarios for evaluation by the IPCC Working Group I. The scenarios cover the emissions of carbon dioxide (CO_2), methane (CH_4), nitrous oxide (N_2O), chlorofluorocarbons (CFCs), carbon monoxide (CO) and nitrogen oxides (NO_x) from the present up to the year 2100. Growth of the economy and population was taken common for all scenarios. Population was assumed to approach 10.5 billion in the second half of the next century. Economic growth was assumed to be 2-3% annually in the coming decade in the OECD countries and 3-5 % in the Eastern European and developing countries. The economic growth levels were assumed to decrease thereafter. In order to reach the required targets, levels of technological development and environmental controls were varied.

In the **Business-as-Usual scenario** (Scenario A) the energy supply is coal intensive and on the demand side only modest efficiency increases are achieved. Carbon monoxide controls are modest, deforestation continues until the tropical forests are depleted and agricultural emissions of methane and nitrous oxide are uncontrolled. For CFCs

the Montreal Protocol is implemented albeit with only partial participation. Note that the aggregation of national projections by IPCC Working Group III gives higher emissions (10 - 20%) of carbon dioxide and methane by 2025.

In **Scenario B** the energy supply mix shifts towards lower carbon fuels, notably natural gas. Large efficiency increases are achieved. Carbon monoxide controls are stringent, deforestation is reversed and the Montreal Protocol implemented with full participation.

In **Scenario C** a shift towards renewables and nuclear energy takes place in the second half of next century. CFCs are now phased out and agricultural emissions limited.

For **Scenario D** a shift to renewables and nuclear in the first half of the next century reduces the emissions of carbon dioxide, initially more or less stabilizing emissions in the industrialized countries. The scenario shows that stringent controls in industrialized countries combined with moderated growth of emissions in developing countries could stabilize atmospheric concentrations. Carbon dioxide emissions are reduced to 50% of 1985 levels by the middle of the next century.

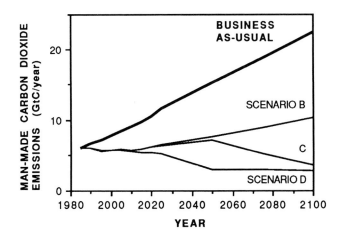 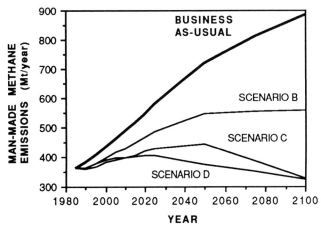

Man-made emissions of carbon dioxide and methane (as examples) to the year 2100, in the four scenarios developed by IPCC Working Group III.

Greenhouse Gases and Other Climate Forcing Agents

by Ulrich Siegenthaler *(Switzerland)* and Eugenio Sanhueza *(Venezuela)*

Abstract

The atmospheric concentrations of several greenhouse gases (CO_2, methane, CFCs and N_2O) have increased in the past centuries due to human activities. CO_2 is produced mainly by combustion of fossil fuels; 10–30% are contributed by tropical deforestation. The increase of methane is due to the cultivation of rice, cattle-rearing and other sources. The CFCs are purely anthropogenic. The time taken for the atmospheric concentrations of CO_2, the CFCs and N_2O to adjust to emission changes is long, 50–200 years. In order to stabilize the atmospheric concentration of the long-lived gases CO_2, CFCs and N_2O at their present levels, a reduction of emissions by 60–80% would be necessary; for methane the necessary reduction would be 15–20%.

The increasing greenhouse gas concentrations have changed the radiation balance; about 61% of the calculated change in radiative forcing since 1765 is due to CO_2, 23% to CH_4, 12% to CFCs and 4% to N_2O. At present, due to many uncertainties, it is difficult to estimate the effect of changes in ozone and aerosols. The concept of a Global Warming Potential (GWP) is being developed, which indicates (relative to CO_2) the radiative effect of the emission of a gas, integrated over a long period of time. This index could prove useful for assessing the climatic consequences of different policy options regarding the future of greenhouse gases .

1.   Introduction

The climate of the Earth depends on the global radiation balance, that is how solar radiation is scattered and absorbed and thermal infrared radiation is emitted and absorbed by the Earth's surface and atmosphere. Of particular importance for the radiation budget are radiatively active gases ("greenhouse gases") and aerosols. There is a natural greenhouse effect which already keeps the Earth's surface warmer by about 33°C than it would

otherwise be. Due to human activities, the concentrations of these gases have changed in the past decades and centuries to a degree that threatens to seriously perturb the balance between incoming solar and emitted infrared radiation and thus to change the global climate.

In this paper, the status of knowledge about the increase of greenhouse gases and their effects on the Earth's radiation balance are summarized. It is essentially a summary of the contents of Sections 1 and 2 of the Scientific Assessment of Climatic Change by the Intergovernmental Panel on Climatic Change (IPCC, 1990), and the reader is referred to that report for more details and for full references. In the following, the global cycles of the climate forcing agents will first be discussed, then their effects on the radiation balance (radiative forcing).

2.   Greenhouse Gases and Aerosols

The atmospheric concentration of several greenhouse gases such as carbon dioxide (CO_2), methane (CH_4) and nitrous oxide (N_2O) has increased since the industrial revolution. as shown by direct atmospheric measurements and - for times before about 1960—by analyses of air trapped in old polar ice. Furthermore, chlorofluorocarbons (CFCs), which did not exist in the unperturbed atmosphere, have been industrially produced and emitted to the atmosphere in the last few decades, contributing to the growing greenhouse effect. Table 1 gives a summary of the concentrations, rates of increase and contributions to the change in greenhouse effect of these gases. Many of these gases have long lifetimes in the atmosphere, so that even after a complete stop of man-made emissions, their concentrations would decrease only slowly, over decades to centuries, towards their pre-industrial values. The atmospheric lifetime of CH_4, N_2O and CFCs, given in Table 1, is defined as the ratio of atmospheric content to the total rate of removal. For CO_2, this definition cannot be applied; CO_2 is not removed from the system, but rather it is redistributed

Table 1: Key Greenhouse Gases influenced by Human Activities (from IPCC, 1990)

Parameter	CO_2	CH_4	CFC-11	CFC-12	N_2O
Pre-industrial atmospheric concentration [1]	280 ppmv	0.8 ppmv	0	0	288 ppbv
Current atmospheric concentration (1990)[1]	353 ppmv	1.72 ppmv	280 pptv	484 pptv	310 ppbv
Current annual rate of atmospheric increase	1.8 ppmv (0.5%)	0.015 ppmv (0.9%)	9.5 pptv (4%)	17 pptv (4%)	0.8 ppbv (0.25%)
Atmospheric lifetime (years) [2]	(50-200)	10	65	130	150
Greenhouse forcing (1765-1990, Wm^{-2})	1.50	0.56 [3]	0.06	0.14	0.10

(1)   ppmv = parts per million by volume;   ppbv = parts per billion by volume;
       pptv = parts per trillion by volume.
(2)   For each gas in the table, except CO_2, the "lifetime" is defined here as the ratio of the atmospheric content to the total
       rate of removal. This time scale also characterizes the rate of adjustment of the atmospheric concentrations if the
       emission rates are changed abruptly.
(3)   The value for the greenhouse forcing of CH_4 includes an estimated indirect effect of 0.14 W m^{-2} due to an increase of
       stratospheric water vapour from the oxidation of methane.

between atmosphere, ocean and land biota. Thus, the time it would take the carbon system to adjust towards a new equilibrium distribution is indicated.

Water vapour is the most important greenhouse gas. Its abundance in the global atmosphere is not directly affected by anthropogenic emissions of water vapour (e.g. by cooling towers), since these are very small compared to the natural fluxes by evaporation, precipitation and transport by winds. Changes in atmospheric water vapour concentration are expected to occur with a global warming; they must be considered as an internal feedback effect of the climate system, rather than an external forcing mechanism.

2.1   Carbon Dioxide

The atmospheric concentration of CO_2 is steadily rising. This has been documented by measurements at monitoring stations over the whole globe (e.g. Beardsmore and Pearman, 1987; Conway et al., 1988; Keeling et al., 1989). Since 1958, when the first continuous CO_2 monitoring programme started at Mauna Loa, Hawaii, and at the South Pole, the annual mean CO_2 concentration has increased

every year. Presently (in 1990), the global mean value is 353 ppmv, that is 25% above the pre-industrial value of 280 ppmv. The CO_2 increase started around 1800 A.D., as documented by analyses of air trapped in old polar ice (cf. Figure 1).

Carbon dioxide is continuously exchanged between atmosphere, ocean and the biota. Before the industrial revolution, these fluxes must have been well in balance, as indicated by a relatively constant atmospheric CO_2 content. This is documented by ice core measurements which indicate that in the period 900 to 1800 AD, the atmospheric concentration stayed constant, within 10 ppmv, at a level of 280 ppmv. Major natural variations occurred parallel to the large climatic shifts between glacial and postglacial time, with glacial CO_2 levels at about 190 ppmv, that is 70 percent of the postglacial level.

The CO_2 increase of the past two centuries has been caused by the growing emissions of CO_2 by fossil fuel combustion and deforestation. Presently, 5.7 ± 0.5 GtC (Gigatonnes of carbon; 1 Gt = 10^9 tonnes = 10^{12}kg) are emitted as CO_2 into the atmosphere annually by use of fossil fuels, plus 0.6–2.5 GtC due to deforestation. The

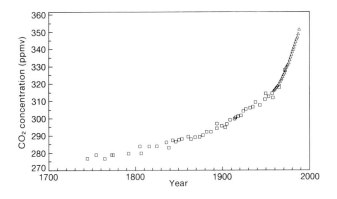

Figure 1: Atmospheric CO_2 increase in the past 250 years, as indicated by measurements on air trapped in old ice from Siple Station, Antarctica (squares; Neftel et al., 1985; Friedli et al., 1986) and by atmospheric measurements at Mauna Loa, Hawaii (triangles; Keeling et al., 1989).

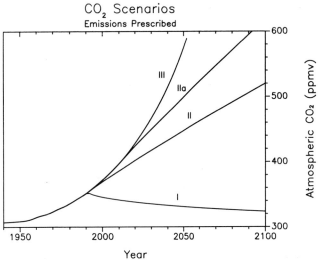

Figure 2: Future atmospheric CO_2 concentrations, calculated by means of a model of the global carbon cycle (Enting and Pearman, 1987), for the following scenarios for the CO_2 production: I, complete production stop in 1990; II, constant production rate from 1990 on; III, production rate grows by 2% per year from 1990 on. In scenario IIa, production rate grows by 2% per year from 1990 to 2010 and then remains constant.

cumulative input of CO_2 into the atmosphere until 1990 is estimated at about 220 GtC (± 10%) from fossil fuels (Marland, 1989), and at 120 GtC (± 30%) from deforestation and land use changes (Houghton and Skole, 1990). Therefore, the total amount of CO_2 produced by human activities corresponds to more than 50% of the pre-industrial atmospheric CO_2 amount (594 GtC).

The observed atmospheric CO_2 increase corresponds to roughly 50% of the estimated emissions, or, more precisely, to 48 ± 8% for the decade 1980–1989. A main sink for anthropogenic CO_2 is the ocean. Using carbon cycle models as well as measurements of the CO_2 partial pressure difference between atmosphere and oceans, the uptake by the ocean is estimated at 2.0 ± 0.8 GtC per year, or 29 ± 12% of the annual emissions of 7.0 ± 1.1 GtC. Thus, the sum of observed increase in the atmosphere and estimated uptake by the ocean is smaller than the estimated emissions by 0–3 GtC per year. This suggests that (1) the uptake of CO_2 by the oceans is underestimated, or (2) the land biota sequester carbon, for instance due to the fertilizing action of the increased CO_2 level, or (3) the amount of CO_2 released from tropical deforestation is at the very low end of current estimates, or some combination of these factors.

It has sometimes been speculated that the current CO_2 growth could be a natural fluctuation. This possibility must be dismissed based on the following arguments: First, the natural variability about the mean of 280 ppmv was small during the 1000 years before the industrial revolution (see above). A concentration as high as the present value of 350 ppm did not occur during the past 150,000 years (i.e. as far back in time as observational data go). Second, the observed CO_2 increase closely parallels the anthropogenic emissions. Since the start of atmospheric monitoring in

1958, the annual atmospheric increase has always been smaller than the fossil CO_2 input. This shows that oceans and biota together must have taken up carbon during all these years. Third, the observed trends of the carbon isotopes ^{13}C and ^{14}C in atmospheric CO_2 agree with those expected due to the emissions of fossil and biospheric CO_2 into the atmosphere. CO_2 from these emissions carries isotopic signals different from atmospheric CO_2, by which its pathway in the various carbon reservoirs can be followed.

Future atmospheric CO_2 concentrations depend primarily on emission rates and on the uptake of CO_2 by the ocean and the land biota. For the sake of illustration, concentrations have been calculated, using a carbon cycle model, for several schematic emission scenarios (IPCC, 1990; Figure 2). If all emissions were completely stopped in 1990 (scenario I), the atmospheric concentrations would slowly decline and by 2100 reach again the level of 1970. If emissions were frozen in the future at their 1990 value (scenario II), CO_2 would still increase and reach a concentration higher than 500 ppmv in 2100 according to the model used. Case IIa in Figure 2 shows what would happen if emissions were allowed to grow by 2% per year till 2010 and then frozen at their 2010 value; comparison of cases II and IIa illustrates the problematic consequences of postponing measures for controlling emissions to a later date. To put it differently, delaying action will need more stringent measures (when they are eventually taken) if a specified concentration is not to be exceeded. Scenario III

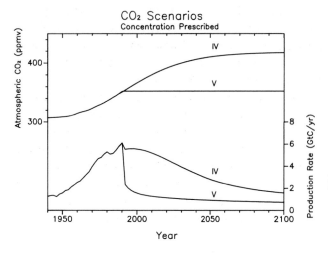

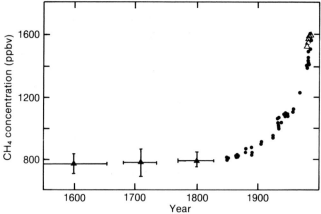

Figure 3: Future CO_2 production rates (lower curves) calculated by means of a carbon cycle model (Enting and Pearman, 1987) so as to yield prescribed atmospheric CO_2 concentrations (upper curves) after 1990. IV, concentration smoothly increasing to 420 ppmv (= 50% above pre-industrial); V, concentration constant after 1990.

Figure 4: Atmospheric methane concentrations in the past few centuries, as measured on air trapped in old polar ice (Pearman and Fraser, 1988). Reprinted by permission from NATURE vol. 332. Copyright © 1988 Macmillan Magazines Ltd.

assumes an exponential growth at a rate of 2% per year from 1990 to 2100. For comparison, the rate of growth after the Second World War till the oil crisis in 1973 was nearly constant at 4.6% per year.

A possible measure to keep future global warming under control would be to prescribe some CO_2 level that must not be exceeded. If, as a drastic assumption, the atmospheric CO_2 concentration should be stabilized at its 1990 level, global emissions would immediately have to be reduced by 60 to 80% (Figure 3, curves V)—obviously an illusory target. (For an immediate stabilization, emissions must first be reduced by an amount equal to the current growth in concentration, i.e. by about 50%, but then very soon more and more as the disequilibrium between atmosphere and ocean, and therefore the uptake by the ocean, decreases.) A less stringent scenario would be to allow an increase to 50% above the pre-industrial concentration (i.e. to 420 ppmv). In that case, global emissions would have to be stabilized immediately at present levels and then to be reduced slowly in the future (Figure 3, curves IV).

In turn, climate changes will affect the oceanic and biospheric carbon cycles. During the millenium preceding the industrial revolution, the concentration was relatively constant, with a variability of less than ± 10 ppmv (compared to an increase of 70 ppmv from 1800 to the present). This indicates that the atmospheric CO_2 concentration is not significantly influenced by minor climatic fluctuations such as the Little Ice Age (lasting roughly from the 16th to the mid-19th Century), during which temperatures were lower and glaciers advanced in many regions of the world. However, the anticipated

climatic and environmental changes may soon become large enough to have an impact on atmospheric CO_2 through changing ocean temperature and circulation or through processes on the continents such as accelerated decomposition of organic matter in soils. It is difficult to assess the possible feedback processes quantitatively, but it seems likely that there could be a positive effect, amplifying the man-made increase.

2.2   Methane

The concentration of CH_4 in the atmosphere is 1.7 ppmv presently, more than double its pre-industrial value, (see Table 1 and Figure 4). Ice core analyses indicate that it was relatively constant during the 2000 years preceding industrialization. During the ice age, the concentration was lower than during postglacial time by about a factor 2. The present CH_4 concentration is higher than at any time in the past 150,000 years. Methane has a number of natural and anthropogenic sources (Cicerone and Oremland, 1988). It is produced by decomposition under anaerobic conditions. The main anthropogenic sources that have increased are cultivation of rice and cattle-breeding; somewhat smaller but still significant sources are due to exploitation of natural gas and coal. biomass burning and land fills. It is difficult to assess the size of the anthropogenic sources of methane, and their estimated sum is somewhat lower than the value needed to balance the CH_4 atmospheric budget.

Methane is removed from the atmosphere by chemical processes. The main sink is reaction with the hydroxyl radical (OH) in the atmosphere. The OH concentration is controlled by a complex series of photochemical reactions involving CH_4, CO, non-methane hydrocarbons, NO_x and tropospheric O_3. Globally, the efficiency of this sink may have decreased during the last century, because of a supposed reduction of the OH concentration in response to increasing levels of trace gases. A minor sink for CH_4 may be destruction in soils. The atmospheric lifetime of CH_4 is estimated at 10 ± 2 years, which is relatively short compared to other anthropogenic greenhouse gases. Hence, CH_4 concentrations react relatively fast to changes in emissions. In order to stabilize the concentration at the present level, an immediate reduction in global man-made emissions by 15–20% would be necessary.

Methane emissions from natural sources might react in response to changing climate. The CH_4 production in tropical natural wetlands and rice paddies depends sensitively on soil moisture and thus on changes in evaporation and precipitation; however, neither the size, nor the sign of the change is presently known. Furthermore, thawing of permafrost in high northern latitudes might lead to a release of methane now trapped in it. This might produce a large emission under a global warming, but the size of this source is very difficult to estimate.

2.3   Chlorofluorocarbons

The CFCs have been comprehensively evaluated and widely discussed because they are depleting the stratospheric ozone (WMO 1985, 1989 a, b). In addition. they are also important greenhouse gases. The 1990 atmospheric concentrations of the two most important CFCs are 280 pptv for CFC–11 (CCl_3F) and 484 pptv for CFC–12 (CCl_2F_2). The CFCs are exclusively of anthropogenic origin. They are used as aerosol propellants, refrigerants, foam blowing agents and solvents in industry and in households. Over the past decades these two and other CFCs have increased more rapidly (percentage-wise) than the other greenhouse gases, currently by 4% per year or more.

CFCs are used by Man, amongst other reasons, because they are chemically inert and stable. This property makes them also resistant to destruction once they have been released into the atmosphere. The only significant natural removal process is photochemical destruction in the stratosphere, which proceeds at a slow rate only. CFC–ll and CFC–12 have atmospheric lifetimes of 65 and of 130 years, respectively, other CFCs have similar or even longer lifetimes.

In recognition of the harmful effect of CFCs on the environment, many governments signed the "Montreal Protocol on Substances that Deplete the Ozone Layer" (UNEP, 1987), in which measures were agreed upon to freeze and reduce the production of a number of CFCs. Current negotiations about more stringent measures let it appear likely that emissions will be reduced considerably or stopped completely in the future. However, atmospheric conceentrations of CFCs will still be significantly high during at least the next century because of their long atmospheric lifetimes.

Certain hydrofluorocarbons (HFCs) and hydro–chlorofluorocarbons (HCFCs) are foreseen as replacements for the long-lived CFCs. These substitute gases are also radiatively active, but they have shorter atmospheric lifetimes than the CFCs, ranging from 1 to 40 years. Therefore, and because it is expected that the release of substitute gases will (initially) be lower than the release of substituted CFCs, the atmospheric increase of HFCs and HCFCs will be much lower for the next several decades than the increase of CFCs without such a substitution. However, if emissions of these substitutes grow, the atmospheric concentrations may become important for the earth's radiation balance in the next century.

2.4   Nitrous Oxide

The present-day concentration of N_2O is 310 ppbv, about 8 percent higher than during pre-industrial times. Nitrous oxide is produced from various biological sources. Soils are the dominant source, another important source are the oceans. A quantitative assessment of the emissions, particularly from soils, is difficult in view of the complexity of the processes (N_2O is produced, but can also be destroyed in soils) and because of the heterogeneity of conditions as well as the paucity of observational data.

The total annual N_2O emissions needed to account for the presently observed rate of increase amount to roughly 150% of the estimated pre-industrial source. However, the estimated size of the anthropogenic sources is smaller than required to explain the current rate of increase. Processes supposed to be responsible for the atmospheric increase of N_2O are production in soils from the application of nitrate and ammonium fertilizers and, less important, combustion of fossil fuels and biomass burning; the estimates are affected by large uncertainties.

The main atmospheric sink for N_2O is photochemical destruction in the stratosphere. The atmospheric lifetime of this gas is estimated at about 150 years. In order to stabilize the atmospheric N_2O concentration at its current level, an immediate reduction by 70–80% of the additional emissions that are in excess of the pre-industrial source would be necessary.

2.5   Ozone

Atmospheric ozone plays an important role in the chemistry of the stratosphere and troposphere, and it also influences the radiation balance. Stratospheric ozone, which shields the surface from harmful ultraviolet radiation, is produced and destroyed photochemically. It is endangered by the CFCs, which efficiently destroy ozone, and also by anthropogenic N_2O, if its increase continues. The Antarctic "ozone hole" in the southern hemisphere spring was an unexpected phenomenon; this dramatic ozone decrease has unequivocally been shown to be caused by the CFCs. In mid-latitudes of the northern hemisphere, a decrease in total stratospheric ozone of 3.4–5.1% during the past 20 years has been observed in winter, but no significant trend in summer (WMO, 1989a, b). Future changes of stratospheric ozone essentially depend on emissions of CFCs and also of CH_4 and N_2O. Assuming that future emissions would be controlled according to the Montreal Protocol, but not reduced more strongly, the predicted reduction of total stratospheric ozone for 2060 is 0–4% in the tropics and 4–12% in high latitudes (excluding the Antarctic ozone hole). If, as expected, the control measures are strengthened such as to eliminate the emissions of a number of CFCs by 2000, then similar ozone levels to those of today are predicted for 2060.

Tropospheric ozone is a secondary product of complex photochemical processes involving, amongst other substances, carbon monoxide (CO), CH_4, reactive nitrogen oxides ($NO_X = NO + NO_2$) and non-methane hydrocarbons (NMHC). CO, NO_X and NMHC are increasing due to human activities, in particular fossil fuel use and biomass burning. The concentration of tropospheric ozone depends in a complex way on the concentration of these precursor gases and on the intensity of sunlight. Tropospheric ozone is short-lived (lifetime: several weeks), and thus its distribution in space and time is rather heterogeneous.

In northern mid-latitudes, where most long-term ozone measurements have been made, annual mean tropospheric concentrations seem to be a factor of 2–3 higher now than a hundred years ago, and they have increased by roughly 1% per year during the past 20 years. In contrast, no increase or even a decrease is indicated by the scanty data series from the southern hemisphere. This is due to different anthropogenic emissions between both hemispheres, and it is mainly related to the concentration of NO_X in the atmosphere. For the radiative forcing, the O_3 change near the tropopause (upper troposphere and lower stratosphere, ca. 8–20 km altitude) is important. Data from these altitudes are insufficient to indicate a clear trend. In the lower stratosphere (below 25 km) O_3 has decreased.

Understanding the cycles of tropospheric ozone and its precursors is essential to understand the hydroxyl radical (OH), which controls the atmospheric lifetimes of CH_4 and the NMHCs. A continued increase of CO, which consumes OH, could lead to a reduction of the OH concentration, which in turn would increase the atmospheric lifetime, and thus the concentration, of CH_4.

2.6   *Aerosols*

Aerosol particles absorb and scatter solar and terrestrial radiation and can thus influence the climate. They also act as cloud condensation nuclei and are climatically important also in this way.

An important anthropogenic source of aerosols is SO_2, produced by combustion of fossil fuels and of biomass. SO_2 is rapidly converted to sulphuric acid, which forms small aerosol droplets. Aerosols have a lifetime of only days to weeks in the troposphere and are thus concentrated over industrialized regions. There are almost no long-term direct observations of aerosols in the atmosphere. Indirect evidence from visibility observations indicates an increasing trend during the past decades over North American and European regions.

The very short lifetime, compared to that of greenhouse gases, means that aerosol concentrations react instantaneously to changes in emissions.

3.   Changes of the Radiative Forcing of Climate

3.1   *The Earth's Radiation Balance*

A primary factor determining the Earth's climate is the balance between incoming solar radiation and outgoing thermal (infrared) radiation, emitted by Earth to space. During the present postglacial epoch, and before man significantly perturbed the atmospheric composition, the absorbed and the emitted radiation were very well in equilibrium on the global scale (although not on a regional scale). Figure 5 is a schematic representation of the global energy balance. Of 100 units of solar radiation impinging on the Earth's atmosphere, about 30 are reflected back to space by clouds and the surface (left-hand side); the albedo (reflectivity) of the planet Earth is 30%. The remaining 70 units are absorbed by the surface and the atmosphere and thus contribute to warming them. The amount of absorbed solar radiation per time thus depends on the intensity of the solar irradiance (the "solar constant")—which seems to vary within narrow limits only, at least over a time scale of a decade, according to direct observation—and on the Earth's albedo, which may vary because of natural or anthropogenic causes.

Corresponding to its temperature, the Earth emits at longer wavelengths, in the infrared (right-hand side in Figure 5). In this part of the spectrum, many atmospheric gases, like water vapour, CO_2, CH_4 etc., are not transparent (as for visible light), but absorb infrared radiation, forming an insulating cover. These radiatively active gases (or greenhouse gases) re-emit in the infrared in all directions; as a consequence, the surface receives an additional 95 units of radiation—more than the absorbed solar radiation. In this way, the surface is heated up to a higher temperature (15°C on a global average) than it would have without the greenhouse gases (about–18°C). Thus, the greenhouse effect has always been active, providing the climate to support Earth's life.

The hatched arrow in Figure 5 marked "convection" includes the fluxes of latent and of sensible heat between surface and atmosphere. These are internal processes of the climate system.

On a global average, 240 W/m^2 of sunlight are absorbed and - in a radiative equilibrium - also emitted to space. If

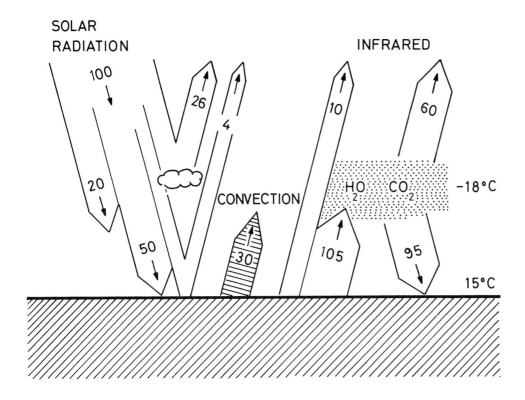

Figure 5: Schematic representation of the greenhouse effect. The total incoming solar radiation is arbitrarily set at 100 units.

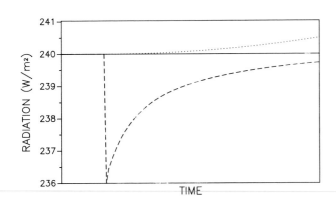

Figure 6: Schematic changes of the global radiation fluxes after a sudden CO_2 doubling. Solid line: absorbed solar radiation; dashed line: infrared radiation emitted to space. Immediately after the CO_2 doubling, the emitted infrared flux decreases by about 4 W m^{-2} (= radiative forcing). Consequently, the surface temperature, and therefore also the infrared flux increases. Dotted curve: due to the warming, ice and snow may melt, so that more solar radiation is absorbed.

the concentrations of greenhouse gases change, this equilibrium is perturbed; this is schematically shown in Figure 6 for an assumed sudden doubling of the CO_2 concentration. Before the perturbation, the absorbed solar flux (solid line) and the infrared flux emitted to space

(dashed line) are equal, at 240 W/m^2. Immediately after the CO_2 doubling, the infrared flux is reduced by about 4 W/m^2, while the absorbed solar flux remains constant (at least initially), so that there is a surplus of absorbed over emitted radiation by 4 W/m^2 (more precisely 4.4 W/m^2); we call this the radiative forcing. The Earth's surface and the troposphere now heat up and the infrared flux increases, until a new radiative balance is reached—but now with a warmer Earth. Based on climate model studies, using different models and different assumptions, the global mean surface temperature will, for a forcing of 4.4 W/m^2, change by 1.5–4.5°C, as discussed by Mitchell and Zeng (this conference). Here, we only consider the direct changes in the radiation balance.

As a consequence of the global warming, several feedback effects may become active. One of them is an increase in the atmospheric concentration of water vapour: the atmospheric water cycle tends to keep the relative humidity constant, therefore the absolute humidity (the water vapour concentration) increases as the temperature increases. Water vapour is the most important greenhouse gas, so this increase will amplify considerably the initial warming. Another feedback effect is a decrease of the Earth's albedo, due to a reduction in snow- and ice-covered areas when climate becomes warmer. This leads to an increase of absorbed solar radiation (dotted curve in Figure 6), which is again a positive feedback. Clouds are an important radiative agent, influencing the solar flux (via albedo) as well as the outgoing infrared flux (via absorption and emission). Changes in cloud cover or properties can thus sensitively affect the Earth's energy balance. All these feedbacks are part of the behaviour of the climate system and are not further considered in this article; the following discussion is restricted to the changes in the direct radiative forcing by greenhouse gases and other forcing agents.

3.2   *Radiative Forcing due to Greenhouse Gases*

The influence of a greenhouse gas on the radiation balance depends on several factors: on its atmospheric concentration, on its absorption strength and on the position in the Earth's infrared spectrum of the wavelengths where it absorbs, and on its atmospheric lifetime. As an example, CO_2 has a strong absorption band at 15 µm wavelength, near the maximum of the thermal emission of the surface. At the same time, its concentration is high enough that the atmosphere is very opaque at the centre of the 15 µm band and the absorption is nearly saturated. Therefore, an atmospheric CO_2 increase has little effect in the centre of this band, but at its edge, the effect is still significant. Consequently, when CO_2 increases, the radiative forcing of CO_2 does not increase proportionally to the CO_2 concentration; rather, it varies approximately as the logarithm of the concentration. On the other hand, the CFCs have very low concentrations and the total atmospheric absorption at their absorption bands is not strong. Therefore, their forcing is proportional to their concentration. The forcing of CH_4 and N_2O, which have concentrations intermediate between those of CO_2 and of the CFCs, varies approximately as the square root of the concentration. These dependencies have been derived by means of radiation transfer models that take into account the molecular optical properties, transfer of radiation in the atmosphere and interference of different gases by overlap of absorption bands (e.g. Ramanathan et al., 1987). A CO_2 doubling results in a radiative forcing of 4.4 W/m^2.

Besides the direct greenhouse forcing by anthropogenic gases, indirect effects occur. By the oxidation of methane, water vapour is produced in the stratosphere. It is assumed that this has led to a significant increase of stratospheric water vapour, although the observations are inadequate to confirm this. The suspected water vapour increase strengthens the greenhouse forcing. Stratospheric ozone absorbs ultraviolet radiation from the sun and thus heats the stratosphere; at the same time, tropospheric and stratospheric ozone absorbs and emits in the infrared. The overall effect of ozone depends on its vertical distribution; ozone changes in the upper troposphere and lower stratosphere are most effective in changing the radiative forcing.

3.3   *Changes in Radiative Forcing up to the Present and in the Future*

The radiative forcing corresponding to the concentration increases of greenhouse gases from 1765 to 1990, as calculated by IPCC (1990), is given in Table 1. In addition to the gases indicated in Table 1, other CFCs have contributed an estimated 0.085 W/m^2. This adds up to a total estimated forcing of 2.45 W/m^2 till now, 61% of which is due to CO_2, 23% to CH_4 (including the supposed effect of a stratospheric water vapour increase; without this, CH_4 would account for 18%), 12% to the CFCs and 4% to N_2O. Thus, CO_2 bears the main responsibility, but the contribution of the other greenhouse gases to the total forcing is comparable to that of CO_2. It should be noted that these numbers cannot be directly interpreted in terms of temperature increase, since the Earth is still warming up and a new radiative equilibrium has not yet been reached. Therefore, not only the total radiative forcing of a gas at the present time is important, but also its past rate of growth, and this has been different for the different gases. See the paper by Mitchell and Zeng (this conference) for more details.

Table 2: *Emissions Scenarios from Working Group III of IPCC*

BUSINESS-AS-USUAL SCENARIO:
The energy supply is coal intensive and on the demand side only modest efficiency increases are achieved. Carbon monoxide controls are modest, deforestation continues until the tropical forests are depleted and agricultural emissions of methane and nitrous oxide are uncontrolled. For CFCs the Montreal Protocol is implemented albeit with only partial participation.

SCENARIO B:
The energy supply mix shifts towards fuels containing less carbon, notably natural gas. Large efficiency increases are achieved. Carbon monoxide controls are stringent, deforestation is reversed and the Montreal Protocol implemented with full participation.

SCENARIO C:
A shift towards renewables and nuclear energy takes place in the second half of next century. CFCs are now phased out and agricultural emissions limited.

SCENARIO D:
A shift to renewable and nuclear energy in the first half of the next century reduces the emissions of carbon dioxide, initially more or less stabilizing emissions in the industrialized countries.

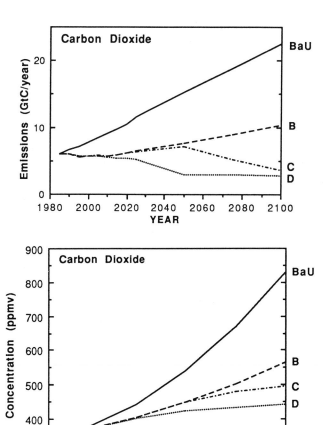

B, C and D assume a progressively stronger control on greenhouse gas emissions. Main aspects of these four scenarios are summarized in Table 2. Figure 7 shows, as examples, the assumed emissions and the resulting atmospheric concentrations of CO_2. Figure 8 shows the radiative forcing calculated for these scenarios, with the contributions of the individual gases. The effects of ozone are neglected. CO_2 remains the main contributor in all cases; for case BaU, its share always exceeds 60% of the total forcing. The relative contribution of the CFC substitutes (summarized as HCFC–22 in Figure 8) becomes significant by the middle of the next century, particularly for cases B, C and D.

3.4 A "Global Warming Potential" Index for Greenhouse Gases

For developing energy policies for the future, a simple way to compare the climatic impact of the emissions of different gases would be useful. This impact depends on the infrared absorption properties of the considered gas, but also on how long a given amount of the gas emitted will reside in the atmosphere, i.e. on its atmospheric lifetime. For example, the lifetimes of CO_2 and of the CFCs are about 10 times longer than that of CH_4 (Table 1), so that control measures for the former will become effective much more slowly than for CH_4.

A Global Warming Potential for a gas is defined here as

$$GWP = \frac{\int_0^{t_1} \Delta F(i)\, c(i)\, dt}{\int_0^{t_1} \Delta F(CO_2)\, c(CO_2)\, dt}$$

Figure 7: Assumed CO_2 emissions (top) for the four energy policy scenarios developed by IPCC Working Group 3, and corresponding atmospheric CO_2 concentrations (bottom) as calculated by means of a carbon cycle model (Siegenthaler and Oeschger, 1987).

The increase of tropospheric O_3 may have contributed an additional 10% to the forcing, although this estimate is uncertain. However, a decrease in lower stratospheric O_3 may have led to a reduction in radiative forcing that may have compensated for the effect of the tropospheric ozone increase. Therefore the net contribution of atmospheric ozone to the change in the greenhouse effect seems to have been small.

Based on assumptions on the future growth of economy and population and on energy policies, four scenarios for future emissions of greenhouse gases until 2100 have been developed by Working Group 3 of the Intergovernmental Panel on Climatic Change. The resulting future changes in atmospheric concentrations as well as their climatic effects have been estimated by means of different models by Working Group 1 of the IPCC (IPCC, 1990). The four scenarios are termed Business-as-Usual (BaU), B, C and D.

where $\Delta F(i)$ is the instantaneous radiative forcing due to an increase of the atmospheric amount of gas i by one mass unit (e.g. 1 kg) (see section 3.2 and Table 3) and $c(i)$ the concentration of the gas remaining in the atmosphere at time t, after the release of mass unit at time 0. The integration time, t_1, is chosen here as 100 years. The GWP thus gives a rough index of the climatic effect, integrated over 100 years, of gas i compared to that of CO_2. The index is only approximate, because the climatic impact is not related directly to the integrated radiative forcing, but also to the distribution in time of the forcing. The GWP thus does not account for transient effects.

The results for the main greenhouse gases are given in Table 3. It is seen that the forcing per unit mass is much larger for the CFCs than for CO_2, amongst other reasons because the 15 μm absorption band of CO_2 is near saturation. As the atmospheric lifetimes of the CFCs and of

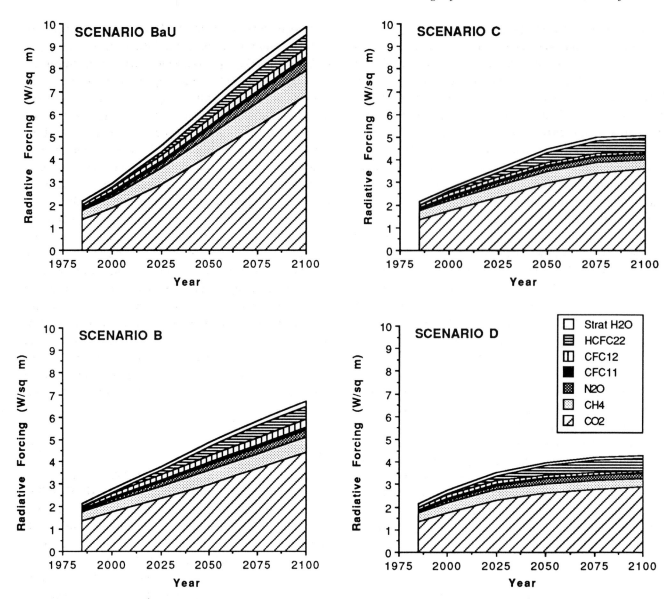

Figure 8: Future changes in radiative forcing due to the increase of greenhouse gases between 1985 and 2100, based on the four policy scenarios developed by IPCC Working Group 3. The contributions of the individual greenhouse gases are indicated.

Table 3: *Radiative forcing (ΔF) due to the addition of 1 mass unit of gas to the present atmosphere, and Global Warming Potential (GWP) for an integration time of 100 years. ΔF and GWP are given relative to CO_2. GWP. E(1990), with E(1990) = global emissions in 1990, provides a measure of the future climatic effect of present emissions (in percent of the total effect).*

	ΔF	GWP	GWP . E(1990)
CO_2	1	1	61%
CH_4	58	21	15%
CFC–11	3970	3500	2%
CFC–12	5750	7300	7%
N_2O	206	290	4%

CO_2 are similar, the GWP of the CFCs per unit mass emitted into the atmosphere is much larger than of CO_2 or CH_4. However, the absolute amounts annually emitted are much larger for the latter than for the CFCs, so the future climatic effect of the 1990 emissions is still largest for CO_2, and approximately the same for CH_4 and all CFCs together.

3.5 Other radiative agents

Besides increasing greenhouse gas concentrations, climate can also be influenced by aerosols—man-made and natural—by changes in albedo and by solar variations. Aerosols can have a direct effect, by scattering or absorbing radiation, or an indirect effect, as cloud condensation nuclei. It is not easy to determine even the sign of the changes due to aerosols. Depending on their optical properties, aerosols can either increase or decrease the albedo (they can be "brighter" or "darker" than the underlying surface). Over industrial regions, aerosols significantly reduce the solar irradiance at the surface. Some of the missing radiation is, however, absorbed in the air and thus warms the lower troposphere. However, on a larger spatial scale, conditions vary very much, and it is not possible to determine the global effect of anthropogenic aerosols. Aerosols can affect the optical density (the "whiteness") of clouds, acting as condensation nuclei. At present, it is hardly possible to determine whether this is an important effect for global climate.

Large volcanic eruptions inject sulphur dioxide (SO_2) into the stratosphere, which is then converted to sulphuric acid droplets. A slight decrease in mean surface temperature has been observed after major eruptions, obviously due to the reduction of solar radiation by the stratospheric aerosol. This aerosol is, however, removed from the stratosphere within a few years. Therefore, only a series of large volcanic eruptions within a short time can noticeably influence the climate on a decadal time scale. In the long run, volcanoes cause small year-to-year fluctuations of the temperature, but no long-term trend.

Desertification, deforestation and urbanization have increased the Earth's albedo. However, the albedo change over the last few decades has produced a radiative forcing of only about -0.03 W/m^2; therefore, it seems that this forcing agent is globally negligible compared with the effect of greenhouse gases.

If the energy output of the sun would vary significantly, this would have an immediate impact on the Earth's climate. Accurate measurements of the "solar constant" (i.e. the solar irradiance at the average distance of the Earth from the sun) have only become possible since 1978 from satellites, and they have shown a decrease by about 0.1% from 1980–86 and a slight increase afterwards. This variation seems to parallel the sunspot cycle, although the observational record is not yet long enough to state this with certainty. The decrease of the mean absorbed solar radiation by 0.1% or 0.2 W m^{-2} can be compared with a calculated change in radiative forcing from 1980–1990, due to the growth of greenhouse gases, of 0.5 W m^{-2}. Thus, the solar variations seem to be small compared to the anthropogenic radiative forcing—provided the satellite record is representative also for longer-term solar changes.

At present, there is no means to assess directly the long-term solar radiation variations. Based on a coincidence of very low sunspot numbers (the "Maunder minimum"), a high atmospheric $14C$ concentration and a cold climate in Europe around 1700, a relatively strong change of solar output correlated with sunspot numbers has been hypothesized. Over longer time scales, there is no convincing correlation between climate and $14C$ concentration measured in tree rings (as an indicator of solar activity). Alternatively, significant solar irradiance variations correlated with sunspot numbers have been postulated, because the trends of temperature and sunspots during the past 100 years were similar. Such a relation is, however, based on the extreme assumption that the recorded temperature increase is entirely due to solar variation, which is not realistic. The ideas about solar variations as a significant radiative forcing mechanism were originally put forward, based on indirect correlations, at a time before accurate observations of the solar constant became available. If the satellite observations are representative for the variability of the solar output also for other periods, then solar variations are too small—at least over a decadal time scale—to have a significant effect on climate.

Acknowledgements

This paper is essentially based on sections 1 and 2 of the Scientific Assessment of Climatic Change (IPCC, 1990). The lead authors of theses sections are R.T. Watson, H. Rodhe, H. Oeschger and U. Siegenthaler (section 1, Greenhouse Gases and Aerosols), and K. Shine, R.G. Derwent, D.J. Wuebbles and J.-J. Morcrette (section 2, Radiative Forcing of Climate). Numerous other scientists contributed to these sections; a full list of contributors is given in IPCC (1990). We profited much from the help of G. Jenkins, J. Ephraums and the IPCC staff who co-ordinated the work on the IPCC report.

References

Beardsmore, D.J. and G.I. Pearman, 1987: Atmospheric carbon dioxide measurements in the Australian region: Data from surface observatories. Tellus 39B, 459–476.

Cicerone, R., and R. Oremland, 1988: Biogeochemical aspects of atmospheric methane, Global Biochem. Cycles, 2, 299–327.

Conway, T.J., P. Tans, L.S. Waterman, K.W. Thoning, K.A. Masarie, and R.H. Gammon, 1988: Atmospheric carbon dioxide measurements in the remote global troposphere, 1981–1984, Tellus, 40B, 81–115.

Enting, I.G., and G.I. Pearman, 1987: Description of a one-dimensional carbon cycle model calibrated using techniques of constrained inversion, Tellus, 39B, 459–476.

Friedli, H., H. Loetscher, H. Oeschger, U. Siegenthaler, and B. Stauffer, 1986: Ice core record of the 13C/12C record of atmospheric CO_2 in the past two centuries, Nature, 324–238.

Houghton, R.A., and D.L. Skole. 1990: Changes in the global carbon cycle between 1700 and 1985, in The Earth Transformed by Human Action, B.L. Turner, ed., Cambridge University Press, in press.

IPCC, 1990: Intergovernmental Panel on Climate Change, Scientific Assessment of Climatic Change, World Meteorological Organization/United Nations Environmental Programme.

Keeling, C.D., R.B. Bacastow, A.F. Carter, S.C. Piper, T.P. Whorf, M. Heimann, W.G. Mook, and H. Roeloffzen, 1989: A three dimensional model of atmospheric CO_2 transport based on observed winds: 1. Analysis of observational data in: Aspects of climate variability in the Pacific and the Western Americas, D.H. Peterson (ed.), Geophysical Monograph, 55, AGU, Washington (USA), 165-236.

Marland, G., 1989: Fossil fuels CO_2 emissions: Three countries account for 50% in 1988. CDIAC Communications, Winter 1989, 1–4, Carbon Dioxide Information Analysis Center, Oak Ridge National Laboratory, USA.

Mitchell, J.F.B., and Q. Zeng, 1990: Climate Change Prediction, this volume.

Neftel, A., E. Moor, H. Oeschger, and B. Stauffer, 1985: Evidence from polar ice cores for the increase in atmospheric CO_2 in the past two centuries, Nature, 315, 45–47.

Pearman, G.I., and P.J. Fraser, 1988: Sources of increased methane, Nature, 332, 489–490.

Ramanathan, V., L. Callis, R. Cess, J. Hansen, I. Isaksen, W. Kuhn, A. Lacis, F. Luther, J. Mahlman, R. Reck and M. Schlesinger, 1987: Climate-chemical interactions and effects of changing atmospheric trace gases. Rev. Geophys., 25, 1441–1482.

Siegenthaler, U., and H. Oeschger, 1987: Biospheric CO_2 emissions during the past 200 years reconstructed by deconvolution of ice core data, Tellus, 39B, 140–154.

UNEP, 1987: Montreal Protocol on Substances that Deplete the Ozone Layer, UNEP conference services number 87–6106.

WMO, 1985: Atmospheric ozone 1985: Assessment of our understanding of the processes controlling its present distribution and change, Global Ozone Research and Monitoring Project, report 16, Geneva.

WMO, 1989a: Report of the NASA/WMO Ozone Trends Panel, 1989, Global Ozone Research and Monitoring Project, Report 18, Geneva.

WMO, 1989b: Scientific assessment of stratospheric ozone: 1989, Global Ozone Research and Monitoring Project, Report 20, Geneva.

Climate Change Prediction

by John F. B. Mitchell (*United Kingdom*) and Zeng Qingcun (*China*)

Abstract

A brief description of climate processes is given. Numerical models of climate are described and an assessment of their ability to simulate climate and climate change is given. Results from simulations with increased atmospheric CO_2 concentrations are given, including the temporal evolution of the warming and estimates of the geographical distribution of changes. The main sources of uncertainty are listed , and possible shortcomings in current research programs are identified.

1. Introduction: The climate system

In order to predict changes in climate, one must first identify and understand the various components of the climate system. These include the atmosphere, the oceans, the cryosphere (land-ice and sea-ice), the biosphere and (over geological timescales) the geosphere. The fundamental process driving climate is radiation. The climate system is heated by solar radiation, which is generally strongest in low latitudes, and cooled by long (thermal or infrared) radiation to space, which is more uniformly distributed with latitude. The resulting temperature contrast between low and high latitudes drives atmospheric and oceanic motions which transport heat polewards and thus tend to reduce the equator-to-pole temperature gradient. The resulting circulations and their interactions with orography and the biosphere determine the earth's climate.

Changes in the radiative heating of the earth will produce changes in its climate. Changes in solar heating can arise from external factors, such as variations in incident solar radiation due to changes in the earth's orbit, or in the earth's reflectivity, for example, due to addition of volcanic aerosols to the atmosphere, and to changes in the reflectivity of the surface such as that caused by deforestation. The longwave cooling can be modified by changes in atmospheric composition, notably by changes in the concentration of greenhouse gases. Other non-radiative factors such as changes in orography can also affect climate, but these occur on geological timescales, and so will not be considered further here. Note that changes in climate will themselves produce changes in the earth's radiation budget: these are known as climate feedbacks and are discussed later in this section.

In this paper, we are primarily interested in predicting changes in climates over the next century or so, hence it may be possible to neglect changes in some of the more slowly varying components of the system. The atmosphere and land surface, including seasonal snow cover, adjust to changes in heating on timescales of a year or less. The seasonal mixed layer of the ocean (typically up to 100 metres deep) and sea-ice respond over periods of up to a decade or so. The oceanic warm water sphere (down to the permanent thermocline, typically at half a kilometre) has a thermal response time of several decades, whereas the deep ocean, extending to several kilometres, has a timescale of centuries to millenia. The major ice-sheets vary over millennia, so apart from the substantial contributions from changes in accumulation and melting on sea-level, they can be regarded as fixed in the present context.

In predicting climate change, it is generally assumed that for each specification of the factors affecting climate (solar heating, atmospheric conditions, orography), the earth's climate will move to a long term steady state, known as the equilibrium climate, in which there is no net heating of the system. Note that this does not preclude considerable variability on interannual or even interdecadal timescales, but there are no longer term trends. If the factors affecting climate are changed (for example, by increasing the concentration of greenhouse gases), then the system will adjust slowly towards a new equilibrium state. The transition period is sometimes referred to as a "transient" or "time-dependent" climate change: the difference between the initial and final states is the "equilibrium climate change".

One possible method of predicting climate changes is to look for periods in the past when the factors affecting climate (solar parameters, greenhouse gas concentrations) were similar to those expected in the near future. One could then use reconstructions of the relevant past climates as a forecast for the future climate. Unfortunately, in the recent geological past, there are no close analogues to an increase in greenhouse gases, and so this method cannot provide reliable forecasts of climate over the next century or so.

The only alternative to the climate analogy approach is the use of physical-mathematical models of climate. Among such models, the most highly developed are general circulation models (GCMs). These are numerical models in which the equations of classical physics (including the laws of motion, and requirements for conservation of heat and mass) are solved in a three-dimensional grid, with a horizontal spacing of 250 to 800 km, depending on the model concerned.

Many processes (for example, those associated with cloud or vertical mixing in the oceans) occur on much smaller scales than can be resolved by the model grid, and so are represented in a simplified or idealized way (parameterisations). These may be based on approximations of the underlying equations, data from observational studies or laboratory experiments, results from numerical experiments using a finer grid, or as a last resort, sensitivity experiments using the GCM itself. In a climate model, an atmospheric component (essentially the same as a weather prediction model) is coupled to a model of the oceans and sea-ice, which may be equally complex.

Climate forecasts are derived in a different way from weather forecasts. A weather prediction model gives a description of the atmosphere's state up to 10 days or so ahead, starting from a detailed description of an initial state of the atmosphere at a given time. Such forecasts describe the movement and development of large weather systems, though they cannot represent very small scale phenomena; for example, individual shower clouds.

To make a climate forecast, the climate model is first run for a few (simulated) decades. The statistics of the model's output will be a description of the model's simulated climate which, if the model is a good one, will bear a close resemblance to the climate of the real atmosphere and ocean. The above exercise is then repeated with increasing concentrations of the greenhouse gases in the model. The differences between the statistics of the two simulations (for example in mean temperature and interannual variability) provide an estimate of the accompanying climate change.

In this paper, reference will be made to atmospheric models run with prescribed sea surface temperature (referred to henceforth as AGCMs), atmospheric models coupled to an oceanic mixed layer, with or without allowance for ocean heat transport (A/MLMs) and atmospheric models coupled to full dynamical ocean models (CGCMs).

As alluded to above, changes in climate can lead to changes in the components of the earth's radiation budget

which may be amplified (positive feedback) or reduced (negative feedback). The simulated strengths of feedbacks (such as those due to cloud) vary from model to model. The relative strengths of the radiative feedbacks in different models have been analysed by substituting the globally averaged changes in the simple energy balance equation

$$\Delta Ts = \Delta Q/\lambda$$

where ΔQ is the applied radiative perturbation (Wm^{-2}), ΔTs is the equilibrium change in globally averaged surface temperature, and λ is the climate sensitivity parameter (Wm^{-2} K^{-1}). For example, doubling CO_2 produces an increase of 4.4 Wm^{-2} in the radiative heating of the troposphere and surface. If the system responded as a radiative black body, this would produce an equilibrium warming of 1.2°C. Warming the atmosphere leads to an increase in atmospheric water content—water vapour is also a greenhouse gas, and estimates based on both observations (Raval and Ramanathan, 1989) and models indicate that it produces a positive feedback, increasing the warming to 1.7°C.

Snow and sea-ice reflect solar radiation back to space. The global warming associated with increases in greenhouse gases would reduce the areal extent of snow and ice, increasing solar absorption, leading to a further increase in temperature. The magnitude of this feedback is generally much smaller than that due to water vapour, though estimates of the strength vary from model to model.

Clouds cool climate through reflecting solar radiation back to space, and warm climate through their greenhouse effect. In our present climate, it appears that the solar effect dominates. Any change in cloud amount, height or cloud radiative properties will alter the net effect of clouds on the earth's radiation budget.

Numerical studies suggest that doubling CO_2 amounts would lead to an equilibrium warming of 2 to 5°C , much of the uncertainty being associated with uncertainties in the strength of cloud radiation feedbacks and associated processes (for example, Mitchell et al, 1989, Cess et al, 1989).

Some scientists have suggested that the climate sensitivity is much smaller than indicated by models. A few, including Idso, and Newell and Dopplick considered the energy balance at the surface only or neglected the water vapour feedback (see Luther and Cess, 1985, for references and a detailed explanation why such approaches are misleading). More recently, Lindzen (1990) and Ellsaesser (1989) have questioned the strength of the water vapour feedback in climate models. Lindzen argues that cumulus convection "short circuits " much of the potential water vapour feedback by transporting heat from the boundary layer to the upper troposphere. This process is undoubtedly represented in current general circulation models and provides, as expected, a strong negative feedback through changes in lapse rate (see for example Schlesinger and Mitchell, 1987). Lindzen and Ellsaesser also argue that as convection penetrates to higher levels in a warmer climate, the compensating subsidence will start from higher and therefore colder and drier levels. This would lead to a reduction in the absolute humidity in the upper troposphere in the decending branch of the Hadley circulation, and hence a local negative water vapour feedback . Since Lindzen has yet not published the details of his argument, it is not possible to assess it quantitatively. Nevertheless, this mechanism is also included in current models (see for example Gregory and Rowntree,1990).

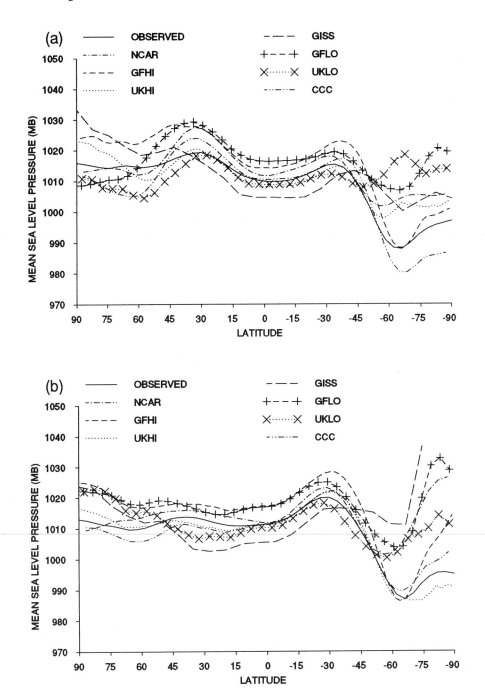

Figure 1: Zonally averaged sea-level pressure (mb) for observed (Schutz and Gates, 1971, 72) and models: (a) December, January, February; (b) June, July, August. NCAR - National Center for Atmospheric Research (15 spectral waves); GISS - Goddard Institute for Space Studies (8° x 10° latitude x longitude); GFLO - Geophysical Fluids Dynamics Laboratory (15 spectral waves); UKLO - United Kingdom Meteorological Office (5° x 7.5°); GFHI - Geophysical Fluid Dynamics Laboratory (30 spectral waves); UKHI - United Kingdom Meteorological Office (2.5° x 3.75°); CCC - Canadian Climate Centre (32 spectral waves). The more recent high resolution runs referred to in the text are GFHI, UKHI and CCC.

Indeed, on doubling CO_2 concentrations, the UK Meteorological Office high resolution model produces decreases in absolute humidity in this region (Senior, 1990, personal communication), but the magnitude and spatial extent of the drying is limited. It appears that other processes including the detrainment of warmer, moister air and lateral mixing of moistened air from the inner tropics and mid-latitudes reduce the extent of the drying through deeper subsidence. The increased drying through subsidence appears to be responsible for the decreases in relative humidity in mid-latitudes and the tropics which leads to the reduction in cloud in the upper troposphere and a strong positive cloud feedback found in many models (Mitchell and Ingram,1990). Thus, if models have exaggerated the positive water vapour feedback, it is also likely that they have underestimated the strength of the positive feedback associate with reductions in cloud amount. In summary, although tropical convection remains one of the major sources of uncertainty, it is unlikely that errors in the parameterization of convection lead to a gross overestimate of climate sensitivity. (Even removing the water vapour feedback completely and globally in a model with a sensitivity of 2.5°C due to doubling CO_2—the IPCC "best guess"—would only reduce the equilibrium warming to about 1.4°C.)

2.   Validation of numerical climate models

In this section, we attempt to answer the question "To what extent should one believe results from climate models?" First, the models are based on a firm physical basis with a minimum of adjustable parameters, as discussed in the previous section. Second, they show considerable skill in reproducing the large-scale features of current climate. Third, they have been shown to be capable of reproducing many features of contemporary climate change, and some of the main features in more recent paleoclimate. These last two points are expanded on below.

2.1   Simulation of present climate

The simplest way of validating a climate model is to prescribe present day "boundary conditions" and compare the long term statistics of the resulting simulation with observational climatologies. This shows that AGCMs and A/MLMs have considerable skill in the portrayal of the large scale distribution of pressure, temperature, wind and precipitation in both summer and winter, although this success is due in part to the constraints on sea surface temperature and sea-ice. There has been a general reduction in the errors in more recent AGCMS as a result of increased horizontal resolution, improvements to the parameterisation of convection, cloudiness and surface processes and the introduction of parameterisations of gravity wave drag. For example, the three most recent

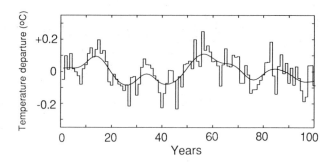

Figure 2: The temporal variation of the deviation of global mean surface air temperature (°C) of the GFDL coupled ocean atmosphere model from its long term average. (Manabe, 1990; personal communication).

simulations considered in the IPCC Working Group I Report show a marked improvement in the simulation of the depth and position of the Antarctic circumpolar surface pressure trough (Figure 1).

Changes in variability of climate are as important as changes in mean climate in the assessment of climate impacts. Hence, the ability of models to simulate the variability of current climate should also be assessed. The daily and interannual variability of temperature and precipitation have been examined but only to a limited extent. There is evidence that variability is overestimated in some models, especially in summer. The daily variability of sea-level pressure can be simulated well, but the eddy kinetic energy in the upper troposphere (indication of the variability of the flow) tends to be underestimated. The level of interannual variability of global mean surface temperature in CGCMs (for example, Figure 2) is comparable to that observed over similar timescales if allowances are made for the estimated trend in the observed data.

On regional scales, there are significant errors in all models. A validation of A/MLMs for five selected regions (typically 4×10^6 km^2) showed errors in area average surface temperature of 2 to 3°C (IPCC,1990). This is small compared with the average seasonal range of temperature of 15°C. Errors in mean precipitation for the same five regions ranged from 20 to 50% of the observed average. All the recent models reproduce the northern summer monsoon rainfall maximum over South East Asia (Figure 3), but in the example shown, the mean rainfall over South East Asia is substantially less than observed as the simulated rainband does not extend far enough to the north and east. Given the large errors in the simulation of present day regional climate, confidence in the simulated changes in regional climate must be low.

Models of the oceanic circulation simulate many of the observed large-scale features of ocean climate, especially in

(a) JJA PRECIPITATION: OBSERVED

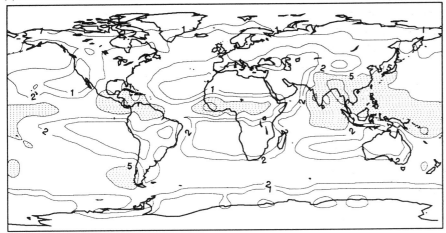

(b) JJA PRECIPITATION: UKHI

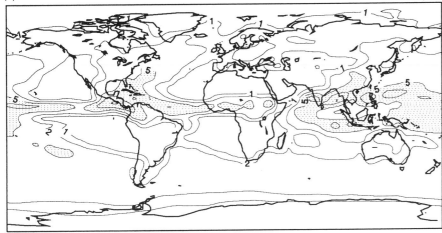

Figure 3: Precipitation (mm/day) for June, July, August: Contours at 1, 2, 5 and 10 mm/day, stippled where greater than 5mm/day: (a) Observed (Jaeger, 1976); (b) Simulated (United Kingdom Meteorological Office high (2.5° x 3.75°) resolution model.

low latitudes, although their solutions are sensitive to resolution and to the parameterisation of sub-grid scale processes such as mixing and convective overturning. It is particularly important that vertical mixing in the ocean is reproduced correctly as this determines the rate at which the ocean responds to increases in greenhouse gases. This can be validated to some extent by simulating the spread of passive tracers, such as that of tritium following the atomic bomb tests in the 1950s and 1960s. Current models are capable of reproducing the main features of the spread of passive tracers (for example, Figure 4), though there may be errors in the detailed changes (for example, in Figure 4b, underestimating the strength of penetration near 30 to 50°N). In CGCMs, it has proved necessary to add empirical adjustments to the ocean surface fluxes in order to reduce errors in the simulated climate.

2.2 Simulation of contemporary climate change and recent paleoclimate

AGCMs have been used to simulate the atmospheric response to prescribed anomalies in sea surface temperature (SST), and have been notably successful in simulating the response to tropical SST anomalies. Large scale positive SST anomalies occur in the eastern tropical Pacific during the occurrence of El Niño , and models have simulated successfully the associated observed anomalies in precipitation and circulation (for example, Fenessy and Shukla, 1988). Other AGCMs have reproduced the relationship between large scale SST distribution and summer rainfall in the Sahel (for example, Folland et al,1990).

A/MLMs have reproduced some of the large scale features of the mid-Holocene climate, only the changes in the earth's orbital parameters being specified, and also the last glacial maximum, with the distribution of land ice and

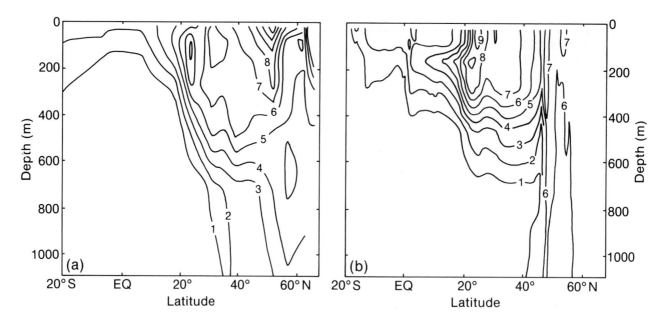

Figure 4: Tritium in the GEOSECS section in the Western North Atlantic approximately one decade after major bomb tests: (a) GEOSECS observations; (b) as predicted by a 12 level model (Sarmiento, 1983) in tritium units.

the prevailing CO_2 concentration being the only prescribed changes (for example, COHMAP members, 1988).

2.3   Summary

The most recent models are able to simulate the large scale features but not the regional (4000–2000 km) scale details of present climate. AGCMs reproduce the atmospheric response to contemporary SST anomalies, and A/MLMs are capable of reproducing the major changes in climate over the last 18000 years provided certain boundary conditions (CO_2 concentrations, orbital parameters, major ice sheets) are prescribed. This, combined with the physical basis on which the models are developed, gives us some confidence in their ability to predict future changes in climate, particularly on larger scales. The least reliable aspects of models are their treatment of sub-grid scale processes, and their ability to simulate variations in climate on a regional scale. Aspects of the simulation of present day climate by CGCMs may be inadequate unless corrective measures, such as adjustments to the surface fluxes, are made.

3.   Simulation of climate change

We now consider the simulation of climate change due to increases in greenhouse gases. Most of our understanding of greenhouse gas induced climate change is based on equilibrium experiments based on mixed layer models. In the last year or so, several studies have been made of the time-dependent response to increases in greenhouse gases using CGCMs. However, results presented here are based largely on the equilibrium experiments which have been

analysed in much greater detail. Furthermore, the rates of increase in the time-dependent coupled experiments do not correspond to the IPCC scenarios of increases in greenhouse gases. Hence, in order to calculate the evolution of temperature resulting from the IPCC scenarios, we have used results from one-dimensional upwelling diffusion models calibrated using the more complex A/MLMs and CGCMs.

3.1   Equilibrium climate sensitivity

The equilibrium sensitivity of mixed layer models to doubling CO_2 ranges from 2 to 5°C. The range is due to uncertainties associated with sub-grid scale para-meterization, particularly those associated with cloud. Some of the more recent studies attempt to allow for changes in the microphysical properties of cloud—these more detailed, but not necessarily more accurate, models give an equilibrium sensitivity of 2 to 4°C.

The observed change in temperature (1860–1990) seems to be consistent with an equilibrium warming of 1 to 3°C, allowing as far as possible for natural variability and assuming other factors such as the effect of sulphate emissions on cloud albedo have not affected the warming due to increases in greenhouse gases. On the basis of modelling studies, the sensitivity of climate due to doubling CO_2 is most likely to lie between 1.5 and 4.5°C, and, in view of the observational evidence, a "best guess" of 2.5°C was chosen to illustrate the IPCC scenarios.

3.2   Equilibrium and transient climate change

When the radiative heating of the earth-atmosphere system is changed by increases in greenhouse gases, the

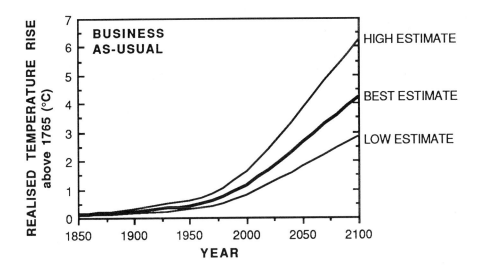

Figure 5: Evolution of global mean warming (above pre-industrial temperatures) assuming IPCC Business as Usual Scenario. The curves correspond to an equilibrium sensitivity to doubling CO_2 of 4.5, 2.5 and 1.5°C respectively. (IPCC, 1990).

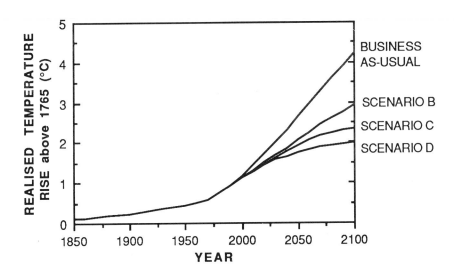

Figure 6: Evolution of global mean warming (above pre-industrial temperatures) assuming an equilibrium sensitivity to doubling CO_2 of 2.5°C for the IPCC scenarios (top to bottom), "Business as Usual", B, C and D. (IPCC,1990).

atmosphere will respond (by warming) immediately. The atmosphere is closely coupled to the oceans so the oceans also have to be warmed; because of their thermal capacity this takes decades or centuries. This exchange of heat between the atmosphere and ocean will act to slow down the temperature rise forced by the greenhouse effect.

Consider the concentration of greenhouse gases in the atmosphere rising to a new level and remaining constant thereafter. The radiative forcing would also rise rapidly to a new level. This increased radiative forcing would cause the atmosphere and oceans to warm, and tend towards a new equilibrium temperature. Commitment to this equilibrium temperature rise would occur as soon as the greenhouse gas concentration had changed. But at any time before equilibrium is reached, the actual temperature will only

have risen by part of the equilibrium temperature change—known as the realised temperature change.

One CGCM (Stouffer et al, 1989) predicts that, for the case of a steady increase in radiative forcing similar to that currently occuring, the realised temperature at any time is about 60–80% of the committed temperature. If the forcing were to be held constant, temperatures would continue to rise slowly, but it is not certain whether it would take decades or centuries for most of the remaining rise to equilibrium to occur.

3.3   *Changes in global mean temperature based on the IPCC scenarios*

The following results are derived from an upwelling—diffusing model (*) assuming that the temperature of surface water sinking in high latitudes does not change. The IPCC "business as usual" scenario, in which effective CO_2 concentrations double over pre-industrial levels by 2020, gives a warming of 1.3 to 2.6°C above pre-industrial levels at 2030, corresponding to a prescribed climate sensitivity of 1.5 to 4.5 °C (Figure 5). The "best guess" sensitivity gives a warming of 1.8°C at the time of doubling, which is reduced to about 1.5°C in the other scenarios (Figure 6). Scenario B assumes an effective doubling of CO_2 by 2040, scenario C by about 2050. Note that the changes in emission scenarios take some time to produce an effect, as a result of the slowness with which both the gas concentrations and the oceans respond. Much of the warming immediately after 1990 can be attributed to the system "catching up" the effect of previous emissions.

3.4   Patterns of climatic change due to an effective doubling of CO_2

Only one AGCM has been used in both an equilibrium experiment coupled to a mixed layer ocean, and in a long time-dependent experiment coupled to a deep ocean model (a CGCM, Stouffer et al 1989) (Figure 7). The equilibrium sensitivity of this model to doubling CO_2 is 4°C. In most regions heat is mixed down to the main thermocline, at about 500m; and the surface temperature change at the time of doubling CO_2 in the time-dependent experiment (Figure 7b) is similar to the equilibrium response (Figure 7a), but reduced by 20 to 40% (Figure 7c). The exceptions are around Antarctica, where there is little or no warming in the time-dependent experiment, and over the northern North Atlantic and North Western Europe, where the reduction is 40 to 60%. In these regions, a deep wind driven circulation (circumpolar ocean) or deep convection (N. Atlantic) mix heating down to several kilometres. The surface warming is greatest in high northern latitudes, and north of 30°S, is generally greater over land than over the ocean at the same latitudes, as found in equilibrium experiments. The warming in northern high latitudes is generally greatest around the sea-ice margins in late autumn and early winter, and around the snow-line over the continents in spring. Over the low latitude and moist areas of the low latitude continents, the warming is small relative to the global mean.

The changes in the hydrological cycle are qualitatively similar in these equilibrium and transient experiments. Hence, the remainder of this section is based on results from equilibrium experiments. Precipitation generally increases in high latitudes throughout the year and in mid-latitudes in summer (Figure 8). Precipitation generally increases in the tropics, these changes are accompanied by shifts in the main tropical rain bands, which vary from model to model, so there is little consistency in results for a particular region.

The warming produces an increase in global mean precipitation and evaporation. Enhanced precipitation increases soil moisture levels in high northern latitudes in winter. Most models produce a drying of the land surface over the northern mid-latitude continents in summer (for example, Figure 9) as a result of earlier snow melt, enhanced evaporation (and in some regions, reduced precipitation).

Predicted changes in the day to day variability of weather are uncertain. However, simulated episodes of high temperature become more frequent simply due to the substantial increases in mean temperature. There is some evidence of an increase in convective precipitation. Numerical experiments show a reduction in mid-latitude storms, but as current models only resolve the larger disturbances, this does not rule out the possibility of smaller but more intense storms. An increase in the maximum intensity of tropical cyclones might be expected on theoretical grounds due to the increased availability of latent heat, but changes in the intensity of tropical disturbances simulated in different A/MLMs are inconsistent.

3.5   Estimating regional climate change: Pre-industrial to 2030 (IPCC "business as usual" scenario)

The reader is reminded of the limited ability of current climate models to simulate regional climate change. In deriving the results below, the following assumptions have been made:

i)   The "best guess" of the magnitude of the global mean equilibrium increase in surface temperature at 2030 is 1.8°C (this is consistent with a climate sensitivity of 2.5°C)

ii)   The patterns of equilibrium and transient climate change are similar (this may be approximately true for the regions considered, but not for the high latitude Southern Ocean, see Figure 7)

iii) The regional changes in temperature, precipitation and soil moisture are proportional to the global mean changes in surface temperature (this will be approximately valid except in regions where the changes are associated with a shift in the position of steep gradients, for example where the snowline retreats, or on the edge of a rainbelt which is displaced).

Although it is hard to justify some of these assumptions on rigorous scientific grounds, the errors involved are substantially smaller than the uncertainties arising from the threefold range in climate sensitivity. The results below (IPCC,1990) are derived using equilibrium results from three high resolution models (see caption to Figure 1) and

---

(*) Upwelling velocity 4 my^{-1}, Diffusivity 0.63 cm^2 s^{-1}

(a) YEARS 60-80 OF TIME-DEPENDENT TEMPERATURE RESPONSE

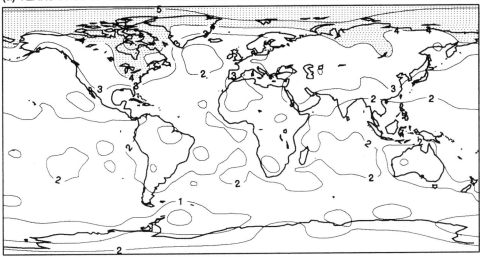

(b) EQUILIBRIUM TEMPERATURE RESPONSE

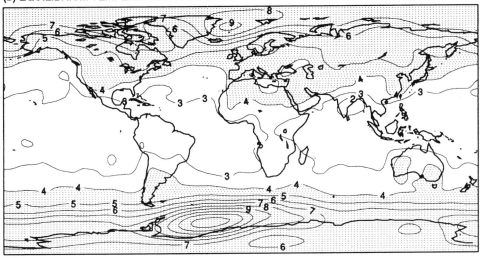

(c) RATIO OF TIME-DEPENDENT RESPONSE TO EQUILIBRIUM RESPONSE

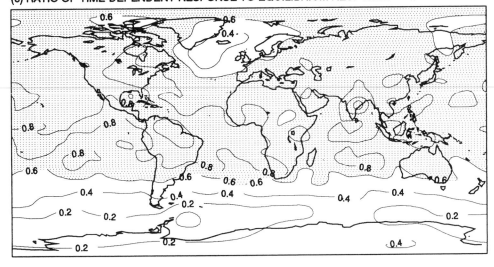

Figure 7: (a) The geographical distribution of the time- dependent response of surface air temperature (°C) in the GFDL coupled ocean-atmosphere model to a 1% per year increase of atmospheric CO_2. Shown is the difference between the 1% per year perturbation and the control run for the 60th–80th year period when the CO_2 approximately doubles. (Stouffer et al 1989); (b) The geographical distribution of the equilibrium response of surface air temperature (°C) in the atmosphere-mixed layer ocean model to a doubling of atmospheric carbon dioxide. (Stouffer et al,1989); (c) The geographical distribution of the ratio of time dependent to equilibrium responses shown above. (Manabe, personal communication).

(c) DJF 2 X CO2 - 1 X CO2 PRECIPITATION: UKHI

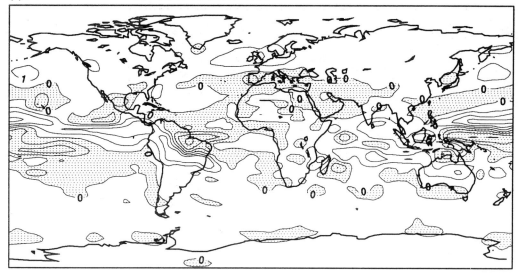

(f) JJA 2 X CO2 - 1 X CO2 PRECIPITATION: UKHI

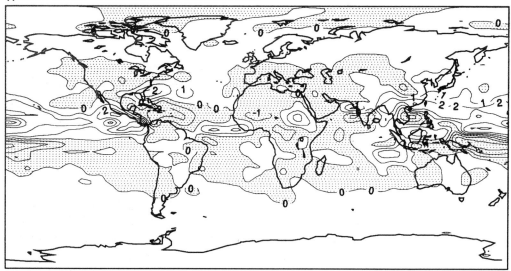

Figure 8: Simulated changes in precipitation rate at equilibrium following a doubling of atmospheric CO_2 concentrations, in the United Kingdom Meteorological Office 2.5° x 3.75° resolution model.(mm/day, contours at 0, ± 1, 2, 5mm/day) Areas of decrease stippled: (a) December, January, February; (b) June, July and August.

scaling them to give the appropriate change in global mean temperature. For a climate sensitivity of 1.5°C, the changes should be reduced by 30%, and for a climate sensitivity of 4.5°C they should be increased by 50%. All the changes below are averages over the region - the range arises because of the use of different models.

Central North America (35–50°N, 85–105°W)
The warming varies from 2 to 4°C in winter and 2 to 3°C in summer. Precipitation increases range from 0 to 15% in winter, whereas there are decreases of 5 to 10% in summer. Soil moisture decreases in summer by 15 to 20%.

South East Asia (5–30°N, 70–105°E)
The warming varies from 1 to 2°C throughout the year. Precipitation changes little in winter and generally increases throughout the region by 5 to 15% in summer. Summer soil moisture increases by 5 to 10%.

Sahel (10–20°N, 20 W–40°E)
The warming ranges from 1 to 2°C. Area mean precipitation increases and area mean soil moisture decreases marginally in summer. However, there are areas of both increase and decrease in both variables throughout the region which differ from model to model.

(d) JJA 2 X CO2 - 1 X CO2 SOIL MOISTURE: CCC

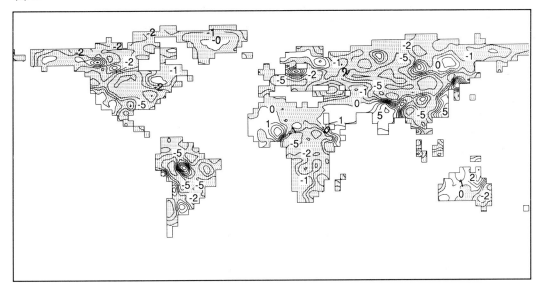

Figure 9: Simulated changes in soil moisture at equilibrium due to doubling atmospheric CO_2 in June, July, August. Contours every cm, areas of decrease are stippled. (Canadian Climate Centre model with a horizontal resolution of 32 spectral waves. Boer, personal communication, 1990)

Southern Europe (35–50°N, 10 W–45°E)

The warming is about 2°C in winter and varies from 2 to 3°C in summer. There is some indication of increased precipitation in winter, but summer precipitation decreases by 5 to 15%, and summer soil moisture by 15 to 25%.

Australia (10–45°S, 110–155°E)

The warming ranges from 1 to 2°C in summer and is about 2°C in winter. Summer precipitation increases by around 10% , but the models do not produce consistent estimates of the changes in soil moisture. The area averages hide large variations at the sub-continental level.

3.6   Summary

Evidence from both modelling and observational studies suggest that the equilibrium global mean warming due to doubling atmospheric CO_2 is most likely to lie in the range 1.5 to 4.5°C. Assuming an effective doubling of CO_2 by 2020 and an equilibrium sensitivity of 2.5°C, simple models calibrated using the more complex GCMs predict a warming of just under 2°C above pre-industrial levels by 2030. The warming is expected to be greatest over the higher northern latitudes in winter and least over the southern ocean throughout the year. Simulated precipitation increases in middle and high latitudes in winter, but soil moisture generally decreases in northern mid-latitudes in summer. The magnitude of changes in precipitation is generally 0 to 15%. Little confidence can be placed in the variations in the changes on a regional scale (less than 2000 km)

4.   Reducing Uncertainties: Current Research Programmes and Their Possible Shortcomings.

The assessment of recent results by the IPCC gives an opportunity to review existing research programmes and to identify areas where further initiatives are required. The major sources of uncertainty in the simulation of climate change arise from:

i)   lack of knowledge concerning the processes leading to the formation and dissipation of clouds, and determining their radiative properties. Furthermore, an improved understanding of the relevant microphysical processes needs to be matched with improved parameterization of these processes in large scale models of climate if the current uncertainties in climate sensitivity are to be reduced. This issue is addressed to some extent by the World Climate Research Programme (WCRP) through its Global Energy and Water Cycle Experiment (GEWEX) and the ongoing International Satellite Cloud Climatology Project (ISCCP), though it is not obvious, for example, that clouds and related processes are the top priority in GEWEX. Thus, it may be necessary to review widely spread programmes such as GEWEX to ensure that the major sources of uncertainty are given due priority.

ii)   lack of knowledge concerning the processes leading to the vertical mixing of heat into the deep ocean, particularly in high latitudes. The World Ocean Circulation Experiment (WOCE) of the WCRP was set up to describe the ocean circulation at all depths during a five year period (1990–1995). The area under consideration was belatedly extended to cover the high latitude southern oceans and does not include the Norwegian Sea. In view of the apparent importance of high latitude oceanic mixing in recent CGCM experiments, there may be a need for additional field experiments in high latitudes to supplement WOCE. The activities of the Joint Global Ocean Flux Study (JGOFS, part of the International Geosphere-Biosphere Programme of the International Council of Scientific Unions) is also relevant to the problem of ocean mixing.

iii)   lack of knowledge concerning the parameterization of tropical convection, which appears to be associated with an uncertainty of a factor of two in the magnitude of the sensitivity in the tropics. This is presumably addressed as part of GEWEX, but may need to be made more explicit.

iv)   lack of knowledge of land surface processes.and the interaction between climate and ecosystems. This gives rise to uncertainties in the.simulation of important hydrological characteristics of the land surface such as soil wetness and evaporation This is addressed explicitly in GEWEX and in the International Satellite Land Surface Climatology Programme (ISLSCP).

Other factors which limit progress are insufficient computing power and lack of scientists with experience in the relevant topics. Note that in this paper we have assumed that changes in greenhouse gases are prescribed, and so additional uncertainties due to the,.effect of changes in climate on greenhouse gas concentrations are ignored.

Acknowledgements

This chapter is based on sections 3 to 6 of the Intergovernmental Panel on Climate Change Working Group I Report (Scientific Assessment of Climate Change). The sections are Processes and Modelling (Lead authors U. Cubasch and R.D. Cess); Validation of Climate Models (Lead authors W.L. Gates, P.R. Rowntree, and Q.-C. Zeng); Equilibrium Climate Change (Lead authors J.F.B. Mitchell, S. Manabe, T. Tokioka and V. Meleshko) and Time - Dependent Greenhouse - Gas - Induced Climate Change (Lead authors F. Bretherton, K. Bryan and J.D. Woods). The reader is referred to the original report for further details of the findings presented above, and a full list of contributors to each section. We are grateful to Howard Cattle, Peter Rowntree and an anonymous reviewer for useful comments on the text.

References

Cess, R.D., G.L. Potter, J.P. Blanchet, G.J. Boer, S.J. Ghan, J.T. Kiehl, H. Le Treut, Z.X. Li, XZ Liang, J.F.B. Mitchell, J.-J. Morcrette, D.A. Randall, M.R. Riches, E. Roeckner, U. Schlese, A. Slingo, K.E. Taylor, W.M. Washington, R.T. Wetherald and I. Yagai, 1989: Interpretation of cloud climate feedback as produced by 14 atmospheric general circulation models. Science, 245, 513–516

COHMAP members, 1988: Climatic changes over the last 18,000 years: Observations and model simulations. Science, 241, 1043–1052

Cubasch, U., E. Maier Reimer, B. Santer, U. Mikolajewicz, E. Roeckner and M.Boettinger; 1990, The response of a global coupled O-AGCM to CO_2 doubling. MPI report, MPI fuer Meteorologie, Hamburg, FRG.

Ellsaesser H.W. 1989. A different view of the climatic effect of CO_2 Updated. Atmosphera, 3, 3–29

Fennessy, M.J. and J. Shukla,1988: Impact of the 1982–3 and 1986–7 Pacific SST anomalies on time mean prediction with the GLAS GCM. WCRP-15, WMO, pp26–44.

Folland, C.K., J.A. Owen, and K. Maskell, 1989: Physical causes and predictability of variations in seasonal rainfall over sub-saharan Africa. IAHS Publ. No. 186, 87–95.

Gates, W.L., 1975: The physical basis of climate and climate modelling; GARP Publication Series No. 16, WMO, Geneva.

Gregory D. and P.R. Rowntree 1990. A mass flux convection scheme with the representation of cloud ensemble characteristics and stability-dependent closure. Monthly Weather Review, 118, 1483–1506.

IPPC ,1990 Report of Working Group I "Scientific Assessment of Climate Change". WMO/UNEP Geneva, pp 365

Jaeger, L. 1976: Monsatskarten des Niederschlags fur der ganze Erde. Bericht Deutscher Wetterdienst, 18, Nr 139, 38pp

Lindzen R.S. 1990 Some coolness concerning global warming Bulletin of the American Meteorological Society, 71, 288–299.

Luther F.M. and R.D. Cess 1985. Review of the recent carbon dioxide-climate controversy. In "Projecting the climatic effects of increasing carbon dioxide", eds M.C. MacCracken and F.M. Luther, pp321–335. US Department of Energy, Washigton DC, 1985. (Available as NTIS, DOE ER-0237 from Nat Tech. Inf. Serv. Springfield Va.)

Mitchell J.F.B. and W.J. Ingram 1990. On CO_2 and climate.Mechanisms of changes in cloud. J of Climate (submitted)

Mitchell, J.F.B., C.A. Senior and W.J Ingram, 1989: CO_2 and climate: A missing feedback?, Nature, 341, 132–134.

Oort, A.H., 1983: Global Atmospheric Circulation Statistics, 1958–1973. NOAA Professional paper 14, US Department of Commerce.

Raval A. and V. Ramanathan, 1989: Observational determination of the greenhouse effect. Nature, 342, 758–761.

Sarmiento, J.L., 1983: A simulation of bomb -tritium entry into the Atlantic Ocean. J. Phys. Oceanograph., 13, 1924–1939.

Schlesinger M.E. and J.F.B. Mitchell, 1987. Climate model simulations of the equilibrium climatic response to increased carbon dioxide. Reviews of Geophysics, 25, 760–798.

Schutz, C. and W.L. Gates. 1971: Global climatic data for surface, 800mb, 400mb: January. Rand, Santa Monica, R-915-ARPA, 173pp.

Stouffer, R.J., S. Manabe, and K. Bryan,: 1989. Interhemispheric asymmetry in climate response to a gradual increase in atmospheric CO_2. Nature, 342, 660–662.

Climate Trends and Variability

by M. J. Coughlan *(Australia)* and B.S. Nyenzi *(Tanzania)*

Introduction

Climate in its most general sense may be defined as the organized summary over time of the observed behaviour of the planetary land, atmosphere and water system. The components of this system interact in many ways and on many time and space scales; to model and predict the system to the maximum extent possible is the principle aim of climate research.

To gain insight on the nature of variability and change within the climate system as a prerequisite to achieving this aim, it is necessary to study its components in a systematic way. A record of temperature or any other climatic variable will typically consist of a complex mixture of variations. Such variations can be separated or filtered out and identified as long term trends, annual and semi-annual cycles, discontinuities, quasi-cyclical inter-annual variability and a residual, random or "noise" component (Figure 1). The process of differentiating and examining the components of the climate system is best carried out in ways that are natural to the system and care must be taken not to introduce spurious features that are by-products of the statistical filtering process or a consequence of inadequate sampling. Many "cycles" in climate records of limited length have been shown to be insubstantial and evanescent with the addition of more data. The climate system can also be differentiated on spatial scales and at the physical component level, for example, one may study the circulation of the Atlantic Ocean or the distribution and processes of rainfall over the Sahel region of Africa.

In this paper we summarize the evidence for trends and variability in the climate record. We draw particular attention to the practical difficulties in attributing trends and specific modes of variability to a single cause against a background of variations intrinsic to the climate system and due to other unknown or at least unspecified forcing mechanisms on the system. These difficulties are especially acute at regional and local levels where the real problems of a change in the climate must be addressed. First, however, we will summarize briefly what has occurred to the earth's climate on the largest scales.

Is the Earth's Climate Changing?

There is no obviously simple answer to this question since it is necessary first to agree on what change is and on the base climatic state from which to measure change. It is generally accepted that climate is inherently variable on all time scales. The appropriate length of the base period for

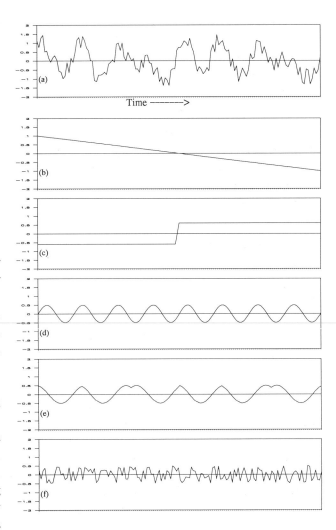

Figure 1: Hypothetical data series (a) and its component parts: (b) linear trend, (c) discontinuity, (d) periodic, (e) quasi-periodic, and (f) random.

determining a climatic state, the date of its beginning and whether one base period is appropriate for all locations, however, are not universally agreed upon. Climatic change then might be said to occur only when the climate system is externally disturbed, e.g. by an increase in volcanic or solar activity, or by a major anthropic influence. Yet it remains a difficult task to track from climatic records alone the trail of impacts on say the global radiation balance through to some variation in global or regional precipitation patterns, and to distinguish the effects of that externally imposed influence from the intrinsic variability of the system. This problem of detecting a forced or induced change in climate is addressed in some detail by Wigley in Chapter 8 of the IPCC report.

The situation is further complicated by suggestions that climate may be capable of moving between significantly different quasi-equilibrium states without any external forcing; that climate is subject, in part at least, to the emerging laws of chaotic systems (Lorenz, 1989).

The Palaeoclimatic Record

The long term climate record derived from palaeological records provides clear evidence of fluctuations in temperature and precipitation on all time scales (Figure 2). Swings in broad-scale average temperatures of around 4–5°C in amplitude reflect the onset and retreat of ice ages; periods lasting thousands of years when large tracts of North America and Northern Europe were covered by year-round ice and snow. Data from Antarctica confirm that such fluctuations were indeed global. It is generally believed that vacillations in the earth's orbit can cause large scale swings in surface temperature (Milankovitch, 1941; Berger, 1980), with ice ages being more likely to occur when radiation from the sun is least over the Northern Hemisphere. However, there is no reason to believe that the Milankovitch theory is the only cause for the onset of glacial conditions (Winograd et al, 1988; Johnson and Wright, 1989 and Winograd and Coplen, 1989).

While the onset and retreat of ice ages at higher latitudes and alternating periods of wetter and drier conditions at lower latitudes characterize the climate record on time scales of tens to hundreds of thousands of years, the record also reveals occasions when significant, large scale fluctuations in temperature and presumably precipitation occurred over much shorter periods; one example of such a period is the so-called Younger Dryas. As the earth was emerging from the last major ice age around 10–11,000 years ago, the upward trend of temperature throughout the North Atlantic region (eg. Atkinson et al., 1987) and some other parts of the globe including the Southern Hemisphere (Salinger, 1989) reversed. This cooling period lasted for about 500 years and was caused, it has been suggested, by

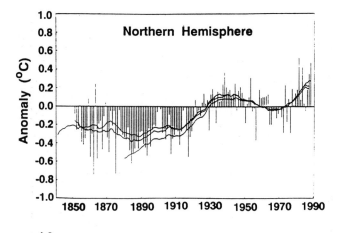

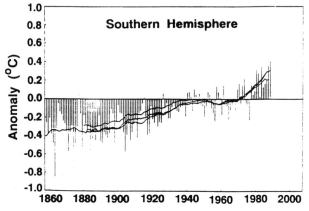

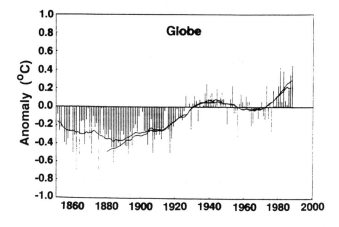

Figure 2: Land air temperatures, relative to 1951-80. Annual values (bars) by Jones (1988) and smoothed curves from Jones (1988), Hansen and Lebedeff (1988) and Vinikov et al. (1990).

changes in the deep Atlantic ocean circulation (Broecker et al, 1985; Broecker and Denton, 1989).

Several recent studies (e.g. Webb et al, 1987; Huntley and Prentice, 1988) have highlighted periods of various lengths since the end of the last ice age when near-surface temperatures were warmer than at present. However the evidence is neither globally consistent nor synchronous. The 300 or so years leading up to the instrumental record (1550–1850) were accompanied by advances in many of the world's glaciers. During this period the middle latitudes of the northern hemisphere, at least, endured particularly

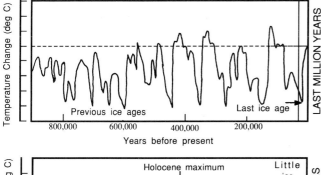

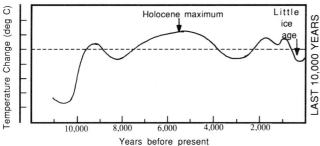

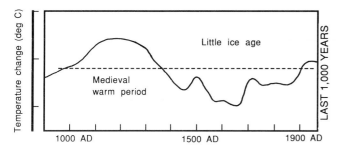

Figure 3: Schematic diagrams of global temperature variations since the Pleistocene on three time-scales. The dashed line nominally represents conditions near the beginning of the twentieth century.

harsh winters. It is important then to realize that the instrumental record begins at a time when significant portions of the earth's surface where the earliest records began may have been emerging from a period that has been frequently designated as the Little Ice Age.

The Global Instrumental Record

Temperature

Over the past five years or so there have been several new attempts to compile composite global records depicting the mean land surface temperature (Figure 3) and mean sea surface temperature (not shown), derived from instrumental observations (Jones et al, 1986a and b, Jones, 1988; Hansen and Lebedeff (1987, 1988); Vinnikov et al, 1990). All conclude that a record of global mean temperature cannot be compiled with confidence from instrumental observations prior to the middle of the 19th century. Since the bulk of the observations used to derive the land surface curves in Figure 3 are common to each study there is a high degree of correspondence between the three analyses. The differences arise from variations in the presumption of biases and their treatment. Such biases can arise from a number of sources, some of which are discussed below.

The consensus from the three studies is that there has been an increase in the global mean of near surface temperature of $0.45 \pm 0.15°C$ since the late nineteenth century. The increase has not been steady and moreover, there are distinct differences between the mean temperature curves for the two hemispheres. The northern hemisphere, which has the greater proportion of the earth's land masses, exhibits two periods of warming—one from around 1920 to the early 1940s and the other from the mid 70s to the present. During the years 1945-1975 the mean northern hemisphere temperature actually trended downward. The southern hemisphere, which is predominantly ocean, exhibited a more steady rise in temperature especially since about the turn of the century, although there too the rise seemed to falter during the years 1945-1975.

Not surprisingly, the curve for the whole northern hemisphere has many of the characteristics of the curve of globally averaged land temperatures while the southern hemisphere curve has many of the characteristics of the curve of globally averaged sea surface temperatures. There are subtle differences between the separate curves for land, sea and hemispheres, however, such as the starting and end points of the hiatus in the upward trend in temperature between the mid-30s and mid-70s, and for which there are as yet no unequivocal explanations.

Precipitation

Coupled with any change in global temperature as a consequence of the enhanced greenhouse effect there will most likely be concurrent impacts on the global hydrological cycle leading to changes in precipitation patterns and perhaps in the global average of precipitation. Precipitation in any of its forms (rain, hail, snow etc.) is not a continuously measurable variable, has large spatial discontinuities and falls virtually unrecorded over the oceans. Consequently, there are no reliable estimates of global or hemispheric periodic accumulations of precipitation. The search for changes in precipitation records, therefore, must remain at the regional or local level, with particular attention being given to the associated systematic variations in atmospheric circulation which would accompany such changes.

Other Variables

As with precipitation, the records of other climatic measures such as wind, and humidity are too patchy or short to derive meaningful global trends. Because of the significance of clouds as a potential regulator of global change, some effort has been made to detect continental-scale changes in cloudiness. The International Satellite

Cloud Climatology Project (ISCCP) has been established with this as one of its objectives in mind, however only one or two years of data have been processed so far. Visual observations of clouds made at ground level provide the only source for detecting long-term trends in cloudiness (e.g. Henderson-Sellers, 1986a, 1986b, 1989; McGuffie and Henderson-Sellers, 1989). There are indications of increases in total cloud cover in some extensive continental regions during this century, however there are many difficulties associated with systematic changes in the way cloud amounts have been observed and recorded. Variations and discontinuities in climatic records which are due to non-climatic factors are a pervasive problem in the search for genuine trends and fluctuations.

The Ideal Climatic Record

While it is of extreme interest to determine whether or not the earth's climate system is changing or will change as a consequence of the enhanced greenhouse effect, the ultimate need is to determine how it might change at a regional and even local level. It is at these latter scales that the impacts of climate change will be most keenly felt and for which many response strategies and actions will be devised. The large pooling of data relating to surface temperature and, for a more restricted period, upper air temperatures has enabled us to construct the records of average global temperature changes outlined above with a relatively high degree of confidence. There is sufficient internal consistency in the global data sets for us to believe that for the most part, the errors in the individual records which make up the data sets can be considered as random noise and hence have not contaminated the global record, at least as far as determining trends or long term fluctuations.

Nonetheless, the existing instrumental records, when coupled with the longer term palaeoclimatic and proxy records, are the measures against which modelling results of past, present and predicted climatic change and variability at the regional and local level are verified or compared. It is essential then that we know what an individual record has measured and to what, in general terms at least, the variations therein can be ascribed.

There are four attributes which together determine the quality of an instrumental record:

- Continuity
- Representativeness
- Length
- Homogeneity

Continuity is not absolutely essential if gaps in the record are short relative to its overall length and data from other nearby locations can be used to estimate the missing records.

Representativeness of the observing site is important for interpreting the context of the record and its fluctuations: observations taken at the bottom of a narrow river valley may be quite misleading if the record is unwittingly extended to the surrounding upland plains.

A long record of a locality's climate would seem to be an obviously important requirement, and for the assessment of climate change it is most desirable. However, climate is a synthesis of many interactive processes involving the sun, the atmosphere, the earth's surface and its oceans, and the value of each record will depend on the time-scale of a process under investigation. Thus a climate record of limited length will still have value if one can be confident of its quality in other respects.

Inevitably, there are trade-offs. Surface instrumental records provide high levels of detail on individual climatic variables for processes up to the decadal scale with mixed spatial resolution. However, there are few good instrumental records longer than 150 years. Proxy-records derived from pollen counts in lake varves, growth-ring variability in trees, coral accretion rates and ice core composition etc., provide much longer records of past climatic fluctuations. These are integrated measures of climate with temporal resolutions that are typically annual and often related more to a growing season than to a full calendar year. In some palaeoclimatic reconstructions, however, the temporal resolution of proxy records may be no better than millenial. Much effort and care is required in analyzing and interpreting the variations observed in proxy-records and associating them directly to specific quantitative changes in temperature and rainfall.

This is not to say that existing instrumental records are free from the difficulties of interpretation inherent in the proxy records. Homogeneity (as distinct from stationarity) is undoubtedly a most desirable attribute of any climatic record, but is usually the most difficult to quantify. The value of a homogeneous instrumental record is best illustrated by the range of factors which can lead to non-homogeneity at an observing site. A long and continuous instrumental record may be degraded by any one or combination of the following factors:

- Site change (even small shifts in location may be critical)
- Instrument change or drift (e.g. lack of standardization and routine calibration)
- Change in immediate surroundings (e.g. growth of nearby trees, erection of buildings)
- Change in observational practice (e.g. change of recording times due to "daylight saving")

There are techniques for assessing homogeneity and the comparison of records between adjacent stations provides the simplest and most reliable check, although, even this procedure assumes that each station is under similar climatic influences and that the effect of any changed observational condition can be diagnosed for each site. Homogeneity is less of a problem if to obtain a spatial average there is an adequate number of sites to reduce the impact of site specific fluctuations, albeit at the expense of lowering the signal-to-noise ratio. The problem is more serious if the inhomogeneity is systematic and affects a significant number of stations in a network under investigation, whether that network is global, regional or local.

One of the principal culprits of inhomogeneity in the instrumental climatic record is man's own activity. The "urban heat-island" effect is probably the most common example. A rapid increase in overnight minimum temperature compared to the concurrent daytime maximum temperature over a period of years is typical of temperature records taken in and around industrialized centres of population. Even small town growth can significantly affect the more sensitive climatic parameters such as temperature, wind and relative humidity. Re-locations of observing sites to airports outside the major centres of population, such as occurred during the 50s and 60s and then the gradual build-up in activity around those sites need to be considered and allowed for. Inadequate documentation of such changes has made the task of identifying inhomogeneities extremely troublesome in many cases, with researchers applying differing methods and levels of correction. Assessments by Karl and Jones (1990) and Jones et al (1990) lead to the conclusion that in global terms, the contribution of urban heating to the global trend over the last 100 years is around 0.05°C, i.e., an order of magnitude less than the observed trend. Systematic changes in measuring sea surface temperature, particularly a change from observing the temperature of a sample of sea water collected in a bucket to that of observing the temperature of the water at a hull intake point, appear to have contributed at least 0.1°C to the 100 year upward trend of uncorrected temperatures at the ocean surface (Folland and Parker, 1989).

Throughout the period from the mid-nineteenth century to the establishment of standards by the International Meteorological Organization and subsequently by the World Meteorological Organization there has been a good deal of variation from nation to nation and even within nations in observational practices and instrumental exposure. For example, the small open-fronted Glaisher Stand was in wide-spread use in some countries during the nineteenth century; this form of instrumental enclosure, while acceptable perhaps in higher latitude countries, is plainly unsuitable for lower latitudes, especially during summer when the sun can shine directly into the exposed enclosure giving unrealistically high readings of air temperature. Others nations favoured larger structures, e.g. the so-called "round-house", an instrumental enclosure about the size and shape of a small garden summer house.

There have been (and still are) several variants on the official Stevenson Screen standard. These differences and unrecorded changes in site render many of the earliest of the instrumental records suspect at least. With the exception of a few of the longest records, work to "clean up" early historical records has only just begun.

The situation improved around the turn of the century as better communications made daily weather forecasting a realizable goal; many nations expanded their observing networks and commenced archiving climate records in a more systematic fashion. Nevertheless, while the number of recording stations has grown enormously during the last 50–100 years the climatic record continues to be plagued by station interruptions and closures, and changes in instrumental and observing practices which are often the result of pressures on meteorological agencies to "improve" their efficiency and minimize costs in data collection processes. It is important to note that the specifications for weather observations and climate observations are similar but not identical. If meteorological agencies responsible for weather forecasting networks are to continue to carry the responsibility for climate observations then they must incorporate the differing requirements into their networks.

Satellites are now providing very high spatial resolution for observing the climate system in its entirety, however the records are still short and sometimes not well-reconciled with ground-based data. There are also homogeneity problems due to changes in instruments, calibrations and the algorithms used to derive standard climatic variables from the raw radiation measurements made by on-board sensors. The recent studies of Spencer and Christie (1990), however, demonstrate the promise of an extensive capability to monitor climate change and variability from space. While the plans by the major space agencies for the remaining years of this decade and into the 21st century demonstrate a commitment to achieving this capability, the need for a well-defined and maintained surface network will be a continuing requirement.

Regional Climate Variability

There have been many attempts to classify climates into various categories as an aid to studying climate (e.g. Köppen, 1884; Thornthwaite, 1931), however, each location on the earth is unique and its climate is determined principally by its latitude, elevation, longitude, distance from the sea and the local orographic features. On the temporal climate scale, which relates typically to averages for periods longer than a month to a season, the regular northward and southward "passage of the sun" provides the strongest driving force for variability in the climate system; its very regularity often causes us to overlook its importance. Yet the the annual cycle of seasonal conditions and the equally familiar diurnal cycle of night and day are

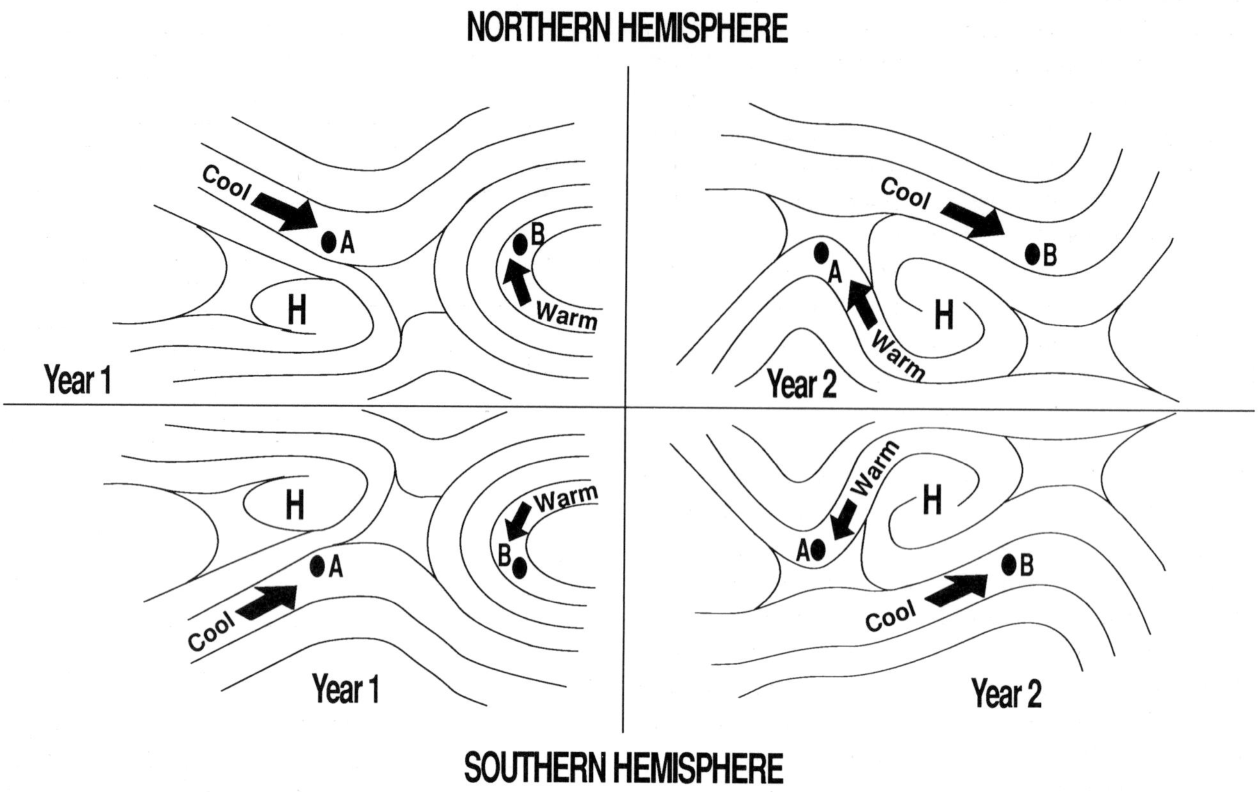

NORTHERN HEMISPHERE

SOUTHERN HEMISPHERE

Figure 4: Schematic illustrations for locations in the northern hemisphere (upper) and southern hemisphere (lower) showing how mean monthly temperatures from one year to the next can vary with shifts in average wind systems.

the strongest regulators of human and indeed all living behaviours.

Despite the overall persistence of the seasonal rhythm in the climatic system, we are aware that the annual climate cycle is not constant. For example, some summers may be hotter than others while one winter may be drier than the last; during one month the wind may blow incessantly from one quarter but during the same month of the following year it may blow from quite a different direction or be much more variable.

The major factors underlying existing local climates, including the role of the the seasonal cycle, are reasonably well known and understood. As noted, factors such as latitudinal position, proximity to the oceans or major mountain ranges, location with respect to the major monsoon regions and so on, provide the general framework for local climates. From year to year, however, for reasons that are not always understood and mostly not yet predictable, there are differences in the shape of this general framework. For example, if the wind patterns over an area are averaged for about one month a relatively smooth pattern emerges, especially if the averaging process is carried out on winds in the atmosphere away from the influences of surface roughness. In the left hand side of Figure 4 at the points marked A, the air moving from lower

latitudes is warm relative to that at the locations marked B where the air has a trajectory out of higher latitudes (upper part of the diagram depicts the northern hemisphere case and the lower part, the southern hemisphere). It may also be wetter at A than at B depending on the season and geographical location. In another year (right hand side of Figure 4) the pattern may be shifted, such that it is not as warm or wet at A and not as cold but wetter at B.

Such inter-annual fluctuations in average wind patterns are a natural part of the climate system and we see their effects in events such as a failure of the monsoon over the Indian sub-continent or Southeast Asia, swings between cold and mild winters in Eurasia and North America, and the inevitably recurring sub-tropical droughts across large tracts of Africa, Australia and South America. Often there are systematic, contemporaneous climatic anomalies in different parts of the globe; these have now come to be known as teleconnection patterns (Wallace and Gutzler, 1981). While we are able to describe with hindsight some of the underlying causes of interannual climate fluctuations and the teleconnection patterns, we are only just beginning to understand the processes sufficiently well enough to attempt to model them with confidence for predictive purposes.

Teleconnection Patterns

The strongest teleconnection pattern known within the climate system on seasonal to decadal time-scales, apart from the annual cycle, is the set of processes known variously as El Niño, the Southern Oscillation or simply ENSO. ENSO involves a set of complex interactions between the tropical oceans and the atmosphere centred on the Pacific and Indian Ocean Basins with a life-cycle typically lasting from 3–6 years (Rasmusson & Carpenter, 1982). ENSO is now known to be at the root of many of the disastrous interannual climate fluctuations affecting tropical and sub-tropical countries and probably also, although not as directly, many of the seasonal variations observed at higher latitudes. Through the Tropical Oceans Global Atmosphere (TOGA) project of the World Climate Research Programme we are approaching, some would say are achieving, a reliable and useful level of skill in modelling and predicting the onset of the various phases of ENSO and hence the ability to make seasonal predictions.

The interannual variability of such phenomena as drought in Northeastern Brazil and winter temperatures and precipitation over the eastern United States, have been associated with teleconnection patterns in sea surface temperature over the North Atlantic, hence the term North Atlantic Oscillation (NAO). This interannual variability has been shown to be modulated by longer period variations on scales from a decade to several decades. As noted earlier, these longer scale fluctuations in the North Atlantic may account for the downward trend in temperature in the Northern Hemisphere in the decades after 1960. The phase of the NAO during the period prior to 1925 is also consistent with the cooler temperatures prevailing at that time.

Persistent systematic changes in ENSO and the NAO, whether occurring abruptly or gradually over many years would cause the climate of a locality to change. While there have been quiescent periods in the ENSO record, there is no evidence that the basic character of the phenomenon is currently undergoing any systematic change (Nicholls, 1988). The longer time scale fluctuations of the NAO and its own inherent variability renders it more difficult to extract or identify systematic changes in its character from the limited instrumental record.

Integrated Measures of Climate

When viewing climate change in the context of its ecological and socio-economic impacts, it is often more important to look at integrated measures of climate variability. For example, will an increase in temperature matter if there is a concurrent increase in rainfall to offset the probable increase in evaporation, and what will be the effect of a change in the wind regime? The examination of changing vegetation types can provide information of this type over periods of a century or more while other proxy

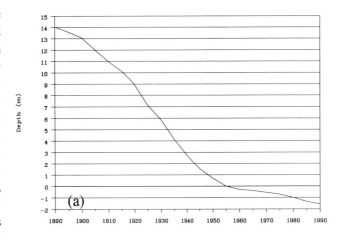

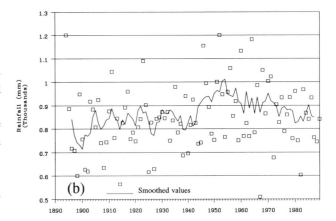

Figure 5: (a) Annual level, Lake Keilambete, (b) Annual rainfall, Lake Keilambete. Lake Keilambete is located in southeastern Australia at approx. 38°S 143°E.

measures such as the variation in tree ring and coral layer growth as noted earlier can provide good estimates of climate change as long as the climate and morphological relationships are well understood. Closed lakes can be used as simple indicators of changes in net water availability at a local or regional level depending on their represent-ativeness. Closed lakes have negligible surface or sub-surface outflow and their only source of water is from precipitation over a well defined catchment only marginally greater than the surface area of the lake itself. Although not well distributed over the globe, a number are to be found in Africa, China, North America and southeastern Australia.

Figure 5a shows that Lake Keilambete, a well documented closed lake site in southeastern Australia, has been falling consistently for the past 100 years or so despite the fact that between 1900 and the mid-1950s rainfall increased by 15–20%, and has since fallen by less than 10%. Mean annual temperature over southeastern Australia has increased by around 0.4°C over the past 56 years and

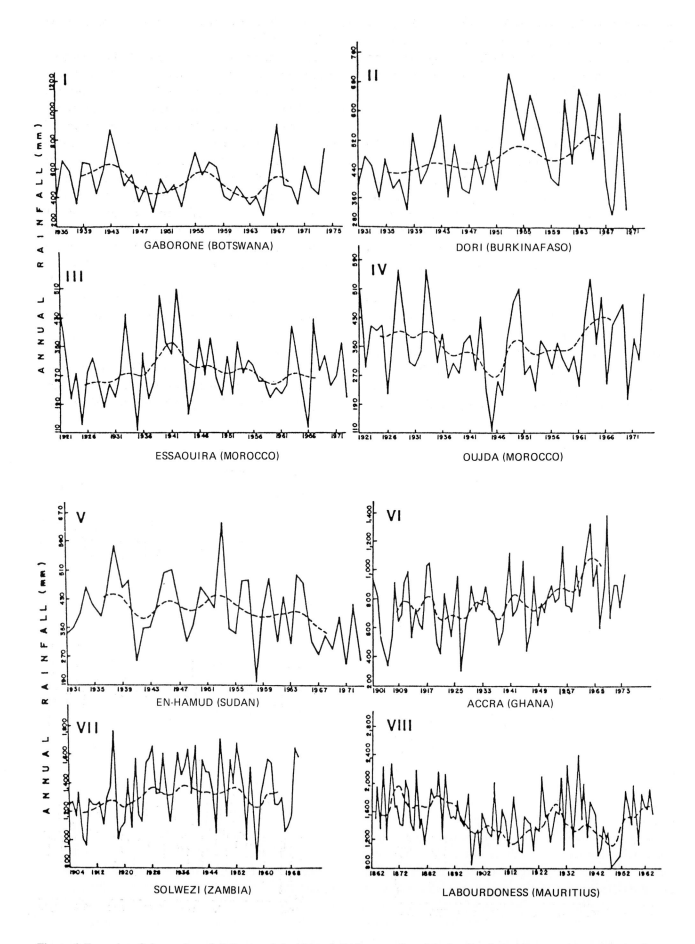

Figure 6: Examples of observed precipitation trends in Africa. Solid lines are the original series, dashed lines are the smoothed series.

by around 0.6–0.8°C since 1945. On the face of it then, a significant increase in rainfall would have been necessary to offset the loss of water from the lake through increased evaporation brought about by the variations in temperature and perhaps wind also.

Regional Variability—African Rainfall

In a review of this length, it is impossible to examine all the regional trends in temperature and precipitation which, because of their magnitude or persistence, are already having socio-economic impacts of varying intensities. The recent fluctuations in precipitation of large expanses of the African Continent, however, stand out as particularly significant, partly as a consequence of their magnitude but perhaps more so as a consequence of their impacts on the nations and communities affected.

Significant work has been done recently on rainfall variability in Africa, for example Ogallo (1979, 1980), Nicholson and Entekhabi (1987), Nyenzi and Nicholson (1989) and many others. Most of these studies have used data collected during the instrumental era and therefore do not represent a period long enough to describe climate change satisfactorily. However, their results provide reasonably reliable indications of recent general trends on the African continent.

Several researchers have speculated about climate change in Africa including Lamb (1978) who suggested that this would occur as a result of the general circulation slowly changing through an equatorward shift of the principal high pressure belts. Other studies on annual rainfall series for certain African regions revealed no established trends (e.g. Landsberg, 1975), although short period trends of both signs, typical of many rainfall records, were evident in some cases. Figure 6 shows the series of some African rainfall stations constructed by Ogallo (1979). These results, although for a short period, show that most of the annual series have a generally oscillatory character without significant overall trends. Note that the short term positive or negative trends observed as a result of the smoothing process which reveal the oscillatory nature of the curves are not in themselves statistically significant.

This oscillatory behaviour has also been observed by other authors in different parts of Africa, for example Rodhe and Virji (1979), Nicholson and Entekhabi (1986) and Nyenzi and Nicholson (1989). Figure 7 shows the

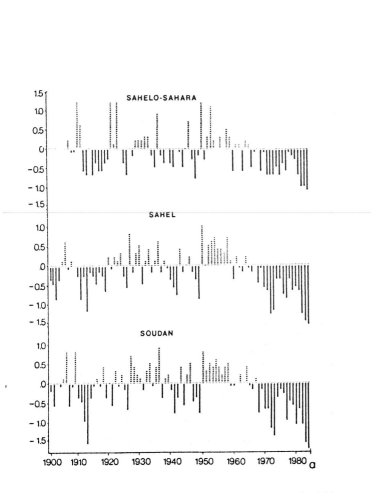

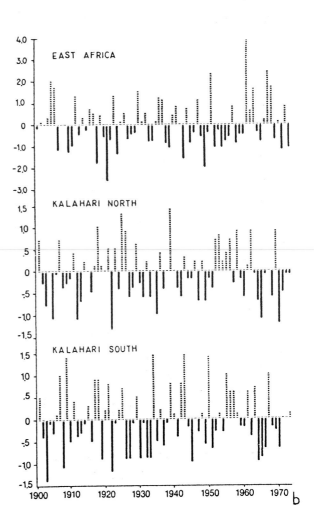

Figure 7: Regional, annual precipitation anomalies over the African continent.

standardized annual rainfall departure series in some regions for the period 1901–1990. The series for the Sub-Sahara region has above normal rainfall in 1930–1940 and 1950–1960; and below normal rainfall in 1910–1920, 1940–1950 and 1968–1990. The negative trend observed after 1968 reflects the severe drought conditions which prevailed in the Sahel in those years. An oscillatory nature is also clearly observed in the rainfall time series for the northern and southern Kalahari and for East Africa. The following quasi-periodic cycles have featured prominently in African rainfall series by different authors: 2.0–2.5, 3.3–4.4, 5–6.5, and 10–12 years. These values have been evident in records from other countries, e.g. Australia (Coughlan, 1976). The first, a quasi-biennial signal is common in many meteorological time series, the second and third are characteristic of harmonics of the ENSO cycle, and the longest period is typical of the North Atlantic Oscillation referred to above. Relationships between the Southern Oscillation and rainfall in Africa have been investigated by Nicholson and Entekhabi (1986) and Nicholson and Nyenzi (1989). Their results suggested a strong influence of the Southern Oscillation on rainfall variability in southern Africa and parts of the equatorial

belt. Coherence with the Southern Oscillation Index was particularly strong in the quasi-biennial range (2.0–2.5 years), especially in the tropics and southern Africa.

In summary, the available records of rainfall over Africa do not reveal convincing evidence of persistent positive or negative trends which would suggest that the climate may be changing over a period comparable in length to the series, typically 50–80 years. However there is strong evidence for quasi-oscillatory, short to medium term trends which appear to be strongly related to well established climatic teleconnections.

Extreme Events

A complementary approach to studying variability is to examine the frequency of extreme events. For many forms of statistical distribution, the frequency of extreme events is very sensitive to relatively small changes in the mean or central tendency. For example, a small increase in monthly mean temperature at a given location may suggest a doubling of the the number of days exceeding the old "very hot day" threshold. The examination of changes in extremes over time should therefore be a sensitive measure of climate change. Unfortunately, there is a trade-off and it

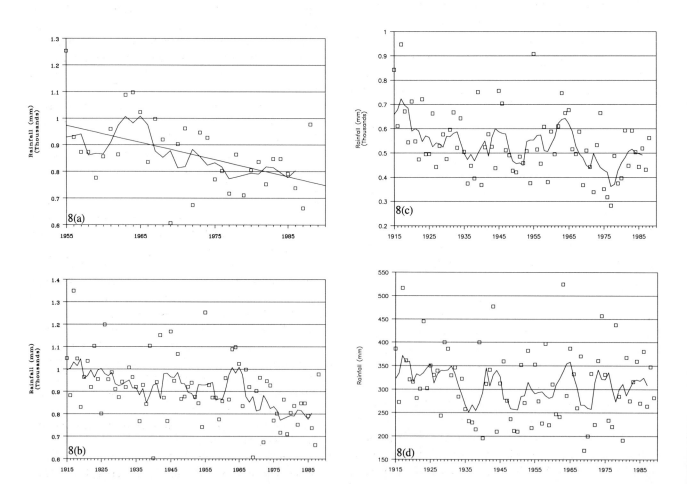

Figure 8: Annual rainfall totals for locations in the far southwest of Western Australia. (a) District since 1955 - data points, smoothed and linear trend. (b) District since 1915 - data points and smoothed. (c) New Norcia since 1915 - data points and smoothed. (d) Merredin since 1915 - data points and smoothed.

is usual that a longer record is required to establish a stable set of statistics for the extremes of a variable than for its basic statistics. Although some work has been done to identify systematic changes in climatic extremes, the general consensus is that the evidence is still too patchy and region specific to draw any meaningful inferences. In an assessment made especially for the IPCC Report, there was very little evidence for changes in the incidence of very wet (flood) or very dry (drought) years over five major regions of the globe: Australia, India, the USA, portions of the USSR and China. Although this assessment was by no means global in extent, it was judged highly unlikely that systematic changes on a global scale could be occurring without any of these diverse regions being affected.

The Task Ahead

While there is an obvious eagerness to demonstrate that the enhanced greenhouse effect is causing the global average temperature near the surface to rise, it is of far more practical importance to determine how such a temperature rise is going to be distributed and what may happen to the statistical distributions of rainfall patterns.

The difficulty of the task ahead and the spatial resolution that may be needed to study impacts of climate change at the regional level should not be underestimated. The existing record of presumably "natural" climate variability provides many salutary lessons in this regard. For the past two decades or so, for example, it has been observed that annual rainfall in the district representing the coastal and hinterland strip south of Perth in the far southwest of Australia has fallen by around 20% (Figure 8a): a regional reduction in rainfall in an important agricultural zone that might be quoted in a typical scenario for a "greenhouse" induced global warming. A careful inspection of the graph reveals that the downward trend commenced in the late 60s. To put the trend in context, one should look at a longer record (Figure 8b), and some nearby records: New Norcia, 100km to the northeast of Perth (Figure 8c) and Merredin, 200km to the east (Figure 8d). From the long-term district record, the most recent period does appear to have been exceptionally dry. However, the downward trend from the late 60s in the New Norcia record, was of much shorter duration and reversed in the late 70s, and there is little trend at all during this period in the Merredin record.

In attempts to put some sense and order into assessing the probable impacts of climate change, some researchers have turned to the construction and use of scenarios. The validity of any scenario for change in climate is partially contingent, at least, on our skill in distinguishing "changes" from "natural variability" in the climatic record of the past; whether or not a change is anthropic in origin or attributable to "natural" events or processes on any given scale. The level of this skill will also have a major bearing on the validity of linking any observed changes in regional climatic records in the future to large scale anthropically linked climatic changes. To put it more succinctly, if we don't know what caused the climatic changes and fluctuations of the past, then it is unlikely that we will be any wiser in our assessments or speculations about the nature of future changes and fluctuations in the climate. Devising scenarios about a system as complex as climate in the absence of such knowledge is therefore fraught with hazards and gambling on or being swayed by one in particular is apt to be dangerous. Doing nothing in the absence of a reliable prediction, however, implies just that - the acceptance of the single scenario of "no change".

Acknowledgements

This paper has drawn heavily on the efforts of those who contributed to Chapter 7 of the IPCC Report on the Scientific Assessment of Climate Change. The lake level data for Figure 5 was kindly provided by Dr Jim Bowler.

References

Atkinson, T.C., Briffa, K.R., and Coope, G.R., 1987: Seasonal temperatures in Britain during the past 22,000 years reconstructed using beetle remains. Nature, 325, 587–592.

Berger, A., 1980: The Milankovitch astronomical theory of palaeoclimates: a modern review, Vistas in Astronomy, 24, 103–122.

Broecker, W.S. and Denton, G.H., 1989: The role of ocean-atmospheric reorganizations in glacial cycles, Geochemica et Cosmochimica Acta, 53, 2465–2501.

Broecker, W.S., Peteet, D. and Rind, D., 1985: Does the ocean-atmosphere system have more than one stable mode of operation? Nature, 315, 21–25.

Coughlan, M.J., 1976: Observed variability in Australian climate records, In Climatic Change and Variability: a Southern Perspective, A.B. Pittock. L. Frakes and J.W. Zillman (eds). Cambridge University Press.

Folland, C.K. and Parker, D.E., 1989: Observed variations of sea surface temperature, Proc. NATO Advanced Research Workshop on Climate-Ocean Interaction, Oxford, UK, 26–30 Sept. 1899. Kluwer Academic Press, pp31–52.

Hansen, J. and Lebedeff, S., 1987: Global trends of measured surface air temperature. J. Geophys. Res., 92, 13345–13372.

Hansen, J. and Lebedeff, S., 1988: Global surface temperatures: Update through 1987, Geophys. Res. Letters., 15, 323–326.

Henderson-Sellers, A., 1986a: Cloud changes in a warmer Europe, Clim. Change. 8, 25–52.

Henderson-Sellers, A., 1986b: Cloud changes in a warmer world, Clim. Change. 9, 267–309.

Henderson-Sellers, A., 1989: North American total cloud amount variations this century, Glob. and Planet. Change, 1, 175–194.

Huntley, B and Prentice, I.C. 1988: July temperatures in Europe from pollen data, 6000 years before present, Science, 241, 687–690.

Johnson, R.G. and Wright, H.E.(Jr.), 1989: Great Basin Calcite and the Pleistocene Time Scale, Science, 246, 262.

Jones, P.D., 1988: Hemispheric surface air temperature variations: recent trends and an update to 1987, J. Climate, 1, 654–660.

Jones, P.D., Raper, S.C.B., Bradley, R.S., Diaz, H.F., Kelly, P.M. and Wigley, T.M.L., 1986a: Northern Hemisphere surface air temperature variations, 1851–1884. J. Climate and Appl. Met., 25, 161–179.

Jones, P.D., Raper, S.C.B., Bradley, R.S., Diaz, H.F., Kelly, P.M. and Wigley, T.M.L., 1986b: Southern Hemisphere surface air temperature variations, 1851–1884. J. Climate and Appl. Met., 25, 1213–1230.

Jones, P.D., Groisman, P.Ya., Coughlan, M., Plummer, N., Wang,W-C., and Karl, T.R., 1990: Assessment of urbanization effects in time series of surface air temperature over land. Nature, 347, 169–172.

Köppen, W., 1884: Zur Frage der 'gestrengen Herren', Zeit. österreich. Gesell. f. Met. (Vienna), 18, 447–458.

Karl, T.R. and Jones, P.D., 1990: Reply to comments on "Urban bias in area-averaged surface temperature trends". Bull. Amer. Met. Soc., 71, (in press).

Lamb, P.J., 1978: The current trend of world climate, CRU RP3, East Anglia University, 25pp.

Landsberg, H.E., 1975: Sahel drought, change of climate or part of climate? Arch. Meteor. Geophys., B23, 193–200

Lorenz, E.N., 1989: Chaos, spontaneous climatic variations, and detection of the greenhouse effect. Proc. US DOE Workshop on Greenhouse Gas-Induced Climatic Change: A critical appraisal of simulations and observations, Amherst, MA, May 8–12, 1989, Elsevier.

McGuffie, K. and Henderson-Sellers, A., 1989: Is Canadian cloudiness increasing?, Atmos.-Ocean, 26, 608–633.

Milankovitch, M.M., 1941: Canon of Insolation and the Ice Age Problem. Beograd, Koniglich Serbische Akademie. (English translation by the Israel Program for Scientific Translations, U.S Department of Commerce, Washington, DC. 1969).

Nicholls, N., 1988: More on early ENSOs: Evidence from Australian documentary sources, Bull. Amer. Meteor. Soc., 69, 4–6.

Nicholson, S.E. and Entekhabi, D., 1986: The quasi-periodic behaviour of rainfall in Africa and its relationship to the Southern Oscillation, Arch. Met. Geoph. Biokl., Series A34, 311–348.

Nicholson, S.E. and Entekhabi, D., 1987: Rainfall variability in Equatorial and Southern Africa: Relationships with sea-surface temperatures along the southwest coast of Africa, J. Appl. Meteor. and Climate, 26, 561–578.

Nicholson, S.E. and Nyenzi, B.S., 1990: Temporal and spatial variability of SSTs in the tropical Atlantic and Indian Oceans. Meteorol. and Atmos. Physics, 42, 1–17.

Nyenzi, B.S. and Nicholson, S.E., 1990: An analysis of the interannual variability of rainfall over East Africa, In Press.

Ogallo, L., 1979: Rainfall variability in Africa, Mon. Wea. Rev., 107, 1133–1139.

Ogallo, L., 1980: Time series of rainfall in East Africa. PhD Thesis, University of Nairobi, Kenya, pp238.

Rasmusson E.M. and Carpenter, T.H., 1982: Variations in tropical sea surface temperature and wind fields associated with the Southern Oscillation/El Niño, Mon.Wea. Rev., 111, 517–528.

Rhode, H. and Virji, H., 1976: Trends and periodicities in East Africa rainfall data, Mon. Wea. Rev., 104, 307–315.

Salinger, M.J., 1989: New Zealand climate: from ice age to present. Environmental Monitorina in New Zealand. 32-40.

Spencer, R.W. and Christie, J.R., 1990: Precise monitoring of global temperatures from satellites. Science, 247, 1558–1562.

Thornthwaite, 1931: The climate of North America, according to a new classification, Geogr. Rev., 21, 633–655.

Vinnikov, K.Ya., Groisman, P.Ya. and Lugina, K.M., 1990: The empirical data on modern global climate changes (temperature and precipitation), J. Climate, 3, 662–667.

Wallace, J.M. and Gutzler, D.S., 1981: Telecommunications in the geopotential height field during the Northern hemisphere winter. Mon. Wea. Rev., 109, 784–812.

Webb, T. III, Bartlein, P.J. and Kutzbach, J.E., 1987: Climatic change in eastern North America during the past 18000 years: comparisons of pollen data with model results. In: The geology of North America, Vol K-3: North Ameica and Adjacent Oceans During the Last Deglaciation, (eds. W.F. Rudddiman and H.E. Wright). Geol. Soc. of America, Boulder, 447–462.

Winograd, I.J., Szabo, B.J., Coplen, T.B. and Riggs, A.C., 1988: A 250,000-year climatic record from great basin vein calcite: implications for Milankovitch theory, Science, 242, 1275–1280.

Winograd, I.J. and Coplen, T.B., 1989: Reply to comments on "Great Basin Calcite and the Pleistocene Time Scale", Science, 246, 262–263.

Climate Change Impact Studies: The IPCC Working Group II Report

by Yu. A. Izrael *(USSR)*

The terms of reference of IPCC Working Group II (WG-II) specified an evaluation of the environmental and socio-economic implications of possible climate changes caused by increased, human-induced greenhouse gas concentrations in the atmosphere. Over 150 specialists from 24 countries participated in preparing the group's report. The final session, which was devoted to discussing and adopting the definitive text of the report and its policy-makers' summary, was attended by delegations from 35 countries and a number of international organizations. The report is thus the result of the group of authors' scientific analysis and broad generalization.

The report is divided into sections describing potential climate change impacts on: agriculture and forestry; natural terrestrial ecosystems; hydrology and water resources; human settlements: energy, transport, and industrial sectors, human health, and air quality; oceans and coastal zones; and seasonal snow cover, ice and permafrost. It must be said that this work was made very difficult from the very beginning by the absence of WG-I's recommendations on the use of climate change scenarios. These were in fact received when the work on WG-II's report was practically completed and no changes could therefore be made to the text. Moreover, the WG-I report contains many uncertainties, particularly as regards the possibility of forecasting precipitation and soil moisture and predicting extreme situations on regional scales. The WG-II report is thus based on various global warming (climate change) scenarios existing in the literature at the time of writing, particularly scenario A (as described by WG-I), i.e. a worst-case situation.

The group's work frequently involved using the paleoclimate analog method, which provides additional means of validating climatic models, making regional assessments, and forecasting changes in precipitation.

On the basis of the work carried out, we can make the generalization that, in most cases, the impacts will be felt most severely in regions already under stress, primarily the developing countries. Moreover, human-induced climate changes caused by uncontrolled greenhouse-gas emissions will accentuate the impacts of natural factors.

Clearly, regional assessments and the extreme situations arising in some regions must therefore play a key role in the process of estimating climate change impacts.

Despite the many uncertainties WG-II made preliminary climate-change impact assessments, which are reflected in its report and its policy-makers' summary.

1. Potential Impacts on Agriculture and Forestry

Assessments based on a possible warming over the next few decades show that certain regions with variable moisture supply will become drier, permitting even greater soil degradation and crop losses. Moist regions will become even more saturated with moisture as a result of more frequent and intense tropical storms. There will be a change in the frequency and nature of extreme impacts on agriculture caused by flooding, persistent drought, forest fires and crop pests.

Estimates of the impacts of doubled CO_2 on crop potential have shown that in the northern mid-latitudes summer droughts will reduce the potential production by 10–30%. The impact of climate change on agriculture in all, or most, food-exporting regions will entail a rise in the average cost of world agricultural production by no less than 10%.

In the tropical and sub-tropical regions, where most developing countries are located, considerable climate changes are predicted. Besides the prolonged drought in the Sahel, Ethiopia, and a number of other African countries, there are other notable examples such as the extreme *El Niño* phenomenon in 1982–1983 which scientists assume to have caused the drought in Brazil, Australia, India and part of Africa during that period. This phenomenon was connected with an unusual warming in the surface waters of the eastern part of the Pacific Ocean. Also noteworthy are the droughts in the USSR (1972, 1975, 1981, etc.), and in the USA and Canada (1988).

The duration of the warm season is increasing in many parts of the world, which may lead to increased production owing to the adaptation of agriculture through the introduction of more late-ripening and, as a rule, higher-yield varieties. A review of farming techniques and, in particular, of sowing times will naturally also lead to increased production through the fuller use of the increased heat resources. This is important for all farming regions in the temperate belt, except for some where precipitation may decrease substantially.

Our estimates show that, globally, agriculture may remain stable at levels which are sufficient for meeting human food requirements, but that the cost of the measures to maintain these levels is unclear. Serious negative impacts may occur in some regions, especially those which are currently most climate-dependent. These may include the semi-arid and subtropical parts of Africa (a 30–70% reduction in crops may be expected in some regions with a warming of up to 3.5°C) and South America, as well as the tropical and equatorial parts of South-east Asia and Central America (crop reduction of up to 7% and 5–25% respectively); and these impacts will primarily affect the developing countries. Crop yields in some countries may increase (Philippines, Japan, etc.).

In the south and south-east of the European USSR, where the annual total precipitation may drop by up to 20% by the end of the 20th century, mainly during winter, the increased frequency of droughts may cause a 10–20% average reduction in cereal crops.

Reduced production may affect regions with a traditionally high level of cereal crop production, such as southern Europe (by 5–36% with a warming of 4°C), south of the USA (by several per cent, but in some cases there will be increased crops), and part of South America and western Australia. On the other hand, the production of cereals in northern Europe may rise. It is thought that in some parts of the world the climatic boundaries for agriculture will shift by 200–300 km per degree of warming.

Climate change may also cause considerable shifts in the main forest zones, which, in the northern hemisphere, may move several hundred kilometres northwards. The boreal forests and forests in the arid and semi-arid zones are particularly sensitive to climate change. Changes in forest systems may also have considerable impacts on both man and animals. Forests are, of course, a most important user of CO_2.

2. Potential Impacts on Natural Terrestrial Ecosystems

The main vegetation zones may be subjected to serious impacts which will affect, in turn, the plant communities and associated animal populations. It is possible that the speed of climate change will be so great that some ecosystems will not be able to adapt to the new climatic conditions. Some species may become extinct, thus reducing biological diversity.

Losses in the vegetation zones will be greatest in the higher latitudes, where the land area classified as polar deserts, tundra and boreal forest is expected to be reduced by about 20%. In the northern part of the Asian USSR, the zonal boundary will shift 500–600 km northwards. The tundra zone may completely disappear from the north of Eurasia.

The socio-economic consequences of these impacts will be particularly badly felt in regions where society's welfare largely depends on the natural ecosystems.

3. Potential Impacts on Hydrology and Water Resources

Many regions can expect an increase in the amount of precipitation, which will lead to changes in agricultural production and natural ecosystems. In other areas, soil moisture may diminish, especially where a moisture deficit in the soil is already experienced, such as in Africa. A 1–2°C rise in air temperature accompanied by a 10% reduction in the amount of precipitation may cause a 40–70% drop in mean annual river runoff, which will substantially affect agriculture, water supplies and hydroelectricity.

From the example of the study of the Sacramento River Basin, it can be seen how upstream water-resource systems depending on snowmelt-generated runoff may be influenced by global warming. Higher air temperatures will increase runoff through snowmelt by 16–81%. At the same time, summer runoff will drop by 30–68% and soil moisture will simultaneously be reduced by 14–36%.

4. Potential Impacts on Human Settlements: the Energy, Transport and Industrial Sectors; Human Health; and Air Quality

Flooding related to the sea-level rise caused by climate change accompanied by a change in precipitation may lead to widespread human migration. The most significant consequences can be expected in highly urbanized areas with high population densities. These may be connected with water-supply difficulties, the influence of rising heat loads, and more favourable conditions for the spread of infections. A change in the amount of precipitation and air temperature may radically alter the distribution of viral diseases by shifting their distribution boundary to higher latitudes.

Consequences for the energy, transport and industrial sectors will depend more on climate-change response strategies than on the climate changes themselves. Global warming is expected to considerably affect water resources and the biomass, which are important energy resources in most developing countries.

The impact on air quality will also largely depend on the response strategy in the area of fossil fuel burning for energy production, industry and transport.

5. Potential Impacts on the World Ocean

The projected global warming will affect the ocean's thermal budget, cause sea-level rise, modify ocean circulation and cause changes to marine ecosystems, ultimately with serious socio-economic consequences.

By the year 2050, a 30–50 cm sea-level rise can be expected and by the year 2100 a 1 m sea-level rise, which will cause erosion of the coastline, increase the salinity of estuaries and increase the tidal range in river mouths and inlets. These impacts will lead to the loss of large areas of productive land, contamination of freshwater sources and the displacement of millions of people from flood-prone areas. Of the areas affected by climate change, the most vulnerable from the human population point of view may be the large areas close to coastlines, such as Bangladesh, the arable Nile delta area, the small islands in the Pacific Ocean and the Caribbean Basin and the small island States such as the Maldives, Tuvalu, Kiribati, etc., as well as many large ports. The comparatively fast sea-level rise may change the ecological situation in offshore areas and present a serious threat to marine fisheries.

The currently projected increases of greenhouse gas concentrations in the atmosphere will lead to a rise in globally-averaged sea-surface temperature by 0.2–2.5°C and changes in practically all components of the energy balance.

The strongest impact of the increased sea-surface temperature and particularly the sea-level rise will be felt in the offshore ecosystems.

Alteration of the energy balance and circulation system in the world ocean will directly affect the productivity of marine ecosystems. Given that 45% of the total gross primary production is concentrated in upwellings in the ocean and coastal areas as well as in the high polar latitudes, the changes in functioning of these ecosystems in global warming conditions will determine the future productivity of the ocean.

Paleoclimate research shows that global warming will reduce the most productive area of the ocean by about 7%. It is expected that the ocean's primary production will, on the whole, drop by 5–10%. As a result, the production zones may be redistributed and the natural habitat of commercially valuable species of fish may change.

Current estimates show that global warming will cause an intensification of biodegradation processes, which will affect the high-latitude regions 30–50% more. The

vegetation period in the sub-Arctic and Arctic ecosystems, such as the Bering Sea, is also expected to change.

6. Potential Impacts on Seasonal Snow Cover, Ice and Permafrost

The continental cryosphere includes seasonal snow cover, glaciers, continental ice sheets, permafrost and the seasonal freeze-thaw layer. These ice masses currently cover about 41 million km^2.

The predicted climate change will quite substantially reduce the area and volume of the terrestrial cryosphere. This will affect not only the quantity of available drinking water, change the sea-level and relief characteristics, but also the life and activity of societies and their respective economic systems, which depend to differing degrees on the existence of the cryosphere.

The reduction of the seasonal snow cover in both space and time may considerably alter the hydrological regime and water resources.

The response of glaciers to climate change will depend on their type and geographical location. Melting of glaciers in the Soviet Arctic archipelagos may result in their disappearance in 150-250 years. In contrast, an assessment of mountain glaciers in the temperate zone of Eurasia indicates that up to 2020 these glaciers will, in general, remain essentially unchanged.

The projected warming during the next few decades would significantly deepen the active layer and initiate a northward retreat of permafrost. It is expected that a 2°C global warming would shift the southern boundary of the climatic zone currently associated with permafrost over most of Siberia north-east by at least 500–700 km. The socio-economic impacts of permafrost degradation will be mixed. Maintenance costs of existing facilities such as buildings, roads and pipelines will rise, with abandonment and relocation needed in some cases. The melting of permafrost would result in the release of methane from previously frozen biological material and from gas hydrates, which will enhance the greenhouse effect.

7. Concluding Remarks

These are the main results of the work of IPCC WG-II. Although the WG-II report on socio-economic and environmental impact assessments reflects the current knowledge and understanding of the problem, many questions are insufficiently covered because of the limited time available and the many uncertainties in the climate change predictions.

Nevertheless, the following conclusion can be drawn: the work of WG-II showed the great importance of climate change for the future development of mankind and the economy, and pointed out the main areas of unfavourable changes and damage, which will necessitate concerted international action.

However, the calculations and assessments made did not point to any catastrophic socio-economic consequences on *global* scales.

The critical regional effects and local extreme situations could not be demonstrated clearly because of the large uncertainties in the predictions and research results using the existing models.

Our next task is to make maximum use of the results obtained:

- In assessing climate changes and their impacts
- In verifying the choice of measures, optimal economic outlay and areas in which to use this outlay
- In determining expenditure priorities (choice between controlling the still insufficiently proved damage and assisting those who are already in need, i.e. fighting hunger and poverty).

This is the responsibility of the scientists and decision makers.

To fill the existing gaps, international scientific and technical co-operation and studies must be promoted to improve climate monitoring, develop climate change prediction, analyse the impact of climate change on crucially important human activities and the environment, and make socio-economic studies. Special attention must be given to the situation in the developing countries.

As a joint WMO/UNEP body with recognized scientific authority, the IPCC is an effective international instrument which is capable of solving this problem successfully; the work of this powerful scientific assembly should be continued.

In conclusion, I express my sincere gratitude to the large group of scientists and specialists from many countries who took part in the work of WG-II, on whose behalf I have presented this report.

Impacts on Hydrology and Water Resources

by H. F. Lins *(USA)*, I. A. Shiklomanov *(USSR)* and E. Z. Stakhiv *(USA)*

Abstract

It is possible that relatively small climate changes could produce large water resource problems in many areas, especially arid and semi-arid regions and those humid areas where demand or pollution has led to water scarcity. It is important to realize, however, that regional details of greenhouse-gas-induced hydrometeorological change are virtually unknown. If meaningful regional estimates of water resources conditions, appropriate for planning and policy formulation, are to be produced, then studies must include estimates on the frequency, intensity and duration of potential future hydrological events. This is especially critical for evaluating effects on agriculture and the design of water resource management systems, and for producing reasonably accurate water supply estimates. At the present time, the capability to produce such specific regional forecasts does not exist. What is currently possible, however, is the capability to assess water resource sensitivities. Such assessments are necessary in all countries, but especially those located in environmentally sensitive arid and semi-arid regions, where the potential for conflicts associated with low water-resource system development and rapidly increasing water demands is high. Studies that produce improved procedures for operating water management systems considering climate uncertainty are also both possible and necessary at the present time.

Introduction

As has been the pattern throughout Earth history, changes in climatic conditions, for whatever reasons, will probably alter land and water resources, their distribution in space and time, the hydrological cycle of water bodies, water quality and, in more recent millenia, water supply systems and requirements for water resources in different regions. Quantitative estimates of the hydrological effects of climate change are essential for understanding and solving potential water-resource problems associated with domestic water use, industry, power generation, agriculture, transportation, future water-resource system planning and management, and protection of the natural environment.

Climate change can be expected to produce changes in soil moisture and water resources. The most important climate variable that may change is regional precipitation, which cannot be predicted accurately. Water supply and use in semi-arid lands are very sensitive to small changes in precipitation and evapotranspiration by vegetation, because the fraction of precipitation that runs off or percolates to groundwater is small. Increased heat will lead to more evapotranspiration, but the increase is expected to be partly offset by reduced plant water use in a CO_2-enriched atmosphere.

Higher temperatures may also have an impact in the transitional winter snow zones. More winter precipitation would be in the form of rain instead of snow, thereby increasing winter season runoff and decreasing spring and summer snowmelt flows. Where the additional winter runoff cannot be stored because of flood control considerations or lack of adequate storage, a loss in usable supply would be the result.

Significantly, the ability to assess hydrological and water-resource impacts of climatic change is predicated on satisfying several important requirements. First, it is necessary to understand how hydrological systems respond to meteorological and climatological conditions. Notably, the current capability for meeting this requirement is only fair. Second, one needs estimates of regional meteorological conditions attendant to a potential future climate. This is the 'water supply' side of the impacts equation. Unfortunately, the ability to produce an accurate regional forecast of climatic conditions decades, years, or even months into the future is currently very poor. Third, estimates of the regional social and economic conditions likely to exist contemporaneously with a future changed climate are necessary. This is the 'water demand' component of the impacts equation. As with climate forecasting, the current ability to accurately forecast regional social and economic conditions is poor. It is noteworthy that studies of the accuracy of socio-economic forecasts in general, and water-use forecasts in particular, have shown that the majority of such forecasts during the past 30 years have been substantially in error (Osborn *et al.*, 1986).

This, then, is the background on which the IPCC subgroup that focused on the hydrological and water-resource impacts did its work. The present report is a condensed version of the material contained in the Hydrology and Water Resources chapter of the IPCC final report. The objectives of the work reported in that chapter were (1) to assess the potential hydrological and water-resource impacts of a greenhouse-gas-induced climatic change; (2) to assess the related potential social and economic impacts; (3) to identify weaknesses in the current knowledge base regarding hydrological impacts of climate change; and (4) to recommend steps to improve this knowledge base. Given the two-year time frame of the IPCC study, the working groups were constrained to draw upon existing published investigations. No new studies were to be undertaken as part of the process.

The reader will note that there has been no attempt made by the authors to critically evaluate the quality of the findings of the reports cited herein. There are two reasons for this. First, in no case are the estimates of effects reviewed herein intended to be unequivocal 'statements of impact.' The consensus view, and that of the authors, is that it is not currently possible to produce impact assessments of climatic change since it is not yet possible to forecast either future regional climatic or socio-economic conditions. Second, the authors were charged in the IPCC process with reviewing the literature that focused on regional effects of climate change without regard to the methods used. The process emphasized geographical inclusivity and not critical evaluation. Thus, the reader is cautioned not to place any degree of confidence in the 'potential' effects that have been reported in the literature and synthesized herein.

Methods

Estimates of the potential hydrological and water-resource effects of climate change have been produced using simulated climatic conditions or scenarios developed by several distinct methods, including (1) hypothetical (or prescribed) scenarios, (2) scenarios obtained from atmospheric general circulation models (GCMs), and (3) scenarios based on historical and paleoclimatic reconstructions.

In addition, during the past ten years hydrologists from many countries have extensively studied the potential hydrological consequences of future anthropogenic climate change. These studies have been based on various methods that can be grouped as follows: (i) analysis of long-term variations in runoff and meteorological elements over past periods using either statistical analyses of the relations between runoff, air temperature, and precipitation, or analyses of the hydrological consequences of past periods of very warm or cold, wet or dry conditions; (ii) use of water balance methods over a long period of time; (iii) use of atmospheric GCMs. In this approach, GCMs with prescribed increases in the concentrations of trace gases in the atmosphere (usually $2 \times CO_2$) are used to obtain direct estimates of changes in the climatic and hydrological characteristics for large regions; and (iv) use of deterministic hydrological models. In this approach, rainfall-runoff models for river basins are employed with climatological data sets, including GCM outputs, to determine changes in hydrological conditions.

The first and second methods have been widely applied to estimate changes in water resources over large areas because a relatively small amount of initial data is required, usually annual runoff, precipitation, and air temperature. Caution should be exercised in extrapolating regression relationships over past years to future periods. One cannot assume that a past inter-annual pattern of meteorological factors will be repeated in the future. It is also true that for the same annual precipitation and temperature, annual runoff can vary widely, depending on the distribution of the meteorological variables within months and seasons.

The results obtained in hydrological simulations based on different GCMs are inconsistent for certain important hydrological conditions and regions. This can be attributed to the low resolution of the current generation of GCMs, and to their simplified description of hydrological processes. Nevertheless, the approach is very promising and studies of this type should be continued.

Deterministic hydrological models have some desirable properties. They allow explicit study of causal relations in the climate/water-resource system for estimating the sensitivity of river basins to changing climatic conditions. In addition, when regional climatic forecasts are available, possible runoff changes in different hydroclimatic environments may be simulated for water planning and management.

Purported Hydrological and Water-Resource Changes in Large Regions and Countries

Numerous studies have been completed recently that purportedly identify potential hydrological and water-resource impacts of climatic warming. Despite the fact that, in the absence of accurate regional forecasts of future climatic and socio-economic conditions, it is not possible to produce a true water-resource impact assessment, these studies have been drawn upon in some instances as justification for the need to develop immediate policy responses. Therefore, a summary of the broad conclusions of many of these investigations, organized by large geographical areas and countries, follows. This summary is not intended to be comprehensive. Rather, it is intended to provide an overview of the type of studies that currently exist in the literature and their level of specificity.

North America

In both the United States and Canada, there is a rapidly growing literature on the likely effects of a greenhouse-induced climatic change on water resources. In analysing the conclusions drawn by different researchers, based generally on the output of general circulation models, the following regional tendencies are presented as purported hydrological effects of global warming for the United States. Some increase in annual runoff and floods is possible in the Pacific north-west. A considerable increase in winter and decrease in summer runoff with insignificant rise in annual runoff may occur in California. In the Colorado and Rio Grande river basins, as well as in the Great Basin, a tendency toward decreasing runoff is often cited. Similarly, decreasing runoff and increasing evaporation are frequently reported for the Great Lakes basin. In the Great Plains, and across the northern and south-eastern states, disparity in model output indicates a high degree of uncertainty exists regarding potential water-resource effects.

Looking very broadly at the impact of climatic warming on the water resources of Canada, the outlook is generally reported to be somewhat favourable. Assuming conditions simulated by a number of general circulation models, with temperature rises generally in the 2°C to 4°C range and precipitation generally increasing across the country from 11% to 54%, runoff could increase in all major regions by 10% to 235% (Ripley, 1987).

Western Europe and Scandinavia

Effects of climate change on water resources have been estimated for western Europe (primarily for the northern regions) by using various scenarios and approaches. A very broad assessment of the potential water-resource response to climate change for over 12 countries within the EEC has been conducted by da Cunha (1989). Information about water resources and water availability for each country, coupled with the scenarios of climatic change simulated by general circulation models for the EEC region, is used to suggest that man-induced climatic change is likely to influence water resources differentially in different parts of the region. If the simulated general circulation model results turn out to be correct, then average precipitation values in the southern part of Spain, Portugal, and Greece could decrease, while increases could occur throughout the rest of the EEC region. The model results also indicate a tendency for reduced precipitation variability in some of the northern parts of the region during the entire year, and for increased variability in Greece and southern Italy during

the summer. Such changes in variability are interpreted by da Cunha (1989) as meaning that higher flood peaks are possible in parts of the United Kingdom, Germany, Denmark, and the Netherlands, while more severe droughts are possible in Greece and southern Italy.

In Norway, Sælthun *et al.* (1990) have produced an initial assessment of the consequences of climatic change for water resources. Using a climate change scenario of a temperature increase of 1.5°C–3.5°C (with the higher temperatures prevailing during winter and in upland areas) and a precipitation increase of 7%–8%, the following hydrological consequences were posited. Annual runoff in mountainous areas and regions of high annual precipitation will likely experience a moderate increase. In lowland and forest inland basins, annual runoff may decrease in response to increased evapotranspiration. The intensity of the spring flood could decrease in most basins, while winter runoff may increase substantially and summer runoff may decrease. Floods could possibly occur more frequently in autumn and winter. The period of snow cover may be shortened by one to three months. One positive potential outcome, however, is that the hydroelectric power production of Norway could increase by 2%–3% under the assumed climate change scenario. This is due in part to an increase in reservoir inflows, and in part to reductions in reservoirs spillovers. The seasonal distribution of runoff may more closely match that of energy consumption, thereby increasing firm power yield.

Union of Soviet Socialist Republics

Quantitative estimates of potential climate change effects on rivers in the USSR have generally been based on paleoclimatic analogs. From these proxy records, predictions of changes in summer and winter temperature and annual precipitation have been developed from which estimates of mean annual runoff changes over the USSR for a 1°C global warming (expected to occur between 2000–2010) have been made. The runoff estimates were derived from precipitation-runoff relationships, or from water balance equations, developed over long time periods and are summarized as follows.

The most serious effects on water resources are possible south of the forest zones of the European USSR and western Siberia. Annual runoff could fall to 10–20% of normal (20–25 mm). In the southernmost steppe regions, in response to a potential marked increase in precipitation, the annual runoff may increase by 10–20%. In the northern regions of the European USSR and Siberia, runoff may increase by 280–320 km^3/yr or 7%, which may compensate for potential future increases in water consumption. Estimates for the more distant future are even less certain; nevertheless, there is evidence to suggest that with a 2°C global warming the annual runoff will possibly increase by 10–20% on all the large rivers of the USSR. This increase could possibly provide an additional 700–800 km³/yr of water for the whole country.

Asia

An analysis of hydrometeorological observational data for Northern China (Liu, 1989) shows that the warmest period over the last 250 years began in 1981. The mean air temperature over the period 1981–1987 was 0.5°C above the normal, while for the same period precipitation was somewhat lower than normal (for Beijing by 4%). Studies of natural climate variations for the past 100 years suggest that the warming in northern China will continue up to the next century. Estimates of the potential influence of increasing atmospheric CO_2 on northern China are not available.

At the same time, it is possible that small climate changes can have considerable hydrological consequences. Simulations using the Hinangchzang hydrological model indicate that in semi-arid regions, with a 10% increase in precipitation and a 4% decrease in evaporation, the runoff increases by 27%. Increasing precipitation by 10% and evaporation by 4% results in an 18% increase in runoff (Liu, 1989).

A number of agencies and institutes within Japan have begun to focus on how climate warming would affect the water resources and related environment systems of that country. To date, the studies have emphasized the empirical description and characterization of effects and problem areas with the aim of identifying how to derive quantitative estimates of likely impacts. Given the critical concern with which the government of Japan views the potential threat imposed by a greenhouse-induced climatic warming, a preliminary study was undertaken with the support of the Japanese Environment Agency (Matsuo *et al.*, 1989). It qualitatively assessed potential effects on a number of environmental systems, including water resources, associated with a possible severe change in climatic conditions. The specific water resources issues addressed included flood control, water use, and water quality.

Looking first at flood control the report notes that current reservoir storage capacity is limited, so much so that the multipurpose use of reservoirs for flood control and water use could become increasingly difficult. An increase in rainfall intensity brought by typhoons or fronts would give rise to serious flooding. Prolonged periods of drought, punctuated by short bursts of intense precipitation, could lead to increased frequency of mudslides. Since many Japanese cities are situated in coastal areas and lowlands, any flood could produce heavy damage. In terms of water use, a number of effects were identified. In general, water demand is expected to increase, while water supply decreases, spreading localized water shortages to broader regions or to the nation as a whole. Reduced precipitation and the associated reduction in discharge can be expected

to reduce both the amount and regularity of hydroelectric power generation. In the particular case of rivers dependent upon snowmelt, the impact could be quite significant. Moreover, contamination due to salt-water intrusion could lead to acute agricultural and municipal water supply problems. Finally, with respect to water quality, the expectation is that river water quality will decline with decreases in minimum flow. Lake quality could be threatened by a prolonged state of thermal stratification, and increasing temperatures may lead to increased eutrophication. It is also thought that declines in the water level and storage capacity of lakes and dams could adversely affect water quality, although the precise mechanism behind such deterioration remains unclear.

New Zealand

In June 1988, the New Zealand government instituted a national climate change programme under the co-ordination of the Ministry for the Environment. Three working groups were formed, one each addressing facts, impacts, and responses, that are consistent with the three working groups of the IPCC. The Impacts Working Group was formed to produce a series of reports addressing the broad range of physical, biological, economic, and social issues likely to be affected by the climatic change estimates provided by the Facts Working Group. One of these impact reports deals with water resources (Griffiths, 1989). All the impact assessments considered two scenarios of climate change, one based on a temporal analog (referred to as S1) and the other on the limiting conditions simulated by general circulation models (S2). The analog scenario is based on the period of maximum warmth 8 000 to 10 000 years ago, when westerlies were weaker and there was more airflow from the north-west. New Zealand temperatures were 1.5°C warmer than at present and westerly winds were lighter, especially in winter. There was a reduced frequency of frontal storms, and global sea-level was 20 cm to 40 cm higher than at present. The second scenario is based on the model-simulated upper limit to greenhouse warming in the New Zealand area of 3°C. It is accompanied by positive-mode Southern Oscillation conditions (*La Nina*) on average. Frequent incursions of tropical air from the north are expected and, although the intensity of the westerlies may decrease, they will still prevail. Moreover, global sea-level may be 30 cm to 60 cm higher than at present.

The major hydrological and water-resource impacts which purportedly would occur in New Zealand can be summarized as follows. In broad terms, along the main axial mountain range on the North and South Islands, little change in runoff is anticipated. Along the eastern margins of both islands notable decreases (40%–80%) could occur, especially in the southern half of the South Island. In most other areas, runoff increases are predicted. Recreational and scenic opportunities may be enhanced in lakes and rivers, and more reservoir storage will assist hydroelectric operations. Wetlands may grow in number and size, and may be more difficult to drain in areas where such drainage is desirable. As temperatures rise, there will be a tendency for consumers to increase water use, thus increased water-use efficiency will be an imperative in those areas where runoff decreases are likely. The simulations suggest that the magnitude and frequency of droughts may increase in eastern New Zealand, resulting in intense competition between instream and offstream uses during low flows. Existing standards for setting minimum flows and for water quality will come under severe pressure for revision. Sea-level rise, as well as flooding and drainage problems in low-lying coastal areas affecting the design performance of local hydraulic structures, may also cause some rivers to adjust to new base levels.

Case Studies of Purported Effects in Critical or Sensitive Environments

In addition to describing the potential water-resource effects of global warming at national to continental scales, a number of more specialized evaluations were included in the IPCC study that focused on critical or sensitive environments. These included purported impacts on selected large water bodies, arid and semi-arid zones, critical agricultural regions, intensively urbanized areas, and regions of snowmelt-generated water supply.

Large Water Bodies
The Great Lakes, with a surface area of 246 000 km^2 and a volume over 8 000 km^3, store 20% of the world's fresh surface water and 95% of the United States' fresh surface water. The drainage area, including the surface area of the Lakes themselves, encompasses nearly 766 000 km^2. Eight States of the United States and two provinces of Canada border the Great Lakes. Two of the Great Lakes are regulated: Lake Superior and Lake Ontario; i.e., their outflows are controlled. Lakes Michigan and Huron are connected by the Straits of Mackinac and their surface water elevations respond synchronously to changes in water supply. Twenty-nine million Americans (12% of the US population) and eight million Canadians (27% of the Canadian population) live within the Great Lakes basin (Cohen *et al.*, 1989). Millions of people benefit directly from the water-resources-related services of the Great Lakes in the form of hydroelectric power, navigation/ transportation of mineral resources including coal and iron and food. In 1975 economic activity in the US portion amounted to $155 billion (1971) while that of Canada was estimated to be $27 billion (1971) (International Joint Commission, 1985). By 1985, the Great Lakes basin accounted for 37% of the US manufacturing output, consisting of transportation equipment, machinery, primary

metals, fabricated metals and food and beverage products. Clearly, therefore, any climatic change that would alter the social and economic conditions associated with the Great Lakes assumes major national and international importance.

A detailed examination of available water supply and users for the Great Lakes basin (in both the USA and Canada) reveals a number of commonly encountered conflicts among competing water uses, even under present climatic conditions. First, it should be recognized that there are two major and different hydrological regions: the Great Lakes proper and the tributary watersheds. Most of the salient conflicting water uses for which the regulation and stabilization of the Great Lakes are important are navigation and hydropower production, which favour high lake levels, versus recreation, reduction of flooding and shoreline erosion which favour low lake levels. The former (hydropower and navigation) affect the basis of an extensive industrial economy in the region, while the latter affect millions of residents who wish to recreate on the shores of the Great Lakes and whose houses and property are affected by storm damage. Since the Lakes themselves are a vast reservoir of water, the projected typical consumptive uses of water (municipal and industrial water supply, thermal cooling) are not expected to impose a significant incremental adverse impact. Tributary watersheds, on the other hand, experience highly variable flows, with significant present constraints on in-stream and off-stream water uses.

Based on the output from three general circulation models, Croley (in press) compared runoff simulations under doubled CO_2 conditions. The analysis used daily models to simulate moisture storage and runoff for 121 watersheds draining into the Great Lakes, over-lake precipitation into each lake, and the heat storage and evaporation for each lake. A 23% to 51% reduction in net basin supplies to all the Great Lakes was suggested by the three models. The largest changes in net basin supply would occur in Lake Erie, a feature that was uniform in all three models. Thermoelectric power cooling is projected to increase significantly, accounting also for much of the increase in consumptive water use. Consumptive uses of water, largely through evaporation, represent a net loss of water to the hydrological system. This fairly large loss was estimated by the US Geological Survey to be 85 cm in 1985, rising to 170 cm in 2000. This loss, however, would account only for a reduction of a few centimetres in lake levels in comparison to a one-half metre drop in lake levels projected from a reduction of net basin supplies due to increased natural evapotranspiration. If the dramatic climate-change scenario were to materialize, the large socio-economic impacts would stem from water-use conflicts and shortages that are currently encountered primarily during naturally fluctuating low lake-level stands.

That is, hydroelectric power production would be significantly affected, particularly in the Lake Ontario and Lake Erie drainage areas. There are substitutes for the loss of hydropower, but most will add to the emission of greenhouse gases and consumption of thermoelectric cooling water. Commercial navigation is a very important component of the Great Lakes economy that would be seriously affected by lower lake levels. Either cargoes would have to decrease to get through the locks, imposing increased transportation costs, or the locks on the Great Lakes would have to be rebuilt. In either case, the economic cost would be high.

The water-resource impacts are also likely to be of great socio-economic consequence in the streams of the watersheds that drain into the Great Lakes. These watersheds are largely unregulated and the municipalities and industries depend on natural streamflow and groundwater. Although the variability of weather and associated shifts in the frequency and magnitude of climate events were not available from the outputs of the GCMs, there is reason to believe that the increased precipitation regime predicted by some of the GCMs will result in greater and more frequent flooding in the tributary watersheds. Along with this trend, it is possible that the frequency, duration and magnitude of droughts might also increase as a consequence of the warming trend. In other words, the cycles of floods and droughts experienced in the current hydrological record could become worse, exacerbating future conditions of higher water demands.

Considerable attention has also been focused on another large water body, the Caspian Sea in the Soviet Union, which is the world's largest enclosed lake, with a surface area of 371 000 km^2. Lake-level variations have a very large amplitude and depend mainly on the river inflow-water surface evaporation ratio (precipitation on the lake surface is small). Over the last 2000 years the amplitude of lake-level variation has been about 10 m. Instrumental observations (since 1837) indicate that lake-level variations ranged between 25.5 m and 26.5 m below sea-level during the period 1837 to 1932. The lake then began to fall and reached a low of 29.10 m below sea-level in 1977. From 1978 to present the lake level has increased, reaching a level of –27.7 m in 1989 (Shiklomanov, 1988). Future lake levels depend on natural inflow variation (80% of the inflow is contributed by runoff from the Volga River). Other controlling factors are man's activities in the river basin, anthropogenic climate changes, and changes in precipitation and evaporation.

To assess future levels of the Caspian Sea, methods for probabilistic prediction have been developed which take into account potential effects of natural and anthropogenic factors, including global warming. Scenarios based on empirical data for all water-balance components obtained since 1880 were used. According to estimates prepared by the USSR State Hydrological Institute, the mean decrease (in km^3 yr^{-1}) in river water inflow due to present and projected human activities is: in 1989, –40; 2000, –55; 2010, –60; 2020, –65. For every year the values of river runoff were computed separately for all the main rivers.

Caspian Sea-level projections, based on water-balance simulations under a variety of potential climate change conditions, were also made using characteristic statistical properties (variability of water-balance components, autoregression and cross-correlation values, etc.) of the observational record. Results of this work suggest that by the end of the century some lake-level lowering may take place due to increasing human activities in the basin. After the year 2000, some stabilization may occur followed by a significant rise in the Sea's level due to predicted increases in inflow and precipitation over its surface (Shiklomanov, 1988; 1989). These results, though preliminary, show the importance and necessity of taking into account anthropogenic climate-change forecasts to estimate the fate of continental reservoirs and to plan future management.

Arid and Semi-Arid Zones

North Africa and the Sahelian zone are both subject to frequent and disastrous droughts, progressive aridization, and encroaching desertification. The Sahel, a vast though narrow belt stretching from West Africa to the Horn of Africa, is the transition zone from the Saharan desert to the hot and semi-arid African savannahs. Analysis of annual precipitation data from stations in the Sahelian zone for the period 1968–1988 indicate that this period was exceptionally dry. Although the current Sahelian drought appears to many scientists and policymakers to be a persistent and unprecedented phenomenon, this drought is far from unique. During the current century, the Sahelian region experienced several dry periods of varying duration, magnitude, and spatial extent (Grove, 1973; Sircoulon, 1976). During the last five centuries, historical accounts and paleoclimatic evidence indicate that there have been rainfall fluctuations of extremely variable duration (from years to several decades) (Nicholson, 1978;1989). Demarée and Nicolis (1990) viewed the Sahelian drought as a fluctuation-induced abrupt transition between a stable state of 'high' rainfall and a stable state of 'low' rainfall.

The major reason for Sahelian droughts is a decrease in annual precipitation. Ojo (1987) analysed precipitation data for 1901–1985 using 60 stations in western Africa. He found that during 1970–79 average precipitation there was 62% of normal and, during 1981–84, about 50% of normal. The Sahelian zone is characterized by a strong sensitivity to hydrometeorological conditions, especially precipitation. This is confirmed by analysis of precipitation and runoff over the most dry and wet five-year periods of time in the Senegal, Niger and Shari River basins, as well as by Lake Chad levels and areas indicated in Sircoulon (1987).

According to these data, with precipitation increasing by 20–30% the runoff rises by 30–50% in the river basins; with precipitation decreasing by 9–24% the runoff is reduced by 15–59%. In addition, between the early 1960s and 1985, the area of Lake Chad shrank by more than 11 times, i.e. from 23 500 km^2 to 2 000 km^2.

Very little research has focused directly on the water-resource impacts of an anthropogenic climate change in North Africa. However, using paleoclimatic reconstructions of long-term natural climatic changes in this region, Soviet climatologists have proposed a somewhat optimistic forecast (Budyko and Izrael, 1987). Accordingly, over the next several decades annual precipitation may increase considerably, as may total moisture and river runoff. Such a forecast tends to agree with the longer-term pattern of aperiodic wet and dry transitions for this area. Given this clear pattern of variability and taking into account the importance of climatic and hydrological data for use in economic analyses of the African continent, there is a clear need to establish (i) systematic networks for meteorological data monitoring, and (ii) improved and expanded surface hydrological data collection networks.

Critical Agricultural Regions

Next to the direct impact on water resources, perhaps the most critical social and economic concern associated with climatic change is the impact on agriculture. This is especially true in marginal farming regions of dryland farming or those that are heavily dependent upon irrigation for the maintenance of agricultural productivity. An example of such a region is the South Platte river basin in the west-central part of the United States. The basin comprises 62 210 km^2 of mountains, plains, foothills and alluvial fans. The western part is dominated by the rugged mountains of the Front Range, which is part of the southern Rocky Mountains in Colorado. Of the 49 250 km^2 of the basin in Colorado, about 20% is above 2 700 metres in altitude, and more than half is above 1 800 metres. A few peaks rise above 4 000 metres.

The South Platte river basin is typical of the agriculturally-based economy of the US Great Plains states, which includes a large livestock industry. Farmers in the Great Plains are particularly vulnerable to climate variability and the drought of the 1930s was devastating, serving as the impetus for a large-scale social migration. The Great Plains has become a productive agricultural area in the US thanks to the large degree of water-resource development. In its natural state, however, it is a marginal agricultural region and is very sensitive to climatic variability.

In general, results of recent studies show a net reduction in crop (wheat send corn) yields, even with the compensating effects of CO_2 on plant photosynthesis. Dryland farming would likely become increasingly risky

and less economical. Moreover, the present water resources base (both surface and groundwater) could not compensate for the relatively large loss of marginal dryland acreage. Navigation and hydropower production on the mainstem reservoirs could be threatened, as could surface water quality. Development of the basin's surface water potential is near capacity. Substantial changes in present water use technologies, particularly for irrigation, would have to be undertaken to meet the more stressful conditions postulated under the modelled climate-change scenario. This encompasses increased water use efficiencies, conjunctive use of surface and ground waters, transfer of water rights and, lastly, major new imports of water from outside the basin.

Another detailed look at the hydrological and water resources impacts of a climatic change in a critical agricultural region has been completed for the Murray-Darling basin in south-eastern Australia (Stewart, 1989). In this basin, which has an area of 1.06 million km^2, about 10 million of the 12.4 million ML (mega litres) of water per annum available under present climatic conditions have already been developed and mean water usage already exceeds 8.6 million ML (Zillman, 1989). The region presently accounts for just under 60% of mean annual water use and includes 75% of the total irrigated area of the continent (1.7 million ha). The Murray-Darling basin must be regarded as extremely vulnerable to the effects of climate variability or change and is of major socio-economic importance to Australia.

Close (1988) used a hydrological model to examine how the River Murray system would have behaved if the current storages and current irrigation developments had existed for the last 94 years. By superimposing predictions of climatic changes assumed to be associated with the greenhouse effect, the impacts on water resources and system operation were estimated. Close concluded that:

i) Tributary flows will increase over almost all of the Murray-Darling basin

ii) Flow increases will be more pronounced in the northern rivers of the basin than in those in the south

iii) Spring, autumn, and summer precipitation will increase while winter rainfall may decrease slightly

iv) Demand for irrigation water per hectare will decrease slightly

v) The water resources available for irrigation will increase which will permit either the area of irrigation to be increased or the security of supply of the current irrigation areas to be improved

vi) Dilution flows will increase and hence salinity in the lower river will decrease.

Intensively Urbanized Areas

Estimating the water resources impacts of climatic change in dynamic and intensively urbanized areas has some unique problems. Perhaps the most significant of these is that such areas frequently undergo major population, commercial and industrial changes over time periods of several decades. The water-resource effects of these changes are likely to be more significant than any that would be attendant to climatic change over the same time frame. Thus, estimates of changes in total population and its spatial distribution, as well as changes in industrial and land-use conditions, are as essential to the determination of future water resources conditions as are estimates of climate change. Unfortunately, experience to date indicates that the estimation of future demographic, industrial, technological, and land-use conditions, when available, is generally unreliable (Osborn *et al.*, 1986).

In the United States, a recent study investigated the sensitivity of hydrological and water-resource conditions in the Delaware River basin, an intensively urbanized basin on the middle Atlantic coast, to potential changes in climate (Ayers *et al.*, 1990). This work evaluated water-resource impacts using estimates of climate change both with and without estimates of social and economic change. The Delaware River basin is an environmentally diverse region encompassing an area of approximately 33 000 km^2 and crossing five major physiographic provinces and three distinct ecosystem types. Runoff processes differ considerably over the basin; in addition, human activities influence the movement and storage of water in the basin; adding considerable complexity to the hydrological response characteristics of the various physiographic provinces. Urbanization in the lower portions of the basin has significantly altered the regional runoff response (Ayers and Leavesley, 1988).

Simulations using a variety of deterministic and stochastic hydrological models illustrate two important characteristics about the sensitivity of basin watersheds to climatic change. First, seasonal differences in the expected effects of global warming on streamflow are observed. Maximum daily streamflows increase with time (more positive than negative trends) in mid-winter months, decrease in spring and summer, and change little in autumn and early winter. These seasonal differences in trend primarily reflect changes in snowfall accumulation and snowmelt. With warming, more winter precipitation falls as rain than as snow, and snowmelt occurs earlier. The warming effect is strongest in the northern part of the basin where snow accumulation currently is significant. Second, natural variability in precipitation masks the effects of increasing temperature. The percentage of simulations that do not show a significant increase or decrease is greater than 58% in any month and averages 84%. Results of the simulations using the daily watershed model (without

reservoirs) indicate that, overall, warming alone would cause a decrease in daily streamflow, specifically the maximum and average daily low and seven-day low flow. Most of this decrease would occur in the warmer months. Where snow accumulation currently is significant (in the northern part of the basin), however, the warming would result in an increase in the February average and maximum daily flow, regardless of precipitation changes. In general, watershed runoff was found to be more sensitive to changes in daily precipitation amounts than to changes in daily temperature or precipitation duration. Detectability of runoff changes was masked by the underlying variability in precipitation.

Regions of Snowmelt-Generated Water Supply

The Sacramento-San Jaoquin Rivers comprise the drainage basin of the Central Valley of California, one of the most productive and diverse agricultural regions in the world. It encompasses several large metropolitan areas, and sustains many others. Over 40% of California's total surface water drains from the Central Valley Basin into the San Francisco Bay area, supplying water for irrigated agriculture, municipal and industrial uses and numerous other recreational and ecological purposes.

Elevations in the basin range from sea-level in the Sacramento-San Joaquin River Delta to mountain peaks of over 4 200 m in the Sierra Nevada range. California's climate is characterized by wet winters and dry summers, with mean annual precipitation decreasing from 190 cm in the northern Sacramento River basin to 15 cm in the southern San Jaoquin River basin. The key feature of the hydrological characteristics of this basin is that a significant proportion of the basin's precipitation falls as snow in the high mountains. Consequently, storage of water in the snowpack controls the seasonal timing of runoff in the Central Valley rivers and has shaped the evolution of strategies for water management and flood protection (Smith and Tirpak, 1989).

An assessment of the possible water resources impacts of a doubling of atmosphere CO_2 has been made using the output from three general circulation models. It should be noted that conditions for the entire Central Valley drainage system are encompassed within one grid cell for each of the three GCMs. There is considerable diversity among the three models, especially with regard to the seasonal prediction of precipitation changes. Nevertheless, based on a series of hydrological impact studies conducted through the US Environmental Protection Agency and reported in Smith and Tirpak (1989), it was purported that total annual runoff in the Sacramento-San Joaquin River basin might be expected to remain near current levels or to increase somewhat under the scenarios simulated by each of the GCMs. However, major changes could likely occur in the seasonality of the runoff. In response to higher temperatures, less precipitation would be expected to fall as snow, and the snowpack that does accumulate could melt earlier in the year. Thus, there could be higher runoff in the winter months and considerably less in the traditional spring snowmelt-runoff season. It has also been estimated that a 3°C temperature increase could shift about one-third of the present spring snowmelt into increased winter runoff. Also, the possibility exists for runoff variability to increase substantially during the winter season. Change in the timing of runoff could have profound influences both on the aquatic and terrestrial ecology and on water resources management, despite the fact that total annual runoff could be slightly higher. First, more liquid precipitation in the winter months could lead to reduced runoff in the late spring and summer, thereby reducing the carryover soil moisture needed by the vegetation. Also, such a reduction in streamflow in the late summer could lead to a degradation in the health of aquatic species.

An analysis of 1990 water-use demands, in conjunction with instream flow constraints and delta outflow requirements and current reservoir operating policies, indicates a potential decrease in mean annual deliveries from 7% to 15%. This would occur because under current operating rules combined with earlier (i.e. winter) snowmelt, the reservoir would not have the capacity to store the early runoff for summer season irrigation withdrawals while simultaneously retaining springtime flood-control capabilities. Future demands for surface water from the Sacramento-San Joaquin system purportedly would increase 30% by the year 2010. 'This demand could not be reliably supplied under the current climate and resource system, and the shortage might be exacerbated under either of the three GCM derived scenarios' (Smith and Tirpak, 1989). Furthermore, the potential magnitude of changes in the seasonality of runoff are such that operational changes alone would probably not significantly improve the system's performance.

Summary

It is possible that relatively small climate changes could produce large water-resource problems in many areas, especially arid and semi-arid regions and those humid areas where demand or pollution has led to water scarcity. It is important to realize, however, that regional details of greenhouse gas-induced hydrometeorological change are virtually unknown. Although in recent years there has been an increased interest placed on understanding how world-wide water resources would be impacted by long-term climate change, little attention has focused on the fact that our current understanding of how water resources respond to short-term climate variability is seriously deficient. Significantly, it is only by first improving our knowledge of the latter shorter-term processes that it will be possible to answer questions associated with the former longer-term process. Thus, increased understanding of relations between climatic variability and hydrological response must be developed. Such work should include the development of methods for translating climate model information into a form that provides relevant data for watershed and water-resource system models.

If meaningful estimates of water resource conditions, appropriate for planning and policy formulation, are to be produced, then studies must include estimates on the frequency, intensity and duration of potential future hydrological events. This is especially critical for evaluating effects on agriculture, the design of water-resource management systems, and for producing reasonably accurate water-supply estimates.

In many instances, it is quite possible that changes in hydrological extremes in response to global warming will be more significant than changes in hydrological mean conditions. Thus, attention must be focused on changes in the frequency and magnitude of floods and droughts in evaluating the societal ramifications of water-resource changes.

Initial water-resource planning and policymaking will continue to be implemented even in the face of uncertainty about global change. Clarification and specification of the useful information content of the various methods for estimating future change must be made available to the management community.

Areas particularly vulnerable to even small changes in climate must be identified world-wide. Vulnerabilities must be ascertained considering both natural and anthropogenic conditions and potential changes.

Intensive assessments of water-resource sensitivities are necessary in all countries, especially those located in environmentally sensitive arid and semi-arid regions, where the potential for conflicts associated with low water-resource system development and rapidly increasing water demands is high.

Studies are needed that produce improved procedures for operating water management systems in consideration of climate uncertainty. A related aspect of this work is the development of design criteria for engineered structures that specifically incorporate estimates of climatic variability and change.

Very little is currently known about the effects of climate change on water quality. Although water quality concerns are becoming increasingly important, the separation of human-induced versus climate-induced changes in water quality is a very difficult problem. Specifically, there is an immediate need to identify those aspects of this problem that hold the most promise for yielding credible evaluations of climatic effects on water quality.

Of course, the most essential need is for more reliable and detailed (in both space and time) estimates of future climatic conditions. These estimates must be regionally specific and provide information on both the frequency and the magnitude of events. At the same time, there is a corresponding need for improved estimates of future socio-economic conditions. Recent experience indicates that social and economic dynamics exert a far greater influence on water resources than climate dynamics.

Conclusions and Recommendations

One of the most important outcomes of the IPCC study has been to elucidate that physical and social scientific understanding is not yet at a level where water-resource impact assessments are practical. Significantly, though, it is now feasible to identify river basin sensitivities to both climate and socio-economic change. Clearly, such identification is essential to the development of reasoned and beneficial planning and policymaking activities.

Currently, in the absence of a true impact-assessment capability, there are a number of beneficial tasks to undertake in order to develop useful planning and policy formulation information. For example, improved analytical tools for characterizing climate effects on water resources and water-resource systems can be developed. Considerable work has already been initiated in this regard. Also, techniques and tools specifically designed for planning and managing water-resource systems assuming a non-stationary climate can be developed. Such methods are critical inasmuch as the traditional procedures for water-resource-planning unrealistically assume stationary climatic conditions. Moreover, water system managers can test and evaluate alternative operating procedures and rules for making their systems more robust (less sensitive) to climate variability and change. Having taken such steps as these, as regional climate and socio-economic forecasts begin to improve, an adequate scientific and technical basis will then exist to define the effects of climate change on the physical hydrological system and also on the issues of water demand and water use.

References

Ayers, M., Wolock, D., McCabe, G., and Hay, L 1990. 'Simulated hydrological effects of climatic change in the Delaware River Basin'. *Proc., Symposium on International and Transboundary Water Resources Issues.* American Water Resources Assoc. Toronto, p.587–594.

Ayers, M. and Leavesley, G. 1988. 'Assessment of the potential effects of climate change on water resources of the Delaware River basin'. US *Geological Survey Open-File Rept. 88–478.* 66p.

Budyko, M.I. and Izrael, Y.A. (eds.). 1987. *Anthropogenic Climatic Changes.* Gidrometeoizdat, 406p. (Eng. translat.: The University of Arizona Press, 1990).

Close, A.F. 1988. 'Potential impact of the greenhouse effect on the water resources of the River Murray'. In Pearman, G. (ed.) *Greenhouse, Planning for Climatic Change.* CSIRO, Australia, p.312–323.

Cohen, S., Welsh, L., and Louie, P. 1989. *Possible Impacts of Climatic Warming Scenarios on Water Resources in the Saskatchewan River Subbasin,* Canadian Climate Centre Rept. No. 89–9, 39p.

Croley, T.E. (in press). 'Laurentian Great Lakes double-CO_2 climate change impact'. *Climatic Change.*

da Cunha, L. 1989. 'Water resources situation and management in the EEC'. *Hydrogeology, 2,* p.57–69.

Demarée, G. and Nicolis, C. 1990. 'Onset of Sahelian drought viewed as a fluctuation-induced transition'. *Q.J.R. Meteorol. Soc.,* 116, p.221–238.

Grove, A.T. 1973. 'A note on the remarkably low rainfall of the Sudan zone in 1913'. *Savanna, 2,* p.133–138.

Griffiths, G.A. 1989. 'Water Resources'. Chapter XX, *New Zealand Report on Impacts of Climate Change.* North Canterbury Catchment Board and Regional Water Board. New Zealand, 21p.

International Joint Commission. 1985. *Great Lakes Diversions and Consumptive Uses.* Report to the Governments of the United States and Canada, Washington and Ottawa.

Liu, C. 1989. *The Study of Climate Change and Water Resources in North China.* Ministry of Water Resources.

Matsuo, T. *et al.* 1989. *Report on the Analysis of the Effects of Climate Change and their Countermeasures in the Field of Civil and Environmental Engineering.* Unpublished manuscript prepared for the Japanese Environment Agency.

Nicholson, S.E. 1978. 'Climatic variations in the Sahel and other African regions during the past five centuries'. J. *Arid Environments, 1,* p. 3–24.

Nicholson, S.E. 1989. 'Long term changes in African rainfall'. *Weather,* 44, p.46–56.

Nobre, C., Shukla, J., and Sellers, P. 1989. 'Climatic impacts of Amazonian development'. *Brazilian Association of Water Resources Bulletin, 39.*

Ojo, 0. 1987. 'Rainfall trends in West Africa, 1901–1985'. In Solomon, 5. *et al.* (eds.) *The Influence of Climate Change and Climatic Variability on the Hydrological Regime and Water Resources.* IAHS Publication No. 168, p.37–43.

Osborn, C., Schefter, J., and Shabman, L. 1986. 'The accuracy of water use forecasts'. *Water Resources Bulletin,* 22, p.101–109.

Ripley, E.A. 1987. 'Climatic change and the hydrological regime'. In Healey, M. and Wallace, R. (eds.) *Canadian Aquatic Resources.* Canadian Bulletin of Fisheries and Aquatic Sciences 215. Dept. of Fisheries and Oceans, p.137–178.

Sælthun, N., Bogen, J., Hartmann Flood, M., Laumann, T., Roald, L., Tvede, A., and Wold, B. 1990. *Climate Change and Water Resources.* NVE Publ. V30, Norwegian Water Resources and Energy Administration.

Sellers, P., Mintz, Y., and Sud, Y. 1986. 'A simple biosphere model (SIB) for use within general circulation models'. J. *Atmospheric Sciences,* 43, p.505–531.

Shiklomanov, I.A. 1988. *Studying Water Resources of Land: Results, Problems, Outlook.* Gidrometeoizdat, Leningrad.

Shiklomanov, I.A. 1989. 'Anthropogenic change of climate, water resources and water management problems'. *Proc., Conference on Climate and Water.* Helsinki.

Sircoulon, J.H. 1976. 'Les données hydropluviométriques de la sécheresse récente en Afrique intertropicale. Comparaison avec les sécheresses "1913 et 1940".' *Cah. Orstom, ser. Hydrol.,* 13, p.75–174.

Sircoulon, J.H. 1987. 'Variation des débits des cours d'eau et des niveaux des lacs en Afrique de l'ouest depuis le début 20 ème siecle'. In Solomon, S. *et al.* (eds.) *The Influence of Climate Change and Climatic Variability on the Hydrological Regime and Water Resources.* IAHS Publication No. 168, p.13–25.

Smith, J. and Tirpak, D. (eds.). 1989. *The Potential Effects of Global Climate Change on the United States.* U S Environmental Protection Agency. Washington, DC, 413p.

Sternberg, H.O. 1987. 'Aggravation of floods in the Amazon River as a consequence of deforestation'. *Geografiska Annaler,* 69A.

Stewart, B.J. 1989. *Potential Impact of Climatic Change on the Water Resources of the Murray-Darling Basin.* Unpublished manuscript, 12p.

Zillman, J.W. 1989. 'Climate variability and change—their implications for the Murray-Darling basin'. 12th Invitation Symp., Murray-Darling Basin: A Resource to be Managed. Australian Academy of Technological Sciences and Engineering, Albury.

Impact of Climate Change on Agriculture: A Critical Assessment

by Suresh K. Sinha *(India)*

Introduction

The Intergovernmental Panel on Climate Change (IPCC, 1990) outlines the following objectives for agriculture and forestry:

1. To identify those systems, sectors and regions of agriculture and forestry that are most sensitive to anticipated changes of climate.
2. To summarise the present knowledge about the potential socio-economic impact of changes of climate on world agriculture and forestry.
3. To consider the adjustments in agriculture and forestry that are most likely to occur (even if no policy options were implemented).
4. To establish research priorities for future assessment of the impact of climatic change on agriculture and forestry.

A systematic and commendable analysis of the above objectives has been presented in the report using published and personal sources of information. However, some issues are controversial, such as (i) impact assessment based on GCM scenarios (ii) direct effects of CO_2 (iii) the vulnerability of specific regions (iv) projections of trade. What follows in this paper is an appraisal of some important issues relating to the impact of climate change on agriculture.

Assumptions such as 'business as usual' have been employed to develop future climate scenarios at the global and regional levels. These scenarios provide the basis for assessing the impact of climate change on agriculture. However, for any meaningful assessment of the impact of climate change on agriculture, the regional or, more specifically, local information about climate variables and resources is necessary. The accuracy of the prediction of regional climate is at present poor (IPCC,1990). The prediction of global climate is relatively less important for assessing the impact of climatic change on agriculture. For example, there has been no significant change in global climate in the past two decades. But, the sub-Saharan region has experienced one of the worst climatic aberrations during this period which disrupted the agriculture of the region. Though animal husbandry, fisheries and forestry are important, this paper focusses attention on crop-based agriculture.

GCM Predictions of Regional Climate

The GCM simulations of present climate ($1xCO_2$) from GFDL and UKMO have been compared with the long term means of temperature and precipitation. As an example, the temperature and rainfall for a site in Ethiopia are shown in Figure 1. Though we do not expect a perfect fit between model estimates and long term averages of the climatic variables, because of the enormous complexity of the atmospheric processes, it is clear that the models will require considerable improvement before they can simulate the present regional climate. Consequently it is difficult to have confidence in $2xCO_2$ climate predictions of the models. The proposal that the ratio of $1xCO_2$ and $2xCO_2$ could be used to estimate future climatic change becomes essentially a numbers game. An important inference from the above is that many of the conclusions about the regional impact of climate change in the IPCC report have to be seen more as contributing towards the development of a methodology for climate impact assessment studies rather than as predictions of such impacts.

Factors Affecting Agriculture

Atmospheric carbon dioxide and climatic factors such as temperature, precipitation, solar radiation and others are important for the growth of vegetation, and hence of agricultural crops. In addition, however, agricultural productivity is dependent upon soil characteristics, and farm inputs such as energy, fertilizer, pesticides and other farm inputs. Therefore, in evaluating the potential effects of

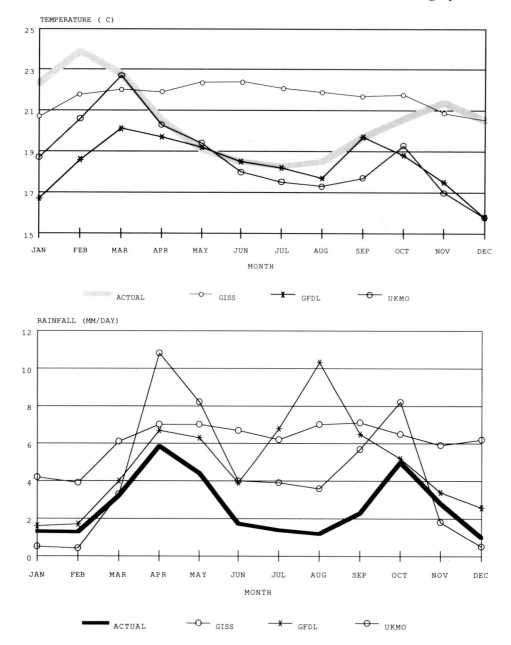

Figure 1: (a) Burgi (Ethiopia) mean temperature; (b) Burgi (Ethiopia) rainfall: Both show actual and simulated (1xCO_2) based on GISS, GFDL and UKMO models.

climatic change on agriculture and subsequently on the economy and trade, we will have to consider the non-climatic factors as well.

Direct Effects of CO_2

Since carbon dioxide is an essential factor in photosynthesis, its increasing levels have received considerable attention in evaluating impacts on agriculture. In most instances the effects of doubling of the concentration of carbon dioxide have been considered to be advantageous for photosynthesis, and hence for crop productivity. The reviews of Kimball (1983) and Cure and Acock (1986), based on surveys of publications on the

effect of CO_2, have been used to predict an improvement in crop productivity substantially in C_3 crops and to a lesser extent in C_4 plants. Therefore, the IPCC Report (1990) gives the impression that the increased concentration of carbon dioxide may lead to improvements in crop yields, as well as their adaptation to stress environments.

The above view is shared by Soviet scientists, because of some additional factors. However, the limitations of the existing literature need to be pointed out lest the direct effects of carbon dioxide lead to complacency. The following points need to be emphasised:

1. Most of the experiments on CO_2 direct effects used 600 ppm or an even higher concentration of carbon dioxide. In a "2xCO_2 climate" resulting from the

increase of CO_2 plus other greenhouse gases to the point where the warming was equivalent to that of doubling of CO_2 alone, the concentration of CO_2 would be only 460 ppm

2. Experiments were conducted in controlled environments, maintaining optimal temperatures for the growth of the experimental plants.
3. Plants were protected from diseases and pests.
4. There are hardly any studies on the interaction of CO_2 effects with other environmental factors, such as temperature, water stress and other greenhouse gases. In fact Cure and Acock (1986) reached the conclusion, that there is just too little quantitative information available to enable us to predict a precise response to CO_2 concentration under well-defined environment conditions. In view of the above, the conclusion of the IPCC Report is not justifiable. The following points need to be considered:

• The past studies, even at 600 ppm or more of CO_2 do not show any effect on the phenology of the plant. (i.e. on how long it takes a plant to complete the vegetation phase, flowering and maturity). The plant phenology will be influenced mainly by temperature, which is a strong determinant of plant yield.

• We can learn something about the response of crops to increasing concentrations of carbon dioxide from the past records. Between 1880–1990, the CO_2 concentration has increased from 280 ppm to 350 ppm, a net increase of 25%. If there has been any effect of this increase, it should be reflected in crop productivity. The per unit area productivity of chickpeas (**Cicer arietinum**) and rapeseed and mustard (**Brassica species**) were analysed for India since 1900. There was a slight increase in productivity in the early part of this century, followed by a decline and subsequently an insignificant change in yield Figure 2). These records show that at the country level no changes in productivity of crops occurred until an improvement in variety or management, or both occurred.

In conclusion, the effects on agriculture of an increased level of CO_2 in the atmosphere can not be simulated adequately by growing crops in glasshouses. Therefore, there is uncertainty with regard to responses of crops to an increased concentration of carbon dioxide.

Effects of Other Factors

In all cereals, pulses and oilseed crops, the utilizable part is the grain or seed. Therefore, it is essential that a plant should enter the reproductive phase if there is to be a harvest of economic yield. However, in tuber and root crops, sugarcane, sugarbeet, jute and others, the delay or absence of flowering is important to obtain economic yield.

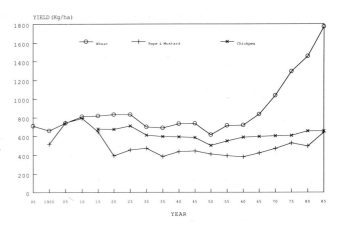

Figure 2: Change in productivity of wheat, rape + mustard and chickpea in India since 1895. Each point is a mean of the predeeding five years.

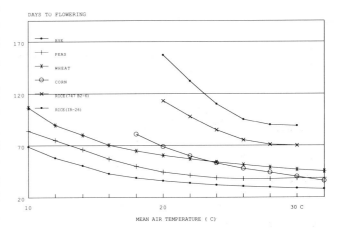

Figure 3: Influence of temperature on the length of the pre-flowering period

Thus response of flowering to various factors will play a crucial role in determining the impact of climate change on a region's agricultural productivity and production. There is no evidence to show that increased levels of carbon dioxide influence the time to flowering. This characteristic will be influenced significantly by temperature, radiation, water availability and the presence of other greenhouse gases.

Effect of Temperature

Day length, temperature and their interaction regulate flowering in plants. In most instances, particularly at low and mid-latitudes, the increase in temperature results in reducing the total duration of the crop, by inducing early flowering and shortening the grain fill period. Several experiments at ICARDA (International Centre for Agriculture Research in Dryland Areas) and elsewhere have shown that the productivities of wheat, barley and

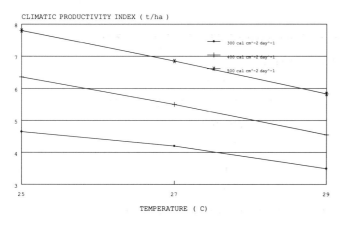

Figure 4: Effect of increasing temperature on productivity of rice at different levels of radiation (Hoshida, 1981)

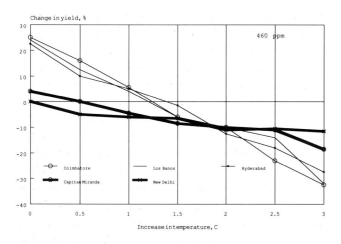

Figure 5: Effect of increased CO_2 and temperature on irrigated wheat yield at Coimbatore (India), New Delhi (India), Los Banos (Philippines), Hyderabad (India) and Capitan Miranda (Paraguay)

chickpeas are reduced by spring planting as compared to winter planting. A broad generalization of the effects of increasing temperature on crop duration is shown in Figure 3. The shorter the crop duration, the lower is the yield per unit area. A rise in temperature will result in decreasing agricultural production in low and mid latitudes. However, in higher latitudes (northern Europe, USSR and North America) where crop duration is limited by low temperature, the crop duration could increase, resulting in improved productivity. Therefore, the question which requires our attention is the interaction between temperature and carbon dioxide. Assuming that there would be some beneficial effects of CO_2, at what temperature would these be cancelled? To this, an additional factor of radiation may be added in the case of rice. An increase in precipitation would be coupled with reduced radiation. Rice is extremely sensitive to reduced radiation and increased temperature (Figure 4) because these conditions cause sterility. Thus a decrease in productivity of rice can be

expected which would not be compensated by increased CO_2 because photosynthate availability does not limit productivity under such conditions. Calculations of the effect of the interaction of temperature and 460 ppm CO_2 at several locations in low and mid latitudes show that the beneficial effects of CO_2 on wheat would be cancelled out by a 1°C increase in temperature (Figure 5). In New Delhi an increase of 3°C could cause a loss of 15–20% in yield.

Effect of Water Availability

Reduced water availability below a threshhold reduces productivity of almost all crops. Though its excess may not necessarily benefit crop production, it may sometimes cause crop losses. Therefore, the occurrence of unpredictable droughts has caused concern around the world. In the general circulation models the simulation of precipitation is not satisfactory. Nonetheless, it is clear that any reduction in precipitation will not only influence the monsoon season crops but also cause shortages for irrigation. In this respect the run-off and water collection could differ very significantly in low latitudes in comparison with the mid-latitudes and high latitudes. For example, in a five year study of a water-shed in north-east India, Singh et al. (1990) observed that for a mean rainfall of 2553 mm, the length of water storage was 152 days but was reduced to 69 days with a rainfall of 2330 mm. This would have a very significant effect on irrigation. There are in fact hardly any studies which have examined the relationship between rainfall, ground water storage, water collection and storage for irrigation and crop production in low and mid-latitudes. These studies are urgently needed in low latitudes because of the nature of precipitation in these areas.

Furthermore, an increase in temperature by even 1 to 3°C would cause enhanced evapotranspiration. Depending upon the changes in precipitation and increased evapotranspiration of certain crops such as rice, the water table is likely to be lowered. Such events have indeed occurred in the North Indian states of Haryana and Punjab during two or more consecutive drought years. Such problems require detailed studies in low and mid latitude to formulate strategies for adapting to climate change.

Effect of Change in Radiation

The relationship between radiation and photosynthesis rate has been studied for several plant species. Depending upon the nature of the species and other reactants the saturation limits for radiation vary. However, there are several studies on wheat and other crops which show a linear relationship between radiation and photosynthesis rate within a certain range. The effects of reduced radiation with increasing temperature are severe on sterility and hence the yield of

Table 1: *Crop disease in temperate and tropical regions*

Crop	Number of diseases reported	
	Temperate	Tropical
Rice	54	500–600
Maize	85	125
Citrus	50	248
Tomato	32	278
Beans	52	250–280

rice. The yields of rice vary considerably in the monsoon season even in experimental plots because of variation in total radiation (Sinha and Swaminathan, 1990). It is thus possible that the productivity of rice would be influenced adversely by increased temperature and reduced radiation, which may not be compensated by increased CO_2 concentration.

Diseases and Pests

The prevalence of pests and diseases is among the major constraints to achieving higher yield. This is particularly true of the tropics where more diseases and pests occur because of higher temperatures and where humidity provides favourable conditions for infestation (Table 1). The increase in temperature and precipitation in the $2xCO_2$ climate could create more conducive conditions for pest infestation. Pests and diseases, which today are unimportant and cause little harm, could become devastating under the changed conditions. Periodic pest and disease-related losses in food production resulted in famine conditions as the following events from history remind us:

- the Irish famine of the 1890s (Potato blight epidemic)
- the 'wheatless days' of 1917 in the United States (Stem rust epidemic)
- the Bengal famine in India in 1943 (*Helminthosporium* brown-spot disease of rice)
- the Southern Corn blight epidemic 1970/71 caused by *Helminthosporium maydis* in the United States
- the rapid shift of brown plant-hopper biotype 1 to biotype 2 1974-76 in the Philippines and in Indonesia causing damage to new rice varieties. The shift in biotypes occurred within two years.

These examples are reminders that temperature and humidity together can sometimes result in massive losses of crops. The occurrence of such situations will be further accentuated by increased CO_2. The leguminous and oilseed crops are particularly vulnerable to insects, and often an entire crop can be lost.

Energy in Agriculture

Agriculture essentially is a mechanism of conversion of solar energy into the products which are directly or indirectly consumed by human beings. However, it is necessary that additional energy be invested through various inputs in agricultural production. The inputs consist of both non-renewable and renewable sources of energy (Sinha, 1986). Highly mechanised and productive agriculture in western Europe, North America and other regions is based on intensive use of commercial energy (Pimentel and Pimentel, 1979). The energy inputs include various farm operations, fertilizers, irrigation, pesticides, transportation, etc. resulting in an input/output ratio. This in fact becomes negative by the time products reach the table because of several other energy costs. However, in many developing countries agriculture is not as energy intensive as in the developed world. Thus, the predictions of the impact of climate change on agriculture based entirely on the western model of agriculture and used by the IPCC report may not be realistic.

Response of Major Crops to Climate Variation

Wheat and rice are the most important staple food around the world. Of the total wheat production, 40.7% occurs in developing countries and 59.3% in developed countries, 94.5% of the total rice production is in the developing countries. Of this, Asia alone produces 92%. Any climatic variation which influences the production of wheat and rice will have a profound effect on global food security, and particularly that of the developing countries. However, Africa is mostly dependent on the production of millets and sorghum among cereals, and root and tuber crops. Rice, millet, sorghum, maize and tubers and root crops (Cassava, sweet potatoes and yam) are mostly grown as monsoon season crops and generally fluctuate with variations in the monsoon. For example, the production of rice in India, Bangladesh, Indonesia etc. is severely influenced by droughts, but is influenced less in China, Korea and Japan where most of it is grown with irrigation. Among the rice producing states in India the maximum adverse impact of drought occurs where the resources of irrigation are limited. However, in northern Japan the productivity of rice is limited by the growing season because of low temperature (Yoshino, 1990). Such a situation may also occur in regions where rice is grown at high altitudes, as in Nepal. If appropriate varieties could be identified, the productivity of rice could increase due to global warming in such places. But, most of the rice is grown in the humid tropical environment where high temperature is a limiting factor for grain yield. Wherever high yields of rice are obtained, as in Korea, Japan, China, parts of India (Punjab and Kashmir) and other places, the temperatures are generally mild (22 to 26°C mean daily temperature). The productivity in such regions is likely to be adversely affected by a rise in temperature. Thus, in summary, the following possibilities can be predicted:

Temperate Regions:
1. The growing season will increase, benefiting the crop yield.
2. Increased CO_2 may also benefit crops.
3. Increases in disease and pest incidence could occur.
4. The demand for irrigation water will increase.

Tropical Regions:
1. The growing season will be reduced, adversely influencing the crop yield.
2. Higher temperature and high humidity would cause higher sterility in rice.
3. Increased CO_2 may benefit dry matter production, but not grain yield.
4. An increase in disease and pests incidence could occur. New pests may emerge.
5. The uncertainty of the monsoon will make crops more vulnerable to water deficits.
6. Cloudiness or low radiation will adversely influence crop yield.

A detailed analysis for coarse cereals and pulses is necessary to predict the effects of climate change on these crops. Therefore, depending upon the location, varieties with appropriate duration and tolerance to diseases, pests and abiotic stresses will have to be found. In addition, a great effort will have to be placed in water management.

Role of Livestock

Livestock constitutes an important component of agriculture. In industrialised countries, livestock farming meets the demands for meat, poultry and dairy produce. However, in most of the developing countries, draught animals provide renewable energy input for agriculture.

Thus it is important in these countries to estimate the impact of climate change on livestock, and identify the mechanisms of their adjustment or to identify breeds which are adaptable to higher temperatures.

In those regions whose grassland productivity is limited by low temperature, as in Iceland, a warmer climate would benefit both the meat and dairy industry. But in tropical regions where an increase in temperature may be combined with an increasing frequency of droughts and there may be an increase in diseases, both the meat and dairy industry may suffer, as studies in Kenya suggest. The draught animal power accounts for an equivalent of 30,000 megawatts annually for agriculture in India. How climate change would influence this renewable source of energy is uncertain and study of this is necessary.

Vulnerability of the Regions

The IPCC report (1990) has identified some regions that are more vulnerable than others to a climate change. This conclusion is based on the following:
1. An FAO assessment of agricultural productivity potential.
2. GCM simulations of climate change.

The vulnerability of the various regions with reference to food security depends in part on the size of the population, the production potential and the consumption patterns. The present per capita consumption of food grains is given in Table 2. Most of the countries of Africa and Asia can be considered vulnerable, if the food consumption pattern in North America and Europe is used as the basis. Since Africa is recognised as potentially one of the most vulnerable regions, an analysis for this region has been made.

The present projection is that the population of Africa in 2025 will be 1.62 billion. The per capita consumption of food grains is 257 kg annually which is the lowest in the world. Even with this consumption, the total food grain requirement will be 416 million tonnes. However, assuming that the consumption should rise to 300 kg per capita annually, similar to Asia, then a total of 486 million tonnes of food grains or energy and protein equivalent would be needed. Are there enough resources in Africa to meet this need? At present it is estimated that the arable and permanent cropped lands are 166 and 18.6 million hectare, respectively. However, only about 100 million hectares are actually cultivated. Between 1984 and 1986, the average productivity of cereals, pulses and root and tuber crops was 1077, 591 and 2717 (as dry weight) kg ha^{-1} respectively. This leads to a total production of about 119 million tonnes, compared with the present requirement of 158 tonnes for the present population at the consumption rate of 260 kg per capita annually. Thus food shortages are common in Africa. These range from severe to marginal

Table 2*: Food grain requirement of different regions of the World in 2025 based on the present level of per capita consumption*

	Population (billion)	Average per capita annual consumption (kg)	Annual food grains requirement (Mt)
Africa	1.62	257	416
South America	0.78	296	231
Asia	4.54	300	1362
North America	0.35	885	310
Europe	0.52	700	364
USSR	0.37	983	364
Oceania	0.04	578	23
World	8.22	373	3070

depending upon the region and national boundaries. Having recognised deficiencies in food availability, what are the future options for Africa?

1. Will African nations continue to import food?
2. How will these nations pay for the imports? And from where are food imports possible, bearing in mind the uncertainties of production in the present day food exporting countries?
3. Would it not be most desirable for the nations of Africa to produce enough to meet their food needs?
4. If increased food production is the most desirable objective, what crop combinations would be most suitable?

Answers to the above questions are necessary for defining the vulnerability of particular regions. The Horn of Africa has been identified as a vulnerable region. An analysis of the resources of this region shows that it could become self-sufficient if all the countries of the region joined together in developing modern but essentially indigenous agriculture. In addition, these nations would have to develop mechanisms for food storage and distribution.

South and South-East Asia are also described as vulnerable to climate change from the point of view of food security. Rice is the most important crop of the region. Within the region there are annual fluctuations in rice production but the overall production remains stable. For example, China in 1985 harvested 10 million tonnes less rice than in 1984, but India harvested about 9 million tonnes more. A similar situation exists for the various smaller nations. Regional cooperation by having a "Rice Pool" of South and South-East Asia could provide a mechanism for ensuring the food security of the region and reducing its vulnerability.

Higher Order Effects on Economy and Society

It is rightly recognised that the effects of climatic changes on crop yields, grasslands and forestry could have profound effects, described as higher order effects, on farm income, rural employment, national food production, food security and exports. Assessments of these would vary from region to region and nation to nation. There is in fact very little work in developing countries on these aspects. Whatever isolated efforts have been made do not really address these problems from a regional and national perspective (Jodha, 1988). The conclusions based on climate change effects in marginal areas can not be extrapolated to the national level because in several countries, such as India, it is the climatic impact on its most productive areas that plays a dominant role in food security.

Though there are several factors for which the higher order effects of climate change are important, only two, rural employment and food grain export, are discussed here. The problem of rural employment is intimately connected with the percentage of population involved in agriculture. In Africa and Asia 70.3 and 62.3 per cent of the population was classified as agricultural population, whereas in the United States of America and Canada the proportion was only 2.7 and 4.0 per cent respectively. Similarly, the population engaged in agriculture varies from 2.2 to 6.4% in some of the West European countries. Thus one of the conclusions of the IPCC report that a 16% decrease in production in Saskatchewan would lead to about a 1% decrease in rural employment is not relevant for the developing countries. A study in some villages of Haryana (India) during the 1987 drought showed that only 25% landless labour got full employment and about 67% got only partial (15 days a month) employment. Therefore, in regions where agriculture is the predominant economic activity, the magnitude of unemployment in the event of climatic anomalies such as droughts would be enormous.

More detailed studies are required in developing countries to give an idea of the enormity of the problem.

The international trade in cereals alone was worth 42, 36 and 30 billion US dollars in 1984, 1985 and 1986 respectively. Most of the exports were from USA, Canada, Argentina, Australia and some European countries. However, the imports were mostly by developing countries. Therefore, it is inevitable that there should be concern for the following reasons:

1. Trade in cereals is important for the economy of the exporting countries.
2. If production is reduced in cereal exporting countries, how could the food needs of the importing countries be met?

Implicit in these questions, as well as in the IPCC report, are the assumptions that food exports from the present day exporting countries will continue. Therefore, there was an effort by the EPA (Environment Protection Agency of USA) and others to estimate the rise in cost of production which would indeed raise the export price. Assuming that the present balance in production remains, would there be resources in many developing countries to buy food, or have they to depend perpetually on food aid? The latter would be the continuation of human indignity.

Food Security

The problem of food security is present around the world both without and with change in climate (Sinha et al., 1988). Can the regions and nations which are identified as vulnerable meet their own food needs? Clearly, the effort should not be towards increased dependence on trade but towards attaining self sufficiency by individual nations or a group of nations. Let us broadly consider Africa where the projected need in 2025 AD is 416 million tonnes of food grains or energy and protein equivalent. The productivity of cereals is less than half of the world average, being 1175 kg ha^{-1} against the world average of 2588 ha^{-1}. The cultivable land is 166 million hectare besides that under permanent crops. Therefore, it is a question of whether 416 million tonnes of food grains or energy and protein equivalent could be produced from 166 million hectare. Almost all FAO studies which have determined the productivity potential are underestimates for various reasons. There are now several studies available to demonstrate that the productivity of sorghum and millets can be enhanced more than two-fold in large parts of Africa. One of the strong points in the agricultural system in Africa is the cultivation of roots and tuber crops, particularly Cassava (Tapioca), sweet potato and yams. At 35% of the dry weight of roots and tubers, these crops have more than double the productivity of cereals. In fact it was argued earlier by deVries et al. (1967) that root crops could play an important role in the production of edible calories in the

tropics. An important advantage of these crops is that they are less or not sensitive to temperature as compared to grain crops. Several products of tapioca and sweet potato flour with legumes have been developed and consumed in India (Subrahamanyan et al., 1954). Furthermore, earlier studies estimating productivity potential have not considered water management as a means for increasing productivity. Therefore, there is an urgent need to realistically reassess the production potential of Africa and other regions which are being described as vulnerable. If such an exercise is not done and appropriate strategies are not developed to increase productivity of different crops, these regions will be left with the following options:

1. Continue to receive food aid, which may be uncertain and prohibitive in financial terms.
2. To clear more forests to increase the area for agriculture.

Conclusions

1. Agriculture is largely dependent on natural resources which can be utilized together with other appropriate inputs. Agricultural development is vastly different in different parts of the world. High commercial energy agriculture is associated with large farms and a small percentage of the population engaged in agriculture. The low commercial energy agriculture in developing countries is associated with small farms and a high percentage of agricultural population. Therefore, it would be too simplistic to describe the impacts of climate change on agriculture for the two systems on the basis of a common scale.
2. Temperature, precipitation and radiation are important factors, besides soil and crop characteristics, which determine crop production. The present GCMs cannot provide reliable estimates of regional climate change to assess the impact of climate change on agriculture.
3. The direct effects of CO_2 are mostly exaggerated because they are based on data obtained for 600 ppm CO_2 or more (and not 460 ppm which could occur in the "2xCO_2" climate) at optimal temperatures. Furthermore, increases in the rate of photosynthesis do not necessarily lead to increases in grain or edible yield. More research is needed on this aspect.
4. In low-latitudes increases in cloudiness associated with increases in precipitation may have adverse effects on rice production at temperatures only 2–3°C above the present normals.
5. There appears to be no evidence to show that average national yields of any crops have increased as a result of the 25% increase in CO_2 concentration in the last 100 years.
6. The vulnerability of regions has not been defined with adequate evidence. More data and a greater

effort are needed to support the statements in the IPCC report.

7. The concept of 'risk pooling' by combining the efforts of the countries of a region has to be promoted. This must include such aspects as a 'Rice Pool' for south and south-east Asia and a 'Coarse Grain Pool' for Africa.

8. The potential of tuber and root crops for producing edible calories has to be assessed for low and mid-latitudes because these crops can provide alternate productive cropping systems.

9. Among other aspects of agriculture, the impact of climate change on livestock needs to be studied in detail, since livestock are quite important for dairy farming, and for energy, transport, etc. in many parts of the world.

10. The emphasis on trade in food grains must be substituted or supplemented by encouragement of self sufficiency in regions of food deficiency.

11. If the productivity potential of the indigenous crops is not increased in food deficit countries, these countries will continue deforestation to increase the area for agriculture.

12. Since agriculture and food security are closely related to population and consumption, there should be an effort to reduce population in developing countries and to reduce consumption in developed countries.

13 Food aid will not be the answer to either present or future food problems. These can only be overcome by "support for production".

Acknowledgement

I would like to thank Mr. B.C. Patil for helping in preparing the manuscript and Professor Yoshino and one anonymous reviewer for their constructive comments.

References

Cure, J.D. and Acock, B., 1986. Crop response to carbon dioxide doubling: A literature survey. Agricultural and Forest Meteorology, 38, 127–145.

Dyke, G.V., George, B.J., Johnston, A.E., Poulton, P.R., and Todd, A.D. 1982. The Broadbalk Wheat Experiment 1968–78: Yields and plant nutrients in crop grown continuously and in rotation. Rothamsted Report for 1982, Part 2, 5–44.

Intergovernmental Panel on Climate Change: Working Group 1 Report 1990, WMO and UNEP, Geneva.

Jodha, N.S., 1988. Potential strategies for adapting to greenhouse warming: Perspectives from the developing countries. In Greenhouse Warming: Abatement and Adaptation. Resources for the Future. Washington D.C., 147–158.

Kimball, B.A. 1983. Carbon dioxide and agricultural yield: An assemblage and analysis of 430 prior observations. Agronomy Jour, 75, 779–788.

Pimentel, D. and Pimentel, M., 1979. Food, Energy and Society, Edward Arnold Ltd.

Singh, A., Varma, A. and Awastihi, R.P., 1990. Research development in land and water resources conservation and prospect of application in North Eastern hill region In National Seminar on Conservation of Land and Water Resources for Food and Environmental Security. Pub. Soil Conservation Society of India. 305–311.

Sinha, S.K., 1986. Energy Balance in agriculture: The developing world. In Global Aspects of Food Production. Eds. M.S. Swaminathan and S.K. Sinha 57–83. Publisher, International Rice Research Institute and Tycooly International.

Sinha, S.K., and Swaminathan, M.S., 1990. Deforestation, climate change and sustainable nutrition security. Climatic Change (in press).

Sinha, S.K., Rao, N.H., and Swaminathan, M.S. 1988. Food security in the changing global climate. Conference Proceedings—The Changing Atmosphere: Implications for Global Security. Toronto, Canada 27–30 June, 1988. WMO No. 710, pp 167–191.

Subrahmanyan, V., Bhatia, D.S., Bains, G.S., Swaminathan, M., and Rao, Y.K.R. 1954. Investigations on grain substitution-1. Production of ground grain from blends of tapioca and groundnut. The Bull. Central Food Technological Research Institute, Mysore (India). III, 180-183.

de Vries, C.A., Ferwerda, J.D., and Flach, M., 1967. Choice of food crops in tropics. Neth. J. Agric. Sci., 15, 214–248.

Yoshino, M. 1990. Impact of Climate Change on Agriculture and Forestry: A Review on the World Climate Impact Studies Programme for Monsoon Asia (in press).

Potential Impacts of Climate Change on Human Settlements; the Energy, Transport and Industrial Sectors; Human Health and Air Quality

by M. Hashimoto and S. Nishioka *(Japan)*

Abstract

Rapid and large-scale climate change is likely to have numerous impacts on human settlement, the energy, transport and industry sectors, human health and air quality, and may affect the mechanism of UV-B change. Their intensity may be magnified by other anthropogenic changes in global, regional or local climate, and combined climatic effects associated with global warming seem certain to have profound and often highly disruptive impacts on human activity. The impacts on developing countries, many of which lack resources for adaptation, may be particularly disruptive. Due to the significant socio-economic aspects associated with impacts on these sectors, there is a strong need for further integrated study and cross integration with impacts on other sectors such as agriculture, water resources and coastal and marine resources.

Introduction

This report represents the results of Section 5 of IPCC Second Working Group, which covers socio-economic impacts of climate change on several human activities. Some likely characteristics of the anticipated change are of special significance to policy-makers.

The various changes induced by global warming may appear in an insidious manner over a time scale ranging from a decade to a century. Some of these changes may happen abruptly providing little lead time for adaptation.

Present natural disasters underscore the vulnerability of humanity to large-scale shifts in climate. Recurring problems such as floods (e.g. Bangladesh), drought (e.g. Africa), severe storms (e.g. Caribbean Sea), land subsidence and exhaustion of fuel wood supply demonstrate the present vulnerability of human society to weather variability and resource depletion. The frequency or intensity of some of these disasters may increase with global warming. Other impacts such as sea-level rise may occur more gradually.

For a number of sectors, especially energy, transport and industry, the impact of response strategies to limit greenhouse emissions may be as significant an impact as the direct impact of the climate change itself. Public policies and shifts in consumer preferences in order to limit greenhouse gas emissions may benefit some industries or technologies while adversely affecting others.

Although a few beneficial aspects can be anticipated from even a rapid global warming, they are likely to be overwhelmed on both a regional and a national scale by a multitude of adverse impacts. Rapid climate change will prove universally damaging to coastal areas and island areas across the world, with the adverse effects differing depending on the vulnerability of particular coasts and the resources of those nations to develop adequate response and adaptation strategies. Climate-change-induced alterations in distribution of water resources, shifts in weather circulation patterns, and changed circumstances for production of biomass and other energy and food sources will have both negative and positive effects within the same regions and even within the same countries.

Due to the large human infrastructure investments based on the assumption of a continuation of the historically observed climate—reservoirs, dams, flood control projects, irrigation systems, water and sewage systems, hydroelectric and other energy systems, etc.—rapid climate change could prove costly to all countries. Many projects with normally long lifetimes may be obsolete long before they would normally be slated for replacement. The unpredictable shifts of the assumptions pose severe difficulties in the planning process of such large infrastructures.

In addition, increases in average annual temperatures may have a major influence on human health, air quality, seasonality and quantity of energy demand, material durability and work patterns. The combination of climate-induced changes in seasonality of energy demand and shifts in fuel use and energy technologies to limit greenhouse emissions could produce significant changes in energy production and transmission.

Although the combined climatic effects due to the increased greenhouse-effect seem certain to have profound and often highly disruptive impacts on human activity, their intensity may also be magnified by other anthropogenic changes in global, regional or local climates. These include a potential intensification of the urban heat-island effect resulting from increased urbanization; the growth of desertification following land degradation; drying of regional climates as a consequence of deforestation; and possible increases in ultraviolet radiation resulting from depletion of the stratospheric ozone layer. This stratospheric ozone depletion, already perceptible over major populated areas of the Earth, may combine with climate warming to exacerbate such problems as air pollution, stress on human health and materials.

A continuation of the current rate of increase of concentrations of greenhouse gases in the atmosphere will ultimately cause a significant rise in global average annual temperatures. Given current emission trends, an effective doubling of pre-industrial concentrations of carbon dioxide appears likely by around 2030. Although considerable uncertainty exists concerning the magnitude and regional implications of climate change following such a growth in greenhouse gas levels, some generalizations can be made.

Thermal expansion of the upper layer of the oceans and possible melting of some glaciers are expected to spur a steady rise in sea-levels with the threat of coastal inundation, shoreline erosion, increased salt-water intrusion into freshwater supplies, and loss of biologically important natural wetlands and socially-valued ocean beaches. Increased sea-surface temperatures are likely to increase the intensity of tropical cyclones, and this increased storm intensity on top of gradually rising sea-levels should significantly raise the vulnerability of coastal communities to severe damage from storm surge.

Although projections of likely changes in the hydrological cycle resulting from global warming are fraught with a high degree of uncertainty, the models indicate that there is a likelihood of an enhanced intensity of that cycle with increased frequency of drought, flooding and severe thunderstorms. These changes in the hydrological cycle together with expected changes in sea-level and storm intensity will have far-reaching effects on human settlements, food production and distribution, water resource and infrastructure planning, shipping on inland waterways, recreational resources and habitats for flora and fauna.

Rapid climate change is likely to have especially adverse effects on developing nations whose food production and distribution systems may be vulnerable to such change and who may lack the resources to respond to threats to coastal areas, water and energy supplies and infrastructure investments. Natural disasters may be intensified in presently vulnerable areas, and problems such as flood, drought, and scarcity of biomass may be accelerated as a result of global warming. Coastal inundation and agricultural changes may produce large populations of environmental refugees. Although these refugees are likely to be generated largely within developing countries, the growth of a population of environmental refugees can be expected to affect not only neighbouring countries but also many others, including those less severely affected by the direct impact of climate change. The growth of large refugee populations may also facilitate the growth of epidemics of various diseases respecting no national boundaries.

Among the most significant impacts in each of the seven areas studied by this group are:

Human Settlement

A principal difficulty in determining the impact of climate change on human habitat is the fact that many other factors, largely independent of climate change, are also important. One can reliably predict that certain developing countries will be extremely vulnerable to climate changes because they are already at the limits of their capacity to cope with climatic events. These include populations in low-lying coastal regions and islands, subsistence farmers, populations in semi-arid grasslands, and the urban poor.

The largest impacts of climate change on humanity may be on human settlements, with the existence of entire nations such as the Maldives, Tuvalu, and Kiribati imperilled by a rise of only a few metres in sea-level and populous river delta and coastal areas of such nations as Egypt, Bangladesh, India, China and Indonesia threatened by inundation from even a moderate global sea-level rise. Coastal areas of such industrialized nations as the United States and Japan will also be threatened, although these nations are expected to have the requisite resources to cope with this challenge. The Netherlands has demonstrated how a small country can effectively marshal resources to deal with such a threat.

Besides flooding of coastal areas, human settlements may be jeopardized by drought, which could impair food supplies and the availability of water resources. Water shortages caused by irregular rainfall may especially affect developing countries as is seen in the case of the Zambezi river basin. Biomass is the principal source of energy for most of the nations of sub-Saharan Africa and changed moisture conditions in some areas, reducing this biomass, could pose grave problems for energy production and construction of shelter.

Although there exist only a handful of city specific studies, they suggest that climate change could prove costly to major urban areas in industrialized nations. A study has projected that an effective CO_2 doubling could produce a major water shortfall for New York City equal to 28 to 42 per cent of the planned supply in the Hudson River Basin, requiring a $3 billion project to skim Hudson River flood waters into additional reservoirs.

Although in the permafrost region global warming may result in expansion of human settlements poleward, thawing of the permafrost may also disrupt infrastructure and transport and adversely affect stability of existing buildings and conditions for future construction.

The gravest effects of climate change may be those on human migration as millions are displaced by shoreline erosion, coastal flooding and agricultural disruption. Many areas to which they flee are likely to have insufficient health and other support services to accommodate the new arrivals. Epidemics may sweep through refugee camps and settlements, spilling over into surrounding communities. In addition, resettlement often causes psychological strains, and this may affect the health and welfare of displaced populations as well as of indigenous people facing cultural changes.

Tasks for the near future

Some important priorities may be defined. First, reliable projections of the implications of climate change on human settlement should relate to specific climate models, none of which can yet provide reliable projections of likely future local climates. Improvement of the grid resolution of the General Circulation Models would seem essential to permit correlation between likely local climate scenarios and potential impacts. Climate-change impact analyses are especially scarce for Latin America, which contains regions highly sensitive to climatic fluctuations associated with such phenomena as *El Niño*.

Second, the complex linkages between urban functions likely to be affected by changed weather conditions and altered urban settlement patterns in developed countries are not well understood, and these interactions may vary considerably in different geographic areas, e.g., central cities, secondary cities, suburbs and rural areas. In developing countries, many largely non-climatic factors, e.g., improvement of agricultural management, increased urbanization, and self-reliance, may produce very different impacts in urban and rural areas.

Third, the relationships between societal change and economic implications of climate change at the scale of cities need to be quantified. Finally, study needs for the effects on building materials and design of buildings have been described by some reseachers. The most difficult task is to correlate analogical studies (assessment of effects from historical and geographical analogies) with the future

climate change projections. This should in addition take into account the effects of policy trends (housing and social policies, energy policies etc). As the settlement policy of refugees sometimes results in another conflict among people, the need to consider such feedback of policies developed to address social and economic problems constitutes a serious difficulty in correctly assessing the expected impacts of climate change on human settlements.

Energy

Among the largest potential impacts of climate change on the developing world are the threats in many areas to biomass, a principal source of energy in most sub-Saharan African nations and many other developing countries. More than 90% of the energy in some African countries is biomass energy (fuel wood). Owing to uncertainties in water resource projections derived from current climate models, it is very difficult to provide reliable regional projections of future moisture conditions in these countries. Drier conditions could be expected in some countries or regions, and in those situations energy resources could be severely impaired. There could be possible compensating effects of faster growth of fuelwood due to higher ambient CO_2. Analysis of this situation should be a top priority for energy planners.

In addition to affecting the regional distribution of water and biomass, climate-related changes in cloud cover, precipitation and wind circulation intensity will affect the distribution of other forms of potential renewable energy such as solar and wind power. Understanding these impacts on hydro-, biomass, solar and wind energy is particularly important because renewable energy sources are playing a significant role in the energy planning of many countries. This could become an increasingly important concern in developing countries, many of which are facing serious economic pressures from the need to import conventional energy resources.

Developing countries, including many in Africa, depend significantly on hydropower. By changing water resource availability, climate change may make some present hydroelectric power facilities obsolete and future energy planning more difficult, although others may benefit from increased runoff.

In developed countries, two Canadian studies of likely impacts of an effective CO_2 doubling have shown widely differing effects on availability of hydroelectric power, with one study showing a potential increase of 93 000 gigawatt hours in Quebec for three drainage basins and another study indicating a loss of 4 165 gigawatt hours of power generation for Canadian hydrogenerating stations on the Great Lakes. Generally, impacts on hydrogeneration would vary among countries and even within regions of the same nation.

Major studies to date of the likely impact of global warming on the energy sector in industrialized countries are confined largely to six countries: Canada, Germany, Japan, the United Kingdom, the United States and the USSR. Generally, they show differing overall aggregate impacts depending on how much energy use is related to residential and office heating and cooling. Climate warming will increase energy consumption for air conditioning and, conversely, lower it for heating.

The aggregate effect, studies suggest, will be to increase energy demand within the United States and Japan while reducing it within Germany. Although aggregate demand is likely to increase within the United States, circumstances will vary very much between particular regions, with some southern utilities requiring large additional increments in capacity and others in northern regions benefiting slightly.

In addition, the energy sector may be affected by response strategies against global warming, such as a policy on stabilizing emissions. This may be among the most significant energy sector impacts in many developed countries, enhancing opportunities for technologies that produce low quantities of greenhouse gases. Controversy on the way to obtain CO_2-free energy has already arisen, with the options of increased reliance on nuclear power or hydropower weighed against related safety and environmental concerns. Energy sector changes in both developing and developed countries may have broad economic impacts affecting regional employment, migration and patterns of living.

Tasks for the near future

Studies of impacts on energy supply and demand are presently limited in terms of types of potential impacts studied, regions studied and consideration of the range of potential impact with the uncertain and highly variable results on regional climate change provided by general circulation models. The rates and patterns of change for regions over several decades need to be more properly represented in the impact assessment. Because technology and energy supply options are changing, more detailed study of capability and vulnerability of new technologies needs to be made. Because water supply and changes in levels of local water bodies appear to be major potential constraints on technologies such as hydropower, detailed study of water basins and potential consumptive uses of water is needed to identify the range of impacts.

Studies are needed of potential effects of climate change on availability of biomass energy, especially in developing countries. Future conditions need to be considered, including population and competition for biomass and land resources, along with possibilities for improvements in efficiency or substitution for utilization of biomass energy.

A survey of the availability of non-traditional forms of energy such as wind or solar power would be helpful. This survey could develop an atlas of availability of such resources in countries around the world together with a preliminary analysis of how changed weather circulation patterns and cloud cover might affect their potential for energy production.

There is a strong need for an in-depth impact study on the energy sector to examine impacts of climate change on all stages from exploration through consumption and even waste disposal.

The potential indirect impacts on energy use patterns due to policy responses to global warming should be analysed, including secondary effects on the natural environment and on regional economies.

Transport

Generally, the impacts of climate change on the transport sector appear likely to be quite modest with two exceptions. Ultimately, the greatest impact of climate change on the transport sector in developed countries would appear to be changes produced by regulatory policies or consumer shifts designed to reduce transport-related emissions of greenhouse gases. Because of the importance of the transport sector as a source of greenhouse gases, it seems certain to be targeted as a major source of potential reductions in greenhouse gas emissions, with potentially added constraints on private automobile traffic, automotive fuel and emissions, and increased use of efficient public transport.

A second large impact on the transport sector concerns inland shipping where changes in water levels of lakes and rivers may seriously affect navigation and the costs of barge and other transport. Studies to date, focused entirely on the Great Lakes Region of Canada and the United States, have shown quite large potential impacts. Climate scenarios have shown a likely drop of lake levels of as much as 2.5 metres resulting from an effective CO_2 doubling. Such changes could increase shipping costs, but the shipping season could be longer than at present due to decreased ice. Lake and river levels may rise in some other regions with potentially enhanced opportunities for shipping.

Generally, impacts on road transport appear likely to be quite modest, except in coastal areas where highways or bridges may be endangered by sea-level rise or in mountainous regions where potentially increased intensity in rainfall might pose the risk of mudslides. Studies in Atlantic Canada and Greater Miami, USA, indicate that highway infrastructure could prove very costly in such exposed coastal areas. Reduced snow and ice and lessened threat of frost heaves should generally produce highway maintenance savings, as suggested by a study of Cleveland, Ohio, USA.

Impacts on railways appear likely to be modest, although heat stress on tracks could increase summertime safety concerns on some railways and reduce operational capability during unusually hot periods. Dislocations due to flooding may increase. Reduced snow and ice conditions in colder regions on the other hand could result in better operating conditions in those areas.

There has been little analysis of likely impacts on ocean transport. The largest effect would appear likely to be some jeopardy to such shipping infrastructure as ports and docking facilities, threatened both by sea-level rise and storm surges. Some climate projections indicate the possibility that tropical cyclone intensity may increase. This could have adverse implications for ocean shipping and infrastructure. On the other hand, decreased sea ice could provide greater access to northern ports and even enable regular use of the Arctic Ocean for shipping. Moderate sea-level rise could also increase the allowable draught for ships using shallow channels.

There is a strong need for analysis of likely impacts of climate change for the transport sector in developing countries, as efficiency of the transport sector is likely to be an essential element in the ability of countries to respond to climate change.

Tasks for the near future

The areas in which near-term research and analysis would seem appropriate include:

- an analysis of likely implications of change in tropical storm patterns, sea ice extent, and iceberg flow on ocean shipping
- an exploration of likely impacts of government regulatory policies to restrain greenhouse emissions on the automotive sector, including likelihood of fuel shifting, increased use of mass transit, or shift to other modes of transport
- an examination of likely climate-induced changes on water levels in inland waterways and potential effects on inland shipping
- an exploration of likely impacts on transportation infrastructure in tropical areas of climate warming and sea-level rise
- an examination of the likely impacts on transport and roadways of a thawing of the permafrost associated with warming in the Arctic and other cold regions
- an assessment of likely climate-induced population migration on traffic and transport patterns
- a refinement of existing transportation models to permit projection of likely shifts between transportation modes as a result of climate change.

Industry

A study in the UK shows that 30.1% of turnover in agriculture, 20.2% in transportation and communication, and 14.7% in construction depend on climatic variability. But studies of likely impacts of climate change on the industrial sector tend to be concentrated heavily in certain sectors such as recreation and only in a handful of industrial countries, principally Australia, Canada, Japan, the United Kingdom and the United States. Very little analysis exists of the likely impacts of climate change on industry in developing countries, although there is some evidence to suggest that developing country industry may be especially vulnerable to climate change. An especially important factor is the likely change in the production map of primary products as a result of climate change.

Changes in the regional and global availability and cost of food and fibre may significantly affect the competitiveness and viability of such derivative industries as food processing, forest and paper products, textiles, and clothing. Climate change may be expected to have impacts on the availability and cost of food, fibre, water and energy which would differ markedly from region to region. But the coarse resolution of current climate models makes it impossible to project reliably any regional shift in distribution of such resources.

Just as the motor vehicle and the energy sectors are likely to be influenced by regulatory decisions and shifts in consumer patterns emanating from concerns about limiting greenhouse emissions, heavy manufacturing may face readjustment to new situations such as transboundary siting constraints and international mechanisms for development and transfer of new technology. Efficiency in use of energy may become an even more significant competitive factor in steel, aluminium and other metal industries and automotive manufacturing. Public concern about limiting greenhouse emissions may also create opportunities for energy conservation or 'clean technology' industries.

Studies of likely impacts of climate change on industry tend to be clustered in the recreational sector, where direct impacts of climate change are more ascertainable. Generally, global warming can be expected to reduce the length of the snow skiing seasons in many areas and to affect the viability of some ski facilities. In some areas, on the other hand, the summer recreation season may be extended. However, in some coastal areas the benefits resulting from a longer summer recreation period may be offset by a loss of economically vital ocean beaches and coastal recreational resources.

With sufficient lead time, industry may be able to adjust to many of the changes accompanying global warming. Shortages of capital in developing countries which may be vulnerable to flood, drought or coastal inundation may, however, constrain such industry's ability to design effective response strategies.

Tasks for the near future

More case studies of likely impacts on particular industries are needed. Many of these studies should be directed at developing-country industries. Central to this analysis is a projection of the likely change in the production map of primary products as a result of climate change. This change in the price and availability of primary products will affect the competitive position of industry in developing countries. Such studies could have major implications for development planning and investment policies in developing countries. Analysis should also be performed of market opportunities created by the new global focus on and response to greenhouse warming.

Mechanisms should be developed for ensuring growth of climate impact expertise within developing countries to cover the whole range of climate impacts. This could be accomplished through the funding of regional climate centres and rotation of scholars from various affected countries to these regional centres.

There is also a strong need to involve industrial strategic planners in the analysis of industry impacts in order to fine-tune the research and ensure that the results of such analysis begin to shape industrial planning.

Human Health

Humans have a great capacity to adapt to climatic conditions. However, adaptations have occurred over many thousands of years. The rate of projected climatic changes suggests that the cost of future adaptation may be significant.

A greater number of heat waves could increase the risk of excess mortality. Increased heat stress in summer is likely to increase heat-related deaths and illnesses. Generally, the increase in heat-related deaths would be likely to exceed the number of deaths avoided by reduced severe cold in winter. Global warming and stratospheric ozone depletion appear likely to worsen air pollution conditions, especially in many heavily populated and polluted urban areas. Climate-change-induced alterations in photochemical reaction rates among chemical pollutants in the atmosphere may increase oxidant levels, adversely affecting human health.

There is a risk that increased ultraviolet radiation resulting from depletion of the stratospheric ozone layer could raise the incidence of skin cancer, cataracts and snow blindness. The skin cancer risks are expected to rise most among fair-skinned Caucasians in high-latitude zones.

Another major effect of global warming may be the movement poleward in both hemispheres of vector-borne diseases carried by mosquitoes and other parasites. Parasitic and viral diseases have the potential for increase and reintroduction in many countries.

Changes in water quality and availability may also affect human health. Drought-induced famine has enormous consequences for human health and survival.

The potential scarcity in some regions of biomass used for cooking, and the growing difficulty in securing safe drinking water due to drought, may increase malnutrition in some developing countries.

Tasks for the near future

The following research would be necessary:

General climate effect

* The effect of global warming on seasonal trends of major causes of morbidity and mortality
* The assessment of the incidence of major causes of death in industrialized and developing countries in the future.

Temperature stress

* The effect of global warming on heat- and cold-wave episodes
* The effect of such episodes on mortality
* Methods of decreasing mortality among high risk groups
* The assessment of capacity of adaptation to hot and cold weather, especially among vulnerable population groups such as the elderly.

Air pollution

* The effect of global climate change on oxidants and organic carcinogens in the atmosphere
* Exposure assessment of these air pollutants
* The incidence of respiratory disease and lung cancer in polluted and non-polluted areas.

Chemical pollution

* The effect of global warming on the world-wide chemical pollution
* Human exposure to these chemical pollutants
* The incidence of morbidity and mortality in acutely or chronically exposed populations.

Water quality

* The effect of global warming on the precipitation in various countries
* The assessment of hygienic quality of water resources in the world.

Vector-borne diseases

* The effect of global warming on geographical abundance of major vector species
* The assessment of incidence of vector-borne diseases in the future

- The improvement of the environment to prevent the breeding of vector species.

UV-B radiation
- The assessment of the elevation of UV-B radiation according to ozone depletion in order to determine the dose received
- Epidemiological association of the rise of incidence of cataracts, non-melanoma and melanoma skin cancer and an increase of UV-B radiation in many countries
- The risk evaluation of immune suppression by UV-B increase on vaccination and infectious diseases.

Air Pollution

SOx, NOx and auto exhaust controls are already being implemented to improve air quality in urban areas in some developed countries. Concerns about possible energy penalties and overall implications of such control measures for greenhouse gas emissions will need to be incorporated in future planning. Moreover, global warming and stratospheric ozone depletion appear likely to aggravate tropospheric ozone problems in polluted urban areas.

The tropospheric temperature rise induced by the greenhouse effect could change homogeneous and heterogeneous reaction rates, solubility in cloud water, emissions from marine, soil and vegetative surfaces and deposition to plant surfaces of various atmospheric gases including water vapour and methane.

A change in water vapour concentration will lead to changes in the concentration of HOx radicals and H_2O_2, which are important for the oxidation of SO_2 and NOx in the atmosphere. The predicted change of the patterns of cloud cover, stability in the lower atmosphere, circulation and precipitation could concentrate or dilute pollutants, and change their distribution pattern and transformation rates in regional or local sectors.

A change in aerosol formation by atmospheric conversion from NOx, and SO_2 and wind-blown dust from arid land could lead to changes in visibility and albedo. Material damage caused by acidic and other types of air pollutants may be aggravated by higher levels of humidity.

Tasks for the near future
- Changes in the frequency and pattern of cloud cover due to the greenhouse effect should be studied in relation to ozone formation and conversion of SO_2 and NOx
- Change in pressure, wind, circulation and precipitation patterns and frequency and intensity of stagnation episodes due to the greenhouse effect should be studied in relation to the change in distribution of air quality, oxidants and acid rain

- Quantification should be made of the effect of atmospheric temperature on photochemical oxidant formation and on acid rain. Particularly the discrepancy between model prediction and the results of photo-chemical smog chamber experiments for different levels of NOx should be pursued
- The temperature dependence of energy use in various sectors other than electricity due to global warming should be studied in relation to its effect upon acid deposition and photochemical ozone formation
- Change in the extent of arid regions should be studied in relation to generation of wind-blown dust and its effect upon air quality, acid rain and planetary albedo
- Change in oxidant formation in remote areas due to the greenhouse effect should be studied, taking the change of emission of precursor gases into account
- The temperature dependence of homogeneous reaction rates, and reaction rates in the aqueous phase of atmospherically important gaseous species should be more accurately quantified
- Further study should be made of the temperature dependence of emissions of important gases from soil, sea surface and plants
- General circulation models and mesoscale meteorology-chemistry models which include the above effects should be linked to predict overall effects of greenhouse gases.

Ultraviolet-B Radiation

Besides the human health implications of increased UV-B radiation already discussed, such radiation may also significantly affect terrestrial vegetation, marine organisms, air quality and materials. Increased ultraviolet-B may adversely affect crop yields. There are some indications that increased solar ultraviolet-B radiation which penetrates into the ocean surface zone where some marine organisms live, may adversely affect marine phytoplankton, potentially reducing marine productivity and affecting the global food supply. Increased ultraviolet-B radiation can also be expected to accelerate degradation of plastics and other coating used outdoors. The enhanced greenhouse effect is expected to decrease stratospheric temperatures and this may affect the state of the stratospheric ozone layer.

Tasks for the near future
- Trends of UV-B radiation should be studied and correlated to ozone trends
- More detailed measurements of the wavelength dependence of UV-B radiation should be made
- Efforts should be made to improve the standardizing of instrumentation and calibrations of UV measurement

- The number of field experiments for impacts of UV-B on agriculture should be increased
- Studies must be initiated to determine the impacts of UV-B on natural ecosystems
- UV-B effects on growth and reproductive cycles of lower plants, such as mosses, fungi and ferns, have yet to be studied
- Increase the effort taken to obtain a better understanding of the effects of multiple stresses and of shifts in competitive balance when plants are given additional UV-B
- An area of world-wide interest may be tropical rice growing regions, where information is limited on how rice will be affected either under enhanced UV-B or under increased temperature and CO_2
- Establish the effect of UV-B stress to aquatic micro-organisms on commercial fisheries
- Determine the threshold of effect and biological action spectra for aquatic organisms
- Determine long-term effects for embryos or larvae exposed to UV-B radiation
- Determine effects on ecosystems, including the Antarctic ecosystems
- Obtain data on the mechanisms of damage, and range of possible adaptation or genetic selection in response to increased UV-B radiation
- Develop predictive three-dimensional models for loss in biomass production with CO_2 increase, including the enhancing effect of temperature increase
- For the ozone formation and destruction problem, the main research areas are: peroxides, photodissociation reactions of compounds that absorb UV-B, and aerosol formation
- Expanded modelling is needed to understand the effects of increased UV-B on tropospheric air quality and model predictions should be verified experimentally
- A comparative study of environmental photodegradation of relevant plastic materials under near-equator and high-latitude conditions
- Quantitative effect of UV-B radiation on the non-plastic materials such as paints and coatings, rubber products, wood and paper products, and textiles
- Set up long-term weathering studies in different geographical regions to study the effects of naturally occurring variations in the UV content of sunlight
- Undertake investigations of surface coating, painting and other means of controlling photodegradation, including a study of environmental impact of such technologies.

Conclusion

Despite the serious limitations of current climate models in projecting the pace, magnitude and regional and local manifestations of global warming, there is a high probability that highly disruptive impacts of the kind summarized above will occur broadly once the projected global warming is fully evident. Many developing countries are already vulnerable due to pressures of poverty, resource depletion and growing population. Disruptions associated with rapid and large-scale climate change may increase this vulnerability. It is essential that response strategies to minimize adverse impacts and limit global emissions be sensitive to the special vulnerability of many developing countries.

A co-operative international program of long-term monitoring and systematic surveillance should be launched immediately to establish a base-line to detect impacts in the future. Criteria should be developed for evaluating the significance of impacts. Such an assessment should facilitate the development of a systematic action plan to ameliorate these impacts. Such a programme can also provide useful quantitative inputs to the development of climate scenarios as developed by Working Group I.

Response strategies for anticipated climate change may also have considerable environmental and economic impacts. It is essential that analysis of such potential impacts be an integral part of the evaluation of response strategy options.

Recommendations for Action

- Assessment of the vulnerability of countries, especially in the developing world, to gain or loss of energy resources such as hydroelectric power, biomass, wind and solar, and an examination of available substitutes under new climate conditions, should be a high priority.
- Research is critically needed into the adaptability of vulnerable human populations, especially the elderly and the sick, to the occurrence of increased heat stress as well as the potential for vector-borne and viral diseases to shift geographically.
- Policy-makers should give priority to the identification of population and agricultural and industrial production at risk in coastal areas subject to inundation from sea-level rise of various magnitudes and to storm surges.
- It is important that developing countries have the capability to assess climate change impacts and to integrate this information into their planning. The world community should assist countries in conducting such assessments and work to create indigenous climate-change impact assessment capabilities in such countries.

Acknowledgements

This report is fully based on the final version of IPCC (Intergovernmental Panel on Climate Change) report. The original report of IPCC Working Group II, Section 5, is titled "Likely impacts of climate change on human settlement, the energy, transport and industrial sectors, human health and air quality, and likely impacts of changes in UV-B", and synthesized by a team with following members:

Co-chairmen:

Dr Michio Hashimoto (Japan), Academician M Styrikovich (USSR) Dr Shuzo Nishioka (Japan, acting)

Lead authors:

Mitsuru Ando, Shuzo Nishioka, Toshiichi Okita, Christian Rouviere (Japan), John Topping, Richard Ball, Ted Williams (USA), Y. Shinyak (USSR)

Expert contributors:

John Connell, Peter Cheng, Peter Roy (Australia), Syed Safiullah (Bangladesh), Roger B. Street (Canada), Ye Ruqiu (China), Haile Lul Tebicke (Ethiopia), Ernst G. Jung, Ewald Tekulve, Fritz Steimle, Gerd Jendritzky, G. Waschke, Karl Gertis, P. Donat-Hader (Germany), Aca Sugandhy (Indonesia), Masami Iriki, Nobuaki Washida, Shaw Nishinomiya (Japan), Richard Odingo (Kenya), J.C. van der Leun (Netherlands), D.S. Wratt (New Zealand), Wolf H. Weihe (Switzerland), Alan Teramura, David Guinnup, J.D. Longstreth, L.D. Grant, Michael Oppenheimer, Michael P. Walsh (USA), Aleksandr F. Yakovlev, Igor Nazarov, Nadezhda F. Vladimirova (USSR), P. J. Waight (WHO)

Editor of Summary:

John Topping (USA)

More than 220 scientific studies, were evaluated and referred to by the lead authors, and a number of comments and suggestions were given by experts through the IPCC lead authors' meetings and peer review processes. The authors express their sincere thanks to those contributors.

References

Ando, M., Hirano, S., and H. Itoh, 1985. Transfer of hexachlorobenzene from mother to new-born baby through placenta and milk. *Archives of Toxicology* 56, 195–200.

Aneja, V.P., J.H. Overton, Jr., L.T. Cuppitt, J.L. Durham and W.E. Wilson (1979). Direct measurements of emission rates of some atmospheric biogenic sulfur compounds. *Tellus,* 31, 174–178.

Aoki, Yoji, Climatic Sensitivity of Downhill Ski Demands, NIES, Tsukuba City, Japan, 1989.

Arthur, Louise M., The Implication of Climate Change for Agriculture in the Prairie Provinces, Environment Canada, *CCD.* 88–01, Downsview, Ontario, Canada, 1988.

Bangladesh Research Institute, *Proceedings on the Bangladesh Floods*, Dhaka, Bangladesh, 1989.

Berz, G., "CIimatic change: impact on international rein–surance", in *Greenhouse: Planning for Climate Change*, G.I. Pearman, Editor, CSIRO Publications, East Melbourne, Victoria, Australia, 1988.

Bidelman, T.F., E.J. Christenson, W.N. Billings, and R. Leonard, 1981. Atmospheric transport of organo-chlorines in the North Atlantic gyre. *Journal of Marine Research* 39, 443–464.

Bird, Eric C.F., "Potential Effects of Sea-level Rise on the Coasts of Australia, Africa, and Asia", in James G. Titus, Ed., *Effects of Changes in Stratospheric Ozone and Global Climate*, Vol. 4: Sea-level Rise, US EPA and UN Environment Programme, October 1986.

Blum, H.F., 1959. *Carcinogenesis by Ultraviolet Light*. Princeton University Press, Princeton, N.J.

Blum, H.F., H.G. Grady, and J.S. Kirby-Smith, 1941. Quantitative introduction of tumors in mice with ultraviolet radiation. *J. of National Cancer Institute of US*, 2, 259–268.

Bolhofer, W.C., "Summary of the US Canada Great Lakes Climate Change Symposium", in *Coping With Climate Change*, John C. Topping, Jr., Editor, Climate Institute, Washington, DC, 1989.

British Meteorological Office Report, February 3, 1989.

Broadus, James M., John D. Millman, Steven F. Edwards, David G. Aubrey, and Frank Gable, "Rising Sea-level and Damming of Rivers: Possible Effects in Egypt and Bangladesh", in James G. Titus, Ed., *Effects of Changes in Stratospheric Ozone and Global Climate*, Vol. 4: Sea-level Rise, US EPA and UN Environment Programme, October 1986.

Brown, H.M., "PIanning for Climate Change in the Arctic—The Impact on Energy Resource Development", Climate Institute *Symposium on the Arctic and Global Change*, Ottawa, Ontario, Canada, October 26, 1989.

CCD 88–07, *Socio-Economic Assessment of the Physical and Ecological Impacts of Climate Change on the Marine Environment of the Atlantic Region of Canada—Phase 1*, Climate Change Digest, Canadian Climate Centre, Downsview, Ontario, Canada (1988). Submitted by Dr Peter Stokoe, Dalhousie University, Halifax, Nova Scotia.

Carter, W.P.L., A.M. Winer, K.R. Darnall and J.N. Pitts, Jr. (1979). Smog chamber studies of temperature effects in photochemical smog. *Environ. Sci. Tech.*, 13, 1094–1100.

Changnon, Stanley A., Steven M. Leffler and Robin Shealey, *Effects of Past Low Lake Levels and Future Climate Related Low Lake Levels on Lake Michigan, Chicago and the Illinois Shoreline*, Illinois State Water Survey, Champaign, Illinois, USA, 1989.

Climate Institute, *Climate Change and the Third World*, Summary Report of the Symposium on the Impact of Climate Change for the Third World: Implications for Economic Development and Financing (March 24–25, 1988, Washington, DC) and Related Symposium for United Nations Missions and Agencies (June 13, 1988, New York, NY).

Cohen, Stewart J., "How Climate Change in the Great Lakes Region May Affect Energy, Hydrology, Shipping and Recreation", in *Preparing for Climate Change*, Climate Institute, Editor, Government Institutes, Inc., Rockville, Maryland, USA, 1988.

Conde, Julien, *South North International Migrations*, OECD, Development Center Papers, 1986.

Dadda, Turkiya Ould, Presentation to Conference on Implications of Climate Change for Africa, Howard University, Washington, DC, USA, May 5, 1989.

Damkaer, D.M. (1987), Possible influences of solar UV radiation in the evolution of marine zooplankton, in Calkins, J., (Editor), *The Role of Solar Ultraviolet Radiation in Marine Ecosystems*, New York, Plenum.

Damkaer, D.M., D.B. Day, G.A. Heron, and E.F. Prentice (1980), Effects of UV-B radiation on near-surface zooplankton of Puget Sound. *Oecologia*, 44, 149–158.

Davis, Devra Lee, Victor Miller and James J. Reisa, "Potential Public Health Consequences of Global Climate Change", in *Preparing for Climate Change*, Climate Institute, Editor, Government Institutes, Inc., Rockville, Maryland, USA, 1988.

Debrah, Lt. Col. Christine (Ret'd), *Address to the Conference on Implications of Climate Change for Africa,* Howard University, Washington, DC, USA, May 5, 1989.

DeFabo, E.C. and F.P. Noonan, "Stratospheric Ozone Depletion, Sunlight and Immune Suppression: A New Connection", in *Coping With Climate Change*, John C. Topping, Jr., Editor, Climate Institute, Washington, DC, USA, 1989.

Dobson, A.P. and E.R. Carper, 1989. Global warming and potential changes in host-parasite and disease-vector relationships. *Consequences of Global Warming for Bio-diversity*, R. Peters (Editor), Yale University Press.

Dobson, Andrew, "Climate Change and Parasitic Diseases of Man and Domestic Livestock in the United States", in *Coping With Climate Change*, John C. Topping, Jr., Editor, Climate Institute, Washington, DC, USA, 1989.

Druyan, Leonard, Presentation, Conference on Implications of Climate Change for Africa, Howard University, Washington, DC, USA, May 5, 1989.

Emanuel, K.A., "The dependence of hurricane intensity on climate", in *Nature*, 326, 483–485, 1987.

Flynn, Timothy J., Stuart G. Walesh, James G. Titus and Michael C. Barth, "Implications of Sea-level Rise for Hazardous Waste Sites in Coastal Floodplains", in *Greenhouse Effect and Sea-level Rise: A Challenge for this Generation*, Michael C. Barth and James G. Titus, Editors, Van Nostrand, Reinhold Co., New York, NY, USA, 1984.

French, Hugh, Presentation to Climate Institute Symposium on the Arctic and Global Change, Ottawa, Canada, October 26, 1989.

Friedman, D.G., "Implications of Climate Change for the Insurance Industry", in *Coping With Climate Change*, John C. Topping, Jr., Editor, Climate Institute, Washington, DC, 1989.

Galloway, R.W., "The potential impact of climate changes on Australian ski fields", in *Greenhouse: Planning for Climate Change*, CSIRO Publications, East Melbourne, Victoria, Australia, 1988.

Gertis, K. and F. Steimle, "Impact of Climate Change on Energy Consumption for Heating and Air Conditioning", *Report to the Federal Minister for Economics, Summary*, 1989.

Gery, M.W., R.D. Edmond and G.Z. Whitten (1987). Tropospheric ultraviolet radiation: Assessment of existing data and effect on ozone formation. *EPA Report 600/3-87/047*, October.

Glantz, M.H., B.G. Brown and M.E. Krenz, Societal Responses to Regional Climate Change: Forecasting by Analogy, ESIG/EPA, 1988.

Goklany, Indur, Personal Communication, February 8, 1990.

Golitsyn, G.S., Institute of Atmospheric Physics, Moscow, USSR, Climate Changes and Related Issues, Presentation August 24, 1989 at *Sundance, Utah Symposium on Global Climate Change.*

Graedel, T.E. and R. McGill, "Degradation of materials in the atmosphere", in *Environmental Pollution Technology*, Vol. 20, No. 11, 1986.

Granger, Orman E., "Implications for Caribbean Societies of Climate Change, Sea-level Rise and Shifts in Storm Patterns", in *Coping With Climate Change*, John C. Topping, Jr., Editor, Climate Institute, Washington, DC, USA, 1989.

Hader, D.P. (1989), Private communication.

Hader, D.P. and M.A. Hader (1988a), Inhibition of motility and phototaxis in the green flagellate, Euglena gracilis, by UV-B radiation, *Arch. Microbiol.*, 150, 20–25.

Hader, D.P. and M.A. Hader (1988b), Ultraviolet-B inhibition of motility in green and dark bleached Euglena gracilis, *Current Microbiol.*, 17, 215–220.

Hader, D.P. and M.A. Hader (1989), Effects of solar UV-B irradiation on photomovement and motility in photosynthetic and colorless flagellates, *in press.*

Han, Mukang, "Global Warming-Induced Sea-level Rise in China: Response and Strategies", *Presentation to World Conference on Preparing for Climate Change*, Cairo, Egypt, December 19, 1989.

Hansen, J., D. Rind, A. Del Genio, A. Lacis, S. Lebedeff, M. Prather, R. Ruedy and T. Karl, "Regional Greenhouse Climate Effects", in *Coping With Climate Change*, John C. Topping, Jr., Editor, Climate Institute, Washington, DC, USA, 1989.

Hassett, G. Mustafa, M.G. Levison, W. and Elastoff, R.M. (1985). Marine lung carcinogenesis following exposure to ambient ozone concentrations, *J. Natl. Cancer Inst.*, 75, 771–777.

Hatakeyama, S., M. Watanabe and N. Washida (1989). Effect of global warming on photochemical smog. *To be presented at the 30th Annual Meeting of Japan Soc. Air Pollut.*

Henderson-Sellers, Ann and Russell Blong, *The Greenhouse Effect: Living in a Warmer Australia*, New South Wales University Press, Kensington, N.S.W., Australia, 1989.

Hiller, R., R.D. Sperduto, and F. Ederer (1983), Epidemiologic associations with cataract in the 1971–1972 National Health Nutrition Survey, *Am. J. Epi.*, 118, 239–249.

Hoffman, John and Michael Gibbs, Future Concentrations of Stratospheric Chlorine and Bromine, US EPA, Washington, DC, August 1988.

Hunter, J.R., S.E. Kaupp, and J.H. Taylor (1982), Assessment of the effects of UV radiation on marine fish larvae, in Calkins, J., (Editor), *The Role of Solar Ultraviolet Radiation in Marine Ecosystems*, New York, Plenum.

Hyman, William, Ted. R. Miller, and J. Christopher Walker, "Impacts of Global Climate Change on Urban Transportation", Transportation Research Record, 1989, *in press.*

ICF Inc., "Potential Impacts of Climate Change on Electric Utilities", study for New York Research and Development Authority, December 1987.

IPCC, Working Group 1. Estimates of Climate Change: A Contribution to the Report of Working Group 2, February 21, 1990.

Iriki, M. and M. Tanaka, 1987. "Accidental hypothermia in Japan", in *Climate and Human Health, World Climate Applications Programme*, World Meteorological Organization.

Izrael, Yuri, Personal Communication, Tokyo, Japan, September 14, 1989.

Jacobson, Jodi I., "Abandoning Homelands", in *State of the World*, Worldwatch Institute, Washington, DC, USA, 1989.

Japan Ministry of Transportation, *Impact on the Harbor Area*, August, 1989.

Jessup, Philip S., "Carbon Dioxide Trends in Canadian Transportation", in *Coping With Climate Change*, John C. Topping, Jr., Editor, Climate Institute, Washington, DC, USA, 1989.

Jodha, N.S., "Potential Strategies for Adapting to Greenhouse Warming: Perspectives from the Developing World", in *Greenhouse Warming: Abatement and Adaptation*, Norman J. Rosenberg, William E. Easterling III, Pierre R. Crosson, Joel Darmstadter, Editors, Resources for the Future, Washington, DC, USA, 1989.

Johansson, E. and L. Granat (1984). Emission of nitric oxide from arable land. *Tellus*, 36B, 25–27.

Jung, E.G., Wie kann man Melanome verhindern, Prävention der Melanome und Fruherkennung der Melanomvorläufer (How to prevent melanoma? Prevention of melanoma and early diagnosis), *DMW Deutsche* Medizinische Wochenschrift. 114 (1989) 393–397.

Kalkstein, Laurence S., "A new approach to evaluate the impact of climate upon human mortality", *Report to Conference on Global Atmospheric Change and Human Health*, NIEHS, 1989b.

Kalkstein, Laurence S., "Potential Impact of Global Warming: Changes in Mortality from Extreme Heat and Cold", in *Coping With Climate Change*, John C. Topping, Jr., Editor, Climate Institute, Washington, DC, USA, 1989.

Kassi, Norma, "Notes for a Panel Discussion on the Social Effects of Global Change", *Climate Institute Symposium on the Arctic and Global Change*, Ottawa, Ontario, Canada, October 27, 1989.

Kavanaugh, M., M. Barth and T. Jaenicke, "An Analysis of the Economic Effects of Regulatory and Non-Regulatory Events Related to the Abandonment of Chlorofluorocarbons as Aerosol Propellants in the United States with a Discussion of Applicability of the Analysis to Other Nations", Aerosol Working Paper Series, Paper 1, ICF Inc., Washington, DC, 1986.

Kennedy, Victor S., "Potential Impact of Climate Change on Chesapeake Bay Animals and Fisheries", in *Coping With Climate Change*, John C. Topping, Jr., Editor, Climate Institute, Washington, DC, 1989.

Kerr, Richard A., "Global Warming Continues in 1989", *Science*, February 2, 1990 at 521.

Khalil, M.A.K. and R.A. Rasmussen, "Carbon Monoxide in the Earth's Atmosphere: Indications of a Global Increase", *Nature* 245, p.332, March 1988.

Klassen, W.J., "Speaking Notes", *Climate Institute Symposium on the Arctic and Global Change*, Ottawa, Ontario, Canada, October 27, 1989.

Kornhauser, A., "Implications of Ultraviolet Light in Skin Cancer and Eye Disorders", in *Coping With Climate Change*, John C. Topping, Jr., Editor, Climate Institute, Washington, DC, USA, 1989.

Lamothe & Periard, "Implications of Climate Change for Downhill Skiing in Quebec", Environment Canada, *CCD* 88–03, Downsview, Ontario, Canada, 1988.

Lashof, Daniel A. and Dennis Tirpak, Editors, *The Draft Stabilization Report of the US Environmental Protection Agency, Policy Options for Stabilizing Global Climate*, Washington DC, USA, February, 1989.

Leatherman, Stephen P., "Beach Response Strategies to Accelerated Sea-Level Rise", in *Coping With Climate Change*, John C. Topping, Jr., Editor, Climate Institute, Washington, DC, USA, 1 989.

Lee, James A., The Environment, Public Health and Human Ecology, Considerations for Economic Development, *World Bank Publication*, Johns Hopkins University Press, 1985.

Last, J.A., Warren, D.L., Pecquet-Goad, E. and Witsohi, H.P. (1987). Modification of lung tumor development in mice by ozone. *J. Natl. Cancer Inst.*, 78, 149–154.

Linder, Kenneth and Michael Gibbs, "The Potential Impacts of Climate Change on Electric Utilities: Project Summary", in *Preparing for Climate Change*, Climate Institute, Editor, Government Institutes, Inc., Rockville, Maryland, USA, 1988.

Linder, Kenneth, "Regional and National Effect of Climate Change on Demands for Electricity", in *Coping with Climate Change*, John C. Topping, Jr., Editor, Climate Institute, Washington, DC, USA, 1989.

Liu, S.C. and M. Trainer (1988), Responses of the tropospheric ozone and odd hydrogen radicals to column ozone change, *J. Atmos. Chem.*, 6, 221.

Lonergan, Steven, "CIimate Change and Transportation in the Canadian Arctic", Climate Institute Symposium on the Arctic and Global Change, Ottawa, Ontario, Canada, October 26, 1989.

Longstreth, Janice, "Overview of the Potential Effects of Climate Change on Human Health", in *Coping With Climate Change*, John C. Topping, Jr., Editor, Climate Institute, Washington, DC, USA, 1 989.

Mabbutt, J.A., "Impact of Carbon Dioxide Warming on Climate and Man in the Semi-Arid Tropics", in *Climatic Change*, Vol. 15, Nos. 1–2, October 1989.

MacCracken, M.C. and G.D. Santer (1975). Development of air pollution model for the San Francisco Bay area. *UCRL* 51920, Vol. 1, Lawrence Livermore Laboratory, Univ. California, Livermore, CA.

MacDonald, Gordon J., "The Greenhouse Effect and Climate Change", presented to Env. & Public Works Committee, January 28, 1987.

Makino, K., 1987. Effects of weather on mortality from major causes using monthly number of deaths for a thirty year period, *Jpn. J. Biometeor.* 25, 69–78.

Manabe, S. and R.T. Wetherald (1967), Thermal equilibrium of the atmosphere with a given distribution of relative humidity, *J. Atmos. Sci.* 21, 361–385.

Marmor, M., 1978. Heat wave mortality in nursing homes. *Environmental Research* 17, 102–115.

Medioli, Alfred, "Climate Change: The Implications for Securities Underwriting", in *Coping With Climate Change*, John C. Topping, Jr., Editor, Climate Institute, Washington, DC, 1989.

Meyers and J. Sathaye, "Electricity in Developing Countries: Trends in Supply and Use Since 1970", Lawrence Berkeley Laboratory, *LBL*–2166, December 1988.

Miller, Ted R., "Impacts of Global Climate Change on Metropolitan Infrastructure", in *Coping With Climate Change*, John C. Topping, Jr., Editor, Climate Institute, Washington, DC, USA, 1 989.

Ministère de l'Equipement du Logement, de l'Aménagement du Territoire et des Transports, *Mutations Economiques et Urbanisation, La Documentation Française*, 1986.

Miura, T., 1987. The influence of seasonal atmospheric factors on human reproduction, *Experientia* 43, 48–54.

Momiyama, M. and K. Katayama, 1972. Deseasonalization of mortality in the world, *Int. J. Biomet.* 16, 329–342.

Morris, R.E., M.W. Gery, M.K. Liu, G.E. Moore, C. Daly, and S.M. Greenfield (1988), Examination of sensitivity of a regional oxidant model to climate variations, *Draft Final Report*, Systems Applications, Inc., San Rafael, CA.

Mortimore, Michael, The Causes, Nature and Rate of Soil Degradation in the Northernmost States of Nigeria and an Assessment of the Role of Fertilizer in Counteracting the Processes of Degradation, *Environment Department Working Paper No. 17*, World Bank, July 1989.

Murday, Maylo, Presentation to Conference on Implications of Climate Change for Africa, Howard University, Washington, DC, USA, May 4, 1989.

Myers, Norman, "Tropical Deforestation and Climatic Change", Environmental Conservation, Vol. 15, No. 4, The Foundation for Environmental Conservation, Switzerland, Winter 1988.

NASA-WMO, Ozone Trends Panel Report, March, 1988.

Nishinomiya, S. and Kato, H. (1990). Potential effects of global warming on the Japanese electric industry—Event tree of impacts on the electric utility industry stemming from climate induced changes in the natural environment, ecosystems and human society. CRIEPI report, *in press.*

Odingo, Richard, Professor, University of Nairobi, Personal Communication, Geneva, Switzerland, IPCC WG2 meeting, November 1, 1989.

OECD, Center of Development, La Migration internationale dans ses relations avec les politiques d'adjustement industriel et agricole, *Comptes rendus du Séminaire de Vienne*, 13–15 Mai 1974, Paris, 1974.

Okita, T. and S. Kanamori (1971). Determination of trace concentration of ammonia using pyridinepyrazolone reagent. *Atmos. Environ.*, 5, 621–628.

Oppenheimer, Michael, "Climate Change and Environmental Pollution and Biological Interactions", in *Climatic Change*, Vol. 15, Nos. 1–2, October 1989.

Parry, M.L. and N.J. Read, The Impact of Climatic Variability on UK Industry, Atmospheric Impacts Group, University of Birmingham, Birmingham, UK, 1988.

Parry, M.L., T.R. Carter, and N.T. Konijn, The Impact of Climate Variations on Agriculture, Reidel Dordrecht, The Netherlands, 1987.

Pearman, G.I., Editor, *Greenhouse: Planning for Climate Change*, CSIRO Australia, Melbourne, 1988.

Peele, B.D., "Insurance and the greenhouse effect", in *Greenhouse: Planning for Climate Change*, G.I. Pearman, Editor, CSIRO Publications, East Melbourne, Victoria, Australia, 1988.

Quinn, Frank H., "Likely Effects of Climate Change on Water Levels in the Great Lakes", in *Preparing for Climate Change*, Climate Institute, Editor, Government Institutes, Inc., Rockville, Maryland, USA, 1988.

Ramanathan, V., "The Greenhouse Theory of Climate Change: A Test by an Inadvertent Global Experiment", *Science*, 15 April 1988.

Raoul, Joseph and Zane M. Goodwin, "Climatic Changes--Impacts on Great Lakes Lake Levels and Navigation", in *Preparing for Climate Change*, Climate Institute, Editor, Government Institutes, Inc., Rockville, Maryland, USA, 1988.

Rapaport, R.A., N.R. Urban, P.D. Capel, J.E. Baker, B.B. Looney, S.J. Eisenreich, and E. Gorham, 1985. "New" DDT inputs to North America: Atmospheric deposition. *Chemosphere* 14, 1167–1173.

Regens, James L., Frederick W. Cubbage and Donald G. Hodges, "Climate Change and US Forest Markets", in *Coping With Climate Change*, John C. Topping, Jr., Editor, Climate Institute, Washington, DC, 1989.

Regier, H.A., J.A. Holmes and J.D. Meisner, "Likely Impact of Climate Change on Fisheries and Wetlands With Emphasis on the Great Lakes", in Preparing for Climate Change, Climate Institute, Editor, Government Institutes, Inc., Rockville, Maryland, USA, 1988.

Roper, Tom, Presentation to Fabian Society Greenhouse Conference, Lorne, Victoria, Australia, May 20, 1989.

Rothlisberg, P., D. Staples, I. Poiner and E. Wolanski, "The possible impact of the greenhouse effect on commercial prawn populations in the Gulf of Carpentaria", in *Greenhouse: Planning for Climate Change*, CSIRO Publications, East Melbourne, Victoria, Australia, 1988.

Rovinsky, F.Y.A., M.I. Afanasjev, L.V. Brutseva, and V.I. Yegorov, 1982. Background environmental pollution of the Eurasian continent. *Environmental Monitoring and Assessment* 2, 379–386.

Roy, Peter and John Connell, "Greenhouse: the Impact of Sea-level Rise on Low Coral Islands in the South Pacific", Research Institute for Asia and the University of Sydney, N.S.W., Australia, 1989.

Safiullah, Syed, "Disaster—Perception Cycle", Remarks presented at *IPCC WG2 Section 5 Lead Authors Meeting*, Tsukuba, Japan, September 18–21, 1989.

Sakai, S., *The Impact of Climatic Variation on Secondary and Tertiary Industry in Japan*, 1987.

Sanderson, Marie, Implications of Climatic Change for Navigation and Power Generation in the Great Lakes, Environment Canada, *CCD 87–03*, Downsview, Ontario, Canada, 1987.

Sathaye, J., A. Ketoff, L. Schippev and S. LeLe, "An End-Use Approach to the Development of Long-Term Energy Demand Scenarios for Developing Countries", Lawrence Berkeley Laboratory, *LBL*-25611, February 1989.

Schneider, Stephen H., "The Greenhouse Effect: Reality or Media Event", in *Coping With Climate Change*, John C. Topping, Jr., Editor, Climate Institute, Washington, D.C., USA, 1989.

Schneider, T., Lee, S.D., Walters, G.J.R. and Grant, L.D. *The Proceedings of the 1988 US-Dutch Symposium on Ozone.*

Schuman, S.H., 1972. Patterns of urban heat-wave deaths and implications for prevention, *Environmental Research 5*, 59–75.

Shands, William and John Hoffman, *The Greenhouse Effect, Climate Change and US Forests*, The Conservation Foundation, Washington, DC, 1987.

Silvey, Joseph F., "Implications of Climate Change for the Environmental Engineering and Construction Industry", in *Coping With Climate Change*, John C. Topping, Jr., Editor, Climate Institute, Washington, DC, 1989.

Simmons, Alan, Serio Diaz-Bricquets and Aprodicio A. Laquian, Social Change and Internal Migration, A Review of Research Findings from Africa, Asia and Latin America, International Development Research Center, 1977.

Singh, Professor Bhawan, The Implications of Climate Change for Natural Resources in Quebec, Environment Canada, *CCD* 88–08, Downsview, Ontario, Canada, 1988.

Smith, Barry, Implications of Climatic Change for Agriculture in Ontario, Environment Canada, *CCD* 87–02, Downsview, Ontario, Canada, 1987.

Smith, J.B. and D.A. Tirpak, Ed., "The Potential Effects of Global Climate Change on The United States", US EPA, 1989.

Smith, Joel B., "An Overview of the EPA Studies of the Potential Impacts of Climate Change on the Great Lakes Region", in *Coping With Climate Change*, John C. Topping, Jr., Editor, Climate Institute, Washington, DC, USA, 1989.

Stark, K.P., "Designing for coastal structures in a greenhouse age", in *Greenhouse: Planning for Climate Change*, CSIRO Publications, East Melbourne, Victoria, Australia, 1988.

Stokoe, Peter, Socio-Economic Assessment of the Physical and Ecological Impacts of Climate Change on the Marine Environment of the Atlantic Region of Canada—Phase 1, Environment Canada, *CCD* 88–07, Downsview, Ontario, Canada, 1988.

Street, Roger, Personal Communication, November 6, 1989.

Sugandhy, Aca, "Preliminary Findings on Potential Impact of Climate Change in Indonesia", *Remarks presented at IPCC WG2 Section 5 Leading Authors Meeting*, Tsukuba, Japan, September 18–21, 1 989.

Sullivan, J.H. and A.H. Teramura (1988), Effects of ultraviolet-B irradiation on seedling growth in the Pinaceae, *Am. J. of Bot.*, 75, 225–230.

Takeuchi, U. and S. Hayashida (1987), Effects of ozone layer depletion upon vegetation, *Kankyogijutsu*, 16, 732–735.

Tanabe, S., R. Tatsukawa, M. Kawano, and H. Hidaka, 1982. Global distribution and atmospheric transport of chlorinated hydrocarbons. *Journal of the Oceanographical Society of Japan* 38, 137–148.

Tebicke, Haile Lul, Personal Communication, *IPCC Lead Authors Meeting*, Tsukuba City, Japan, September 19, 1989.

Tebicke, Haile Lul, "CImate Change, Its Likely Impact on the Energy Sector in Africa", Remarks presented at *IPCC WG2 Section 5 Lead Authors Meeting*, Tsukuba, Japan, September 18–21, 1989.

Tevini, M., and A.H. Teramura (1989), "UV-B effects on terrestrial plants", *to be published in Photochem. and Photobiol.*

Tickell, Sir Crispin, "Environmental Refugees", National Environment Research Council Annual Lecture, at the Royal Society, London, UK, June 5, 1989.

Timmerman, Peter, "Everything Else Will Not Remain Equal, the Challenge of Social Science Research in the Face of a Global Climate Warming", in *Impacts of Climate Change on the Great Lakes*, US National Climate Program Office and Canadian Climate Centre, January, 1989.

Tingey, D.T., M. Manning, L.C. Grothasu and W.F. Burus (1980). The influence of light and temperature on monoterpene emission rates from slash pine. *Plant Physiol.*, 65, 797j.

Titus, Jim, "Cause and Effects of Sea-level Rise", in *Preparing for Climate Change*, Climate Institute, Editor, Government Institutes, Inc., Rockville, Maryland, USA, 1988.

Tokyo Electric Company (1988). *Summary of prediction of electricity demand in 1988–1997* (in Japanese).

Topping, John C., Jr. and John P. Bond, *Impact of Climate Change on Fisheries and Wildlife in North America*, Climate Institute, Washington, DC, 1988.

Topping, John C., Jr., "Where Are We Headed in Responding to Climate Change", *International Environment Reporter*, Bureau of National Affairs, Inc., October 1988.

Tsyban, A.V., Volodkovich, Yu L., Venttsel, M.V., Pfeifere, M. Yu., Zagryaznenie i tsirkulyatsiya toksicheskikh zagryaznyayuschikh veschestv v ekosisteme Baltijskogo morya V knige: Issledovanie ekosistemy Baltijskogo morya. Vypusk 2. Leningrad. Gidrometeoizdat, 1985, 244–257.

US Environmental Protection Agency (1987), Chapter 12. An assessment of the effects of ultraviolet-B radiation on aquatic organisms, in Hoffman, J.S. (Editor), Chapter 7. *Assessing the Risks of Trace Gases that Can Modify the Stratosphere*, US *EPA 400/1–87/001C.*

US Environmental Protection Agency (1987), Ultraviolet Radiation and Melanoma—With a Special Focus on Assessing the Risks of Ozone Depletion, Vol. IV, J.D. Longstreth (Editor), *EPA 4001–87/00 1 D.*

US Environmental Protection Agency, 1988. The potential effects of global climate change on the United States, *Draft Report to Congress.*

US Environmental Protection Agency, Regulatory Impact Analysis: Protection of Stratospheric Ozone, June 15, 1988.

US Environmental Protection Agency (1988). The Potential Effects of Global Climate Change on the United States. *Draft Report to Congress Vol.1: Regional Studies* (eds., J.B. Smith and D.A. Tirpak), Office of Policy, Planning and Evaluation, October.

UNEP (1989a). Synthesis Report of the Four Assessment Panels Reports, the Open-ended Working Group of the Parties to the Montreal Protocol.

UNEP (1989b). Van der Leun, J.C., J.P. Longstreth, D. English, F. Kerdel-Vegas, and Takisawa Y., Draft of UNEP Report on the Environmental Effects of Ozone Depletion, ch.3, 1–18 (1989).

United Nations, 1986. Determinants of Mortality Change and Differentials in Developing Countries. *Population Studies* No. 94, United Nations, New York.

Victoria Ministry for Planning and Environment, *The Greenhouse Challenge*, Melbourne, Victoria, Australia, June 1989.

Viterito, Arthur, "Implications of Urbanization for Local and Regional Temperatures in the United States", in *Coping With Climate Change*, John C. Topping, Jr., Editor, Climate Institute, Washington, DC, USA, 1989.

Vladimirova, Nadezhda, Personal Communication, Nalchik, USSR, February 28, 1990.

Walker, J. Christopher, Ted R. Miller, G. Thomas Kingsley and William A. Hyman, "Impact of Global Climate Change on Urban Infrastructure", *The Potential Effects of Global Climate Change on the United States*, Appendix H.

Wall, G., Implications of Climatic Change for Tourism and Recreation in Ontario, Environment Canada, *CCD* 88–05, Downsview, Ontario, Canada, 1988.

Walsh, Michael P., "Global Trends in Motor Vehicle Pollution Control—a 1988 Perspective", SAE Technical Paper Series, Detroit, Michigan, USA, February 27, 1989.

Walsh, Michael P., "Global Warming: The Implications for Alternative Fuels", *SAE Technical Paper Series*, Washington, DC, USA, May 2, 1989.

Walsh, Michael P., Personal Communication, October 12, 1989.

Walsh, Michael P., "The Global Importance of Motor Vehicles in the Climate Modification Problem", in *International Environment Reporter*, BNA, Washington DC, USA, May 1989, pp. 261–267.

Washida, N., S. Hatakeyama and O. Kajimoto (1985). Studies on the rate constants of free radical reactions and related spectroscopic and thermochemical parameters. *Res. Report Natl. Inst. Environ. Stud.*, R–85.

Weihe, W.H., 1988. Climate change and human health, World Congress of Climate and Development, Hamburg.

Whitten, G.Z. and W.N. Gerry (1986), The interaction of photochemical processes in the stratosphere and troposphere, in Titus, J.G. (Editor), *Effects of Changes in Stratospheric Ozone and Global Climate*, US EPA and UNEP, Washington, DC.

WHO. *Draft of WHO's Expert Group* "Potential Health Effects of Global Climate Change", 1989.

Williams, G.D.V., R.A. Fautley, K.H. Jones, R.B. Stewart, E.E. Wheaton, Estimating Effects of Climatic Change for Agriculture in Saskatchewan, Canada, *CCD*.

Winget, Carl, "Forest Management Strategies to Address Climatic Change", in Preparing for Climate Change, Climate Institute, Editor, Government Institutes, Inc., Rockville, Maryland, USA, 1988.

WMO/NASA, *Ozone Trends Panel Report*, 1988.

Worrest, R.C. (1982), Review of literature concerning the impact of UV-B radiation upon marine organisms, pp. 429–457 in *The Role of Solar Ultraviolet Radiation in Marine Ecosystems*, J. Calkins (Editor), Plenum Publishing Corp., New York.

Wratt, D.S. (1989). Impacts of climate change on the atmosphere. Private communication.

The Response Strategies Working Group of the IPCC

by Frederick M. Bernthal *(USA)*

I am honored to be here at the Second World Climate Conference to report for Working Group III of the IPCC, the Response Strategies Working Group (RSWG).

Throughout our labor to define our task, organize our work, and structure our First Assessment Report, there was one constant amid all the fluidity: we knew that our report had to be ready in time for this Conference. There were moments when meeting that deadline seemed impossible. But we have produced an assessment that the world community has already found to be very useful - even as we all acknowledge that much more work is needed.

Beyond the contributions of its members, RSWG was greatly aided in this endeavor by the support of Messrs. Tolba and Obasi, the leaders of our two sponsoring bodies, by the wise counsel of IPCC Chairman Bolin, and by frequent consultation with the chairmen of WG's 1 and 2 and the Special Committee - Dr. Houghton, Dr. Izrael and Mssr. Ripert. Special praise and credit must go to the IPCC Secretariat, Dr. Sundararaman and his staff, for their tireless support and thoroughly professional work.

As you consider what WG3 accomplished and what remains to be done, please bear in mind two points: 1) all of the activities took place in a mere 18 months; and 2) the three Working Groups necessarily were working concurrently to develop their respective reports, even though a sequential process would have been more logical. Thus, the reader will notice some inconsistencies and discontinuities across the three reports which inevitably arise from such a parallel process.

When WG3 first met in Washington in January 1989, its task was to define its objectives and organize its work in such a way as to assure a useful product within the time allotted. In June of this year, at its third plenary meeting, RSWG approved the Policy Makers Summary of its work. The report represents an ambitious but still limited and provisional effort to survey climate change response options. In recognition of the constraints imposed by time, the identification of areas for future work is an integral part of each section of the report.

At its first meeting, the RSWG established four subgroups to examine the possible response options. Two subgroups, the Energy and Industry Subgroup (EIS) and the Agriculture, Forestry and Others Subgroup (AFOS), explored the possibilities for limitation and mitigation of potential climate change by steps that would reduce greenhouse gas emissions, or enhance GHG sinks. The other two subgroups, the Coastal Zones Management Subgroup (CZMS) and the Resource Use and Management Subgroup (RUMS), dealt with strategies for adapting to impacts of potential climate change such as sea level rise and changing distributions of flora and fauna.

One of RSWG's first tasks was to develop preliminary scenarios of possible future greenhouse gas emissions as a basis for all three IPCC Working Groups to use in conducting long range analyses. At the direction of the first RSWG plenary, an experts group from the US and the Netherlands developed three possible scenarios of future emissions corresponding to (1) the equivalent of a CO_2 doubling from pre-industrial levels by the year 2030; (2) a CO_2 equivalent doubling by 2060; and (3) a doubling by 2090, with stabilization thereafter.

Subsequently, the group developed two additional emissions scenarios corresponding to emissions projections in which atmospheric concentrations of greenhouse gases are stabilized at a level less than a CO_2 equivalent doubling. These emissions scenarios have been complemented by the work of the EIS and AFOS subgroups based on individual country studies of likely long-term greenhouse gas emissions trends.

These scenarios are purely hypothetical, reflecting the state of knowledge and best estimates as of a given date. To underscore that point, it should be noted that in the year since the original scenarios were drawn up the world community has already rendered obsolete one set of assumptions on which they were based. The June 1990 meeting of the Parties to the Montreal Protocol speeded up the timetable for phaseout of CFCs, just after the RSWG had completed its report.

The RSWG was fortunate in being able to call on a host of dedicated experts from around the world to contribute to the analyses and to lead its various groups. As you can see, the lead countries for the RSWG report represent a cross-section of world geography, climate, and economic development.

The Energy and Industry Subgroup (EIS), co-chaired by Japan and China, considered energy uses in the industrial, transportation, and residential sectors that produce CO_2, methane, nitrous oxide and other gases. The EIS developed estimates of future greenhouse gas emissions from these sectors, and defined technological and policy options to reduce emissions of these gases.

The subgroup called Agriculture, Forestry, and Other Human Activities (AFOS), was co-chaired by the Federal Republic of Germany and Zimbabwe. Of interest to AFOS were methane emissions from livestock, rice, biomass, and waste sources, CO_2 emissions from deforestation, CO_2 uptake from reforestation, and nitrous oxide emissions from the use of fertilizers. AFOS also developed estimates of future greenhouse gas emissions from agriculture, forestry, and other sectors not studied by the EIS group.

The Coastal Zones Management Subgroup, co-chaired by Australia and New Zealand, dealt with the impacts of potential sea level rise and the potential effects of storms and other extreme events on coastal regions. The work of CZMS included reviewing information on technology, practices, and administrative mechanisms relevant to adaptation to the impacts of potential climate change on coastal and low-lying areas.

The Resource Use and Management Subgroup was co-chaired by Canada, France, and India. Its mandate covered natural resources including fisheries, wildlife, forests, biological diversity, and water, as well as adaptation to the impacts of climate change on agriculture and animal husbandry. The RUMS considered possible strategies for either reducing the potential negative impacts or enhancing the possible positive impacts of climate change on food security, water availability, and natural ecosystems.

In a parallel effort, the WG III also looked at five categories of implementation measures to carry out the response strategies. The examination of each topic was led by two or three countries, acting as joint topic coordinators. The topics were:

1) Public Education and Information - to promote awareness of the issue as a global concern, and build support for response measures. Coordinators: US and China

2) Technology Development and Transfer - to promote the development of new technologies to mitigate or adapt to climate change, and also to foster the international transfer of technologies relevant to climate change, a subject of special interest to developing countries. Coordinators: Japan and India

3) Economic (Market) Measures - to assure that response strategies are economically viable and cost effective. Coordinators: Australia and New Zealand;

4) Financial Measures - to enhance the ability of countries, in particular, developing countries, to address climate change. Coordinators: France and Netherlands

5) Legal and Institutional Measures, including possible elements of a framework climate convention. Coordinators: UK, Canada, and Malta.

Each of these groups held numerous workshops attended by many experts from a wide range of countries. The chairs of the four RSWG subgroups and the coordinators of the five RSWG topic areas took the responsibility for completing their individual reports. Although the topic papers and the reports of the subgroups represent the best collective judgment of the hundreds of expert contributors, the underlying reports have not been fully examined or formally approved by the RSWG Plenary. Only the Policy Makers Summary was approved in RSWG plenary session by all countries in attendance.

The primary task of the RSWG was, in the broad sense, technical, not political. The charge of IPCC to RSWG was to lay out as fully and fairly as possible a set of realistic response policy options and the factual basis for those options. Consistent with that charge, it was not the purpose of the RSWG to select or recommend specific actions to be undertaken by governments. The essential purpose of the RSWG report is to highlight for national and international policymakers and negotiators major policy issues.

Among the main findings of the Policymakers Summary:

- Climate change is a global issue requiring a global response effort which may have considerable impact on humankind and individual societies

- Industrialized and developing countries have a common responsibility in dealing with problems arising from climate change, making international cooperation essential. Although countries differ in their relative ability to limit greenhouse gas emissions or to adapt to climate change, it is timely for all countries to consider what measures would be feasible to adapt their economies to limit emissions

- Continuing economic development will increasingly have to take into account the issues of climate change. Thus, it is imperative that the right balance between economic and environmental objectives be struck

- Limitation and adaptation strategies must be considered as an integrated package and should complement each other to minimize net costs

- The potentially serious consequences of climate change give sufficient reasons to begin by adopting response strategies that can be justified immediately even in the face of significant uncertainties
- A well-informed population is essential to promote awareness of the issues and provide guidance on positive practices. The social, economic and cultural diversity of nations will require tailored approaches.

Because the large projected increase in world population will be a major factor in causing the projected increase in global greenhouse gases, the RSWG report notes that it is essential that global climate change strategies include strategies and measures to deal with the rate of growth of world population.

The reports of both Working Groups 1 and 2 identified varying levels of confidence with regard to specific predictions about future climate change and its impacts. The RSWG report recognizes the challenge to policy makers confronted with the uncertainties regarding the causes, magnitude, timing, and regional impacts of climate change, on the one hand, and with uncertainties in the economic costs and benefits associated with climate change response strategies on the other.

In considering specific response strategies, the RSWG recognized that response in the short term should emphasize measures which can be justified on the basis of benefits other than those related to climate change.

Short term limitation measures which meet that criterion include: reducing emissions through improved energy efficiency; use of cleaner energy sources and technologies; increasing sinks through improved forest management and expansion of forested areas; phasing out CFCs; and reducing emissions of greenhouse gases from non-industrial sectors through improved agricultural technology and waste management practices.

Prudent adaptation measures for the near-term include: developing emergency and disaster preparedness policies and programs; developing comprehensive management plans for areas at risk from sea level rise; and improving the efficiency of natural resource use, for example from research on crops adaptable to the potential climate changes.

Recognizing that other measures may be required in the long term, the RSWG noted that actions should be undertaken now to provide the requisite information for making such decisions in the future. These investments in the long-term knowledge base include increased research to reduce scientific and economic uncertainties and the development of new technologies in the fields of energy, industry, forestry, and agriculture.

Finally, because effective responses to climate change may require an unprecedented degree of international cooperation, the RSWG noted that the international community should embark upon the negotiation of a framework convention on climate change as soon as possible after completion of the IPCC first assessment report.

To assist in that effort, RSWG was directed to develop possible elements of a framework climate convention. The product is one I have often referred to as a "roadmap" for the Convention negotiations. I commend that map to your attention. We concluded that such a Convention should, at a minimum, contain general principles and should be framed so as to gain the adherence of the largest possible number and most suitably balanced range of countries consistent with timely action.

As this group is aware, it has recently been decided that the first session of negotiations on a framework climate convention will be convened in Washington in early February of next year. This will be exactly one year after the IPCC met in Washington for its third plenary session. In comparison with the typical timetable for international deliberations on a global scale, this is rapid progress indeed, and demonstrates the concern the world community attaches to this issue.

Let me turn now to one piece of unfinished business not covered in the RSWG report. In November 1989, the Noordwijk Environment Ministers meeting requested the IPCC to analyze the feasibility of achieving: 1) targets and timetables for limitation of CO_2 emissions, including, e.g. a 20 percent reduction of CO_2 emission levels by the year 2005, and 2) a net increase of world forest cover of 12 million hectares per year in the beginning of the next century.

Having completed their contributions to the IPCC's First Assessment Report, the EIS and AFOS subgroups of RSWG are returning to the business at hand, including a response to the "Noordwijk remits."

In addition to holding two workshops on this topic, the Energy and Industry Subgroup has reviewed case studies for some 21 countries on the feasibility of various targets for greenhouse gas emissions.

The EIS has identified an array of significant issues that all countries will need to consider in determining the feasibility of targets on greenhouse gases, if such targets are to be decided. These issues fall into three categories -- technical, accounting, and impact. Resolving these difficult issues will require improvements in inventory and monitoring of greenhouse gas emissions.

Technical issues include fundamental questions such as:
- whether the target applies comprehensively to all greenhouse gases or selectively;
- choosing the base and target years,
- whether emission inventories for the various sectors and/or gases are adequate to allow targets to be set and measured.

The accounting aspects include:

- measuring and verifying any possible targets, and how to measure the full, net effects of actions on the fuel cycle, for example;
- monitoring the performance of a country or groups of countries, given the targets or strategies that they might adopt;
- a global economy produces further issues to be addressed, such as: 1) a country may benefit from imported products with "energy value added," for which an exporting country has incurred increased emissions; 2) the attribution of emissions from international fuel storage; 3) energy trade through transboundary electricity grids.

The impact aspects include:

- determining what is technically and economically feasible;
- defining the time frame and discount rate for assessing costs and benefits;
- assessing the overall economic implications of diverse policy measures;
- determining the ramifications for international and domestic energy markets.

EIS has also drawn up a list of current and proposed studies to further assess emissions from the energy and industry sectors, as well as a work plan to pursue further analytical work that it is anticipated will be required to support the convention negotiations.

For its part, AFOS has begun work to assess the implications and feasibility of a net annual increase in world forest area of 12 million hectares. AFOS's work will add additional perspectives to the preliminary estimates already performed by Working Group I, which estimated the potential effect of reforestation in terms of net carbon emissions.

Basing their calculations on an assumption of 10 million hectares planted to forests each year for a period of 40 years, Working Group I estimated that the accumulated uptake of carbon in these forests would be equivalent to some 5-10 percent of the emission due to fossil fuel burning projected to occur during that time frame under current trends.

Studies are underway in a number of countries and by FAO on inventories, costs and risks, and impact of the Noordwijk target on a range of sectors. Some preliminary national studies have raised serious doubts about the feasibility of this target, since they indicate that such ambitious efforts to offset carbon emissions through reforestation could entail significant tradeoffs, for example with food security for an increasing world population.

Such issues are independent of the monetary cost of establishing and maintaining the reforested lands. Further analysis is needed on many aspects of agricultural impact,

including the projected population of various regions, and the long-term strategies to assure food and fiber security in a varying climate.

Since the Noordwijk target is couched in terms of net increase in forest cover, calculations must begin from a known base. The work of AFOS has been hampered by the lack of an international consensus on baseline data for current levels of forest cover and rates of land use conversion. FAO and others have reported on the initiation of the Forest Resources Assessment 1990 Project that will ultimately provide, through the coordination of national inventories, a continuous forest resources monitoring. Such an inventory is essential to assess accurately the full impact of a net annual global forest increase of 12 million hectares per year.

With regard to international instruments on forestry, two options have been proposed: 1) a forestry convention, or 2) a forestry protocol to a climate convention. There is not yet consensus among countries as to which option would be preferable. AFOS plans to convene this fall a preliminary meeting of a small planning group to prepare for an international experts meeting to be held in Thailand early in 1991. The meeting in Thailand, which may be an AFOS meeting, will discuss the Noordwijk remit and also the development of key elements that might be contained in a world forest conservation agreement.

The current AFOS workplan calls for submission of its assessment of the Noordwijk remits in time to enable the negotiators of a forestry agreement and participants in the 1992 UN Conference on Environment and Development to take account of the conclusions of AFOS in their deliberations.

With the Response Strategies Working Group having arrived at this milestone, what lies ahead? Although the IPCC Fourth Plenary, in Sundsvall in August did not have time to take up the issue of its future work plan, it was never assumed that the work of the IPCC could possibly be finished in only 18 months, and indeed the substantive work submitted by the RSWG really represents only one year of effort.

One issue of vital importance, which time constraints did not permit to be included in the first assessment report, is consideration of the economic costs associated with the various options. The EIS and AFOS are attempting to address one aspect of this question insofar as it relates to the Noordwijk targets. Their work to date suggests how difficult the task is. However, world political leaders quite rightly will demand such analyses for the whole range of policy response options, in order to evaluate and compare various strategies for mitigation and adaptation. Continuing work after the IPCC's first assessment report will help policy makers make informed decisions as they work toward international agreement on appropriate and feasible response options.

The work of RSWG to date has been valuable not only in building consensus, but also in identifying areas of dispute, and highlighting issues requiring further analysis. While the spotlight of world political attention should and will inevitably shift to the arena of negotiations, the success of these negotiations will depend in large measure on the continuing policy analysis of the RSWG and the many international groups that have contributed to its efforts.

I regret that my duties at home will not permit me to be present for much of this Second World Climate Conference. I wish you well as you undertake your important deliberations.

The Greenhouse Marathon: Proposal for a Global Strategy †

by Pier Vellinga and Rob Swart *(The Netherlands)*

The race to prevent global warming is like the Marathon that is annually run in New York by amateurs. Suddenly, we are aware that we have to run this Marathon. We don't know what to expect but with the outcome of the IPCC report, the starting shot has been fired. And, this Marathon is not a race of forty kilometers but a race for, at least, forty years.

To make this Marathon successful, we need:

- *A Goal*: A common vision of long term goals and objectives is absolutely necessary to give direction to our efforts and especially to keep us motivated.
- *A Route*: We must know the way and the milestones to prevent us from losing our way.
- *A Good Start*: The first ten years are the toughest. The changes to be made are the greatest. In terms of our energy economy: first, we must get rid of our excess weight and, then, we must develop muscles. It is precisely in this starting phase that the doubts over the necessity of these efforts will be the greatest.

This paper discusses the greenhouse effect along these lines. It begins with the formulation of a long term goal, a route to arrive there and, finally, the start.

The Long Term Goal

Regarding a long term goal, the Noordwijk Declaration on Climate Change, (Ministry of Housing, Physical Planning and Environment,1989) states:

"For the long term safeguarding of our planet and maintaining its ecological balance, joint effort and action should aim at limiting or reducing emissions and increasing sinks for greenhouse gases to a level consistent with the natural capacity of the planet.

Such a level should be reached within a time frame sufficient to allow ecosystems to adapt naturally to climate change, to ensure that food production is not threatened and permit economic activity to develop in a sustainable and environmentally sound manner."

In order to define a long term goal we must first look where we will arrive, if we continue with our present form of development and if we ignore the greenhouse effect and ignore the limits of resources and the limits of the ecological bearing capacity of our Earth.

In the "business-as-usual scenario" developed by IPCC (IPCC, 1990), the use of fossil fuels will grow by a factor of four in the coming century. Similar growth rates are anticipated for the emissions of the other greenhouse gases. As a consequence, the atmospheric concentration of the greenhouse gases will be four times higher than before the industrial revolution.

The resulting global average temperature rise as calculated by IPCC is shown in Figure 2. With unchanged policy, we must expect that the temperature will rise by 3-6°C and the sea level will rise by 30 cm to 1.00 m in the coming century (Figure 3).

The speed of these changes is two to six times greater than the rate of change of climate in the past ten thousand years. And we can be sure that the ecosystems like forests and coastal regions cannot adapt without large scale damage and socio-economic disruptions.

This paper assumes that we want to prevent such disruptive changes. But, before defining our goals, we should further examine the results of the model calculations.

The curved lines of temperature rise and sea level changes suggest that changes will happen gradually and smoothly. And, that is precisely not the case, as explained in the IPCC report.

Through the local variations in the change of climate the risks are much higher than suggested by the smoothly curving graphs of the average global temperature change.

---

† This is a proposal by the authors to the international community dealing with climate change, and it does not necessarily reflect the position of the Netherlands' Government.

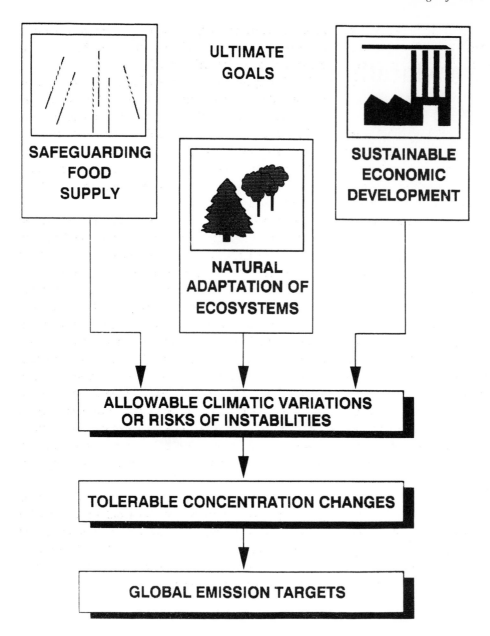

Figure 1: Long term environmental and development goals as a basis for short term policy targets.

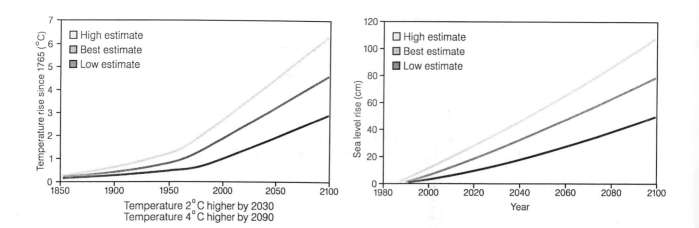

Figure 2: Temperature Rise, IPCC Business as Usual Scenario   **Figure 3**: Sea Level rise, IPCC Business as Usual Scenario

In several places the temperature rise could be two or three times as great, while in other places, almost nothing may change. The same holds true for sea level rise. When ocean currents change even slightly then there can be great impacts on temperature and precipitation on a continental scale.

Even more frightening is the danger of the reinforcement of the greenhouse effect by non-linear interaction of increased warming and emissions (so called feedbacks). One example of this is the temperature-albedo feedback. As a consequence of higher temperatures the snow and the ice cover in high latitudes will melt. As a result, less solar radiation will be reflected from the earth's surface and the warming will be increased. In addition, the higher temperatures in high latitude land areas will lead to additional methane emissions from thawing ground, thus enhancing the greenhouse effect.

There are many feedback mechanisms, some positive with a strengthening effect on the process of warming and some negative with a dampening effect. However, one thing is certain: The greater the disturbance of the climate system, the greater the risk of instability and catastrophes.

The "business-as-usual" scenario shows one outcome if we don't take any greenhouse action. The associated risks are great. But, on the other hand, to reduce emissions by 60 percent overnight, even if this were possible, would also have unacceptable consequences. So, which risks with respect to impacts and feedbacks are we willing to take?

The IPCC did not made any commitments on this issue. Also, there is some skepticism among the specialists about the development of criteria for tolerable climate change. The arguments over the criteria are similar to the discussions we have had over tolerable radioactive radiation. There too, it appeared that we needed and used a criterion for tolerable radiation, although there was no consensus on the dose-effect relationships.

That is why the results of recent research, coordinated by the Stockholm Environment Institute (Rijsberman and Swart, 1990), for the World Meteorological Organisation (WMO) and the United Nations Environment Programme (UNEP) are explained here.

Through this research, a long term target was formulated for climate change. This target is presented here in the form of a traffic light (Figure 4).

The green light (bottom) indicates a changing climate in which there is a global average temperature rise limited to 0.1°C per decade with a maximum total change of 1°C and a sea level rise of 2 cm per decade with a maximum of 20 cms. This rate of change is likely to allow natural adaptation of ecosystems. The orange light indicates the transitional area in which the damage increases and the risk of instability increases too. The red light implies that we are on the way to rapid, unpredictable and non-linear responses that would lead to extensive ecosystem damages. We must expect that in many places in the world there will be a crisis in the local food supply and ecosystems and the corresponding disruption of socio-economic systems and a loss of several islands. Most threatening is the risk of instability of the climate system, which is unacceptably high.

The goal of our efforts must be, therefore, to go for the green light, and in any case, to fully avoid the red light. To avoid the red light means that we want to limit the global average temperature rise to well below 2°C with respect to the pre-industrial level and that we want to limit the sea-level rise to well below 50 cms.

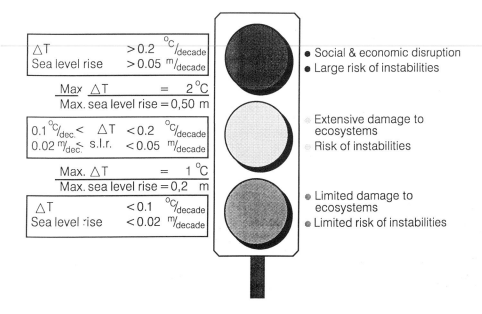

Figure 4: Targets for Climate Change based on Stockholm Environment Institute Studies (Rijsberman and Swart 1990)

The Route

When this is our goal, which route should we choose? In terms of concentrations of greenhouse gases, this goal means that we have to keep the atmospheric concentrations of greenhouse gases well below a doubling as compared to the pre-industrial levels. In terms of emissions this means that we have to stabilise our emissions globally in the short term and reduce them to less than one half of present levels by the end of the next century. To show how this compares with unchanged policy, this paper now refers to the various IPCC scenarios for fossil fuels and CO_2 emissions.

In Figure 5, the horizontal axis shows the years from 1985 to the year 2100. The vertical axis gives CO_2 emissions from fossil fuels. The upper line represents the emissions with unchanged policy (business as usual).

The "2060-doubling" line denotes that during the whole of the next century there will be a sort of no-regrets-policy, i.e. a policy with measures that are beneficial in their own right even without the greenhouse problem. With the "2090-doubling" line such a "no regrets policy" is modified and strengthened to a precautionary policy during the next century. A precautionary policy directly addresses the greenhouse problem and goes beyond the "no-regrets" measures.

Finally, the "EPA line" and the "RIVM line" indicate a policy by which the "no-regrets-policy" already becomes precautionary policy well before the year 2000.

This shows that only the last two scenarios, the "EPA" and the "RIVM" lines can reach our goal and keep the atmospheric concentrations well below doubling and keep climate change within the area of acceptable risks.

Continuing with the "well below doubling route", the total global CO_2 emissions and the manner in which these emissions could be divided between industrialised and developing countries is shown in Figure 6.

The global sharing of emissions is based on the principles as outlined in the IPCC-RSWG report (IPCC, Response Strategies Working Group Report, 1990):

> "Industrialized and developing countries have a common but varied responsibility in dealing with the problem of climate change and its adverse effects. The former should take the lead in two ways:
>
> i. A major part of emissions affecting the atmosphere at present originates in industrialized countries where the scope for change is the greatest. Industrialised countries should adopt domestic measures to limit climate change by adapting their own economies in line with future agreements to limit emissions.
>
> ii. To co-operate with developing countries in international action, without standing in the way of the latter's development by contributing additional financial resources, by appropriate transfer of

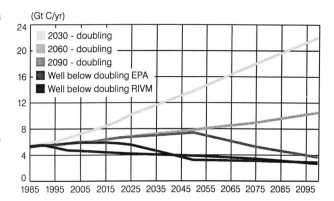

Figure 5: IPCC scenarios for the emission of CO_2 from fossil fuel combustion and cement production

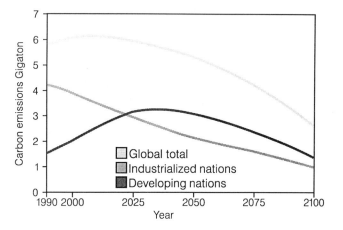

Figure 6: A strategy for the limitation and reduction of global fossil fuel carbon emissions for industrialized and developing nations

technology, by engaging in close co-operation in scientific observation, analysis and research and finally by means of technical co-operation geared to forestalling and managing environmental problems."

.... "Emissions from developing countries are growing in order to meet their development requirements and thus, over time, are likely to represent an increasingly significant percentage of global emissions. As the greenhouse gas emissions in developing countries are increasing with their population and economic growth, rapid transfer on a preferential basis to developing countries, of technologies which help them monitor, limit or adapt to climate change without hindering their economic development is an urgent requirement. Developing countries should, within the limits feasible take measures to suitably adapt their economies. Recognising the poverty that prevails among the populations of developing countries, it is natural that achieving economic growth is given priority by them. Narrowing the gap between the industrialised and developing world would provide a basis for a full partnership of all nations in the world and would assist developing countries in dealing with the climate change issue ...."

In Figure 6 the total global CO_2 emissions are allowed to increase a bit, as shown by the upper line. The CO_2 emissions in the industrialised countries will have to be reduced by 1–2 percent per year throughout the next century. With respect to the CO_2 emissions in the developing countries, this scenario indicates that these would increase by about 2 percent per year, thus a bit less than foreseen without active policy. These emissions are expected to stabilise in the year 2025 and will remain stable till 2050 and thereafter technology and technology transfer would have made such progress that in developing countries emissions could also be reduced by 1–2 percent per year.

These IPCC scenarios were produced by the United States and the Netherlands specialized governmental agencies and their institutes. The scenarios assume that there will be a reasonable economic growth. Such growth is assumed to be twice as high in the developing countries as in the developed countries. The emission reduction of 1–2 percent per year should be achieved in a growing economy by improving energy efficiency by nearly 2 percent per year and by lowering the carbon intensity of energy supply by 1-2 percent per year. In these scenarios there is a shift from coal to gas and from fossil fuels in general to renewable energy sources like biomass, wind energy, water power and especially solar energy. The amount of nuclear energy is relatively low because of the safety problems, the treatment of waste and also because of the costs.

This paper does not go into the details of these scenarios. The main point is that these scenarios and the underlying goals can only be achieved if we can implement an unprecedented and long lasting increase in energy efficiency on a global scale and if we can further develop and use renewable energy sources.

The scenarios illustrated above only deal with CO_2 emissions from fossil fuels. Although these CO_2 emissions are the most important cause of the increased greenhouse effect, there are also CO_2 emissions from deforestation and CH_4 and N_2O, CFCs and O_3 precursors. With respect to these gases, the IPCC "well below" scenario indicates that:

- The CO_2 emissions from deforestation should be reduced by reversing deforestation trends into a net forest growth by the year 2000 with further net forest growth thereafter
- The emissions of methane and nitrous oxide should be stabilised at present levels throughout the next century

- The emissions of CFCs should be phased out in line with the Montreal Protocol as revised in London in June 1990
- The emissions of O_3 precursors, CO, VOC, etc. should be reduced by a factor of 2.

The measures necessary to make these kind of goals achievable are described in the IPCC report in terms of options.

These options can be divided into five different categories of measures:

- technological development and transfer
- economic instruments like energy tax and tradeable emissions permits
- financial instruments such as funding mechanisms to support implementation measures by developing countries
- legal instruments like a UN Climate Change Convention and associated protocols and the strengthening of the global co-operations systems
- Information and education.

These categories are further described in the IPCC Report.

The Start

This brings us to the starting point of the Marathon. From a technical point of view, this phase is not the hardest phase. Every industrialised country can and will do something about an initial reduction of the growth of emissions, for example, by energy efficiency improvement. However, from a psychological point of view, this phase is the most difficult. Every signal from the scientific world which casts doubts on the certainty and effects of global warming, results in further doubts about the necessity for measures, especially with the groups who feel themselves to be most affected by these measures.

To get a feeling about this sort of reaction, you should ask a representative of the tobacco industry whether smoking is bad for your health. The answer to this question will make it clear that we may float in a sea of controversy for a very long time before the necessary measures can be taken. When environmental and health arguments are in competition with economic instincts, the uncertainties are always stressed. But happily enough, the comparison with the example of smoking also illustrates that great changes can be achieved without passing laws.

Given this situation, we must be aware of the relation which exists between certainty about the consequences of the greenhouse effect and the effectiveness of measures, and especially how this relation changes over time. in Figure 7, the horizontal axis shows the time and the vertical axis shows the degree of certainty regarding the consequences of the increasing greenhouse effect and the effectiveness of measures.

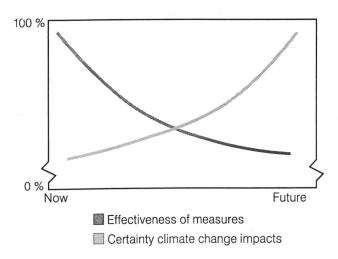

Figure 7: Timing of measures, the policy dilemma

The certainty with respect to the consequences of climate change is still rather low. At the same time, the effects of measures is the greatest. The problem is that if we wait for more certainty, the effect of our measures decreases considerably. In the event that we have a certain final goal to achieve, this means that postponing measures now, lingering at the start of the Marathon, brings with it a corresponding need to run much harder in a later phase, indicating that it will be much more expensive to reach the goal in time.

In the world, we are now witnessing a careful start at managing emissions. It is not astonishing that nations are moving cautiously. The interests are very great. For example, a stabilisation of CO_2 emissions in industrialized nations implies an enormous adjustment of the existing growth scenarios for energy production and consumption.

However, we must realize that, besides the greenhouse effect, there are other reasons why such growth projections should be adjusted. We know that the fossil fuel resources are limited and sooner or later the market mechanism would cause drastic price increases. Moreover, our energy supply system is very vulnerable. The current situation in the Gulf shows how much importance we attach to political stability in the area that supplies us with our energy. The interests are very great, but the potential gains are at least as high, both for the greenhouse effect and for the short and long term safety of energy supply.

This brings us to the last starting problem. With the beginning of the negotiations of a World Climate Treaty, there are many international conferences which, in the eyes of the public and also for the participants, do not always achieve as much as desirable.

Let us not misunderstand this situation. The rumour around these conferences and the clatter of swords is typical for the start of a major race. Although we have heard the starting shot the rules of the game have still to be established. What are the criteria for the sharing of global emissions and to what extent do rich countries have to pay for measures taken by poorer countries?

Conclusions

The rate of global warming and potential climate instabilities are the main threat of the accumulating greenhouse gases. For an adequate response we need a long term goal based on risk assessment and maximum tolerable climate change. The long term goal and the short term emissions control strategies presented here may be a good basis for the further development of a global strategy. Given the seriousness of the issue, it is not a question of whether to act, but how to act. We are now seriously making progress. In the turbulent days leading up to the Second World Climate Conference in October, 1990, most OECD countries set targets for the control of carbon dioxide emissions from the energy sector, varying from stabilization to considerable reductions of emissions. During the coming years the policies to reach these targets have to be specified and implemented. They will lead to a reversal of past trends, when carbon dioxide emissions and economic growth were closely linked.

As indicated before, stabilization of emissions and reduction up to 25% of current levels are just a first step on the way to stabilizing atmospheric concentrations. But a good start is half the race.

References

Ministry of Housing, Physical Planning and Environment: The Noordwijk Declaration on Climate Change, Leidschendam, the Netherlands, 1989.

IPCC Expert Group on Emissions Scenarios: Emission Scenarios for the Response Strategies Working Group of the Intergovernmental Panel on Climate Change, Washington, 1990.

Rijsberman, F. and Swart, R.J. eds.: Targets and Indicators of Climate Change, Stockholm Environment Institute, 1990.

IPCC, Response Strategies Working Group, Policy Makers Summary, Geneva, May 1990.

Adaptation Measures

by K. Madhava Sarma *(India)*

1. Introduction

The potential impacts of global warming are many and could affect the Earth's ecosystems, water resources, air quality, agriculture, forestry, fisheries, human settlements, energy, transport, industry, and human health.

Food security could be threatened in areas of high vulnerability, such as semi-arid regions. Major forest-type zones and species ranges could shift significantly, hundreds of kilometres northward. Species could disappear, leading to a reduction in biological diversity. Both increased precipitation in some areas and reduced precipitation in others will change patterns of agriculture and ecosystems and will have significant effects on water supply, hydro-electric power generation and on the frequency of extreme events, such as floods and droughts.

In the areas vulnerable to sea level rise, inundation and increased frequency of storm surges could lead to large-scale migration. Increased heat stress, more air-pollution, movement poleward of vector-borne diseases and changes in water quality and availability could affect human health. The response strategies could affect industry, energy and transport. The bio-physical properties of estuaries and wetlands could be changed due to sea level rise and could, along with changes of the ocean circulation, affect fisheries and the productivity of the oceans. The general decrease in areal coverage and duration of snow cover will change hydrology and have adverse effects on snow-dependent activities such as transportation and recreation.

These impact predictions are very uncertain. The increases in greenhouse gases emissions, increases in temperatures, the impacts of such temperature increases on precipitation, soil moisture, wind speed, and the frequencies and magnitudes of extreme events such as floods, droughts and storms are all subject to uncertainty regarding the direction of change as well as their rates, magnitudes and timing. These uncertainties increase in attempts to estimate changes on a regional or seasonal basis.

The effects of climate change are long-term, global and largely irreversible. The uncertainty in prediction implies the possibility, however small, of the worst predictions coming true. Prudence would, therefore, dictate that humankind should plan to limit the impacts and to adapt to the impacts which cannot be avoided.

Adaptive and limitation responses should be considered as a single package in order to ensure cost-effectiveness and mutual consistency. A shift away from fossil fuels towards biomass-based energy should not lead to deforestation. More man-made plantations to absorb CO_2 should not result in the reduction of areas needed for preservation of bio-diversity. If adaptation measures are difficult or impossible to implement, as for example, may be the case for islands, limitation strategies will be unavoidable.

Taking into account the uncertainties regarding the direction, magnitude and timing of impacts, the responses should satisfy several criteria including:

- Flexibility: to suit any impact that actually materializes
- Economically justifiable based on other benefits
- Timing: to ensure that the response is neither too early nor too late
- Feasibility: taking into account the legal, institutional, political, social and financial factors
- Compatibility with other measures.

The adaptive measures may be classified in three categories:

- those that augment the information and knowledge base to permit a reasoned judgement on the desirability of various measures
- measures that are economically justified even now
- measures that are costly and that can be taken up after the uncertainties are reduced.

Some of the suggested options for various sectors are discussed below. This is only a menu of options and the decisions on the options have to be taken by individual countries after examining the economic and environmental suitability of each of the options.

2.   General Measures

The following measures could be taken in any sector or country.

- Improvement of information and knowledge base
- Research programmes on all aspects of climate change
- Cataloguing information on resource use and management practices (agriculture, forestry, animal husbandry, fisheries, etc.) under a variety of climatic and other conditions, so that analogues can be found easily by each area to suit its new climatic conditions
- Inventory of resources, land, water, flora, fauna
- Assessment of sensitivities of resources to climate change
- Strengthen and/or create institutions to improve management of resources
- Research, development and promotion for more efficient agriculture, forestry, animal husbandry, fisheries, etc., through use of modern science, e.g., better soil and water conservation, development of more (salinity-drought-pest) resistant varieties of cultivars, conservation of forests.

3.   Water Resources

Water is the most important element in the economy of a country and any impact of climate change on water resources will affect every aspect of a country's life. Every country is continuously striving to manage its water resources, given even the present uncertainties in the supply, increasing demands and growing threats of pollution.

Given the uncertainties of the impacts of climate change on water resources, the following options have been suggested:

- Determining the flexibility and vulnerability of current water supply systems
- Improving the flexibility of the systems by inter-system co-operation and optimization
- Better measurement, monitoring and forecasting
- Water conservation through better irrigation scheduling, lining of canals, drip irrigation, recycling, use of drought resistant varieties, better tillage practices, more efficient water-using appliances, appropriate pricing
- Better flood management
- Disaster management preparedness.

4.   Land Use and Ecosystems

These are already under pressure from increasing population. The increasing demand for food, fuel, fibre, forest products and living and recreational space removes land from the unmanaged ecosystems and leads to deforestation, loss of bio-diversity, soil erosion, ground water depletion and more severe impacts from floods and droughts. If the response to climate change is population migration and there is no readily available land for such migration, stresses will increase.

The suggested adaptation measures are listed below:

- Inventories, research on the response of various species to climate change, on the location of conservation areas for various uses and on the needed conservation corridors
- Development and dissemination of new technologies for better land use, ex-situ conservation, relocation of species in new areas, more efficient use of forest products, increased efficiency of industrial plantations through genetic selection, breeding and propagation, etc
- Strengthening of conservation of vulnerable areas
- Establishing and enlarging conservation corridors between protected natural areas.

5.   Food Security

In spite of vastly increased food production in the world through modern technologies, more than 500 million people in the world, concentrated in the developing countries, do not have access to enough food. The food security situation, determined by the availability of food and the ability to acquire it by dependable long-term access through local production or through power to purchase, is getting worse due to the increasing populations and deteriorating natural resources of most of the developing countries. Protection and regeneration of natural resources, economic growth and equity are essential to ensure food security, even without climate change. The solutions are not simply technological, but will need appropriate policies in developing countries to promote equity and greater international support through additional transfer of resources by the developed countries to promote alleviation of poverty.

The climate change and the consequent changes in precipitation could intensify the problem, particularly in areas vulnerable at present, such as the arid and semi-arid lands. Sea level rise could inundate low-lying coastal plains with fertile alluvial soils, where a high proportion of rice production is located. Fisheries could be affected through the impacts on lakes and physical and chemical changes in the sea. Higher temperatures could increase the range and severity of plant and animal pests and diseases. Erratic precipitation and more frequent floods and droughts will affect the yields. the adverse effects would be particularly noticeable in the poorer sections of the society and reduce their access to food supplies.

The possible response options include:

- Promote economic growth and agricultural development in the developing countries
- Better soil and water management, food storage, protection of germ plasm resources
- Coastal protection, where necessary
- Modified tillage practices, improved drought-heat-pest resistant cultivars
- Better animal husbandry practices.

6.   Sea Level Rise

A large portion of the world population lives in coastal areas and on islands. These areas have several assests including some of the largest cities of the world, ports, industries, fertile lands, fisheries and tourism facilities. A sea level rise of one metre by 2100 would have a severe impact on all coastal areas, particularly in the developing countries. More frequent recurrence of storm surges would pose an intolerable burden on the coastal populations. The impacts of shoreline erosion, inundation of wetlands and mangroves, increases in the salinity of aquifers, and drowning of coral reefs would cause severe ecological damage. The small islands are particularly vulnerable.

The adaptive responses fall into three categories:

- Retreat i.e., abandon vulnerable areas and resettle inhabitants
- Accommodation - adjust to sea level rise
- Protection - for vulnerable areas, particularly those with high population density, economic value or precious natural resources.

(a)   Retreat

Preventing development can reduce future expenditures for adaptation. India, Sri Lanka, USA, Australia and a number of island nations have already such regulations. Where the area that could be inundated is high, the timing of such regulations would be difficult. Spreading of awareness among all the occupants of vulnerable areas could encourage private initiatives and solutions. This is no solution, however, for small island states and poorer countries.

(b)   Accommodation

This response requires advance planning. Many coastal structures could be elevated for protection from inundation. Building codes could be devised for protection from surging waters and high winds. Drainage could be modified. Storm warning and preparedness plans could be strengthened. Where salt water damages agricultural lands, salt-tolerant crops may be introduced. Some agricultural lands could be converted to agricultural uses. Human activities that degrade coastal resources may be banned.

E.g., filling wetlands, damming rivers, mining coral and beach sands and cutting mangroves. Undeveloped land can be set apart to accommodate re-establishment of wetlands and mangroves. Exploitation of ground water could be controlled to prevent salinity intrusion.

(c)   Protection

This response involves defensive measures against inundation, tidal flooding, shore erosion, salinity intrusion and loss of natural resources. These measures could be structures such as dikes, levees, flood walls, sea walls, and tidal barriers. Other measures include beach nourishment, dune building, and pollution control.

(d)   Economic Implications

While it is very difficult to estimate the economic implications correctly, it is possible to make some broad statements about the various options. In densely populated and productive areas, retreat may not be viable in general and, in the case of small islands, be impossible. Accommodation could be extremely costly but could be justifiable in particular situations. Protection would be economically viable, in particular cases, but could be unaffordable for the poorer countries. One estimate quoted in the Intergovernmental Panel on Climate Change report gives the total additional costs to be more than 500 billion dollars with a major part to be incurred in developing countries.

(e)   Social and Cultural Implications

The life styles of people living on coasts are distinct. The adaptive measures may, therefore, impact heavily on the social and cultural mores of millions of people. Relocation of people, whether planned or unplanned, could impose a traumatic burden on those relocated. The impacts on the people originally living in the relocation areas could be very high and cause great conflicts, even within the same country. Where such relocation crosses national boundaries, international conflicts could occur.

7.   Mechanisms for implementing adaptation measures

Adequate implementation of the measures outlined would need establishment and strengthening of several mechanisms through which the measures can be brought into force. These are outlined below.

(a)   Public Information and Education

The objectives are:

- to promote awareness and knowledge of climate change issues
- to encourage positive practices to adapt, and

- to encourage universal participation in the development of measures.

National and international focal points should be developed for each of the issues.

(b)   Technology Development and Transfer

These are needed in the areas of energy, natural resources, and monitoring. Impediments such as lack of financial resources, institutions, trained man-power, restrictive trade practices and over-protection of intellectual property rights should be removed.

(c)   Financial Mechanisms

These are needed in view of the special vulnerability of developing countries and their lack of resources. The resources should be focussed on activities which promote economic growth even while implementing the response measures. The areas could include:

- energy efficiency and alternative sources of energy
- rational forest management practices and agricultural techniques
- research and development in relevant areas
- awareness, education and training.

While existing institutions could be used, new mechanisms are essential. the principles of operation of these mechanisms should be acceptable to all, including the developing countries who will bear the brunt of the suffering due to climate change.

8.   Concluding Remarks

There is considerable optimism among many that adaptation to climate change is feasible, since it is as old as the human species. At every location, human beings and other living things adapted successfully to varying temperatures, precipitation, droughts, floods and storms. Because of economic and technological progress, the in-built adaptability of humankind has improved. The world has many climatic zones and as long as the new climate in an area has existing analogues anywhere in the world, people in that area can learn and adapt. This capacity to adapt, however, is very uneven and varies from country to country. Broadly speaking, the capacity of a country for adaptation is reflected in its development status. The conditions conducive to adaptation - awareness, knowledge, technology, resources, institutional capab- ilities, appropriate social and legal systems - are also the ones which promote development. In the poorer countries, the natural resources are already under stress. These countries have not been able to effectively counter the stresses and the burden of climate change will be even more unmanageable. The resources of the developing countries are inadequate even for their programmes for modest economic growth, public health and environmental needs, which are necessary to generate resources and to increase their adaptive capacity. Hence cost-effectiveness and good cost benefit ratios of adaptive measures are not sufficient to ensure that the developing countries are able to take adaptive measures by diverting resources from their other needs. Such a diversion would reduce their capacity to adapt - a "vicious circle", indeed, unless the total resources available to them are increased through global action.

Limitation Strategies

by Keiichi Yokobori *(Japan)*

Introduction

This paper discusses the options for attempting to limit emissions of greenhouse gases (GHGs), primarily carbon dioxide (CO_2), methane (CH_4), and nitrous oxide (N_2O). It also reflects mainly the deliberations made at the Response Strategies Working Group (RSWG) of the IPCC. It represents only a first tentative step towards the goal of identifying the control strategies compatible with the concept of sustainable development.

GHG Emission Patterns

Various human activities contribute to emissions of GHGs. The most important single source, the energy sector, accounts for 46% of the increased radiative forcing due to anthropogenic sources in the 80s. Industry accounts for 24%, forestry 18%, agriculture 9% and other sources 3%.

Table 1 presents current and future scenarios of GHG emissions and different emission sources. For 2025, two different scenarios are shown: the AFOS/EIS reference case developed by country submissions and the Task A high-emission scenario. Further, Table 2 shows more detailed CO_2 emission patterns by region based on the Energy and Industry Sub-group (EIS) reference case scenario. The latter indicates regional differences, e.g., in *per capita* levels and carbon intensity in energy used.

Future Greenhouse Gas Emissions

Greenhouse gas emissions depend on many factors, which include population size and growth, the level and growth of the economy and, GHG intensity. The GHG intensity reflects, *inter alia*, economic structure, technologies used, climatic and geographic conditions, managerial practices, behavioural attitudes and infrastructure and the mix of energy produced, transported and used. Uncertainties inherent in these factors make it difficult to predict future GHG emission levels.

Based on national studies assuming the absence of further policy measures ('reference scenarios') and other studies submitted to the IPCC/RSWG/EIS process, CO_2 emissions are estimated to increase to 7 gigatons of carbon (GtC) by 2000 and over 12 GtC by 2025, as presented in Table 1. Global emissions rise faster than those in the high-emission scenario provided by the RSWG Scenario Task Force. Methane emissions from coal mining, oil and natural gas systems and waste management systems (i.e. landfills, wastewater treatment facilities) may increase by 183% by 2025.

Technical Options and Policy Measures

There are a number of technical and policy measures options available to reduce greenhouse gas emissions. Technical options include not only the installation of new capital stock with lower emission characteristics (or the modification of existing stock) but also the managerial and behavioural changes. Policy measures are the actions, procedures and instruments which governments adopt to bring about additional or accelerated uptake of the technical measures beyond that in a reference scenario.

Many factors affect the implementation of technical options and adoption of policy measures. These include the technical, economic and market potential of technologies, development status and timescale for implementation of technologies, implications for other GHGs, interaction between measures, resource costs and private costs, macro-economic and micro-economic effects, implications for other policy goals, and social consequences, policy robustness and political and public acceptability, effectiveness and limitations of policy instruments.

Technical, Economic and Market Potential

Understanding the distinction between technical, economic and market potential is important for developing realistic response strategies. The technical potential of a specific technology is its capacity to reduce potential greenhouse gas emissions, irrespective of the costs involved. However, the economic potential may be significantly less. This

Table 1: Anthropogenic greenhouse gas emissions from the RSWG scenarios

	AFOS/EIS reference case		2030 high-emission scenario	
	1985	2050	1985	2025
CO_2 emissions (Petagrams C/Yr)				
Energy	5.1	12.4	5.1	9.9
Deforestation	1.7††	2.6	0.7†	1.4
Cement	0.1	0.2	0.1	0.2
Total	6.9	15.2	5.9	11.5
CH_4 emission (Teragrams CH_4/Yr)†††				
Coal mining	44	126	35	85
Natural gas	22	59	45	74
Rice	110	149	110	149
Enteric fermentation	75	125	74	125
Animal wastes	37	59	–	–
Landfills	30	60	40	71
Subtotal	318	577	304	503
Biomass burning			55	77
Natural			179	179
Total			541	761
N_2O (Teragrams N/Yr)			12.5	16.4
CO (Teragrams C/Yr)			537	781
NOx (Teragrams N/Yr)			51	68
CFCs (Gigagrams/Yr)				
CFC–11			278	245
CFC–12			362	303
HCFC			97	1,340
Halon 1211			1.4	18.6
Halon 1301			2.1	7.4

†   Assuming low biomass per hectare and deforestation rates.

††   Midrange estimates for deforestation and biomass consistent with preferred value from Working Group I.

†††   Differences in the 1985 emission figures are due to differences in definitions and qualifying the emissions from these particular sources.

occurs where there are positive resource costs when evaluated at social discount rates—allowing for second-round effects. Finally, the market potential might be even less, due to market imperfections and the use of higher discount rates. The challenge for policy-makers is to enhance the market uptake of appropriate options, taking full account of all the intersections, second-round effects, costs and benefits. There are, in general, extensive information and data available on the technical potential of the many technological options. However, the economic and market potential of the options depends on specific circumstances (national, local and even sectoral) in which the option is to be applied. Therefore, the quantification of these potentials requires further studies including country specific case studies.

Technical Options and Time Frame

The most relevant categories of technologies to reduce greenhouse gas emissions are efficiency improvements and conservation of substances that add GHGs to the atmosphere, such as fossil energy; substitution by substances which have no or lower greenhouse gas emissions, such as nuclear and renewable energy sources; improved fertilizer formulations and reduction of greenhouse gas emissions by removal, recirculation or fixation.

Technical options must be viewed in terms of the time frame in which they can be effective. This distinction facilitates the formulation of a technological strategy. For example, first-wave or near-term technologies are those which are or will be ready for introduction and/or

Table 2: *Gross CO$_2$ emissions (GtC), per capita carbon emissions and carbon intensity from the energy sector* [†] *(from the reference scenario).*

	1985	%	2000	%	2025	%
Global total	5.15	(100)	7.30	(100)	12.43	(100)
Developed	3.83	(74)	4.95	(68)	6.95	(65)
North America	1.34	(26)	1.71	(23)	2.37	(19)
Western Europe	0.85	(16)	0.98	(13)	1.19	(10)
OECD Pacific	0.31	(6)	0.48	(7)	0.62	(5)
Europe (centrally planned economies)	1.33	(26)	1.78	(24)	2.77	(22)
Developing	1.33	(26)	2.35	(32)	5.48	(44)
Africa	0.17	(3)	0.28	(4)	0.80	(6)
Asia(centrally planned economies)	0.54	(10)	0.88	(12)	1.80	(14)
Latin America	0.22	(4)	0.31	(4)	0.65	(5)
Middle East	0.13	(3)	0.31	(4)	0.67	(5)
South and East Asia	0.27	(5)	0.56	(8)	1.55	(12)

	1985		2000		2025	
	PC [††]	CI [†††]	PC	CI	PC	CI
Global total	1.06	15.7	1.22	15.8	1.56	16.0
Developed	3.12	16.3	3.65	15.9	4.65	15.6
North America	5.08	15.7	5.75	15.8	7.12	16.7
Western Europe	2.14	15.6	2.29	15.1	2.69	14.6
OECD Pacific	2.14	16.1	3.01	16.1	3.68	14.8
Europe (centrally planned economies)	3.19	17.5	3.78	16.9	5.02	16.4
Developing	0.36	14.2	0.51	14.8	0.84	15.5
Africa	.29	12.3	0.32	13.2	0.54	15.2
Asia (centrally planned economies)	0.47	17.3	0.68	18.8	1.15	19.6
Latin America	0.55	11.5	0.61	11.4	0.91	11.8
Middle East	1.20	16.7	1.79	16.1	2.41	15.5
South and East Asia	0.19	12.3	0.32	14.3	0.64	15.6

†   This table presents regional CO$_2$ emissions and does not include CFCs, CH$_4$, O$_3$, N$_2$O, or sinks. Climate change critically depends on all GHGs from all economic sectors. *This table should be interpreted with care.*

†† PC - *Per capita* carbon emissions in tonnes carbon per person.

††† CI - Carbon intensity in tonnes carbon per gigajoule

demonstration up to the year 2005 and beyond. Second-wave technologies are available, but not yet clearly economic, and thus would mainly be implemented in the medium-term time frame of 2005 to 2030. They could be introduced sooner if they were close to being economic, or particularly beneficial to the environment. Third-wave technologies are not yet available, but may emerge in the long term or post-2030 as a result of research and development.

A phased approach to technology development and introduction into the market is offered as an important strategy to be considered for concerted national and international collaborative actions.

Illustrative Sectoral Potentials

Extensive information available to the IPCC/RSWG process indicates the varying technical potentials of many technological options.

For example:

- In the transportation sector, high vehicle efficiency improvements could be achieved (e.g. 50% improvement over the average vehicle on the road)
- In the electricity generation sector, efficiency improvements of 15 to 20% could be achieved for retrofits and repowering of coal plants and up to 65% for new generation v average existing coal plants and fuel substitution could achieve a 30% (for oil to natural gas) to 40% (for coal to natural gas) reduction in emissions of CO_2
- In the building sector, new homes could be roughly twice as energy-efficient and new commercial buildings up to 75% as energy-efficient as existing buildings; retrofitting existing homes could give an average 25% improvement and existing commercial buildings around 50%
- In the industry sector, the technical potential for efficiency improvements ranges from around 15% in some sub-sectors to over 40% in others (i.e. best available technology v. the stock average)
- In the agriculture sector, methane emissions might be reduced by 10 to 30% from flooded rice cultivation and by 25 to 75% per unit of product in dairy and meat production in the long term
- In the forestry and other sectors, methane emissions might be reduced by 30 to 90% for landfills through gas collection and flaring and up to 100% for wastewater treatment in the long run.

The constraints on achieving the technical potential in these sectors can be generally categorized as:

- Capital costs of more efficient technologies *vis-à-vis* the cost of energy
- Artificial prices resulting from subsidies, regulatory barriers and other market distortions
- Lack of infrastructure
- Remaining performance drawbacks of alternative technologies
- Replacement rates
- Lack of relevant data and information
- Reaching the large number of individual decision-makers involved.

The challenge to policy-makers is to enhance the uptake of technological options and behavioural and operational changes as well as to address the broader issues outside the sector in question in order to capture more of the potential that exists.

The pool of *policy measures* is broadly similar for all nations. The RSWG discussed the wide array of policy implementation measures. They include public education/information, economic (market) measures, technology development and transfer, financial measures and legal measures. Of these the first three are considered as most pertinent in addressing obstacles commonly confronting all countries.

Information measures include all efforts to better inform the public about greenhouse gas emissions and the means available for their reduction. These programmes should be aimed at particular sections of the population, and should emphasize the present and potential future costs and benefits of such measures. *Economic measures* include the broad areas of taxes, charges, subsidies, and policies which include incorporating environmental cost into end-use prices (both imposition and removal). Such measures may also be used to complement regulations, making them more effective in meeting environmental goals. Economic measures may also be used to support the research, development, demonstration, or application of technologies. *Technological development and transfers* would also require adequate and trained human resources.

Costs

Full assessment of the costs of emission control strategies is essential. Such evaluation must include not only the resource costs of technical options but also the costs of government policy implementation, macro-economic and micro-economic effects, other environmental and social costs and benefits, and private and non-monetary costs. Some preliminary indications from the individual country studies suggest that emission abatement potential at low or negative resource cost, when tested at social discount rate, might amount to around 20% of global emissions in reference case scenarios by 2020. It is primarily attributable to the accelerated implementation of energy efficiency and conservation measures. The more ambitious the strategy, the higher the associated costs. Achieving a 20% reduction from current emission levels would require major changes in global markets, plans, and infrastructure, and intervention by governments. Maintaining this emission reduction goal would require continued technological improvements, changes in the structure of the global economy, and in the proportion of carbon intensive fuels utilized over the remainder of the next century.

Response Case Studies

Only preliminary analysis of response policy options is possible due to the insufficient number of case studies, differences of criteria in selecting policy options and lack of other cases for comparison. But it indicates the important role of industrialized countries in the total global emissions in the near term. The technical potential is large but differs greatly between regions and countries.

Response Strategies

Climate change offers an unprecedented challenge. Many uncertainties remain about both the impacts of climate change itself and our response to it. Preliminary assessments of a large number of options suggest differences among these options in terms of their economic and social feasibility, and the timing of implementation. This will call for the development of flexible and phased response strategies. The underlying theme of any strategy must be economic efficiency, achieving the maximum benefit at minimum cost. Some are found to be justified in their own right. A balance should be made among alternative abatement options in the energy, forestry and agricultural sectors, and adaptation options and other policy goals where applicable at both national and international levels. Ways must be found to account for consequences for other countries and inter-generational issues when making policy decisions. Encouragement of accelerated implementation of energy efficiency measures (as regards both demand and supply) should be a major common focus of initial energy policy responses. This will need to be supported by enhanced R&D, and encouragement for the increased use of natural gas and low-cost renewable, or less greenhouse-gas-producing, energy technologies.

On policy instruments, the highest initial priority must be to review existing policies and remove inappropriate conflicts with the goals of climate change policy. New initiatives will, however, be required. The international implications of some policy instruments (e.g. trade and competitiveness issues associated with carbon taxes, energy efficiency standards, and emission targets) will need to be resolved quickly if effective responses are not to be hampered.

Options

Generally, the following options have been identified, namely to:

1. Take steps now to attempt to control the emissions of greenhouse gases and prevent the destruction and improve the effectiveness of sinks. Governments may wish to consider setting targets for CO_2 and other greenhouse gases
2. Adopt a flexible and progressive approach, based on the best available scientific, economic and technological knowledge
3. Draw up specific policy objectives and implement wide-ranging, comprehensive programmes covering all greenhouse gases

4. Start implementing strategies which have multiple social, economic and environmental benefits, are cost-effective, are compatible with sustainable development and make use of market forces in the best way possible
5. Intensify international multilateral and bilateral co-operation in developing new strategies
6. Increase public awareness of the need for external environmental costs to be reflected in end-use prices, markets and policy decisions to the extent that they can be determined
7. Increase public awareness of efficient technologies and products and alternatives, through public education and information (e.g. labelling)
8. Strengthen relevant research and development and international collaboration in technological, economic and sectoral policy analysis
9. Encourage the participation of industry, the general public, and NGOs in the development and implementation of strategies to limit greenhouse gas emissions.

Short-term options include:

- Improving diffusion of efficient and alternative technology and practices which are technically and commercially proven
- Improving efficiency of mass-produced goods including motor vehicles and electrical appliances and equipment and buildings (e.g. through improved standards)
- Developing, diffusing and transferring technologies to limit greenhouse gas emissions.

Long-term options include:

- Accelerating work to improve the long-term potential of efficiency in the production and use of energy, fertilizer use, and forest management; encouraging a relatively greater reliance on sources and technologies with no or lower greenhouse gas emissions; and enhancing natural and man-made means to sequester greenhouse gases
- Further reviewing, developing and deploying of policy instruments including public information and standards that will encourage sustainable energy choices by producers and consumers without jeopardizing energy and food security and economic growth
- Developing methodologies to evaluate the trade-off between limitation and adaptation strategies and establishing changes in infrastructure needed to limit or adapt to climate change.

Report of the IPCC Special Committee on the Participation of Developing Countries

by J. Ripert *(France)*

The IPCC created a Special Committee on the Participation of Developing Countries in June 1989, following the report of an *ad hoc* subgroup of representatives of developing countries. The latter had been chaired by the Vice-Chairman of the IPCC, Dr Al-Gain, who had been invited to examine ways to remedy the insufficient participation of developing countries in IPCC activities. The special committee started with a core membership of representatives of ten countries (five developed and five developing) but now functions as an open-ended group. More than 50 people attended its last session, of whom 40 were from developing countries. The committee's purpose, based on the recommendations of Dr Al-Gain's *ad hoc* subgroup, was to promote, as quickly as possible, the full participation of the developing countries in IPCC activities. The key words here are obviously *quickly* and *full participation*. I will not comment on the former of these now, but will return to it at the end of this statement. As regards the latter, the special committee defined 'full participation' as including, and I quote, 'the development of national capacity to address the issues of concern such as the appreciation of the scientific basis for climate change, the potential impacts of such change and evaluations of practical response strategies for national applications'. Briefly, the aim was much more than to ensure the physical presence of as large a number as possible of representatives of developing countries at meetings, even if this objective required special attention in itself.

The committee analysed the factors inhibiting this full participation, which it grouped into the following categories: insufficient information, insufficient communication, limited human resources, institutional difficulties, and limited financial resources. It should perhaps be emphasized that these negative factors are endemic to the developing world and they most frequently slow down the development process. Whilst protection of the global environment from the impacts of human-induced change in the composition of the atmosphere is a growing concern in the industrialized countries, the human activities which cause these changes, such as the increased production and consumption of energy and food, are a priority for the developing countries. Any international response strategy to the adverse impacts of the greenhouse effect must address this apparent conflict of interests. This will be possible if it is recognized that all members of the international community share responsibility for the action to be taken. This responsibility must be exercised differently through time, and, as can be seen from the report of the special committee, the degree of participation may increase as the development gap between countries in the north and the south is reduced. It is necessary to alleviate the poverty which is prevalent in large sections of the population in the developing countries in order to create the basis for a true partnership between all countries. These are of course generalizations, but they simply show that the struggle to master a very important aspect of the future of our planet cannot be dissociated from other efforts which the international community must make to favour a general process of development, to ensure an equitable rise in standards of living and to equalize opportunities between peoples.

By no means could the IPCC, and still less its special committee, cover fully, or even mention all aspects of, these problems. The United Nations Conference on Environment and Development, to be held in Brazil in 1992, will, I think, provide a more appropriate framework for such discussions. For its part, the special committee has focused its attention on the factors specifically affecting the participation of developing countries in the development and implementation of an international strategy to combat adverse climate-change impacts. I now wish to draw your attention to some of its recommendations.

In the light of what I have just said, it will be understood why the representatives of developing countries on the special committee repeatedly stressed the importance they attached to the conditions for access to technology as well as the mobilization of additional financial resources. These questions have been discussed in depth within the special subgroups established by WG-III. The special committee did not wish to cover this aspect explicitly in order to avoid duplication of effort. However, none of its members underestimated the importance of these questions. Indeed, international negotiations on the protection of the ozone layer have shown quite clearly the need to deal with them if a global co-operation project is to become a reality. We can welcome the progress achieved in the field of ozone layer protection at the meeting in London in July 1990 which augurs well for the effort which should be made regarding the global warming problem. In discussing and formulating its recommendations, the special committee wished to distinguish three time-scales:

- Immediate action to finalize the IPCC's first report
- Follow-up action to be continued at least until the conclusion of negotiations on the development of a framework convention and possible protocols, and
- Action forming part of longer-term programmes which would gain from being integrated in a new institutional framework such as that which could be set up by the convention.

The problem of greatest immediate concern to the special committee was the funding of the participation of a larger number of experts from developing countries in the meetings of the IPCC and its bureau, working groups and subgroups. A welcome fact is that many governments and international or non-governmental organizations have responded to the IPCC appeals quickly (i.e. as from 1989 and 1990) with generous contributions. It was thus possible in 1989 to devote SFR 400 000 to funding the participation of 80 experts, in other words double the figure initially allocated. In 1990 and to date, it has been possible to raise the number of experts whose travel and participation have been funded in this way to about 200. The success of this effort is also illustrated by the fact that only 11 developing countries were represented at the first plenary session of the IPCC, 22 at the second, 33 at the third and 45 at the fourth. If I have spoken of this achievement, it is because these results must be consolidated and amplified. The IPCC report stressed the need to ensure the participation of experts in its future work as well as in meetings directly linked with the negotiation process. It made a new appeal to the governments of developed countries to contribute to this effort more, in good time, either through the IPCC trust fund or through bilateral channels. This may seem a trivial matter to many people here, but not to me. Similarly, in order to make best use of the precious time of so few experts and of insufficient resources, the schedules for future work on climate need to be better harmonized. This point has already been stressed by representatives of developing countries. The problem will be particularly acute in 1991, during which a negotiation process will be started up which should not only concern representatives of governments but also mobilize experts, technicians, and scientists, whilst some of the work must be covered in depth within the IPCC.

Harmonization of this process is a problem which requires great attention being given to the communication and information channels. We cannot but admire the IPCC's success in ensuring such broad-based consultation between scientists, engineers, and government representatives on such varied subjects and in so little time, but we remember that it was achieved thanks to much improvisation and numerous entries in the appointments diaries of many eminent people. This has not always permitted the participation in the best possible way of all those wishing to take part. This process may have appeared insufficiently transparent to many participants, particularly those from the developing countries. We must all give more attention to these criticisms if we want this collective enterprise to involve everyone in the future, and not only those who are the most alert.

In order to facilitate the awakening of political attention in the developing countries, the special committee recommended that seminars be organized within these countries themselves to draw the attention of the national leaders to concerns related to the greenhouse effect. I am happy to point out that several governments and non-governmental organizations as well as the IPCC itself have organized such seminars or are in the process of doing so. One point requires attention here. The mutlilateral or bilateral organizations have traditionally carried out extensive activities aimed at increasing the awareness of national leaders about certain problems. But their programmes are usually very specific and are legitimately drawn up on the basis of their own terms of reference. However, the study of climate change impacts and the preparation of adaptation or response strategies have, or should have, a multi-sectoral nature. In particular, the international organizations should adapt themselves to responding to this need for 'transversality', just as our governments are called upon to do in their own internal structures.

In the longer term, the special committee identified areas for co-operation which are listed in its report and will be of no surprise to you. They cover the efficient use of energy resources, development of new, clean and renewable sources of energy, rational use of forest products, proper forest management, development and transfer of clean, reliable technologies in fields such as construction, housing, manufacturing industries and transport, reinforcement of the capacity of developing countries to implement data collection and processing programmes related to climate change, as well as training of the necessary personnel for such programmes, reinforcement of the participation of developing countries in major international programmes such as the World Climate Programme and the International Geosphere/Biosphere Programme. As you can see, this covers the whole agenda of international work to respond to the greenhouse effect. The group wanted to stress that there are activities in all these fields through which it would be possible to prepare for, or facilitate, a more extensive and efficient participation of the developing countries.

Two additional remarks. The special committee highlighted the advantage of developing activities or programmes not only at national level but also, and I stress this, at regional or sub-regional level. In many cases, the nature of the problems creates an affinity between countries. Moreover, in many cases, governments do not have the capacity to cover all aspects of such vast problems. Co-operation is necessary. I am all the more willing to speak of this as a Frenchman, because, in France, we cannot presume with our resources alone to follow an overall strategy to study, and still less to lead, the action needed in all these fields. We must develop many new activities within the European Economic Community. What is true for the Member Countries of the European Community is, to my mind, all the more true for the less developed countries, such as those in the South Pacific or many regions in Africa or Latin America. The international community must therefore propose carrying out programmes which facilitate regional and sub-regional co-operation.

The special committee also emphasized the urgency of the problem, and I thus return to my initial point regarding the need to translate the proposals made in the committee's report into concrete action now. I hope that the Climate Convention, which is to be negotiated shortly, will recognize the importance and give the necessary impulsion to the current co-operation instruments and, if necessary, propose others. However, the IPCC wanted to draw the attention of everyone to the fact that the action needed to promote the participation of the developing countries does not have to wait for the outcome of new negotiations. Co-operation programmes have been drawn up in various fields as pointed out by Dr Obasi and Dr Tolba. These programmes must receive a fresh impetus. The time for action in this field is today, not tomorrow.

The World Climate Programme: Achievements and Challenges

by James P. Bruce *(Canada)*

The World Climate Programme is a truly remarkable international endeavour. Looking back over the decade or so since the start of the World Climate Programme, one is driven to three conclusions. First, we have *learned a great deal* over this period about global climate, the way the climate system works and the factors that affect it. Secondly, the steady increase in our knowledge has been achieved through an unparalleled web of *cooperative effort*—cooperation between scientists, between nations and between international organizations through the WCP. Thirdly, there remain significant *gaps in knowledge* about climatic conditions in some parts of the world, and about some aspects of the way the climate system functions, and how it is disturbed. Redoubled efforts and even closer cooperation will be needed over the next decade to begin to meet the expectations of the public and of policy makers for greater knowledge and better predictions.

The Roots of the WCP

Since the very beginning of the work of the International Meteorological Organization in 1873, the measurement and understanding of climate has been a preoccupation of international meteorology. Standardization of observations, collection and analysis of climatic data stem from the work of the first Technical Commission for Climatology. For many years, first under IMO, later WMO, the patient work went on of collecting observations, describing the world's climate, its means and extremes, determining the "normals" of precipitation, temperature, wind, relative humidity, sunshine duration and atmospheric pressure - those climatological parameters widely observed around the globe.

But during the 60s and 70s several revolutions occurred which changed the very nature of climatology and began to forge a science that was "too important to leave entirely to the climatologists". In the early 60s the *observational revolution* was characterized by the establishment of the World Weather Watch, with its extensive satellite and ground-based observing systems, its rapid world-wide communications (the Global Telecommunications System, GTS) and its network of computerized World and Regional Meteorological Centres which could quickly compile regional and global climatic statistics and analyses, in addition to their operational weather forecasting functions. A second aspect of the observational revolution, that was to profoundly affect climatology in the 80s was the establishment of programmes to measure the chemical changes in the global atmosphere, particularly WMO's Background Air Pollution Monitoring programme and Global Ozone Observing System, both established in the late 1960s. It should be an encouragement to all of those who conduct the painstaking systematic measurement programmes on the atmosphere and earth system, to note the fame (if not fortune) that has fallen to Dr. Keeling who started and continued the CO_2 measurement programme at Mauna Loa, Hawaii and Dr Farnam who first measured the Antarctic ozone hole at Halley Bay, two distinguished contributors to the global atmospheric chemistry observation programmes. BAPMoN and GOOS are now combined in the Global Atmosphere Watch (GAW).

A second revolution was the *computer revolution*. What was originally seen by climatologists as a handy tool for quality control, analysis and storage of climatological information on punched cards, has rapidly evolved into a capability to mathematically simulate the atmosphere, and even the long-term climate system, to permit daily weather predictions, and projections of future climates. This in turn has led to ever-increasing demands for reliable data on many components of the climate system consisting of the atmosphere, biosphere, oceans, ice, lands, hydrological and geochemical cycles, and greater understanding of the processes linking these components. Only if they are understood and measured well can they be modelled well.

A third revolution was generated by the output of previous major efforts to assemble research resources internationally, in the *Global Atmospheric Research Program (GARP)*. GARP was the highly successful forerunner of the World Climate Research Programme and mobilized support from both the operational meteorological services and academic scientists in a truly global research effort. WMO and ICSU began developing the concept of GARP in 1963 following a UN General Assembly resolution on Peaceful Uses of Outer Space. The main objective was to improve understanding of the dynamics of the global atmosphere, and so to extend the period for which useful weather forecasts could be issued. This very ambitious programme was carried out over the 15 year period 1967 to 1982, and comprised a number of major field projects, starting with GATE (GARP Atlantic Tropical Experiment), 1974, involving some 70 countries, 40 research vessels, special observation aircraft, unprecedented numbers of surface based observations and use of satellite imagery. The Global Weather Experiment (GWE) followed in 1979 to measure and study intensively the whole global atmosphere over a full year; with special sub-projects, the Summer MONEX, the Winter MONEX and the West African Monsoon Experiment, on monsoonal circulations. ALPEX (the Alpine Experiment) in 1982 was designed to measure and better understand the role of mountains in cyclogenesis and mountain wind systems. Together these projects provided unparalleled data sets on the world's atmosphere which are still being used to permit major advances in prediction of the weather, up to 7 days in advance, with significant skill.

Finally there was a revolution in perception of the *influence of human activities on climate*. The total energy output of human activities is so small compared to the solar energy received by the atmosphere, that the general view up to 25 or 30 years ago was that human activities could not affect global climate. The effects of major changes in vegetative cover and large water projects were recognized to affect local and regional climates to some extent. There had been, though, a few scientists, especially Arrhenius (1896), who recognized that increases in carbon dioxide will result in global warming. There was also a general understanding of the role of the radiatively active gases, and heat transports by the atmosphere and oceans, in maintaining a habitable climate over much of the earth. Revelle pointed out in 1955 that "Human beings are now carrying out a large-scale geophysical experiment"—by increasing CO_2 concentrations through industrial activities. By the 70s the documentation by Keeling and others showed the steady, apparently inexorable increase of CO_2 at about 0.5% per year, and subsequently the rapidly rising trends of other greenhouse gases, methane (CH_4), nitrous oxide (N_2O), and chlorofluorocarbons (CFCs) were confirmed. The output of early general circulation models (GCMs) demonstrated the possible effects of a doubled CO_2 concentration or the equivalent radiative effect of all the gases combined. These together focussed world-wide attention on the concept that human actions were indeed forcing a major change in the atmosphere, and hence in climate, on a global scale.

First World Climate Conference

The First World Climate Conference in February 1979 was held in the midst of these four revolutions. In the last two days of this "historic Conference" (Kellogg, 1987) the terms of reference of the proposed World Climate Programme were debated. This was done through a number of small working groups. The Conference statement called on all nations to work together to understand climate and climate change. The statement was cautious about the issue of greenhouse gases and climate change, reflecting the knowledge and consensus of the time. It made the point that international cooperation on climate studies would only be effective in a world at peace, which might lead us to hope for even greater cooperation in the 1990s with lessening of cold war tensions.

WMO Congress Decisions 1979

The Congress of WMO, meeting in June 1979 after the First World Climate Conference, took the important decision to launch the World Climate Programme. From the start it was recognized that scientists of all nations must participate, and that many disciplines and international organizations must be involved if understanding of the global climate and its effect were to be advanced. The structure of the WCP adopted by Congress was similar to that recommended by the Working Groups at the First World Climate Conference. There were to be four main components.

- Climate Data Programme
- Climate Applications Programme
- Climate Impact Studies Programme
- Climate Change and Variability Research Program (later shortened to Climate Research Program)

It was agreed that "WMO must take the lead within the framework of the WCP in promoting studies of climate change and variability and their effects on the natural environment and mankind". WMO's role was to provide overall coordination and to manage the Data and Applications Programmes. The inter-disciplinary nature of the WCP was emphasized from the start and national governments and a wide range of UN organizations and agencies were asked to participate. UNEP took the lead on Impact Studies, and IFAD, FAO, UNESCO and WHO were also asked to participate. Collaboration between WMO and the non-governmental ICSU was seen as the cornerstone of the World Climate Research Programme (Climate Change and Variability Research).

In the wonderful light of hindsight, more than a decade later, how well has this structure served us? What have been its achievements? What are the shortcomings and how can they be remedied? It is with some hesitation that I attempt to answer these questions, for within the Secretariats of the various WCP components, and amongst national scientific contributors have been some of the most distinguished scientists, and dedicated managers, of the world's scientific community. It is probably salutory, however, to take stock periodically of how we've done, and how we might improve matters, and the Second World Climate Conference is intended to do just that. I will only be providing the first assessment and commentary on the WCP at this Conference. Others will deal with specific components and there will be opportunity for discussion of the Programme and where it should be going.

Achievements of the World Climate Programme

Data

Under this component four major projects have been developed, Climate Computer (CLICOM), Data Rescue (DARE), Climate System Monitoring (CSM), and most recently the Climate Change Detection Project (CCDP). The first two of these have been particularly aimed at the vital task of assisting developing countries in management, preservation and use of climatic data in their own territories. By the end of 1989 some 120 CLICOM, personal computer based data management systems, had been installed in 80 countries, and 40 more are on the way in 1990. The DARE project has the aim of committing to permanent microfilm form the original written manuscript records which in some cases date back more than 100 years and are in danger of being lost. This work has been completed in most African countries and is expected to move to Latin America and the Caribbean in the early nineties. Both CLICOM and DARE have been supported by financial and technical contributions from a number of industrialized countries, and through UNDP projects, and have had the very positive effect of ensuring that many developing countries now have ready access to the full climatic data-base from their lands and are able to exchange climatic data internationally with greater ease. These two projects must be extended to all regions with a need, and CLICOM software should continue to be upgraded.

Under the Climate System Monitoring Project, WMO with some financial support from UNEP and major input from the WWW's World and Regional Meteorological Centres has published monthly, with summaries every two years, since 1984, analyses of the recent dynamical behaviour and anomalies in the general circulation and climate system. Documentation of "El Elño Southern Oscillation" events, monsoon anomalies and similar features of the climate system has been provided in these widely used publications. In 1989 the WMO Commission for Climatology launched the Climate Change Detection Project to provide for protection of baseline observation sites, and cooperative global compilation of authoritative climatic data for analyses of trends and anomalies. The Leaders of the Group of 7 western economic powers at their 1989 Summit in Paris specifically endorsed this important initiative of WMO. Much more effort must be devoted by WMO and its Members to this activity.

Applications

Reliable climate data are essential to the safe and efficient design of many socio-economic activities. Statistics of climatic means, extremes and variability are necessary inputs in planning of food production, water resources management, energy systems, construction, urban development, tourism and recreation and many other human activities. The implementation of CLICOM and the increased availability of application methods is providing major advances, in many countries, of user-oriented climatological services.

During the first decade of the WCAP, priority has been given to development and exchange of application techniques in the fields of:

energy: efficient and optimal system design, as well as assessment of solar and wind energy potentials;

food: crop-weather modelling and further development of agrometeorological applications;

water: applications aiming at improved water management and design flood determinations;

urban and building: with focus on tropical regions, e.g. through the TRUCE (Tropical Urban Climate Experiment).

The concept of "operational climatology" has begun to influence many national services, wherein information from climatological data bases, together with near-real time data, are being used to provide operational guidance, in for example, energy system operation, and transportation planning and operation.

Research

The World Climate Research Programme, as a successor to GARP, seeks to understand and predict the global climate system and to assess the impact of greenhouse gases and other perturbations on this complex system. The main thrust is to improve the physical understanding and measurements that are incorporated in the climate system models, used to predict changes in global climate due to changes in the radiation balance, and for prediction of climate a month or season ahead. A number of individual projects have been launched or are in final planning stages. The greatest achievements so far have been much better definition of global cloud fields through the International Satellite Cloud Climatology Project (ISCCP), and greater understanding of the coupling of the tropical oceans and the global atmosphere (TOGA), which has permitted increasingly confident seasonal predictions of regional scale anomalies under certain conditions, such as a strong "El-Niño Southern Oscillation". The two most important studies just getting underway are: (1) WOCE (World Ocean Circulation Experiment) to determine poleward fluxes of heat through ocean currents and to permit construction of much more realistic ocean models to underline the climate models and (2) GEWEX (Global Energy and Water Experiment), to greatly improve the capability of modelling the global hydrologic cycle and the exchange of energy between the various component of the climate system. An important subcomponent of GEWEX is the International Precipitation Climatology Project to obtain through satellite, weather radar and surface measurements, a much improved assessment of regional and global precipitation especially over oceanic regions.

The WCRP is overseen by the Joint Scientific Committee of 12 distinguished scientists, established jointly by WMO and the International Council of Scientific Unions (ICSU). Much of the oceanographic work is planned and coordinated by the Intergovernmental Oceanographic Commission (IOC) of UNESCO and the ICSU-IOC Committee on Climate Changes and the Ocean (CCCO). The WCRP, mainly a physical sciences programme, is closely coordinated with ICSU's International Geosphere-Biosphere Program which is concerned with the biological and chemical systems of land and sea that interact within the climate system. Together the WCRP and IGBP programmes are often referred to as the "Global Change Research" programme.

Impacts

From the beginning of the WCP in 1979, UNEP has coordinated studies of impacts of climate change on various ecosystems and human activities in different regions of the world under the World Climate Impact Studies Programme. An international Scientific Advisory Committee was established in 1981 to advise UNEP on this work. The WCIP sponsored a number of ground-breaking individual evaluations on both national and regional scales, and pioneered the development and publication of appropriate methodologies for impact studies. Important assessments of the global impact of sea level rise by groups in the Netherlands and United Kingdom were initiated under this program. In the past few years, emphasis has been placed on networks to ensure good contacts between scientists, and transfer of technology in this field. The efforts of Working Group 2 of the IPCC in assessing socio-economic

effects of climate change owe much to the studies carried out under the WCIP.

The WCP has also provided the organizational framework for an *important assessment* of the greenhouse gas—climate change issue. In 1985 scientists from 29 countries were convened in Villach, Austria, by WMO, UNEP and ICSU, to try to provide a new world-wide consensus on the existing knowledge of climate change. The meeting's discussions were based on a thorough analysis of the scientific situation by ICSU's Scientific Committee on Problems of the Environment, supported through the WCIP, and chaired by Dr. Bert Bolin. The Conference statement concluded that "While some warming of the climate now appears inevitable, the rate and degree of future warming could be profoundly affected by governmental policies on energy conservation, use of fossil fuels, and the emission of greenhouse gases".

In summary, the WCP's first decade has left a legacy of solid achievements, especially in computerization of climate data management in many countries (CLICOM), Climate System Monitoring, climate applications to water problems, the conceptualization of the WCRP and early results of the ISCCP and TOGA, establishment of methodologies and capabilities in socio-economic impact assessment, and in the Villach assessment (1985) of the greenhouse gas—climate change issue. But the situation in 1990 has changed significantly—the effective and powerful Intergovernmental Panel on Climate Change has been established by WMO and UNEP; and these two organizations, WMO and UNEP, have been requested by the UN General Assembly to prepare almost immediately for negotiation by national representatives of a global Framework Convention dealing with climate change. In these circumstances what should be the role and mission of the WCP in the balance of this century, and how should it be organized to fulfill that mission effectively?

Future of the World Climate Programme
The Role
To propose an appropriate continuing purpose for the WCP requires making some assumptions about the future of the IPCC, and the structures that will be created for a Framework Convention on Climate Change. My assumptions about the manner in which these institutional arrangements may evolve in a complementary manner are outlined in the following paragraphs.

The IPCC was established and has performed well, as an assessment organization. With a continuing IPCC there would be no need for periodic ad hoc assessment conferences under the World Climate Programme, such as Villach '85. Rather, periodic assessments of scientific knowledge of greenhouses gases and climate change and the socio-economic impact of climate change would be well performed by the IPCC. In addition the IPCC has the capability of taking matters one step further and recommending alternative policy responses for consideration of the nations of the world. It is presumed that IPCC will perform such periodic scientific assessment functions in response to needs expressed by its parent bodies WMO and UNEP, by the UN General Assembly or by the governments involved in negotiating the international convention or specific protocols to it. For example, it is highly likely that a further assessment of the effects, sources and potential for control of methane and nitrous oxide will be needed before protocols to address these greenhouse gases can be developed.

International policy research is not well provided for under present organizational arrangements. Yet studies based on risk assessments, socio-economic analyses of global warming preventive options, energy policy, and other types of policy analyses will be sorely needed, and soon. It is assumed that one function of a continuing IPCC will be to further develop the policy studies initiated in Working Group 3 (Response Strategies Working Group). If this responsibility is not assigned to IPCC, an additional stream to cover this important field may be needed in the framework of the World Climate Programme. For present purposes it is assumed that IPCC will have continued responsibility for policy analysis activities, in addition to a responsibility for periodic scientific assessments.

The countries negotiating a Convention on climate change will probably wish to establish some negotiation machinery, an executive, working groups, a secretariat, etc. Later the Parties to this Convention will probably hold periodic Conferences to oversee the administration of the Convention. To undertake this task the Parties will probably wish to request that certain assessments of scientific knowledge be done and such requests could be referred to the IPCC. On the other hand, the requirement may be for further research or data collection in some parts of the world or on some aspect of the climate problem. In this case the matter should be referred by the Parties to the World Climate Programme through WMO and its partners UNEP, ICSU, UNESCO, WHO, FAO and others.

In short, it is essential to the success of both a climate Convention, and of IPCC, that internationally coordinated data collection and research on climate and its impact is continued within the World Climate Programme and in the International Geosphere-Biosphere Programme. WMO also has a continuing role, even more vital now than in 1979, of standardizing, harmonizing and exchanging globally, data and information on the climate system and its behaviour.

The World Climate Programme of the Nineties

In the world of the 1990s, in which national leaders frequently discuss issues of climate change, and the public media in almost all countries rarely let a week go by without reporting the latest pronouncements on global warming, the WCP must rapidly become much more relevant and visible. If the essential increased support of nations for the important efforts of the WCP is to be realized, the WCP must be much more effective than in the past in delivering information necessary to policy development, and in a form useful to the public and government officials. In this context, the following proposals are offered.

a)   *Climate System Monitoring and Data*

The *Climate Change Detection Project* should be designed to not only protect observing stations world-wide with long-period, high quality records, but should ensure timely annual reporting of data from a network of such stations. Climate Change Detection should not be limited to surface temperature observations. For example, observations of the temperature of the lower stratosphere and of the height of the tropical region tropopause may be even better indicators of climate trends and changes. In addition, hydrologic data such as river runoff and glacier trends, and data on changes in net radiation should be included. Effort to improve observational programmes and data collection for climate change detection should be closely tied to existing and expanded efforts in WCP Data. Support to improve the observation network and data management through CLICOM and DARE should be expanded. Finally it would be valuable to combine the climate change detection data with the *Climate System Monitoring* reports and information from the *Global Atmosphere Watch* on changes in greenhouse gas concentrations. Reports should be issued annually preferably within three months of the end of the year. This will require a major effort on the part of contributing Member countries of WMO, the World Meteorological Centres and the WMO Secretariat. There is a great need for an annual, authoritative report on the state and behaviour of the global climate and WMO is in a uniquely favoured position to undertake such a project within the framework of the World Climate Programme. This report would be a key vehicle for provision of regular information to government and should be a basis for imaginative public information campaigns. The management of this combined activity (CSM Data and CCDP) should become the first stream in a revamped WCP.

b)   *Applications*

Application of climatic statistics to design and operation of a myriad of human activities remains an important part of the information infrastructure of all countries.

Unfortunately, in spite of efforts under the WCP and in Technical Cooperation projects, many countries in the developing regions do not have the application technology capability to support appropriate statistical reports and actions. Current effort should be increased to develop applications software packages to run on the CLICOM computers, and to form a linkage with hydrologic application through HOMS (Hydrological Operational Multi-purpose System). This suggests that an applications programme, directed primarily at developing regions through technical cooperation and technology transfer, could well be the focus of a second main stream of a re-shaped WCP. All countries, including developing countries, must increase their capability to assess their own climates and effects of climate changes, and to apply information on climate trends and variability to all of their economic activities, especially agriculture and water management, if they are to develop efficiently. The second stream of WCP should have this as its most important goal.

A summary of needs for strengthening climate data applications programmes in Commonwealth countries, especially island nations and deltaic regions, appears in a recent report of the Commonwealth Secretariat (1989). Such an analysis should be expanded to cover other developing countries as well and be made available to prospective donors.

c)   Research

The far-sighted design of the World Climate Research Programme, and its linkages with the IGBP, have stood the test of time. However, worldwide, extensive support for the various projects has often been slow in coming, although early estimates are that contributions to World Ocean Circulation Experiment could exceed $3 billion including three dedicated satellite missions. The WCRP is also not as well known in the scientific community as it should be, so that from time to time national, bilateral or multilateral efforts are proposed, or even launched, outside of the WCRP framework with dangers of overlap and duplication of precious scientific effort. Overlap and duplication is probably very healthy in intellectual and theoretical work, but can be counter-productive when large and costly observation systems are deployed. Recent efforts to describe the goals and methods of the WCRP to a more general audience through the pamphlet Global Climate Change, will assist in ensuring recognition and support for WCRP. Countries should be encouraged to establish national committees responsible for coordination of climate research, and stronger liaison established between the JSC and joint programme office on the one hand and such committees on the other. This would help provide a broader base for inputs to planning of WCRP and for dissemination of results. In addition, the JSC Members should consider participating actively in conferences of scientists of

different disciplines from all regions of the world, to help information exchange about WCRP. The WCRP as the third main stream of the WCP, should be recognized as one of the two main pillars, along with IGBP, of studies of "Global Change" encompassing all of the relevant earth and biological sciences. The continued partnership of WMO and ICSU in providing the Secretariat for WCRP will remain an essential feature of its success. However, since the ocean programmes are an integral part of climate system science, IOC-UNESCO should consider seconding a staff member to the joint Programme Office, and other means should be found to ensure even more active participation of the oceanographic community in the Planning and work of the WCRP.

d)   Adaptation Strategies

Studies of socio-economic *impacts* of climate change have resulted in a much greater public and governmental realization of the manner in which climate change and variability underly and impinge on almost all economic and social activities. However, until more reliable, and fine-scale projections of *future climates* are produced, from WCRP efforts, these studies will not become much more definitive. Efforts to concentrate on ensuring close contact between scientists working on impact assessment and technology transfer in this field should be further strengthened. At the same time assessments of impacts and development of strategies to *adapt* to climate change and sea level rise are closely linked. It is highly likely that any global Convention on climate will address the preventive strategies to decrease the sources and increase the sinks of greenhouse gases, rather than adaptation to climate change. There will then remain a continuing need for development and exchange of information on adaptation techniques—the planting of different tree species, shoreline management and protection, resilient water, land and agricultural policies, and so on. Thus it is proposed that the fourth stream of the World Climate Programme be designated *Adaptation Strategies* and UNEP be asked to coordinate efforts in these fields.

e)   Coordination

WMO should continue to be the lead organization for the WCP but needs to strengthen mechanisms for active involvement of both ICSU Unions and the UN organizations, particularly UNEP, UNESCO (IOC), WHO, FAO, IAEA. To this end the heads of these organizations should continue to be convened at least annually by the Secretary General of WMO. Their agenda should be based upon input from the chairmen of the various senior advisory and coordinating committees, such as JSC, SAC and CCCO. The WMO Secretariat must provide vigorous support to these coordinating activities and should report regularly on progress in the WCP to governing bodies of the participating organizations and to the Designated Officials on Environmental Matters.

Summary

The designers of the World Climate Programme in 1979 built exceedingly well. Much of the present understanding of the global climate system, its state and its behaviour, is due to the efforts of scientists of more than 160 countries working in the framework of the WCP, a truly remarkable achievement in international, inter-disciplinary and inter-agency cooperation. In 1990, at the start of a new era in global cooperation on climate and climate change, we can see the need for three kinds of institutions: a global Convention to take appropriate action based on existing knowledge; a mechanism for periodic assessment of knowledge and policy implications, the IPCC; and a continuing but even more vigorous scientific programme of systematic measurements and research, the WCP phase 2.

To summarize the proposals for this second phase - WCP2, it is proposed that there be four main streams:

1. *Climate System Monitoring and Data*—including Climate Change Detection.
2. *Applications*—involving all countries but with a strong component of support to developing countries.
3. *Climate Research*—with more active participation of national committees.
4. *Adaptation Strategies*—for international exchange of technologies for adaptation to climate change, making use of impact study results.

These are *my* suggestions, but soon you will have *your* say. May we have, during these few days here in Geneva, the courage and wisdom of our colleagues in 1979 in designing an inspirational and effective framework for the World Climate Programme of the 90s.

References

Arrhenius, S. 1896, On the influence of carbonic acid in the air upon temperature of the ground. Phil. Mag. 41, 237-71.

Commonwealth Secretariat 1989, Climate change: meeting the challenge. Report by a Commonwealth Group of Experts, London, 131 pp.

ICSU-WMO 1990 Global Climate Change. Geneva, 35 pp.

Kellogg, W. W. 1987, Mankind's impact on climate: the evolution of awareness. Climate Change 10, 111.

Revelle, R. and Suess, S.E. 1957, Carbon dioxide exchange between the atmosphere and ocean and the question of an increase of atmospheric CO_2 during the past decades. Tellus 9, 18–27.

WMO 1979, Proceedings of the World Climate Conference. Geneva, Feb. 1979, WMO #537.

WMO 1986, Report of the International Conference on the assessment of the role of carbon dioxide and other greenhouse gases in climate variations and associated impacts. ICSU, UNEP, WMO, Villach, Austria, Oct.1985, WMO #661, 78pp.

WMO 1989, The global atmosphere watch. Geneva, Fact Sheet #3.

Modern Data and Applications:
World Climate Data Programme, World Climate Applications Programme

by Victor G. Boldirev *(WMO)*

1. Introduction

The First World Climate Conference in 1979, when outlining the World Climate Programme, considered, as the main thrusts of the programme, improvement of the acquisition and availability of climatic data and the application of knowledge of climate.

Planning and implementation of the World Climate Data Programme (WCDP) and World Climate Applications Programme (WCAP) in 1979–1983 were guided by the "Outline Plan and Basis for the World Climate Programme 1980–1983" adopted by the Eighth Congress of the WMO in 1979.

Later, the World Climate Data Programme and World Climate Applications Programme were outlined in the First WMO Long-term Plan (for the period 1984–1987) and in the Second WMO Long-term Plan (for the period 1988–1991). In these plans, specific formulations of objectives and projects of the programmes are somewhat different. However, the basic purposes, tasks and priorities of the programmes remained essentially the same throughout the whole period of the implementation of the WCP.

Climate data have been maintained for a long time and regularly published by meteorological services of a number of countries and also under the aegis of WMO as so-called "World Weather Records", containing for some selected stations monthly values of some basic meteorological variables (surface temperature, precipitation, surface and sea level pressure) on a decadal basis. Publication of data for the decade 1971–1980 has recently started.

Promotion of climate applications, as one of the major concerns of WMO throughout virtually all the history of the Organization, was regularly called for by resolutions of the WMO Executive Council and by the World Meteorological Congresses. In Resolution 25 (Cg-VIII)—"Climatic Change"—Congress recognized that "Man's activities seem likely to have an increasing impact on global climate" and requested "to review data requirements for the assessment of climatic change and the mechanisms necessary for effective data exchange". Congress also encouraged "further studies of climatic change and the consequential natural, social and environmental effects and .... the exchange of results of such studies".

2. Objectives, Tasks and Implementation of the WCDP and WCAP

2.1 World Climate Data Programme

The main long-term objectives of the World Climate Data Programme may be summarized as follows:

i) assistance to countries in improving their national data management systems and building climate data banks

ii) consolidation of the requirements for climate data observations and exchange, and promotion of fulfilment of these requirements

iii) provision of reference information on climate data and station networks

iv) development of a global climate monitoring system, including establishment of climate baseline data sets.

We will now examine the various projects which have been developed to address these objectives.

i) Assistance to countries to improve climate data management and user services

Within this activity WMO is implementing a project called Data Rescue (DARE), designed to save rapidly deteriorating meteorological records in some developing countries of the world by transferring these records to a microfilm/microfiche format. Presently, the project is being implemented in RA I (Africa) through support of donor countries, especially Belgium, and UNEP. By mid-1990, the project was fully implemented in 9 countries and underway in 10 others. The WMO Commission for Climatology has recommended that this be expanded to include both Asia and Central and South America. Both the digital and microfiche database are provided to the participating country. The country is also provided with training and equipment necessary to continue this process into the future. The data processing is being co-ordinated by the international centre established for this purpose in Brussels (Belgium).

CLICOM provides WMO Members with systems for a personal computer that can perform all the functions of a complete traditional climate data centre. The system is designed to perform data entry, quality control, storage and retrieval, data inventories and basic climatological products. Essentially, this WMO system provides all countries with an easily useable complete data management system.

The CLICOM system therefore includes hardware (personal computer and peripherals), software packages, and training. It is planned to expand continuously CLICOM software packages and integrate into them the outcome of other activities within WCDP and the World Climate Applications Programme.

The CLICOM system has been installed (as of March 1990) in about 90 countries with about 120 systems in operation. The goal is by 1991 to have this system operational in over 100 member countries and by 1997 to have established climate data management systems in all member countries. This is an ambitious but. at the same time, a realistic goal.

ii)   Conslidation of climate data requirements and improvement of data exchange

Every country and nearly every province, state and local community has a meteorological data network of some sort. These data can, and indeed do, provide a great benefit to the local areas, but generally are not available and are not consistent with other data to be useful for global or even, in many cases, national climate studies. It is WMO's role to co-ordinate both the acquisition and analysis of global climate data. For this purpose, WMO co-ordinates the transmission and exchange of climate data, for instance from the network of stations producing CLIMAT messages, via the WWW Global Telecommunications System. The CLIMAT network consists of over 2,000 stations transmitting monthly summaries of data to be used specifically for climate purposes. These data are archived at the World Data Centres for acquisition by scientists and researchers. This data network is still the best available and the most easily accessed for global studies. WMO is constantly striving to improve this network and the protection and expansion of the Reference Climatological Station Network. The reference network. which is a subset of the CLIMAT network. consists of climate stations in controlled environments with excellent observers and instruments. These stations are intended to reduce the effects of man's influence on the long term record and to serve as a baseline for other stations to be compared against.

Criteria for selection of reference climatological stations and procedures to improve the quality of data sets were developed by WMO. What is needed is international recognition at the governmental level of the need to protect existing stations and expand their network.

Besides meteorological data, information on atmospheric composition, the state of the ocean, data on the cryosphere and land surface - that is, data on all components of the global climate system - are obtained by various operational observing systems and research programmes.

As examples of such systems, one could mention the WMO Global Atmosphere Watch (GAW), which provides data on atmospheric composition and the joint IOC-WMO Integrated Global Ocean Services System (IGOSS).

The GAW, approved in June 1989 by the WMO Executive Council, co-ordinates WMO data-gathering activities begun in the 1950s under the Global Ozone Observing System (GO$_3$OS) and the Background Air Pollution Monitoring Network (BAPMoN) with other smaller measurement networks.

The GAW serves as an early warning system to detect further changes in atmospheric concentrations of greenhouse gases. changes in the ozone layer and in the long-range transport of pollutants, including acidity and toxicity of rain and changes in the atmospheric burden of aerosols (soil and dust particles).

At present, BAPMoN (supported by UNEP) consists of 196 stations, 159 of which have the capability of carrying out sampling for measurements of precipitation chemistry (rainfall and snow). Other measurements recorded in the observation programme are: turbidity (transparency or clarity of the air) at 90 BAPMoN stations; suspended particulate matter at 84; carbon dioxide at 23; surface ozone at 22; methane at 7; and chlorofluorocarbons at 5. The atmosphere is sampled according to agreed criteria by specially trained staff using recommended instruments and standard procedures.

Regarding the ozone layer, as early as 1957 WMO established international procedures to assure standardized and co-ordinated ozone observations, their publication and related research. Since then, continous measurements of atmospheric ozone have been carried out by the Global Ozone Observing system It now has approximately 140 stations around the world—complemented during the last decade by satellite remote-sensing. It is the only network in the world that provides information on total ozone content, its vertical distribution and changes.

The IGOSS is a worldwide operational service system which provides physical oceanographic data and information for various marine users (operational since 1975). In 1989, a total of 38,615 Bathythermograph Report messages and 5,444 Temperature Salinity and Currents Report messages were transmitted via the Global Telecommunication System. The daily average was 121 messages. Data are collected by IGOSS World, Specialized and National Oceanographic Centres for preparation of products required for national and international activities. In 1989, approximately 200 ships (both research and merchant) from 23 countries contributed to the collection of IGOSS data.

iii)   Implementation of the climate data referral system (INFOCLIMA)

Information on climate data is contained in the INFOCLIMA Catalogue of Climate System Data Sets; an expanded and updated version of this Catalogue was printed and distributed in 1989. A version of the Catalogue on diskette allowing for direct digital access was completed.

The 1989 version contains more than one thousand data set descriptions from 268 data centres in 112 countries. INFOCLIMA is now cross referenced on three other international referral systems. The Inventory covers both national networks of synoptic and climatological stations and stations operated for other applications. It is expected that ultimately the inventory will contain information on about 40,000 stations and will be available on computer-compatible media. Two of the planned six volumes of the inventory have been published.

iv)   Development of global/regional climate data sets, and of a climate system monitoring capability

These two activities are very closely linked since for climate monitoring and assessment of climate change and variability, quality controlled reference data sets are crucial. Many data centres affiliated with WMO, ICSU, IOC, UNEP, FAO and Unesco develop their own data sets. Co-ordination of all these activities is necessary; the process of such co-ordination started recently.

The Climate System Monitoring Project, implemented by WMO with the support of UNEP, has provided member states of WMO with consistent analysis and anomaly information, which for many Members is the only global climate information available.

Monthly, biennial and annual summaries of the climate and anomalies are prepared on a regular basis. The WMO publication "Climate System Monitoring" is distributed monthly to over 1500 scientists and institutions to help experts worldwide to assess the climate. The publication covers a variety of events which are placed in historical context. Meteorologists and hydrologists may study these climate anomalies to detect climate changes or trends.

Three biennial reviews were published covering the period from 1982 to 1988, and the fourth one for 1989–1990 is planned to be published in 1991.

As a natural development of ongoing projects, a new project has emerged in the WCDP: the Climate Change Detection Project, endorsed by the WMO Executive Council. The concept is to utilize all available facilities and expertise, to undertake construction of a highly controlled climate database and provide the mechanisms necessary to use this database in the detection of climate change.

The project should be a worldwide effort, primarily through meteorological services, to collect more climate data with well-documented station information (metadata) and to process them using uniform (objective) procedures so that eventually more reliable analysis could be made of climate trends and variability.

These data should be utilized along with climate model outputs to produce on a regular basis an authoritative WMO assessment of the state of the world's climate.

At the end of November 1990, a meeting of experts to prepare the detailed plan of project implementation will be held near Toronto, at the invitation of Canada. The meeting will bring together experts in climate data processing and analysis, including analysis by climate models.

2.2   *World Climate Applications Programme*

The main long-term objectives of the World Climate Applications Programme serve the purpose of assisting Members to strengthen their national institutional capabilities to apply climate knowledge.

The World Climate Applications Programme at present consists of nine projects primarily because applications in each individual major socio-economic sector (agriculture, water management, energy, etc.) are covered by separate projects.

In the area of climate applications of energy, the main thrusts were:

- education and training of meteorologists and users on how information on climate can be used in the decision making process with regard to energy matters; for this purpose, expert missions appeared rather effective and more than 30 such missions were undertaken since the inception of the project in 1981

- development of methodologies for operational applications of climate knowledge; some results of this activity were summarized by a Technical Conference on the Operational Aspects of Energy-Meteorology held in Quito. Ecuador. in 1987 and are included in a number of WCAP publications

- climatological aspects of utilization of solar, wind and other renewable sources of energy; the results on solar and wind energy sources are summarized in WMO Technical Notes.

As regards climate application to agriculture and food production, the WCAP relies upon the WMO programme on Agricultural Meteorology. Here, as in the energy area, education and training to promote the application of agroclimatic data in agricultural activities and methodology development aspects are of primary importance. The promotion of agroclimatic data applications was pursued through expert missions (about twenty) and through regional agroclimatic surveys of which one in South East Asia was completed, and another one, in South America, is near completion.

Methodology development is supported through organization of various workshops on the utilization of agrometeorological information for planning and operation in agriculture and through dissemination of information on experience gained in using "crop-climate" models of various levels of complexity (several tens of such models were reported to be used).

The project on climate applications in water resources management is being implemented in co-ordination with the WMO Hydrology and Water Resources Programme. Thus an effective input is ensured from operational hydrology to water-related aspects of studies of climate and the use of climate information for water resources activities.

The following areas of activities are related to climate applications:

- studies of hydrological data in the context of climate variability and and change
- application of climatic information in the planning, design and operation of water-resources systems
- studies of the influence of climate change and variations on water resources

The project on urban and building climatology was initiated in 1983. A plan of actions in this area was adopted by the major international Conference on Urban Climatology and its Applications (Mexico, November 1984).

The project supported various studies which led to the preparation of numerous guidance documents on:

- calculation of climate parameters used for building purposes
- application of climatological data in building
- impacts of climate on building
- use of meteorological information for construction in architecture, building and civil engineering
- design for urban planning in different climates.

On the basis of special urban building climatology studies and recommendations made by the Mexico and Kyoto conferences, a Tropical Urban Climate Experiment (TRUCE) was initiated. The project is intended to fill (at least partly) a significant gap in urban studies in tropical areas (in 1979–1987 only 55 studies were undertaken for these areas in comparison with 637 studies for mid-latitudes).

Specifically, the project will provide for:

- a global data base on the characteristics of tropical urban climates

- development of models to simulate the urban climate system and predict its status
- information exchange and training.

Within the WCAP, other areas of climate applications are included such as human health, transport, tourism and recreation, etc. However, the main feature of the programme at present is an attempt to assist, while covering as wide a range of applications as possible, national climatological services in obtaining and using modern climate application methodologies.

For this purpose, the programme promotes:

- development of CLICOM-compatible software on various climate application methodologies
- development (again in a CLICOM compatible format) of the Climate Applications Referral System (CARS), thus helping to obtain information on available methodologies.

A new development in applied climatology that is included in the WCAP, and should be even better reflected, is the operational use of climate data and information for activities or processes that are happening now or that will happen in the near future (operational applied climatology concept). Operational climatological services are increasingly being demanded and may include, for example:

- to-date accumulations of climate data to put current climatic or weather events into historical perspective
- climate indices based on near-real-time data
- climate forecasts or prediction on various time scales (weeks to seasons).

Assistance to Members, especially to developing countries, should be provided in developing national climate application services through computerized systems (CLICOM), applications software packages, including software to apply climate information. The Climate Applications Referral System and linkages with other data and information systems such as HOMS, should continue to be distributed to all CLICOM users and should be updated regularly.

Utilizing climate information for application purposes is now being expanded and facilitated as new technologies became more easily available to climatological services. These technologies include:

- improved and cheaper access to routine real-time climate information
- enhanced capability to obtain quality-controlled data and information in near-real time from central processing locations
- better collection of data provided by remote sensing (satellite, microwave profilers, Doppler radar, etc.)

- development of networks of microcomputer systems (e.g. CLICOM) with advanced hardware (including digital archiving means) and user-interactive software.

In the future, the WCDP should place more emphasis on climate monitoring and climate change detection and, at the same time. should be able to perform necessary climate data service functions. The programme may, in coming years, include the Climate Change Detection and Climate System Monitoring projects as leading ones. Service functions would cover:

- preservation of climate data for past periods (Data Rescue)
- promoting expansion of climatological station networks, with particular emphasis on protection and expansion of the climatological reference station network
- computer systems for climate data management (CLICOM, INFOCLIMA)
- development of climate data bases including preparation of data sets and data exchange.

The WCAP should continue to promote applications of information on current climate to various socio-economic sectors. An overall emphasis on sustainable development, involving improvement of human well-being and wise use of natural resources and environmental protection, will certainly influence the development of the programme. Here again, as in the case of the WCDP, concern related to climate change will create additional demands for climate applications, especially in the light of potentially adverse socio-economic impacts of climate change.

Many of the activities during the IDNDR (International Decade on Natural Disaster Reduction) will directly or indirectly relate to climate applications, i.e., will require input of both climate information and knowledge. There will be an increasing need for climate information in land use planning, especially for human settlements, as well as a need for near real time climate information, i.e. "operational climatology", in disaster reduction schemes and for other applications.

There is a growing awareness by many users of the economic and social benefits of applying climate information, both in operational and long-term planning processes.

The programme should concentrate upon support to further development of methodologies both to apply information on current and "recent" climate and to assess climate effects on socio-economic activities.

Climate application methodologies in many cases were made and are being developed using accumulated information, both climate and socio-economic. Strictly speaking, before developing any methodology a study of climate impact on a particular area of human activity or on environment should be carried out. If a significant climate impact is identified, it should be worthwhile to apply climate information as necessary.

In the case when climate and/or socio-economic or environmental patterns change substantially, previously developed climate application methodologies may become irrelevant. Therefore, with respect to projected global climate change, methods of assessment of the impact of climate change should be brought together with the "future climate" statistics—for instance, statistics from climate models—to develop methodologies suitable for future climate conditions, or at least, to evaluate to what extent it would be possible to use currently available methods and guides. Linkage and co-ordination of WCAP with WCIP will have to be strengthened.

Evidently, an ultimate goal of WCAP, as well as WCIP, should be the provision of a set of options to decision makers for adaptation to current climate variations and future climate change. More than ten years of implementation of the WCDP and WCAP shows that these programmes are and can continue to be a viable structure capable of responding to the needs for systematic monitoring of climate and climate applications for the purpose of the policy decision making process.

3. Summary

Both the World Climate Data and World Climate Applications Programmes achieved those objectives which were formulated in 1979 and then pursued through the WMO First and Second Long-Term Plans. However, demand always exceeded delivery. This became especially evident when projections of a future climate change began to be a matter of particular concern of governments, international agencies and the general public.

Reliable climate data and proven climate applications methodologies, provided by meteorology in general and specifically by meteorological services, are needed now more than ever before.

Overview of the World Climate Research Programme

by Pierre Morel *(WMO/ICSU)*

Abstract

Much more precise information about the nature, amplitude and timing of regional climate impacts will be needed to plan and implement adaptation to climate changes that are expected in the future. This information is not yet available, as can be readily deduced from the still wide range of uncertainty on estimates of the global warming at the surface of the Earth, 1.5 to 4.5 degrees Celsius, which would result from a permanent doubling of the pre-industrial amount of carbon dioxide in the atmosphere. The World Climate Research Programme (WCRP) is the scientific initiative sponsored by the International Council of Scientific Unions and the World Meteorological Organization to study climate variations on all time-scales, from seasons to centuries. These variations result from the interplay of the various components of the Earth's climate system: the global atmosphere, the world ocean, continental and sea ice, and land surface hydrology. Major international research programmes, organized by the WCRP, are under way or planned for the near future: the study of the Tropical Oceans and Global Atmosphere (TOGA) coupled system, the World Ocean Circulation Experiment (WOCE) and the Global Energy and Water Cycle Experiment (GEWEX). These research programmes are the keys to the required advances in our knowledge of climate processes. However, considerable lead time, ten to twenty years, will be needed to develop the necessary means to observe the Earth system and advance our theoretical understanding to the point where reliable predictions of regional climatic impacts will be feasible.

Introduction

The World Climate Research Programme (WCRP) is a joint undertaking of the World Meteorological Organization (WMO) and the International Council of Scientific Unions (ICSU).

For WMO, the WCRP is the research arm of the World Climate Programme, launched in 1980. Its objectives are to determine the extent to which climate can be predicted and the extent of man's influence on climate, by means of computer models, detailed field studies and global observations from ground-based stations, ships and satellites. The expected result of the WCRP is a capability for climate prediction on all time-scales from seasons or years to decades or centuries.

For ICSU, the WCRP, together with the International Geosphere-Biosphere Programme (IGBP), provides the overall scientific strategy for the study of global change. The two programmes aim to understand the mechanisms of the total Earth system, the WCRP being focused on dynamical and physical aspects while the IGBP is focused on chemical, biological and ecological aspects.

Specifically, the task of the WCRP is to understand, in *quantitative terms*, the interactions between four major components of the Earth climate system.

- The *global atmosphere* which is the most rapidly variable component of the climate system and also the most energetically active, the heat engine that drives the global water cycle and the ocean circulation
- The *world ocean,* which interacts with the overlying atmosphere over periods of months to years, while the deeper ocean circulation responds over periods of decades to centuries and can sequester heat and chemicals (e.g. carbon) over similarly long periods
- The *cryosphere*, which comprises continental ice sheets and ice caps, mountain glaciers, snow and sea-ice
- The *land surface* of continents, including the flow in rivers and underground which controls water storage and evaporation.

The World Climate Research Programme addresses climate variations on all time-scales. Because the Second World

Climate Conference will be emphasizing the issue of climate change caused by human activities, the following discussion will highlight the role of the WCRP in understanding the response of the Earth's climate to the increase in the greenhouse effect. This choice does not reflect on the scientific and practical importance of other WCRP research activities aiming at understanding and predicting transient (natural) variations in the Earth's climate, such as the *El Niño*/Southern Oscillation phenomenon. The work carried out under the WCRP Tropical Ocean and Global Atmosphere (TOGA) programme will be presented in one of the three WCRP scientific lectures later this week. Let us only note that significant progress has been made toward developing a capability for operational prediction of *El Niño* events several months in advance (World Climate Research Programme, 1990), a very significant advance for resource managers and civil protection planners in countries affected by these anomalous weather/climate events.

The Role of Science in the Overall Response to Global Change

It is often stated that policy-makers are aware of the uncertainties of climate predictions but that they can already take action to protect the Earth's climate on the basis of available scientific evidence. This is certainly true when the proposed actions recommend themselves by obvious benefits for the conservation of the global environment and immediate economic feasibility. It is doubtful, however, that the actions envisaged at this time could stop further man-induced modification of the Earth's climate or, alternatively, provide adequate protection against changes to come. Far more precise and reliable scientific information will be needed for effective adaptation to climate change.

In the first place, it is obvious that checking the rise in the concentration of carbon dioxide and other greenhouse gases, with the possible exception of man-made chlorofluorocarbons and similar artificial chemicals, would require a major adaptation of man's activities and the acceptance of serious economic consequences. Modifying industrial and agricultural practices to minimize the expenditure of energy world-wide, changing the way of life of individual consumers, and re-examining the justification for population-planning in the light of the need to preserve non-renewable resources may eventually be forced upon humanity but these will certainly not be easy steps to take. It is quite out of the question that such transformations should be accomplished by administrative decision, without the informed consent and effective support of all citizens. There is no chance that such a consensus could be achieved in the face of existing scientific uncertainties and continuing controversy among scientists. Only further

scientific advances could reduce uncertainties in our knowledge of the nature, amplitude and timing of future climate change, and would provide a solid scientific foundation for wide public support of strong preventive actions.

To illustrate this point, one may quote the question asked by W. T. Brookes (1989) in a widely read business magazine: 'Is the Earth really on the verge of environmental collapse? Should wrenching changes be made in the world's industry to contain CO_2 build-up? Or could we be witnessing the 1990s' version of earlier scares?'. In the face of such (legitimate) queries, policy-makers have the most urgent need for unquestionable evidence that scientists can determine the sensitivity of the Earth's climate to greenhouse gases and can detect unambiguously the physical portents of changes to come. Until the international community of scientists and responsible environmental, meteorological, oceanographic and space agencies have collected the necessary observations and developed the necessary theoretical knowledge, this need will remain unfulfilled.

In the second place, one should remember that the radiative effect of the increase in the amount of carbon dioxide (CO_2) and other greenhouse gases since the beginning of the industrial era has already reached about half the value which corresponds to doubling the pre-industrial concentration of CO_2. Furthermore, it is also clear that the current trend toward even higher concentrations will continue for some time. About two-thirds of the man-made increase in the greenhouse effect is attributable to carbon dioxide released by the combustion of fossil fuel which, in 1985, provided 88.4% of the world-wide commercial production of energy. According to the lowest scenario considered by the 1989 Congress of the World Energy Conference, fossil fuels would still amount to 82% of the total commercial energy production by year 2020, while the total consumption of fossil fuels would rise by 40% above the 1985 level (Barnaby, 1989). Similar considerations apply to methane emissions. Despite courageous measures that may be taken to slow the ongoing accumulation of greenhouse gases in the air, a considerable departure from the undisturbed radiative energy balance of the Earth has already been caused by past emissions and will continue to increase in the near future.

As a consequence, a significant change in the Earth's climate is unavoidable. However, the consequences of such a change need not be unmanageable, especially if suitable warning can be given for each region of the Earth, concerning the kind of climate change that will occur, how much and when. Present scientific knowledge allows estimation of only the global or, at best, very large-scale climatic effects of the increase in greenhouse gases. It is the objective of climate research, complemented by world-wide monitoring of the Earth's atmosphere, land and oceans, to achieve meaningful predictions of regional as well as global climate changes, as required for planning specific responses to adjust to expected changes. Obtaining this information may require one or two decades of serious scientific work.

Period 1990–1995

The sensitivity of the Earth's climate to the increase in greenhouse gases was first estimated by considering the response of the atmosphere only, discounting for simplicity the vast thermal inertia and complex dynamics of the ocean. The first international assessment of the influence of carbon dioxide on climate (World Climate Programme, 1981) concluded, already ten years ago, that the *equilibrium response* of climate to doubling the pre-industrial concentration of carbon dioxide would be a general warming at the surface of the Earth of the order of 1.5 to 4.5 degrees Celsius for the mean temperature averaged over the whole globe and all seasons. Even larger uncertainties apply to similar estimates of regional climate changes. Since this early estimation, the range of uncertainty has not been materially reduced. Indeed, the latest assessment conducted under the aegis of the Intergovernmental Panel on Climate Change led to the same error margin (WMO/UNEP, 1990). The roots of this uncertainty are the complex adjustments which occur in the atmosphere as a result of the increase in the greenhouse effect. These adjustments are especially difficult to determine because the various elementary processes which take place in the atmosphere and at the Earth's surface constitute looped 'feedback mechanisms' that may either amplify or reduce the primary response to radiative forcing. The most powerful positive (amplifying) feedback is the increase in the amount of atmospheric water vapour, which is itself a greenhouse gas. The next most important, and unfortunately poorly known, feedback is the change in the amount, distribution and optical properties of clouds. It is significant that low-altitude liquid water clouds cool the planet, while high-altitude ice-clouds have an opposite warming effect (their own 'greenhouse effect' more than compensates for increased backscatter of sunlight). Our insufficient knowledge of these cloud-related phenomena is the main cause of uncertainty about the *global mean sensitivity* of climate to a specified departure from the pre-industrial composition of the air.

With some luck, one may be able to capture the essence of the cloud-radiation feedback by a combination of measurements of the Earth's radiation budget at the top of the atmosphere (obtained from satellites), radiation observations at the surface and improved atmospheric circulation models, without having to account for the complex microphysics of clouds. Several front-line teams are actively working to improve the formulation of clouds and radiation transfer in atmospheric models, using global observations which are now becoming available from the Earth Radiation Budget Experiment and the WCRP International Satellite Cloud Climatology Project. With a degree of optimism, one may foresee significant progress in the next five years and a considerable narrowing of the range of uncertainty of the computed *global-mean equilibrium response* of the atmosphere to a specified amount of carbon dioxide.

Period 1995–2000

The Earth's climate has changed only minimally during the last fifty years, undoubtedly less than could be inferred from computations of the response of the atmosphere only to the observed increase in the concentration of greenhouse gases. One reason for this weak climatic response to the increasing greenhouse effect is that the behaviour of the Earth system results from the interplay of a fast component, the atmosphere, and several slow components including the deep ocean and polar ice sheets. The gross (global mean) response of the ocean is essentially determined by two factors: the net (annual and global) mean flux of heat which is received by the ocean, and the capacity of the ocean to absorb this extra energy and transport it into its deep interior. The penetration of heat is not a simple diffusion process, but rather the result of water sinking in middle and high latitudes and being transported downward and equatorward. In the final analysis, the problem of estimating the heat capacity of the seas for a *transient warming* occurring over a period of several decades calls for a detailed understanding of the global ocean circulation.

A major global oceanographic programme, the WCRP World Ocean Circulation Experiment (WOCE), has been undertaken to acquire, over a period of about five years (1990–1995), a complete description of the global ocean, similar to the global maps of the atmospheric circulation drawn every day by meteorologists. This unprecedented effort will take up a large fraction of the world oceanographic resources, research vessels and laboratories, and relies on the successful implementation of several satellite missions designed for this purpose. It will also require a substantial upgrading of computing facilities and the development of much-improved mathematical models of the three-dimensional dynamics of the global ocean circulation. From the scientific interpretation of WOCE observations and the use of realistic (topography and eddy-motion resolving) ocean models, one can expect, in the period 1995–2000, a major advance in the ability to accurately predict the penetration of heat into the ocean and thus the *rate of transient climate change* and the *rate of sea-level rise* associated with the thermal expansion of the oceans.

Period 2000–2005

The regional climatic impacts of a change in the global atmospheric circulation regime are actually the results of transient weather phenomena. There is little scope for assessing such regional effects with current climate models which cannot resolve the relatively small scale of weather events, nor describe water transport processes realistically. Far more complete, more penetrating and more accurate information is needed to understand and model these processes, all of which involve the problems of atmospheric radiation, evaporation and rainfall, water vapour transport and cloud formation.

To develop the necessary understanding of these processes world-wide, the WCRP has undertaken a major new atmospheric research programme, the Global Energy and Water Cycle Experiment (GEWEX). GEWEX has actually begun already, as a multi-disciplinary effort of atmospheric scientists, meteorologists and hydrologists, to organize theoretical (model) studies and experimental (field) studies, to develop and test appropriate data and information systems encompassing the required range of observations and to work out with the space agencies the definition of the future international Earth Observing System. The lead time for such major developments is of the order of a decade. Global data from the GEWEX observing systems can thus be expected to flow at the turn of the century, leading to good progress toward prediction of *regional climate changes* in the time frame 2000–2005.

Period 2005–2010

The knowledge of world ocean dynamics derived from WOCE (and continuing global ocean observations), together with the knowledge of global atmospheric fluxes of energy and water obtained from GEWEX, will allow, in the first decade of the twenty-first century, integration of the results of parallel research conducted by the International Geosphere-Biosphere Programme (IGBP) into a comprehensive representation of the Earth's system. Such coupled physical, chemical and biological models are in fact necessary to explore the more subtle second-order climate feedback processes, involving changes in global biogeochemical cycles, which may well eventually dominate the future of our planet. One should remember that the primary cause of the return of glacial ages is believed to be a relatively small, apparently inconsequential, variation in the regional and seasonal distribution of sunlight. The primary energetic effect, much too small to cause significant climate change directly, must have been greatly amplified by (yet unknown) dynamical-biogeochemical processes which produced a marked change in the atmospheric concentration of carbon dioxide and methane, the very chemical species which are being manipulated by humanity. We must learn for example

whether or not the predicted slowing-down of the meridional circulation of the Atlantic Ocean could significantly reduce deep sea-water formation and effectively insulate the deep ocean from the atmosphere, with a corresponding loss of the buffering role of the ocean and reduced oceanic uptake of anthropogenic CO_2. Another important issue to be tackled in the same time-frame is the mass balance of the Greenland and Antarctic ice-sheets, with the objective of determining the extent of the compensation between melting and increase in snowfall and snow accumulation under a warmer climate, with potentially large and yet very uncertain consequences on mean sea-level rise or fall.

Resource Requirements for Climate Research

Climate research can provide a valuable service to society by permitting reliable predictions of the amplitude and timing of expected global climate change and consequent regional effects. These achievements will be possible only with the resolute support of all nations to sustain the operation of necessary observing systems and to develop scientific research on global change.

Unprecedented co-operation of the world scientific community and task-sharing by research institutions and responsible national administrations are needed, together with long-term national commitments to support these efforts: the very nature of climate processes calls for long-term sustained scientific attention. The World Climate Research Programme offers a blueprint for organizing this scientific effort.

Inasmuch as global climate change is recognized as a problem concerning all nations, establishing an effective global climate monitoring and prediction system must also be recognized as a responsibility of all nations. The foremost source of basic information on climatic processes and climate variations is the operational World Weather Watch observing system and associated hydrological and environmental monitoring networks (Global Atmospheric Watch). The maintenance and improvement of these global observing systems is a prerequisite for any serious endeavour to understand and predict climate change. However, these operational observing systems are not comprehensive enough: essential pieces of information are still lacking to understand the role of clouds in the radiation budget and global hydrological cycle or changes in biological activity on land and in the sea. New space observing systems will be essential to acquire this knowledge; the support of nations which can develop and conduct such space missions is essential to the success of global climate research.

Furthermore, we have seen that the key to predicting the rate of change of the Earth's climate is to be found in the knowledge of the global ocean circulation and heat storage.

A major new initiative is needed to carry out the necessary systematic observations of the world ocean and polar ice, based in part on satellite surveys but also on *in-situ* measurements of ocean properties and ice characteristics. This initiative requires not only resources to support long-term systematic measurements at sea and on polar ice, but also international goodwill to facilitate access to national Exclusive Economic Zones when critically needed.

Last but not least, effective prediction of climate change, both on the large scale characteristic of global phenomena and on the regional scale of interest to national planners, will require a major development of climate models. The first order of priority is to incorporate a creditable formulation of the world ocean circulation and processes, by means of considerably more powerful computers than are now available to climate scientists. The second order of priority is to further refine the formulation of atmospheric processes to emulate the level of verisimilitude achieved by operational weather-forecasting models. Both requirements will place heavy demands on numerical computing facilities and also require the dedicated efforts of strongly organized teams of scientists and data-processing specialists.

Conclusion

The evolution of the Earth climate over the next century will be a combination of ongoing natural variations and progressive change forced by anthropogenic modifications of the environment. The problem of discriminating the forced warming trend from the background of natural climate variability cannot be divorced from the detailed investigation of natural transient climate variations on all time-scales. This broad-gauge approach is the guiding principle of the WCRP. We believe it is the best scientific strategy and we hope it will prove effective to deliver the specific and reliable information needed to master the consequences of global climate change.

References

Barnaby, F., 1989: World energy prospects, *Ambio*, 18, pp. 459–460.

Brookes, W. T., 1989: The global warming panic, *Forbes Magazine*, 25 December 1989, pp. 96–102.

World Climate Programme, 1981: On the assessment of the role of CO_2 in climate variations and their impacts, report of WMO/UNEP/ICSU meeting of experts in Villach, Austria. WMO, Geneva.

World Climate Research Programme, 1988A: Concept of the Global Energy and Water Cycle Experiment, WCRP-5. WMO, Geneva.

World Climate Research Programme, 1988B: World Ocean Circulation Experiment Implementation Plan, WCRP-11 and 12. WMO, Geneva.

World Climate Research Programme, 1990: Report of the second session of the Intergovernmental TOGA Board. WMO, Geneva

WMO/UNEP Intergovernmental Panel on Climate Change, 1990: Policymakers' Summary of the Scientific Assessment of Climate Change. WMO, Geneva.

World Climate Impact Studies Programme

by James C.I. Dooge *(Ireland)*

Abstract

This paper reviews the problems tackled and the progress made under the WCIP during the decade 1980-1989. A prime deficiency in 1980 was the lack of a basic methodology of climate impact assessment. This has been largely remedied under the WCIP. The programme was also responsible for the provision of the basic documents for the 1985 Villach Assessment Conference and for the follow-up to that Conference in the area of policy responses. The WCIP was less successful in relation to climate impacts in individual climate-sensitive sectors. Regarding the co-ordination of climate impact studies, progress was made towards the end of the decade when attempts at broad co-ordination were replaced by the promotion of networking among active groups.

1. Planning of the WCIP

The impact on human society of climate variation and change was fully taken into account in the planning and the launching of the World Climate Programme. Of the papers presented to the First World Climate Conference in 1979, a considerable number were concerned with the effect of climate change on individual sectors of human activity such as the major world food production systems, water resources, human health, land use, forestry, and fisheries, (WMO, 1979). One of the five working groups that met during the second week of the 1979 Conference was devoted to a discussion of the Impacts of Climate Variability and Change on Society. The report of that working group became the basis of the World Climate Impact Studies Programme, (WMO, 1980).

The initial Plan of Action for the WCIP for 1980–1983 was drawn up by an ad hoc group of experts in February, 1980 and endorsed by the Governing Council of UNEP in May, 1980. This original Plan of Action consisted of four major programme areas as follows:

i)   Reduction of the vulnerability of food systems to climate

ii)   Anticipation of impacts of man-induced climatic changes

iii) Improvement in the science of climate impact studies

iv)   Identification of climate-sensitive sectors of human activity

This Plan of Action for 1980–1983 was estimated to cost about 22 million US dollars, of which 4.1 million dollars would come from UNEP funds, 6.2 million dollars from other international organisations, and 11.6 million dollars from national programmes. In the event, the actual expenditure for these four years on climate impact studies was substantially less than that recommended in all three categories and did not exceed one half of the above recommended figures.

In May 1980, the Governing Council of UNEP established a Scientific Advisory Committee (SAC) to provide scientific guidance for the WCIP. At its first meeting in February 1981, SAC revised the original plan of action to take account of further proposals received in the year which had elapsed since that Plan was drawn up. In particular the SAC recommended that a fifth programme area:

v)   stimulation and coordination of climate impact studies,

be added to the four programme areas of the original Plan of Action. SAC also recommended as appropriate for the programme an expenditure by UNEP of 3.1 million U.S. dollars over the three year period 1981–1983.

Following the completion of this first phase of the programme, subsequent programmes were prepared for each two year period from 1984–85 to 1990–91. The programme for the present biennium 1990–91 has been grouped into the following programme areas:

i)   Greenhouse gases and climate change

ii)   Methodology of climate impact assessment

iii) Co-ordination of activities in climate impact studies

iv)   Monitoring of climate change and climate impacts.

The general level of expenditure of UNEP in the course of the programme was about 0.5 million U.S. dollars per year. In the current year (1990) a number of ongoing activities within UNEP have been brought under the WCIP and in consequence the budget for 1990 is closer to one million dollars. The problems tackled and the results achieved under the WCIP between 1980 and 1989 are reviewed in the following sections of this paper.

2. Methodology of Impact Assessment

Prior to the decade of the World Climate Programme the subject of climatic impact was largely discussed on a purely sectoral basis. This is reflected in the review papers presented to the First World Climate Conference, (WMO, 1979). These were divided according to economic sector (water resources, health, agriculture, land use, forestry, fisheries, marine resources and offshore development) and in the case of agriculture divided according to climatic region (temperate, tropical moist, semi-arid, Far East, Latin America). By 1979, analysis of climatic impacts had reached the stage of distinguishing between primary impacts (both physical and biological) and secondary or indirect impacts, and, in some cases, tertiary impacts. However, the task of studying the full social and economic impacts and of examining the interactions involved had not been tackled. The need to do so was recognised at the First World Climate Conference in the report of the Working Group on the Impacts of Climatic Change and Variability on Society which stated (WMO, 1979, p.744):

> "A full assessment of climatic impact must trace its consequences well into the economic and social fabric of society and examine the whole complexity of linkages and feedbacks in climatic impacts on the biosphere and on human activities. In this connection, analyses of sensitivity of climate-society interactions are among the most important tasks to be undertaken."

The 1979 Working Group also emphasised that the character of climate impact in any given region would depend both on the nature of the climate fluctuation and on the nature of the society.

A number of specific studies on impact methodology were financed under the World Climate Impact Programme as part of the effort to improve and synthetise impact methodology and make existing knowledge in this field more widely available. The first of these specific studies was the project on Climate Impact Assessment of the Scientific Committee on Problems of the Environment of ICSU (SCOPE) which involved the cooperation of over 300 individuals from 36 countries. The final publication (Kates et al., 1985) was an extensive review (625 pages) of the existing state of knowledge in this area grouped into four parts: four overview chapters, five chapters on biophysical impacts, eight chapters on social and economic impacts and five chapters on integrated assessment.

A second project in the area of methodology supported by WCIP was the IIASA project on "Integrated Approaches to Climate Impacts". A feature of this work was that individual reports on geographically widespread marginal areas were based on a common set of impact assessments and the studies were thus compatible with one another. The project resulted in a substantial publication (Parry et al., 1988) which included contributions from 76 authors in 17 countries. Volume 1 (876 pages) dealt with climate impact assessment of marginal areas of agricultural production in cool temperate and in cold regions. Volume 2 (764 pages) dealt with climate impact assessment on marginal agricultural production in semi-arid regions.

A third project in the area of methodology was concerned with the utilisation of satellite data and their incorporation in climate impact studies. The studies on the evaluation of retrospective satellite data and the workshops under phases I and II of the International Satellite Land-surface Climatology Project (ISLSCP) were supported under WCIP. This work has been described in a form suitable for non-experts in a publication sponsored by WCIP (Becker et al., 1989).

At its first meeting in February 1981, the Scientific Advisory Committee (SAC) recommended that UNEP should play an active role in the gathering and diffusion of existing knowledge on climate impact assessment with particular reference to impact assessment methodology. However, this intention was hampered by the scattered nature of the existing information and the necessity to put this information on a firm basis. The publication of the SCOPE volume in 1985 and of the two IIASA volumes in 1988 did not resolve the basic difficulty, since these comprehensive treatments were not suitable for absorption by non-specialists in the area of climate impact assessment. An important development aimed at resolving this difficulty was the preparation of special publications in a form suitable for use by researchers, government officials and policy makers. W. E. Riebsame prepared a guide of 82 pages entitled "Assessing the Social Implications of Climate Fluctuations" (Riebsame, 1988), based largely on the SCOPE publication (Kates et al., 1985). With the encouragement of SAC, this publication was widely disseminated and was the basis of training seminars in India and Nepal in 1988, and in Egypt in 1989. Information on the UNEP/IIASA project on the implications of climate variations in agriculture was also disseminated during these seminars. A shorter and more digestible version of the two volume publication arising from the IIASA study has been prepared (Parry, 1990).

3. Impact of Greenhouse Gases

At its first meeting in February 1981, SAC identified two tasks for the WCIP arising from the carbon dioxide issue. The first was the elaboration and improvement of the international assessment of the socio-economic impacts of climatic changes induced by carbon dioxide. The second was the coordination of national programmes and international efforts in this impact area. In January 1983, SAC agreed to support the SCOPE/IMI project for the development of a scientific basis for assessing the impact of increased atmospheric carbon dioxide on the interaction between climate and the biosphere. The third meeting of the SAC in 1984 was scheduled for Stockholm in order to facilitate a scientific review by the members of SAC of the SCOPE/IMI project on biospheric aspects of the carbon dioxide issue. The report on this project provided the major input to the important Villach Assessment Conference in 1985 and the final version was published in the SCOPE series of publications (Bolin et al., 1986; see also WMO 1986). At its fourth meeting in 1985, SAC recommended that the SCOPE/IMI background report for the Villach Conference should be supplemented by a working paper on the socio-economic aspects of climatic change induced by carbon dioxide. This was produced as part of a programme of IIASA by Clark (1985).

Following the Villach Assessment Conference of 1985, SAC recommended that priority be given under the WCIP to projects aimed at encouraging better regional cooperation between national programmes in order to promote a coordinated response to common problems in the region arising from climate variation and change. Efforts have been made to establish projects in a number of regions, especially among developing countries, but the procedure has proved slow and difficult. The most advanced of these projects on regional assessments is that relating to the socio-economic impact and policy responses resulting from climate change in South East Asia which involves the participation of Indonesia, Malaysia and Thailand. A second regional climate study in Brazil has concentrated on case studies in different parts of that large country of the impact on various economic sectors of the climate factors of particular importance to the region. Other suggestions for regional climate studies on which extensive discussions have taken place are proposed studies in Africa, the USSR and Vietnam.

A number of studies of particular impacts of increased greenhouse gases were undertaken under the WCIP following the Villach Assessment Meeting of 1985. At its fifth meeting in March 1986, SAC recommended the preparation of a state of the art report on the impact of sea-level rise on coastal and estuarine regions with special emphasis on heavily populated areas and regions considered especially vulnerable to sea-level rise. It was suggested that such a study could focus on the socio-

economic implications of rising sea-levels under conditions of climatic change. A publication by the Climate Research Unit of the University of East Anglia on "The Greenhouse Effect on Sea-Level Rise and Tropical Storms" resulted from this project (Warrick and Wigley 1990).

In November 1987, UNEP agreed with the Government of the Netherlands a memorandum of understanding for the execution of a project on Impact of Sea-Level Rise on Society (ISOS) to be carried out by Delft Hydraulics Laboratory. A report on the first phase of ISOS was issued in May, 1989 with the title of "Criteria for assessing vulnerability to sea-level rise: A global inventory to high risk areas." (UNEP, 1989a). Since then a number of site specific studies of sea-level rise have been considered. The continuation of the project has been transferred from the WCIP to the Oceans and Coastal Areas Programme (OCA) of UNEP.

Though the latter programme (OCA) does not operate under the WCIP, a mention of its activities on sea-level rise and related matters is relevant at this point. In 1987, task teams on the Implications of Climatic Changes were established for each of the six regions covered by the Regional Seas Programme of UNEP. Additional task teams were established later for a number of other regions. These teams were asked to prepare regional overviews and case studies on the possible impact of expected climate changes on the coastal and marine ecological systems and also on the socio-economic structures and activities of their respective regions. The first phase of the work of these teams has been completed and reports have been published (e.g. Sestini et al. 1989, Pernetta and Sestini 1989), together with a bibliography on the subject (UNEP 1989b). Work has now commenced on the second objective of the programme which is to assist Governments in the identification and implementation of suitable policy options and response measures which may mitigate the negative consequences of the impact.

A number of activities in the area of policy responses to climate change induced by increased emissions of greenhouse gases have been stimulated and supported under the WCIP. The support of these endeavours has lead to a deeper and more systematic understanding of the socio-economic impact and the actual and potential societal responses. Activities supported included a three day task force meeting at IIASA in 1986 (Chen and Parry, 1987), the workshops organised by the Beijer Institute at Villach in October 1987 and at Bellagio in November 1987 (Jaeger, 1988), and the report prepared by the Beijer Institute for the UNEP General Assembly on "The full range of responses to anticipated climatic change" (Beijer Institute, 1989). UNEP is also supporting the further work, (organised by the Stockholm Environment Institute) of three AGGG working groups—established as a follow-up to the Villach and Bellagio workshops. These three working groups deal with the analysis of limitation strategies, indicators of climatic change, and assessment of response strategies.

4. Climate-sensitive Sectors

Just under half of the expenditure proposed in the original Plan of Action for 1980–83 was for the programme area on the reduction of the vulnerability of the food systems to climate. This was intended to provide a climatic element for the study entitled "Food systems and society" established by the United Nations Research Institute for Social Development (UNRISD) to study food systems and food security under modern conditions particularly at times of crisis. A good deal of information was amassed in this project in relation to the vulnerability of food systems in Eastern India, but there was a serious delay in the production of report material suitable for assimilation by persons interested in applying the techniques to their own regions. A publication in two volumes covering the overall research programme on "Food and Society" will also appear in 1990 (UNRISD 1990).

At its second meeting in January 1983, SAC recommended that an approach be made to the Interagency Group on Agricultural Biometeorology (IGAB) with proposals that socio-economic consequences of climate variability and change be included for study within the forthcoming survey to be carried out in the humid tropics of South America. This proposal was accepted by IGAB whose project commenced in January 1985 in Brazil, Ecuador and Peru with the help of local experts recruited as consultants by UNEP.

In the initial WCIP Plan of Action for 1980–83, fisheries and water resources were identified as strong candidates for investigation under the programme area on the identification of climate-sensitive sectors of human activity. Because of financial constraints, no money became available for projects in these sectors. The WCIP proved somewhat more effective in relation to the problem of drought. It was agreed that Ethiopia could serve as a suitable case study and at its fifth meeting in March 1986, SAC recommended as a pilot project the proposal submitted by the Relief and Rehabilitation Commission of Ethiopia and the National Meteorological Service for building up a national capacity to evaluate drought response and vulnerability in different ecological zones in Ethiopia. The fifth meeting of SAC also recommended support for the International Symposium and Workshop on Drought proposed by the Centre for Agricultural Meteorology and Climatology of the University of Nebraska. As a follow-up to this 1986 Symposium, a management training seminar on drought preparedness was held on Botswana in September, 1989 and further seminars are planned for Asia, South America, and West Africa. At its sixth meeting in April 1988, SAC supported the concept

of the development of a drought early warning system in Ethiopa in cooperation with other UN Agencies who were operating early warning systems for other purposes.

Attempts were made to identify suitable projects in relation to other climate-sensitive economic sectors such as health, energy, fisheries, etc. At its second meeting in January 1983, SAC supported the proposal of WMO to hold a symposium on human biometeorology in cooperation with WHO and UNEP. The objectives of the conference were to bring together meteorologists, physicians, biologists and architects from both the developed and the developing worlds to discuss the principles and mechanisms of the interaction between the atmosphere and the human system, and to discuss and encourage appropriate operational activities by meteorological and health services. The conference was held in Leningrad in September 1986 and the Proceedings published in two volumes (WCAP 1987).

In addition to the workshops dealing with the general aspects of Climate Impact or with individual economic sectors, there have been a number of workshops sponsored by WCIP dealing with specific types of climate variation. Thus in 1985 a workshop was organised and financed by UNEP and by the Environmental and Societal Impacts Group (ESIG) of NCAR dealing with the topic of "Economic and Societal Impacts associated with the 1982–83 Worldwide Climate Anomaly". At this workshop, twelve case studies were presented dealing with the impact of the 1982–83 El Nino-Southern Oscillation phenomena on conditions in Eastern and Southern Africa, West African Sahel, Australia, Brazil, Chile, China, Western Europe, India, Indonesia, Japan, the United States and the USSR (Glantz et al., 1987). Following this Workshop an ad-hoc working group on ENSO phenomena was set up under WCIP. The group met at Bangkok in January, 1988 to discuss weather anomalies associated with ENSO events in 1982 and 1987 and met at Macau in September 1990 to discuss the effect of climate change on ENSO events.

5. Co-ordination of national impact studies

At its first meeting in February 1981, SAC expressed the view that an important part of the World Climate Impact Studies Programme was the task of stimulating, promoting and coordinating Climate Impact Studies by both international and national organisations. The committee considered this task so important that it added such co-ordination as a fifth programme area to the four programme areas in the original Plan of Action. A number of international conferences dealing with the greenhouse gas issue and climate generally were supported by UNEP under the WCIP. These included Villach (1985), Toronto (1988), Hamburg (1988), New Delhi (1989), Tokyo (1989), Noordwijk (1989) and Cairo (1989). Conferences supported in 1990 include Penang (March), Nairobi (May), Sao Paolo (June), Bangkok (September) and the present Second World Climate Conference.

In January 1983, SAC considered a compendium of fifty Climate Impact projects prepared by the WCIP secretariat and recommended that it should be distributed to governments, institutions and relevant individual scientists for comment and for amplification as soon as possible. There was a good response to this circulation of the first version of the inventory and considerable additional material was provided. A second version of the inventory containing this additional information was circulated in 1985. At its sixth meeting in April 1988, SAC recommended that a further attempt be made to update the inventory which then contained information on 181 projects, 59 of which related to economic and social impacts. The response to the resulting questionnaire was very limited and at its meeting in January 1990, SAC recommended that the issuing of a further version of the inventory be postponed pending the further development of an international network of Climate Impact activities.

At its third meeting in February 1985, SAC concerned itself with the relationship between the WCIP and National Climate Impact Activities. In following up the recommendations of that meeting, SAC devoted two days of its fifth meeting in Warsaw in March 1986 to the consideration of National Climate Programmes. Although invitations were extended to a number of national climate programmes, the attendance at the meeting was disappointing. Information was made available to the meeting, either directly by participants representing National Climate Programmes or through members of the SAC, in regard to the position on Climate Impact Studies in Canada, China, Egypt, Ireland, Italy, Japan, New Zealand, Poland and the USA. The secretariat reported that responses to a questionnaire concerning National Climate Programmes indicated that interest in Climate Impact Studies was limited even in those countries that had an active National Climate Programme. In discussing the results of this enquiry, SAC agreed that there was a need for a coordinating committee within each country to provide a focus for climate related activities and research, and also recommended the establishment of an international network of Climate Impact Studies Programmes based on existing active programmes in this area into which other national programmes could be added as their climate impact component developed.

At its fifth meeting in March 1986, SAC recommended that an international network be established on the basis of existing active Climate Impact Programmes. With the support of WCIP, an International Networkshop on Climate-related Impacts was held at the National Centre for Atmospheric Research in Boulder in March 1989 involving 24 participants from twelve countries. Besides presentations on special topics in relation to climate impacts, there were presentations on the National Climate Programme Offices in Canada, Hungary, Italy, Japan and the United States. This international networkshop made a number of interesting recommendations concerning national climate-related impact networks and the relationships between such national networks, national climate programmes, and the international networkshop.

In January 1990, SAC noted with satisfaction various developments under the WCIP in regard to national networking and national newsletters and recommended that UNEP should build on these developments in order to increase its effort at coordinating Climate Impact Studies developed outside the WCIP. It also recommended that international co-ordination should be extended in accordance with the recommendations made at the Boulder International networkshop in 1989 and in particular that UNEP should devote resources towards the establishment of a widely representative international network devoted to Climate-Related Impact Assessment in all its dimensions including multidisciplinary research and the publication of an international newsletter reflecting activities throughout the world.

At its seventh meeting in January 1990, SAC recommended that in place of a further issue of the UNEP World Inventory of Climate Impact Studies that consideration be given to extending the US National bibliography on Climate and Society edited by the National Hazard Research and Applications Centre on behalf of NOOA and to transform it into an annotated bibliography of Research on Climate and Society in all countries.

There is a lack at national level of factually based information on the greenhouse gas climate issue in a form suitable for the general public. In order to remedy this, UNEP is producing and distributing visual and textual materials about this issue and other climate problems. It is a matter of satisfaction that the first publication in the UNEP/GEMS Environment Library Series was concerned with the Greenhouse Gases and the second with the Ozone Layer. The fourth in this series will be on El Nino and the Southern Oscillation (UNEP 1990) and it is understood that a future issue will deal with the impacts of climate variation and change. The UNEP Industry and Environment office plans to produce a publication on Climate Change and Industry. The various written material is being supplemented by slides and tapes, videos, T.V. and radio programmes.

6. Future of Climate Impact Studies

The Second World Climate Conference offers an opportunity to evaluate what has been accomplished under the World Climate Impact Studies Programme over the past ten years and to determine whether its objectives should be modified and how these objectives should be promoted in the coming decade. The situation of the WCIP in relation to the other components of the WCP should be carefully considered. The restriction of the World Climate Research Programme to physically based sciences creates an uncertainty in relation to other valid areas of research in the biological and social sciences of importance in impact studies. There is also the special problem of the blurred distinction between the subject matter of the WCIP and the WCAP with its special activities in the area of food, energy, water, health, urbanisation, etc. The possibility of overlap and of gaps between the two programmes and the nature of the cooperation between them should be thoroughly explored before deciding on structures and procedures for the future.

In the programme area of Climate Impact Methodology there has been a marked change during the past decade. The development of methodology, much of it stimulated and financed under the WCIP, enabled Working Group II of the IPCC to carry out its work with an effectiveness that would have been impossible ten years ago. The lack of basic methodology in 1980 inevitably led to a slower start for the WCIP than for the other components of the WCP. The systematic compilations of methodology sponsored under the WCIP have largely remedied the defect in this regard. A start has been made on the processing of this material in a form suitable for use in training workshops and for a wide distribution to those in many disciplines and organisations concerned with socio-economic impact. In the future, the emphasis should be on an even wider dissemination of such material.

In the programme area of the assessment of the impact of the enhanced greenhouse effect, there was good cooperation with other international organisations on the global impact issue and a good start has been made under the WCIP in regard to the integrated assessment of socio-economic impacts in different regions, particularly among developing countries. As in the case of methodology, a foundation exists for fruitful work in the coming decade in relation to regional integrated assessments and the study of more specific aspects of climate-sensitive sectors and economies. A good deal of attention should be paid to building up regional cooperation. The lack of reliable scenarios of climate change at a regional level is a serious handicap and the effect of this deficiency must be taken into account in planning and implementing regional activities. The actual structures and the mechanisms of cooperation necessary in this connection will depend to a large extent on the form of the World Climate Programme in the future and of any related activities under the Intergovernmental Panel for Climate Change (IPCC) undertaken as a follow up to its first assessment.

In the programme area of climate-sensitive sectors, the degree of achievement under the first decade of the WCIP

was disappointing. A start was made in the area of climate impacts on the vulnerability of food systems. The integration of climate impact studies into existing programmes of international organisations (which was strongly recommended by the First World Climate Conference and in the plan for the WCIP) proved more difficult and time consuming than had been expected. In other sectors (water resources, energy, and health), direct activity proposed under the WCIP was hampered by financial restrictions. Coordination with the WCAP and programmes such as WCP-Water were intermittent and largely ineffective. A distinct improvement in coordination both in programme planning and project implementation would be required in order to eliminate these weaknesses in future activities under the World Climate Programme.

In the programme area of the coordination of National Climate Impact activities and public information, the situation improved greatly after a slow start and seemed to be moving along the right lines at the end of the decade. Using the experience of the first decade of the WCIP, it should be possible through further networking and through the planned encouragement of National Climate Impact Programmes to provide the stimulation for effective activity at the national level coordinated under WCIP. The detailed manner in which this would be done could be discussed at the Second International Networkshop proposed by the WCIP for mid-1991.

A key problem for the future success in the field of Climate-Related Impacts is the level of direct expenditure by the sponsoring agency. At the meeting of experts which drew up the original plan of action in February 1980 and at the first meeting of the SAC in February 1981, the opinion was clearly expressed that an effective programme of World Climate Impacts Studies would require a total expenditure of over 5 million US dollars per year of which just over 1 million dollars per year should be expenditure by UNEP as the sponsoring agency. In the event the expenditure by UNEP only reached a level of 0.25 million dollars in 1983 and 0.5 million dollars in 1984 and subsequent years. This low level of expenditure, and the added disadvantage that the secretariat available to implement the programme was less than appropriate even for this reduced expenditure, led to the WCIP being substantially less effective than it otherwise might have been. A small amount of additional expenditure in the future would allow the secretariat of the WCIP to spend more time on the coordination of national activities and could even result in the long run in a reduced expenditure by the sponsoring agency.

The fact that some of the comments made in the above paragraphs and earlier in the paper are somewhat critical should not be taken as an indication that the achievements of the WCIP have not been worthwhile. The critical comments have been made so that some of the weaknesses of the programme in the past decade can be avoided in the future. Looking back to the state of knowledge and the degree of cooperation in regard to Climate Impact Studies at the time of the first World Climate Conference, all those concerned (the sponsoring agencies and governments, the scientific community and the general public) can all be well satisfied with what has been achieved over the past decade for an expenditure that was extremely moderate in view of the difficulty of the subject area and the global importance of the issues involved.

Publications Related to WCIP Activities

Becker, F., Bolle, H -J., Rowntree, P.R., 1989. The International Satellite Land-surface Climatology Project. ISLSCP Report No.10. UNEP and ISLSCP. 100pp.

Beijer Institute, 1989. The Full Range of Responses to Anticipaated Climate Change. Stockholm, 182pp.

Bolin, B., Döös, B.R., Jager, J., and Warrick, R.A., 1986. The Greenhouse Effect, Climatic Change and Ecosystems. SCOPE, 29, Wiley, Chichester. 541 pages.

Bolle, H. -J. and Rasool, S.I., (ed.), 1985. Development of the Implementation Plan for the International Satellite Land-Surface Climatology Project (ISLSCP)—Phase I. WCP Publication 94. Geneva.

Chen, R.C., and Parry, M.L., 1987. Policy-oriented Impact Assessment of Climate. IIASA Research Report, RR–87–7. Reprinted as Climate Impacts and Public Policy by UNEP and IIASA. March 1988. 54pp.

Clark, W.C., 1985. On the practical implications of the greenhouse Question. WP–85–43. IIASA, Laxenburg.

Glantz, M., Katz, R., and Krenz, M. (Editors): 1987. The Societal Impacts Associated with the 1982–83 Worldwide Climate Anomalies, UNEP and NCAR, 105pp.

Jaeger, J., 1988. Developing Policies for Responding to Climate Change, WCIP–1, WMO/TD No. 225, Geneva. 53pp.

Kates, R.W., Ausubel, J.H., and Berberian, M., 1985. Climate Impact Assessment, SCOPE 27, Wiley, Chichester, 625 pp.

Parry, M.L., Carter, T.R., and Konijn, N.T., (editors), 1988. The Impact of Climatic Variations on Agriculture, Kluwer, Dordrecht, Vol. 1, 876pp. Vol. 2, 764pp.

Parry, M., 1990. Climate Change and World Agriculture. Earthscan. London. 150pp.

Pernetta, J., and Sestini, G., 1989. The Maldives and the impact of expected climatic changes. UNEP Regional Seas Reports and Studies 104. Nairobi. 84 pages.

Riebsame, W.E., 1988. Assessing the Social Implications of Climate Change, Natural Hazards Center, University of Colorado, 82pp.

Sestini, G., Jeftic, L., and Milliman, J.D., 1989. Implications of expected climate changes in the Mediterranean region: an overview. UNEP Regional Seas Reports and Studies No. 103. Nairobi. 46 pages.

UNEP, 1983. Inventory of Climate Impact Projects. WCIP, Nairobi. revised and enlarged. 1985 and 1988. 117pp.

UNEP, 1987a, The Greenhouse Gases, UNEP/GEMS Environment Library No. 1, Nairobi, 40pp.

UNEP, 1987b, The Ozone Layer, UNEP/GEMS Environment Library No. 2, Nairobi, 36pp.

UNEP, 1989(a), Criteria for Assessing Vulnerability to Sea Level Rise, UNEP and Delft Hydraulics Laboratory, 54pp.

UNEP, 1989 (b), Bibliography on effects of climatic change and related topics. UNEP Regional Seas Directories and Bibliographies. Nairobi.

UNEP/ESIG, 1989. Report on International Workshop on Climate-Related Impacts. United Nations Environmental Programme and Environmental and Societal Impact Group of NCAR, March 1989.

UNEP, 1990. El Nino and Southern Oscillation, GEMS Environment Library. (In preparation)

UNRISD, 1990, Food and Society.

Wallen, C.C., 1984. Present Century Climate Fluctuations in the Northern Hemisphere and Examples of their Impact. WCP Publication 87. Geneva. 91 pages.

Warrick, R.A., and Wigley, T..L.M., (editors) 1990, Climate and Sea Level Change: observations, projections, implications. Cambridge University Press (in press).

WCAP, 1987, Climate and Human Health, Proceedings of Leningrad International Symposium, September 1986. WCAP Publications No. 1 (274 pages) and No. 2 (203 pp).

WMO, 1979, Proceedings of the World Climate Conference, 13–23 February, 1979, WMO Publication No. 537, Geneva, 791pp.

WMO, 1980. Outline Plan and Basis for the World Climate Programme 1980–1983. WMO Publication No. 540, Geneva, 64pp.

WMO, 1984. Report of the Study Conference on Sensitivity of Ecosystems and Society to Climate Change. Villach, September 1983. WCP Publication 83. Geneva. 34 pages.

WMO, 1986, Report of the International Conference on the Assessment of the Role of Carbon Dioxide and of other Greenhouse Gases in Climate Variations and Associated Impacts, WMO Publication No. 661. 78pp.

Global Climate, Energy and Water Cycle

by Gordon A. McBean *(Canada)*

Abstract

Interactions of energy and water are main driving factors in determining the earth's climate. Climate variability is determined by the interactions of components of the climate system. The earth's climate is a highly nonlinear system with at least three major feedback processes: water vapour feedback, snow-ice albedo feedback and cloud feedback. In order to adequately predict the response of the climate system to external forcing we must improve our knowledge of the feedback processes. We also must be able to reduce the uncertainties of predictions of the precipitation which greatly exceed those for temperature.

The Global Energy and Water Cycle Experiment (GEWEX) and Programme of the World Climate Research Programme is now underway to directly address these issues. The GEWEX Programme will include a series of process studies of land surfaces and hydrology and of clouds and radiation, global data collection projects for clouds, radiation and precipitation and numerical modelling, leading to a global observing experiment in the late 1990s. Only through major commitments of nations will it be possible to make major reductions in these uncertainties about climate change.

1. Introduction

To even the most casual observer, it is apparent that water, in its many forms, is an important element for the climate of our planet. Complex interactions between energy and water provide richness to the variations of our climate and, at the same time, make complete understanding and prediction of the climate system a far distant goal. In addition to its role in climate variations, water is a critical component of the whole earth ecosystem; it provides the basis for life and its relative abundance as freshwater is a key to many of our social and economic activities. In this paper, some of these complexities will be discussed and the meagre level of our understanding explored. In recognition of the theme of the Conference on the changing climate due to increasing concentrations of greenhouse gases, the presentation will emphasize the role of water-energy interactions in amplifying climate changes, initiated by greenhouse gases. In concluding, the Global Energy and Water Cycle Experiment (GEWEX) and Programme will be described as the vehicle to enhance our understanding and lead to the possibility of reliable climate prediction on a regional basis.

2. Background

Although the Sun is the source of energy for the climate system, most of the energy for atmospheric motions comes indirectly, through radiative, sensible and latent heat transfers from the earth's surface. As we examine Figure 1, we note continuously the role of water in the radiative budget of the planet. The average irradiance at the top of the atmosphere is 342 Wm^{-2} (indicated as 100 units; all numbers are approximate; Ramanathan et al., 1989). The atmosphere absorbs 68 Wm^{-2} of solar energy and receives 90 Wm^{-2} due to evaporation and subsequent condensation to precipitation. Net long-wave emission from the earth's surface is the result of 390 Wm^{-2} of upward irradiance minus 327 Wm^{-2} of downward irradiance from the atmosphere. Most of the earth's losses are to the atmosphere which in turn emits long-wave radiation to space to balance its budget. The atmosphere loses about 106 Wm^{-2} of radiative energy which is balanced by latent and sensible heat transfer from the earth's surface. Clearly, water is very important in the global energy budget. Clouds play a major role in controlling heat input to the earth's surface as well as the radiative heat loss to space. Low clouds generally cool the planet by reflecting more solar radiation back to space, although they also trap radiation below them. High clouds generally warm the planet since cirrus clouds reflect little solar radiation while trapping earth's radiation. Although the balance is very delicate, our best measurements from space now indicate that the present distribution of cloudiness cools our planet (Ramanathan et al., 1989). Water vapour, as a gaseous component of a cloud-free atmosphere, also plays a major role. There are

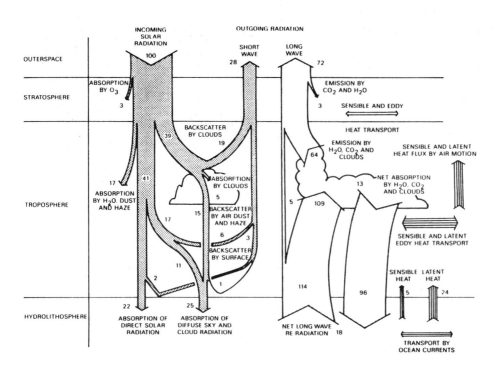

Figure 1: Schematic diagram of the earth's radiation budget. (from EOS Science Steering Committee Report, Volume II)

very large exchanges of long-wave radiation between the atmosphere and the surface, but the net flux is smaller than the latent heat flux. Almost half of the net energy loss from the earth's surface is by evaporation of water. The heat engine of the earth is a "wet engine", more like a refrigerator or a heat pump, rather than a "dry engine", that is characteristic of the other planets.

One of the most important aspects of our climate system is the greenhouse effect. Although shortwave solar radiation passes through a cloud-free atmosphere with relatively little attenuation, long-wave radiation from the earth's surface is partially absorbed in the atmosphere and then re-radiated to space at a lower temperature. The result is that the lower atmosphere and earth's surface are warmer than they would otherwise be. This is the natural greenhouse effect and is due to the presence in the atmosphere of water vapour (the most important greenhouse gas; about 1% of the atmosphere by volume), carbon dioxide (about 0.04%), methane and nitrous oxide. Without it, the average temperature of the earth's surface would be -18°C; instead of the observed 15°C, an increase of 33°C. Note that the increase takes the average temperature from well below freezing to well above; this is critical for the formation and maintenance of life. It is interesting to compare the earth with its nearest neighbouring planets. The atmosphere of Venus is dominated by carbon dioxide and has a very large greenhouse effect. Venus's surface temperature would be -46°C in the absence of the greenhouse effect; it is

observed to be 477°C; a greenhouse effect of 523°C. Mars, on the other hand, has very little greenhouse effect because of its very thin atmosphere. The observed surface temperature is only raised to -47°C, a 10°C greenhouse effect. Further, if we go back to the early formative years of planet earth, we see that the increase in atmospheric water vapour, with outgassing from the planet, resulted in the condensation to liquid water, rain-out and a quasi-equilibrium (Walker, 1977). Because Venus was warmer, saturation was not achieved and there was a "runaway greenhouse effect" resulting in a very hot climate. For Mars, the saturation occurred at much lower atmospheric concentrations and the phase change was directly from gas to ice. The characteristics of water have made earth a distinct planet.

The global water budget is simply the balance of global precipitation and global evaporation (both about 16×10^6 m^3 s^{-1} or a little more than 1 m of water per square metre per year averaged over the globe). 1 m of evaporation per year corresponds to 79 Wm^{-2}. The distribution of precipitation is the controlling influence on vegetation, on the extent of snow and ice and on freshwater input to oceans. River runoff plays a major role in the terrestrial water budget, is a significant contributor to the total fresh water input of oceans, and dominates the Arctic Ocean. River runoff to the Arctic basin is estimated to be about 3 times larger than the precipitation and must be largely balanced by ice export.

The salinity budget of oceans is closely linked to the fresh water budget as the total salt content of a basin is determined by the balance of fresh water input (sum of precipitation plus runoff minus evaporation) and the net influx of salt across oceanic boundaries of the basin. Oceanic precipitation is difficult to estimate due to sparsity of observations and biases in ship and island station observations and is generally the major source of uncertainty in freshwater budgets. For example, North Pacific precipitation estimates from three published sources disagree more than evaporation estimates and preclude determining whether the North Pacific is gaining or losing freshwater (McBean, 1989).

While existing observing systems and atmospheric models are very useful for short range (a few days) weather prediction, they are rather primitive in their ability to observe and predict water and its phase changes and transport in the atmosphere. As noted above, quantitative estimates of rainfall over the ocean have large uncertainties. The vertical redistribution of latent heating is extremely important, but neither observations nor methods of properly including it in models are available. Heat exchange, both sensible and latent, between the ocean and the atmosphere is known only in a gross climatological sense. The situation over land is even worse because of the heterogeneity of the surface and the changing coverage of vegetation with time—seasonally and otherwise. It is possible that the hydrologic transport aspects of the climate system may be sensitive to relatively small readjustments in the dynamics of the meridional cells. The total meridional heat transport is determined by the requirement to balance radiative exchanges at the top of the atmosphere. However, if the atmosphere has the option of moving heat either as sensible heat or in the form of latent heat, it is under less constraint. Further, since the water can be in the form of clouds and these will feed back upon the radiation balance, a multitude of interactions and non-linear feedbacks are possible. Hence, it is clear that precise understanding of climate variability requires understanding of the hydrologic cycle.

3. Climate Variability

The climate of the world is clearly not a static system; there is a wide range of time scales (Figure 2). The atmosphere contains only a small amount of water, equivalent to about one week's precipitation over the globe. Residence times for whole ocean basins and major ice sheets (Antarctica and Greenland) are up to thousands of years. The response time of the upper ocean is tens of years. The residence times for soils and lakes and rivers will be very dependent on depth and local conditions. Residence times for components of the water cycle vary considerably: (all very approximate) rivers, 2 weeks; lakes, 10 years; soil

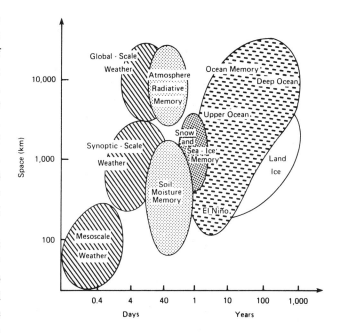

Figure 2: Space-time scales of different climatic processes (adapted from Dickinson, 1986)

moisture, 2–50 weeks; biological waters, a few weeks; ground waters, up to 10^4 years. The amount of freshwater is only 2.6% of the total and over 89% of that is tied up in ice caps, glaciers and icebergs or more than 800 m deep in the ground. The portion of the freshwater that is readily useful to humans and the land ecosystem is a minute fraction of the total and generally has the shorter residence times (days to few years). Thus, the important cycle of freshwater is a dynamic system and constantly evolving.

We are primarily interested in the possibility of climate change over the next one-to-two hundred years. For this period, the oceans provide the dominating factor for longer-term adjustment of the climate system. It is useful to think of the climate system as being divided into two components: the fast and slow climate systems (WCRP, 1990a). The fast climate system is the atmosphere, the upper ocean (that part subject to an annual cycle of vertical mixing due to wind) and the transient processes at land surfaces. The fast climate system is active, driven by the atmospheric engine, and comes to statistical equilibrium in a few years. The slow climate system consists mainly of the deeper ocean and perennial land ice, with a response time of decades to centuries. The major interactions between the fast and slow climate systems take place in a limited number of areas where heat is transferred by up- or downwelling of ocean water and at high latitudes where cold, dense water sinks to great depths. To a first approximation, the magnitude of climate response to external forcing, such as increasing greenhouse gas concentrations or changes in solar irradiance at the top of the atmosphere, will be determined by the fast climate system; the decadal rate of change will be determined by the slow climate system.

The past history of the earth's climate provides some clues as to possible changes in the future. Over the past two million years, glacial and interglacial periods have alternated at intervals of about 100,000 years. Global surface temperatures typically varied by at least 5°C through the Pleistocene ice ages, with middle-to-high latitude variations as great as 10–15°C. At the time of the last glacial maximum, about 18,000 years before present (BP), ice sheets covered Canada and large areas of northern Europe and Asia. Sea level was 120 m lower than present. The glaciation ended between 10,000 and 15,000 BP. About 10,500 BP, there was an abrupt reversal of the warming trend, the Younger Dryas cold episode, that lasted for about 500 years and ended suddenly. It is generally believed that an important cause of the glaciations was variations in seasonal radiation received at the top of the atmosphere due to Milankovitch orbital variations. Actual changes in solar radiation were only about 0.2%. Such a small change in the solar radiation flux must have been considerably amplified by the earth's climate system in order to induce transitions from ice age to interglacial periods.

During the period 5000–6000 BP, the world-wide summer temperatures were likely higher than present. From the late tenth to early thirteenth centuries, temperatures were exceptionally warm around the North Atlantic, but not in China. A pronounced global climatic feature was the Little Ice Age between 150 and 450 years ago. There is no agreed explanation for the Little Ice Age and the warming that has occurred since 1850 may be, in part, recovery from this cool period, due to natural reasons. It is important to recognize that natural climatic variations may mask anthropogenically-induced changes.

4.   Climate System Feedbacks

The earth's climate is a highly nonlinear system with a multitude of interacting components and its response to external or internal forcing is not always clear. The existence of many feedbacks, both positive and negative, makes it difficult to understand and model the climate's response to changing concentrations of greenhouse gases. An instantaneous doubling of CO_2 results in a 4 Wm^{-2} increase in the radiative forcing on the troposphere-surface system. Effectively the climate system absorbs 4 Wm^{-2} additional energy and warms up in order to increase outgoing radiation and re-establish radiative equilibrium. If no feedback or other complicating processes were involved, the global mean surface temperature would warm by about 1.2°C. However, there are at least three important feedback processes that change the climate system's response: water vapour feedback; snow-ice albedo feedback; and cloud

feedback. Note that each involves a component of the water cycle interacting with energy fluxes.

Water vapour feedback is the most straightforward. If the temperature increases due to extra radiative heating caused by increased CO_2 concentration, air can hold more water vapour and the warmer surface will evaporate more. Since water vapour is a greenhouse gas, increasing the concentration of one greenhouse gas (CO_2) causes an increase in another (H_2O), resulting in a positive or amplifying feedback mechanism. Computations show that the direct effect is to increase the warming from 1.2°C to 1.7°C (Cess et al., 1989; Cess, 1989). The increased water vapour will also absorb more solar radiation, resulting in a further warming to 1.9°C, an amplification factor of 1.6. A second positive feedback mechanism is snow-ice albedo feedback. A warmer earth with less snow will absorb more solar radiation. This process likely contributes to the poleward enhancement of warming seen in global climate models. The actual magnitude of the amplification is difficult to compute in isolation.

Feedback mechanisms related to clouds are extremely complex and depend upon cloud amount, cloud altitude, and cloud water content. Reduced cloud amount decreases the greenhouse effect due to clouds and acts as a negative feedback. However, solar radiation reaching the surface will increase, a positive feedback. If clouds move to higher altitude, they have a stronger greenhouse effect, a positive feedback. Changing the cloud water content and whether it is solid or liquid lead to very complicated interactions, with some models indicating positive and others, negative feedbacks. At present, it is not possible to compute the strengths of these feedbacks with any confidence.

5.   Prediction of Climate Change

Since these feedbacks all interact, the most appropriate way to understand the climate system response is through use of very sophisticated global climate models. Although simplified models, such as one or two-dimensional energy balance or radiative-convective models, have their purposes, it is only fully three-dimensional coupled models, with the atmosphere, oceans and ice all represented, that can represent the feedback processes and determine the climate system's sensitivity to changes in greenhouse gas concentrations.

At the present time, global climate models are all deficient in some aspects. Most have very simplified treatment of the water cycle, particularly clouds and radiation, precipitation and land-surface hydrology. Existing ocean climate models are still of very coarse resolution and have highly-parameterized mixing. The approach adopted by most global climate modellers has been to represent the ocean as a mixed layer of typically 50m depth that exchanges heat and mass with the atmosphere but has no interacting currents nor deep circulations. In essence, most global climate models presently being used are of the fast climate system. Thus, they can simulate the climate in equilibrium with some fixed forcing, such as doubled atmospheric concentration of greenhouse gases, but cannot simulate the evolution of the climate as a time-dependent response to varying greenhouse gas concentrations.

Before examining results of global climate change models, it is appropriate to consider a simplified conceptual model (Figure 3). Depending on economic and pollution control policies of governments, emissions of greenhouse gases to the atmosphere will likely continue to increase. Some emissions will be absorbed by the oceans or the terrestrial biosphere. Presently about 50% of the

anthropogenic emissions of CO_2 are absorbed, presumably in the ocean. The Joint Global Ocean Fluxes Study (JGOFS) and the International Global Atmospheric Chemistry (IGAC) Programme and other projects will make contributions towards understanding these biogeochemical cycles. Given an atmospheric concentration of a greenhouse gas, it is possible to compute with reasonable confidence the additional radiative forcing at the top of the troposphere. As discussed above, this is 4 Wm^{-2} for doubled carbon dioxide concentration. However, the total radiative forcing will depend on the climate system feedbacks or amplifiers. Here, we have a large uncertainty. The time-dependent response of the climate system to this additional radiative forcing will depend on the response of the slowest component, i.e., the oceans.

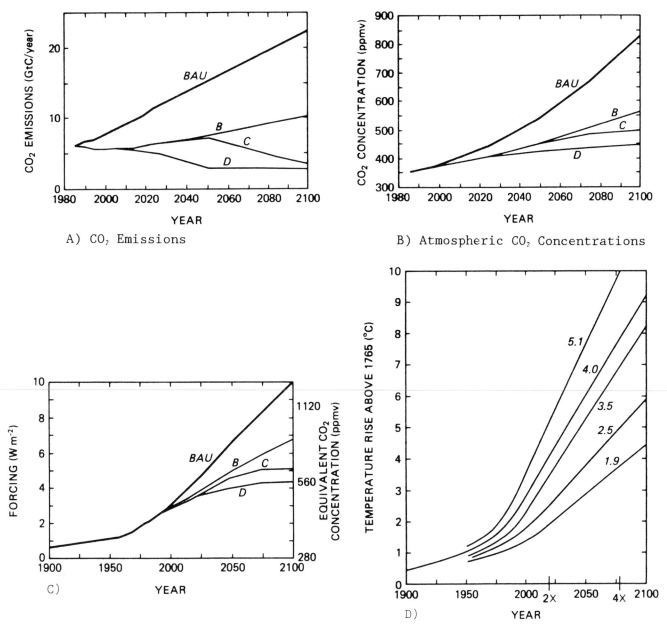

A) CO_2 Emissions

B) Atmospheric CO_2 Concentrations

C) YEAR

D) YEAR

Figure 3: Schematic diagram of the factors determining the change in climate due to increasing greenhouse concentrations. a) CO_2 emissions; b) Atmospheric CO_2 concentrations; c) Radiative forcing (all adapted from IPCC, 1990; and d) Temperature change of fast climate system (no deep ocean) for various amplificiations.

This will result in a delay and possible reduction in climate change.

The fast climate system essentially determines the amplifiers or feedbacks while the slow climate system determines the response time. Since global climate models with a mixed-layer ocean are models of the fast climate system, they provide the best means of determining the net strength of climate system feedbacks.

The IPCC WG1 Report (IPCC, 1990) has summarized the results from global climate models with mixed layers for doubled carbon dioxide equilibrium experiments. The changes in the global mean surface temperature ranged from 1.9 to 5.2°C. Since the additional radiative forcing of 4 Wm^{-2} would cause a surface temperature change of 1.2°C, these results indicate feedback multipliers of 1.6 to 4.3. It should be noted that the extremes of this range, 1.9 and 5.2°C change, are both due to the same model (the United Kingdom Meteorological Office model) with different parameterizations of cloud and radiative properties. Three higher resolution models gave global temperature changes of 3.5, 4.0 and 3.5°C. The IPCC WG1 Report concluded "near the earth's surface, the global warming lies between +1.5°C and +4.5°C, with a "best guess" of 2.5°C". These results demonstrate the uncertainty in global warming due to doubled atmospheric carbon dioxide concentrations. The geographical distributions of the temperature changes, with respect to the global mean, also vary significantly from model to model (IPCC, 1990). The dominant factor in the uncertainty is water-related feedback processes.

6. Changes to water cycle

Global climate models also show that the precipitation will change, between 3 and 15% in terms of global mean. Regional variations are large. The socio-economic impacts of changes in precipitation may well be greater than impacts of a warmer climate. In many respects, the most important weather element in terms of its impact on human activities is precipitation. Excess rain and snow can create floods and avalanches while droughts result from the excess of evaporation and runoff over precipitation. Improvements in our ability to predict the occurrence and intensity of precipitation, over periods of days to decades, will have immediate and significant economic and social benefits.

The most important parameter of practical interest is the difference between precipitation and evaporation, referred to as available water. In many places, the available water is a small difference between two much larger numbers. This is also represented by the runoff ratio (= runoff/ precipitation). For wet climates the ratio may be about 0.5, but for drier climates the ratio is near zero. Suppose we have a climate where the mean annual precipitation is 500

mm and the mean annual evaporation is 450 mm; the available water for runoff and replenishing soil moisture will be 50 mm, with a runoff ratio of 0.1. With climate change, the precipitation may increase by, for example, 5% (to 525 mm) and the evaporation by 10% (to 495 mm). The result is a decrease in available water to 30 mm, a change of 40%. Changes to available water, in regions where there are already water supply problems, may be catastrophic.

Manabe and Wetherald (1987), Meehl and Washington (1988) and others have examined the large-scale changes in soil moisture accompanying climate change. Due to the simplicity of the treatment of the hydrological cycle in such models the results can only be considered illustrative of what may happen. For doubled CO_2, with a variable clouds version of their model, Manabe and Wetherald found that soil moisture in the Great Plains of North America for July and August was about half of the values for the control case. The decrease in soil moisture was due to a combination of factors. Although winter rainfall increased, spring and summer rainfall decreased due to changes in atmospheric circulation. Snowmelt, which is a major source of spring moisture in the control case, occurred earlier with the warmer climate and resulted in more spring runoff. Evaporation increased through the winter and spring due to higher temperatures and increased absorbed solar radiation. Changes in snow storage/runoff relationships will be very important for many regions. Over most of the Canadian prairies, river flow, used for irrigation, domestic and industrial consumption and power generation, results primarily from the melting of winter snow accumulations on the Rocky Mountains to the west. Over the prairies themselves, local precipitation and evaporation are in near equilibrium (Hare and Thomas, 1979). It is essential that research be carried out to improve our understanding and prediction of these critical aspects of water budgets.

7. The Global Energy and Water Cycle Experiment (GEWEX) and Programme

In view of the uncertainties and the potentially devastating impacts of climate change and its effects on the water cycle, it is essential that further studies be undertaken. The studies need to cover a range of activities aimed at: improving our understanding of processes; enhancing our ability to model the global climate system, in particular, the important interactions of energy and water; and providing global data fields for model development and validation. The Global Energy and Water Cycle Experiment (GEWEX) and Programme (WCRP, 1988) is aimed at the heart of these climate change and variability questions. It is clear that the main controls on climate change due to increased greenhouse gas radiative forcing are the diabatic feedback processes that amplify or reduce the direct greenhouse forcing. The GEWEX Programme is a component of the World Climate Research Programme (WCRP, 1984) and refers to an ongoing set of activities leading up to and including a 5-year duration global experiment, to start in the late 1990s, and continuing research that will follow thereafter.

Scientific objectives of GEWEX are:

i) to determine the hydrological cycle and energy fluxes by means of global measurements of observable atmospheric and surface properties

ii) to model the global hydrological cycle and its impacts on the atmosphere and ocean

iii) to develop the ability to predict the variations of global and regional hydrological processes and water resources, and their response to environmental change

These scientific objectives are clearly aimed at understanding and predicting the energetics of the atmospheric heat engine, which drives the whole earth system. Improved knowledge of clouds and radiation, precipitation and evaporation, including their regional variations and trends, is essential to reduce the uncertainty in predicting climate change by increased greenhouse warming. The GEWEX Programme is the core component of the WCRP aimed at understanding the fast climate system.

GEWEX has a fourth objective, of a more technical nature, namely:

iv) to foster the development of observing techniques and data management and assimilation systems, suitable for operational applications to long-range weather forecasts, hydrology and climate predictions.

Plans and activities for the GEWEX Programme are being developed by the Scientific Steering Group for GEWEX, under the overall guidance of the Joint Scientific Committee for the World Climate Research Programme. A complementary programme, under the auspices of the International Geosphere Biosphere Programme, is the Biospheric Aspects of the Hydrological Cycle. Close collaboration between the two projects has been developed. This will include joint field studies. GEWEX will also cooperate with other projects of the WCRP and the IGBP.

8. Elements of the GEWEX Programme

The GEWEX Programme is developing into a broad set of programme elements addressing key problems of our observing, understanding and modelling of the cycles of water and energy and their complex interactions, within the climate system. Some of these activities are now underway, some predated GEWEX, and others are being developed.

Process Studies

There are two main areas of emphasis: cloud and radiation interactions; and land-surface processes. An example of a cloud-radiation study is the First ISCCP Regional Experiment (FIRE) which was conducted off the coast of California to investigate the extensive stratocumulus cloud layers over the oceans. Other studies have investigated cirrus clouds and clouds over the Northwest Pacific Ocean. Within the Tropical Ocean Global Atmosphere (TOGA) Programme (WCRP, 1985, 1990b), there will be the Coupled Ocean Atmosphere Response Experiment (COARE) which will investigate the processes in the important cloud-radiation linkages in tropical regions. GEWEX will join with other projects to investigate other cloud systems. Initial planning has been made for a multi-region cloud study.

Present techniques to estimate or measure areal evaporation are quite weak and there is a major problem with scale integration. The GEWEX strategy is to combine intensive field experiments, mesoscale simulations and parameterizations in climate models. In 1986, the first Hydrological-Atmospheric Pilot Experiment (HAPEX) was conducted in southwestern France to investigate atmospheric-land surface interactions on a 100 km scale (Andre et al, 1986). The International Satellite Land-Surface Climatology Project (ISLSCP) was organized to devise algorithms to relate land-surface characteristics and satellite observations, to develop techniques of estimating evaporation, radiation, etc., from space and to create global data bases of land-surface characteristics. A variety of small scale experiments were conducted, leading to the First ISLSCP Field Experiment (FIFE) in 1987/9 in Kansas. Experiments in the Soviet Union are underway and future experiments are planned for China, Niger, the boreal forest of North America and Spain.

GEWEX Modelling

The GEWEX Programme will lead to improvements in the representation of the clouds, radiation and other diabatic processes and land-surface hydrology in climate models. An intercomparison of radiation codes in climate models; sensitivity studies on cloud-radiation parameterizations in climate models; and development of algorithms for determining the surface and top of the atmosphere radiation budgets are examples of projects now underway. A project to improve atmospheric boundary-layer parameterizations for climate models has begun by comparing existing schemes with observations and with higher-resolution models. This project will be extended to include surface and sub-surface parameterizations.

The aim of the GEWEX Continental-Scale Project is to stimulate the development of continental-scale models, dealing with large river basins, that are driven by precipitation and evaporation data derived from operational analyses of meteorological fields and intensive ground- and space-based measurements. The GEWEX Continental-Scale Project is expected to be conducted during the mid-1990s in the Mississippi River basin.

With the emphasis on the water cycle and the advent of new data sources, it is important that new multivariate four-dimensional data assimilation techniques be developed. Data retrieval techniques are needed for optimum information extraction from satellite radiance measurements in all wavelength ranges. Rainfall and non-conventional surface observations should be more fully incorporated into data assimilation systems.

Global Climatological Data Projects

The initial projects are aimed at producing from existing space and ground-based observations climatological quantities that would not otherwise be available and are seen as interim solutions, pending the availability of more powerful or more complete observing schemes that we expect to be part of the earth observing system of the late 1990s.

The International Satellite Cloud Climatology Project (ISCCP) (Schiffer and Rossow, 1985) has been preparing monthly cloud climatologies using a combination of geostationary and polar orbiting satellite data since 1983. These data provide the first globally consistent data set on cloud coverage that can be used for climate studies. The global climate modelling community is now undertaking a systematic intercomparison with data from ISCCP. Using the radiance data accumulated in the ISCCP, a project is now beginning to produce global radiation budgets for the top of the atmosphere and the earth's surface.

The Global Precipitation Climatology Project (GPCP) has recently begun to provide global estimates of precipitation. The Project will assemble all land precipitation data, satellite estimates of precipitation over the oceans and other data-sparse regions, and weather radar data into a global precipitation data set. The Project works with the Global Runoff Data Project to produce global data sets of these two critical elements of the hydrological cycle.

9. Global Earth Observing System for GEWEX

The Global Earth Observing System for GEWEX will be a combination of traditional observations, from the World Weather Watch (WWW) and the Integrated Global Ocean Surveillance System (IGOSS), future meteorological satellites and special experimental satellite missions, and special ground and oceanic measurements. Much of the data must be processed in real time using advanced multivariate four dimensional data assimilation schemes to produce optimum global data sets, including derived products.

The ideal geostationary component for supporting GEWEX would include five essentially identical satellites, each equipped with:

*   a multi-spectral (visible and infrared), high-resolution imaging radiometer to observe cloud developments and cloud motions within several atmospheric layers
*   a medium resolution, infrared/microwave spectrometer or atmospheric sounder to observe the time-dependent, three-dimensional distribution of temperature and moisture

This geostationary component is essential for determining the basic atmospheric variables as well as providing reasonably frequent observations of rapidly variable atmospheric processes.

Orbiting satellites can be in two principal configurations: namely, low-inclination orbits and polar orbits. Global coverage must be strived for but it is recognized that certain observations are more critical in some latitude bands and others may be more effectively made at the same or different latitudes.

The following observational systems are important for GEWEX:

i)   Atmospheric temperature and moisture sounders with improved, higher accuracy, better vertical resolution sounders of both temperature and moisture. These sounders should use high-spectral resolution sensors for a range of infrared and microwave channels

ii) Tropospheric wind measurement from doppler-lidar profilers

iii) Precipitation measurement system including: multiple-frequency passive microwave radiometers; an improved AVHRR instrument to provide cloud patterns and cloud-top temperatures; and a one or preferably two frequency rain radar with suitable beamwidth and scanning capability to cover a 200 km swath along the satellite track and to resolve the vertical distribution of rain

iv) Clouds and radiation: advanced visible and infrared imaging radiometers for cloud patterns and cloud-top temperatures; scanning radiometers similar to ERBE instruments; and, preferably, a backscatter lidar to determine the three-dimensional distribution of cloud tops, semi-transparent clouds and aerosols

v)   Land surface, vegetation and hydrology using a combination of remote sensing and in situ observations.

10. Summary

The principal uncertainties in our understanding and constraints on our ability to predict climate evolution relate to the water cycle and its interactions with the radiative fluxes. These factors determine the response of the fast climate system to changes in radiative forcing due to increasing greenhouse gas concentrations. The Global Energy and Water Cycle Experiment and Programme is specifically aimed at resolving these uncertainties.

References

Andre, J.C., J.P. Goutorbe and A. Perrier, 1986: HAPEX-MOBILHY: a hydrologic atmospheric experiment for the study of water budget and evaporation flux at the climatic scale. Bull. Amer. Meteor. Soc. 67, 138–144.

Cess, R.D., G.L. Potter, J.P. Blanchet, G.J. Boer, S.J. Ghan, J.T. Kiehl, H. Le Treut, Z.X. Li, X.Z. Liang, J.F.B. Mitchell, J.-J. Morcrette, D.A. Randall, M.R. Riches, E. Roeckner, U. Schlese, A. Slingo, K.E. Taylor, W.M. Washington, R.T. Wetherald and I. Yagai, 1989: Interpretation of cloud climate feedback as produced by 14 atmospheric general circulation models. Science 245, 513–516.

Cess, R.D., 1989: Gauging water vapour feedback. Nature 342, 736–737.

Dickinson, R.E., 1986: Impact of human activities on climate—a framework, Sustainable Development of the Biosphere, W. Clark and R. Munn (eds.), 252–289.

EOS Science Steering Committee Report, 1988: From Pattern to Process: The Strategy of the Earth Observing System, NASA, Washington, Vol. II, 140 pp.

Hare, F.K., and M.K. Thomas, 1979: Climate Canada, Wiley, Toronto, 223 pp.

IPCC, 1990: Scientific Assessment of Climate Change. WMO-UNEP, Cambridge Univ. Press.

Manabe, S., and R.T. Wetherald, 1987: Large-scale changes of soil wetness induced by an increase in atmospheric carbon dioxide. J. Atmos. Sci. 44, 1211–1235.

Meehl, G.A., and W.M. Washington, 1988: A comparison of soil moisture sensitivity in two global climate models. J. Atmos. Sci. 45, 1476–1492.

McBean, G.A., 1989: Global energy and water fluxes, Weather 44, 285–291.

Ramanathan, V., B.R. Barkstrom and E.F. Harrison, 1989: Climate and the earth's radiation budget, Physics Today 42, 22–34.

Schiffer, R.A., and W.B. Rossow, 1985: ISCCP global radiance data set, a new resource for climate research. Bull. Amer. Meteor. Soc. 66, 1498–1505.

Walker, J.C.G., 1977: Evolution of the Atmosphere, MacMillan, New York, 318 pp.

WCRP, 1984: Scientific Plan for the World Climate Research Programme. WCRP No. 2, WMO/TD 6, Geneva.

WCRP, 1985: Scientific Plan for the Tropical Ocean and Global Atmosphere Programme. WCRP No. 3, WMO/TD 64, Geneva.

WCRP, 1988: Concept of the Global Energy and Water Cycle Experiment. WCRP-5, WMO/TD 215, Geneva.

WCRP 1990a: Global Change. WMO-ICSU, Geneva, 35pp.

WCRP, 1990b: Scientific Plan for the Tropical Ocean and Global Atmosphere Programme. WCRP No. 3, Addendum, Geneva.

Remote Sensing and Global Climate Change: Water Cycle and Energy Budget

by Hartmut Grassl *(Germany)*

Introduction

Water is the dominant substance for climate because it strongly interacts with the energy budget of the Earth. It covers on average 80 percent of our globe either as liquid (up to 65 percent) or solid (up to 20 percent); its vapour is by far the dominant absorbing gas, responsible for more than two thirds of the greenhouse effect of the atmosphere; it contributes most to the shortwave albedo of the Earth by strong reflection from snow surfaces and by backsacttering from water and ice clouds; it transports four fifths of the energy needed by the radiatively cooled atmosphere by evaporation at the surface and subsequent condensation in the troposphere; its partitioning into liquid and solid determines global mean sea level; it supports life on Earth.

The present inability to detect or to foresee quantitatively anthropogenic climate change is mainly a consequence of the rather poor understanding of parts of the global water and energy cycle. Since we need global and continuous observations of all water and energy cycle components, remote sensing from space has to play a major role. It has already given many new insights into phenomena and processes like the global net cooling by clouds and the cell structure of the convective planetary boundary layer, however, often still only qualitatively and most of the data gathered are not evaluated. All global climate models have to be validated by global data sets coming increasingly from remote sensing from space.

The next section summarizes satellite remote sensing achievements with respect to water cycle components and net radiation budget at the top of the atmosphere and also points to the main gaps. A logical approach seems to be the start from water vapour column content to water vapour profiles, clouds, ocean surface temperature, ocean topography, sea level, sea ice, snow cover and ice caps, before the final section discusses the radiation budget at the top of the atmosphere and the clouds' role in this budget.

Water Vapour Column Content

The strong temperature dependence of maximum water vapour content confines the high water vapour column content to tropical air masses where in extreme cases 60 to 70 kg m^{-2} may be reached. Any remote sensing technique reaching an accuracy of about 10 percent like radiosondes and better areal coverage is a step forward. At present the best way for measuring global water vapour column content over the oceans is the use of passive microwave imager data because they penetrate through clouds. Over land surfaces the strong variability of surface emissivity in the microwave region of the electromagnetic spectrum hampers water vapour measurements.

Up to now, no routine algorithm using either microwave or thermal infrared data exists. This is due to the cloud contamination of thermal infrared data (see Thiermann et al., 1990; Chesters et al., 1983) and the experimental character and calibration problems of the microwave radiometer onboard the Nimbus 7 satellite (Staelin et al., 1976; Chang et al., 1979; Prabhakara et al., 1985; Katsaros et al., 1981). With the Special Sensor Microwave/Imager (SSM/I) onboard DMSP* satellites a new microwave imager, with global coverage in at least three days, exists. As shown by Schlüssel and Emery (1990) global fields of water vapour column content over ice free oceans can be retrieved under nearly all weather conditions with an rms error of 1.5 kg m^{-2}, which is in most cases well below a 5 percent relative error in the tropics. Figure 1 shows a 24 day mean in August 1987 and clearly demonstrates the strong latitudinal dependence of water vapour column content over the oceans, the anomaly caused by the Asian monsoon with values above 50 kg m^{-2} up to 40°N and the dry areas off the west coasts of all continents.

---

* Defense Meteorological Satellite Program

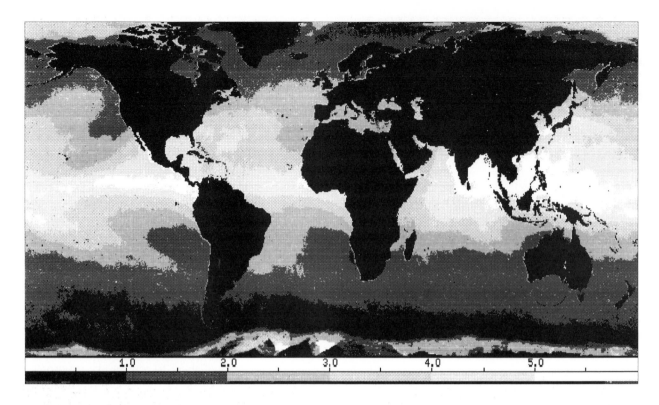

Figure 1: Water vapour column content in g m^{-2} or of 10 kg m^{-2} derived from the SSM/I onboard a DMSP satellite for August 1987; the algorithm applied was taken from Schlüssel and Emery (1990).

Water Vapour Profiles

The vertical profile of water vapour density often shows strong structures, especially a strong decrease at the top of the planetary boundary layer, whose magnitude determines stratocumulus cloud deck lifetime through the entrainment of dry air from above (Kuo and Schubert, 1988). By passive remote sensing only a few (up to 3) independent pieces of information on the water vapour profile can be retrieved. This drawback of all passive sounders is due to the broad contribution function of any radiometer, i.e. the measured radiance at the satellite still gets contributions from height levels ± 1 scale-height distant from the level of maximum contribution. The water vapour scale-height is typically 2 km. Therefore, only very precise multichannel measurements allow the retrieval of a rough water vapour profile because of the strongly overlapping contribution functions. This information is in principle available from the HIRS (High Resolution Infrared Radiation Sounder) and the AVHRR (Advanced Very High Resolution Radiometer) onboard the NOAA satellite series. However, it is hampered by clouds in the most interesting parts like frontal zones. It is possible to derive the relative humidity of some broad layers as shown by Smith and Zhou (1982) for HIRS and by Schlüssel (1989) for a combination of HIRS + AVHRR. The improvement in comparison to radiosoundings is the resolution of mesoscale features, since a profile is available roughly every 30 km. Another step forward is the possibility to detect the water vapour

content in the planetary boundary layer over oceans also with cloud cover using SSM/I data. According to Schulz (1990) this opens the possibility for better estimates of latent heat fluxes over the oceans from satellite data alone. In conclusion, water vapour profile retrievals are far from satisfying the needs. Improvements of passive techniques will not suffice. The first strong improvement could come from a combination of passive radiometers with a backscattering lidar (Flood et al., 1990) or an imaging spectrometer (Fischer, 1988) giving cloud heights to 100 m and thus more precise water vapour density at cloud top.

Clouds

There is no doubt that one of the most impressive achievements of remote sensing is the routine observation of clouds from geostationary satellites used for nowcasting and short-term weather forecasting. However, if climatologists ask for the total cloud cover over an ocean area or the frequency of occurrence of cirrus they still do not get data sufficiently accurate to start a trend analysis.

The International Satellite Cloud Climatology Project (ISCCP) in the framework of the World Climate Research Programme (WCRP) tries to give cloud cover and type in an objective way. However, a comparison of 5 different cloud climatologies for part of the South Atlantic (Bakan, 1990) with predominant stratocumulus revealed such strong differences in monthly mean total cloud cover, that we are

Table 1: RMS-Differences between monthly mean total cloud cover (in percent cloud cover); subtropical South Atlantic, July 1983.

	ESA-CDS	ESA-CDS (1100 GMT)	ISCCP	ISCCP (1200 GMT)	Saunders (1985)	COADS
HCCSA (1200 GMT)	10	11	15	13	18	13
ESA-CDS		4	19	18	20	12
ESA-CDS (1100 GMT)			21	20	23	12
ISCCP				8	15	17
ISCCP (1200 GMT)					10	17
Saunders (1984)						17

HCCSA   - Hamburg Cloud Climatology of the South Atlantic
ESA-CDS - ESA Climate Data Set
ISCCP   - International Satellite Cloud Climatology Project
COADS   - Comprehensive Ocean Atmosphere Data Set, Surface Observations
Saunders - Satellite data presented by Saunders (1985), J. Clim. Appl. Meteor., 24, 117-127.
All data are interpolated to 5° x 5° regions for intercomparison and all satellite climatologies use METEOSAT imaging data.

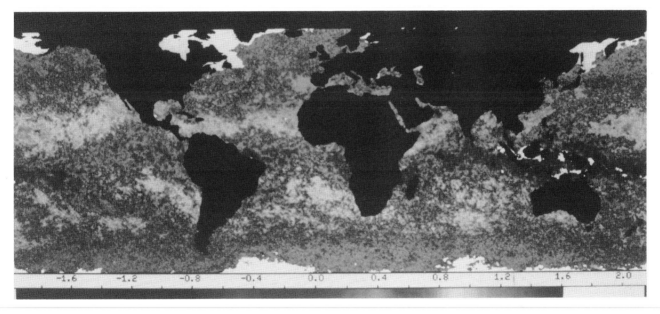

Figure 2: The skin effect: difference between two sea surface temperature algorithms calibrated with the bulk ocean temperature (measured for example, by drifting buoys or a bucket) and the ocean skin temperature (as measured by a radiometer) for March 1–March 15,1984; modified after Schlüssel et al. (1989).

far from having solved cloud cover determination (see Table 1). Not only in this case, the question of comparability of conventional and remote sensing data arises.

Sea Surface Parameters

Remote sensing of ocean surfaces can in principle deliver many important variables for the understanding of the global water and energy cycles: for example, sea surface temperature, ocean topography, mean sea level, sea ice extent and age.

The only remotely sensed geophysical parameter that has since 1981 routinely reached an absolute accuracy comparable to conventional sensors is bulk surface temperature (see, for example, McClain et al., 1985; Schlüssel et al., 1990), although a 0.5 K absolute error still does not allow a trend analysis. This drawback is due to:

- firstly the unresolved bulk-skin temperature difference (see Figure 2) ranging from a tenth of a Kelvin to approximately one Kelvin
- secondly to water vapour and temperature structure in

the lower troposphere, if strongly deviating from mean conditions and not fully accounted for by the split-window alogrithm[†] as well as

- thirdly by strong volcanic aerosol layers in the stratosphere causing too low surface temperature retrievals.

For an improvement one has to calibrate the satellite sensors with shipborne or airborne radiometers, since the satellite sensor receives surface radiation only from the uppermost micrometers of the ocean. Thus it measures skin temperature and the skin-effect is of the same order of magnitude as the accuracy desired by climatology for anomaly fields and trends, for instance 0.3 K absolute in tropical ocean areas. Using more sophisticated algorithms accounting for aerosol effects and including information on temperature and moisture profiles from sounders onboard the NOAA satellites and new satellite radiometers like the Along Track Scanning Radiometer onboard the ESA Remote Sensing Satellite (ERS-1) a bulk ocean surface temperature trend analysis will be possible soon.

The oceanic circulation determines, besides tides and wind-driven waves, the ocean topography. Radar altimeters like those onboard SEASAT in 1978 or GEOSAT allowed the detection of relative differences in surface elevation of about 5 cm and thus give insight into the ocean eddy

distribution and the variability of ocean currents, which cause surface elevation changes of more than 5 cm up to 1.5 m in western boundary currents like the Gulf Stream. However, only the improvement in our knowledge of the geoid, expected with the future missions like the French-American TOPEX-POSEIDON satellite and the Radar Altimeter onboard the ERS-1 (ESA Remote Sensing Satellite No. 1), will give a continuous global ocean current observation from 1991 onwards. Radar altimeter data are seen as a backbone for the World Ocean Circulation Experiment (WOCE), of the World Climate Research Programme (WCRP).

From tide gauge stations measuring for decades we know that global mean sea level has been rising at a mean rate of 15 ± 5 cm per century. If satellite radar altimeters should contribute to the trend analysis and the short term variability detection, they have to reach the millimeter range accuracy for space-time averages. This is only possible with a continuous correction of the atmospheric masking. The above mentioned missions will allow this correction.

The determination of sea ice extent from space relies mainly on passive microwave radiometers and uses the strongly different emissivity of sea ice in comparison to water in the microwave part of the spectrum. Using the SMMR data Comiso and Sullivan (1986) as well as Gloersen and Campbell (1988) derive ice extent and age for the Antarctic or the Arctic, respectively. Thus there exists a first trend analysis (since the first passive microwave radiometer onboard NIMBUS 7 was put into space in 1978) by Gloersen and Campbell (1988) for both

---

[†] This algorithm uses two spectral intervals in the thermal infrared window at roughly 11 and 12 micrometer wavelength available from NOM 7, NOAA 9 and NOAA 11 satellites within the TIROS-N series of NOAA (National Oceanographic and Atmospheric Administration).

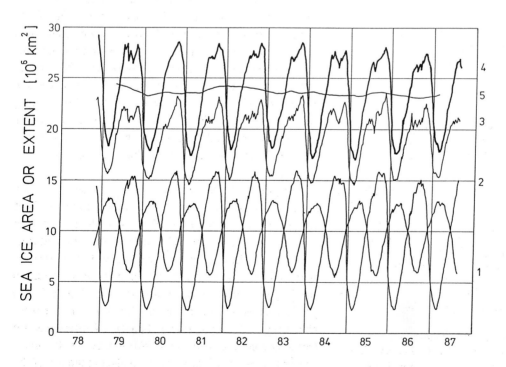

Figure 3: Sea ice area (Arctic, Antarctic, global) and sea ice extent (global including one year running mean) during 1978 to 1987; taken from Gloersen and Campbell (1988).

the Antarctic and Arctic. In the 9-year period from 1979 to 1987, global sea-ice area as well as maxima and minima of sea-ice area declined but the changes were not significant. However, maximum ice extent which includes leads and polynas within the sea ice area became 5 percent smaller and this change was highly significant (see Figure 3).

This astonishing result was commented by Gloersen and Campbell in the following way: We are faced with the paradoxical fact that the same decrease does not occur in the record for global sea ice area. The global sea ice area variations are not exactly tracked by those of global sea ice extent. Since the global sea ice area decreased slightly, if at all, while a significant decrease occurred in the maxima of the global sea ice extent, a decrease occurred in the amount of open water within either or both of the Arctic and Antarctic ice packs.

Microwave radiometers are also capable of distinguishing between seasonal sea ice and multi-year sea ice. Therefore, a trend analysis of the relative contributions to sea ice area could be made also. The SSM/I onboard the DMSP satellite allows an even finer distinction between newly formed, young and multi-year ice and respective ice coverage as demonstrated in Figure 4 from Taurat (1990). The use of the 85.5 GHz channel seems to be at least as good as the 37 GHz channel.

Ice and Snow on Continents

While global sea level changes by only millimeters per year, the elevation of an ice cap may change by

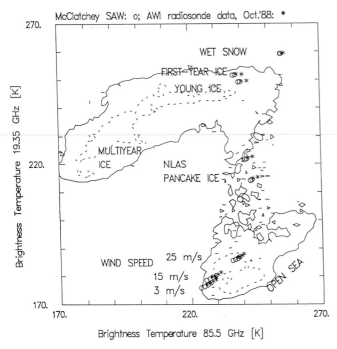

Figure 4: Two-dimensional histogram of brightness temperatures over Arctic sea ice from SSM/I for January 16, 1988 together with radiative transfer simulations for different ice types; taken from Taurat (1990).

centimeters, decimeters or locally even meters per year. Therefore, satellite radar altimeters are already sufficiently accurate for a trend analysis of those parts of ice caps which have a rather high mean accumulation rate like Southern Greenland. Such a trend analysis by Zwally (1989) using the SEASAT altimeter and GEOSAT altimeter data revealed a growing ice cap. If the average 23 cm accumulation per year observed during the last decade were extrapolated to Greenland as a whole, it is equivalent to 0.3 mm per year sea level reduction. The observed sea level rise must therefore be due to higher rates of thermal expansion of sea-water and of melting of alpine glaciers. A trend analysis of the Antarctic ice cap still does not exist due to the low accumulation rate there.

Since the seasonal snow cover is a climate variable strongly related to winter temperature, the observed temperature increase especially since 1975 should be visible in a seasonal snow cover decline. As reported in WMO/UNEP (1990) snow cover data are available on a weekly basis using satellite imagery. According to these data, more reliable since 1975, the snow cover extent of the eighties was lower by 13% in Eurasia during autumn, and by 9% during spring compared to the seventies. The overall decrease of total northern hemisphere snow cover since 1970 amounts to roughly $3 \cdot 10^6$ km^2 for $30 \cdot 10^6$ km^2 total area during winter.

Net Radiation Budget at the Top of the Atmosphere

The imbalance of the net radiation at the top of the atmosphere is the cause for atmospheric and oceanic circulation. It can only be measured from space. With the advent of a dedicated satellite experiment, the Earth Radiation Budget Experiment (ERBE) measuring short and longwave radiation from two sun-synchronous and one drifting satellite (Barkstrom et al., 1986), the global climate system research has reached a new level. Scientists are now able to answer the following two questions: When during the course of a year does which part of our globe gain or lose energy? Where do clouds heat or cool the planet and is the overall effect already measurable? The answers are as follows.

In April 1985, for instance, the atmosphere-surface system lost up to 165 W m^{-2} over the Weddell Sea while it gained up to 125 W m^{-2} off south-western Mexico and over South-East Asian areas. The most prominent features, however, are the energy loss over the Eastern Sahara for nearly all months and the restriction of strong yearly mean energy gain to ocean areas clearly shown in Figure 5. The yearly average does not exceed +120 Wm^{-2} in the main gain area in the equatorial Indian and Western Pacific Ocean and not fall below -100 Wm^{-2} in both polar areas. The yearly course of the radiation budget at the top of the atmosphere in Figure 6 first of all shows the 7% change of

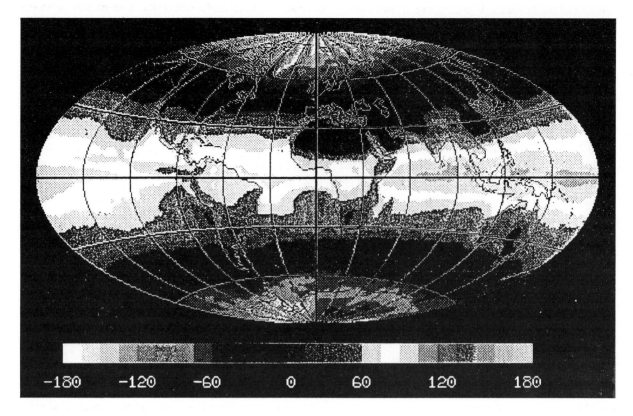

Figure 5: The average radiation budget of the Earth at the top of the atmosphere as measured from February 1985 to January 1986 by the Earth Radiation Budget Experiment (ERBE) radiometers onboard one dedicated and two NOAA satellites (Wm^{-2}); Rieland (1990).

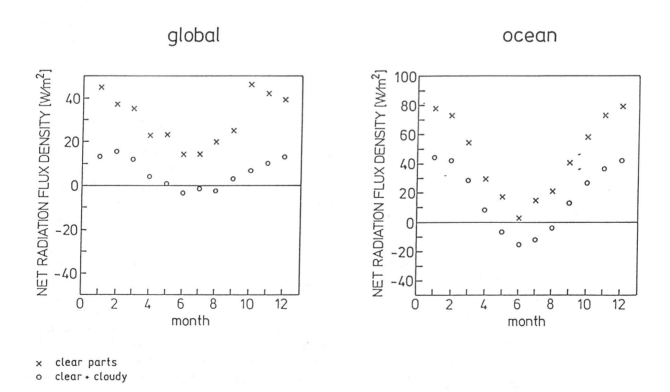

Figure 6: The annual course of the radiation budget for the entire globe and the ocean area, derived from the ERBE data set by Rieland (1990) both for cloudless parts and the entire area.

insolation caused by the ellipticity of the Earth's orbit with the minimum in summer of the northern hemisphere. The enhanced variation for the ocean areas is mainly due to the larger ocean area and its low albedo in the southern hemisphere.

The global mean effect of clouds was a cooling in April 1985 (Ramanathan et al., 1989), which was the result of strong cooling by optically thick low clouds especially over ocean areas and the heating by high, optically rather thin clouds. The main contribution to the cooling stems from optically thick mid-latitude clouds. Clouds cool throughout the year (Rieland, 1990). However, they cool more strongly during the northern hemisphere winter. The water vapour feedback, a main cause for the warming of the globe after a trace gas increase, could also be derived from the ERBE data (Raval and Ramanathan, 1989). This has enabled the test of climate models. All 19 general circulation models tested (WMO/UNEP, 1990) showed a remarkable agreement with the satellite derived climate sensitivity parameter $\lambda = 0.5$ K/(Wm^{-2}) for cloudless parts of the atmosphere/surface system.

Final Remarks

Remote sensing with satellites has given first global data sets of water cycle components. It still lacks precision in many cases, for instance for cloud cover. However, it has allowed first trend estimates for sea ice and snow cover, constrains climate model results by the measured energy budget at the top of the atmosphere, and has validated the water vapour feedback in climate models. With every new year of remotely sensed data, the basis for trend estimates and better validation of climate models increases strongly due to continuously improved alogrithms, better satellite sensors, the combination of satellite sensors and—most important—the combination of in-situ and remote sensing data.

References

Bakan, S. (1990): A METEOSAT cloud climatology for the subtropical South Atlantic. Proceedings 8th METEOSAT Scientific Users' Meeting, ESOC, Darmstadt, in press.

Barkstrom, B.R., and ERBE Science Team (1986): First data from the Earth Radiation Budget Experiment. Bull. Am. Met. Soc., 67, 818–824.

Chang, A.T., T.T. Wilheit (1979): Remote sensing of atmospheric water vapour, liquid water and wind speed at the ocean surface by passive microwave techniques from the Nimbus-5 satellite. Radio Science, 14, 793–802.

Chesters, D., L.W. Uccelini, W.D. Robinson (1983): Low level water vapour fields from the VAS split window channels. J. of Clim. Appl. Met., 22, 725–743.

Comiso, J.C., and J.W. Sullivan (1986): Satellite microwave and in situ observations of the Weddell Sea ice cover and its marginal ice zone. J.G.R, 91, C8, 9663–9681.

Fischer, J. (1988): High resolution spectroscopy for remote sensing of physical cloud properties and water vapour. in: Current Problems in Atmospheric Radiation. Ed.: Lenoble and Geleyn, Deepak Publishing, 151–154.

Flood, C., H. Grassl, G. Megie (1990): Backscatter Lidar: The potential of a space-borne lidar for operational meteorology, climatology and environment research. ESA-SP-1121, 27 pages.

Gloersen, P., and W.J. Campbell (1988): Variations in the Arctic, Antarctic and global sea ice covers during 1978-1987 as observed with the Nimbus 7 Scanning Multi-channel Microwave Radiometer. J.G.R., 93, C9, 10666–10674.

Katsaros, K.B., P.K. Taylor, J.C. Alishouse, R.B. Lipes (1981): Quality of Seasat SMMR atmospheric water determinations. in: J.F.R. Gowers (Ed.) Oceanography from Space, Plenum, 691–710, New York.

Kuo, H.-C., and W.H. Schubert (1988): Stability of cloud-topped boundary layers. Quart. J. Roy. Met. Soc., 114, 887–916.

McClain, E.P., W.G. Pichel, C.C. Walton (1985): Comparative performance of AVHRR-based multichannel sea surface temperatures. J.G.R., 90, 11587–11601.

Prabhakara, C.D., A. Short, B.E. Vollmer (1985): El Niño and atmospheric water vapour: Observations from Nimbus–7 SMMR. J. Clim. Appl. Met., 24. 1311–1324.

Ramanathan, V., R.D. Cess, E.F. Harrison, P. Minnis, B.R. Barkstrom, E. Ahmad, D. Hartman (1989): Cloud-radiative forcing and climate: results from the Earth Radiation Budget Experiment. Science, 243, 57–63.

Raval, A., and V. Ramanathan (1989): Observational determination of the greenhouse effect. Nature, 342, 758–761.

Rieland, M. (1990): The seasonal variations of the Earth radiative budget from satellite data. Private communication.

Saunders, R. (1985): Monthly mean cloudiness observed from METEOSAT-2. J. of Clim. a. Appl. Met., 24, 117–127.

Schlüssel, P. (1989): Satellite-derived low-level atmospheric water vapour content from synergy of AVHRR with HIRS. Int. J. Rem. Sens., 10, 705–721.

Schlüssel, P., W.J. Emery, H. Grassl, T. Mammen (1990): On the bulk-skin temperature and its impact on satellite remote sensing of sea surface temperature. J.G.R., 95, Paper 90 JC 00515.

Schlüssel. P., and W.J. Emery (1990): Atmospheric water vapour over oceans from SSM/I measurements. Int. J. Rem. Sens., 11. 753–766.

Schulz, J. (1990): Water vapour in the atmospheric boundary layer over oceans from SSM/I-measurements. Unpublished manuscript.

Smith, W.L., F.X. Zhou (1982): Rapid extraction of layer relative humidity, geopotential thickness and atmospheric stability from satellite sounding radiometer data. Appl. Optics, 21, 924–928.

Staelin, D.H., K.F. Kiinzi, R.L. Pettyjohn, R.K.L. Poon, R.W., Wilcox, J.W. Waters (1976): Remote sensing of atmospheric water vapor and liquid water with the Nimbus-5 microwave spectrometer. J. Appl. Met., 15, 1204–1214.

Taurat, D. (1990): Klassifizierung von Meereis mit SSM/I-Messungen. Thesis, University of Hamburg, Meteorology Department.

Thiermann, V., P. Schlüssel, E. Ruprecht (1990): Determination of total precipitable water over sea from IR satellite data. Submitted to J.G.R.

WMO/UNEP (1990): Climate change: the IPCC Scientific Assessment. Ed.: Houghton, Jenkins, Ephraums, Cambridge University press, 365 pages.

Zwally, H.J. (1989): Growth of Greenland ice sheet: interpretation. Science, 246, 1589–1591.

World Ocean Circulation and Climate Change: Research Programmes and a Global Observing System

by D. James Baker *(USA)*

Abstract

This paper provides an overview of the state of understanding of the role of the ocean in climate change. The focus is on long-term variability and the results to be expected from the World Ocean Circulation Experiment of the World Climate Research Programme and the Joint Global Ocean Flux Study of the International Geosphere-Biosphere Programme. An emphasis is placed on the need for a global observational system that would systematically monitor ocean changes and provide data for prediction of climate change.

1.   Introduction

The first World Climate Conference in Geneva in 1979 led to the establishment of the World Climate Programme and the World Climate Research Programme (WCRP). In recognition of the importance of the interaction between the ocean and atmosphere in the dynamics of climate, the WCRP soon established two major ocean-related programmes: the Tropical Ocean and Global Atmosphere (TOGA) programme and the World Ocean Circulation Experiment (WOCE).

This year, 1990, is the mid-point of the TOGA programme and the first year of the field programme for WOCE. Other programmes related to climate have also begun. It is now a logical time to assess the state of understanding the role of the ocean in climate change and to consider what will be required after these research programmes to provide an observation system for describing and modeling the ocean as part of climate change.

2.   Ocean Circulation, Heat Flux and Climate

With its large heat capacity and mobility the ocean absorbs and redistributes heat from the sun and the atmosphere and takes up and emits carbon dioxide and other greenhouse gases. The ocean thus has an intimate part to play in the dynamics of climate. We need to understand this role so that we can make the best predictions of climate change due to natural fluctuations, human perturbations, and their interaction.

In the most local sense, the ocean buffers and impacts coastal climate and sea level. The large heat capacity of water dampens diurnal and seasonal temperature swings, leading to more temperate coastal climates. On a global scale, the ocean stores and transports heat and acts as a heat source that forces atmospheric convection and winds. Sea surface temperature (SST) patterns and their associated heat and moisture fluxes affect weather events like storms and hurricanes and are a key element of climate change.

SST anomalies in the tropical Pacific Ocean are an integral element of *El Niño's* effect on climate, currently being studied by the WCRP's Tropical Ocean and Global Atmosphere (TOGA) programme. Shifts in the Gulf Stream and Kuroshio have each been correlated with changes in SST and regional climate. The patterns of SST are produced by a complex feedback process between ocean currents and mixing and atmospheric winds, heat and moisture fluxes. Knowledge of all of these processes is necessary for understanding climate change and the effect of greenhouse warming.

From direct measurements of radiation impinging on the Earth we can calculate how much heat must be transported towards the poles for radiative equilibrium. In the subtropics, the ocean and the atmosphere appear to be about equally important in this respect. The effect of ocean currents on heat transport can be seen in the sea surface temperature, but more than the surface is involved. The broad-scale currents are deep and a global-scale convective cell returns cold water from the polar regions to the tropics.

In the North Atlantic, the heat flux produced by a combination of the Gulf Stream carrying warm water north and cold water sinking in the Greenland Sea and subsequently drawing warmer water from the south leads to a warmer western Europe. Despite our general notions of these overall patterns, we know little about the exact magnitude and variability of ocean heat flux. The situation is only somewhat better for the atmosphere: atmospheric models are much more advanced than ocean models, but even today, the best general circulation models of the atmosphere differ in their estimates of heat transport from observational analyses by as much as a factor of two (Stone and Risbey, 1990).

On geological time scales the ocean and the atmosphere are tightly linked. At the termination of the last glaciation, about 11,000 years ago, geological evidence points to some dramatic changes leading to a short term reversal of the warming and a thousand-year-long cold spell in the Northern Hemisphere. One explanation (Broecker and Denton, 1989) is that melting fresh water from glaciers diluted the salinity of surface waters of the North Atlantic, reducing the density and preventing them from sinking. As a consequence, warm water no longer flowed northward and coastal regions all around the North Atlantic Ocean were chilled. The change was rapid, possibly occurring in less than 100 years. This phenomena shows that all the ocean is involved when it comes to climatic processes on the longer time scales. And we note that the phenomena was discovered from a paleoclimate record, emphasizing the importance of long-term data records for understanding climate.

In order to develop these general notions into realistic models of the coupled system, we need a full description of the state of the ocean and its variability. However, the description of the global ocean circulation is far from being adequate; in vast regions of the ocean, particularly in the southern hemisphere, data sets are very sparse. Changes in ocean climatology have only been observed in a few areas.

The formation and evolution of oceanic water masses determines the extent to which the ocean can store heat resulting, for example, from greenhouse warming. These processes and the role of sea ice and other fresh water fluxes need to be accurately represented in ocean models. The oceanic data to be collected during the field phase of the World Ocean Circulation Experiment of the WCRP will both characterize the ocean circulation for the first time and thus provide an extensive data set for testing and improving models of these physical processes.

The ocean is a source of and a sink for carbon dioxide and other greenhouse gases. We need to understand these oceanic sources and sinks and how they affect and vary with climate change. The magnitude of these interactions is crucial to understanding global biogeochemical cycles, yet our understanding of them is poor. The Joint Global Ocean Flux Study (JGOFS) of the IGBP is aimed at understanding the coupling of circulation, mixing, and biogeochemical cycles.

Because of its role in the climate system and biogeochemical cycles, the ocean is fully implicated in the response of the Earth to increased greenhouse gases. The report of Working Group 1 of the Intergovernmental Panel on Climate Change (McBean et al, 1990) notes that if the atmosphere and upper ocean alone were responding to the increase in greenhouse heating, then the surface of the earth should already be one to two degrees Celsius warmer than the temperatures of the nineteenth century. But the data, although not comprehensive, do not yet show such an unambiguous increase. The response of the Earth's atmosphere to increased greenhouse heating may be tempered by the uptake of heat and carbon dioxide by the ocean.

Understanding and modeling quantitatively the global circulation of the ocean is an essential element in determination of the timing and extent of global warming in the Earth's atmosphere, cryosphere, and ocean. In the absence of a convincing ocean model, climate predictions will have unsolvable uncertainties. Fortunately, models are beginning to become realistic enough to begin to show the effect of the oceans on greenhouse warming. Stouffer, Manabe, and Bryan (1989) have developed a coupled ocean-atmosphere model for study of effects of increasing carbon dioxide which explicitly incorporates the effect of ocean heat transport and deep water formation. In the model's circumpolar ocean of the Southern Hemisphere, a region of deep vertical mixing, the surface air temperature increases very slowly. In the Northern Hemisphere of the model, the surface air warms faster than in the south; and the warming increases with latitude.

In spite of the recognition of the importance of the ocean in all of these aspects of climate dynamics, progress has been slow in developing an adequate understanding. The measurement problems are formidable: the ocean, unlike the atmosphere or space, is opaque to electromagnetic radiation. Like the atmosphere, the ocean is a turbulent, unstable fluid with energy in an enormous range of space and time scales. For example, a major feature of ocean circulation, mesoscale eddies, are relatively small, on the order of 10 to 50 kilometers, so measurements must be closely spaced. Measurements must be made in a corrosive medium and often at very high pressure. Chemical measurements must be made with high precision. Biological populations have a patchy distribution, requiring closely spaced stations and satellite measurements.

Modeling is equally difficult: we do not have many guiding measurements and many of the controlling processes are not well understood. The small scale required for resolving these processes leads to large computer time requirements. But now with new techniques for measurement, both in situ and from space, many of the difficulties can be overcome. And state-of-the-art computers are approaching the speed and capacity necessary for eddy-resolving ocean modeling. The programmes to be discussed in the next section show how we are making progress in these areas.

3. The World Ocean Circulation Experiment

On the decadal and longer time scales, we must study the entire ocean, top to bottom, and its driving forces which include winds, precipitation, and radiation. The World Climate Research Programme established the World Ocean Circulation Experiment (WOCE) in recognition of this need. The primary goal of WOCE is to develop models useful for predicting climate change on decadal time scales and to collect the data necessary to test them.

The data from WOCE will help us to determine and to understand the large-scale fluxes of heat and fresh water on a global scale. Major improvements are expected in the accuracy of numerical models of ocean circulation for coupling to atmospheric models for better simulation of climate. The WOCE data will also help to determine the design of a long-term observing system.

The success of WOCE requires the application of large observational resources because the ocean circulation as we know it varies on many time and space scales. Eddy-resolving oceans models (e.g. Semtner and Chervin, 1988) show the broad pattern of global currents. Because of the small horizontal scale and the relatively rapid time period of the eddies, it has not been possible to achieve a synoptic observation of the state of the ocean. The general features of the model calculation, particularly the eddy field, are corroborated by satellite and in situ measurements of currents, but many important details are not yet accurate.

To provide a clear understanding of the circulation, WOCE will merge many different space and time scales. As a consequence, WOCE is one of the largest oceanographic programmes ever mounted. More than 50 countries will be involved, and studies will be carried out in every major ocean. The programme will include measurements from several satellites, dozens of ships, hundreds of moored instruments, and thousands of surface drifters, mid-water floats, and expendable instruments to take a comprehensive, several-year "snapshot" of the physical properties of the ocean.

WOCE investigators will use the data set to probe the dynamical balance of the global circulation and its response to changing surface conditions of winds, temperature, and rainfall. Information will be collected on the components of ocean variability on time scales of months to years and space scales of kilometers to global. The rates and nature of formation and circulation of water masses that influence the

climate system on time scales from ten to one hundred years will be measured.

Over the five to seven years of the programme, there will be a major effort to determine air-sea fluxes globally by combining marine meteorological and satellite data. An upper-ocean measurements programme will determine the annual and interannual oceanic response to atmospheric forcing. When the experiment is complete, WOCE investigators will have carried out measurements at close to 25,000 precision hydrographic stations, more than three times the total number available from all expeditions before. WOCE scientists will also make intensive use of historical oceanographic data to assess the longer-term variability of ocean circulation. With the complete data set, they hope to achieve the first full global description of both surface and deep ocean circulation.

One of the major reasons that WOCE can be envisioned at this time is the availability of new in situ and satellite-borne instruments. For example, new "pop-up" buoys are being used that automatically track temperature and other properties at a given depth in the ocean, then come to the surface and report their information by satellite to shore-based stations.

In early 1991, the in situ programmes will be supported by the launch of the European Space Agency's ocean satellite ERS-1. Instruments carried by ERS-1 will provide global wind measurements by scatterometer, a device that probes the surface with radar and uses the scattered radiation to determine the state of the surface, surface topography measurements by altimeter for ocean circulation, and sea surface temperature measurements. In 1992, the joint US/French precision altimeter mission TOPEX/ POSEIDON will begin to provide more accurate measurements of surface topography than available before. A NASA scatterometer, also for global wind measurements, is scheduled for flight on the Japanese Advanced Earth Observation Satellite (ADEOS) in 1995.

Development of models is also a key element of WOCE. Modeling experts are now working with a variety of eddy-resolving models on different scales. With the guidance from the global data sets, modelers are expected to be able for the first time to provide understanding of the ocean circulation and its interaction with the atmosphere on a global scale.

The interaction of data and models occurs in three ways: the first is specification of initial conditions, so that the model starts off with the right state. Second is assimilation of data to achieve the best possible forecast with the most recent data available. Third is to obtain the best possible estimate of an ocean field based on limited observations. As more data become available, ocean modelers will be able to ascertain how well they are mimicking the real oceans.

WOCE systems are to be activated for a limited time, 5 to 7 years depending on availability of resources, extending into the mid-1990s. In addition to providing a snapshot of the ocean over this period, the WOCE programme also will focus on the longer time scales by identifying those oceanographic parameters, indices and fields that are essential for continuing measurements in a climate observing system. This work will help to provide a basis for the development of a global observing system for long-term systematic measurements of the ocean.

4.   Chemical and Biological Effects

Changes in climate and ocean circulation affect the ocean's chemistry and biological ecosystems. The reverse is also true: changes in ocean ecosystems can affect climate. The most direct effect occurs in the upper layers: phytoplankton blooms increase the absorption of light and thus concentrate the heating effect of solar radiation.

But perhaps even more important is the fact that the oceanic reservoir is the largest pool of carbon in the world cycle, and thus has a major part in determining the concentration of carbon dioxide in the atmosphere through physical processes of mixing and circulation, chemical processes, and biological processes. The uptake and recirculation of carbon by ocean ecosystems and the eventual storage of much of it in the deep sea is a primary element in the global carbon cycle.

If we are to understand the cycles of the chemical elements, we must understand their uptake and reactions with the ocean and its ecosystems. This need has led to the development of the Joint Global Ocean Flux Study (JGOFS), which is a core project of the International Geosphere-Biosphere Programme, and to a newly emerging programme, Global Ocean Ecosystems Dynamics (GLOBEC). The goal of JGOFS is to determine and understand on a global scale the processes controlling the time-varying fluxes of carbon and associated biogenic elements in the ocean, and to evaluate the related exchanges with the atmosphere, sea floor, and continental boundaries. GLOBEC is concerned with animal ecosystems which mediate these biogeochemical transformations and the impact of global climate change on them in terms of the living resources they represent.

JGOFS scientists are using newly developed in situ techniques for direct measurement of fluxes of biogenic material in the water column by sediment traps and high-precision methods for detection of trace species in very small amounts. Satellite ocean color measurements and shipboard pigment studies will be used to determine the state and variability of global biological productivity. The first JGOFS field experiment was held in the North Atlantic in the spring and summer of 1989. Noting the special importance of monitoring oceanic carbon dioxide, JGOFS

scientists have initiated time series stations and a global survey as part of the WOCE global coverage.

Plans are less firm for the ocean color measurement by satellite, but NASA is currently making arrangements for a private company to build and fly an ocean color instrument in return for NASA's agreement to buy the data. If these arrangements hold up, a new satellite ocean color instrument will be available in the early 1990s. Without flight of the NASA instrument, the next scheduled satellite ocean color instrument is on the Japanese ADEOS, scheduled for 1995. For the late 1990s, an enhanced color imager is planned to fly as a part of the international Earth Observing System (EOS).

5.   A World Ocean Observing System: Requirements and Elements

Ocean circulation, chemical transports, and the dynamics of ecosystems all have variability on periods well beyond the seven-year "snapshot" of WOCE or the lifetimes of the other global research programmes. Thus it is essential that measurements of the relevant parameters continue beyond the intensive field periods so that we can continue to learn about the state of natural variability. To do this, we will need a network of systematic global long-term observations of physical, chemical, and biological parameters. Such a network must be based on scientific requirements, provide data on a routine basis, and have continuing support; in other words it should be an "operational" programme.

The transition from research programmes to operational measurements must be a focus for the 1990s and into the 21st century. The transition will require enhancement of existing systems based on the scientific needs, development of new technology for cost-effective measurements, and proper archives and timely distribution of data. The interaction between the operational measurements and operational climate forecast models is essential. If necessary, new institutional arrangements may be required.

But above all, the system must be driven by scientific requirements. TOGA in the Pacific Ocean is an excellent example of how a focused oceanographic programme can lead to a longer-term effort. Research efforts demonstrated that a long-term measurement system was needed to provide the data necessary for improving the understanding of El Niño and to begin regular prediction efforts. As a consequence, ocean measurements in the Pacific Ocean grew from sparse coverage in 1985 to basin-wide observations using systems developed under TOGA sponsorship.

The initial requirements identified in the planning for WOCE are aimed at global ocean heat transport and storage which is caused by the oceanic circulation, sinking of cold surface water and upwelling of deep water, and periodic (decadal) surveys of transient tracers. The Global Energy and Water Cycle Experiment (GEWEX) will require global measurements of temperature, salinity, and air-sea fluxes. The JGOFS plans call for global CO_2 measurements and ocean color monitoring. The design of a long-term monitoring system will also be driven by other requirements, such as those from commercial shipping and national defense. For the purpose of the discussion here, we will assume that these requirements largely overlap.

For convenience, we can divide up the variables to be measured into five components (Stommel, 1990): temperature and salinity, or a "hydrographic component"; nutrients, dissolved oxygen content, carbon dioxide and geochemical tracers, or a "chemical component"; changes of state of water, ice, precipitation, and evaporation, or a "water phase component"; and the external forcing of wind stress, air-sea heat flux, insolation, and heat flux from the seabed, or a "flux component"; and concentration of plankton in the surface layers, or a "biological component."

Specific measurement programmes that would contribute to each of these components include (McBean et al, 1990):

- Upper ocean monitoring programme to determine the time and space dependent distribution of heat and fresh water, sea level, CO_2 content, air-sea fluxes, and plankton distribution
- Systematic deep ocean measurements at suitable time and space intervals to determine the state of the ocean circulation, physical and chemical properties, transient tracers and CO_2
- Satellite observations of the ocean surface temperature, wind and topography, precipitation, sea-ice concentration, and chlorophyll content (ocean color) by an international array of space platforms in suitable orbits around the earth

Several global measurement and data management systems have been established on the basis of operational agency and commercial needs. Perhaps the most comprehensive of these is that co-ordinated through the International Global Ocean Services System (IGOSS), jointly sponsored by the Intergovernmental Oceanographic Commission and the World Meteorological Organization. Ocean temperature and salinity data are provided through IGOSS on a regular basis from many parts of world ocean. Sea level data is co-ordinated through the Global Sea Level Observing System (GLOSS). WMO and IOC also encourage and support the use of drifting buoys for regular observations of air pressure, air and sea surface temperature. The Voluntary Observing Ships (VOS) scheme of the WWW include ships transmitting meteorological and surface oceanographic observations in real time over the GTS. Under separate auspices, regular satellite observations of the surface temperature and ice cover are provided from the operational environmental satellites.

Long-term regional monitoring efforts for biological oceanography have been carried out by a number of organizations around the world, including the International Council for the Exploration of the Sea (ICES) in the North Atlantic, and the California Cooperative Fisheries Investigation (CALCOFI) programme off the coast of California. The UK continuous plankton recorder has been operated in various parts of the North Atlantic Ocean. Programmes supported by Japan monitor physical, chemical, and biological conditions in the western North Pacific Ocean. Ocean color was monitored for several years by the Coastal Zone Color Scanner on the Nimbus-7 satellite.

Most of the measurements made today, however, are on a temporal and spatial scale that is too coarse or too regionally focused for documenting the processes important to climate change. For example the sea surface distribution of one-degree squares containing temperature observations from 1900 to 1978 (Levitus, 1982) shows that aside from a heavy concentration in the western boundary current region, the global coverage, particularly in the southern hemisphere and deep ocean is poor.

Many of the measurements which are necessary for understanding climate are not being made at all. Although regular measurements of wind speed and air temperature are being made from volunteer ships, accurate measurements of air-sea fluxes of heat, mass, and momentum are not being made. In addition, much of the data that are collected are not delivered to users through the GTS. More rapid data delivery is essential for climate modeling.

New technology will help solve some of these problems. A comprehensive global system will include direct measurements by floats, current meters, and water sampling, and indirect measurements from acoustics and satellites (Munk and Wunsch, 1982). Many improvements in direct techniques are being supported by TOGA, WOCE and other programmes. Telemetering moorings and autonomous vehicles are already showing their worth, and will be a key element of the system.

Munk and Forbes (1989) have described a scheme for acoustic measurement of global ocean temperature change that could be part of a long-term monitoring system. In this case, a sound source from Heard Island in the South Indian Ocean produces signals that can be heard in much of the world ocean. The time for sound to traverse the different ray paths is a function of the average temperature along that path. The sensitivity of such a measurement is great, on the order of millidegrees Celsius. The signals expected from greenhouse warming estimates should be detectable.

Plans now in place in a number of countries indicate that the satellite measurements from research missions in the early 1990s will continue and be expanded in the late 1990s and early 21st century. Ocean surface waves, temperature, topography, color, surface winds, precipitation, and ice cover are expected to be measured by a variety of satellites. It will be essential to have in place a mechanism for delivery and exchange of such data.

As we look to the development of a global ocean observing system, it is logical to take the World Weather Watch (WWW) as an analogy. The WWW has been a remarkably successful demonstration of international cooperation. The WWW is a carefully integrated union of national weather services acting at their own expense and using their own facilities to accomplish a specified portion of the global weather task (WMO, 1988).

A start has been made towards the establishment of an analogous ocean observing system through IGOSS and the WWW, but we are far from having the system we need. This is partly due to the fact that lack of knowledge of ocean processes has delayed the establishment of requirements, and partly due to the fact that up to now, governments have not had compelling reasons to support climate observations within the ocean. Furthermore, it is difficult to sustain the long-term support of programmes whose payback only comes after decades of observations and analysis. The lack of interest has led to underfunding the implementation of systematic ocean observations by operational agencies.

The funds for many of the existing observational systems have come from a combination of operational programmes and research programmes. There needs to be a transition from this research mode to a more operational mode, but an operational mode that has science requirements as a driver. Scientists must be an integral part of the observing system, to assure quality and to interact as users.

Data from a global observing system must be made available in a timely fashion to researchers and to the modelers for predictions. This "real-time" requirement is less stringent than that of the weather forecast, because the inherent times are longer. But it has been generally recognized that if data are not transmitted as soon as they are taken, they are often lost. Data must also be archived properly so that accurate climatological benchmarks can be established. Against such benchmarks could be measured the slow temperature change caused by greenhouse warming, change of transports of currents, spreading of pollutants, and changes in primary productivity (Stommel, 1990). Archives such as this will be basic when it comes to testing numerical climate models of the ocean in the future.

What would such a global ocean observations system cost? Since the design of the comprehensive system is not yet in place, costs cannot be estimated with any certainty. But we might expect that the costs of the system would be at least that of continuing a programme like WOCE which is already global in scope. The costs will also involve data storage, handling, and transfer facilities, related computer facilities for modeling and analysis. These may be just as large as the instrumentation and field work costs. One way to reduce costs is to develop autonomous instrument technology that can replace the intensive research programmes.

Planning for a global observational system is now vigorously under way. The JSC and CCCO have jointly established an Ocean Observing System Development Panel (OOSDP). The Panel is charged with developing the overall design concepts for a global ocean observing system and is working closely with the IOC Ad Hoc Group on Ocean Observing Systems to ensure interaction with existing programmes.

But even with the best laid plans, if the institutional support is missing, the plans will not materialize. We have learned from the WWW that the operational parts of the system must be operated and funded by national agencies. It is clear that the same is true for the ocean. Adequately funded national organizations that interface effectively with international organizations and programmes are required. Today there are few of these, but support is growing. Stronger national agencies will help to develop better ties to both the academic and private user community. Stronger national agencies will also help on an international scale by providing the necessary base for coordination by WMO, IOC, and other relevant groups. It will also be necessary to look for additional sources of funding; organizations such as the World Bank could be involved.

Climate prediction requires more than just an observational system and scientific results from TOGA, WOCE, JGOFS and all the other research programmes. Climate dynamics on long time scales involves the coupling of the ocean, atmosphere, and biogeochemical cycles. Regular or operational climate predictions require routine assimilation of data into operational climate forecast models. This can be done at national centers, and pilot studies with existing data and models should be started as soon as practicable. In the end, there may be a need for an international ocean or climate analog to the successful European Center for Medium Range Weather Forecasting. With an ongoing observing system and adequate operational models, we have every reason to believe that governments will be able to make informed decisions about how to cope with the inevitable future global change.

Acknowledgements

I would like to thank the following for helpful reviews:
D. Adamec, N. Andersen, P. Brewer, L. Brown, A. Clarke, P. Dexter, D. Haidvogel, G. Kullenberg, J. Ledwell, A. Leetmaa, S. Levitus, R. Molinari, P. Morel, Y. Nagata, P. Niiler, W. Nowlin, J. Price, M. Reeve, P. Schlosser, T. Spence, R. Stewart, L. Talley, B. Taft,

R. Tipper, A. Tolkachev, B. Voituriez, R. White, W. White, J. Willebrand, G. Withee, W. Woodward, and C. Wunsch.

References

Broecker, W. S. and Denton, G. H. 1989. The role of ocean-atmosphere reorganizations in glacial cycles. *Geochimica et Cosmochimica Acta* 53: 2465–2501

Levitus, S. 1982. *Climatological Atlas of the World Ocean.* US Department of Commerce National Oceanic and Atmospheric Administration, 173 pp.

Manabe, S. and Stouffer, R. J. 1988. Two Stable Equilibria of a Coupled Ocean-Atmosphere Model. *J. Climate.* 1: 841–866.

McBean, G., McCarthy, J., Morel, P., Browning, K. 1990. Narrowing the Uncertainties: A Scientific Action Plan for Improved Prediction of Global Climate Change. Prepared for the Intergovernmental Panel on Climate Change Report on *Scientific Assessment of Climate Change.*

Munk, W. H. and Forbes, A. M. G. 1989. Global Ocean Warming: An Acoustic Measure? *J. Phys. Oceans.* 19: 1765-1778.

Munk, W. and Wunsch, C. 1982. Observing-the Ocean in the 1990s. *Phil Trans. R. Soc. Lond.* A 307: 439-464.

Semtner, A. J. Jr. and Chervin, R. M. 1988. A simulation of the global ocean circulation with resolved eddies. *J. Geophys. Res.* 93: 15,502-15,522.

Stommel, H. 1990. General principles for the design of a World Ocean Observing System (WOOS). Draft Manuscript.

Stone, P. H. and Risbey, J.S. 1990. On the Limitations of General Circulation Climate Models. MIT Center for Global Change Science, Report No. 2, Cambridge, Massachusetts, USA.

Stouffer, R. J., Manabe, S. and Bryan, K 1989. Interhemispheric asymmetry in climate response to a gradual increase of atmospheric CO_2. *Nature* 342: 660–662.

World Meteorological Organization (WMO). 1988. The World Weather Watch. WMO Publication No. 709. Geneva.

Short Term Climate Variability and Predictions

by J. Shukla *(USA)*

Abstract

This paper first describes the nature of short term variability of the coupled atmosphere-ocean-biosphere system as shown by analysis and diagnosis of observations during the past 100 years. By "short term" we mean those fluctuations of the coupled climate system whose time scales range from 10 days to 1000 days. We have deliberately excluded any discussion of short range weather forecasting (less than 10 days) and decadal changes (more than 1000 days).

We next present a discussion of the present status of our knowledge of the predictability of short term fluctuations of the coupled climate system. Based on a large number of observational and modeling studies using complex models of this system, we suggest that most of the major short term climate fluctuations observed during the past 100 years of reliable data are consequences of the interactions among the different components of the climate system. For example, interactions between the atmosphere and the biosphere play an important role in the maintenance of prolonged drought conditions over the land areas. Interactions between the oceans and the atmosphere produce large and significant changes in the locations and the intensities of the large scale rain belts and also produce large changes in the global atmospheric circulation patterns.

We then present a brief description of TOGA (Tropical Oceans and Global Atmosphere), which is a 10 year program (1985–1995) launched by the World Climate Research Programme (WCRP) to monitor and model the interactions between the tropical oceans and the global atmosphere. Its ultimate objective is the design and development of an ocean-atmosphere observing system for operational climate prediction using advanced models of the coupled climate system.

We also point out that the natural variability of regional climate is so large that the uncertainty of predicted climate change due to such factors as increase of greenhouse gases would be significant, unless the climate models can realistically simulate the interannual variability of the coupled climate system. For example, decadal mean global temperature can be significantly affected by the number and intensity of El-Niño events, which can be produced by interactions between the atmosphere and the oceans. Better simulation and prediction of short-term climate variability will increase our confidence in climate models used to predict climate change.

Finally, we make a recommendation that as a natural extension of the earlier and ongoing programs like the Global Atmospheric Research Program and TOGA, WCRP should now initiate a comprehensive Global Climate Prediction Program to investigate the feasibility of operational prediction of monthly, seasonal and interannual variability of regional climatic anomalies over the globe. Such a Global Climate Prediction Program would utilize realistic models of the atmosphere, oceans and biosphere including snow and sea ice. Most of the important ingredients of a global climate prediction program are already in place. The ongoing program of TOGA would need to be expanded to cover global oceans; the ongoing WCRP project of hydrological and atmospheric pilot experiments (HAPEX) would provide better treatment of the biosphere; the ongoing WCRP radiation projects would help improve the treatment of clouds and radiative processes; and the ongoing projects on sea ice research would help improve the atmosphere-ocean-sea ice interactions in the climate system model.

1. Introduction

Before coming to the topic of short-term climate variability and prediction, let us begin by asking a more fundamental question. What determines the mean climate of any region of earth's atmosphere-ocean-biosphere system?

The primary energy source for atmospheric motions is the radiation heating of the warm equatorial regions and the cooling of the cold polar regions. The actual rates of heating and cooling are determined by astronomical variables (the earth's distance from the sun, the periods of rotation of the earth around its axis and of the earth's orbit around the sun) and the planetary variables (size, shape and mass of the earth; chemical composition of the atmosphere, ocean and biosphere; and distribution of land, ocean, mountains, and vegetation). In addition to these fixed parameters, the amount of heat transported around by atmospheric and oceanic currents, which we would refer to as the dynamical variables, also plays an important role in determining the mean climate of any region on the earth.

The mean climate of the earth, therefore, is an equilibrium resulting from the various factors described above. This mean climate contains strong spatial and temporal gradients of pressure, temperature, salinity, velocity and water vapor. These gradients combined with the rotation of the earth give rise to day-to-day fluctuations of "weather" in the atmosphere and oceans which are routinely measured and diagnosed in order to predict their future evolution. It is the statistical average of these day to day weather fluctuations which gives rise to the weekly, monthly, seasonal and annual average climates, whose variability from one year to another is referred to as the interannual variability. It is no surprise, therefore, that superimposed on a well defined seasonal cycle - which itself has a rich space-time structure - there are large weekly, monthly, seasonal and interannual variations in the earth's climate system. For convenience of discussion in this paper, we shall make the following somewhat arbitrary classification of various time scales:

Time/Scale	Qualitative Description
0–10 days	Hourly & Daily Changes
10–100 days	Monthly, Intraseasonal & Seasonal Changes
100–1,000 days	Annual & Interannual Changes
1,000–10,000 days	Decadal & Interdecadal Changes
10,000-100,000 days	Centennial & Beyond

For discussion in this paper I have chosen to include the variations in the range of 10 days to 1000 days to define short-term climate variability. This means that we will not discuss weather prediction, and we will not discuss decadal and longer climate change. As a partial justification for this choice of time scales, it should be noted that the needs of water, energy and agriculture, and, in fact the entire socio-economic fabric of the global community are affected significantly by climatic fluctuations on these time scales.

Understanding and predicting short-term climate variability are also important, for such variability can be helpful in verifying climate models which are used for the prediction of climate change. Just as the numerical weather

prediction models were useful for simulation and prediction of short-term climate variability, likewise, models with realistic simulation of short term climate variability will enhance our confidence in predictions of long-term climate change. It should be noted that most of the climate models that have been used so far to predict the climate change due to the increase in greenhouse gases have not been sufficiently validated in terms of the simulation and prediction of short-term climate variability.

In addition to the two factors of societal importance and validation of climate models, there is yet another important reason for our discussion of short-term climate variability. A large body of modeling and empirical studies, and our current understanding of the mechanisms that govern interannual changes, suggest that there is a scientific basis, and indeed some hope, for making useful predictions of climate variations on seasonal and interannual time scales.

It is, of course, true that day to day atmospheric fluctuations are not predictable after a few weeks because of the chaotic nature of atmospheric motions; however, it is now recognized that there is predictability in the midst of chaos. The interactions between the atmosphere and ocean, and atmosphere and biosphere produce long-period variations in the coupled system which enhance the predictability of the coupled system for months to years. These interactions are found to be much stronger in the tropics than in the extratropics, and therefore, the predictability of the short-term climate variability is also much higher for the tropics. We will come to this point in a later section.

In this paper, we shall address the following aspects of the short-term climate variability:

- Examples of short-term climate variability
- Mechanisms of short-term climate variability
- Predictability of short-term climate variability
- Tropical Oceans Global Atmosphere (TOGA)
- A proposal to initiate an international program for prediction of global short-term climate variability.

2. Examples of Short-term Climate Variability

During the past 100 years of global observations of the earth's climate, there are many examples of significant short-term climate variability, such as the El-Niño-Southern Oscillation (ENSO), the monsoons, the tropical droughts and heat waves/severe cold winters in the extra-tropical regions. It will be pointed out in the next section that these regional short-term climate anomalies are manifestations of regional and global scale atmosphere-biosphere and atmosphere-ocean interactions. The interrelationships among El-Niño-Southern Oscillation and monsoons are the most remarkable examples of interannual changes in the coupled climate system which affect global circulation and rainfall. A comprehensive summary of mechanisms of air-sea interaction and worldwide climate anomalies associated with the 1982–83 El-Niño has been presented by Rasmusson and Wallace (1983).

It was noted by Walker (1924) that "when pressure is high in the Pacific Ocean, it tends to be low in the Indian Ocean from Africa to Australia", and for this recurrent pattern of planetary scale atmosphere fluctuations he coined the term "Southern Oscillation." It was later suggested by Bjerknes (1966) that Walker's Southern Oscillation is but one component of a coupled ocean-atmosphere climate system, the other being the sea surface temperature fluctuations in the tropical Pacific Ocean. It is the interaction between the ocean (ocean warming off the South American coast being referred to as El-Niño) and the atmosphere that is responsible for such fluctuations and produces short-term climate anomalies in different regions of the globe. Figure 1 shows fluctuations of surface pressure over Darwin, Australia (dashed line). The anomalies (departures from climatological mean) are first smoothed by a 12 month running mean and then divided by the standard deviation, and then smoothed again by a 12 month running mean. Darwin pressure has been chosen to illustrate the fluctuations in Southern Oscillation. The solid curve in this figure represents the sea surface temperature (SST) anomalies for the eastern equatorial Pacific. It can

be seen that both variables, surface pressure and SST, show long-period (2–5 years) fluctuations, and it is also remarkable that the two are highly correlated. These coupled fluctuations of tropical ocean and atmosphere are referred to as ENSO (El-Niño-Southern Oscillation).

Interannual variations in seasonal mean rainfall averaged over the whole of India also show very large interannual variability. Figure 2 shows the summer (June, July, August, September) monsoon rainfall averaged for more than 300 stations over the whole of India except the northern and the eastern hilly regions. It can be seen that even after averaging over a large spatial region and the whole monsoon season, there are significant fluctuations in monsoon rainfall from year to year, and from decade to decade. The solid line shows the 30 year running mean of the seasonal mean rainfall and the dashed line shows the standard deviation of rainfall for each 30 year period. It is remarkable that the 30 year mean as well as the variability within a thirty year mean show such large changes from one to the other thirty year period. It is unlikely that such large-scale, long-period fluctuations could be explained as a mere consequence of different sampling of high-frequency, small-scale, rain-producing disturbances. It is more likely that such fluctuations are produced by planetary scale, long-period fluctuations of the coupled ocean-land-atmosphere system.

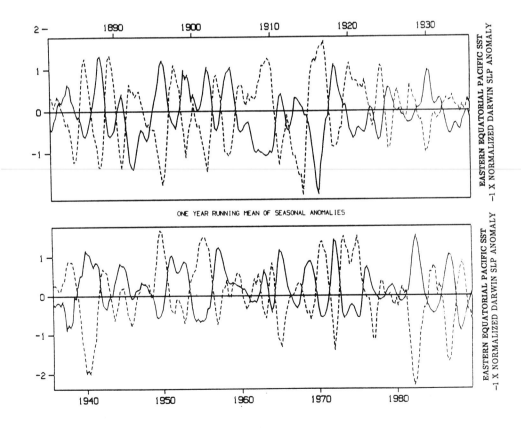

Figure 1: Darwin sea level pressure anomaly with sign reversed (dashed line) and eastern equatorial Pacific SST anomaly. Each anomaly is normalized by its own standard.

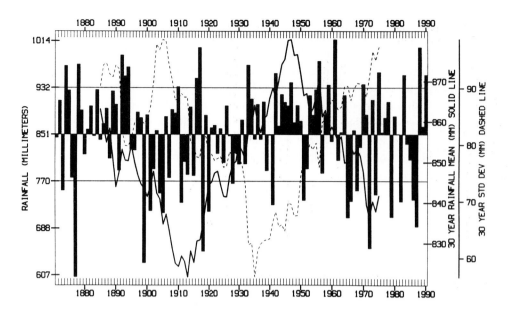

Figure 2: Summer monsoon rainfall over India (solid bars), 30 year running mean rainfall (solid line) and standard deviation for successive 30 year periods (dashed line).

These selected examples of short-term climate variability suggest that quite large changes in our climate system can, and do, occur which are not necessarily either due to external or anthropogenic factors. We will refer to such fluctuations as the natural variability of our climate system. For lack of a better definition of natural variability, we would define it as the climate variability that would occur if the planet were never inhabited by the human species. This provides a baseline for detecting and predicting changes in climate and climate variability due to human influences. According to this definition, the examples described above will be categorized as being part of the natural variability of the climate system.

3. Mechanisms of short-term climate variability

The mechanisms responsible for the short-term climate variability can be described conceptually in two categories (Shukla, 1981).

- Internal dynamics of the individual components of the climate system
- Interactions among the various components of the climate system.

Internal Dynamics:

Even if the external forcings of the solar radiation and boundary conditions at the earth's surface were constant in time, the regional atmospheric circulation will exhibit short-term climate variability due to the combined effects of the hydrodynamical instabilities of the climate system and nonlinear interactions among various scales of motion. Although the distribution of oceans, continents, and mountains are fixed with time, their interactions with fluctuating winds can produce short-term climate variability. The occurrence of nearly zonal or persistent non-zonal regional circulation regimes ("blocking") are possible examples of anomalies due to the internal dynamics of the atmosphere. Likewise, the internal dynamical instabilities of ocean currents can produce variability in the ocean circulation and possibly the overlying atmosphere. We know that the spectrum of the atmospheric observations is red. A certain amount of interannual variability will be produced solely due to the unpredictable weather and, therefore, that will remain unpredictable too.

Interactions (Atmosphere - Ocean; Atmosphere - Biosphere; Atmosphere- Cryosphere):

We suggest that all the major events of short-term climate variability observed during the past 100 years - a period for which reasonably reliable instrumental measurements of climate variables exist - are due to the interactions among the atmosphere, ocean, biosphere and cryosphere components of the climate system.

Atmosphere-Ocean Interactions:

Changes in SST produce changes in evaporation, sensible heat flux and low level moisture convergence which in turn produce changes in atmospheric heating. The anomalous atmospheric heating produces changes in atmospheric circulation which in turn produce changes in wind stress and heat flux at the ocean surface. If this air-sea coupling has a positive feedback, it can produce long-lived anomalies of SST and the associated atmospheric

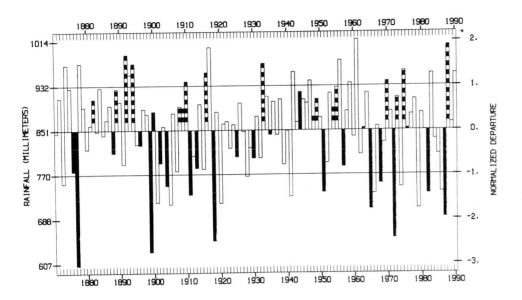

Figure 3: Summer monsoon rainfall over India. Solid and hatched bars denote the years when the eastern equatorial Pacific SST anomaly was rising and falling respectively.

circulation. Because of the differing rotational forces in the tropics and the extra-tropics, and because tropical ocean temperatures are warmer, even a small change in SST in the tropics can produce much larger changes in moisture convergence and heating than similar SST changes in the extratropics. This is the main mechanism for the occurrence of tropical droughts and floods which are manifestations of spatial and temporal shifts of mean climatological maxima of rainfall. The tropical atmosphere and oceans also do not have strong dynamical advections (as they do in midlatitudes) and therefore changes in atmospheric heating and surface wind stress can produce significant changes in atmospheric circulation and SST respectively. This is the primary reason why short-term climate variability of the tropical oceans and atmosphere are so strongly linked, and also why there is hope for predicting this coupled variability. In Figure 1 it was seen that the tropical SST and surface pressure fluctuations were highly correlated. Many researchers have further shown that in association with ENSO, several large regions of the globe experience droughts and heavy floods. For example, Figure 3 shows the Indian monsoon rainfall fluctuations (as in Figure 2) except that the years in which the tropical Pacific SST was rising and falling during the monsoon season are represented by black and hatched bars respectively. It is again remarkable that most of the severe droughts and floods over India occur during the anomalous warming (El-Niño) and cooling of the equatorial Pacific ocean respectively.

Atmosphere-Biosphere Interaction:

Changes in vegetation produce changes in albedo, surface roughness and soil moisture. These changes in turn produce changes in ground temperature, evaporation and sensible heat flux. Changes in horizontal gradients of ground temperature produce changes in convergence of moisture, and changes in vertical gradients of temperature along with moisture convergence produce changes in convection and rainfall, which in turn changes the soil moisture. The nature and degree of this interaction again depend on the character of the dynamical circulation regime where the land surface changes are taking place. The occurrence of prolonged droughts in sub-tropical regions (where the atmospheric dynamics are relatively weak) and even the tendency of heat waves to persist in the extra-tropical regions can be explained, at least in part, by such atmosphere-biosphere interactions. The West African Sahel has experienced persistent drought conditions for more than 20 years with significant interannual changes during these 20 years. Figure 4 shows fluctuations of rainfall over the west of the Sahel during the period 1940-1990. It can be seen that Sahel rainfall, like the Indian monsoon rainfall shown in Figure 2, displays large year to year changes in seasonal mean rainfall. However, in addition there is a significant shift in the mean rainfall after 1968. While such shifts in seasonal and annual mean values are not uncommon for regionally averaged climatic parameters, this is a rather unique case because there is not a single year during the past 20 years when the seasonal rainfall was significantly above the climatological normal. Atmosphere-ocean interactions over the global oceans, as well as local atmosphere-biosphere interactions have been suggested as

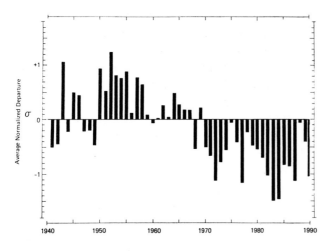

Figure 4: Rainfall index for sub-Saharan West Africa. Anomaly is normalized by standard deviation (courtesy of P.J. Lamb).

possible causes for these changes. Since monsoon rainfall over India as well as China also showed a notable shift (reduction) from the decade of 1950s to the decade of 1970s, it is reasonable to assume that this decadal shift in rainfall was perhaps due to some planetary scale circulation changes. However, it is quite likely that the local atmosphere-biosphere interactions, exacerbated by human activities leading to changes in the land-surface properties, could contribute towards the continuation of the reduced rainfall regime.

4.   Predictability of short-term climate variability

It is well known that the instantaneous weather conditions are not predictable beyond a few days. It is also well understood that this lack of predictability is due to dynamical instabilities and non-linear interactions which amplify even very small initial uncertainties which may be either due to inadequate observations or imperfect equations for physical principles (Lorenz, 1965). However, it should be noted that lack of deterministic predictability of day-to-day weather beyond a few weeks does not necessarily mean that space and time averages for a month or season or beyond are also not predictable. In fact, we would like to propose that the large body of observational, theoretical and modeling results collectively suggest that there is indeed a scientific basis for the predictability of space-time averaged short-term climate variability. The primary scientific reasons for such an optimistic view on the predictability of short-term climate can be summarized as follows:

> The space time spectra of atmospheric observations show that most of the variance in the interannual variability is accounted for by long-period, large scale fluctuations which are intrinsically more predictable than the day-to-day small-scale weather

systems, and it is these relatively longer period large scale variations which are important for the prediction of short-term climate variability. In addition, atmosphere-ocean and atmosphere-biosphere interactions produce predictable changes in the coupled climate system. The atmospheric circulation anomalies are likely to be more predictable for those time scales for which the boundary forcings due to the anomalies in SST and soil moisture can also be predicted. For time periods beyond the persistence of boundary forcings, we must be able to predict the evolution of the boundary forcings themselves. Recent developments in the modeling of the coupled system suggests that for the particular example of ENSO, the coupled tropical ocean-atmosphere climate system is theoretically predictable up to 1–2 years (Goswami and Shukla, 1990).

However, while considering the predictability of any climate signal, we must also consider the possible sources of climate noise which would tend to introduce a lack of confidence in the predictions. The following table gives some simple examples of possible signals we may wish to predict and important sources of noise which will make the predictions unreliable.

SIGNAL	NOISE
Daily Weather	Thunderstorms
1–10 Day Mean	Cyclones
Monthly Mean Climate	Blocking
Seasonal Mean Climate	30–60 Day Oscillations
Interannual (1–3 Year) Climate (ENSO)	Coupled Air-Sea Instabilities
Decadal Climate Change	ENSO

It should be noted that predictions of monthly and seasonal mean climate anomalies (from a given initial state) will be affected by the presence or absence of blocking regime, and the amplitude and phase of 30–60 day oscillations. Similarly, the predictability of ENSO will be determined by the instabilities of the coupled ocean-atmosphere system much like the predictability of short and medium range weather is influenced by baroclinic instability in the atmosphere. It should also be noted that the predictability of decadal climate changes (e.g., greenhouse warming) will be strongly influenced by the intensity and frequency of ENSO events. Pan and Oort (1983) have shown that interannual changes in the global mean temperature of the entire atmospheric column are highly correlated with SST anomalies in the eastern equatorial Pacific.

This is particularly relevant for the ongoing controversy over the detection and prediction of greenhouse warming. Based on the observational record of the global mean temperature, it is not possible to conclude that the climate change has been detected. The observed changes are entirely within the range of the natural variability of the coupled climate system. Likewise, the present climate models - the ones used for predicting effects of increased greenhouse gases - have not been adequately validated against the actual observed climate variability during the past. Therefore, there is no strong basis to accept the model predicted changes in global climate. The present climate models show large systematic errors in simulating the mean climate. It is assumed that although the simulated mean climate of the models is wrong, the differences between the simulated climate for the future (with increased greenhouse gases) and the simulated climate for the present can be accepted because the errors in simulating the present climate get removed when we subtract one model simulation from the other. This assumption is generally correct if the models were used to simulate the direct response of a strong external forcing. The simulation of the effects of increased greenhouse gases falls in an altogether different category. The direct radiative forcing due to increased greenhouse gases is quite small (3–4 Watts m^{-2} compared to the mean value of about 300 Watts m^{-2}), and the corresponding increase in surface temperature will also be quite small. Thus, the model predicted climate changes due to greenhouse gases are entirely due to a number of positive feedbacks as formulated in the model parameterizations. Therefore, in order to accept the model results, it is quite important that the models are validated against some known climate variations in the past before we can have confidence in the model predictions for the future. It is in this context that it is considered particularly important that the climate models are validated for their ability to simulate and predict the short-term climate variability including the frequency and intensity of El Niño events.

5. Tropical Ocean Global Atmosphere (TOGA)

Recognizing the role of tropical SST anomalies in forcing global scale atmospheric circulation anomalies, and that the tropical SST anomalies are deterministically forced by atmospheric circulation anomalies, the World Meteorological Organization (WMO) and the International Council of Scientific Unions (ICSU) established a scientific steering group for the international TOGA program which is an element of the World Climate Research Programme (WCRP).

The scientific objectives of TOGA are (WMO, 1985):

1. To gain a description of the tropical oceans and the global atmosphere as a time-dependent system, in order to determine the extent to which this system is predictable on time scales of months to years, and to understand the mechanisms and processes underlying that predictability

2. To study the feasibility of modeling the coupled ocean-atmosphere system for the purpose of predicting its variations on time scales of months to years

3. To provide the scientific background for designing an observing and data transmission system for operational prediction if this capability is demonstrated by coupled ocean-atmosphere models.

TOGA was conceived as a ten year program (1985–1995) of data collection and modeling research. It is hoped that by use of advanced four-dimensional data assimilation techinques, all the surface and sub-surface ocean observations can be synthesized to produce an internally consistent basin-wide synoptic description of tropical oceans. Likewise, an internally consistent homogeneous data set for the four-dimensional structure of the global atmosphere can also be produced. Current efforts of tropical ocean data assimilation in USA and France have already begun to produce basin scale synoptic maps of ocean circulation. This is a major breakthrough for dynamical oceanography.

The ongoing modeling efforts using atmospheric models with prescribed SST, ocean models with prescribed wind stress and heat flux, and coupled ocean-atmosphere models have produced the highly promising result that interactions between the tropical oceans and the global atmosphere enhance the predictability of short-term climate variability. A description of the accomplishments in the first five years of the US TOGA program and challenges for the future are summarized in a recent report (TOGA, 1990).

It should be recognized, however, that although the largest predictable part of the interannual variability of climate arises from ENSO, and more generally TOGA phenomena, to successfully carry out the prediction of short-term climate variability on seasonal and longer time scales will require adequate treatment of the other important interactive components of the climate system, including the extra-tropical oceans, land surface processes (biosphere) and variations in snow cover and sea ice.

6. A Proposal

In order to exploit the scientific advances in understanding the dynamics of the coupled Tropical Ocean/Global Atmosphere system as well as relevant results of other studies by WCRP and Climate and Global Change Programs, serious consideration should be given to the initiation of an international program for the prediction of global short-term climate variability.

The overall objectives of this program might be to:

- Provide real-time predictions of variations of the earth climate system on time scales of seasons to several years
- Validate predictive models of global climate change by demonstrating skillful forecasts of short-term variations of the coupled ocean-land-atmosphere climate system

A transition from TOGA and other ongoing WCRP programs to this project will require a transition from ocean models that focus entirely on tropical oceans to global atmospheric models with fully interactive land-surface processes including snow cover, and global oceans including sea ice. It is likely that in the initial phase, such a program may need to take into account only the thermodynamics of the global upper ocean and the fast dynamics of tropical ocean basins. The information base and the results from this program will be highly valuable in putting quantitative confidence limits on predictions of climate change on decadal time scales.

The program might include three main components, in addition to development of global observing systems foreseen in the framework of the World Weather Watch (WWW), World Ocean Circulation Experiment (WOCE) and Global Energy and Water Cycle Experiment (GEWEX). These components are:

(i) A global atmosphere-ocean-land climate data analysis and prediction component, based on one or several dedicated central facilities for data acquisition, analysis, quality control and climate forecasts

(ii) An operational global observing system to provide the required data inputs for atmosphere, surface and upper ocean, sea ice, snow cover and soil moisture

(iii) A research component to address outstanding problems and new scientific issues which may arise in the course of the program.

It is recognized that for entirely fundamental scientific reasons (differences in the rotational force, dynamical instabilities and non-adiabatic heat sources), the potential predictability of the tropical atmosphere and oceans is much higher than that of the extratropics. Therefore, initially, the greatest beneficiaries of any organized, internationally coordinated effort in short-term climate prediction will be the tropical countries. However, the tropical countries do not have, at this time, the required resources of trained scientific personnel and computation-communication facilities to exploit this gift from nature for the well being of their respective societies. Therefore, we

conclude with a suggestion that the nations of the world join together to exploit the recent scientific advances by initiating an international program on the prediction of short-term climate variability.

References

Bjerknes, J.: 1966, A possible response of the atmospheric circulation to equatorial anomalies of ocean temperature. Tellus, 18, 820–829.

Goswami, B.N. and Shukla, J.: 1990, Predictability of a Coupled Ocean-Atmosphere Model. J. of Climate, 3, 2–23.

Lorenz, E. N.: 1965, A study of the predictability of a 28-variable atmospheric model. Tellus, 17, 321–333.

Pan, Y.H. and Oort, A.H.: 1983, Global climate variations connected with sea surface temperature anomalies in the eastern equatorial Pacific Ocean for the 1958–73 period. Mon. Wea. Rev., 111, 1244–1258.

Rasmusson, E.M. and Wallace, J.M.: 1983, Meteorological aspects of the El Nino/Southern Oscillation. Science, 222, 1195–1202.

Shukla, J.: 1981, Dynamical predictability of monthly means. J. Atmos. Sci., 38, 2547–2572.

Tropical Ocean Global Atmosphere (TOGA): 1990, A review of progress and future opportunities. National Research Council, National Academy Press Washington, D.C.

Walker, G.T.: 1924, Correlations in seasonal variation of weather. Mem. India Met. Dept., 24, 333–345.

World Meteorological Organization (WMO): 1985, Scientific plan for the Tropical Ocean Global Atmosphere Programme. WCRP Publication Series No. 3, WMO/TD 64, Geneva.

Paleodata, Paleoclimates and the Greenhouse Effect

by Hans Oeschger *(Switzerland)*

Abstract

Data from the past provide information on the history of climate per se, as well as on the related changes – both causes and effects – in the entire Earth system; and they are necessary for determining natural baselines for environmental parameters (such as atmospheric greenhouse gas concentrations), against which human impacts can be measured. Paleoclimate reconstructions (derived from the oxygen and hydrogen isotopic composition of polar ice) and atmospheric trace gas data (from bubbles trapped in ice) provide a record of natural climatic variability and greenhouse gas fluctuations over the past 160 000 years. Data covering the most recent glacial cycle, the last deglaciation and the Holocene are of particular interest in assessing the potential environmental and climatic impacts of human-induced rises in greenhouse gas concentration.

Studies of the composition of the gases occluded in Antarctic ice enabled researchers to determine the preindustrial concentrations of CO_2 (ca. 280 ppmv), CH_4 (ca. 0.8 ppmv) and N_2O (ca. 0.29 ppmv) as well as their increases during the period before direct atmospheric measurements began. Whereas in the present interglacial the concentrations of these gases have been relatively constant, CO_2 and CH_4 varied in step with major surface temperature changes recorded in the deep ice core drilled at Vostok Station, Antarctica, as well as in Greenland ice cores. This implies that greenhouse forcing played a critical role in the climate changes of the last glacial cycle. Interpretation of Vostok ice core data suggests that about half of the warming during the last glacial-interglacial transition can be attributed to the greenhouse gases CO_2 (40%) and CH_4 (10%).

The greenhouse gas/climate relationship is still under investigation; changes in ocean circulation probably affected climate and the atmospheric concentration of CO_2 (changes in the ocean's biospheric processes) simultaneously. The continental sources of CH_4 then responded rapidly to the climatic shifts. Still, we are far from a detailed understanding of the greenhouse gas/climate interactions of the past.

The records of the climatic change in marine sediments and the $d^{18}O$ and d^2H records of the last 100,000 to 160,000 years in ice cores support the Milankovitch theory of ice ages. However, the Greenland ice cores also show rapid transitions between a cold and a mild climate state during the glacial period, which are probably due to the switching on and off of North Atlantic Deep Water formation. The existence of two (and perhaps more) modes of operation of the Earth system is of relevance with reference to the anticipated climate change due to human forcing.

1. Introduction

The ultimate goal of global change research is to develop an integrated Earth system model with predictive capabilities that will link the full range of physical, chemical and biological interactions that comprise the total system. Even optimistic estimates suggest that this goal is still decades from becoming a reality (Woods, personal communication). Earth system models are based, in part, on well known, fundamental equations of physics and chemistry; but many complex processes, particularly interactions involving the biosphere, must be simulated by parameterization. Most of the models that have been developed to predict future climate, for example, only consider purely physical feedbacks, such as the role of atmospheric water vapor, surface albedo and clouds. However, other feedbacks in which biological systems play an important role (e.g., the atmospheric concentration of greenhouse gases) might also have significant climatic impacts, but are generally difficult to quantify.

A reconstruction of the evolution of Earth history is necessary to accomplish the goals of Global Change studies, which include:

- Disentangling human-induced change from what would have been the natural behavior of the system, had it not been disturbed. To accomplish this requires an assessment of climatic and atmospheric greenhouse gas concentration trends prior to human impacts

- Evaluating Earth system models. This requires a reconstruction of the forcings (e.g., solar variations, volcanic dust loading of the atmosphere) and responses (e.g., changes in temperature, precipitation, circulation patterns, vegetation) of the Earth system
- Assessing the range of natural system variability and determining the primary causes for the fluctuations. What was the relative importance of internally forced stochastic-deterministic and externally forced processes in the past?
- Understanding the complex interactions between the biosphere and the physical climate system, one of which is the biospheric feedback on atmospheric greenhouse gas concentrations.

In the remainder of this paper, the following topics will be introduced and discussed:

a) Historical and natural archives of the past and examples of pertinent data and measurements (Section 2)
b) The greenhouse effect and recent trends in trace gas concentrations (Section 3)
c) Paleoclimate and greenhouse gas variations (Section 4). Glacial cycles (data from ocean sediments and polar ice cores taken in Greenland in Antarctica)
d) The greenhouse gas-climate relationship (Section 5)

The paper then concludes with a summary of the state of present knowledge concerning these subjects (Section 6) and with an assessment of important directions for future paleoresearch as related to the issue of climate change (Section 7).

2. Data from Natural Archives of the Past

Physical, chemical and biological processes can be traced back in time through studies of a variety of variables recorded in natural archives. Two types of paleo-environmental information are available: historical, written accounts that provided proxy or direct descriptive information and the proxy information that is stored in natural archives.

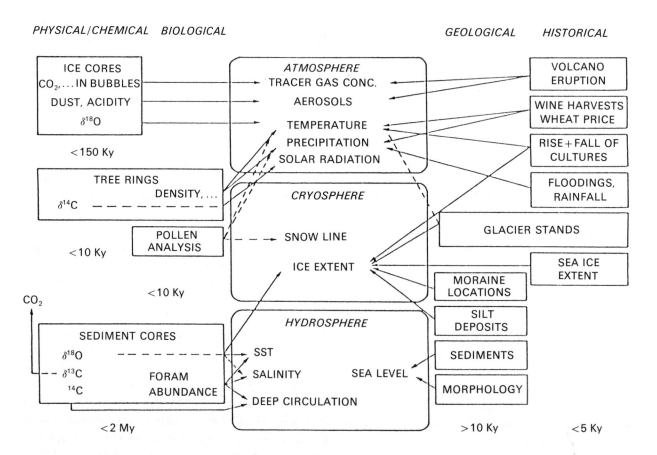

Figure 1: Summary of a selection of proxy data methods, including the various natural archive sources, the proxy data that can be obtained and examples of the types of paleoenvironmental interpretations that can be made using these paleodata (Stocker, 1989).

2.1   Types of Natural Archives

Natural archives of the history of the Earth system are diverse; and each has its own contribution to make in reconstructing paleoenvironmental conditions and in improving our understanding of how the global system functions now and how it functioned in the past. The utility of a selection of historical information and each of the different types of proxy records (including ice cores, tree rings, sediment cores, geological deposits, and pollen analysis) for the determination of past states of the atmosphere, cryosphere and hydrosphere is summarized in Figure 1 (Stocker, 1989). Proxy data also provide a great deal of information about the biosphere.

2.2   Derived environmental information

Paleoclimate reconstructions

Stable isotopes of oxygen and hydrogen atoms in the water molecule are typically used to infer paleoclimate, particularly paleotemperature. The long-term trend in climate over the past 160 000 years is clearly documented in the $d^{18}O$ and dD records of Antarctic (Jouzel et al., 1987) and Greenland ice cores (Dansgaard et al., 1984) and in the ice volume changes reconstructed from the $d^{18}O$ of foraminifera fossils in ocean sediments. Reconstructions of local or regional patterns of temperature and/or precipitation can also be derived from pollen (Birks and Birks, 1980) and tree-ring analysis (Cook and Kaiterus, 1990), and from studies of mountain glaciers and ice caps (Thompson et al., 1988).

Global forcing mechanisms and natural variability

Evidence that orbital, solar and greenhouse forcings are responsible for climate changes of the Quaternary can be found in natural archives. Periodicities of ice volume changes derived from $d^{18}O$ measurements of foraminifera fossils in ocean sediments were found to correspond to those of the orbital parameters eccentricity (100 ky), obliquity (41 ky) and precession (23 ky and 19 ky) (Hays et al., 1976). This was the first ground-truth evidence that glacial cycles are driven by changes in the Earth's orbit relative to the sun. In addition, cosmogenic isotopes (^{14}C in tree cellulose and ^{10}Be in ice cores) provide indirect information on changes in the solar energy flux supplied to the Earth; and trace gas measurements on bubbles of atmospheric air trapped in ice cores can be used to assess the magnitude of greenhouse forcing of climate change. The role of volcanically-forced changes in atmospheric turbidity is addressed through studies of ash layers in ice cores and sedimentary sequences.

Paleoenvironmental reconstructions and evolution of biogeochemical cycles

Paleodata are also necessary to document the state of the Earth system at various times in the past, in terms of such characteristics as oceanic and atmospheric circulation patterns, lake levels, vegetation patterns and landscapes, atmospheric aerosol loading, or sea-surface temperature, to name but a few. Information from the entire suite of paleoarchives is needed to obtain a clear, comprehensive picture of the Earth environment at specific times in the past on a global basis. Without these basic data, it is impossible to assess possible changes in biogeochemical cycles over time.

Utility of the polar ice archive

With respect to the issue of greenhouse forcing of climate change, ice is the most valuable paleoarchive, because it is directly coupled to the atmosphere (Oeschger and Langway, 1989). Ice cores can be analyzed for the composition of atmospheric gases trapped in bubbles in solid ice (e.g., total gas, CO_2, CH_4, N_2O, CH_3Cl, light hydrocarbons, O_2/Ar); for stable isotopes in gas bubbles and ice; and for cosmogenic isotopes, inorganic and organic chemistry, particulates, conductivity, and physical properties (Table 1). In ice older than about 15 ky, the resolution of the record limits dating accuracy using current state-of-the-art methods.

3.   Greenhouse Effect and Recent Trends in Greenhouse Gases

The climate of the Earth is affected by changes in radiative forcing due to several sources (known as radiative forcing agents). These include the concentrations of radiatively active (greenhouse) gases, solar radiation, aerosols, and albedo. In addition to their direct radiative effect on climate, many gases produce indirect effects on global radiative forcing (IPCC, 1990a).

3.1   The Greenhouse Effect

Polyatomic molecules in the atmosphere absorb infrared radiation. Thus the presence of such molecules increases the fraction of the long-wave radiation emitted by the Earth that is absorbed by the atmosphere. This "greenhouse effect" causes a warming of the lower atmosphere and Earth surface, and a compensating cooling of the upper atmosphere. Without greenhouse gases in the atmosphere, the Earth would be a frozen, lifeless planet.

Calculations show that the excess CO_2 that has accumulated in the atmosphere since the beginning of the last century has increased the radiative heating of the Earth's surface by an average of about 1.3 Watt/m^2 (enhanced greenhouse effect); when the other major greenhouse gases are also included, this value rises to 2.2 Watt/m^2 (WCRP, 1990). Doubling the concentration of CO_2 from 300 to 600 ppmv would yield a net increase in radiative heating of about 4 Watt/m^2 (Ramanathan, 1988).

Table 1: Short summary of climatic information and corresponding ice core signal

Atmosphere	Ice core
Temperature	D/H, $^{18}O/^{16}O$
Precipitation	d/h, $^{18}/O^{16}O$, ^{10}Be
Humidity	D/H, $^{18}O/^{16}O$
Aerosols	
• natural (continents, sea, volcanoes, biosphere)	chemicals A1, Ca, Na, H^+, SO_4, NO_3
• man made	SO_4, NO_3, Pb, radioactive fallout
Circulation	Particles
Gases: Natural and man made	O_2, N_2, CO_2, CH_4, N_2O

3.2   Preindustrial CO_2, CH_4 and N_2O Baselines and Anthropogenic Inputs

Trends in CO_2, CH_4 and N_2O since 1750 are presented in Figure 2 (IPCC, 1990b) and discussed separately below.

Carbon dioxide (CO_2)

Precise, continuous measurements of the atmospheric CO_2 content have been performed since 1958 at two stations: South Pole Station and Mauna Loa, Hawaii (Keeling et al., 1982). For the past 30 years, CO_2 has been increasing at the rate of about 1 ppmv/yr (0.4%/yr), with a current rate of 1.8 ppmv/yr (0.5%/yr; IPCC (1990a)). Atmospheric measurements prior to 1958 are afflicted with uncertainties for a variety of reasons, but during the last decade, the analysis of air bubbles trapped in ice sheets and glaciers has allowed an accurate reconstruction of the atmospheric CO_2 increase since the beginning of the last century (Figure 2a).

Results from Antarctic ice cores indicate a preindustrial atmospheric CO_2 concentration of about 280 ppmv (Neftel et al., 1985; Raynaud and Barnola, 1985; Pearman et al., 1986). Additional data suggest that this value varied by less than 10 ppmv throughout the millennium prior to about 1800 A.D. The increase in atmospheric CO_2 plotted in Figure 2a, for the period 1740 A.D. (ca. 280 ppmv) to present (> 350 ppmv), therefore, can be unambiguously attributed to human activities, primarily fossil fuel burning and minor effects of agricultural soil organic matter oxidation and deforestation (Oeschger and Siegenthaler, 1988). This monotonous increase is astonishing, in light of the fact that the human contribution to atmospheric CO_2 is minor relative to the 30 times larger natural CO_2 exchange fluxes. Why has the natural CO_2 system not had a larger buffer effect on anthropogenic inputs? Given the present CO_2 balance, the magnitude of natural exchange fluxes to and from the atmosphere is such that if the average annual CO_2 flux out of the atmosphere were only about 1-2% lower than the influx, the atmospheric concentration would

have increased similarly to the observed anthropogenic trend following industrialization.

Because of the slow response time of the ocean reservoir, even if all anthropogenic emissions of CO_2 were immediately stopped, the atmospheric CO_2 concentration would not approach its preindustrial level for many centuries. By some model estimates, an immediate > 60% reduction in global anthropogenic emissions would be necessary to stabilize atmospheric CO_2 at the current level of 353 ppmv (IPCC, 1990a).

Methane (CH_4)

For the past 20-30 years atmospheric CH_4 has been increasing at a rate of about 1%/yr, or 16 ppb/yr (Figure 2b; Rasmussen and Khalil, 1981). Ice core data show that its increasing trend began only about 200 years ago, with the result that there is at least twice as much methane in the troposphere now as at any time in the past 16 000 years (see section 4). It has been observed that the methane trend parallels world population growth. Part of the observed increase in CH_4 could be due to a reduction in the oxidation capacity of the atmosphere caused by a net depletion of OH radicals (which is the major CH_4 sink), but the largest fraction of the enhanced emissions is attributed to human activities, primarily rice cultivation, animal husbandry, fossil fuel recovery and burning, and landfills. These human sources account for about half of present CH_4 emissions, with the remainder attributed to natural sources. particularly wetlands.

In order to stabilize the atmospheric methane concentration at the current level, an immediate 15-20% reduction in global man-made emissions would be necessary (IPCC, 1990a).

Nitrous oxide (N_2O)

Nitrous oxide has also increased in the past 30 years, although at a much slower rate that CO_2 and CH_4 (Figure 2c). Atmospheric N_2O is presently increasing by

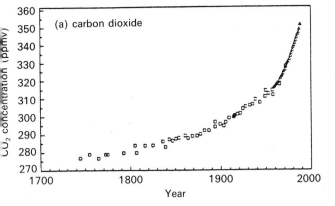

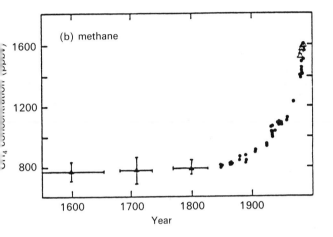

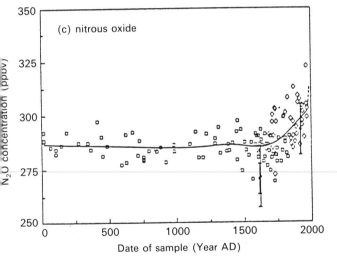

Figure 2: Trends in atmospheric greenhouse gas concentrations prior to and following industrialization: a, CO_2 since 1740 AD, based on air trapped in ice from Siple Station, Antarctica (squares; Neftel et al., 1985; Friedi et al., 1986) and by direct atmospheric measurements at Mauna Loa, Hawaii (crosses; Keeling et al., 1989); b, CH_4 since 1600 AD (Etheridge et al., 1988; Pearman and Fraser 1989); and c, N_2O for the past 2 ky (squares, Khalil and Rasmussen, 1988; diamonds, Etheridge et al., 1988; filled circles, Zardini et al., 1989). All figures from IPCC (1990a).

about 0.8 ppbv/yr, which is a 0.25% annual increase. Quantification of the various natural and anthropogenic sources is uncetain, and it is difficult to account for the annual increase based on known sources. The total source needed to account for the observed annual atmospheric increase is 10-17.5 Tg N/yr against a flux of N_2O from known sources of only 4.4-10.5 Tg N/yr (IPCC, 1990a). These data suggest that there are missing sources of N_2O, or that the strengths of some of the identified sources have been underestimated.

In order to stabilize the N_2O concentration at the present-day level, an immediate reduction of 70-80% of the enhanced flux since industrialization would be required (IPCC, 1990a).

3.3 The Key Question to be Answered

To what extent do changes in the concentration of atmospheric greenhouse gases force climate change?

4. Paleoclimate and Atmospheric Gas Concentration Data

4.1 Principal Ice Core information

Paleoclimate indicators

The isotopic fractionation of H and O during evaporation and condensation of water lead to a linear relationship between temperature and the depletion of heavier isotopes in water (Dansgaard, 1964; Johnsen et al., 1988). Thus $d^{18}O$ and dD measurements on polar ice cores allow local surface temperature reconstructions back to 160 ky BP, at present.

Atmospheric Greenhouse Gas Concentrations

Natural ice contains air bubbles that were entrapped during the change from snow to ice, which takes place at depths between 80-120 m in polar ice sheets, over periods of a few decades to centuries (which represents the age distribution of the gas trapped in bubbles) depending on temperature and accumulation rates (Stauffer et al., 1985). Because both CO_2 and CH_4 are well-mixed in the atmosphere, measurements of these gases in ice cores can generally be taken as global values. In the absence of melt layer (which lead to a CO_2 enrichment), concentrations of bubble gases are indicative of atmospheric values.

4.2 Natural Climatic Variability and Atmospheric Greenhouse Gas Fluctuations

Paleoclimate reconstructions (derived from the oxygen and hydrogen isotopic composition of polar ice) and atmospheric trace gas data (from air bubbles trapped in ice) currently provide a record of natural climatic variability and greenhouse gas fluctuations over the past 160,000 years, with plans to extend the record (although at lower resolution) to about 300 ky BP. It is hoped that this level will be reached in Greenland by 1992, in conjunction with the US Greenland Ice Sheet Project (GISP II) and the European Science Foundation Greenland Ice Core Project (GRIP). Data covering the most recent glacial cycle, the last deglaciation and the Holocene are of particular interest in assessing the potential environmental and climatic impacts of human-induced rises in greenhouse gas concentrations.

Glacial-Interglacial Climatic Cycles and Atmospheric Gases

Long ice cores drilled in Antarctica and Greenland both provide clear evidence for climatic fluctuations associated with the 100,000-yr glacial cycles that have dominated the Earth's climate throughout Pleistocene and Holocene (Recent) time. It is well established –both on theoretical and empirical grounds– that the effect of the sum of changes in the Earth's orbital parameters, referred to as orbital forcing, is the major cause of the glacial cycles. This (Milankovitch) theory states that it is the interplay of changes in eccentricity (with a period of 100 ky), axial tilt, or obliquity (41 ky period), and precession (23 ky and 19 ky periods), which modulate July insolation at 65°N, the parameter believed to be the main control on the mass balance of ice sheets during a glacial - interglacial period.

The record of the ice ages was first described for Indian Ocean sediments (Hays et al., 1976), but has since been seen clearly in Greenland and Antarctic isotope and trace gas records from ice cores. Figure 3 presents ΔTa (Jouzel et al., 1987), CO_2 (Barnola et al.,1987) and CH_4 (Chappelaz et al., 1990) data spanning the past 160ky for a core drilled at Vostok Station. The previous and most recent deglaciations are marked by large, abrupt increases in methane (>300 ppbv), surface temperature (>9°C) and carbon dioxide (>100ppmv). The transition to full glacial conditions during the last ice age began about 130 ky ago, but the glacial maximum did not occur until about 18 ky BP. Greenland cores show similar events, but are unique in the amount of variability during the glacial period, which will be discussed in the following section. Relatively rapid terminations of glacial conditions and more gradual interglacial-glacial transitions are characteristic of glacial cycles, and are due to feedback mechanisms between insolation, ice sheet mass balance, albedo, etc.

Rapid Climatic Oscillations During the Previous Glacial Period

In addition to the relatively slow glacial-interglacial climatic variations that have been documented in Antarctic ice and marine sediment cores, supporting the Milankovitch theory, evidence for rapid climatic oscillations has been detected both in the Dye 3 and Camp Century ice cores (Dansgaard et al., 1982), during the time interval 80–30 ky

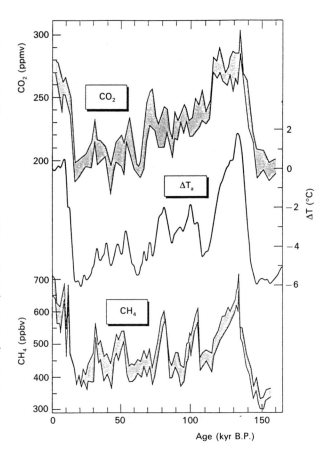

Figure 3: Vostok, Antarctica, ice core records for methane (top curve; Chappellaz et al., 1990), deviations in isotope temperature above the inversion layer (as a difference from the modern value of ca. -40°C; middle curve; Jouzel et al., 1987) and carbon dioxide (Barnola et al., 1987). The CH_4 and CO_2 curves have been adjusted to fit the timescale given (taking into consideration gas occlusion time).

BP (i.e., Stage 3 of the marine oxygen-isotope record; Figure 4).

Between 40-30 ky BP, for example, rapid CO_2 variations with periods of the order of 1000 years and an amplitude of about 50 ppm (about half the glacial-interglacial amplitude) are recorded in the Greenland cores (Stauffer et al., 1984). The $d^{18}O$ values alternated between two levels corresponding to approximately −35.5 and −32 per mil (Dye 3) and −42 and −38 per mil (Camp Century), Figure 4. Corresponding temperature changes are of the order of 5°-6°C, and dust concentrations changed drastically, as well (Dansgaard et al., 1984; Langway et al., 1984). The changes between a cold and mild climatic conditions are also reflected in ^{10}Be contents. In fact, all of the parameters measured in Greenland ice cores fall into distinct, narrow bands for the two states, suggesting that, during the last glacial period, the Earth system oscillated between two quasi-stable systems states (Oeschger et al., 1985).

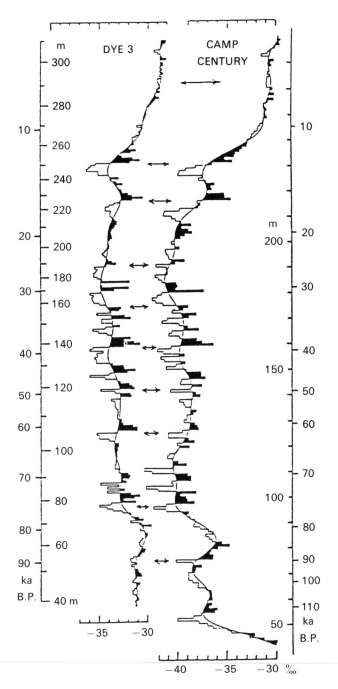

Figure 4: $d^{18}O$ profiles for Greenland ice cores showing rapid bi-modal fluctuations during the last glacial period (Oeschger et al., 1984)

It has been hypothesized the main driver for the climatic oscillations is changing circulation in the North Atlantic Ocean (Broecker et al., 1985). When water vapor is exported from one ocean basin drainage to another, salt is left behind by evaporation, increasing the salinity and density of the surface water mass, thus enhancing deep-water formation. This, in turn, affects ocean circulation, sea-surface temperature and the amount of heat available to warm the air currents that regulate the climate of the North Atlantic region, particularly the European subcontinent.

The short duration of the bimodal North Atlantic oscillations precludes orbital forcing mechanisms and suggests the possibility of oscillation internal to the Earth system, resulting from complex Earth system feedback mechanisms (Broecker et al., 1985). Broecker and Denton (1989) point out that the ocean-atmosphere system is susceptible to mode switches, because of the nonlinear coupling of these two reservoirs. Changes in water vapor transport, for example, can lead to changes in both the rate and pattern of ocean circulation (Broecker and Denton, 1989).

In contrast with the climate signal recorded in Greenland ice cores during the last glacial period, however, Antarctic records lack a pronounced bimodal signal. A series of detailed CO_2 measurements performed on the section of ice core from Byrd Station, Antarctica, that should have recorded rapid CO_2 variations showed no such signal, although indications of smaller variations were observed. One explanation for the discrepancy might be a smoothing of the CO_2 variations due to the longer occlusion time for air in the Byrd core relative to the Greenland cores. Drilling of a new, high-resolution Antarctic core should provide the data necessary to test this hypothesis (Neftel et al., 1988). CO_2 and CH_4 signals are expected to be globally in-phase, because they are well-mixed in the atmosphere on yearly time scales but it is possible that the transfer of the North Atlantic climate signal to the Antarctic might have resulted in a smoothing of the signal, such that the fewer $d^{18}O$ oscillations observed in the Vostok core during the last glacial period (Figure 3) represent clusters of several Greenland oscillations.

The Last Glacial-Postglacial Transition

The transition from glacial to postglacial conditions in Greenland and Antarctic occurred at the same time. The timing of the transition can be inferred from changes in the CO_2 concentration. However, in Antarctica the transition is more monotonous than in Greenland, where strong climatic variations are superimposed on the general warming trend.

In Greenland (Dye 3 and Camp Century), initial warming occurred about 13 ky BP (Figure 5). After a warm period of about 200 years, a strong cooling followed which lasted about 1000 years. At about 10 ky BP, within less than 100 yrs, the final transition to the Holocene followed. This climatic sequence, even in the details, matches the $d^{18}O$ record in lake marl from Lake Gerzen (Gerzensee, Switzerland) and many other European records that have been radiocarbon-dated. The ^{14}C-ages cited for the transition in the different archives differ somewhat from the absolute ages. In Figure 5 it is shown that, shortly before the initial warming (13 ky BP), the CO_2 concentration was still low (ca.200 ppmv). During the Bølling-Allerod warm period, CO_2 was about 300 ppmv, and there are indications of lower values (250 ppmv) during the Younger Dryas cold

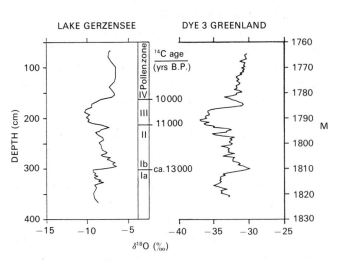

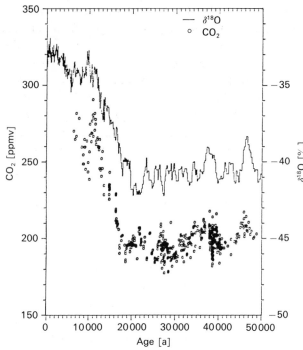

Figure 5: Comparison of a section of the $d^{18}O$ profile from the Dye 3 ice core (right) with the $d^{18}O$ record in lake carbonate from Gerzensee, Switzerland (adapted from Oeschger et al., 1984). The profiles span late glacial and early postglacial times (original timescale based on ^{14}C dating of the Lake Gerzen carbonates (Eicher, 1980)).

Figure 6: Variations in CO_2 and $d^{18}O$ at Byrd Station, Antarctica, for the last 50 000 years (unpublished data). The 10% drop in CO_2 at the end of the last deglaciation (10 ky BP) may have been caused by vegetative regrowth following the melting of continental ice sheets.

period. However, the high values (>300 ppmv) at the beginning of the Holocene indicate that, during the postglacial, meltlayers resulted in CO_2 values higher that the atmospheric concentrations.

In the Antarctic ice cores (e.g., Figure 6), the increases in stable isotopes and CO_2 occurred parallel and, within the uncertainties, in phase. In all the Antarctic cores a slight cooling in the upper part of the increase is indicated. In the Vostok core it is of the order of 2°C. Given the uncertainties in the time scales, this cooling probably corresponds to the Younger Dryas cold period observed in the Northern Hemisphere. This correspondence is supported by a significant decrease of the CH_4 concentrations occurring in the Greenland cores as well as in the Vostok core (for the Vostok comparison, refer to Figure 3).

The problem of assessing the possible causes for the Younger Dryas is currently attracting considerable interest among the scientific community. The short duration of the cold event makes it difficult to tie it to cycles of orbital forcing; and it has been hypothesized that the oscillation is related to changes in the heat input into the North Atlantic caused, specifically, by the interruption of deep water formation in the Norwegian Sea. Interestingly, the Younger Dryas does not seem to be a unique event, but rather the last of a series of similar events during a major part of the last glaciation (see discussion of rapid oscillations in Greenland ice cores, above). The importance of ocean circulation in influencing the state of the climate system

has been recognized and theoretically studied since the 1960's (e.g., Stommel, 1961; Manabe and Stouffer, 1988); and the possibility of more than one stable climatic state was shown by these researchers. New research will focus on two fundamental questions:

- Was the Younger Dryas a unique event related to the deglaciation of North America, or rather the last of a series of similar events during a major part of the last glaciation?
- How was the signal of North Atlantic climate change spread over the globe? Candidates are via changes in ocean circulation and by reduction in the greenhouse effect (especially if not only CH_4, but also CO_2 showed a significant decrease).

The Holocene Climate and the Carbon Balance

Was the preindustrial earth system in equilibrium? Records of the last 10 ky are smoother and the climate appears to be more stable than during glacial times. These differences suggest that the dynamics of the climate system may be dependent on the state of the climate at any given time.

An interesting observation concerns the carbon balance: Measurements on the Byrd Station core, Antarctica, show a decrease in CO_2 from about 280 ppmv to 250 ppmv at the beginning of the Holocene (Figure 6), during a period of about 4 ky (Neftel et al., 1988). A probable explanation for this rapid decrease in CO_2 is the regrowth of soils and plants on the continents in areas which had been covered by ice during the preceding glacial period.

Assuming that ocean chemistry was determined by processes similar to those operating in the present-day ocean, the 10% decrease in atmospheric CO_2 (requiring a 10% decrease in the partial pressure of CO_2 in the surface ocean) corresponds to an extraction of about 400 GT of carbon from the ocean-atmosphere system during this 4000-yr period. In this long lasting change in the CO_2 system, interaction with the ocean sediments also played a role.

5. The Greenhouse Gas/Climate Relationship

On the time scale of 10-100 ky, insolation changes linked to the earth's orbital variations are recognized as a major forcing of the climate system (Hays et al., 1976; Imbrie et al., 1984). However, ice core records indicate that greenhouse gases such as CO_2, CH_4 and N_2O also changed significantly on the same and shorter time scales (Khalil and Rasmussen, 1989a), and modeling experiments and statistical analyses on ice core data suggest that changes in atmospheric greenhouse gases can have a large effect on climate (Broccoli and Manabe, 1987; Lorius et al., in press).

5.1 Evaluation of the Climatic Effect of Greenhouse Gases

The greenhouse gas/climate relationship is still under investigation; however, data suggest that changes in ocean circulation can simultaneously affect both climate and the atmospheric concentration of CO_2. Methane and N_2O would then respond rapidly to climatic shifts. Still, many of the causes of changes in greenhouse gas concentrations remain poorly quantified and the climatic effect of greenhouse gases is only beginning to be estimated.

Contemporary ΔT and Greenhouse Gas Trends

Are data from the past useful in explaining present trends in global T and the likely effects of increasing atmospheric concentrations of greenhouse gases?

Phase Relationships between Climate Change and Variations in Greenhouse Gases

Within the resolution of the available data, the last glacial-interglacial transition was accompanied by in-phase changes in CO_2 and CH_4. This suggests that greenhouse forcing has played a critical role in the climate changes of the last two glacial cycles; however, it is possible that changes in ocean circulation could simultaneously affect both climate and the atmospheric concentration of CO_2.

Thus cause and effect relationships between trends in atmospheric greenhouse gas concentrations and climate changes remain a topic of intense study.

It also remains to be proven whether atmospheric CO_2 and CH_4 changes enhanced the rapid climatic changes that have been detected in the Greenland ice core record. In other words, how significant are fast-acting Earth system feedback mechanisms in modulating climate changes forced by other mechanisms, such as external forcing factors like changes in the Earth's orbit? And what will human impact amount to? These remain critical unknowns.

Estimates of Global Climate Forcing by Greenhouse Gases

Best estimates of the relative importance of the various trace gases to the enhanced greenhouse effect during the period 1765–1990 are 61% for carbon dioxide, 23% for methane, 12% for CFCs as a whole, and 4% for nitrous oxide (IPCC, 1990a).

It has been estimated from Vostik ice core data that CO_2 (ca. 40%) and CH_4 (ca. 10%) explain about half of the last glacial-interglacial warming (Chappellaz et al., 1990); and GCM experiments (Broccoli and Manabe, 1987) and statistical analyses of the Vostok data (Lorius et al., in press) show that half of the forcing for glacial cycles can be attributed to changes in the atmospheric concentration of greenhouse gases.

Although summer insolation signals (Milankovich forcing) at mid-latitudes in the Northern and Southern Hemisphere were out of phase, glacier records from these regions show a simultaneous termination of the last Ice Age (Broecker and Denton, 1989). GCM modeling shows that greenhouse gas variations can partially explain the synchronous response of the hemisphere to orbitally-forced climatic shifts.

5.2 Climate Feedbacks and Other Controls on Greenhouse Gases

Feedbacks on the atmospheric concentration of greenhouse gases

High-resolution data for the Little Ice Age reveal climate-induced feedbacks on the atmospheric concentrations of N_2O and CH_4 (Khalil and Rasmussen, 1989). Both CH_4 and N_2O decreased during the Little Ice Age (1450-1750 A.D.), at rates of up to -30 ppbv CH_4/100yrs and -9 ppbv N_2O/100 yrs (Figure 7; Khalil and Rasmussen, 1989), probably in response to lowered soil emissions under cooler climatic conditions. These rates of natural change are of the same order of magnitude or faster than current increases of CH_4 and N_2O attributed to anthropogenic sources.

Other known positive feedbacks include changes in ocean temperature. Changes in ocean circulation, rates of gas exchange between ocean and atmosphere caused by changing wind patterns, modification of biogeochemical cycles, etc. all have possible feedbacks on the carbon cycle.

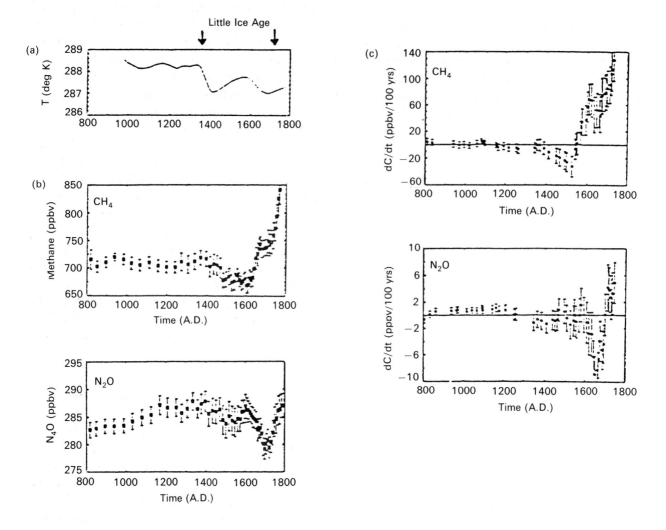

Figure 7: High-resolution ice core data reflecting the imbalance of sources and sinks during the Little Ice Age: a, temperature, b, absolute CH_4 and N_2O concentrations; and c, rates of change for atmospheric CH_4 and N_2O in ppbv/century. Error bars represent ± 90% confidence limits (Khalil and Rasmussen, 1989)

The role of the oceanic and terrestrial biosphere

The oceanic biological pump is important in regulating the atmospheric concentration of CO_2. Without marine biological activity, the partial pressure of CO_2 in surface water would dictate an equilibrium atmospheric CO_2 concentration of ca. 450 ppmv (conversely, if all nutrients were consumed, atmospheric CO_2 would be ca. 150 ppmv, which is lower than the concentration during the last glacial maximum 18ky BP). The rain of dead organic particles to the seafloor represents a continuous export flux of carbon out of the surface ocean, which is balanced by an equal upward transport of dissolved carbon by water motion. In polar regions and strong upwelling zones, where production is not limited by N or P, however, the balance could be upset, with consequent effects on the concentration of

atmospheric CO_2 (IPCC, 1990a). Ocean warming could accelerate the breakdown of dissolved organic carbon (DOC, the largest carbon sink in the ocean) to CO_2, enhancing the atmospheric concentration of CO_2 (IPCC, 1990a).

Feedbacks related to effects of soil water content, temperature, nutrient supply, and atmospheric CO_2 and ozone levels on the terrestrial biosphere could have significant impacts on the carbon cycle and the magnitude of the greenhouse effect resulting from a given temperature increase. However, most of these feedbacks are understood only quantitatively; therefore it is impossible to assess the net effect of all feedbacks between the terrestrial biosphere and climate at the present time.

6. Conclusions to Date

Studies of natural archives have enabled us to reconstruct atmospheric concentrations of CO_2 and CH_4 throughout the past 160,000 years, with detailed information on natural preindustrial background concentrations and increases in these gases and N_2O due to human activities. This information is crucial for an improved understanding of the sources and sinks of these gases, as well as for estimating the enhanced greenhouse climate-forcing effect of gases emitted through human activities.

Throughout the last glacial cycle, there is a close correlation between surface temperature, CO_2 and methane in ice cores from both Greenland and Antarctica. Whereas methane probably merely reflects a changing terrestrial production rate through time, with the possibility of minor changes in the atmospheric sink, the CO_2 changes are related to vigour of the oceanic biological pump. These changes probably are caused by variations in ocean circulation, which may, simultaneously, have influenced climate, at least in certain regions of the globe. The explanation of these variations and their phase relations is a great challenge for experiments with coupled Earth system models.

The last glacial-postglacial transition in temperature (stable isotopes in ice) was of similar size in Greenland and Antarctica and occurred during the same period, as deduced from the change of the CO_2 concentrations: At the beginning of the transition the CO_2 concentrations in the cores from both ice sheets were still low, whereas, at the end of the transition, they had reached typical postglacial levels.

However, there are also considerable differences between the temperature records from ice cores in the Northern and Southern Hemispheres. Whereas the Greenland ice cores show about a dozen changes between a cold and a mild climate state during the period 80-25 ky BP, only 4 such changes appear in the Antarctic records. One explanation is changes in the formation of North Atlantic Deep Water (NADW). When the deep water formation stopped, the influx of heat from surface ocean currents also came to an end, with the consequence of a cooling. After a few thousand years, NADW formation resumed, leading to a mild climate state. These changes in the North Atlantic region probably spread throughout the globe, with the few transitions in the Antarctic cores possibly reflecting clusters of the northern hemisphere events.

Stable isotope profiles of the last glacial-interglacial transitions in the two hemispheres also show significant differences. Two warming steps are observed in the Northern Hemisphere: the first, at 13 ky BP, and the second, at 10 ky BP, are separated by the Younger Dryas cold interval that lasted from 11-10 ky BP. In contrast, the Southern Hemisphere records are much smoother, though

in all the three Antarctic ice cores a short cooling period that likely corresponds with the North Atlantic Younger Dryas signal is indicated. The question remaining to be answered, how the cold Northern Hemisphere signal spread into the Southern Hemisphere, is very intriguing and represents a great challenge to modellers.

The behavior and subsequent effects of fluctuations in the atmospheric concentrations of greenhouse gases on climate, the biosphere, etc. are an example of the intimate coupling between the biosphere and the physical climate system. Although the importance of internal feedbacks among different components of the Earth system is beginning to be appreciated, these mechanisms remain ignored, to a large extent, by contemporary Earth system models. Data on past changes show the onset and increasing impact of human activities on the atmospheric concentration of greenhouse gases, and make it clear that this and other important feedback effects must be incorporated into models of Earth system behavior if we are to develop the capacity for accurate predictions of future changes.

7. Future Contributions of Paleoresearch to the Greenhouse Issue

7.1 Focussed Research of the ICSU-IGBP Core Project "Past Global Changes" (PAGES)

PAGES will enlist the intellectual efforts of scientists and technicians in many nations, employing the full spectrum of techniques to recover and interpret the past record of global changes. It will also require a significant increase in the level of human resources that are now being applied to the problem. The four key research activities of the project, which are extracted from IGBP Report No. 12 (IGBP, 1990), are described below.

Solar and Orbital Forcing and Response

Solar luminosity and Earth orbital parameters are the primary external controls on the amount of solar energy received by the Earth system. Research will be conducted to determine the role of external forcing mechanisms in fixing the climate of the late Quaternary including that of the last 2000 yrs, and to identify the feedbacks in the natural Earth system that respond to these forcing mechanisms to bring about major climatic changes.

Fundamental Earth System Processes

Our clearest insights into the fundamental biogeochemistry of the globe are based on the findings from polar ice cores that associate glacial-interglacial changes in surface temperature with concurrent changes in CO_2, CH_4 and aerosols. However, as discussed above, the nature of these records still precludes an unambiguous distinction between cause and effect. Greater knowledge is needed of the natural processes and feedbacks that establish the fundamental chemical cycling of the geosphere/biosphere system; the mass balance of ice sheets, and subsequent changes in sea level; the relationship between changes in biogeochemical cycles and changes in the physical climate system; how the magnitude of human-induced changes in chemistry compares with the range of naturally-induced changes of the past; and the extent to which the activities of man have disturbed the natural cycling of chemical constituents through the major reservoirs of the Earth system. The goal of this activity is to develop a better understanding of the internal dynamics of the coupled Earth system that are relevant to global environmental changes, including climate change.

Specific tasks in the PAGES projects will address this research activity initially in three subject areas: Atmospheric Trace-Gas Composition and Climate; Global Impacts of Volcanic Activity; and the Role of Ice-Sheet Mass Balance in Global Sea-Level Change.

Rapid and Abrupt Global Changes

The abrupt climatic changes of decade to century scale that have been identified in paleorecords such as the Younger Dryas event and features found in Pleistocene reconstructions of temperatures from Greenland ice cores, may pertain directly to modern concerns about rapid climate change. Such changes are possible analogues of the greenhouse induced warming now anticipated in the first decades of the next century. They reveal the natural potential for rapid change, including possible abrupt changes in ocean circulation and in the physical climate system. They are also our only source of information on the stability and natural regulation of the climate system. This activity will identify abrupt environmental changes in the record of the past and establish their characteristics (e.g., duration, global extent and causes and effects, including impacts on the biosphere).

It will be necessary to determine the internal causes of each event and to distinguish natural forcing (i.e. variation of solar radiation, solar flux, volcanic dust loading of the atmosphere and stratosphere) from possible human influence (such as changes in albedo, hydrology, the evapotranspiration balance, aerosol loading, and atmospheric trace gas concentrations), and to distinguish either type of forcing from what may be stochastic changes within the system.

Multi-Proxy Mapping

The mapping activity will support and link the other activities of the PAGES project, provide the raw data from which parameters necessary for the validation of general circulation models can be quantitively derived and allow assessment of the extent to which the past represents a useful analogue for future global changes. The activity is designed around two complementary streams of temporal emphasis:

- Stream I research is focused on the recovery, calibration, correlation, and interpretation of high-resolution (seasonal, annual, decadal) historical and natural proxy records; whereas
- Stream II emphasis is placed on the integration of long records (including reliable land-sea correlations), which will provide the data necessary to complete the studies of global changes over the last glacial cycle.

7.2   Concluding Remarks

The recovery and interpretation of historical and proxy data, including the development of tools and techniques, has been traditionally done through individual or single-laboratory efforts, employing an often specialized technique to examine typically regional or continental records spanning a limited temporal domain. The emergence of an integrated Earth system science calls for a much fuller knowledge of the past, in both space and time, and for data sets that are drawn as composites from different efforts and disparate techniques.

8.   Acknowledgments

Thanks are due to Anne Arquit, who contributed to the conception of the paper, composed initial drafts of some sections and assisted with editing. I also thank Walter Zimmerli for typing and editorial work. The manuscript was reviewed by Claude Lorius, who made valuable suggestions for improvement and provided one of the figures.

References

Birks, H.B.J., and H.H. Birks, 1980, Quaternary Paleoecology, Edward Arnold (Publishers) Ltd., London, 289pp.

Broccoli, A.J., and S. Manabe, 1987, The influence of continental ice, atmospheric CO_2, and land albedo on the climate of the last glacial maximum, Climate Dynamics, 1, 87-99.

Broecker, W.S., and G.H. Denton, 1989, The role of ocean-atmosphere reorganizations in glacial cycles, Geochim. Cosmochim. Acta, 53, 2465-2501.

Broecker, W.S., D. Petoet, and D. Rind, 1985, Does the ocean-atmosphere system have more than one stable mode of operation? Nature, 315, 21-25.

Chappellaz, J., J.M. Barnola, D. Raynaud, Y.S. Korotkevich, and C. Lorius, 1990, Ice-core record of atmospheric methane over the past 160,000 years, Nature.

Cook, E.R., and L.A. Kaitierus (eds.), 1990, Methods of Dendrochronology: Applications in the Environmnetal Sciences, Kluwer Academic Publisher, London, 394 pp.

Dansgaard, W., 1964, Stable isotopes in precipitation, Tellus, 16, 436-468.

Dansgaard, W., S.J. Johnsen, H.B. Clausen, D. Dahl-Jensen, N. Gundestrup, C.U. Hammer, and H. Oeschger, 1984, North Atlantic climatic oscillations revealed by deep Greenland ice cores, in J.E. Hansen and T. Takahashi (eds.), Geophysical Monograph 29: Climate Processes and Climate Sensitivity, A.G.U., Washington, D.C., 288-298.

Dansgaard, W., H.B. Clausen, N. Gundestrup, C.U. Hammer, S.F. Johnsen, P.M. Kristinsdottier, and N. Reeh, 1982, A new Greenland deep ice core, Science, 218, 1273-1277.

Eicher, U., 1980, Pollen- und Saurestoffisotopenanalysen an spatglazialen Profilen vom Gerzensee, Faulenseemoos und vom Regenmoos ob Boltigen, Mitt. Naturforsch. Ges. Bern, 37, 65-80.

Etheridge, D.M., G.I. Pearman, and F. de Silva, 1988, Atmospheric trace-gas variations as revealed by air trapped in an ice core from Law Dome, Antarctica, Ann. Glaciology, 10, 28-33.

Friedli, H., H. Loetscher, H. Oeschger, U. Siegenthaler, and B. Stauffer, 1986, Ice core record of the $^{13}C/^{12}C$ record of atmospheric CO_2 in the past two centuries, Nature, 324, 237-238.

Hays, J.D., J. Imbrie, and N.J. Shackleton, 1976, Variations in the earth's orbit, pacemaker of the Ice Ages, Science, 194, 1121-1132.

IGBP, 1990, The international Geosphere-Biosphere Program: A Study of Global Change. The initial Core Projects, IGBP Report No. 12, ICSU (IGBP), Stockholm.

Imbrie, J., J.D. Hays, D. Martinson, A. Mcintyre, A. Mix, J. Morley, N. Pisias, W. Prell, and N.J. Shackleton, 1984, The orbital theory of Pleistocene climate: support from a revised chronology of the marine $d^{18}O$ record, pp.269-305 in A.L. Berger et al. (eds.), Milankovitch and Climate, Part 1, D. Reidel, New York.

IPPC, 1990a, Scientific Assessment of Climate Change, Intergovernmental Panel on Climate Change Working Group 1 Report, pp. 365.

IPCC, 1990b, Policymakers Summary of the Scientific Assessment of Climate Change, Intergovernmental Panel on Climate Change Working Group 1 Report.

Johnsen, S.J., W. Dansgaard, and J. White, 1988, The origin of Arctic precipitation under glacial and interglacial conditions, Tellus.

Jouzel, J.C., Lorius, J.R. Petit, C. Genthon, N.I. Barkov, V.M. Kotlyakov, and V.N. Petrov, 1987, Vostok ice core: A continuous isotope temperature record over the last climatic cycle (160,000 years), Nature, 329, 403-409.

Keeling, C.D., R.B. Bacastow, A.F. Carter, S.C. Piper, T.P. Whorf, M. Heimann, W.G. Mook, and H. Roeloffzen, 1989, A three dimensional model of atmospheric CO2 transport based on observed winds: 1. Analysis of observational data, in D.H. Peterson (ed.), Geophysical Monograph 55: Aspects of Climate Variability in the Pacific and the Western Americas, A.G.U., Washington, D.C., 165-236.

Keeling, C.D., R.B. Bacastow, and T.P. Whorf, 1982, Measurements of the concentration of carbon dioxide at Mauna Loa Observatory, Hawaii, in W.C. Clark (ed.), Carbon Dioxide Review, Oxford University Press, New York, 377-385.

Khalil, M.A.K., and R.A. Rasmussen, 1989a, Climate-induced feedbacks for the global cycles of methane and nitrous oxide, Tellus, 41 B, 554-559.

Khalil, M.A.K., and R.A. Rasmussen, 1988, Nitrous oxide: trends and global mass balance over the last 3,000 years, Ann. glaciol., 10, 73-79.

Langway, C.C. Jr., H. Oeschger, and W. Dansgaard, 1984, The Greenland ice sheet in perspective, in C.C. Langway et al. (eds.), Geophysical Monograph: The Greenland ice Sheet Program, A.G.U., Washington, D.C.

Lorius, C.J., 1989, Polar ice cores and climate, pp. 77-103 in A. Berger et al. (eds.), Climate and Geo-Sciences, Kluwer Academic Pub.

Lorius, C., J. Jouzel, D. Raynaud, J. Hansen, and H. Le Treut, in press, The ice-core record of atmospheric methane over the past 160,000 years, Nature, 345, 127-131.

Manabe, S., and R.J. Stouffer, 1988, Two stable equilibria of a coupled ocean-atmosphere model, Jour. of Climate, 1, 841-866.

Neftel, A., H. Oeschger, T. Staffelbach, and B. Stauffer, 1988, CO_2 record in the Byrd ice core 50,000-5,000 years BP, Nature, 331, 609-611.

Neftel, A., E. Moor, H. Oeschger, and B. Stauffer, 1985, Evidence from polar ice cores for the increase in atmospheric CO_2 in the past two centuries, Nature, 315, 45-47.

Oeschger, H., and C.C. Langway, Jr. (eds.), 1989, The Environmental Record in Glaciers and Ice Sheets, John Wiley & Sons, New York, 400 pp.

Oeschger, H., and U. Siegenthaler, 1988, How has the atmospheric concentration of CO_2 changed? pp. 5-23 in F.S. Rowland and I.S.A. Isaksen (eds.), The Changing Atmosphere, John Wiley & Sons Ltd., New York.

Oeschger, H., A. Neftel, T. Staffelbach, and B. Stauffer, 1988, The dilemma of the rapid variations in CO_2 in Greenland ice cores (abstract), Ann. Glaciol., 10, 215-216.

Oeschger, H, B. Stauffer, R. Finkel, and C.C. Langway, Jr., 1985, Variations of the CO_2 concentration of occluded air and anions and dust in polar ice cores, in E.T. Sundquist and W.S. Broecker (eds.), Geophysical Monograph 32: The Carbon Cycle and Atmospheric CO_2: Natural Variations Archean to Present, A.G.U., Washington, D.C., 132-142.

Oeschger, H., J. Beer, U. Siegenthaler, B. Stauffer, W. Dansgaard, and C.C. Langway, 1984, Late glacial climate history from ice cores, in J.E. Hansen and T. Takahashi (eds.), Geophysical Monograph 29: Climate Processes and Climate Sensitivity, A.G.U., Washington, D.C., 299-313.

Pearman, G.I., and P.J. Fraser, 1988, Sources of increased methane, Nature, 332, 489-490.

Pearman, G.I., D. Etheridge, F. de Silva, P.J. Silva, 1986, Evidence of changing concentrations of atmospheric CO_2, N_2O, and CH_4 from air bubbles in Antarctic ice, Nature, 320, 248-250.

Ramanathan, V., 1988, The Greenhouse theory of climate change: a test by an inadvertent global experiment, Science, 240, 293-299.

Rasmussen, R.A., and M.A.K. Khalil, 1981. Increase in the concentration of atmospheric methane, Atmos. Envir., 15, 883-886.

Raynaud, D., and J.M. Barnola, 1985, An Antarctic ice core reveals atmospheric CO_2 variations over the past few centuries, Nature, 315. 309-311.

Siegenthaler, U., and H. Oeschger, 1987, Biospheric CO_2 emissions during the past 200 years reconstructed by deconvolution of ice core data, Tellus, 39B, 140-154.

Stauffer, B., E. Lochbronner, H. Oeschger, and J. Schwander, 1988, Methane concentration in the glacial atmosphere was only half that of the preindustrial Holocene, Nature, 332, 812-814.

Stauffer, B., J. Schwander, and H. Oeschger, 1985, Enclosure of air during metamorphosis of dry firn to ice, Ann. Glaciol., 6, 108-112.

Stauffer, B., H. Hofer, H. Oeschger, J. Schwander, and U. Siegenthaler, 1984, Atmospheric CO_2 concentration during the last glaciation, Ann. Glaciol., 5, 160-164.

Stocker, T.F., 1989, Review of Climate Fluctuations on the Century Timescale, CRG Report No. 89-7, McGill University, Montreal, 69 pp.

Stommel, H., 1961, Thermohaline convenction with two stable regimes of flow, Tellus, 13, 224-230.

Thompson, L.G., W. Xiaoling, E. Mosley-Thompson, and X. Ziehu, 1988, Climatic records from the Dunde Ice Cap, China, Ann. Glaciol., 10, 1-5.

WCRP, 1990, Global Climate Change: A Scientific Review, WCRP, Geneva, 35 pp.

Zardini, D., D. Raynaud, D. Scharffe, and W. Swiler, 1989, N_2O measurements of air extracted from Antarctic ice cores: Implications on atmospheric N_2O back to the last glacial-interglacial transition, Jour, Atmos. Chem., 8, 189-201.

Climate Prediction Based on Past and Current Analogues

by M. I. Budyko *(USSR)*

1. Introduction

In the early 1970s Soviet scientists concluded that global warming was inevitable because of the increasing concentration of atmospheric carbon dioxide (CO_2) due to fossil fuel combustion (Budyko et al., 1978). A prediction of expected changes in mean air temperatures in the Northern Hemisphere published in the USSR in 1972 was confirmed by meteorological observations in the 1970s and 1980s. It was much more difficult to determine expected changes in regional air temperature and precipitation, which could have significant effects on human activities and the biosphere. In order to determine such regional changes, two main methods are used at present: climate models (mainly in the USA and Western Europe) and the palaeoanalogue method (predominantly in studies of Soviet scientists, Budyko *et al.*, 1978).

2. Palaeoanalogue Method

Given estimates of the mean annual temperature at the end of the 20th Century over the Northern Hemisphere, the problem is to estimate the geographical distribution of anomalies of the main meteorological variables (air temperature and precipitation) for several stages of the expected rise of global mean temperature between 2000 and 2050. For this purpose, Soviet scientists have used palaeoclimatic evidence for three past warm epochs, including the Holocene (5–6000 YBP), the last interglacial (125 000 YBP) and the Pliocene (3–4 million YBP) climatic optima. During these warm periods temperatures were, respectively, 1, 2, and 3–4°C warmer than in the 19th century. Maps of the mean air temperature (January and July) and annual precipitation anomalies have been compiled for the greater part of the continents of the Northern Hemisphere. It is assumed in the palaeoanalogue method that these maps can be used to estimate climatic conditions in 2000, 2025 and 2050.

The palaeoanalogue method has been described in a number of studies published mainly in the Russian language since 1978. The most comprehensive studies are described by Budyko and Izrael (1987). The method has been widely used by the participants of the Soviet/American cooperative project on estimating future climate conditions. The results were presented in the joint reports of the Soviet and American experts between 1981 and 1990.

Because of their great economic and social importance, the estimates of future climatic conditions must be checked for their reliability. One of the effective ways of checking them is to compare the results of two independent methods.

3. Mean Air Temperature

An analysis of the observed temperature data of the last 100 years showed that from the end of the 19th Century until the early 1970s anthropogenic warming was to a large extent hidden by the temperature changes due to natural factors.

Both observational data and theoretical calculations indicate that in the last quarter of the 20th Century the influence of the increased greenhouse effect on the mean air temperature trend is larger than natural variability. Taking these results into consideration, it can be hypothesized that the mean air temperature rise of 0.3 °C/decade during 1975–1989 was mainly a result of anthropogenic factors.

A detailed calculation made in "Anthropogenic Climatic Changes" showed that the mean air temperature will increase by 0.32°C/decade during the period 1975–2000. This rate is very similar to the rate observed between 1975 and 1989. For the more distant future the above-mentioned book concludes that the rate of increase of mean air temperature will be similar to that of 1975–1985.

Table 1 compares the results of theoretical (model)

Table 1*: The increase of the mean air temperature (°C)
above the value at the end of the 19th Century*

	1975	2000	2025	2050
Theoretical calculation	0.5	1.3	2.5	3-4
Extrapolation of the observed trend 1975-1985	0.5	1.25	2.0	2.75

calculations of the mean air temperature changes for the period 2000–2050 with the results obtained by linear extrapolation of the trend between 1975 and 1985. At least until 2025 the results are very similar.

4.   The Causes of Climatic Change

The question of the causes of climatic change during the warm epochs of the past mentioned above is clearly important. It is not, however, the main question to be answered in order to justify the use of the analogue method to forecast future climatic change. The basic requirement to be met by the analogues is that the main climatic variables change similarly during the period of global warming at the end of the 20th Century, in the first half of the 21st Century and in the warm epochs of the past. This condition would obviously be fulfilled if the causes of the past warmings were the same as the anthropogenic global warming, namely the increasing concentrations of greenhouse gases in the atmosphere. However, if the past warmings were mainly due to other factors, further studies are necessary in order to decide whether the past warm epochs can be used as analogues of future climate.

In the model studies of Wetherald and Manabe (1975) and the empirical studies of, for example, Budyko and Yefimova (1984) similar features of climatic change were found as a result of different forcing factors. However, some differences were found in the changes of meteorological regime between the warming that occurred in the 1930s and that of the 1980s (Budyko, 1988). These differences were obviously a result of the fact that the warming periods were forced by different factors. The first was probably due to the increased transparency of the lower layers of the stratosphere. The second warming was probably a result of the increased atmospheric concentrations of greenhouse gases. However, this does not mean that the climate during the first warming period cannot can be regarded as a satisfactory analogue of the climate of the second warming period. Since the requirements for accuracy of the analogues used for estimating future climate are limited, it is quite possible that these differences are not important. Using the

empirical approach to choosing satisfactory analogues of future climates, the distribution of relative values of the latitudinal mean surface air temperature anomalies during the warm epochs of the past in the Northern Hemisphere can be compared. These relative values are equal to the departures of the past temperatures from the modern ones divided by the values of the anomalies of the whole hemisphere. These distributions describe fundamental characteristics of global climatic change. Their similarities or differences are important for deciding whether the warm epochs of the past can be taken as analogues of future warming.

The distributions of these anomalies for the three warm epochs mentioned above show that the features of the latitudinal distribution of the air temperature practically coincide for the three epochs. This means that these three epochs can be used as the analogues of future climates if (a) the warming of at least one of them occurred as a result of the increased atmospheric concentration of greenhouse gases, or (b) the reaction of the climate system to the factors which caused the global warming is similar to that caused by an increased concentration of greenhouse gases.

The answer to the question of what caused the climate to be warmer than that of today during the Tertiary period, including the Pliocene, is certain. Both empirical studies (Budyko et al., 1985) and model calculations (Manabe and Bryan, 1985) confirmed the hypothesis of Arrhenius that the pre-Quaternary climates were warm due to a high atmospheric CO_2 concentration.

With respect to the climatic changes during the Pleistocene (including two of the above-mentioned warm epochs), the problem is more complex. It was discovered long ago that the causes of climatic change in the Pliocene and Pleistocene were different to some extent. Consequently, there could have been certain differences in the responses of the climate system during each of these epochs (Budyko, 1980). Most important for solving the problem of whether it is possible to use the Pleistocene warming periods as analogues of future climate is the comparison of latitudinal and seasonal distributions of air temperature anomalies during the three warm epochs of the past. Some of the explanations of the Pleistocene climatic changes are discussed below.

The analysis of the air bubbles from ancient ice showed that during the last 160,000 years the atmospheric CO_2 concentration changed between 200 and 300 ppm. These changes were closely related to temperature variations, with a correlation coefficient of 0.9. It is widely accepted that the surface air temperature depends on the atmospheric CO_2 concentration. Therefore, it is possible to calculate that the above changes of CO_2 concentration would lead, all other climate factors remaining the same, to a change of the mean surface air temperature of about 2°C. Many investigators believe that a decrease of the CO_2

concentration was associated with the summer reduction of incoming radiation in the middle and high latitudes of the Northern Hemisphere due to the changes of the earth`s orbital parameters. When this occurred, the glacial areas expanded considerably on the continents and over the oceans and this caused an increase of the earth`s albedo and a decrease of the global average temperature.

Given that during the last 160000 years the global average temperature changed by about 6°C (from -4°C during the last glacial epoch to 2°C at the last interglacial optimum), it can be concluded that the positive feedback of the earth`s albedo resulted in an additional air temperature increase of not more than 4°C. This estimate is in good agreement with the calculations of how the changes of the areas covered by snow and ice, as well as the changes of the areas with different vegetation cover, which occurred in the Pleistocene, influenced the albedo. It should be noted that the direct effect of the Earth`s position with regard to the Sun on the incoming global solar radiation and, consequently, on the global average surface temperature is insignificant.

If there is no change of the mean solar radiation arriving at the top of the atmosphere, astronomic factors lead to a redistribution of this solar radiation in different latitudes and in different seasons. If these factors have a direct influence on the air temperature, their changes are followed by similar changes of the temperature. However, this hypothesis contradicts the available empirical data. Thus, unlike the incoming radiation change, the Northern and Southern Hemisphere air temperatures varied synchronously in the Pleistocene.

The same contradiction applies to the radiation and temperature changes during summer and winter in the middle and high latitudes. Despite the changes in the incoming solar radiation, i.e., when the difference between the maximum summer radiation and the minimum winter radiation increased, the difference between mean summer and winter temperatures in both hemispheres decreased.

The most probable explanation for these contradictions is that the CO_2 concentration fluctuations played the leading role in the Pleistocene climate changes, with the Earth`s albedo variations amplifying this main factor. Astronomic factors, which triggered the Pleistocene climate change, had an insignificant effect on the air temperature compared with the CO_2 impact.

It is interesting to see how this question was treated in a recent study that analyzed the CO_2 concentration and air temperature changes derived from ice core data for the last 160,000 years from the Vostok station in Antarctica (Lorius et al., 1989). According to this analysis, the fluctuations of the CO_2 concentration were responsible for more than 50 percent of the air temperature changes in the vicinity of the Vostok station.

Bearing in mind that in the high latitudes where the Vostok station is situated, the impact of astronomical factors on climate is much larger than in the middle and low latitudes, it can be concluded that the mean air temperature variations over most of the Earth's surface in the Pleistocene depended mainly on the CO_2 concentration changes, which were more effective due to positive feedbacks. The same conclusion has been drawn in the climate modelling studies of Manabe and Broccoli (1985) and is also confirmed by empirical studies.

Why were the Pleistocene warm epochs 1–2°C warmer than recent centuries although the CO_2 concentration was similar (about 280 ppm) during these periods? It seems that there are two reasons. The dependence of the global surface albedo on the incoming solar radiation in summer in the high latitudes is different in the two hemispheres. It is more noticeable in the Northern Hemisphere, where the shifts of the snow-ice cover boundary are less influenced by thermal inertia, since they occur mainly over land, while in the Southern Hemisphere the boundary is over the ocean surface, which has a larger thermal inertia. The second reason is the feedback of the summer warming in the high latitudes of the Northern Hemisphere with the precipitation increase in the arid regions of the middle and low latitudes, which resulted in a global albedo decrease and a mean air temperature increase. For these reasons the indirect influence of the astronomic factors on climate as well as the effect of the increased CO_2 concentration led to a more even distribution of the air temperature variations compared with the distribution of solar radiation changes.

5. Warm Epochs of the Past as the Analogues of Future Climates

To use past climates as analogues of future climate, one needs to know how the anomalies of climate variables in different latitudes depend on the mean global temperature rise. If, for different levels of warming, the anomalies are directly proportional to the mean global air temperature changes, the problem of using climate analogues is easier, since the data on different epochs can be used for mutual testing. If the relationship is not linear, the possibility for such testing is limited.

Comparison of the mean latitudinal values showed that within the latitudinal band between the equator and 80°N, the air temperature changes corresponding to different levels of warming are in good agreement (Budyko et al., 1989).

The comparison of the reconstructions of the winter and summer air temperature increase for the three mentioned epochs with the air temperature of the 19th Century showed that they are quite similar. All three reconstructions reveal a distinct spatial variability of air temperature anomalies. The higher the latitude, the greater is the anomaly both in summer and winter. Winter reconstructions show greater anomalies in the regions where climate is more continental (Budyko and Izrael, 1987; Budyko et al., 1989).

The great similarity of the reconstructions of the air temperature anomalies for these three warm epochs of the past allows the conclusion to be drawn that regional changes of the temperature anomalies are approximately proportional to the mean global air temperature increase. The regional distributions of the surface air temperature anomalies due to the above-mentioned causes of climatic change are analogous.

Another conclusion allowed by the similarity of the reconstructions is that empirical methods for estimating spatial temperature distributions during global warming are reliable. Thus, it is possible to average three winter reconstructions and three summer reconstructions to obtain a more precise pattern of temperature distribution during the warming. This produces a more a reliable result than the use of three individual reconstructions developed by three independent empirical methods of estimating temperature changes.

Such averaging was carried out in two studies (Budyko and Izrael, 1987, Budyko et al., 1989). In both studies two reconstructions of the mean summer and mean winter temperature changes in the Northern Hemisphere for a global temperature increase of 1°C were developed.

6. The Reliability of Information on Future Climate

The problem of the accuracy of the estimates of future climate change has been discussed in studies by Soviet authors (e.g., Budyko and Izrael, 1987). Some of the conclusions from these studies are presented here. It should be noted that the accuracy of climate predictions is significantly limited by factors that do not depend on the level of development of atmospheric sciences. The first factor is the reliability of available data on the future atmospheric chemical composition, including information on the effects of economic activities. In this regard, the future rate of fossil fuel combustion is very important.

Comparison of the most reliable estimates of the atmospheric greenhouse gas concentrations in the next few decades suggests that the estimates are more or less precise for the next 20 years. Beyond this time, the reliability of the forecasts decreases and forecasts for the second half of the 21st Century are unreliable and should rather be regarded as scenarios of future changes in the atmospheric chemical composition.

There is a probability that at the end of the 20th century changes of the mean global air temperature could occur due to unpredictable natural factors, such as explosive volcanic eruptions. In some years, such changes could be only slightly less than the global temperature increase due to

human activities. In the early 21st Century the temperature variability due to natural factors will be much smaller than the anthropogenic global warming. However, in the more distant future the role of other factors limiting the accuracy of the forecast will, as mentioned above, become more significant.

Numerical experiments suggest that the most reliable forecast of future climate change will be for the period 2000–2020, while the forecasts for 1990–2000 and 2030–2050 will be less precise. There are reasons to suppose that for most of the period between 1990 and 2050 the only accurate estimates of air temperature and precipitation anomalies in certain regions will be for the sign of the anomaly (if it is not close to zero) and the order of its magnitude. Under more favourable conditions the accuracy could be slightly higher.

In this connection, the question arises, whether the existing methods of predicting future climatic changes guarantee that these rather modest requirements for accuracy can be met.

The only quantitative forecast that can be tested against observational data is the prediction of mean air temperature variations in the Northern Hemisphere (Budyko, 1980). The comparison of the predicted values with the observational data shows that the observed mean air temperature anomaly for the past 10 years differs from the predicted value by about 15%. The high accuracy of the forecast indicates that unpredictable natural variability had a comparatively insignificant influence on the temperature anomaly of the last 10 years.

However, such a level of accuracy can hardly be achieved in the forecasts of regional changes of temperature and precipitation. The numbers presented in Tables 2 and 3 give an idea of the probable accuracy of forecasts of meteorological regime changes in different regions. Tables 2 and 3 show the correlation coefficients between the relative values of the temperature anomaly in the Northern Hemisphere for the three warm epochs of the past.

The coefficients of correlation between the data on mean latitude air temperature distribution during the three warm epochs confirm the conclusion that the climate system

Table 2: The correlation between empirical data on the mean latitudinal temperature changes in past warm epochs

| Time interval | Temperature | | | | | |
| | Winter | | | Summer | | |
	1	2	3	1	2	3
1. Holocene	1			1		
2. Eemian	0.92	1		0.99	1	
3. Pliocene	0.92	0.95	1	0.97	0.97	1

Table 3: *The correlation between empirical data on the temperature and precipitation changes in 12 regions in three past warm epochs*

Time interval	Temperature						Precipitation		
	Winter			Summer					
	1	2	3	1	2	3	1	2	3
1. Holocene	1			1			1		
2. Eemian	0.5	1		0.6	1		0.99	-	
3. Pliocene	0.3	0.7	1	0.9	0.8	1	0.7	-	1

reacts similarly to different forcing factors.

As the data on the spatial distribution of temperature and precipitation during these warm epochs are not very accurate, it is understandable that the correlation coefficients in Table 3 are noticeably lower than those in Table 2. The lowest correlation coefficients were obtained for the temperature distribution during the Holocene winter and can be explained by the insufficient sensitivity of the available methods of estimating palaeotemperatures during a comparatively insignificant warming in winter in the middle and high latitudes. Nevertheless, it should be noted that in all of the cases, correlation coefficients are positive and (except for the winter of the Holocene optimum) high enough to confirm that the accuracy of the information about past epochs is sufficient for its use for analogues of future climates. A similar conclusion can be obtained from the analysis of recent changes in temperature and precipitation based on observational data.

REFERENCES

Budyko, M.I., 1980: The Earth's Climate: Past and Future. Leningrad, Gidrometeoizdat, 352 pp, English Translation, Academic Press, 1982.

Budyko, M.I., 1988: The climate of the end of the 20th Century. Meteorologiya, 10, 5–24.

Budyko, M.I. and Izrael, Yu. A. (eds.), 1987: Anthropogenic climate changes. Leningrad, Gidrometeoizdat, 402pp. English Translation, University of Arizona Press, 1990.

Budyko M.I. et al., 1978: Future climatic changes. Izv.Akad.Nauk USSR, ser. geogr., N6, p. 5–10.

Budyko, M.I., Ronov A.B., and Yanshin, A.A., 1985: The history of the atmosphere. Leningrad, Gidrometeoizdat, 209 pp. English translation, 1987.

Budyko, M.I. and Yefimova, N.A., 1984: Annual variations of the meteorological elements as a model of climate change. Meteorologiya i Gidrologiya, N 1, pp. 5–10.

Budyko, M.I., Yefimova, N.A. and Lokshine, I. Yu., 1989: The expected anthropogenic changes of global climate. Izv.Acad.Nauk USSR, ser. geogr., N5, pp 45–55.

Lorius, C. et al., 1990: The ice core record of atmospheric methane over the past 160,000 years. Nature, 345, pp. 127–131.

Manabe, S. and Broccoli, A.J., 1985: A comparison of climate model sensitivity with data from the last glacial maximum. J. Atm. Sci., 42, N 23, pp. 2643-2651.

Manabe, S. and Bryan, K., 1985: CO_2 induced change in a coupled ocean-atmosphere model and its paleoclimatic implications. J. Geophys. Res., 90, C11, pp. 11689–11707.

Wetherald, R.T. and Manabe S., 1975: The effect of changing the solar constant on the climate of a general circulation model. J. Atmos.Sci., 32, pp. 2044–2059.

Detection of the Enhanced Greenhouse Effect on Climate †

by T. M. L. Wigley and S. C. B. Raper *(United Kingdom)*

Abstract

Possible causes of climatic change over the past few centuries are reviewed. The problem of detecting the enhanced greenhouse effect on climate is introduced, and similarities between this problem and those of explaining past climatic change and estimating the climate sensitivity are noted. The record of global-mean temperature changes over the past 100 years or so is then used to estimate the climate sensitivity empirically. Changes in global-mean temperature are simulated by forcing a climate model with observed greenhouse-gas concentration changes and the results are compared with the observed temperature data. The simulations are carried out for various values of the climate sensitivity parameter (defined here as the equilibrium global-mean temperature change for a CO_2 doubling, ΔT_{2x}) and the ΔT_{2x} value that gives the best fit (i.e., minimum mean square error) is identified. If uncertainties in the model parameters and the observed temperature data and the effects of passive internal variability are accounted for, ΔT_{2x} is found to lie in the range 0.73–4.25°C. This is consistent with the results of General Circulation Models, which suggest that ΔT_{2x} lies in the range 1.5–4.5°C, but it does allow a probability of 10–20 per cent that ΔT_{2x} is less than 1.5°C. This possibility is discussed in the light of the hypothesized global-scale cooling effect of Man-made sulphate aerosol concentration changes resulting from fossil-fuel derived SO_2 emissions. Even a minor aerosol effect would require an upward revision of the empirical ΔT_{2x} estimates bringing them more in line with GCM results. We then consider whether or not better empirical estimates of ΔT_{2x} might be possible in the future, specifically in the year 2005. Even if future warming were large (0.6°C over 1990–2005) this would still allow small values of ΔT_{2x} (<1.5°C) to be consistent with observations. Conversely, even if there were a slight cooling over 1990–2005, the value of ΔT_{2x} could still be above 2.5°C. Finally, we note that while detection of the enhanced greenhouse effect on climate is an important issue, detection per se need not lead to a better knowledge of ΔT_{2x}. It is the better quantification of ΔT_{2x} which is crucial if we are to narrow uncertainties in projections of future climatic change to acceptable and useful limits.

Introduction

Global-mean temperature has risen by about 0.5°C over the past 100 years. We know this through the careful analysis of both land data (Jones et al., 1986a, b; Jones, 1988; Hansen and Lebedeff, 1987, 1988) and marine data (Folland et al., 1984; Jones et al., 1986c, 1991). Recent reviews of global-scale changes in temperature have been given by Jones and Wigley (1990) and Folland et al. (1990).

The cause of this warming is still uncertain. There are a number of possible causal factors which may be involved to greater or lesser degrees. These may conveniently be divided into external and internal factors (see, e.g., Robock, 1978). External factors are those that are either external to the planet (such as changes in solar output), or derived from sources external to the climate system (such as large volcanic eruptions, which modify the planet's radiation balance by injecting material into the stratosphere). External factors also include those arising from Man's activity—for example, the enhanced greenhouse effect of gases such as CO_2, CH_4, N_2O, O_3 and the halocarbons, and the possible effects of SO_2 emissions and attendant sulphate aerosol concentration changes.

Internal factors are those within the climate system, which we define here as the atmosphere, the oceans and those parts of the cryosphere that respond on time-scales up to centuries. Internally-generated variability frequently involves interactions between different components of the climate system. In Wigley and Raper (1990a, 1991), we distinguish between passive and active internal variability. The former has its primary driving force in the atmosphere, and arises through the modulation of atmospheric variability by the ocean in a strictly passive mode. Because the ocean has a very large heat capacity, or "thermal inertia", relative to the atmosphere, the effects of random variations in the atmosphere's radiation balance are significantly modified by the ocean. The overall effect is to amplify slower, lower-frequency climate variations relative to high-frequency variations. This process, first elaborated by Hasselmann (1976), acts independently of changes in ocean circulation. In contrast, active internal variability, which may also be stimulated by the atmosphere, refers to climate changes that arise directly from changes in the ocean circulation (either the vertical, thermohaline circulation and/or the horizontal current system).

Attempts to explain recent changes in global-mean temperature have been reviewed by Wigley et al. (1985, 1986) and, more recently, by Jones and Wigley (1990) and Hansen and Lacis (1990). There are three main difficulties. The first is our inadequate knowledge of past changes in the various possible causal factors. The second is our inadequate understanding of how the climate system behaves naturally, i.e., the precise magnitude and character of natural variability. The third is our uncertain knowledge of the magnitude of the sensitivity of the climate system to external forcing changes. We discuss these further in the next two Sections.

Possible Causes of Recent Climatic Change

Of the possible external forcing factors, we consider solar variability first. Reliable information regarding changes in solar output (or "irradiance") has only come in the past decade, through satellite data (e.g., Hickey et al., 1988; Willson and Hudson, 1988). Because these data are so recent and span such a short time period, we cannot yet quantify variations in solar output beyond those inter-annual to decadal time scale variations directly associated with the 11-year sunspot cycle (Foukal and Lean, 1990). Sunspot-cycle-related effects on global-mean temperature are very small having a total range of only a few hundredths of a degree Celsius over the past century (Wigley and Raper, 1990b).

It is possible that these variations are superimposed on larger, lower-frequency effects, as suggested, for example, by Vinnikov and Groisman (1981, 1982), Gilliland (1982), Gilliland and Schneider (1984), Reid (1987, 1990) and others. While there is no convincing, direct observational evidence for such changes this century, there is indirect evidence of 100-year time scale solar variability in the paleoclimatic record. The Holocene period (i.e., the last 10,000 years) is punctuated by a number of intervals of

alpine glacier advance and retreat (Röthlisberger, 1986; Grove, 1988), the last of which has been referred to as the Little Ice Age. Solar irradiance changes, associated with fluctuations in the level of carbon-14 in the atmosphere, are thought by some to be the most likely cause of these "neoglaciations" (Eddy, 1977; Wigley and Kelly, 1990).

The main climatic effects of volcanic eruptions appear to be short-term, lasting for at most a few years (Kelly and Sear, 1984; Sear et al., 1986; Bradley, 1988). Various authors have contended that lower frequency effects (i.e., on decadal and longer time scales) are both possible and important (e.g., Hammer et al., 1980; Hansen et al., 1981; Porter, 1986; Mass and Portman, 1989; Oerlemans, 1988). However, the evidence is equivocal and there is no single generally-accepted record of past volcanic forcing. The different proxy records of volcanic forcing that are available (Lamb, 1970; Mitchell, 1970; Oliver, 1976; Bryson and Dittberner, 1976; Pivovarova, 1977; Bryson and Goodman, 1980; Hammer et al., 1980; Simkin et al., 1981) show large differences (Bradley and Jones, 1985, Table 3.2). It may be that this is because most are poor indicators of the actual forcing, but deciding a priori which is best is difficult, and any such judgement would be controversial.

Of the potential Man-made causes of recent climatic change, one that has received considerable attention recently is SO_2 emissions. These have undoubtedly raised sulphate aerosol concentrations in the Northern Hemisphere (Schwartz, 1988), changes which may well have led to a cooling effect either directly, through clear-sky radiative effects (Charlson et al., 1990; Wigley, 1991a), or indirectly, through changes in the albedo of marine stratiform clouds (Twomey et al., 1984; Charlson et al., 1987; Wigley, 1989). The magnitude of this cooling effect is difficult to determine theoretically and impossible to quantify observationally. Estimates made to date suggest that both direct and indirect effects could, individually, have caused a Northern Hemisphere negative forcing of order $1 Wm^{-2}$ over the past 80–100 years, but the precise magnitudes of these effects are highly uncertain.

In fact, the only past forcing record that is reasonably well established and quantified is that due to the increasing concentrations of greenhouse gases (the most recent review is that of Shine et al., 1990). These have caused an overall positive radiative forcing at the top of the troposphere of around $2.5 Wm^{-2}$ since the late eighteenth century, roughly equivalent to a 1% increase in solar output.

Just how much internal variability has contributed to global-mean temperature change over the past 100 years or so is even more uncertain. On a century time scale, passive internal variability could cause trends of up to 0.3°C, either warming or cooling (Wigley and Raper, 1990a, 1991). This aspect of internal variability will be considered further below. The possible importance of active internal variability, through changes in the thermohaline circulation, was first suggested by simple modelling studies (e.g., Watts, 1985) and has been re-emphasized by the observations of noticeable temperature changes below the thermocline (Levitus, 1989). Wigley and Raper (1987) have estimated that changes in global-mean temperature of up to 0.2°C over 30 years might possibly arise through thermohaline circulation changes, and that such changes could explain the hiatus in global-mean warming over 1940–75, while recent coupled ocean/atmospheric general circulation modelling studies (e.g., Stouffer et al., 1989) suggest that the thermohaline circulation may vary markedly under the influence of external forcing. However, while the reality and potential importance of thermohaline circulation changes has been demonstrated by observation and theory, we have no time series of past changes and no way to quantify the global-scale effects over the past century.

The Climate Sensitivity

The other main hurdle in trying to explain the past temperature record is our inadequate knowledge of the sensitivity of the global climate system to external forcing. Climate sensitivity uncertainties are also the primary reason for uncertainties in predicting future changes in global-mean temperature due to the enhanced greenhouse effect.

The climate sensitivity is determined largely by the strength of feedbacks within the climate system. The classic example of a feedback mechanism is the ice-albedo feedback, whereby a warming would reduce snow and ice cover, which would reduce the amount of solar radiation reflected back into space, which would, in turn, amplify the initial warming. There are a number of other feedbacks, most of which are still relatively poorly understood. For recent reviews, see Cubasch and Cess (1990) and Cess et al. (1990).

The effect of these feedbacks can be studied using a simple type of climate model called an energy-balance model. We briefly describe this type of model here for a number of reasons: since the concept of energy balance is central to the issue of explaining and predicting climate change, because it allows us to show how the climate sensitivity is related to feedback effects, and because we will make use of such a model later in this paper. Before doing so, however, we give the popular definition of climate sensitivity in the greenhouse context. In this arena, sensitivity is frequently defined as the equilibrium (i.e., eventual) global-mean warming that would occur if the atmospheric CO_2 concentration were doubled. We denote this by ΔT_{2x}. ΔT_{2x} is thought to lie in the range 1.5-4.5°C, with a best guess value of 2.5°C (Mitchell et al., 1990).

The link between forcing and global-mean temperature changes can be quantified using an energy-balance climate model, which can be written in the form:

$$C(d\Delta T/dt) = \Delta Q - \Delta F - \lambda \Delta T \qquad (1)$$

This equation describes how changes in global-mean temperature ($\Delta T(t)$), multiplied by the heat capacity of the ocean's mixed layer (C) to give a heat change term on the left-hand side of the equation, are determined by a balance, on the right-hand side of the equation, between the time-dependent external forcing ($\Delta Q(t)$), the flux of heat from the upper, mixed-layer of the ocean into the deeper ocean ($\Delta F(t)$) and the feedback term ($\lambda \Delta T$). All important feedbacks are effectively linear functions of temperature change and they are lumped together in the term $\lambda \Delta T$. The proportionality constant, λ, is called the feedback parameter. This single number (units: $Wm^{-2}/°C$) covers the effects of all of the feedbacks in the climate system, both positive and negative. Uncertainties regarding these feedback processes are reflected in the uncertain magnitude of λ.

Climate sensitivity, as defined by ΔT_{2x}, is related to the inverse of the feedback parameter. This can be seen using equ. (1) by considering the equilibrium warming due to a forcing change ΔQ. In equilibrium, both $C(d\Delta T/dt)$ and ΔF must be zero, so the equilibrium temperature change is:

$$\Delta Te = \Delta Q/\lambda \qquad (2)$$

For $2xCO_2$, $\Delta Q \cong \Delta Q_{2x}$ 1 4.4Wm^{-2} (Hansen et al., 1988; Shine et al., 1990), so that:

$$\Delta T_{2x} \cong 1\ 4.4\ \lambda^{-1} \qquad (3)$$

Thus, λ^{-1} is directly proportional to our earlier definition of climate sensitivity: indeed, λ^{-1} is also often referred to as the climate sensitivity parameter. It is, in fact, a slightly more fundamental property of the climate system. (Note that the symbol λ as used by Cubasch and Cess, 1990, and Cess et al., 1990, is the same as our λ^{-1}. This is confusing terminology, but both usages are quite common.)

Detection of the Enhanced Greenhouse Effect

The problem of explaining the past record of global-mean temperature fluctuations is closely related to that of estimating the climate sensitivity and both are clearly relevant to the issue of detecting the enhanced greenhouse effect. Both problems are made more difficult by the considerable "noise" of natural climatic variability. At the global-mean scale, this noise has considerably obscured the "signal" of the evolving greenhouse effect, as will be shown below. An alternative approach to detection, which partially circumvents the global-scale noise problem, is to consider not just global-mean temperature, but more variables and/or smaller spatial scales. This multivariate approach to detection, which requires identifying a greenhouse-gas-specific signature (or "fingerprint") of climatic change in the past record, is generally referred to as the fingerprint method. In the Intergovernmental Panel on Climate Change's (IPCC) review of the detection issue (Wigley and Barnett, 1990), it is considered to be the way in which the detection problem is likely to be solved first. At present (late 1990), however, we cannot claim to have positively identified the enhanced greenhouse "signal" in any of the past records of climatic change in a way that clearly links cause and effect.

An important and little realised aspect of the detection issue is that, even if detection is achieved by the fingerprint method at some future date, this does not necessarily mean that the magnitude of the enhanced greenhouse effect (i.e., the climate sensitivity) will automatically be well quantified. In other words, even if a specific greenhouse-gas-induced climate change pattern were identified with high confidence by some pattern recognition method, such as spatial correlation, there could still be considerable uncertainty in quantifying the strength of the signal. Spatial correlation methods, by their very nature, rely heavily on the relative characteristics of the pattern, and so may be insensitive to the assumed climate sensitivity. This leads to the question: can we estimate the climate sensitivity from the past record of global-mean temperature changes? The remainder of this paper addresses this question.

Estimating the Climate Sensitivity

Our approach to estimating the climate sensitivity is to use a relatively simple type of energy-balance model of the climate system in which heat transport into the deeper ocean (which determines the ΔF term in equ. (1)) is parameterized as an upwelling-diffusion process. As noted by IPCC, such models, in spite of their numerous shortcomings, constitute our best tool for studying how global-mean temperature might respond to time-dependent external forcing (Bretherton et al., 1990). Their main strength is that they allow many of the uncertainties to be quantified. All of the key projections of future global-mean warming and sea level rise given by IPCC were made with the model used here.

In order to estimate the climate sensitivity, as characterized by ΔT_{2x}, we first use the model to predict the global-mean temperature change that should have occurred over 1861–1990 in response to greenhouse-gas forcing for a range of ΔT_{2x} values. We then compare the various modelled temperatures with the observations, quantifying

differences between them in terms of the mean square error. The best fit ΔT_{2x} value is the one with the least mean square error. A range of possible best fit values can then be obtained by considering uncertainties in the model parameters and in the temperature record, and by accounting for some of the effects of internal variability.

The model we use is that of Wigley and Raper (1987, 1990a, 1991). This model couples an energy-balance model for the ocean mixed layer (equ. (1)) to a deep ocean in which vertical heat transport is parameterized as an upwelling-diffusion process. The model explicitly considers land-sea and hemisphere-hemisphere heat exchange processes. The main parameters that control the magnitude of the model's response to external forcing are (with their ranges of uncertainty)

ΔT_{2x}: the climate sensitivity (1.5–4.5°C)
K: the vertical diffusivity (0.5–2.0cm^2/sec)
h: the mixed-layer depth (70–110m)
π: the sinking water temperature change ratio (0.0–0.4)

The upwelling rate, w, is also an important model parameter, but it is linked to the diffusivity through the length scale of the steady-state vertical temperature profile in the ocean. For further details on the model parameters and on their most appropriate values, see the Appendix.

A simple comparison of modelled and observed changes was given in the IPCC Working Group 1 report (Wigley and Barnett, 1990, Figure 8.1—reproduced here as Figure 1). Two K (0.63 and 1.27cm^2/sec) and two π (0 and 1) values were used to show the sensitivity of the results to these parameters. The forcing used by Wigley and Barnett was that given by IPCC WG1 Section 2 (Shine et al., 1990), beginning in 1765, and the observed temperature curve in Figure 1 is that given by IPCC WG1 Section 7 (Folland et al., 1990). By eye, the best fit value for ΔT_{2x} from Figure 1 is around 1.5°C. The effect of K and π uncertainties is minimal. This is because the modelled temperature changes are insensitive to K and π at low ΔT_{2x}.

It would be naive, however, to accept this single empirical estimate of the climate sensitivity as a useful number without first evaluating its uncertainties further. We will therefore attempt to provide confidence limits on ΔT_{2x}, by deriving a range of values which is consistent both with the observed data and with the many uncertainties surrounding its interpretation. For this analysis, we use the same IPCC forcing as used by Wigley and Barnett (1990). The temperature data we use differ slightly from those used by Wigley and Barnett in that the influence of the El Niño/Southern Oscillation effect (ENSO) has been factored out using the method of Jones (1989). The temperature data have also been updated using an estimated value for 1990. The results to 1990 are not

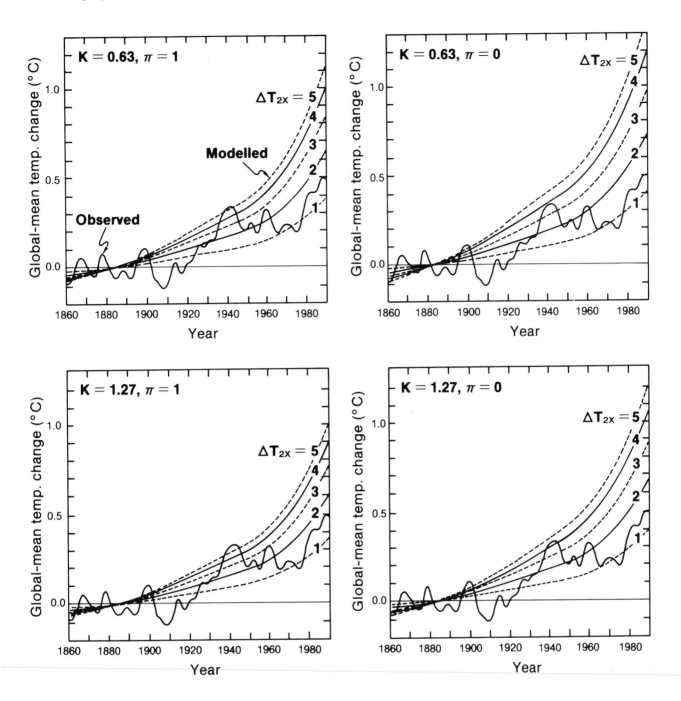

Figure 1: Observed global-mean temperature changes (1861–1989) compared with predicted values. The observed changes are as given in Section 7 of the IPCC WG1 report (Folland et al., 1990). The data have been smoothed to show the decadal and longer time scale trends more clearly. Predictions are based on observed concentration changes and concentration/forcing relationships as given in Section 2 of the IPCC WG1 report (Shine et al., 1990) and have been calculated using the upwelling-diffusion climate model of Wigley and Raper (1987). To provide a common reference level, modelled and observed data have been adjusted to have zero mean over 1861–1900. To illustrate the sensitivity to model parameters, model results are shown for ΔT_{2x} = 1,2,3,4 and 5°C (all panels) and for four K,π combinations. The top left panel uses the values recommended in Section 6 of the IPCC WG1 report (Bretherton et al., 1990) (K = 0.63cm²/sec, p = 1). As noted in the Appendix, these values are not necessarily the best possible choice, but, since sensitivity to K is relatively small and sensitivity to p is small for small ΔT_{2x}, the best-fit ΔT_{2x} depends little on the choice of K and π.

sensitive to the presence or absence of ENSO. In the next section, however, where future temperature changes relative to 1990 are considered, it is more desirable to eliminate the effects of ENSO.

First, we calculate the "best guess" value of ΔT_{2x} more rigorously. In Figure 2 we show how the modelled temperature changes vary with ΔT_{2x} for the "best guess" set of model parameters (K = 1cm²/sec, h = 90m, π = 0.2,

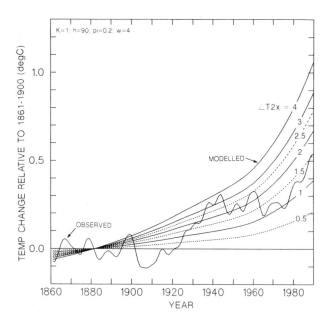

Figure 2: As Figure 1, but using updated temperature data with the ENSO effect factored out, using the best-estimate values of model parameters recommended by the present study (K = 1cm^2/sec, π = 0.2, h = 90m, w = 4m/yr), and a different set of ΔT_{2x} values. The ΔT_{2x} value which gives the least-mean-square-error, i.e., the best fit between modelled and observed data, is ΔT_{2x} = 1.43°C.

w = 4m/yr: see Appendix). The value of ΔT_{2x} that minimizes the mean square error is ΔT_{2x} = 1.43°C. The corresponding warming to 1990 is 0.52°C above the 1861–1900 reference level. This value is slightly above the IPCC "best guess" value of 0.45°C based on similar data (Folland et al., 1990). However, there are a number of ways in which one could define the warming amount. A value of 0.52°C is the best possible estimate of the warming to 1990 within the present model-fitting framework.

Next, we consider uncertainties in the estimate of ΔT_{2x} arising from uncertainties in K, h and π by repeating the analysis for various values of these model parameters changing one parameter at a time. For $0.5 \leq K \leq 2.0$cm^2/sec, the optimum (i.e., least mean square error) value for ΔT_{2x} lies between 1.36 and 1.53°C; for $70 \leq h \leq 110$m, ΔT_{2x} lies between 1.41°C and 1.44°C; while for $0.0 \leq \pi \leq 0.4$, ΔT_{2x} also lies between 1.40°C and 1.48°C. The combined effect of these uncertainties can be obtained by carrying out further model simulations using the K,h,π combinations that maximize and minimize the best fit ΔT_{2x}. This gives $1.32 \leq \Delta T_{2x} \leq 1.62$°C. This small range of possibilities shows that the best fit value is insensitive to these model uncertainties when ΔT_{2x} is relatively small. However, as we show below, the sensitivity increases for higher values of ΔT_{2x}.

Our next step is to consider how natural climatic variability might affect the implied value of ΔT_{2x}. Natural

variability includes both external and internal factors. As noted in the Introduction, however, the only external factor that can be reasonably quantified is greenhouse gas forcing. It is true that other authors have attempted to factor out the effects of solar irradiance changes and/or changes in volcanic activity (e.g., Hansen et al., 1981; Vinnikov and Groisman, 1981, 1982; Gilliland, 1982; Gilliland and Schneider, 1984—these and other works have been reviewed by Wigley et al., 1985, 1986) but all of these studies use different (and all equivocal) records of past forcings. There is no way at present to reliably and consensually quantify decadal and longer time scale solar and volcanic forcing effects. Schönwiese (1991) has considered a wide range of proposed forcing histories for these factors, but his is a purely statistical analysis which has its own problems.

Of the internal sources of variability, active internal variability cannot be accounted for because the history of ocean circulation changes is virtually unknown. We are left, therefore, with passive internal variability. Since this form of variability can be quantified using the same model that we are employing here, its influence on the best fit ΔT_{2x} can easily be determined. We first consider this influence alone, using best-guess values of the model parameters.

Consideration of passive internal variability as the sole source of natural variability is not necessarily a severe restriction. This is because all of the statistical characteristics of the observed temperature record can be simulated using just greenhouse-gas forcing and passive internal variability, and because the magnitude of the low-frequency aspects of this form of variability depend on the climate sensitivity. In these two regards, model-generated passive internal variability also characterizes the statistical and physical characteristics of externally-forced variability, within their uncertainties.

In our earlier work (Wigley and Raper, 1990a, 1991) we derived confidence limits for the natural trend over various time intervals. By considering the 90% confidence band for the internally-generated global-mean temperature trend over 1881–1990 we can therefore deduce the 90% confidence interval for ΔT_{2x}. Because other aspects of natural variability may also have contributed to the century time-scale trend, the true 90% confidence band may be wider than determined here.

Accounting for internal variability requires an iterative procedure because the confidence limits for the magnitude of internal variability trends depend on ΔT_{2x}. This iterative process is illustrated by the following example. Using the best guess values for K, h and π, we first note that the lower confidence limit for the natural, internally-generated trend is about -0.2°C. A natural trend of -0.2°C means that a greenhouse-gas-induced warming of 0.72°C could have occurred, offset by -0.2°C to give the observed warming.

The implied ΔT_{2x} value is about 2.1°C (see Figure 2). However, for ΔT_{2x} = 2.1°C the lower confidence limit for internal variability differs a little from - 2°C. We therefore compute the 90% trend limit corresponding to ΔT_{2x} = 2.1°C, add it to 0.52°C and re-evaluate the best fit ΔT_{2x}. The process is repeated until a stable answer is reached. The final result obtained is that the 90% limits on ΔT_{2x} are 0.95°C to 2.39°C. Accounting for this form of natural variability therefore considerably broadens the range of possible ΔT_{2x} values that is consistent with the observations. If natural variability had contributed positively to past global warming, the climate sensitivity could be very low, but if the natural trend had been a cooling, the climate sensitivity could be relatively high.

The final element of uncertainty that we account for, is that in the observational record. IPCC WG1 Section 7 (Folland et al., 1990) estimates the past global warming amount to lie in the range 0.3–0.6°C, with a best guess of 0.45°C. Using the data with ENSO factored out and our model-based method for its quantification, the best estimate of the warming is 0.52°C. Based on the IPCC uncertainty of ±0.15°C, we consider the uncertainty in our estimate to be ±0.1°C. We therefore calculate the ΔT_{2x} range implied by a warming in the range 0.42–0.62°C. This uncertainty aspect, considered alone, gives $1.10 \le \Delta T_{2x} \le 1.80$°C.

We now combine the effects of all of these uncertainties. We do this, as before, by carrying out further model simulations using combinations of the extreme values of the elements considered individually above. The combined effect of K, h and π uncertainties has already been given, viz. $1.32 \le \Delta T_{2x} \le 1.62$°C. Including the effect of natural, internally-generated variability widens this to $0.95 \le \Delta T_{2x} \le 3.19$°C. Finally, including the effect of a ±0.1°C uncertainty in the observed warming leads to $0.73 \le \Delta T_{2x} \le 4.25$°C. These results are summarized in Figure 3.

The above results show much larger changes in ΔT_{2x} at the upper compared with the lower bound of the range of possible values, manifest in the asymmetrical distribution of the confidence limits about the base value (Figure 3). This is because the sensitivity to uncertainties is greater for higher ΔT_{2x}. To illustrate this further, consider the effect of uncertainties in model parameters for the case where the effects of other uncertainties are included (rows 7 and 10 in Figure 3). The range of possible ΔT_{2x} values changes from $0.73 \le \Delta T_{2x} \le 3.00$ to $0.73 \le \Delta T_{2x} \le 4.25$, demonstrating a much larger effect at the high sensitivity end.

The final range of possible ΔT_{2x} values ($0.73 \le \Delta T_{2x} \le 4.25$°C), because it is based on low probability values of a number of factors probably represents a 95% or higher confidence interval (within the limitations of this analysis). It should be compared with the estimated range for ΔT_{2x} based on General Circulation Model (GCM) experiments, viz. $1.5 \le \Delta T_{2x} \le 4.5$°C. Because the two ranges overlap, we can say with confidence that the GCMs and the

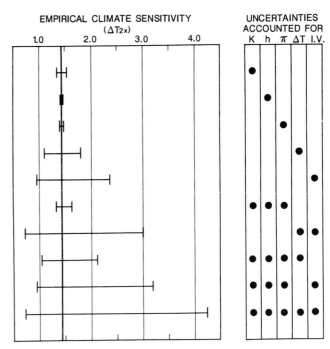

Figure 3: Effect of model, data and natural variability uncertainties on the best-fit estimate of climate sensitivity. The bold vertical line at ΔT_{2x} = 1.43°C is the base value using best-estimate values for the model parameters and the observed warming. In the columns headed "Uncertainties accounted for", ΔT denotes uncertainties in the observed warming and I.V. denotes uncertainties due to internally-generated natural variability.

observations give consistent estimates of the climate sensitivity. However, the observations do tend to point to lower values than the models—they even indicate a reasonable possibility, around 10–20 per cent, that the climate sensitivity is below the GCM-based lower limit of ΔT_{2x} = 1.5°C. This possibility will be discussed further below.

Because we have not accounted for the possible effects of ocean circulation changes and external forcing factors like solar irradiance changes and/or volcanic activity, the range of possible ΔT_{2x} values for any given confidence interval must actually be wider than indicated above. Our opinion is that solar irradiance effects over the past 100+ years have been negligible (see Wigley and Raper, 1990b, and Kelly and Wigley, 1990, for the bases of this opinion). We also believe long-term (>10 year time scale) volcanic effects to be minor, based on observational analyses (Kelly and Sear, 1984; Sear et al., 1986) and on modelling studies (Wigley, 1991b). However, the possibility that one or both of these factors has contributed to low-frequency, global-scale temperature changes since 1861 still remains as an open issue which is unlikely to be resolved adequately in the near future. Similarly, the effects of past ocean circulation changes on global-mean temperature are likely to remain unknown.

One external forcing factor that could, if it is a real effect, lead to an underestimation of the climate sensitivity is the effect of sulphate aerosol changes resulting from Man-made SO_2 emissions. Since SO_2 emissions have occurred predominantly in the Northern Hemisphere, an upper limit to any SO_2-derived negative forcing can be estimated on the basis of inter-hemispheric temperature contrasts. Wigley (1989) places a limit of $-1.5 Wm^{-2}$ to the forcing over the past 100 years or so in this way. Globally, this represents an offsetting of the greenhouse-gas forcing change by about one third, and, not surprisingly, this allows a much larger ΔT_{2x} than would be implied in the absence of an SO_2 effect (see Wigley, 1989, Figure 1). As an example, if ΔT_{2x} were estimated to be 1.6°C with no SO_2 effect, it could be as high as 3.0°C if the SO_2 effect amounted to $-1 Wm^{-2}$ in the Northern Hemisphere. A hemispheric forcing disparity of this magnitude would cause a relative warming of the Southern Hemisphere compared to the Northern Hemisphere, but by an amount that would likely be obscured by natural variability. Although the SO_2 effect is of uncertain magnitude, even if it is small it would be sufficient to bring the empirical and model-based estimates of ΔT_{2x} into closer accord.

Estimating the Climate Sensitivity in the Year 2005

Given the wide range of uncertainty in the empirical ΔT_{2x} value (for any given confidence band, the range is probably greater than given above because we have not directly accounted for ocean circulation changes or the effects of non-greenhouse external forcing factors), will future changes in global-mean temperature allow us to narrow the range? In particular, what if the globe failed to warm or even cooled over the next 15 years? Would this disprove the enhanced greenhouse effect hypothesis? To answer this question, we imagine ourselves in the year 2005 repeating the above analysis. For future forcing over 1990–2005 we use IPCC scenario "B" (Shine et al., 1990). Uncertainties in future forcing over such a short period are small, and have practically no effect on the results presented below. We consider four scenarios for temperature changes to the year 2005: first, a global temperature cooling of 0.2°C between 1990 and 2005; second, no temperature change between 1990 and 2005; third, a warming of 0.2°C; and fourth, a warming of 0.6°C. The latter two numbers bracket the predictions for future global-mean warming under the enhanced greenhouse effect (see, e.g., Wigley and Barnett, 1990, Figure 8.5). Since it is likely that solar irradiance changes will be monitored continuously through to 2005 (and beyond) we also tacitly assume that these will either have no effect on global-mean temperature, or that their

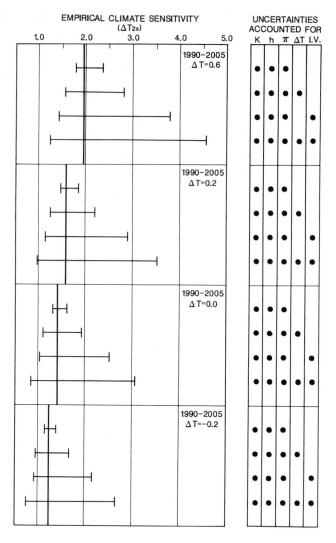

Figure 4: As Figure 3, but for the year 2005 for four different scenarios for the warming between now and 2005 (viz. $\Delta T(1990–2005) = 0.6°C, 0.2°C, 0.0°C$ and $-0.2°C$). The bold vertical lines at 1.97°C, 1.59°C, 1.41°C and 1.24°C are the base value estimates of ΔT_{2x} based on best-estimate model parameter values.

effects have been factored out before performing the empirical sensitivity analysis.

The results are shown in Figure 4. For the cases of negative or zero future temperature change, $\Delta T(1990 – 2005) = -0.2°C$ or $0.0°C$, the base-case estimates for ΔT_{2x} (i.e., using best guess model parameters, ignoring internally-generated natural variability and ignoring data uncertainties) are 1.24 and 1.41°C, respectively. When natural variability and both model and data uncertainties are accounted for, as described in the previous section, the ranges of possible values are $0.74 \leq \Delta T_{2x} \leq 2.67$ for a cooling of 0.2°C and $0.84 \leq \Delta T_{2x} \leq 3.08°C$, for no change. Thus, even these relatively extreme possibilities still admit values of ΔT_{2x} higher than the IPCC best guess value (which is 2.5°C). For ΔT(1990–2005) = 0.2°C, the base-case ΔT_{2x} is 1.59°C and the range is 0.99°C to 3.52°C; while for ΔT(1990–2005) = 0.6°C, the base-case value is 1.97°C with a range of 1.24°C to 4.57°C. In all three cases, the range of possible ΔT_{2x} values is narrowed by the passage of time, only slightly for ΔT(1990–2005) = 0.6°C but more so for smaller values of ΔT(1990–2005).

For the scenarios with no warming between now and 2005, one might expect the lower limit value of ΔT_{2x} to be less than the current lower limit, but this is not the case (compare 0.74°C and 0.86°C with 0.73°C). This somewhat paradoxical result arises because, for any given forcing history, the transient global-mean temperature response becomes increasingly sensitive to ΔT_{2x} as time goes by. This can be seen in Figs. 1 and 2, which show the modelled temperature changes for different ΔT_{2x} diverging from each other. The effect on ΔT_{2x} of a specified temperature departure from the assumed base value, which is essentially the way natural variability effects are accounted for, is therefore less in 2005 than in 1990. The magnitudes of the trend limits for natural variability, however, hardly change at all for 1861 to 2005, compared with 1861 to 1990, so the net effect is a smaller range of ΔT_{2x} values about the base-case estimate.

It is also worth noting that the best estimates for the overall warming to 2005 cannot be obtained simply by adding the various ΔT(1990–2005) values to the 0.52°C warming to 1990. Instead, as before, the best estimate warming is that obtained by using the particular simulated warming curve for the ΔT_{2x} value that minimizes the overall mean square error. For ΔT(1990-2005) = 0.0°C, therefore, the warming to 2005, relative to the 1861-1900 reference period is not 0.52+0.00 but 0.71°C. For ΔT(1990–2005) = -0.2°C, the best fit 2005 temperature is 0.64°C (still higher than 0.52°C), while for ΔT(1990–2005) = 0.2°C and 0.6°C, the best fit 2005 temperatures are 0.78°C and 0.92°C above the reference level.

Independent of the results for ΔT_{2x}, it is still likely that future warming will increase our confidence in the reality of a substantial enhanced greenhouse effect, since continued warming must eventually lead to confident detection on the grounds of unprecedented change (see Wigley and Barnett, 1990, Section 8.4). However, as the above analysis shows, this form of detection need not appreciably affect our ability to determine ΔT_{2x} empirically. Better estimation of ΔT_{2x} is going to require the use of more sophisticated methods than those employed here.

The above results have an important implication in that they illustrate how difficult it would be to disprove the enhanced greenhouse effect hypothesis. Having already experienced a substantial warming over the past century, it would take many years of cooling and/or little change before we could confidently assert that the climate sensitivity was negligibly small (ΔT_{2x} less than 0.5°C, say).

Conclusions

Our main conclusions are as follows. First, if the past global-mean warming of 0.52°C is interpreted solely as an enhanced greenhouse effect, then the implied ΔT_{2x} lies in a relatively narrow range, 1.32–1.62°C, even if model uncertainties are accounted for. This range is near or below the low end of the range suggested by GCM-based studies. Internally-generated natural variability and data uncertainties widen the range of possible ΔT_{2x} values substantially, to 0.73–4.25°C. The climate sensitivity may therefore be anywhere from well below the GCM-based lower limit (which is 1.5°C) to close to the GCM-based upper limit (4.5°C).

Our analysis suggests that there is a probability of about 10–20 per cent that the climate sensitivity is less than 1.5°C. This conclusion, however, depends quite critically on whether or not changes in sulphate aerosol concentrations, arising from Man-made emissions of SO_2, have offset the enhanced greenhouse effect over the past century. Quite a small SO_2-derived negative forcing, of order -1Wm^{-2} in the Northern Hemisphere, would raise the lowest possible ΔT_{2x} noticeably and ensure that the observed global-mean warming was in excellent agreement with General Circulation Model estimates of the climate sensitivity.

Because of these uncertainties, the observations do not allow us to narrow the range of possible ΔT_{2x} values beyond that based on GCM studies. Nor could we confidently say that the lower bound for climate sensitivity was less than the model-based lower limit of 1.5°C. Furthermore, no matter what global-mean temperature changes occur over the next few decades, we are unlikely to be able to reduce these ΔT_{2x} uncertainties substantially, unless more sophisticated empirical methods for estimating ΔT_{2x} can be developed. Even if there were no change or a slight cooling in global-mean temperature between now and 2005, this would still admit a value of the climate sensitivity above the present best guess value of ΔT_{2x} = 2.5°C. "Detection" of the enhanced greenhouse effect is not a panacea in this regard, since detection per se need not improve our knowledge of climate sensitivity. It is this knowledge which is crucial if we are to narrow uncertainties in projections of future climatic change.

Acknowledgement

Part of the work presented here was undertaken as a background contribution to the IPCC Working Group 1 "detection" chapter (Wigley and Barnett, 1990), and was supported by the US Department of Energy, Atmospheric and Climate Research Division, under grant no.FG02-89ER69017-A000.

Appendix:
Model Parameter Values

Uncertainties in predictions of global-mean temperature changes based on an upwelling-diffusion energy-balance climate model are determined by the model structure (i.e., how well such a simple model represents the complexity of physical processes which go to make up the climate system), and by attendant uncertainties in the model parameters. Structural deficiencies are usually assumed to be accounted for by using a range of model parameter values - i.e., it is these deficiencies that are partly responsible for model parameter uncertainties.

In the text, we noted that the upwelling rate, w, and diffusivity, K, are linked. This linkage arises through the steady-state solution of the model which has an exponential term of the form $\exp(-wz/K)$, with K/w as a length scale. This length scale can be estimated by fitting an exponential-decay curve to observed vertical temperature data in the ocean (e.g., those of Levitus, 1982), but the value depends on over what depth the fit is made. For the thermocline, K/w $\approx$ 500m, but larger values obtain if the fitting depth is greater. Based on the strength of the present thermohaline circulation, w is usually taken to be 4m/yr (Hoffert et al., 1980). The implied K is 0.63cm^2/sec. This is the basic (w,K) pair used by IPCC (Bretherton et al., 1990). We choose a slightly higher K value as our base case (and best estimate), viz. 1cm^2/sec, since tracer experiments point to a value noticeably larger than 0.63cm^2/sec, but we retain w = 4m/yr as the base value. This combination is still consistent with the observed vertical profile of global-mean ocean temperature. When K is varied in our analyses, w is varied too so as to maintain constancy of K/w.

The value of the p parameter, defined as the ratio of sinking water temperature change to global-mean temperature change, is the subject of some controversy. Bretherton et al. (1990), representing IPCC, assume a value of 1, but we believe this to be a poor choice, particularly for analyses of past temperature changes. Although it is likely that high latitudes will warm at least as fast as the global-mean, this does not mean that the temperature of the sinking waters that form North Atlantic Deep Water and Antarctic Bottom Water will change to the same extent. The temperature of these waters is primarily controlled by relatively small scale processes associated with sea ice melting and freezing, so any changes in them are likely to be small and decoupled from broader-scale high-latitude temperature changes (see Hoffert et al., 1980, and Harvey and Schneider, 1985, for further discussion). Furthermore, ocean GCM experiments show quite small changes in bottom water temperature even in response to large forcing changes (Stouffer et al., 1989; Mikolajewicz et al., 1990) supporting the view that the best choice for p should be a value near zero. The range chosen here, 0.0-0.4, is a subjective choice based on the above arguments and results. As shown in the main text, however, the simulated temperature changes are relatively insensitive to π and a value of π as high as 1 would not affect our results appreciably.

References

Bradley, R.S., 1988: The explosive volcanic eruption signal in Northern Hemisphere continental temperature records. Climatic Change 12, 221–243.

Bradley, R.S. and Jones, P.D., 1985: Data base for isolating the effects of increasing carbon dioxide concentration. Pp. 29–53 in Detecting the Climatic Effects of Increasing Carbon Dioxide, DOE/ER-0235, (M.C. MacCracken and F.M. Luther, Eds.). US Dept. of Energy, Carbon Dioxide Research Division.

Bretherton, F.P., Bryan, K. and Woods, J.D., 1990: Time-dependent greenhouse-gas-induced climate change. Pp. 173–193 in Climate Change: The IPCC Scientific Assessment, (J.T. Houghton, G.J. Jenkins and J.J. Ephraums, Eds.). Cambridge University Press.

Bryson, R.A. and Dittberner, G.J., 1976: A non-equilibrium model of hemispheric mean surface temperature. Journal of the Atmospheric Sciences 33, 3094–3106.

Bryson, R.A. and Goodman, B.M., 1980: Volcanic activity and climatic changes. Science 207, 1041–1044.

Cess, R.D., Potter, G.L. and 30 others, 1990: Intercomparison and interpretation of climate feedback processes in 19 atmospheric general circulation models. Journal of Geophysical Research 95, 16601–16615.

Charlson, R.J., Langner, J. and Rodhe, H., 1990: Perturbation of the Northern Hemisphere radiative balance by backscattering of anthropogenic sulfate aerosols. Tellus (in press).

Charlson, R.J., Lovelock, J.E., Andreae, M.O. and Warren, S.G., 1987: Oceanic phytoplankton, atmospheric sulphur, cloud albedo and climate. Nature 326, 655–661.

Cubasch, U. and Cess, R.D., 1990: Processes and modelling. Pp. 69–91 in Climate Change: The IPCC Scientific Assessment, (J.T. Houghton, G.J. Jenkins and J.J. Ephraums, Eds.). Cambridge University Press.

Eddy, J.A., 1977: Climate and the changing Sun. Climatic Change 1, 173–190.

Folland, C.K., Karl, T.R. and Vinnikov, K.Ya., 1990: Observed climate variations and change. Pp. 195–238 in Climate Change: The IPCC Scientific Assessment, (J.T. Houghton, G.J. Jenkins and J.J. Ephraums, Eds.). Cambridge University Press.

Folland, C.K., Parker, D.E. and Kates, F.E., 1984: Worldwide marine temperature fluctuations 1856–1981. Nature 310, 670–673.

Foukal, P. and Lean, J., 1990: An empirical model of total solar irradiance variation between 1874 and 1988. Science 247, 556–558.

Gilliland, R.L., 1982: Solar, volcanic and CO_2 forcing of recent climate change. Climate Change 4, 111–131.

Gilliland, R.L. and Schneider, S.H., 1984: Volcanic, CO_2 and solar forcing of the Northern and Southern Hemisphere surface air temperature. Nature 310, 38–41.

Grove, J.M., 1988: The Little Ice Age. Methuen, London.

Hammer, C.U., Clausen, H.B. and Dansgaard, W., 1980: Greenland ice sheet evidence of post-glacial volcanism and its climatic impact. Nature 288, 230–235.

Hansen, J.E., Fung, I., Rind, D., Lebedeff, S., Ruedy, R., Russell, G. and Stone, P., 1988: Global climate changes as forecast by Goddard Institute for Space Studies three-dimensional model. Journal of Geophysical Research 93, 9341–9364.

Hansen, J.E., Johnson, D., Lacis, A., Lebedeff, S., Lee, P., Rind, D. and Russell, G., 1981: Climate impact of increasing atmospheric carbon dioxide. Science 218, 957–966.

Hansen, J.E. and Lacis, A.A., 1990: Sun and dust versus greenhouse gases: an assessment of their relative roles in global climate change. Nature 346, 713–719.

Hansen, J.E. and Lebedeff, S., 1987: Global trends of measured surface air temperature. Journal of Geophysical Research 92, 13345–13372.

Hansen, J.E. and Lebedeff, S., 1988: Global surface temperatures: update through 1987. Geophysical Research Letters 15, 323–326.

Harvey, L.D. and Schneider, S.H., 1985: Transient climate response to external forcing on 10^0–10^4 year time scales Part 1: Experiments with globally averaged, coupled, atmospheric and ocean energy balance models. Journal of Geophysical Research 90, 2191–2205.

Hasselmann, K., 1976: Stochastic climate models, 1, Theory. Tellus 28, 473–485.

Hickey, J.R., Alton, B.M., Kyle, H.L. and Hoyt, D., 1988: Total solar irradiance measurements by ERB/Nimbus-7. A review of nine years. Space Science Reviews 48, 321–342.

Hoffert, M.I., Callegari, A.J. and Hsieh, C.-T., 1980: The role of deep sea heat storage in the secular response to climate forcing. Journal of Geophsyical Research 85, 6667–6679.

Jones, P.D., 1988: Hemispheric surface air temperature variations: recent trends and an update to 1987. Journal of Climate 1, 654–660.

Jones, P.D., 1989: The influence of ENSO on global temperatures. Climate Monitor 17, 80–90.

Jones, P.D., Raper, S.C.B., Bradley, R.S., Diaz, H.F., Kelly, P.M. and Wigley, T.M.L., 1986a: Northern Hemisphere surface air temperature variations: 1851–1984. Journal of Climate and Applied Meteorology 25, 161–179.

Jones, P.D., Raper, S.C.B. and Wigley, T.M.L., 1986b: Southern Hemisphere surface air temperature variations: 1851-1984. Journal of Climate and Applied Meteorology 25, 1213–1230.

Jones, P.D. and Wigley, T.M.L., 1990: Global warming trends. Scientific American 263, 84–91.

Jones, P.D., Wigley, T.M.L. and Farmer, G., 1991: Marine and land temperature data sets: a comparison and a look at recent trends. In, Greenhouse-Gas-Induced Climatic Change: A Critical Appraisal of Simulations and Observations, (M.E.Schlesinger, Ed.). Elsevier Science Publishers, Amsterdam (in press).

Jones, P.D., Wigley, T.M.L. and Wright, P.B., 1986c: Global temperature variations, 1861–1984. Nature 322, 430–434.

Kelly, P.M. and Sear, C.B., 1984: Climatic impact of explosive volcanic eruptions. Nature 311, 740–743.

Kelly, P.M. and Wigley, T.M.L., 1990: The contribution of solar forcing trends in global mean temperature. Nature 347, 460–462.

Lamb, H.H., 1970: Volcanic dust in the atmosphere. Philosophical Transactions of the Royal Society of London A266, 426–533.

Levitus, S., 1982: Climatological Atlas of the World Oceans. NOAA Professional Paper, 13. US Government Printing Office, Washington, D.C.

Levitus, S., 1989: Interpentadal variability of salinity in the upper 150m of the North Atlantic Ocean, 1970-1974 versus 1955–1959. Journal of Geophysical Research 94, 9679–9685.

Mass, C.F. and Portman, D.A., 1989: Major volcanic eruptions and climate: A critical evaluation. Journal of Climate 2, 566–593.

Mikolajewicz, U., Santer, B.D. and Maier-Reimer, E., 1990: Ocean response to greenhouse warming. Nature 345, 589–593.

Mitchell, J.F.B., Manabe, S., Tokioko, T. and Meleshko, V., 1990: Equilibrium climate change - and its implications for the future. Pp. 131-172 in Climate Change: The IPCC Scientific Assessment, (J.T. Houghton, G.J. Jenkins and J.J. Ephraums, Eds.). Cambridge University Press.

Mitchell, J.M., Jr., 1970: A preliminary evaluation of atmospheric pollution as a cause of the global temperature fluctuation of the past century. Pp. 139–155 in, Global Effects of Environmental Pollution, (S.F. Singer, Ed.). Springer Verlag, New York/D. Reidel, Dordrecht.

Oerlemans, J., 1988: Simulation of historic glacier variations with a simple climate-glacier model. Journal of Glaciology 34, 333–341.

Oliver, R.C., 1976: On the response of hemispheric mean temperature to stratospheric dust: an empirical approach. Journal of Applied Meteorology 15, 933–950.

Pivovarova, Z.I., 1977: Radiation Characteristics of Climate of the USSR. Gidrometeoizat, Leningrad (in Russian).

Porter, S.C., 1986: Pattern and forcing of Northern Hemisphere glacier variations during the last millenium. Quaternary Research 26, 27–48.

Reid, G.C., 1987: Influence of solar variability on global sea surface temperature. Nature 329, 142–143.

Reid, G.C., 1990: Total solar irradiance variations and the global sea-surface temperature record. Journal of Geophysical Research (in press).

Robock, A., 1978: Internally and externally caused climate change. Journal of the Atmospheric Sciences 35, 1111–1121.

Röthlisberger, F., 1986: 10,000 Jahre Gletschergeschichte der Erde. Verlag Sauerländer, Aarau.

Schönwiese, C.-D., 1991: Multivariate statistical assessments of greenhouse-induced climatic change and comparison with the results from general circulation models. In, Greenhouse-Gas-Induced Climatic Change: A Critical Appraisal of Simulations and Observations, (M.E. Schlesinger, Ed.). Elsevier Science Publishers, Amsterdam (in press).

Schwartz, S.E., 1988: Are global cloud albedo and climate controlled by marine phytoplankton? Nature 336, 441–445.

Sear, C.B., Kelly, P.M., Jones, P.D. and Goodess, C.M., 1987: Global surface-temperature responses to major volcanic eruptions. Nature 330, 365–367.

Shine, K.P., Derwent, R.G., Wuebbles, D.J., Morcrette, J.-J., 1990: Radiative forcing of climate. Pp. 41-68 in Climate Change: The IPCC Scientific Assessment, (J.T. Houghton, G.J. Jenkins and J.J. Ephraums, Eds.). Cambridge University Press.

Simkin, T., Siebert, L., McClelland, L., Bridge, D., Newhall, C. and Latter, J.H., 1981: Volcanoes of the World. Hutchinson Ross, Stroudsbourg, Pennsylvania.

Stouffer, R.J., Manabe, S. and Bryan, K., 1989: Interhemispheric asymmetry in climate response to a gradual increase of atmospheric CO_2. Nature 342, 660–662.

Twomey, S.A., Piepgrass, M. and Wolfe, T.L., 1984: An assessment of the impact of pollution on global cloud albedo. Tellus 36B, 356–366.

Vinnikov, K.Ya. and Groisman, P.Ya., 1981: The empirical analysis of CO_2 influence on the modern changes of the mean annual Northern Hemisphere surface air temperature. Meteorologiya i Gidrologiya 1981(11), 30–43.

Vinnikov, K.Ya. and Groisman, P.Ya., 1982: An empirical study of climate sensitivity. Atmospheric and Oceanic Physics 18(11), 1157–1167.

Watts, R.G., 1985: Global climate variation due to fluctuations in the rate of deep water formation. Journal of Geophysical Research 90, 8067–8070.

Wigley, T.M.L., 1989: Possible climate change due to SO_2-derived cloud condensation nuclei. Nature 339, 365–367.

Wigley, T.M.L., 1991a: Could reducing fossil-fuel emissions cause global warming? Nature 349, 503–506.

Wigley, T.M.L., 1991b: Climate variability on the 10-100 year time scale: observations and possible causes. In, Global Changes of the Past, (R.S. Bradley, Ed.). (In press.)

Wigley, T.M.L., Angell, J.K. and Jones, P.D., 1985: Analysis of the temperature record. Pp. 55–90 in Detecting the Climatic Effects of Increasing Carbon Dioxide, DOE/ER-0235 (M.C. MacCracken and F.M. Luther, Eds.). US Dept. of Energy, Carbon Dioxide Research Division.

Wigley, T.M.L. and Barnett, T.P., 1990: Detection of the greenhouse effect in the observations. Pp. 239–255 in Climate Change: The IPCC Scientific Assessment, (J.T. Houghton, G.J. Jenkins and J.J. Ephraums, Eds.). Cambridge University Press.

Wigley, T.M.L., Jones, P.D. and Kelly, P.M., 1986: Empirical climate studies: warm world scenarios and the detection of climate change induced by radiatively active gases. Pp. 271–323 in The Greenhouse Effect, Climatic Change, and Ecosystems, (B. Bolin, B.R. Döös, J. Jäger and R.A. Warrick, Eds.). SCOPE series, John Wiley and Sons, Chichester.

Wigley, T.M.L. and Kelly, P.M., 1990: Holocene climate change, ^{14}C wiggles and variations in solar irradiance. Philosophical Transactions of the Royal Society of London A330, 547–560.

Wigley, T.M.L. and Raper, S.C.B., 1987: Thermal expansion of sea water associated with global warming. Nature 330, 127–131.

Wigley, T.M.L. and Raper, S.C.B., 1990a: Natural variability of the climate system and detection of the greenhouse effect. Nature 344, 324–327.

Wigley, T.M.L. and Raper, S.C.B., 1990b: Climatic change due to solar irradiance changes. Geophysical Research Letters 17, 2169–2172.

Wigley, T.M.L. and Raper, S.C.B., 1991: Internally generated natural variability of global-mean temperatures. In, Greenhouse-Gas-Induced Climatic Change: A Critical Appraisal of Simulations and Observations, (M.E. Schlesinger, Ed.). Elsevier Science Publishers, Amsterdam (in press).

Willson, R.C. and Hudson, H.S., 1988: Solar luminosity variations in solar cycle 21. Nature 332, 810–812.

Climate, Water and Development

by J. Sircoulon *(France)*

1. Abstract

There is increasing awareness of the significant relationships between climate, continental waters and environment. The effects of climatic hazards which regularly affect numerous regions of the planet are now better understood. The countries which are not ready for strict water management become highly sensitive to climatic fluctuations, while the arid tropical countries are faced with survival problems. With regard to the foreseeable population growth, the water resource element seems to be fragile and its quantity as well as its quality worsens, while the hydrometric and pluviometric networks decline or become unsuitable.

Over the last ten years, numerous studies have been carried out in order to understand better the relations between the climate-water cycle and water balance and to simulate the hydrological effects of possible climatic changes. However, despite the improvements made in models, satellite techniques, data banks and field studies, the results remain insufficient and sometimes contradictory.

In the future, it will be necessary to study more thoroughly the hydrological processes related to the biosphere (IGBP), to make closer links between GCMs and the hydrological models (GEWEX), to create genuine planetary observatories and to transfer technologies from northern to southern countries.

The international programmes which have been launched recently should address the major challenge raised by water in the 21st Century.

2. Introduction

In his introductory report at the First World Climate Conference, White (WMO, 1979) mentioned the main reasons which led to the organization of that significant scientific event. He emphasized the disastrous consequences of the climatic catastrophes of the 1970s on human activities and the extent to which man remained vulnerable, despite technological progress. As examples White mentioned the serious effects of the 1976 drought in Western Europe and the dramatic drought which affected the African Sahel throughout the 1970s.

It is obvious that the concerns shown at that time have kept on increasing and that the increasing atmospheric concentrations of greenhouse gases give rise to further concerns. For example, the drought which affected France in 1989–90 was worse than the previous one, and that of the Sahel, which is still going on, exhibits an exceptionally lasting character and is of unrivalled seriousness. (Sircoulon, 1990 a,b).

Over recent years, the continuing increase in the atmospheric concentration of greenhouse gases has been accompanied by a gradual increase in temperature on the surface of the planet. The results of GCMs for a doubling of CO_2 have largely concentrated on the increase in temperature and its effects on the melting of ice bodies, which contributes to a considerable increase in the sea level. On the other hand, the possible influence of the increasing greenhouse effect on the water cycle and balance (rainfall, evaporation, runoff) is less often mentioned. There is much greater model uncertainty concerning these aspects. However, the close interaction between climate and water and the vulnerability of water resources (the available and exploitable part) are now taken into account much more frequently than a few years ago. The year 1989 was undoubtedly a significant turning point in that the politicians of numerous countries became aware that the natural resources and the environment of our planet must be preserved. The Scientific Conference on Climate and Water held in Helsinki in September 1989 was, in this context, an important contribution.

With regard to the future crises (Duplessy and Morel, 1990; Brown, 1990), the demographic situation seems to be particularly alarming, due to the difference observed between a constantly increasing water demand and a water resource which is necessarily limited in its amount and in its distribution. Currently, only 5 to 10% of the running waters theoretically available are really consumed but water is not necessarily found where it is needed. Moreover, the non-renewable ground waters of very vast arid zones (Margat, 1990) are running dry. A decrease in the amount of the resource and degradation (pollution) of its quality will be major challenges to be faced by mankind in the 21st Century. Indeed, an increase of temperature in the future could have some advantages by increasing the circulation of water vapour, leading to significant increases of rainfall in certain places. However, we will see that for the arid zones the improvement is much debated (Mabbutt, 1989).

3. The hydrological consequences of some recent climatic anomalies

Through history interest in weather and climate has increased (Le Roy Ladurie, 1983). Although the weather has always been a favourite theme of discussion, meteorology and weather extremes have been mentioned increasingly in the media over the past ten years.

There have been many natural disasters in recent years

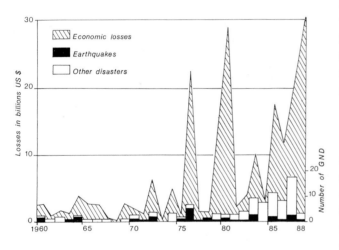

Figure 1: Great natural disasters and loss extent. Source: Munich Reinsurance Company.

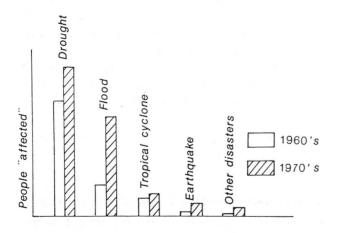

Figure 2: Proportion of people affected per year by natural disasters. Source: Wijkman and Timberlake (1984).

(see Figures 1 and 2), including hydrological extreme values. Examples of these extremes are: the 1982–83 El Niño phenomenon and the related anomalies in other regions (Cadet and Garnier, 1988); the 1979–83 drought in the Brazilian Nordeste (Molinier and Cadier, 1985); the floods in Bangladesh in 1987 and in particular in 1988 when 62% of the country was affected; and the drought in the USA in 1988. Two instances are worth detailed study since they are especially illustrative of hydrological aspects.

3.1. The 1988–1990 drought in France

In fact the depletion in water storage was less alarming than the degradation in the quality and the mismatch between locations of supply and demand. In France, the annual rainfall amounts to 750mm on average and the whole national water availability (as runoff) is 33 times higher than the requirements. Each inhabitant has about 3000 m^3 of water renewed each year. However, in mid-September 1990 ("hot September"), after a 24 month continuous drought, the situation became more than worrying with water limitations in 57 departments, 11000 km of dried up rivers, nuclear power stations not running at full load, shipping stopped in canals, previous records of low water surpassed, and aquatic ecosystems in danger (see Figures 3 and 4). The situation (Merillon, 1990) was worse than in 1976 (Sircoulon, 1989b) and can be compared to 1949. 1921 remains the driest year in the rainfall records.

This shows that even in an advanced temperate zone country, prolonged and intense water deficits can upset numerous economic sectors and accentuate the deficiencies of an unsuited water management or its inadequate distribution. If climatic variability can have such effects, the effects of a significant climatic change must receive serious attention. In the months to come, France will change its water laws.

3.2 The drought in the Sahel since 1968

This example is vastly more serious than the previous one. The drought extended from Mauritania to the Sudan, whose social and economic equilibrium is in any case fragile due to the "arid climate". This zone whose annual rainfall ranges from 100 to 750mm extends over several millions of km^2 and receives rainfall which is highly irregular in space and time within the 2 to 3 month rainfall period. In the great river basins of the zone, the rainfall deficit leads to a still higher runoff deficit. The nine countries of Western Africa affected by drought had a population amounting to 18 million inhabitants in 1958, to about 40 million in 1988, and it could amount to 105 million in 2025 (after the United Nations ST/ESA/SER-A/106/addl).

The drought is exceptional for three reasons:

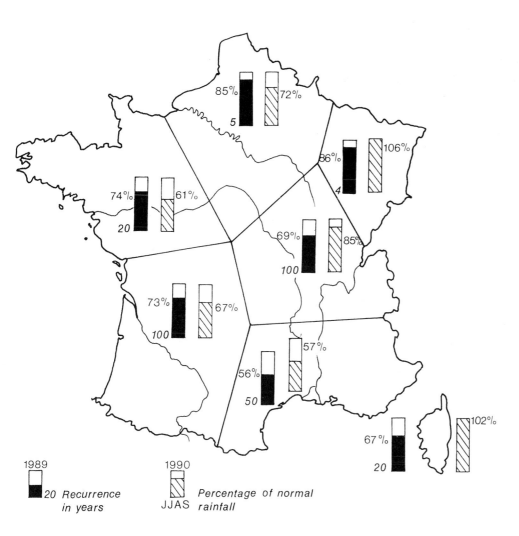

Figure 3: Precipitation 1989 and 1990. Source: Ministère de l'Environnement - France.

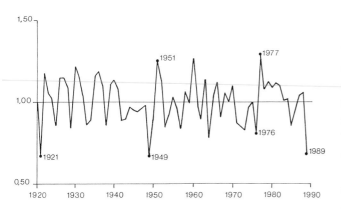

Figure 4: Index of hydroelectricity power capacity. Source: Duband, 1990.

- its persistence; it has been observed since 1968 or 1970 according to region
- its extent; it affects several millions of km² and spread extensively towards the humid tropical zones in 1972-73 and 1983–84

- its seriousness; the rainfall deficits often reach 20 to 40%, and even more.

The long records of discharge (see, for example, Figure 5) of the large tropical rivers show a slump in maximal floods, a longer period of low waters with often a suspension of runoff. The total water yield (see Figure 6) in the francophone Sahelian zone amounted, on average, to only 79.10^9 m³ per year for the 1970–1988 period, or a deficit of 43% as compared to the previous period (Sircoulon, 1990b). The case of the Sahel drought raises crucial questions about the causes of the persistent deficit, stationarity of climate, hydropluviometric standards to be used for water management, and most importantly the possible development of this extensive zone in the coming decades if the drought continues.

4. Instrumental Data and Hydrological Variability

The existence of a more or less pronounced natural variability of hydrological parameters (Beran, 1984) has always been accepted, but the common standpoint was that

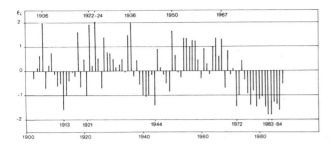

Figure 5: Annual runoff departures (in number of standard deviations) in Senegal at Bakel.

the data of the recent past were adequate for any development project. The stationary character of the rainfall and discharge series was accepted. Several authors, such as Hare (WMO, 1979) and Askew (1987), pointed out that it is necessary to distinguish between climatic variability (anomaly) and climatic change, and to show the significance of the time scales used. For example, a climatic change could lead to a different seasonal distribution of runoff (e.g., more humid winters and warmer summers), while the annual volume of runoff is not considerably modified (Gleick, 1987). Thus, a mere annual study would not reveal the evolution of a hydrological regime.

Has there been a significant statistical change in the rainfall or discharge instrumental series (which seldom go beyond a century) in the Sahelian region during the last

hundred years, when the global temperatures are observed to have increased?

Bradley et al. (1987) observed that in the northern hemisphere, rainfall has tended to increase in the temperate latitudes and to decrease in the tropical latitudes (Figure 7), which is in keeping with the rainfall indices defined by Nicholson (1979), see Figure 8 updated to recent years, or Lamb (1982) for the Sahel. By using the data gathered for the Frend project (Gustard, 1989), Arnell and Raynaud (1989) showed (as had already been noted by Wigley and Jones (1985)) that in England and Wales floods tended to be more frequent due to more humid winters and springs and to drier periods in the summers. Generally, few significant changes have been observed up to now, (Duband, 1990). It has proved difficult to separate the effects of climate change from other natural causes or indirect human actions and it is obvious that the instrumental series, with problems of homogeneity of data and quality of measurements, have to be analyzed very critically. The direct human influence through the withdrawals from rivers is very apparent (see, for instance, Shiklomanov, 1990).

With a view to a probable climatic change, Liebscher (1989) extensively reviewed the available information and suggested necessary future analyses. He showed to what extent it was useful to extend the instrumental data through palaeoclimatology and he emphasized the difficulties due to the increasing influence of man on the natural environment. Liebscher also briefly reviewed the various projects in WCP-WATER concerning this field, such as the

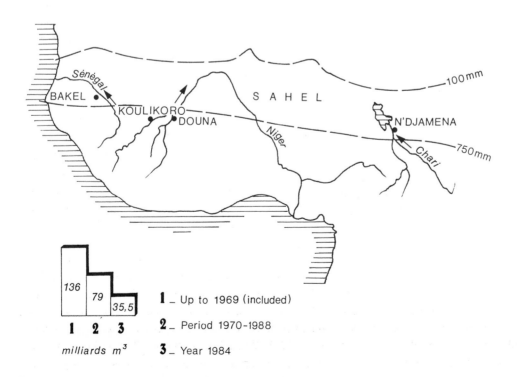

Figure 6: Annual water yields average in $10^9 m^3$.

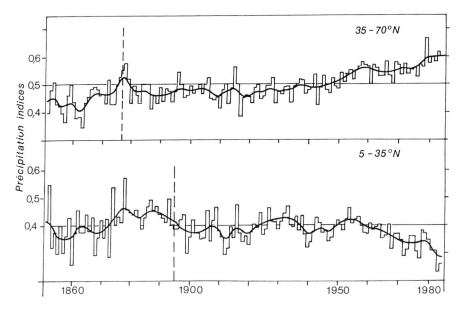

Figure 7: Changes in area-averaged precipitation of the northern hemisphere. Source: Bradley et al. (1987).

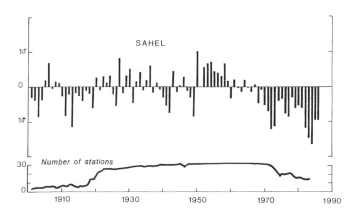

Figure 8: Standardized annual rainfall anomalies.

A2 project (WMO, 1988c), which aims at comparing the results drawn from the analysis of long hydrological series. The inauguration of the Global Runoff Data Center (GRDC) at Koblenz should facilitate such studies (WMO, 1989 a).

However the discontinuity (change in the mean) in the rainfall series of the Sahel (Todorov, 1985; Carbonnel and Hubert, 1985; Snijders, 1986; Hubert et al, 1989) shows the fragility of:

- The 30 year rainfall normals which are used (WMO, 1989 c) in order to assure comparability between data collected at various stations and to give climatological standards, but which become locally unsuited for a non- stationary climate. The transition from the 1931–1960 normal to the 1961–1990 one will upset the rainfall indexes (for instance, in Senegal, the new normal will be lower than the previous one by about 40%, thus decreasing from 500 to 300mm or from 350 to 200mm..). Furthermore, because of station moves and closures, the comparison between 1931–1960 and 1961–1990 is often biased.

- The hydrological standards to be selected for hydraulic systems. (A conference held by the Comite Interafricain d'Etudes Hydrauliques (CIEH) dealt with this problem at Ouagadougou in 1987.) The question is roughly the following one: must one choose a pre-1970 standard with the possibility of a non-profitable infrastructure or post-1970 standard which would considerably reduce the cost of dams but may lead to insufficient storage in the case of a return to higher rainfall?

In Sub-Saharan Africa, a number of water resource projects were constructed in the 1960s and 1970s based on the instrumental series available then. Therefore, the observations made in the 1950–1965 period, which was a humid phase, were given a great significance. Numerous impoundments have proved to be too large. A famous example is the hydroelectric dam situated at Kossou in the Ivory Coast which entered service in 1971 and was designed around an average annual yield of $5.5.10^9$ m^3 (production: 530 GWh). However, over the fifteen following years, the annual yields have amounted only to $2.7.10^9$m^3 on average (or -51%) for a production of 115 GWh (or -78%). Another famous example is the Aswan dam which has suffered an exceptional series of dry years since 1978. While it was almost filled up in 1978 with a live storage of more than 100.10^9m^3, its level has gradually decreased since that year. The live storage amounted only

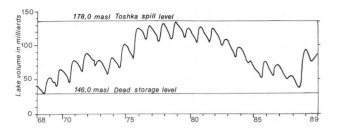

Figure 9: Lake Nasser storage (1968-1989).

to 17.10^9m^3 at its maximum level at the end of 1987 and only the catastrophic flood observed in Khartoum in August 1988 prevented the dam turbines from being stopped (see Figure 9).

Although our present knowledge does not allow us to define strict rules to re-examine existing water resource schemes, it is, nevertheless possible to determine the resilience of water resources to the climatic changes (WMO Bulletin, 1989).

5.   From Climatic Evolution to Hydrological Evolution

Even if there were no significant changes of greenhouse gases, we would still be unable to forecast the climatic changes over the next centuries. Even if, for the case of a doubling of CO_2, we could calculate the precise increase in temperature and its distribution over the planet, it would remain highly problematic to evaluate accurately the changes of regional hydrological parameters.

It is recognized that the warming of the lower atmosphere leads to an acceleration of the water cycle (water transfers between the ocean, the atmosphere and the continental surfaces) and to an increase in the water vapour content of air (of about 6% per extra degree) and to a more intense potential evaporation.

If CO_2 were to double, the rainfall and evaporation which come into equilibrium in the global water balance, should increase from 7 to 15% according to the models (Santer et al., 1990) but these globally averaged values do not give any information about the regional distribution.

In his review of the Conference on Water and Climate held at Helsinki in 1989, Beran summarized a number of processes and interactions observed in the hydrological cycle as well as the role of other gases (methane, nitrous oxide, sulphur derivatives).

In the present state of the art it is not possible to determine how runoff will be modified (annual amount, seasonal distribution, extreme values), since it is necessary first to evaluate rainfall and then to calculate evaporation and antecedent conditions for runoff, which depend on showers, their intensity and their space-time distribution, as well as on vegetation, the soil moisture and the surface

conditions in general. All these factors are likely to change considerably with the climate.

Over the last few years, several authors have studied the methods used to test the sensitivity of the hydrological cycle, e.g., Dooge (1986), Beran (1986), Beran and Arnell (1989), Gleick (1986, 87, 89), Bach (1989), and WMO (1987 and 1988b).

The evaluation of the implications of climatic change for water can be divided into three stages:

- development of quantitative scenarios for the key climatic variables
- simulation of the hydrological cycle in a given basin using the scenario values
- evaluation of the impacts of the changes on the constituent elements of water resource management or of the various water uses.

The following paragraphs review briefly a few methods used.

The literature describing general circulation models (GCMs) and their limitations is massive. Good reviews are given by Dickinson (1986), WMO (1987) and, above all, the IPCC WGI report (1990). Bach (1989) showed one way that GCM results may be applied to Europe. An example of the limitations and uncertainties (results differing from one model to the other) is given by the US Environment Agency (1988) and Beran and Arnell (1989). Santer et al. (1990) also studied the differences between five GCMs (GISS, NCAR, GFDL, UKMO and OSU).

Researchers often use the GCM outputs as inputs to the hydrological models, which range from simple regression models to complex conceptual models. Beran (1986) reviewed these different models, such as the Sacramento model used by Nemec and Schaake (1982). Bultot et al (1988) applied a conceptual model to three Belgian basins and their response to climate change. Klemes (1985) recommended that researchers who use such models need to obtain a good physical basis and to test the model's adequacy for a range of conditions.

The palaeoclimatic analogues (empirical approach) approach is very often used in the USSR. Budyko (1989) compared the situation observed in the Holocene Optimum (considered to be the climate analogue for 2000), in the Last Interglacial and in the Pliocene to the current period.

In the spatial analogues the current climate for site A, is replaced by the climate observed in the site B.

In arbitrary scenarios arbitrary variations in the primary climatic data are determined. For Western Europe, comparisons between the 1934–1953 warm period and the 1901–1920 cold period were made by

Lough et al. (1983), quoted by Beran et al. (1989), for rainfall and by Palutikof (1987) for discharge. These studies find an increase in rainfall and runoff in northern Europe and a decrease in the south. But if the influence of increasing CO_2 concentrations on plant stomata is taken into account, this results in a considerable decrease in evapotranspiration, which would lead to higher runoff in the whole region.

The IPCC WGII report (1990) contains numerous examples of the potential impacts of climatic change on hydrology and water resources.

Generally, most of the hydrological applications of the climatic simulations made up to now were for the temperate countries of Europe or the USA and show that global warming is accompanied by more humid winters and drier summers.

On the other hand, there are fewer studies of the tropical zone where, however, 80% of the population of the planet are found and where the problems related to water are the most crucial.

The models agree quite well in forecasting modest increases in temperature in low latitudes ranging only from 0.5° to 2°C, which may appear insignificant. However, the sensitivity of these regions to variations in rainfall or evaporation is considerable.

The humid tropics could experience an increase in rainfall and in precipitation intensity (WMO, 1988b) along with more pronounced seasonal water deficiencies. The arid zone would be highly sensitive to the soil moisture or runoff changes, as found by Nemec and Schaake (1982). Depending on the GCM used, both water surpluses or water deficits are found (Mabbutt, 1989; Santer et al., 1990), but the latter appear to be the more common. This disturbing finding for the arid tropical zones is particularly worrying given the current effects of the scarcity of rainfall. A priority research effort should therefore be directed towards these regions.

6. A Few Observations and Perspectives

Because of the global importance of climatic variations and the serious phenomena which cannot be easily reversed, multi-disciplinary scientific efforts have been developed in recent years. The results of the work carried out on climatic change and the land-ocean-atmosphere interactions emphasized both the deficiencies of the research and the types of research which must be conducted by the end of this century.

The general physical, chemical and especially the biological phenomena involved in the terrestrial system where we are living are at the heart of the International Geosphere-Biosphere Programme (IGBP). One of the core projects of this programme concentrates on the influence of soil and vegetation on the hydrological cycle (IGBP, 1990). This project needs to be closely related to the WCRP and to its major programme "Global Energy and Water Cycle Experiment" (GEWEX), whose details are currently being worked out (WMO, 1988a) and which is described by McBean in these proceedings.

In order to understand better the phenomena involved in the changes of the water cycle and of its global balance, and to evaluate and forecast the possible transformations, it is necessary to model better and therefore to measure better.

In order to model better, it is necessary to better couple GCMs with improved descriptions of hydrological processes. It is obvious that the current size of the GCM grids (several hundred km) is not suited for a detailed hydrological evaluation. In order to obtain more precise data for hydrological models, it will be necessary to improve the GCM grid (requiring more computer power), to improve the description of processes over continental surfaces, which in turn requires a clearer view of the phenomena (macrohydrology, Shuttleworth, 1988), to make large-scale precise measurements (case of the experiment HAPEX- MOBILHY, Andre, 1990) on a 100x100km square and to launch new experiments such as HAPEX-Niger scheduled in 1992–1993 following the current operation EPSAT-Niger.

Only remote sensing can provide a better understanding of the space-time behaviour of potential evapotranspiration, of rainfall, of nebulosity, of surface temperatures in oceans or continents etc. Thus, the METEOSAT satellite permits a general and continuing view of the African continent or the observation of the evolution of the African monsoon in the Gulf of Guinea, while the SPOT satellite will give precise detail for the 100 km square of HAPEX-Niger.

One can only welcome the successes of the American project EOS in providing a general view of the Earth, but we must keep in mind that the ground validation remains essential as well as satellite calibration and continuity of observation of the same variables.

As Schaake emphasized data are the core of hydrology (WMO Bulletin, 1990). Two aspects must be considered: the equipment (material) and the measurements (networks and archiving of data) themselves.

Concerning the equipment, obvious progress has been made through data tele-transmission (Pouyaud and Le Barbe, 1987).

Concerning measurements, the situation is rather gloomy. Improvements in the network management are feasible (Olivry, 1986) but only if the regionalization of hydrologic parameters can be undertaken. Efforts are being made in the data archiving (INFOCLIMA), data storage (CLICOM), and their collation and dissemination (GRDC, GPDC, CSM). Generally, the meteorological data networks have been worsening since the 1960s or 1970s (for

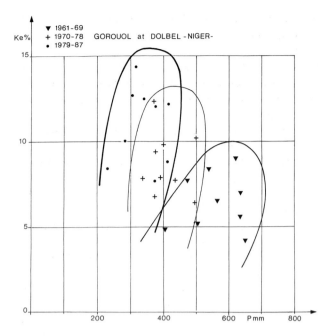

Figure 10: Annual runoff coefficients versus annual precipitation.

instance, WMO 1989b) and this is very apparent since the 1979 GARP experiment. This gives rise to a situation where it is now difficult to establish the new normals (WMO, 1990) or certain rainfall indexes (cf. Lamb or Nicholson for the Sahel). There are also crucial problems concerning the length of the series, their quality, and their origin (significance of METADATA or "Data about Data"), as well as the "naturalization" of data discharge (Beran et al., 1989), since man is increasingly disturbing the natural runoff regime.

The problem of measurement is particularly significant, since data supplied for modelling and the climatic follow-up depend on it in the course of the instrumental period. The measurement, validation, storage and circulation are the major imperatives (Actes Planète Terre, 1989).

However, the measurements do not concern only hydrometeorological or oceanographic data but also all those which aim at better knowing the continental surfaces (vegetation, soil moisture, surface states). The observation of surface conditions (Casenave and Valentin, 1989), can thus give essential information on the evolution of runoff and floods. In the Sahelian basins, (Sircoulon, 1990b), the increase of the annual runoff coefficient—due to soils becoming more impervious—may mitigate the effects of an annual precipitation decrease (see Figure 10 for a basin of 7500 km^2 in Niger.)

Finally another recommendation seems to be significant: given the worldwide problem of the future climatic change and the probable changes of water resources, it is necessary to pool field experiments, data, methodologies and

financial resources better. This requires the establishment of close international collaboration.

For this purpose, it is essential that the northern countries which would be affected by these climate changes co-operate with the southern countries which may be affected much more seriously. A close collaboration is also necessary between the countries affected by the same climatic hazards. In this respect, the creation of the "Sahara-Sahel Observatory" which involves 20 African countries affected by aridity (Northern, Western and Eastern Africa) is particularly stimulating.

Conclusion

- Water is of vital importance and even without the mismatch between water resources and water demand, due to population growth, will be crucial in many arid countries in the near future.
- Water is strongly involved in climate processes but further research remains necessary.
- There is currently evidence of a change in hydrologic variability and a growing concern about design and management of water resources systems.
- The knowledge of the impact of future climatic changes on surface waters remains modest in numerous countries and the application of model calculations to large river basins or to large area grids (GCM) is still problematic.
- The initiation of new international programmes, such as IGBP or GEWEX, to improve the understanding of physical and biological processes for different scales is very important.
- Hydrological observations are vitally important for assessing climatic variability and the data systems must be enhanced.
- It is essential to associate developed and developing countries in international research and data programmes, as well as field experiments.

References

Actes du Colloque Planète Terre, Paris 12 et 13 juin 1989, Ministere de la Recherche et de la Technologie, 347 pp.

ANDRE, J.C.: 1990, Interaction entre climat, processus de surface continentale, cycle de l'eau et variation de l'humidité des sols, La Météorologie, VIIème série, n°33, 10–19.

ARNELL, N., REYNAUD, N.: 1989, Estimating the impacts of climate change on river flows ; some examples from Britain, Conference on Climate and Water, Helsinki, National IHP Committee of Finland, 426–436.

ASKEW, A.J., 1987: Climate change and water resources, IAHS, Publ. n°168, 421–430.

BACH, W.: 1989, Projected climatic changes and impacts in Europe due to increased CO_2, Conference on Climate and Water, Helsinki, National IHP Committee of Finland, 31-50.

BERAN, M.A.: 1984, Climate change: New problems for water resources and hydrology, UNESCO/WMO/IAHS, Sc. 84/W5/53.

BERAN, M.A.: 1986, The water resources impact of future climatic change and variability, Effects of changes in stratospheric ozone and climate: Volume 1: Overview, J.G. Titus Ed., 299–330, Washington.

BERAN, M.A., ARNELL, N.W.: 1989, Effect of climatic change on quantitative aspects of United Kingdom Water Resources, Institute of Hydrology.

BRADLEY, R.S., DIAZ, H.F., EISCHEID, J.K., JONES, P.D., KELLY, P.M., and GOODESS, C.M.: 1987, Precipitation fluctuations over Northern Hemisphere land areas since the mid 19th century, Science, 237, 171–175.

BROWN, L.R.: 1990, State of the world, World Watch Institute, 385 pp.

BUDYKO, M.I.: 1989, Climatic conditions of the future. Conference on Climate and Water, Helsinki, National IHP Committe of Finland, 9–30.

BULTOT, F., COPPENS, A., DUPRIEZ, G.L., GELLENS, D., MEULEN-BERGHS, F.: 1988, Repercussions of a CO_2 doubling on the water cycle and on the water balance. A case study for Belgium, Journal of Hydrology, 99, 319–347.

CADET, D., GARNIER, R.: 1988,: L'oscillation australe et se relations avec les anomalies climatiques globales, La Météorologie, VIIème, n°21, 4–18.

CARBONNEL, J.P., HUBERT, P.: 1985, Sur la sécheresse au Sahel d'Afrique de l'Ouest. Une rupture climatique dans les séries pluviométriques du Burkina Faso, C.R. Acad. Sci., Paris, 301, Ser. II, 13, 941–944.

CASENAVE, A., VALENTIN, C.: 1989, Les états de surface de la zone sahélienne, influence sur l'infiltration, Col. Didactiques, ORSTOM, Paris, 229pp.

DICKINSON, R.E.: 1986, How will climate change ? The climate system and modelling of future climate, B. BOLIN, B. DOOS, J. JAGER and R.A. WARRICK Eds. The Greenhouse Effect, Climate Change and Ecosystems, SCOPE 29, John Wiley, Chichester, 207–270.

DOOGE, J.C.I.,: 1986, Effects of CO_2 increase on hydrology and water resources, Paper presented at the Commission of the European Communities Symposium on CO_2 and other greenhouse gases: climatic and associated impacts, Brussels, November 4 Ed. R. Fantechi and A. Ghazi, Kluwer Pub.

DUBAND, D.: 1990, Reflexions sur l'utilisation des longues series d'observations climatologiques dans le cadre de l'étudu climat et de son évolution, La Houille Blanche, 303, 6–1990, 399–407

DUPLESSY, J.C., MOREL, P.: 1990, Gros temps sur la planète, ED. Odile Jakob, Paris 297pp.

GLEICK, P.H.: 1986, Methods for evaluating the regional hydrologic impacts of global climate changes, J. of Hydrology, 88, 97–116.

GLEICK, P.H.: 1987, Global climatic changes and regional hydrology: impacts and responses, IAHS, Publ. n°168, 389–402.

GLEICK, P.H.: 1989, Climate change, hydrology and water resources, Reviews of Geophysics, 27, 3, 329–344.

GUSTARD, A. et al.: 1989, Flow regimes from experimental and network data (FREND), Institute of Hydrology, Wallingford, 2 Vol.

HUBERT, P., CARBONNEL, J.P., CHAOUCHE, A.: 1989, Segmentation des séries hydrométéorologiques—Application à des séries de précipitations et de débits de l'Afrique de l'Ouest, J. of Hydrology, 110, 349–367.

IGBP (1990)—Global Change, IGBP, Report n°12, A study of Global Change: the initial core projects.

IPCC WGI (1990)—Scientific assessment of climatic change, WMO and UNEP.

IPCC WGII (1990) Potential impacts of climatic changes, WMO and UNEP.

KLEMES, V.: 1985, Sensitivity of water resource systems to climate variations, WCP-98, 136pp.

LAMB, P.J.: 1982, Persistence of subsaharan drought, Nature, 299, n°5878, 46–48.

LE ROY LADURIE, E.: 1983, Histoire du climat depuis l'an mil, Ed. Champs, Flammarion, Paris.

LIEBSCHER, H.: 1989, Effects of climate variability and change on fresh water bodies, Conference on Climate and Water, Helsinki, National IHP Committee of Finland, 365–391.

LOUGH, J.M., WIGLEY, T.M.L., PALUTIKOF, J.P.: 1983, Climate and climate impacts scenarios for Europe in a warmer world, J. Clim. Appl. Meteor. 22, 1673–1684.

MABBUTT, J.A. 1989, Impacts of carbon dioxide warming on climate and man in the semi-arid tropics, Climate Change, 15, 191–221.

MARGAT, J.: 1990, Les gisements d'eau souterraine, La Recherche, Spécial l'EAU, n°221, mai, 590–596.

MERILLON, Y.: 1990, La sécheresse de 1989 (en France). Bilan, Secret. d'Etat chargé de l'Environnement, Service de l'Eau. DEPPR/SE-AEYM-NL.

MOLINIER, M., CADIER, E.: 1985, Les sécheresses du Nordeste brésilien, Cah. ORSTOM, sér. Hydrol., Vol. XXI, n°4, 1984-85, 23–49.

NEMEC, J.; SCHAAKE, J. 1982, Sensitivity of water resources systems to climate variation, Hydro. Sciences J., Vol. 27, n°3, 327-343.

NICHOLSON, S.E.: 1979, Revised rainfall series for the West African subtropics, Mon. Wea. Rev., 107, 473–487.

OLIVRY, J.C. 1986, Possibilités d'allégement des réseaux hydrométriques dans les pays en voie de développement après réalisation de synthèses hydrologiques régionales, Publ. AISH, n°158, Budapest.

PALUTIKOF, J.P.: 1987, Some possible impacts of greenhouse gas induced climatic change on water resources in England and Wales IAHS Publ. n°168, 585–596.

POUYAUD, B., LE BARBE, L.: 1987, La télé-transmission satellitaire une méthode de gestion économique des réseaux hydrologiques dans les pays en voie de développement, Laboratoire d'Hydrologie, ORSTOM, Montpellier.

SANTER, B.D., WIGLEY, T.M.L., SCHLESINGER, M.E., MITCHELL, J.F.B.: 1990, Developing climate scenario from equilibrium GCM results, Max-Planck- Institut fur Meteorologie, Report n°47, Hamburg, 29pp.

SHIKLOMANOV, I.A.: 1989, Climate and water resources, Hyd. Sciences J., vol.34, n°5, 495–530.

SHIKLOMANOV, I.A.: 1990, The world water resources : How much do we really know about them? IHD/IHP 25-year commemorative symposium, UNESCO, Paris, 79pp.

SHUTTLEWORTH, W.J.: 1988, Macrohydrology—The new challenge for process hydrology, J. of Hydrology, 100, 31–56.

SIRCOULON, J.: 1989a, Impact des changements climatiques sur les ressources en eau de surface en Afrique de l'Ouest et Centrale, Proceedings of the Sahel Forum, UNESCO/IWRA Ed., Ouagadougou, 964–975.

SIRCOULON, J.: 1989b, Effets des sécheresses sur l'hydrologie de surface, La Houille Blanche, 298, n°7/8–1989, 505–515.

SIRCOULON, J.: l990a, Aspects hydrologiques des fluctuations climatiques en Afrique de l'Ouest et Centrale, Third WMO Symposium on meteorological aspects of droughts, report n°36, WMO/TD n°353, 205–212.

SIRCOULON, J.: l990b, Impact possible des changements climatiques à venir sur les ressources en eau des régions arides et semi-arides, WCAP-12, WMO/TD-N°389, 97pp.

SNIJDERS, T.A.: 1986, Interstation correlations and non stationarity of Burkina Faso rainfall, J. of Climatol. and Applied Meteo, 25, 524–531.

TODOROV, A.V.: 1985, Sahel : The changing rainfall regime and the "normals" used for its assessment, J. of Climate and Applied Meteorology, Vol.24, N°2, 97–107.

US Environmental Protection Agency, The potential effects of global climate change on the United States, Draft report to Congress, Office of Policy, Planning and evaluation, J. SMITH and D. TIRPAK Ed., October 1988.

WIGLEY, T.M.L., JONES, P.D.: 1985, Influences of precipitation changes and direct CO_2 effects on streamflow, Nature, 314, 149–152.

WIJKMAN, A., TIMBERLAKE, L.: 1984, Natural disasters; acts of God or acts of man?, Earthscan, London, 145pp.

WMO, 1979, Proceedings of the World Climate Conference, N° 537, GENEVA.

WMO, 1987, Water resources and climatic changes: Sensitivity of water-resources systems to climate change and variability, WCAP-4, WMO/TD-n°247, 50pp.

WMO, 1988a, Concept of the global energy and water cycle experiment, Report of the JSC study Group on GEWEX Montreal and Pasadena, WCRP-5, WMO/TD-N°215, 126pp.

WMO, 1988b, Developing policies for responding to climatic change, WCIP-l, WMO/TD-N°225, 53pp.

WMO, 1988c, Analyzing long time series of hydrological data with respect to climate variability, WCAP-3, WMO/TD-N°224.

WMO, 1989a, The global water runoff data project, Workshop on the global runoff data set and grid estimation, Koblenz, WCRP-22, WMO/TD-N°302.

WMO, 1989b, Statistics on regional networks of climatological stations, WMO Region I-Africa, WCDP-7, WMO/TD-N°305, 32pp

WMO, 1989c, Calculation of monthly and annual 30-year standard normals, WCDP-N°10, WMO-TD/N°341, llpp.

WMO, 1990, Report af the expert group on global baselines data sets, Asheville, USA, WCDP-ll, WMO/TD-N°359, 95pp.

WMO Bulletin, Volume 38, N°3, July 1989, 256–258.

WMO Bulletin, Volume 39, N°2, April 1990, 98–105.

Drought Issues for the 1990s

by Michael H. Glantz *(USA)* and Workineh Degefu *(Ethiopia)*

Abstract

The authors seek to draw attention to several contemporary drought-related topics that should be of prime interest to societies everywhere. These topics can shed light on how societies might better mitigate the impacts of climate variability (especially drought) and climate change. The paper is divided into three parts. Part I focuses on "what drought is" and "what drought does". The importance of drought definitions to drought planning is discussed along with the linkages between drought and famine and drought and desertification. The concept of "drought follows the plow" is presented in order to underscore the point that not all climate-related problems are the direct result of climate variability. Part II focuses on "What we can do about drought." The issue of climate information as a resource is discussed with suggestions about what one might include in a national climate resources inventory. In addition, a neglected climate characteristic-seasonality or the natural annual rhythm of the seasons--is discussed, as is the importance of a drought early warning system and a famine early warning system. Part III addresses some aspects of "drought and the future," focusing on climate change issues as they relate to drought and to a climate-related impacts research approach referred to as "forecasting by analogy," that is, forecasting society's ability to cope with the consequences of environmental change.

Introduction

Drought is perhaps one of the most widely investigated topics in research on climate-society interactions, because it directly affects societies through changes in the abundance and/or availability of food and fiber, water resources and energy supply. It also has direct implications for the quality of the Earth's environment.

There are innumerable publications on drought, many of which are collections of presentations at symposia on the topic (e.g., Wilhite and Easterling, 1987; Palmer and Denny, 1971). In fact, several bibliographies already exist that catalog publications on the various aspects of drought in specific geographical locations (e.g., OECD, 1977 to present). We are not seeking here to catalog all of the successes in coping with drought nor all of the missed opportunities in coping with the societal impacts of drought, but to draw attention to several contemporary drought-related topics that should be of prime interest to societies everywhere, as they may shed additional light on how societies might better improve their interactions with climate variability (especially drought) and climate change.

The paper is divided into three parts. Part I, following a brief introductory section, discusses the importance of drought definitions and the linkages between drought and famine and drought and desertification. The concept of "drought follows the plow" is presented in order to underscore the point that not all climate-related problems are the direct result of climatic variations. Part II focuses on what we can do about drought. The issue of climate information as a resource is discussed with suggested items that might be included in a list of such resources. Also, a neglected climate characteristic - seasonality or the natural annual rhythm of the seasons - is discussed, as is the importance of a drought early warning system (DEWS). This is followed by a section on famine early warning systems of which the output of DEWS is an input. The final part of the paper, "Drought and the Future," focuses on climate change issues as they relate to drought and to an approach referred to as "forecasting by analogy", that is, forecasting society's ability to cope with the consequences of environmental change.

PART I
Drought: What it is and What it does

Droughts, like floods, famines, disease and pestilence, plague societies at all levels of economic development. In this regard, little has changed over the thousands of years of history of human settlements and sedentary agricultural practices. No country is immune from the impacts of drought on its food and water supplies. Developing countries, such as Ethiopia and the Sudan, for example, have been ravaged by recurrent drought throughout the past few decades. The Brazilian Nordeste was plagued in the early 1980s with an extremely devastating multiyear drought. But industrialized countries have also not been immune to them. Two years ago, droughts adversely affected different parts of the United States and Canada; with droughts occurring both in the winter and in the summer.

For industrialized countries, however, drought is for the most part an economic problem. It has led to default on bank loans by farmers, to a need for government subsidies to see farmers through hard times, and temporarily to higher prices in the marketplace. While local impacts may be quite severe, national impacts are often much less severe. However, droughts in industrialized countries are not life-threatening situations. In the rest of the world drought can often mean the difference between life and death.

Nevertheless, even if we cannot stop droughts from occurring or cannot fulfill occasional political or scientific promises to drought-proof an area, there are ways either to protect the more vulnerable countries or to prepare them to be better able to cope when such situations recur (e.g., Moremi, 1987). These countries, such as Botswana and Vietnam, need, at the least, drought preparedness training, drought-technology and transfer of drought-coping mechanisms. Such transfers must be supported by those countries that can afford to do so. Drought-related hunger and drought-related famines can be dealt with effectively, if the national as well as international will to do so is there.

To draw the conclusion, however, that progress in dealing with drought has not occurred would be erroneous and misleading. Technologies and techniques have been developed to enable countries to cope better with the vagaries of the weather. Lessons about the successes and failures of drought-coping mechanisms in one region can at the least provide useful insights for other drought-plagued regions (e.g., Magalhaes, 1990). Yet, many of these approaches have not been transferred from one region to another. When such transfers do occur, drought coping mechanisms for the most part need to be adapted to local conditions, meshing with a region's history, economic conditions, and culture.

Drought means different things to different people. People will act according to what they perceive drought to be (Saarinen, 1966). Even though their perceptions might prove to be erroneous, actions taken based on those perceptions and the consequences of those actions will be real. Thus, how decision makers define drought directly affects the types of strategies and tactics that governments might pursue to mitigate their impacts (Wilhite and Glantz, 1985).

Drought Definitions

Maintaining a distinction among different types of droughts (and their impacts) can be very important for developing drought-coping mechanisms. In each case, the scientific community must state clearly what it means by drought and must identify direct and indirect impacts of drought on the environment and society so that the responses of decisionmakers will be more appropriate for resolving specific drought-related problems.

A meteorological drought can be defined as a reduction in rainfall of a specified (or desired) amount such as some percentage reduction of a long-term average. This amount can vary according to the human activities for which rainfall measurements are being used, whether for livestock management or agricultural production, for example. Yet, the reduced amount of rain that does fall might fall at the most beneficial time in the crop production cycle, averting crop failure. Such a situation might be defined as a meteorological drought but need not be viewed as drought by policymakers.

Agricultural droughts derive from the fact that different crops have different moisture requirements. Some crops require large quantities of water while others, such as millet and sorghum are considered relatively drought-resistant. Thus, a rainfall-related crop failure can be referred to as an agricultural drought.

A devastating drought in the Canadian Prairie Provinces in 1974 provides an interesting example (Glantz, 1979). Spring wheat production had been reduced by about 20% because of a cold and wet spring, a hot and dry summer and a wet harvest period. The newspapers during the summer carried pleas from farmers for drought relief from government agencies. However, the drought occurred at a time of worldwide grain production problems and shortages, leading to the highest prices ever in the international marketplace. Thus, farmers received more for the reduced amount of grain than they had for the larger harvests during normal weather periods. Today, 1974 is considered by the Canadian farmers as having been a relatively normal as opposed to a drought year. Meteorologically (and agriculturally), however, 1974 was clearly a devastating drought year.

A similar example can be found in a Brazilian context. Many observers of the drought-prone Brazilian Nordeste believed that the region had been plagued by a 5-year drought between 1979 and 1984. This was measured by agricultural and social indicators. Brazilian scientists, however, suggested that according to meteorological information only 3 years of drought had occurred. Decisions were being made as if a 5-year extended meteorological drought had occurred.

Hydrologic drought can be defined as a reduction of streamflow below a specified level for a given period of time. Hydrologic droughts in sub-Saharan Africa, for example, have had devastating social and economic implications in the last few decades because of the high dependence on flood recession farming and livestock grazing.

It is important to note that there are scores of specific definitions tailored to local, regional or national needs (for a discussion of drought definitions and their implications for action, see Wilhite and Glantz, 1985). The term drought is frequently used in other ways as well; a green drought, a paper drought, etc. A green drought is a phrase used in a recent Brazilian climate-society report (Magalhães, 1990) to describe a situation in which there was enough moisture to produce green vegetative matter but not enough to produce the fruits of that vegetation (i.e., grain). A "paper" drought is one in which bureaucratic reactions to a meteorological drought had more impact on human responses and the economic situation than did the drought itself, which might not have adversely affected agricultural production or human activities in general.

Drought and Famine

In many parts of the world, the major concern about drought comes from the fear that it could lead to severe food shortages and famine. While droughts are often blamed for causing famines, the historical record shows that drought alone seldom leads directly to famine (e.g., Sen, 1981; Glantz, 1989); and famines are not always preceded by severe drought. Agricultural drought often combines with and exacerbates existing societal problems to generate food shortages. For example, drought at a time of high prices for stored grain, or drought plus high prices for fertilizers, drought plus locust infestation, drought plus internal wars, and so on, create food shortages that can turn into famine-like conditions (see World Situation Reports for various years, prepared by the US Department of Agriculture's Economic Research Service). Drought, depending on its severity, can add additional people to the roll of the chronically hungry.

Drought and Desertification

Desertification has received considerable attention from the research and the policymaking communities since the early 1970s, especially as a result of the UN Conference on Desertification held in Nairobi in 1977 (UNCOD, 1977). Scores of definitions of desertification exist (e.g., see Glantz and Orlovsky, 1983) and have in fact made it a difficult concept to deal with because it has come to mean so many different things to people with different perspectives.

Desertification, according to one set of definitions, can generally be considered as the creation of desert-like conditions where they had not existed in the recent past. A degradation of the biological potential of a region is also viewed as a characteristic of desertification. These

processes of degradation can take place in high rainfall areas as well as in low ones. It can result from deforestation, fuelwood gathering, poor land management, livestock overgrazing, and inappropriate irrigation practices, among other human activities. It can also result from natural processes such as natural decadal-scale fluctuations away from and toward aridity on a regional scale. Desertification can lead to changes in albedo (reflectivity) of the earth's surface which can in turn adversely alter local or regional rainfall processes (e.g., Charney et al., 1975).

The adverse impacts on the environment of land-degrading activities are often hidden during periods of favorable rainfall, only to be exacerbated as well as highlighted during periods of extended drought. Droughts are frequently blamed for the destruction of the land's biological potential, although they may have only served to worsen an already deteriorating situation (e.g, Garcia, 1981; Glantz, 1976).

Drought Follows the Plow

The concept of "drought follows the plow" is a play on the words of another concept that was prominent in the late 1800s - rain follows the plow. This concept, originally applied to the US Great Plains, and later to many other areas in Africa and the USSR, suggested that the development of human settlements often led to the enhancement of regional precipitation because the planting of trees as well as the plowing up of grasslands increased evapotranspiration and thus the moisture content of the atmosphere. The view continues to draw support today. For example, attempts to develop treebelts across the southern and northern fringes of the Sahara desert or across eastern China are supposed to stimulate processes that would enhance rainfall and improve the biological potential in an otherwise arid or semiarid region. Anthes (1984), like many others before him, has suggested that the judicious placement of treebelts could positively affect atmospheric rain-producing processes.

The opposing notion that "drought follows the plow" is based on the premise that the best rainfed agricultural land is already in production, excluding those areas that are denied by government decree or use to settlers. This suggests that any need for additional agricultural lands means that people will have to move into areas that are less productive than where they are. These new lands will most likely be marginal in terms of rainfall, soil fertility or topography.

Usually, new settlers will attempt to grow crops such as the ones they grew in the more humid productive areas from which they originally came. However, these marginal areas are often not suited to the same type of crops or production techniques. Thus, yields may decline and, more important, where the newly cultivated lands are more marginal in terms of rainfall, the crops may suffer from an increase in frequency, intensity and duration of drought, however defined. The tendency would be to blame climate for the crop failure. However, the farmers had moved into marginal areas, setting themselves up for drought conditions, conditions which are a function of the moisture requirements of the particular crop. The point is that we will most likely hear about more droughts in the future than we did in the past, not so much because the global climate may be changing but because people are moving into marginal areas, for which their agricultural practices are inappropriate.

There are some very stark examples of how different land management practices can reduce (or enhance) the impacts of drought on the land's surface. Satellite imagery supplies some examples of such condition for southern Africa as well as the Middle East (Glantz, 1977). Another excellent example was supplied some years ago by a satellite photograph of the southwestern United States (Figure 1).

This image clearly shows a 200-mile dust storm that does not cross over to New Mexico, illustrating how a difference in state water laws, in combination with other geological factors, can heighten the level of vulnerability of soils to drought conditions (Kessler, *et al.*, 1978). The

State of Texas allows those who own the land to use it and the groundwater under it as they wish. Thus, a heavy dependence on irrigation for agricultural production (e.g., cotton) developed. In New Mexico, however, the groundwater is owned by the state and permission to use it must be obtained from the State Engineer. In this instance, the State Engineer considered the vegetative cover in eastern New Mexico to be more suitable to livestock grazing than to irrigated agricultural development. When the drought occurred in the early months of 1977 and was accompanied by persistent strong winds, the topsoil of western Texas blew away while that in eastern New Mexico remained in place. Similar situations have been found in Brazil, Ethiopia, Somalia, the West African Sahel, and Australia (Glantz, 1991). Information about these kinds of human responses to social, political, economic or environmental conditions can be useful to those involved in land-use planning in yet-to-be developed areas.

Droughts not only affect agricultural activities but also industrial and municipal water supplies. As rural-urban migration continues to increase urban populations in developing areas, these areas, too, become increasingly vulnerable to the impacts of meteorological droughts (urbanization in industrialized countries is also at a high

Figure 1: Satellite photograph of the southwestern United States showing a 200-mile duststorm.

level). In future years droughts of the same intensity in the same locations will have increasingly devastating impacts because of demographic (if not climatic) changes.

PART II
Drought: What we can do about it

Climate Information as a Resource

The idea that climate is a resource is not a new one. Almost a century ago geographers, among others, believed that climatic conditions in the mid-latitudes generated the preconditions for industrialization. They also argued that human populations in the tropics were plagued by adverse climatic conditions (high temperatures and humidity) which hindered development activities (e.g., Huntington, 1915). This view of climate was that of a boundary condition about which societies could do little. They were dealt their climate by nature and there was little they could do to mitigate its effects on human populations. Many of these early views have since been discarded as having been either wittingly or unwittingly (implicitly) racist.

If we break down the concept of climate into its components, however, such as rainfall, seasonality, temperature, it is much easier to disclose those elements of climate that are rigid boundary conditions from those that are merely constraints. It also enables us to demonstrate that societies can counteract some of the adverse impacts of specific components of climate on human activities. About fifty years ago one author, viewing the temperature component of climate as the major constraint to economic development in the tropical countries, discussed the importance to economic development of what he termed the "air-conditioning revolution". The hope was that this technological innovation would create islands of temperate climate in the midst of the hot and humid tropics (Markham, 1944).

With regard to the precipitation component of climate, there has always been a high interest in the potential value of a weekly, monthly and seasonal forecast of precipitation. As forecasts improve in their reliability, their potential value will be realized. Even in areas plagued by droughts, floods or freezes, information about those climate-related hazards can become a resource, if received in a timely fashion and in a format that can be acted upon by appropriate decisionmakers.

Clearly, in any kind of geographic setting, information about climate in general, and about drought, specifically, can be an asset to those who know how to use it to society's advantage. Although almost everyone in the world is familiar with drought and its impacts on society and economy, many sectors of society still need to be taught about what climate-related information is important to them and how to use that information most effectively. But what kind of information about drought, other than the obvious (e.g., precipitation amounts and timing, temperature), is important to individuals as well as national decisionmakers?

One important aspect of climate information is its use in identifying the natural resource base of a country. For example, the determination of agroclimatic resources, length of the growing periods (LGP), agroecological zonation (AEZ) and land suitability classification for crops, livestock, etc., can be done through the use of climate information. The identification of a country's natural resource base, its conservation and rational use are effective tools in minimizing the impacts of drought.

Thus, it is important to encourage each government to undertake a climate-related resource inventory. Such an inventory would require innovative thinking, that is, it would identify not only traditional views of climate as a resource but also non-conventional aspects. The former would include the usual assessments of rainfall, temperature, evaporation, wind direction and speed, soil moisture and so on, on all time and space scales. The non-traditional, generally overlooked, aspects of climate-as-a-resource include but are not limited to the following:

(1) Resource consultants, available on request from most UN agencies and from many non-governmental organizations

(2) Reports prepared by foreign scholars but which are often unavailable in the regions or countries about which they were written. In addition, many of these reports, books, and articles have never been translated into the lingua franca of the country about which they were written. The lack of scientific translations puts great pressures on local scientists and decisionmakers to develop a fluent capability in foreign languages in the absence of which they remain deprived of this existing wealth of information

(3) Fellowships and the like can also be considered as climate-related resources. Studies of how climate variability and change affect societies are also resources, as are international meetings about climate-related factors

(4) Climate information about other regions in the world is also an important aspect of a climate-as-a-resource inventory. One can see, for example, that although no clear connection might be evident between the occurrence of an El Niño-Southern Oscillation (ENSO) event and anomalous climatic conditions in distant locations, it would be useful to know how such ENSO events affect agricultural production, and therefore market and export activities of a nation's trade competitors

(5) Information about the effects of seasonality and their relationship to drought

(6)   A drought early warning system (DEWS) would be a special climate-information related resource in sub-Saharan Africa.

The last two of these "resources," information about the impacts of seasonality and DEWS/FEWS, will be considered in some detail.

Seasonality

Seasonality is an important characteristic of regional climate. The natural annual rhythm of the seasons and its occasional disruption by extreme climatic events are more related to the success or failure of human activities than is generally realized (e.g., Sahn, 1989). The rhythm of the seasons is something to which every society has had to adjust its climate-related activities. More specifically, one should assess this seasonal rhythm and its implications for rural poverty in the Third World. To date, interest in climate impacts on societies, and more specifically on rural poverty, has focused on spectacular or unusual events such as droughts and floods. Yet, such events may only serve to distort socio-economic relationships that were already established by the less spectacular but perhaps equally important natural rhythm of the seasons. This has been a neglected aspect of research in economic development, with only a few notable exceptions.

For example, Chambers *et al.* (1979, p.3) summarized how seasonality relates to various aspects of rural poverty in the Third World. They noted that "...besides climate, seasonal patterns are also found in labour demand in agriculture and pastoralism, in vital events, in migration, in energy balance, in nutrition, in tropical diseases, in the condition of women and children, in the economics of agriculture and in social relations, and in government interventions."

Not all people in a given area are adversely affected by seasonality. In fact, some segments of society, such as rich peasants and grain merchants, benefit from seasonality at the expense of the larger number of rural poor. Sen (1981, p.43), whose observations were based on case studies in India and the Sahelian zone of West Africa, noted that "...it is by no means clear there has ever occurred a famine in which all groups in a country have suffered from starvation, since different groups typically do have very different commanding powers over food...".

Take the following simple scenario as an example. At the end of a successful harvest the farmer may have sufficient grain to pay back his debts (of borrowed seeds, tools, etc.), to store grain for the dry season and to fulfill other social obligations in the following months. However, as the next harvest season approaches, his grain supplies are most likely at their end and he may have to borrow once again. If the harvest is again a good one, he will be in a position to pay back his debts once more. If, however, a drought reduces the harvest, the farmer will have to begin the year with a considerably reduced supply of grain assuring that by the time of spring planting he will have to go deeper into debt in order to obtain seed for planting and food for his family. If the next harvest is again poor, he will go deeper into debt and before the harvesting season the family might experience severe hunger. The seasonal rhythym in this scenario tends to reinforce a social relationship between the local "haves" and "have-nots", while extreme meteorological events such as drought tend to exacerbate an already distorted social relationship. The following table (Table 1) summarizes the possible societal responses to the cycle of the seasons during favorable and drought-plagued periods.

Thus, understanding seasonality and the relationship of drought and seasonality on rural poverty is essential for establishing an effective system for mitigating climate-related societal impacts.

Drought Early Warning Systems (DEWS)

A reliable Drought Early Warning System would be designed to forewarn governments, those directly affected by drought episodes, as well as those in the donor community, about impending climate-related crop production problems and potential food shortages (NMSA, 1988). Such early warning information can increase the efficiency and effectiveness of the management of a country's food supply, thereby freeing up scarce resources for use in other economic development activities.

DEWS is based on the gathering and interpretation of both quantitative and qualitative scientific information. The physical parameters of a DEWS often include rainfall measurements, crop area planted, phenological and biological observation, crop yields, rangeland conditions, and seasonal forecasts linked to global phenomena such as ENSO events in the eastern and central equatorial Pacific. Other kinds of meteorological forecasts, from daily to 10-day to monthly to seasonal, and climate probability statistics, are of great value to decision-makers in climate-affected activities. Societal parameters might include types of crops being grown, and so forth.

A distinction must also be made between quantitative and qualitative indicators of drought conditions. Quantitative measures are those that can be systematically monitored and measured, such as those cited above. Qualitative indicators include anecdotal information such as the sighting in the marketplace of "famine" foods (less preferred food resorted to only in time of severe shortages) or a sharp increase in the sale of livestock, or the sighting of unusual population movement by truck drivers. Qualitative indicators are often neglected in early warning systems because of the difficulties in calibrating such information. Nevertheless, managers of a good, effective DEWS should be very open-minded, looking for any early

Table 1*: Possible Societal Responses to the Cycle of the Seasons During Favourable and Drought-plagued Periods*

(Usual) Seasonal Impact	Prolonged Drought Impact
Post-Harvest (early dry season) • food available • food prices decline • migrant laborers return to villages • morbidity declines • mortality declines • nutritional status improves "Post-harverst food availability largely determines the size and distribution of village calorie supplies, not just at the time but until the next harvest" (Schofield, 1974, p. 23). **Dry Season** (late dry season) • food becomes less available • food prices increase • nutritional status (especially women, children) delines • drinking water becomes scarce • dry season irrigation becomes more • important	• food availability declines • food prices continue to rise • domestic food self-sufficiency jeopardized • families borrow from kin/friends • disposal of assets for money • migrants do not return • additional family members migrate • nutritiional intake deteriorates • morbidity stays high or increases
Wet Season (early wet season) • "hunger season" begins • wild foed use begins • gathering added to agricultural labour • high food prices • poor families borrow agricultural inputs • distress borrowing/distress sales • draft animals in relatively weak conditions • diseases more prevalent • morbidity increase "During the wet season itself when seasonal food shortages peaked, hardship could be partially alleviated vy participation in communal work parties and short-distance migration making use of the variation in the onset of the rains (and hence in the timing of planting, weeding, and harvest)" (Watts, 1983, p. 49). **Pre-Harvest** (late wet season) • food prices are at seasonal high • food intake is lowest (especially women and children) • morbidity declines • body weights decline • "hunger season" peaks "Peak-season labour inputs often coincide with seasonal food shortages, as on-farm grain stocks are running low before prices begin to be pulled down by the impending harverst" (Schofield, 1974, p. 23).	• food prices go even higher • little work available in rural areas to earn cash • distress sales increase: livestock, stored gain, household goods • food unattinable (due to lack of availability, high price) • food-gathering activities intensify • nutritional status declines • seeds eaten (reduces future production) • eat plants/leaves, etc., not usually eaten • children's illnesses increase • mobidity increases • mortality increases • irrigation becomes limited as streamflow is reduced • call on wider networks; reliance on more distant kin, and on national and international agencies

sign that drought-related food and water problems might be just beginning. Secondary indicators of DEWS, such as the sighting of "famine" foods in the market place, are often used as primary indicators for a famine early warning system (FEWS).

Other aspects of a DEWS await critical review as well. For example, how early is early with respect to a warning system that is supposed to spark timely reaction within a country as well as internationally? Clearly, any drought-prone country that is in the midst of an internal war has a high probability of drought-related food shortages. But, for use as a catalyst to action this would be a difficult indicator to use in order to prompt governments or their agents to act. On the other hand, by the time agencies are able to count the number of people who have abandoned their villages and entered food distribution centers or refugee camps it is most likely too late to avoid famine conditions. Some activities require long lead times to respond to or compensate for drought-related impacts.

It is important to identify the users of the early warning information, given that different users have different needs. Political leaders need such information for one set of reasons (e.g., to alert the donor community to impending food shortages) while rural farmers need it for another (e.g., to develop tactics for planting and for coping with food shortages at the household level). Donor agencies need this information for yet a different set of reasons. A single output in a particular format may not serve all the users implying that output from a DEWS should be tailored to the varying needs of the different target audiences.

Other constraints to the development of an efficient, optimal DEWS include the paucity of relevant data in many drought-prone regions of the world. Not all countries, for example, have optimal meteorological networks. Many do not have a long time series on which to base reliable and credible analyses. While educated guesswork is often required, there is a need to strengthen national capabilities to generate basic informational inputs into a DEWS.

Even if lengthy reliable time series were available, drought forecasts based on those time series would still just be probabilistic forecasts, that is, they are probabilistic statements about a future state of the atmosphere; they are not guarantees. Although several of the forecasts produced by a DEWS may not actually be realized, that should not, by itself, be taken to mean that the DEWS does not work. A number of these forecasts will have to be made before one can ascertain whether a DEWS is, in fact, reliable, and to what level.

Another major problem with DEWS is the lack of continued support by appropriate government agencies, once a drought has seemingly ended. Droughts (but not necessarily famines) are recurrent phenomena. It is

important that societies strengthen their DEWS during those periods when drought is not threatening so that society can better prepare to deal with them when droughts recur (Wilhite and Easterling, 1987). An important aspect of this is the fact that although droughts have identifiable beginnings (in retrospect) and definable ends, their societal impacts often continue for years. For example, debts incurred to purchase grains during drought-related food shortages will require repayment long after the drought itself has ended.

Famine Early Warning Systems (FEWS)

Today, there is considerable interest in FEWS. Drought, however, is only one element of such a system. Others include nutritional status, grain in storage, grain and livestock prices, consumption of "famine" foods, and so forth. The output of a DEWS (which would take into account the different kinds of droughts referred to earlier) would serve as one of the inputs into the FEWS.

While famines may be defined in scores of ways, how we view them determines how, when, and even whether we might identify a famine. Some people, such as those responsible for releasing grain from warehouses in donor countries, continue to view famines as events. They require hard evidence or quantitative indicators to release emergency grain shipments. Their indicators include the number of deaths, the number of people in an emergency feeding center, or whether mass migrations are taking place in the countryside.

Others, such as those who deal with nutritional problems, often see famine in a totally different light. They consider famine to be a process and include pre-famine conditions in their definitions. They tend to sound alarms early in the process, so that the last stages of the famine process - mass starvation and death - can be averted. They consider different indicators to be important, including reduced crop yields, increased prices for grain, declining nutritional status of infants, reduced rainfall, the sale of personal items (jewelry, cooking utensils), and the increasing dependence on "famine" foods consumed only under duress.

Thus, when someone in the field suggests that a famine is emerging, and someone in Washington, Paris, or London wants a "body count," they are not communicating about the same phenomenon. The first issue that must be resolved is the need to broaden the definition of what constitutes a famine. For an early warning system, the broader the definition, the better the possibility of an early identification and an early response.

In summary, governments as well as individuals remain concerned about the potential devastating effects that extended drought can have on food production. Many efforts are underway in most countries to try to understand the causes of droughts as well as their consequences. The focus of attention now appears to be on seeking an improved understanding of how these adverse climate anomalies affect human activities. Research on these topics should be of concern to all countries and not just those that are directly affected. Forearmed with the results of such research efforts, societies will be able to shift the balance from a recurrent need for emergency relief activities to the enactment of appropriate drought mitigation policies. Many countries have come a long way in coping with drought situations. Their experiences should be shared with other countries that have not as yet reached such a level of resilience.

Part III
Droughts and The Future

Drought and Climate Change

There has been considerable discussion in the past decade and a half about climate change in general and more specifically the possibility of a global warming of the atmosphere. While the theory behind the implications of the buildup of radiatively active trace gases in the atmosphere, popularly referred to as the greenhouse effect, is certain, the ultimate effects on, for example, cloud cover or absolute amount of temperature increase, and the impacts on the viability of ecosystems and societies, are not.

At this time it is not possible to identify with an acceptable degree of reliability those parts of the globe that will undergo an increase in the frequency, intensity, duration and location of drought (or those areas that might benefit from enhanced precipitation). General circulation models of the atmosphere do not agree on the spatial and temporal changes that might occur in the redistribution of and changes in regional precipitation regimes. Nevertheless, there are actions that societies can take in order to improve their ability to cope with droughts today that will also be of value in the mitigation of possible climate-change-related droughts in the future.

Some countries have even embarked on ambitious programs to drought-proof their drought-prone regions, such as Canada's grain-growing Prairie Provinces (Anderson & Assoc., Ltd., 1981). But apparently attempts at drought-proofing this region in the early 1980s have yet to be perfected, as witnessed by the severe impacts of the 1988 drought in Canada (Canadian Wheat Board Report, 1990). Ethiopia, among other countries, has also chosen to create a national strategy to prepare for and mitigate the impacts of natural hazards (NMSA, 1988). There are many national and regional examples of similar efforts at mitigating the impacts of drought on environment and on human activites. As Wilhite has suggested, "governments must provide the basic guidelines and be the policy innovators in this area if vulnerability to future droughts

and other extreme climatic events is to be lessened" (Donald Wilhite, personal communication, 1990).

This has not been an easy task because many people still believe that there is little or nothing that societies can do to stop droughts from occurring. To these people meteorological and even agricultural droughts are natural phenomena over which humans have no control. We can, however, mitigate the impacts of droughts by improving our understanding of the way in which they affect our lives directly and indirectly. Better and more timely information can arm decision-makers with the tools needed to predict them some time in advance. Forewarned means, in theory, forearmed. A recent approach, referred to as forecasting by analogy, is an attempt to understand how well society might be prepared to cope with climate change in the future.

Forecasting by Analogy

To understand how societies might best respond to a yet-unknown change in regional as well as global climate regimes, it would be necessary to know how societies have been affected by and coped with the effects of extreme meteorological events (EMEs), such as droughts, that have occurred in the recent past. Current scientific wisdom suggests that with a global warming of the atmosphere there will be an increase in climate variability which includes an alteration in the intensity, frequency, and duration of EMEs.

Although the climate in the future might not be like that of the recent past, one can assume that, barring unforseeable shocks to social systems, such as the energy crises of the 1970s or the rapid democratization of Eastern European countries, societal institutions in the near future will be like those of the recent past. By identifying societal strengths and weaknesses in past responses to EMEs, societies can act in a much more informed manner to eradicate the weaknesses and capitalize on existing strengths. Only then can they better prepare for the implications of an uncertain climate future (e.g., Glantz, 1988).

Forecasting by analogy (societal responses, not future states of regional climate) can be viewed as a win/win situation. Focusing on coping with EMEs in general, and drought in particular, will produce improved responses to EMEs, whether or not the climate of the future is different from that of the recent past. In other words, such research would result in an improved understanding of the interactions between climate variability and society.

Climate Change, Global Change and Multidisciplinary Research

Climate change is in fact only one of several environmental changes of concern to societies today. Deforestation, desertification, loss of biodiversity are just some of the more visible important environmental changes occurring worldwide (Price, 1989). Many of these changes can adversely affect climate on the regional and local levels. Land use, for example, affects climate on these spatial scales. Therefore, to best understand droughts and to improve our ability to cope with them and to mitigate their impacts we must look at a broad range of factors, not just meteorological ones.

Fifteen years ago there was considerable discussion about the "hydra-headed" crises facing industrialized and developing societies. For each crisis isolated and attacked by society, it appeared that two or more new environmental problems emerged. Coping with the food problem led to an increasing dependence on irrigation. This, in turn, led to an increase in the use of fertilizers and pesticides. These led to increased groundwater depletion as well as contamination, and so on. In this regard one can effectively argue that multidisciplinary research is important not only to improving our scientific understanding of environmental problems but also to strengthen the disciplinary bases on which multi-disciplinary research depends.

Clearly, the focus on global change will require multidisciplinary, multinational efforts. The interest of the meteorological community in global change issues might best be centered on those changes in environment and society that affect atmospheric processes. These include, but are not limited to desertification, deforestation, energy consumption, landuse patterns, and the like. It was climate change that prompted concern about global environmental change and we suggest that climate change issues remain as the focal point for global change issues for our community's contribution to global change programs.

References

Anthes, R.A., 1984. Enhancement of convective precipitation by mesoscale variations in vegetative covering in semiarid regions. Journal of Climate and Applied Meteorology, 23, 541–54.

Anderson & Assoc., Ltd., 1981. Draft Outline of the Proposed Saskatchewan Drought Proofing Studies. Report to the Government of Saskatchewan. Regina: Sask: Government of Saskatchewan.

Canadian Wheat Board, 1990. Annual Report 1989/90. Winnipeg, Manitoba: The Canadian Wheat Board.

Chambers, R., R. Longhurst, D. Bradley, and R. Feachem, 1979. Seasonal Dimensions to Rural Poverty: Analysis and Practical Implications. Brighton, England: Institute of Development Studies, University of Sussex.

Charney, J.G., P.H. Stone, and W.J.Quirk, 1975. Drought in the Sahara: A biophysical feedback mechanism. Science, 187(4175), 434–35.

Garcia, R.V., 1981. Nature Pleads Not Guilty. Oxford: Pergamon Press.

Glantz, M.H. (ed.), 1976. Politics of a Natural Disaster: The Sahel Drought. New York: Praeger.

Glantz, M.H., 1977. Dealing with a global problem. In Desertification: Environmental Degradation in and around Arid Lands, ed. M.H. Glantz, 1–15. Boulder, CO: Westview Press.

Glantz, M.H., 1979. Saskatchewan Spring Wheat Production 1974: A preliminary assessment of a reliable long-range forecast. Environment Canada Climatological Studies No. 33.

Glantz, M.H. (ed.), 1988. Societal Responses to Regional Climate Change: Forecasting by Analogy. Boulder, CO: Westview Press.

Glantz, M.H., 1989. Drought, famine and the seasons in sub-Saharan Africa. In African Food Systems in Crisis, ed. R. Huss-Ashmore and S.H. Katz, 45–71. New York: Gordon and Breach.

Glantz, M.H., 1991. Drought Follows the Plow. Tucson: University of Arizona Press.

Glantz, M.H. and N. Orlovsky, 1983. Desertification: A review of the concept. Desertification Control Bulletin, 9, 15–21.

Huntington, E., 1915. Civilization and Climate. Reprinted 1971. Hamden, CT: The Shoe String Press.

Kessler, E., D.Y. Alexander, and J.F. Rarick, 1978. Duststorms from the US High Plains in late winter 1977 - search for cause and implications. Proceedings of the Oklahoma Academy of Science, 58, 116–28.

Magalhães, A.R., 1990. Impactos Sociais e Económicos de Variacõones Climáticos e Responstas Governmetais no Brasil. Fortaleza: Secretaria de Planejamento e Coordenacão do Cear'a.

Markham, S.F., 1944. Climate and the Energy of Nations. London: Oxford University Press.

Moremi, T.C., 1987. Drought planning and response: Botswana experience. In Planning for Drought: Toward a Reduction of Social Vulnerability, ed. D.A. Wilhite and W.E. Easterling, 445–452. Boulder, CO: Westview Press.

NMSA (National Meteorological Service Agency), 1988. Report of the workshop on drought early warning in Ethiopia, Addis Ababa, 29 November - 2 December 1988.

Palmer, W.C. and L.M. Denny, 1971. Drought Bibliography. NOAA-ETM-EDS-20. Silver Springs, MD: Environmental Data Service.

Price, M.F., 1989. Global change: Defining the ill-defined. Environment, 31, 18-20 and 42-44.

OECD (Organization for Economic Cooperation and Development), 1977-present. Elements for a bibliography of the Sahel Drought. Paris: OECD.

Saarinen, T.F., 1966. Perception of the Drought Hazard on the Great Plains. Chicago, IL: University of Chicago, Department of Geography.

Sahn, D.E. (ed.), 1989. Seasonal Variability in Third World Agriculture. Baltimore, MD: The Johns Hopkins University Press.

Schofield, S., 1974. Seasonal factors affecting nutrition in different age groups and especially of pre-school children. Journal of Development Studies, 11, 22–40.

Sen, A., 1981. Poverty and Famines: An Essay on Entitlements and Deprivation. Oxford: Clarendon Press.

UNCOD (UN Conference on Desertification), 1977. Desertification: Its Causes and Consequences. Oxford: Pergamon Press.

Watts, M., 1983. The political economy of climate hazards: A village perspective on drought and peasant economy in a semiarid region of West Africa. Cahiers d'Etudes Africaines, 89–90, 37–72.

Wilhite, D.A. and W.E. Easterling, 1987. Planning for Drought: Toward a Reduction of Social Vulnerability. Boulder, CO: Westview Press.

Wilhite, D.A. and M.H. Glantz, 1985. Understanding the drought phenomenon: the role of definitions. Water International, 10, 111–20.

Agriculture and Food Systems

by M. S. Swaminathan *(IUCN)*

1. The Changing Scenario

The unfavourable weather of 1972 and 1974 in the middle latitudes and the failure of monsoons in the sub-Saharan countries provided the backdrop to the discussions on the impact of climate on global food security at the first World Climate Conference in 1979. The eighties witnessed both uncomfortable food gluts in some parts of the world and acute food scarcity in others, particularly in the Sahelian region of Africa. According to the State of Food and Agriculture Survey of FAO (1989), the years 1987 and 1988 witnessed a marked turnaround in the world agricultural and food security situations. Some important agricultural commodity markets shifted from having a global surplus to a situation of relative scarcity, and international prices increased significantly, after having fallen to their lowest levels in many years. FAO's March 1990 Food Outlook predicts that the cereal supply situation will remain tight in 1990/1991. Even assuming normal weather, 1990 production is unlikely to be large enough to meet trend consumption in 1990/1991 and allow stock replenishment. With stocks at their lowest level for many years, adverse weather would have serious consequences (Figure 1).

The past 10 years have witnessed great progress in methods of both monsoon forecasting and climate impact assessment. In addition, several basic shifts have occurred in our approach to the analysis of the inter-relationships between climate and food production systems. First, while in the past the focus was on the *impact of climate on human activity*, the current concern is more on the i*mpact of human activity on climate*. Second, while studies on the possible impact of a cooling trend in the world's climate on crops like maize and soybeans attracted interest in the seventies (Thompson, 1975), the interest now is more on interactions among CO_2 concentration in the atmosphere, temperature, precipitation and biological productivity. The feed-back linkages among climate change, crop yields and spatial shifts of crop potential are being investigated in several countries and in several major crops (Parry, Carter and Konijn, 1989). There have been apprehensions about a possible increase in the warming of the El-Niño current, thought to be a major contributory factor to droughts in Brazil, India, Australia and the Sahelian region of Africa during 1982–1983, in case of a rise in world temperature.

Third, with an increase in human population, precipitation patterns are assuming great significance not only for stability of crop production but also for drinking water security for human and animal populations. In spite of a growing awareness of the need to stabilise human population, a global annual population growth of 1.6 to 1.7% is still occurring. This would lead to a net increase of 90 million more people to feed each year. The course of fertility decline during the current decade will largely determine whether the world's ultimate population reaches 10 billion, 15 billion or some other plateau. Demo-

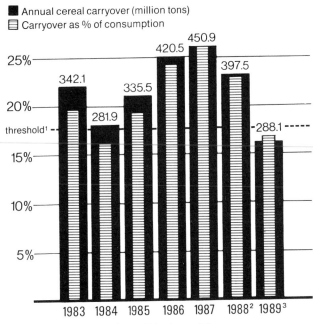

Figure 1: World carry-over of cereal stocks, 1983-1989 (million tons). Source: FAO (1989).

Table 1*: Changes per person staple food production, imports and consumption in developing countries 1970–1986/1987*

	Production (1970–1987)	**Imports**	**Consumption**
	Average annual percent change		
Low-income food-deficit countries	0.81	2.89	1.39
Africa	-1.25	4.82	0.37
Near East	-1.23	4.82	0.78
Far East	1.22	0.75	1.59
Latin America	-0.66	3.53	0.88
Oceania	-0.86	2.71	2.99
Developing countries	0.71	4.32	1.24

* 69 countries with GNP/per caput of up to $940 in 1987, net cereal importers on average during 1983/84–1987/88
Source: Food and Agriculture Organization of the United Nations. AGROSTAT

graphically the industrialised and developing countries will differ not only in the absolute size of the population but also in the age composition. For example, in India about 70% of the population will be below the age of 35 in the year 2000, while in the United States a similar percentage will be above the age of 50. In many developing countries, over 60% of the population will be rural, depending for their livelihood security on crop husbandry, animal husbandry, fisheries, forestry and agro-industries. While at present about 75% of the world's population lives in developing countries, the proportion will increase to 79% in 2000, to 81% in 2010 and to 83% in 2025, according to UN data. FAO (1989) predicts that although agricultural production will continue to increase in most developing countries, there will be little improvement on a per capita basis.

About a half a billion people are believed to be at nutritional risk because they have limited access to a balanced diet. FAO (1989) estimates that the number of people with consumption levels below the critical threshold value is likely to increase from the present 510 million, to 630 million in 2000. A recent report from the World Bank defined poverty as the inability to attain a minimum standard of living (World Bank, 1990). By this standard, it has been estimated that 1.1 billion people are struggling to survive on less than $370/year. About three quarters of these poor, about 800 million people, live in Asia, primarily in Bangladesh, China and India. The largest absolute number of people affected by potentially critical shortage of land is also in Southeast Asia. It is in this context that the further complication likely to arise in relation to national and global food security systems will have to be viewed. Finally, an important development between 1979 and 1990

is the addition of the dimensions of ecological sustainability and equity to the goal of enhancing productivity and profitability in research and development programmes designed to improve major farming systems. Equity is now defined in two time dimensions: (a) intra-generational equity safeguarding the interests of those living today, and (b) inter-generational equity safeguarding the interests of the generations yet to be born. Agricultural scientists are thus now faced with the task of combining ecological sustainability, economic viability and intra- and inter-generational equity in technology development and dissemination.

2. Nature of Food Security Challenge

During the seventies, the major food security challenge was quantitative adequacy or physical access to food supplies. During the eighties also, physical access continued to be an important problem in some years and in some regions. For example, the 1988 harvest worldwide was about 5% less than the 1985 harvest. The Sahelian food famine of the mid-eighties also underlined the precarious food situation in many African countries where *per capita* food production was declining steadily (Table 1).

On the other hand, the spread of modern technologies, particularly irrigation, fertilizer and high yielding seeds of cereals helped the densely populated countries of South and South-East Asia to keep growth rates in food production above that of the population. In such countries the food security challenge became one of economic access to food or one of lack of entitlements, to use the terminology of Amartya Sen (1981). Sen (1987) has also emphasized that a public distribution system geared to the needs of the vulnerable sections of the community can bring the essentials of livelihood within easy reach of people whose lives may remain otherwise relatively untouched by the progress of real national income (Table 2 and 3).

Swaminathan (1987) has stressed that while dealing with food systems the following three evolutionary steps should be recognized.

a) *Food Self-sufficiency* which implies adequate supplies in the market

b) *Food Security* which involves both physical and *economic* access to food, and

c) *Nutrition Security* which implies physical and economic access to *balanced diets and safe drinking water* for all people at all times.

Many developing countries including India which have experienced a "green revolution" are still in the first stage of this evolutionary process.

Kates et al (1989) have calculated that the biosphere at current production levels could support about 6 billion people, if the diet of all human beings were vegetarian. If 15% of calories were derived from animal products, planet earth could support about 4 billion people. If 25% of calories are derived from animal products, the total population which could be supported would fall to 3 billion. Fish production is the most efficient from the point of view of conversion of plant calories into animal protein. In this context, FAO's (1989) report that the world fish catch levelled off in 1987, after ten years of steady growth is a matter for concern. The world harvest of fish in 1987 was 92.7 million tons, as compared to 92.4 million tons in 1986. El Niño warm currents in the Southeastern Pacific substantially reduced catches of some pelagic species in South America. In contrast to this trend, consumer demand for fish is likely to increase by 28.4 million tons by the year 2000. Therefore, an integrated approach to coastal and inland aquaculture and capture fisheries will be important.

3. Challenges Ahead

Sustainable advances in biological productivity are essential for meeting the needs of both enhanced agricultural production and greater agricultural diversification. For this purpose we need new agricultural technologies capable of raising the population-carrying capacity of land and water. Such techniques will have to be tailored to the following major land use systems:

a) *Mountain ecosystems*, where damage upstream would have serious repercussions on downstream agriculture (e.g. Himalayas and the Indo-Gangetic agricultural region)

b) *Coastal ecosystems*, where it will be necessary to promote the integrated management of land and sea surfaces,

c) *Sustainable intensification areas* (could also be referred to as "green revolution" areas) which, with appropriate support from maintenance and anticipatory research, can sustain intensive output of crops and livestock at high and rising levels of productivity

d) *Semi-arid areas*, where inadequate or unreliable rainfall coupled with over-exploitation leads to chronic land degradation

e) *Arid areas*, belonging to both the hot and cold desert categories, where sylvi-pastoral and sylvi-horticultural systems of land use are ideal, and

f) *Island ecosystems* where changes in sea-levels could have important implications.

Unsustainable management of the resource and environmental systems associated with such regions is having serious repercussions, as is evident from the following visible signs in many developing countries.

- Precipitous drying up of drinking water resources
- Vanishing forests, flora and fauna
- Intensifying drought and floods
- Loss of grazing lands and growing degradation of land
- Deterioration of the quality of air and water
- Explosive growth of rural and urban unemployment
- Mushrooming of urban slums

It is the poor and the marginalised who suffer most from such environmental breakdown.

The nexus among people, resources, environment, technology and agricultural development is thus a closed one. FAO's estimates of total harvested land of different potentials is given in Figure 2. FAO's study, World Agriculture: Toward 2000, concluded from an analysis of 93 developing countries, excluding China, that nearly 60% of harvested land in 1982–84 belonged to the high potential category. Unirrigated arid and semi-arid areas of developing countries accounted in 1983–88 for only about 9% of total cereal production and 6% of root and tuber production. Thus for achieving the production goals of the future, it will be essential to maintain and further enhance the production potential of "sustainable intensification areas" and upgrade degraded land. In this context, it will be useful to consider briefly the present state of global land and water resources management.

i) *Land*: It is estimated that 30 to 50% of the earth's land area is degraded due to improper management. In particular, both the conversion of forest land for agriculture and exploitative agricultural practices

Table 2: *Opulence, Life and Death*

	GNP Per Head (Dollars) 1985	Life Expectancy at Birth 1985	Under-5 Mortality Rate (Per Thousand) 1985
Oman	6,730	54	172
South Africa	2,010	55	104
Brazil	1,640	65	91
Sri Lanka	380	70	50
China	310	69	48

Source: World Development Report, 1987; The State of the World's Children, 1987

Table 3: *Mortality rates of children under the age of 5 (U5MRs) per 1,000 live births, 1987*

Per capita GNP (US$) 1986	Lowest U5MRs		Highest U5MRs	
300–400	China	45	Afghanistan	304
	Sri Lanka	45	Sudan	184
			Haiti	174
401–600	Guyana	39	Angola	288
			Yemen Democratic Republic	202
			Yemen	195
			Ghana	149
601–800	Dominican Republic	84	Nigeria	177
	Papua New Guinea	85	Bolivia	176
			Côte d'Ivoire	145
801–1000	Jamaica	23	Cameroon	156
			Botswana	95
1000+	Hong Kong	10	Gabon	172
	Singapore	12	Peru	126
	Costa Rica	23	Saudi Arabia	102
	Trinidad and Tobago	24		
	Chile	26		
	Mauritius	30		

Source: Based on United Nations Children's Fund. State of the World's Children 1989

have led to an increase in soil erosion over the past 25 years. The rate of soil erosion is almost imperceptible (1mm of soil loss in a storm amounts to 15 tonnes per hectare) and significantly exceeds its floor renewal rate (2.5 cm/500 years and at best 1 tonne per hectare per year). The rate of soil erosion in temperate countries is 10 to 20 times the soil renewal rate, while in the tropics it is almost 20 to 40 times.

FAO's estimates suggest that approximately 6 million ha/year are becoming unfit for agriculture. In some areas, the productivity of eroded soils cannot be restored even at enormous costs (equivalent to the application of 2000 tonnes of quality soil per hectare, or 50 tonnes dry rotted cattle manure per hectare).

Soil loss also leads to nutrient depletion. One tonne of good agricultural soil may contain a total of

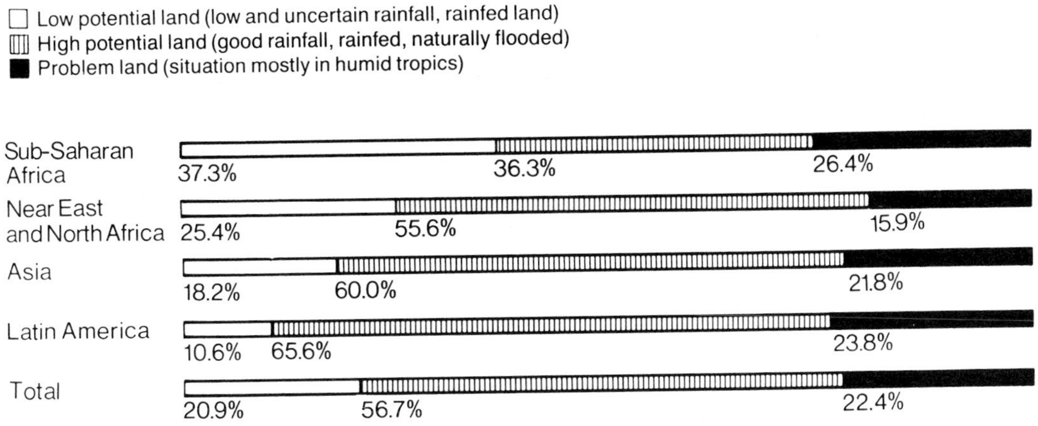

□ Low potential land (low and uncertain rainfall, rainfed land)
▥ High potential land (good rainfall, rainfed, naturally flooded)
■ Problem land (situation mostly in humid tropics)

Sub-Saharan
Africa 37.3% 36.3% 26.4%

Near East
and North Africa 25.4% 55.6% 15.9%

Asia 18.2% 60.0% 21.8%

Latin America 10.6% 65.6% 23.8%

Total 20.9% 56.7% 22.4%

Figure 2: Shares of total harvested land of different potentials, 1982-1984 in 93 developing countries. Source: FAO, World Agriculture: Toward 2000.

Table 4*: Current Status of Threatened Species*

	EX	E	V	R	I	Globally Threatened Taxa
Plants	384	3325	3022	6749	5598	19,078
Fish	23	81	135	83	21	343
Amphibians	2	9	9	20	10	50
Reptiles	21	37	39	41	32	170
Invertebrates	98	221	234	188	614	1,355
Birds	113	111	67	122	624	1,037
Mammals	83	172	141	37	64	497

Key: EX = Extinct (post–1960), E = Endangered, V = Vulnerable, R = Rare, I = Indeterminate
Source: Reid and Miller, 1989; WCMC, unpublished data, Jan. 1989

4 kg. of nitrogen, 1 kg of phosphorous, 20 kg of potassium and 2 kg of calcium. Further, soil erosion results in a loss of organic matter which plays a pivotal role in improving infiltration, water retention, soil structure and cation exchange capacity.

Organic matter and soil micro-flora and micro-fauna (including earthworms) are interdependent in maintaining soil quality and in promoting recycling of nutrients and degradation of wastes.

ii) **Water**: The World Commission on Environment and Development in its report, "Our Common Future" (1987) has drawn attention to the serious state of global water resources.

Global water use has doubled between 1940 and 1980 and it is expected to double again by 2000, with two thirds of the projected water use going to agriculture (WCED, 1987). Yet 80 countries, with 40 percent of the world's population, experience serious water shortage even now. There will be growing competition for water for irrigation, industry and domestic use. River water disputes will multiply between nations and within nations. No easy solution is in sight unless solar desalination of sea water becomes an economic proposition.

Concurrently steps to increase national efforts in saving and sharing water in the rainfed areas are important. The most serious component of potential changes in climate is precipitation. Efforts in enhancing the efficiency of water harvesting and on-farm water management are therefore essential.

In areas of intensive agriculture, problems of salinity, sodicity and water logging as well as the incidence of malaria, schistosomiasis and other waterborne diseases are becoming important. Ground water resources are being adversely affected qualitatively by the excessive use of mineral fertilizers and pesticides. Most countries are yet to develop policies for regulating ground water use in accordance with the recharge capacity of the aquifer.

Figure 3: Population experiencing a fuelwood deficit, 1980 and 2000 (millions). Source: FAO. 1983.

1980	Latin America	Africa	Near East and North Africa	Asia and the Pacific	Total	
Total population	26	55		31	112	Acute
Rural population	18	49		29	96	Scarcity
Total population	201	146	104	832	1283	
Rural population	143	131	69	710	1052	Deficit
2000						
Total population	512	535	268	1671	2986	Acute scarcity
Rural population	342	464	158	1434	2398	or deficit

Source: FAO, Fuelwood Supplies in Developing Countries, 1983

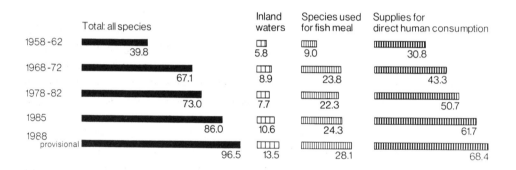

Figure 4: Annual landings (million tons) of aquatic resources (excluding mammals and seaweeds).

iii) *Living aquatic resources*: The world catch of marine living resources is now growing only 1 to 2 percent annually and is approaching its estimated maximum sustainable yield of 100 million tonnes (Figure 4). Since as much carbon is fixed in the sea as on land, it is important that coastal zones are subjected to an ocean capability analysis, on the model of land classification systems currently in use.

iv) *Biological diversity*: Biological diversity is the foundation upon which the edifice of sustainable advances in biological productivity can be built. Recent advances in molecular biology and genetic engineering, which render the transfer of genes across sexual barriers possible, have further enhanced the economic and ecological value of global biological wealth. Unfortunately, serious losses are now occurring at all the three levels in which biological diversity manifests itself, namely intra-species and ecosystem levels, due largely to the destruction of habitats rich in genetic resources, such as tropical rain forests. Recent estimates put the rate of deforestation in the tropics at 17 million ha or 1% per annum (TFAP, 1990). The disappearance of forests also reduces the extent of carbon absorption on the earth. The carbon emission-absorption balance is thus upset.

Many plant and animal species are now under threat of extinction (Table 4). Some scientists predict that if present trends in habitat destruction continue, at least 25 percent of the world's species will be lost in the next several decades. Such a loss of bio-diversity has profound implications for development. Biological resources are renewable; forests, fisheries, wildlife and crops reproduce themselves and even increase when managed properly. Further, the highly diverse natural ecosystems which support this wealth of species also maintain hydrological cycles, regulate climate, build soils, cycle essential nutrients, absorb and break down pollutants and provide for recreation, research and richer quality of life.

Marine protected areas are yet to receive the same attention as their counterparts on land. The area of sea and seabed is more than two and a half times as great as the total area of land masses of the world but less than one percent of that marine area is currently within established protected areas. This compares with about 3 percent of area which is protected in the terrestrial environment. Conservation of biological diversity under wetlands and aquatic conditions should receive greater support.

Conserving biological diversity, therefore, is urgent because diversity provides the raw material for human communities to adapt to change. The loss of each additional gene, species or habitat reduces the available options.

Loss of the biological potential of the soil, adverse changes in water availability and quality, increasing biotic and abiotic stresses and the biological impoverishment of the earth are all occurring at a time when on the one hand, human population is expanding and on the other, the pathways of economic and industrial development chosen so far have built-in potential for climatic alterations.

4. Energy and Fuelwood

FAO's (1989) estimate of fuelwood shortages is given in Figure 3. In India, it has been calculated that a minimum of 1.4 percent growth rate in energy availability will be needed for achieving a 1% growth rate in GNP. However, the energy-mix made available is itself often the cause of both short-term and long-term environmental damage. The deficit of firewood is growing day by day. At the present level of consumption of forest produce and on the current productivity of forests, India needs a minimum of 0.47 ha of forest land for every individual. The existing forest area on this basis would be adequate only for a population of 150 million (in contrast to the present population of 850 million). The task of improving the productivity of forests and mobilising alternative sources of energy is thus urgent. A similar situation obtains with regard to fodder for animals. The population of domesticated animals is increasing, while the area under grasslands and pastures is shrinking. Thus, we will need land-saving agricultural technologies, grain-saving animal rearing methods and energy and cost saving methods of enhancing biological productivity, if we are to face the challenges on the food production and distribution front.

5. Challenges on the Economic Scene

The financial and technical resources needed to promote ecologically sustainable agricultural and food production systems and to meet possible future changes in temperature, precipitation and sea levels are enormous. For example, the participants of the Keystone Dialogue on Plant Genetic Resources held at Madras, India, in January 1990, concluded that a minimum of US $500 million per year of new money will be needed to undertake the tasks

essential for conserving for posterity a sample of the genetic variability existing in crop plants through *ex situ* conservation techniques. The tasks associated with sustainable agriculture such as upgrading degraded land, conserving water and *in situ* protection of biological diversity need considerable additional resources. It is hence unfortunate that there is today a net outflow of resources from developing to developed countries. (WCED, 1987). For developing countries as a whole, external debt increased 4 percent in real terms in 1987, reaching US $1,218,000 million by the end of the year. High debt-servicing payments coupled with the low level of commercial lending and new investment, resulted in growing net transfers of resources from the poor nations to the rich (by World Bank estimates, no less than US $43,000 million in 1988, compared to US $38,100 million in 1987).

International agricultural trade is in disarray. The policies adopted by industrialised countries with regard to trade and domestic subsidies have contributed to surplus production and subsidized exports of agricultural products. Developing countries, already burdened by the debt crisis, fluctuations in exchange rates, economic recession and volatile oil and commodity markets, are also experiencing a deterioration of terms of trade.

Even the slogan "trade and not aid" is losing its meaning. Trade barriers are growing. The scope of intellectual property rights is expanding, while there is little effort to give economic recognition to the informal innovation system which is the very foundation of agricultural evolution and of the conservation of biological diversity. The GATT-TRIPS (Trade related intellectual property rights) negotiations should give serious attention to methods of recognising informal innovations (Keystone Dialogue, 1990).

The concept of "dependence" now being introduced in the revised draft of the UPOV (International Union for the Protection of New Varieties of Plants) should recognise the dependence not only on varieties covered by the PBR (Plant Breeding Rights) System but also on land races and the genetic material arising from informal innovation. A better common future for all will not be possible without a better common present.

6. Implication of Potential Changes in Climate

The impacts on agriculture could be of two major types. First, by altering production adversely in the main food-producing areas, climate change could enhance food scarcities. The location of main food-producing regions could change. Second, there could be profound impacts on the physiological mechanisms regulating plant and animal productivity. The greatest impact is likely to come from changes in precipitation patterns.

Spatial Impact

North America continues to be the principal grain surplus area today. The heavy dependence on North America for world grain reserves has increased the sensitivity of world food supply to the weather and climate of that region. If unfavourable growing conditions occur simultaneously in the major mid-latitude regions of North America, the USSR and Australia, the global food security system will be under severe stress. Fortunately, detailed studies are now in progress on the potential effects of global climate change on US agriculture.

Oram (1985) has drawn attention to four vulnerable producer groups who may be affected severely by adverse changes in temperature and precipitation. The first is located in humid tropics, in lowland areas of Asia and in the Pacific and Caribbean. These areas, normally prone to excessive rains and flooding, may be less severely affected by climatic change. The second group located in the arid and semi-arid areas of the tropics in Africa and South Asia and in the Mediterranean climate of West Asia and North Africa will be extremely vulnerable. A third group comprising farmers at high altitudes may experience both favourable and unfavourable effects. The fourth group consisting of farmers located at the cold margins at higher latitudes may also experience diverse effects. Parry, Porter and Carter (1990) feel that overall levels of production can be maintained through a combination of shifts of agricultural zones and adjustments in technology and management. They have identified the countries in the lower middle and lower latitudes of Africa, South America and elsewhere as the areas most at risk from the effects of climate change.

During the last 10 years, special attention has been given to the agriculture of countries in sub-Saharan Africa. This region is ecologically diverse and covers an area of 22,245 sq. kms. An estimated 30% of the area can sustain production of rainfed crops. The World Bank in a long-term perspective study titled "Sub-Saharan Africa: From Crisis to Sustainable Growth" (World Bank, 1990) has proposed a series of measures which can enhance productivity and reduce vulnerability to ecological and economic factors. An important recommendation related to arresting soil erosion. Soil erosion, widespread in all areas of sub-Saharan Africa, is perhaps most serious in Ethiopia, where top soil losses of up to 290 metric tons a hectare have been reported for steep slopes. The report lays specific stress on developing farmer's associations and recognising the role of women.

Physiological Impact

The Goddard Institute of Space Studies (GISS) and the University of Birmingham, UK, initiated in 1989 a three-year study of the impact of climate change on global agricultural output and food trade. To date, only three comprehensive regional or national assessments of the consequences of climate change for agriculture have been completed. The International Institute for Applied Systems Analysis has conducted several case studies with support from the United Nations Environment Programme (Parry *et al.* 1989).

Generally, it is assumed that increased atmospheric CO_2 would enhance growth rates of certain types of crop plants and that changes in temperature and precipitation would affect livestock, crops, pests and soils. They will also affect ground water replenishment patterns and evapotranspiration rates. Normally, increased CO_2 in the atmosphere can help to increase the rate of photosynthesis, if water and nutrients do not become limiting factors. C3 and C4 plants (i.e. those which have a 3-carbon or 4-carbon path for photosynthesis) respond differently. C3 crops like wheat, barley, rice and potatoes could respond positively to CO_2 enrichment. For wheat and barley, yield increases of as much as 40% have been suggested (Cure, 1985).

Sinha and Swaminathan (1990) examined the integrated impact of a rise in temperature and in CO_2 concentration on the yield of rice and wheat in India. The study showed that for rice, increasing mean daily temperature decreases the period from transplantation to maturity. Such a reduction in duration is often accompanied by decreasing crop yield. There are, however, genotypic differences in per-day yield potential. Breeders can consciously select strains with a high per-day productivity. Increasing levels of CO_2 increase photosynthetic rate and hence dry matter production but an increase in temperature reduces crop duration and thereby yield.

In the case of wheat, there will be an adverse impact on yield if mean temperatures rise by 1 to 2°C. For each 0.5°C increase in temperature there would be a reduction of crop duration of 7 days, which in turn would reduce yield by 0.45 t per ha. According to Parry *et al* (1990), a 1°C increase in mean annual temperature would tend to advance the thermal limit of cereal cropping in mid-latitude northern hemisphere regions by about 150–200 km, and to raise the altitudinal limit to arable agriculture by about 150–200 m.

For India as a whole, rice may become even more important than now in the national food security system, since rice can give high yields under a wider range of growing conditions than wheat. International Testing Networks like those operated by the International Rice Research Institute (IRRI) and the International Centre for Maize and Wheat Research (CIMMYT) provide scope for identifying genotypes suitable for diverse growing conditions (Table 5).

Warrick (1988) has investigated the sensitivity of crop yields to changes in climate variables using several approaches - crop impact analysis, marginal-spatial analysis and agricultural systems analysis. Such investigations suggest that for the core mid-latitude cereal

Table 5: Rice - Varietal Response to Climatic Stresses in IRTP Trials

Drought tolerance
- From rainfed lowland: IR 13240–39–3–3–P1
  IR19274–26–2–3–1
  IR21015–80–3–3–1–2

- From upland nurseries: ADT31
  CR222–MW10
  CR289–1208
  IR9729–76–3

Submergence tolerance FR13A
  IR26933–16–2–3
  IR31338–30–2–1
  IR31429–14–2–3
  IR31432–8–6

Low temperature tolerance JKAU9KO450–126–10
  No. 11 (Takanenishiki)
  Suweon 303, Akiyudaka

Source: International Rice Testing Program (IRTP).
  IRRI Annual Report, 1988

regions, an average warming of 2°C may decrease potential yields by 3 to 17 percent.

It is obvious that more studies are needed at the micro-level in order to understand fully the inter-relationships among CO_2 concentration, temperature and precipitation with reference to their impact on crop yields. The position relating to world food security was characterised by FAO's Committee on World Food Security, at its Session held in Rome from 26–30 March 1990, as "critically dependent on the outcome of the 1990 crops". This underlines the urgency of intensifying on-going research on crop-weather interaction.

Implications of Sea-level Rise
The magnitude of a likely rise in sea level is not yet clear, because of inadequate knowledge of the processes of ice-melt and thermal expansion of ocean water. The definition of sea-level at shoreline positions also presents difficulties, as a result of variations arising from tectonic or isostatic movement of the land mass. Several projections indicate a rate of sea-level rise of 0.5 cm per year or 50 cm per 100 years over the next half century.

Rise of mean sea-level has an immediate and direct effect on the ecosystems of the intertidal zone. It is likely that species with specific tolerances within the tidal spectrum will migrate landwards. Several studies have been conducted on the potential effect of sea-level rise on mangrove swamps (Ellison, 1989). Mangroves generally withstand tropical storms. However, if the intensity and frequency of tropical storms increase as a result of global warming, mangrove ecosystems will be under greater stress.

Coral reefs, sea grasses and mangroves are being destroyed in the coastal areas of several countries due to expansion of tourism, coastal aquaculture, industrial exploitation and pollution. It will be difficult to adapt to new situations in the future if the existing genetic variability in coastal fauna and flora is lost. Hence efforts to conserve the mangrove genetic resources are needed.

7. Consolidating Gains and Adapting to Change

Improving Consumption

Considering the delicate balance between current food production and consumption, every effort will have to be made to improve the productivity, profitability, stability and sustainability of major farming systems. This will have to be done through integrated packages of technology, services and public policies. Agriculture is by and large a location-specific vocation and hence these packages will have to be tailored to match the prevailing agro-ecological, socio-economic and socio-cultural situations.

An area of widespread concern is the persistence of chronic hunger in many countries of the world, in spite of progress in improving agricultural production. The twin deficiencies of past developmental efforts are, first, failure to eliminate chronic hunger (over 600 million children, women and men still go to bed hungry each day) and secondly, failure to arrest the degradation of the environmental capital stocks upon which sustained advances in productivity improvement depend. Most often, hunger arises from inadequate purchasing power. This in turn is due to lack of opportunities for gainful employment.

Experience from employment guarantee programmes organized in India and elsewhere, as well as from the programmes sponsored by the World Food Programme, have shown that such intervention methods are very helpful in mitigating chronic hunger. What is needed urgently is a Social Security Programme for the poorest billion of the human population consisting of the following three components:

a) **Work**: Employment guarantee for ecological security, which integrates food for work and similar programmes with measures to strengthen the ecological foundations of a productive agricultural system

b) **Wage**: Enforcement of minimum wages, with special attention to women workers, since when women have independent access to income, child nutrition is better, and

c) **Welfare**: Steps to ensure that all children have access to better nutrition, health care and education.

A Social Security system for the economically underprivileged sections of the human family involving concurrent attention to work, wage and welfare will help to eliminate chronic hunger, reduce infant mortality and create conditions favourable for the stabilisation of human population and promotion of ecological security. This will call for a bold and imaginative initiative on the part of the affluent nations, since development which is not equitable will not in the long run be sustainable. The World Summit for Children held at the United Nations issued on September 30, 1990, a charter for the very young. If implemented by all nations, the global capacity to meet disasters arising from future changes in climate would be greatly enhanced owing to the availability of the necessary social infrastructure for adaptation to change.

Improving Production

During the decade of the eighties, there has been rapid progress in several areas of science and technology as applied to the improvement of terrestrial and aquatic farming systems. Advances in biotechnology and micro-electronics have enhanced our agricultural capability and improved our early warning and monitoring systems. The new technologies, if integrated in an appropriate manner with traditional technologies, can help to improve the efficiency of management of land and water, energy and post-harvest operations. Improved post-harvest techno-logies can not only help to save what is produced but also to prepare value-added products from every part of the plant and animal biomass. Biomass refineries can be established in rural areas. This will help to provide more jobs in the off-farm sector. What is needed is the stepping up of agricultural research and extension efforts in developing countries through the establishment of **Technology Blending Centres** for deriving maximum benefit from existing and emerging technologies.

Adapting to Change

Every effort should be made to arrest deforestation and to promote upgrading of degraded lands through agro-forestry and other appropriate forms of land-use. This will help to increase carbon fixation on the earth. Along with efforts to prevent the continued increase of CO_2 in the atmosphere, it is necessary to initiate anticipatory measures which can help us to enhance our ability to adapt to new climatic patterns. The following are some of the steps needed urgently.

a) Establishment of Crop-Weather Watch Groups

Special groups consisting of meteorologists, agricultural research and extension workers, dev-elopmental administrators and mass media representatives should be established in every agro-ecological region of a country to continuously monitor the weather situation and analyse its implications for crop growth and pest incidence. Such **crop-weather watch groups** can benefit at the national level from the World Climate Programme of WMO and the early warning systems of FAO. This will help to equip farmers and fisher families with location-specific information to increase their preparedness for floods, tropical cyclones and drought. Mass media and computer-aided extension techniques can help to take the latest information derived from weather satellites to every village.

b) Contingency Planning

Computer simulation models facilitate the preparation of contingency plans to meet different weather probabilities. Alternative land and water use plans can be prepared to adapt to different precipitation patterns. For example, if it is too late to sow a traditional late duration **Sorghum**, an early maturing pearl millet or grain legume crop can be taken. For implementing such contingency plans at the field level, reserves of seeds of the alternative crops will be needed. Fortunately, modern plant breeding has helped to incorporate genes for photo- and thermo-insensitivity in many crop varieties. Thus there is built-in resilience concerning sowing dates. It will be essential that the crop-weather watch groups identify suitable alternative cropping patterns. Just as grain reserves are important for food security, seed reserves are essential for the security of crop production.

c) Ocean Capability Analysis

Over 40% of the global population live in coastal areas. We should initiate in each country an ocean capability analysis through an integrated coastal systems research program (CSR) involving concurrent attention to capture and culture fisheries and coastal forestry and agroforestry. While land capability classifications exist we have yet to initiate similar studies in the coastal zones. We must derive greater benefit from the enormous amount of carbon fixed in the oceans.

d) Genetic Resource Centres for Adaptation to Climate Changes

Gene pools occur in nature for adaptation to drought, floods and sea level changes. Unfortunately, gene erosion, through a variety of causes, is leading to the loss of valuable genetic material. Every effort should be made to establish specialised genetic resources centres for collecting and conserving species and genotypes with the desirable genes. For example, adaptation to sea level rise will be facilitated by the conservation of mangrove species, sea grasses and deep-water or floating rices. A centre for the conservation of genetic resources of plants for adaptation to sea level rise is being established in the State of Tamil Nadu in India. The Centre for Research on Sustainable Agricultural and Rural Development, Madras and the International Tropical Timber Organization (ITTO) are organising in Madras in January 1991 an International Workshop for organising a global network of genetic resource centres for adaptation to climate change. These centres will identify and maintain candidate genes for use in recombinant DNA experiments.

e) Establishment of Genetic Enhancement Centres

Tools of molecular biology and recombinant DNA techniques make the transfer of genes across sexual barriers possible. For example, theoretically the capacity to withstand sea-water conditions can be transferred from mangroves to other species. The "elongation" gene from rice which enables it to grow to the needed height for remaining above flood water levels, can be transferred to other crops. To achieve such goals, it will be necessary to establish **Genetic Enhancement Centres** consisting of experts in molecular biology and genetic engineering. An integrated global grid of Genetic Resources Conservation Centres and Genetic Enhancement Centres will be of immense help in the breeding of crop varieties possessing greater tolerance to drought, floods and sea water intrusions.

f) National Studies

Adams *et al* (1990) have undertaken a detailed analysis of the likely impact of global climate change on the agriculture of the United States of America using the NASA/Goddard Institute of Space Studies (GISS) and the Princeton Geophysical Fluid Dynamics Laboratory (GFDL) models. They found that climate change could lead to a slight reduction in total US cropped acreage. For GISS scenarios, aggregate acreage is reduced due to increased yields. For the GFDL scenarios, acreage is reduced owing to shifts in production between regions and associated changes in resource availability. Because of the possible change in domestic and foreign production assuming the GFDL climate scenario, the role of the USA in agricultural exports may change. Also, patterns of agriculture in the USA are likely to shift as a result of changes in regional crop yields and in crop irrigation requirements. Such potential changes have implications both for public policy and research priorities. **It will be desirable that as a follow-up of the Second World Climate Conference, an International Research Network on Climate Change and Food Security** is organised to conduct at critical locations around the world studies of the kind undertaken by Adams *et al* (1990) in the United States.

g) Specific Measures for Reducing Ggriculture's Contribution to the Build-up of Greenhouse Gases

Reilly and Bucklin (1989) have estimated the contribution of the farm sector to increases in carbon dioxide, methane, nitrous oxide and other greenhouse gases in the atmosphere (Figure 5). It would be useful to standardize methods of reducing the contribution of agriculture to greenhouse gas accumulation in the

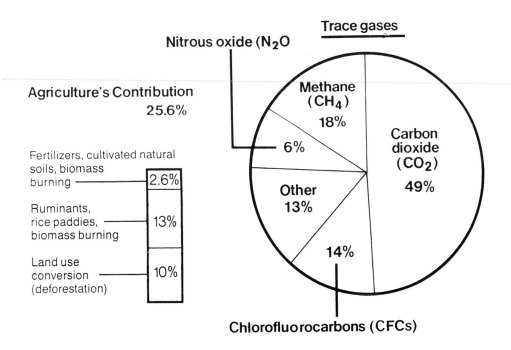

Figure 5: Contributions to increased radiative forcing in the 1990's. Source: Reilly and Bucklin (1989).

atmosphere. Some of these measures are indicated below.

i) CO_2 emissions

Developing countries contribute 10 to 20% of global CO_2 emissions due to deforestation. This contribution can be reduced or eliminated through the restoration of damaged ecosystems and through extensive afforestation.

ii) Methane

Irrigated rice fields, swamps and bogs are believed to be important sources of methane. There is a need for more extensive studies of methane fluxes in rice growing countries. Microbiological and irrigation management methods can help to reduce the release of methane. Methane from the decomposition of organic refuse can be captured and utilized in biogas plants.

iii) Nitrous oxide

Nitrogenous fertilizer as well as the exposure of soils after deforestation could contribute to the emission of nitrous oxide. We need more research on the agricultural sources of nitrous oxide and of methods to reduce such emissions. At the same time, research on the use of biofertilizers, green manure crops and other substitutes for mineral fertilizers needs to be stepped up.

iv) Chlorofluorocarbons (CFC)

Besides contributing to global heating, CFCs also deplete the stratospheric ozone layer which shields the earth from ultraviolet rays. In tropical and sub-tropical areas, more facilities for refrigerated food storage are essential for reducing the spoilage of perishable commodities like meat, milk, fish, vegetables and fruits. There is therefore need for the intensification of research on alternative methods of refrigerated storage of food products, which will not damage the ozone layer. Solar refrigeration methods will be particularly valuable in the tropics, provided they become more economical and efficient.

Every country should develop an integrated package of greenhouse gas free agricultural technologies which will minimise and, where possible, eliminate the contributions of farm operations to the increase of CO_2, CH_4 and NO_2 in the atmosphere and find substitutes for CFC based refrigeration methods.

Conclusions

National and global food and nutrition security systems can be insulated from potential adverse impacts of alterations in temperature, precipitation, sea levels and incidence of UV radiation through well planned anticipatory and participatory research and development programs. This will

call for mutually reinforcing packages of technologies, services and public policies. The organization of an International Research Network on Climate Change and Food Security would help to promote the needed degree of multi-disciplinary and multi-institutional cooperation. The speedy development and effective dissemination of avoidance and adaptation strategies will then become possible.

The major components of such a global research network which can be jointly sponsored by WMO, UNEP, FAO and CGIAR should be the following.

a) Estimation and monitoring of the emission of greenhouse gases as a result of farm operations under different agro-ecological and technological conditions

b) Standardisation of techniques both for minimising agriculture's contributions to the accumulation of greenhouse gases and for withstanding additional biotic and abiotic stresses on crops and farm animals arising from climate change

c) Promotion of research and training in the field of restoration ecology and in the afforestation of degraded forests and organisation of demonstration and training programmes for spreading sustainable management procedures for tropical rain forests

d) Anticipatory research and development measures in coastal areas to avoid or minimize the adverse impact of potential changes in sea levels

e) Standardisation of methods for deriving benefit from higher CO_2 concentration in the stratosphere

f) Standardisation of post-harvest technologies for perishable commodities like vegetables, fruits and animal products based on non-CFC dependent refrigeration methods, and

g) Stimulating policy research designed to strengthen the public policy back-up for both avoidance and adaptation.

References

Adams, R.M., Rosenzwerg, C., Pearl, R.M., Ritchie, J.J., McCarl, B.A., Olyer, D.J., Curry, B.R., Jones, J.W., Boote, K.J., and Allen, H.L., 1990. Global Climate change and US Agriculture. Nature Vol. 345: 219–224 pp.

Cure K.D., 1985. Carbon dioxide doubling response: a crop survey. In: Strain, B.R. and Cure, J.D. (Eds.) Direct effects of increasing carbon dioxide on vegetation. US Department of Energy, Washington. 19–116 pp.

Ellison, Joanna C., 1989. The Effect of Sea-Level Rise on Mangrove Swamps. Review for the Commonwealth Secretariat Expert Group on Climatic Change and Sea-Level Rise. 22 pp.

FAO, 1989. The State of Food and Agriculture Food and Agriculture Organization of the United Nations, Rome. 171 pp.

Human Development Report, 1990. United Nations Development Programme and Oxford University Press, New York, 189 pp.

IRRI, 1988: International Rice Research Institute, Annual Report, Los Banos, Philippines, 646 pp.

Kates, R.W., Chen, R.S., Downing, T.E., Kasperson, J.X., Messer, E. and Millman, S.R., 1989. The Hunger Report: Updated 1989. The Alan Shawn Feinstein World Hunger Program, Brown University, Providence, R.I., USA.

Keystone Dialogue, 1990. International Dialogue on Plant Genetic Resources. Proceedings of Madras Plenary Session, Keystone Center, Colorado, USA.

Oram, P.A., 1985. Sensitivity of Agricultural Production to climatic Change. Climatic Change 7: 129–152 pp.

Parry, M.L., Carter, T. and Konijn, N. (Editors), 1989. The Impact of Climatic Variations on Agriculture. Volume 1—Assessments in Cool Temperate and Cold Regions. Volume 2—Assessments in Semi-Arid Regions. Kluwer Academic Publications.

Parry, M.L., Porter, J.H. and T.R. Carter, 1990. Climatic change and its implications for agriculture. Outlook on Agriculture 19: 9–15.

Reilly, J. and Bucklin, R., 1989. Climate Change and Agriculture. World Agriculture Situation and Outlook Report. Washington, D.C.: USDA/ERS, WAS-55.

Reid, W., and K. Miller, 1989: Programme for the Conservation of biodiversity. World Resources Institute, Washington, D.C.

Sen. Amartya, 1981. Poverty and Famines. Oxford, Clarendon Press, 79 pp.

Sen. Amartya, 1987. Food and Freedom. Sir John Crawford Memorial Lecture, Washington, D.C., 25 pp.

Sinha, S.K. and Swaminathan, M.S., 1990. Climate Change and Food Production. Climate Change (under publication).

Swaminathan, M.S., 1987. Building a National Nutrition Security System. In Global Aspects of Food Production, Tycooly Press.

TFAP (1990). Taking Stock: The Tropical Forestry Action Plan after five years. World Resources Institute. Washington, D.C., USA.

Thompson, L.M., 1975. Weather Variability, Climatic Change and Grain Production. Science 188: 535–541 pp.

Warrick, R.A., 1988. Carbon Dioxide, Climatic Change and Agriculture. The Geographical Journal 154: 221–233 pp.

World Bank, 1990. Sub-Saharan Africa. From Crisis to Sustainable Growth. The World Bank, Washington, D.C., 300 pp.

World Commission on Environment and Development. Our Common Future, 1987. Oxford University Press. 400 pp.

World Development Report, 1987. The State of the World's Children. World Bank.

World Development Report: Poverty, 1990. Published for the World Bank by Oxford University Press. 260 pp.

World Food Council, 1989. The Cyprus Initiative against Hunger in the World. 23 pp. (Mimeographed).

World Resources (1990–91)—1990 A Guide to the Global Environment, Oxford University Press, Oxford & New York. 388 pp.

The Potential Effect of Climate Changes on Agriculture

by Martin Parry *(United Kingdom)* and Zhang Jiachen *(China)*

Abstract

The paper reviews current knowledge of the likely effects of CO_2-induced changes of climate on agriculture and food production. Attention is first directed to the types of climate change, particularly reductions in soil water availability, likely to be most critical for agriculture. Secondly, a number of types of effects on agriculture are considered: 'direct' effects of elevated CO_2, shifts of thermal and moisture limits to cropping, effects on drought, heat stress and other extremes, effects on pests, weeds and diseases, and effects on soil fertility. Thirdly, a summary is presented of likely overall effects on crop and livestock production. The conclusion is that, while global levels of food production can probably be maintained in the face of climate change, the cost of this could be substantial. There could occur severe negative effects at the regional level. Increases in productive potential at higher latitudes are not likely to open up large new areas for production. The gains in productive potential here are unlikely to balance possible reductions in potential in some major grain-exporting regions at mid-latitudes.

1. Introduction

1.1 Objectives and Approach

The purpose of this paper is fourfold: firstly, to identify those systems, sectors and regions of agriculture that are most sensitive to anticipated changes of climate; secondly, to summarize present knowledge about the potential socio-economic impact of changes of climate on world agriculture; thirdly, to consider the adjustments in agriculture that are most likely to occur; finally, to establish research priorities for future assessments of impact.

2. Critical Types of Climatic Change

The potentially most important changes of climate for agriculture include: changes in climatic extremes, warming in the high latitudes, poleward advance of monsoon rainfall, and reduced soil water availability (particularly in mid latitudes in midsummer, and at low latitudes).

2.1 Climatic Extremes

It is not clear whether changes in the variability of temperature will occur as a result of climate change. However, even if variability remains unaltered an increase in average temperatures would result in the increased frequency of temperatures above particular thresholds. Changes in the frequency and distribution of precipitation are less predictable, but the combination of elevated temperatures and drought or flood probably constitutes the greatest risk to agriculture in many regions from global climate change.

2.2 Warming in High Latitudes

There is relatively strong agreement among GCM predictions that greenhouse-gas-induced warming will be greater at higher latitudes (IPCC, 1990). This will reduce temperature constraints on high-latitude agriculture and increase the competition for land here (Parry and Duinker, 1990). Warming at low latitudes, although less pronounced, is also likely to have a significant impact on agriculture.

2.3 Poleward Advance of Monsoon Rainfall

In a warmer world the intertropical convergence zones and polar frontal zones might advance further poleward as a result of an enhanced ocean-continent pressure gradient. If this were to occur then total rainfall could increase in some regions of monsoon Africa, monsoon Asia and Australia, though there is currently little agreement on which regions these might be (IPCC, 1990). Rainfall could also be more intense in its occurrence, so flooding and erosion could increase.

2.4   Reduced Soil Water Availability

Probably the most important consequences for agriculture would stem from higher potential evapotranspiration, primarily due to the higher temperatures of the air and the land surface. Even in the tropics, where temperature increases are expected to be smaller than elsewhere and where precipitation might increase, the increased rate of loss of moisture from plants and soil would be considerable (Parry, 1990; Rind *et al.*, 1989). It may be somewhat reduced by greater air humidity and increased cloudiness during the rainy seasons, but could be pronounced in the dry seasons.

3.   Types of Effect

There are three ways in which increases in greenhouse gases (GHG) may be important for agriculture. Firstly, increased atmospheric carbon dioxide (CO_2) concentrations can have a direct effect on the growth rate of crop plants and weeds. Secondly, GHG-induced changes of climate may alter levels of temperature (as well as temperature gradient), rainfall and sunshine and this can influence plant and animal productivity. Finally, rises in sea-level may lead to loss of farmland by inundation and to increasing salinity of groundwater in coastal areas. These three types of potential impact will be considered in turn.

3.1   Effects of CO2 Enrichment

3.1.1   Effects on photosynthesis

CO_2 is vital for photosynthesis, and the evidence is that increases in CO_2 concentration would increase the rate of plant growth (Cure, 1985; Cure and Acock, 1986). There are, however, important differences between the photosynthetic mechanisms of different crop plants and hence in their response to increasing CO_2. Plant species with the C_3 photosynthetic pathway (e.g. wheat, rice and soybean) tend to respond more positively to increased CO_2 because it tends to suppress rates of photorespiration.

However, C_4 plants (e.g. maize, sorghum, sugarcane and millet) are less responsive to increased CO_2 levels. Since these are largely tropical crops, and most widely grown in Africa, there is thus the suggestion that CO_2 enrichment will benefit temperate and humid tropical agriculture more than that in the semi-arid tropics. Thus, if the effects of climate changes on agriculture in some parts of the semi-arid tropics are negative, then these may not be partially compensated by the beneficial effects of CO_2 enrichment as they might in other regions. In addition we should note that, although C_4 crops account for only about one-fifth of the the world's food production, maize alone accounts for 14 per cent of all production and about three-quarters of all traded grain. It is the major grain used to make up food deficits in famine-prone regions, and any reduction in its output could affect access to food in these areas (Morison, 1990).

C_3 crops in temperate and subtropical regions could also benefit from reduced weed infestation. Fourteen of the world's 17 most troublesome terrestrial weed species are C_4 plants in C_3 crops. The difference in response to increased CO_2 may make such weeds less competitive. In contrast, C_3 weeds in C_4 crops, particularly in tropical regions, could become more of a problem, although the final outcome will depend on the relative response of crops and weeds to climate changes as well (Morison, 1990).

Many of the pasture and forage grasses of the world are C_4 plants, including important prairie grasses in North America and central Asia and in the tropics and subtropics. The carrying capacity of the world's major rangelands are thus unlikely to benefit substantially from CO_2 enrichment (Morison, 1990). Much, of course, will depend on the parallel effects of climate changes on the yield potential of these different crops.

The actual amount of increase in usable yield rather than of total plant matter that might occur as a result of increased photosynthetic rate is also problematic. In controlled environment studies, where temperature, nutrients and moisture are optimal, the yield increase can be substantial, averaging 36 per cent for C_3 cereals such as wheat, rice, barley and sunflower under a doubling of ambient CO_2 concentration. Few studies have yet been published, however, of the effects of increasing CO_2 in combination with changes of temperature and rainfall.

Little is also known about possible changes in yield quality under increased CO_2. The nitrogen content of plants is likely to decrease, while the carbon content increases, implying reduced protein levels and reduced nutritional levels for livestock and humans. This, however, may also reduce the nutritional value of plants for pests, so that they need to consume more to obtain their required protein intake.

3.1.2   Effects on water use by plants

Just as important may be the effect that increased CO_2 has on the closure of stomata. This tends to reduce the water requirements of plants by reducing transpiration (per unit leaf area), thus improving what is termed water-use efficiency (the ratio of crop biomass accumulation to the water used in evapotranspiration). A doubling of ambient CO_2 concentration causes about a 40 per cent decrease in stomatal aperture in both C_3 and C_4 plants which may reduce transpiration by 23–46 per cent (Morison, 1987; Cure and Acock, 1986). This might well help plants in environments where moisture currently limits growth, such as in semi-arid regions, but there remain many uncertainties, such as to what extent the greater leaf area of plants (resulting from increased CO_2) will balance the reduced transpiration per unit leaf area (Allen *et al.*, 1985; Gifford, 1988).

In summary, we can expect that a doubling of atmospheric CO_2 concentrations from 330 to 660 ppmv might cause a 10 to 50 per cent increase in growth and yield of C_3 crops (such as wheat, rice and soybean) and a 0 to 10 per cent increase for C_4 crops (such as maize and sugarcane) (Warrick *et al.*, 1986). Much depends, however, on the prevailing growing conditions. Our present knowledge is based on experiments mainly in field chambers and has not yet included extensive study of response in the field under sub-optimal conditions. Thus, although there are indications that, overall, the effects of increased CO_2 could be distinctly beneficial and could partly compensate for some of the negative effects of CO_2-induced changes of climate, we cannot at present be sure that this will be so.

3.2 Effects of Changes of Climate
3.2.1 Changes in thermal limits to agriculture
Increases in temperature can be expected to lengthen the growing season in areas where agricultural potential is currently limited by insufficient warmth, resulting in a poleward shift of thermal limits of agriculture. The consequent extension of potential will be most pronounced in the northern hemisphere because of the greater extent here of temperate agriculture at higher latitudes. There may, however, be important regional variations in our ability to exploit this shift. For example, the greater potential for exploitation of northern soils in Siberia than on the Canadian Shield may mean relatively greater increases in potential in northern Asia than in northern N. America (Parry, 1990).

A number of estimations have been made concerning the northward shift in productive potential in mid-latitude northern hemisphere countries. These relate to changes in the climatic limits for specific crops under a variety of climatic scenarios, and are therefore not readily compatible (Newman, 1980; Blasing and Solomon, 1983; Rosenzweig, 1985; Williams and Oakes, 1978; Parry and Carter, 1988; Parry *et al.*, 1989). They suggest, however, that a 1°C increase in mean annual temperature would tend to advance the thermal limit of cereal cropping in the mid-latitude northern hemisphere by about 150–200 km, and to raise the altitudinal limit to arable agriculture by about 150–200m.

While warming may extend the margin of potential cropping and grazing in mid-latitude regions, it may reduce yield potential in the core areas of current production, because higher temperatures encourage more rapid maturation of plants and shorten the period of grain filling (Parry and Duinker, 1990). An important additional effect, especially in temperate mid-latitudes, is likely to be the reduction of winter chilling (vernalization). Many temperate crops require a period of low temperatures in winter to either initiate or accelerate the flowering process. Low vernalization results in low flower bud initiation and, ultimately, reduced yields. A 1°C warming has been estimated to reduce effective winter chilling by between 10 and 30 per cent, thus contributing to a poleward shift of temperate crops in southern mid-latitudes (Salinger, 1989).

Increases in temperature are also likely to affect the crop calendar in low-latitude regions, particularly where more than one crop is harvested each year. For example, in Sri Lanka and Thailand a 1°C warming would probably require a substantial re-arrangement of the current crop calendar, which is finely tuned to present climatic conditions (Kaida and Surarerks, 1984; Yoshino, 1984).

3.2.2 Shifts of moisture limits to agriculture
There is much less agreement between GCM-based projections concerning GHG-induced changes in precipitation than there is about temperature—not only concerning changes of magnitude, but also of spatial pattern and distribution through the year. For this reason it is difficult to identify potential shifts in the moisture limits to agriculture. This is particularly so because relatively small changes in the seasonal distribution of rainfall can have disproportionately large effects on the viability of agriculture in tropical areas, largely through changes in growing period when moisture is sufficient and thus through the timing of critical episodes such as planting, etc. However, recent surveys for the IPCC have made a preliminary identification of those regions where there is some agreement amongst 2 x CO_2 experiments with general circulation models concerning an overall reduction in crop-water availability (Parry, 1990; Parry and Duinker, 1990). It should be emphasized that coincidence of results for these regions is not statistically significant. The regions are:

a) Decreases of soil water in December, January and February:

> Africa: north-east Africa, southern Africa
> Asia: western Arabian Peninsula; south-east Asia
> Australasia: eastern Australia
> N. America: southern USA
> S. America: Argentine Pampas

b) Decreases in soil water in June, July and August:

> Africa: north Africa; west Africa
> Europe: parts of western Europe
> Asia: north and central China; parts of Soviet central Asia and Siberia
> N. America: southern USA and Central America
> S. America: eastern Brazil
> Australasia: western Australia

3.2.3 Regions affected by drought, heat stress and other extremes

Probably most important for agriculture, but about which least is known, are the possible changes in climatic extremes, such as the magnitude and frequency of drought, storms, heat waves and severe frosts (Rind *et al.*, 1989). Some modelling evidence suggests that hurricane intensities will increase with climatic warming (Emanuel, 1987). This has important implications for agriculture in low latitudes, particularly in coastal regions.

Since crop yields often exhibit a non-linear response to heat or cold stress, changes in the probability of extreme temperature events can be significant (Mearns *et al.*, 1984; Parry, 1976). In addition, even assuming no change in the standard deviation of temperature maxima and minima, we should note that the frequency of hot and cold days can be markedly altered by changes in mean monthly temperature. To illustrate, under a $2 \times CO_2$ equilibrium climate the number of days in which temperatures would fall below freezing would decrease from a current average of 39 to 20 in Atlanta, Georgia (USA), while the number of days above 90°F would increase from 17 to 53 (EPA, 1989). The frequency and extent of area over which losses of agricultural output could result from heat stress, particularly in tropical regions, is therefore likely to increase significantly. Unfortunately, no studies have yet been made of this. However, the apparently small increases in mean annual temperatures in tropical regions (c. 1 to 2°C under a $2 \times CO_2$ climate) could sufficiently increase heat stress on temperate crops such as wheat so that these are no longer suited to such areas. Important wheat-producing areas such as N. India could be affected in this way (Parry and Duinker, 1990).

There is a distinct possibility that, as a result of high rates of evapotranspiration, some regions in the tropics and subtropics could be characterized by a higher frequency of drought or a similar frequency of more intense drought than at present. Current uncertainties about how regional patterns of rainfall will alter mean that no useful prediction of this can at present be made. However, it is clear in some regions that relatively small decreases in water availability can readily produce drought conditions. In India, for example, lower-than-average rainfall in 1987 reduced food grains production from 152 to 134 million tonnes (mt), lowering food buffer stocks from 23 to 9 mt. Changes in the risk and intensity of drought, especially in currently drought-prone regions, represent potentially the most serious impact of climatic change on agriculture at both the global and the regional level.

3.2.4 Effects on the distribution of agricultural pests and diseases

Studies suggest that temperature increases may extend the geographic range of some insect pests currently limited by temperature (EPA, 1989; Hill and Dymock, 1989). As with crops, such effects would probably be greatest at higher latitudes. The number of generations per year produced by multivoltine (i.e. multigenerational) pests would increase, with earlier establishment of pest populations in the growing season and increased abundance during more susceptible stages of growth. An important unknown, however, is the effect that changes in precipitation amount and air humidity may have on the insect pests themselves and on their predators, parasites and diseases. Climate change may significantly influence interspecific interactions between pests and their predators and parasites.

Under a warmer climate at mid-latitudes there would be an increase in the overwintering range and population density of a number of important agricultural pests, such as the potato leafhopper which is a serious pest of soybeans and other crops in the USA (EPA, 1989). Assuming planting dates did not change, warmer temperatures would lead to invasions earlier in the growing season and probably lead to greater damage to crops. In the US Corn Belt increased damage to soybeans is also expected due to earlier infestation by the corn earworm.

Examination of the effect of climatic warming on the distribution of livestock diseases suggests that those at present limited to tropical countries, such as Rift Valley fever and African Swine fever, may spread into the mid-latitudes. For example, the horn fly, which currently causes losses of $730.3 million in the US beef and dairy cattle industries, might extend its range under a warmer climate leading to reduced gain in beef cattle and a significant reduction in milk production (Drummond, 1987; EPA, 1989).

In cool temperate regions, where insect pests and diseases are not generally serious at present, damage is likely to increase under warmer conditions. In Iceland, for example, potato blight currently does little damage to potato crops, being limited by the low summer temperatures. However, under a $2 \times CO_2$ climate that may be 4°C warmer than at present, crop losses to disease may increase to 15% (Bergthorsson *et al.*, 1988).

Most agricultural diseases have greater potential to reach severe levels under warmer and more humid conditions (Beresford and Fullerton, 1989). Under warmer and more humid conditions cereals would be more prone to diseases such as Septoria. In addition, increases in population levels of disease vectors may well lead to increased epidemics of the diseases they carry. To illustrate, increases in infestations of the Bird Cherry aphid (*Rhopalosiphum padli*) or Grain aphid (Sitobian avenae) could lead to increased incidence of Barley Yellow Dwarf Virus in cereals.

3.3 Effects of Sea-Level Rise on Agriculture

GHG-induced warming is expected to lead to rises in sea-level as a result of thermal expansion of the oceans and partial melting of glaciers and ice caps, and this in turn is expected to affect agriculture, mainly through the inundation of low-lying farmland but also through the increased salinity of coastal groundwater. The current projection of sea-level rise above present levels is 20cm ±10 by c 2030, and 30 cm ±15 by 2050 (Warrick and Oerlemans, 1990).

Preliminary surveys of proneness to inundation have been based on a study of existing contoured topographic maps, in conjunction with knowledge of the local 'wave climate' that varies between different coastlines. They have identified 27 countries as being especially vulnerable to sea-level rise, on the basis of the extent of land liable to inundation, the population at risk and the capability to take protective measures (UNEP, 1989). It should be emphasized, however, that these surveys assume a much larger rise in sea-levels than is at present estimated to occur within the next century under current trends of increase of GHG concentrations. On an ascending scale of vulnerability (1 to 10) experts identified the following most vulnerable countries or regions: 10, Bangladesh; 9, Egypt, Thailand; 8, China; 7, western Denmark; 6, Louisiana; 4, Indonesia .

The most severe impacts are likely to stem directly from inundation. SE Asia would be most affected because of the extreme vulnerability of several large and heavily populated deltaic regions. For example, with a 1.5 m sea-level rise, about 15 per cent of all land (and about one-fifth of all farmland) in Bangladesh would be inundated and a further 6 per cent would become more prone to frequent flooding (UNEP, 1989). Altogether 21 per cent of agricultural production could be lost. In Egypt, it is estimated that 17 per cent of national agricultural production and 20 per cent of all farmland, especially the most productive farmland, would be lost as a result of a 1.5 m sea-level rise. Island nations, particularly low-lying coral atolls, have the most to lose. The Maldive Islands in the Indian Ocean would have one-half of their land area inundated with a 2 m rise in sea-level (UNEP, 1989).

In addition to direct farmland loss from inundation, it is likely that agriculture would experience increased costs from saltwater intrusion into surface water and groundwater in coastal regions. Deeper tidal penetration would increase the risk of flooding and rates of abstraction of groundwater might need to be reduced to prevent re-charge of aquifers with sea water.

Further indirect impacts would be likely as a result of the need to relocate both farming populations and production in other regions. In Bangladesh, for example, about one-fifth of the nation's population would be displaced as a result of the farmland loss estimated for a 1.5 m sea-level rise. It is important to emphasize, however, that the IPCC estimates of sea-level rise are much lower than this (about 0.5 m by 2090 under the IPCC Business-As-Usual case).

3.4 Summary of Plant, Pest and Sea-Level Effects

Potential impacts on yields vary greatly according to types of climate change and types of agriculture. In general, there is much uncertainty about how agricultural potential may be affected.

In the northern mid-latitudes where summer drying may reduce productive potential (e.g. in the US Great Plains and Corn Belt, Canadian Prairies, southern Europe, south European USSR) yield potential is estimated to fall by c. 10–30% under an equilibrium 2 x CO_2 climate (Parry, 1990). However, towards the northern edge of current corn-producing regions (e.g. the northern edge of the Canadian Prairies, northern Europe, northern USSR and Japan, southern Chile and Argentina), warming may enhance productive potential, particularly when combined with beneficial direct CO_2 effects. Much of this potential may not, however, be exploitable owing to limits placed by inappropriate soils and difficult terrain, and on balance it seems that the advantages of warming at higher latitudes would not compensate for reduced potential in current major cereal-producing regions.

Effects at lower latitudes are much more difficult to estimate because production potential is largely a function of the amount and distribution of precipitation and because there is little agreement about how precipitation may be affected by GHG warming. Because of these uncertainties the tendency has been to assert that worthwhile study must await improved projection of changes in precipitation. Consequently very few estimates are currently available of how yields might respond to a range of possible changes of climate in low-latitude regions. The only comprehensive national estimates available are for Australia, where increases in cereal and grassland productivity might occur (except in western Australia) if warming is accompanied by increase in summer rainfall (Pearman, 1988).

The impacts described above relate to possible changes in potential productivity or yield. It should be emphasized that such potential effects are those estimated assuming present-day management and technology. They are not the estimated future actual effects, which will depend on how farmers and governments respond to altered potential through changes in management and technology. The likely effects on actual agricultural output and on other measures of economic performance such as profitability and employment levels are considered in the next section.

4. Effects on Production and Land Use

To date (1990) six national case studies have been made of the potential impact of climatic changes on agricultural production (in Canada, Iceland, Finland, the USSR, Japan and the United States) (Parry *et al.*, 1988; Smit, 1989; EPA, 1989). These studies are based on results from model experiments of yield responses to altered climate and the effects that altered yields might have on production.

Other countries have conducted national reviews of effects of climate change, basing these on existing knowledge rather than on new research. The most comprehensive of these are for Australia and New Zealand (Pearman, 1988; Salinger *et al.*, 1990). Brief surveys have also been completed in the UK and the Federal Republic of Germany (DOE, 1988; SCEGB, 1989). Several other national assessments are currently in progress but not yet complete. This section provides a summary of results from the most detailed of these surveys. These provide us with an array of assessments for three world regions: the northern and southern mid-latitude grain belts and northern regions at the current margin of the grain belt. We shall take these regions in turn. Unless otherwise stated the estimated effects are for climates described by 2 x CO_2 GCM experiments. No national assessments have been completed using climates described by transient response GCM experiments. Some of the estimates relate to the effect only of altered climate, others to the combined effect of altered climate and the direct effect of increased atmospheric CO_2.

4.1 Effects on Production in the Northern and Southern Mid-Latitude Grain Belts

4.1.1 United States
Increased temperatures and reduced crop water availability projected under the GISS and GDFL 2 x CO_2 climate experiments are estimated to lead to a decrease of yields of all the major unirrigated crops (EPA, 1989). The largest reductions are projected for the south and south-east. In the most northern areas, however, where temperature is currently a constraint on growth, yields of unirrigated maize and soybeans could increase as higher temperatures increase the length of the available growing season. When the direct effects of increased CO_2 are considered, it is evident that yields may increase more generally in northern areas but still decrease in the south where problems of heat stress would increase and where rainfall may decrease.

Production of most crops is estimated to be reduced because of yield decreases and limited availability of suitable land. The largest reductions are in sorghum (–20%), corn (–13%) and rice (–11%), with an estimated fall in net value of agricultural output of $33 billion. If this occurred, consumers would face slightly higher prices, although supplies are estimated to meet current and projected demand. However, exports of agricultural commodities could decline by up to 70%, and this could have a substantial effect on the pattern of world food trade.

4.1.2 Canada
On the Canadian prairies, where growing season temperatures under the GISS 2 x CO_2 equilibrium climate would be about 3.5°C higher than today, average potential yields could decrease 10 to 30%, wind erosion potential could increase by about a quarter and the frequency of drought could increase thirteen-fold (if there was no change in precipitation). Spring wheat yields in Saskatchewan are estimated to fall by 28% (Williams *et al.*, 1988). Since Saskatchewan at present produces 18% of all the world's traded wheat, such a reduction could well have global implications.

Assuming (unrealistically) that the present-day relationship between production and profit in Saskatchewan holds in the future, average farm household income is estimated to fall by 12%, resulting in a reduction in expenditure by agriculture of Can$277m on the goods and services provided by other sectors, leading to a Can$250m (6%) reduction in provincial GDP in sectors other than agriculture and a 1% loss of jobs.

In Ontario precipitation increases of up to 50% would be more than offset by increases in evapotranspiration with consequent increased moisture stress on crops (Smit, 1987). Maize and soybean would thus become very risky in the southern part of the province. In the north, where maize and soybean cannot currently be grown commercially because of inadequate warmth, cultivation may become profitable but this is not expected to compensate for reduced potential further south and, if there were no adjustment of current land use and farming systems, the overall cost in lost production is reckoned at Can$100m to 170m.

4.1.3 Japan
Under a warming of 3.0-3.5°C and a 5% increase in annual precipitation (the GISS 2 x CO_2 climate), rice yields are expected to increase in the north (Hokkaido) by c. 5%, and in the north-central region (Tohoku) by c. 2%, if appropriate technological adjustments are made (Yoshino *et al.*, 1988). The average increase for the country overall is c. 2-5%. Cultivation limits for rice would rise about 500 m and advance c. 100 km north in Hokkaido. Yields of maize and soybeans are both estimated to increase by about 4%. Sugar cane yields in the most southern part of Japan could decrease if rainfall were reduced. The northern economic limit of citrus fruits would shift from southern Japan to northern Honshu Island (Yoshino, personal communication, 1989). Net primary productivity of natural vegetation is expected to increase by c. 15% in the north, c. 7% in the centre and south of Japan (Yoshino *et al.*, 1988).

4.1.4 Australia and New Zealand
In Australia and New Zealand national assessments have been based on a thorough review of existing knowledge

and on use of expert judgement rather than on model experiments (Pearman, 1988; Salinger *et al.*, 1990). Overall, it is reckoned that wheat production in Australia could increase under a 2 x CO_2 climate, assuming a quite simple scenario of increased summer rainfall, decreased winter rainfall and a general warming of 3°C. Increases are expected in all states except Western Australia, where more aridity might cause a significant reduction in output (Pittock, 1989).

More generally, the major impact of production would probably be on the drier frontiers of arable cropping. For example, increases in rainfall in subtropical northern Australia could result in increased sorghum production at the expense of wheat. Increased heat stress might shift livestock farming and wool production southward with sheep possibly replacing arable farming in some southern regions.

Many areas currently under fruit production would no longer be suitable under a 3°C warming, and would need to shift southwards or to higher elevations in order to maintain present levels of production. All of these changes would also be affected by changes in the distribution of diseases and pests.

4.2 Effects on Production in Northern Marginal Regions

Some of the most pronounced effects on agriculture would be likely to occur in high-latitude regions because GHG-induced warming is projected to be greatest here and because this warming could remove current thermal constraints on farming. Inappropriate terrain and soils are, however, likely to limit the increase in extent of the farmed area and, in global terms, production increases would probably be small (Parry, 1990). A summary of available information is given below.

4.2.1 Iceland

With mean annual temperatures increased by 4.0°C and precipitation 15% above the present average (consistent with the GISS 2 x CO_2 climate), the onset of the growing season of grass in Iceland would be brought forward by almost 50 days, hay yields on improved pastures would increase by about two-thirds and herbage on unimproved rangelands by about a half (Bergthorsson *et al.*, 1988). The numbers of sheep that could be carried on the pastures would be raised by about 250% and on the rangelands by two-thirds if the average carcass weight of sheep and lambs is maintained as at present. At a guess, output of Icelandic agriculture could probably double with a warming of 4°C.

4.2.2 Finland

Assuming in Finland an increase in summer warmth by about a third and precipitation by about half (consistent with the GISS 2 x CO_2 climate), barley and spring wheat yields increase about 10% in the south of the country but slightly more in the north (due to relatively greater warming and lower present-day yields) (Kettunen *et al.*, 1988). The area under grain production in Finland might increase at the expense of grass and livestock production as a consequence of raised profitability, with the greatest extension being in winter crops such as wheat rather than spring crops such as barley or oats.

4.2.3 Northern USSR

The only other region for which an integrated impact assessment has been completed is in the north European USSR. In the Leningrad and Cherdyn regions, under climates that are 2.2°C/2.7°C warmer during the growing season and 36%/50% wetter (consistent with the GISS 2 x CO_2 climate), winter rye yields are estimated to decrease by about a quarter due to faster growth and increased heat stress under the higher temperatures (Pitovranov *et al.*, 1988). However, crops such as winter wheat and maize, which are currently low-yielding because of the relatively short growing season in these regions, are better able to exploit the higher temperatures and exhibit yield increases in Cherdyn of up to 28% and 6% respectively with a 1°C warming.

The differential yield responses described above are reflected in substantial changes in production costs incurred in meeting production targets. Thus, while production costs for winter wheat and maize in the Central Region around Moscow are estimated to be reduced by 22% and 6% under a 1°C warming and with no change in precipitation, they increase for most other crops, particularly quick-maturing spring-sown ones which are the dominant crops today. This would suggest that quite major switches of land use would result and the land allocation models used in the study indicate that, to optimize land use by minimizing production costs, winter wheat and maize would extend their area by 29% and 5% while barley, oats and potatoes would decrease in extent.

4.2.4 Summary of potential effects in mid-latitude regions

The effects of possible climatic changes on regional and national production have not yet been investigated in any great detail, nor for more than a few case studies. The effects are strongly dependent on the many adjustments in agricultural technology and management that undoubtedly will occur in response to any climatic change. So numerous and varied are these potential adjustments that it is extraordinarily difficult to evaluate their ultimate effect on aggregate production. In this section we have therefore considered the effects on production that are likely to stem directly from changes in yield, unmodified by altered technology and management. Adjustments in technology will be considered, briefly, in the last section of this paper.

In summary, it seems that overall output from the major present-day grain-producing regions could well decrease under the warming and possible drying expected in these regions. In the USA grain production may be reduced by 10–20% and, while production would still be sufficient for domestic needs, the amount for export would probably decline. Production may also decrease in the Canadian prairies and in the southern Soviet Union. In Europe production of grains might increase in the UK and the Netherlands if rainfall increases sufficiently, but may fall in southern Europe substantially if there are significant decreases in rainfall as currently estimated in most GCM 2 x CO_2 experiments (Parry, 1990). Output could increase in Australia if there is a sufficient increase in summer rainfall to compensate for higher temperatures.

Production could increase in regions currently near the low-temperature limit of grain growing: in the northern hemisphere in the northern Prairies, Scandinavia, north European Soviet Union, and in the southern hemisphere in southern New Zealand, and southern parts of Argentina and Chile. But it is reasonably clear that, because of the limited area unconstrained by inappropriate soils and terrain, increased high-latitude output will probably not compensate for reduced output at mid-latitudes. The implications of this for global food supply and food security are considered in the next section.

5.   Implications for Global Food Security

Although, on average, global food supply currently exceeds demand by about 10 to 20 per cent, its year-to-year variation (which is about + or –10 per cent) can reduce supply in certain years to levels where it is barely sufficient to meet requirements. In addition, there are major regional variations in the balance between supply and demand, with perhaps a billion people (about 15 per cent of the world's population) not having secure access to sufficient quantity or quality of food to lead fully productive lives. For this reason the working group on food security at the 1988 Toronto Conference on The Changing Atmosphere concluded that: 'While averaged *global* food supplies may not be seriously threatened, *unless appropriate action is taken to anticipate climate change and adapt to it*, serious regional and year-to-year food shortages may result, with particular impact on vulnerable groups' (Parry and Sinha, 1988). Statements such as this are, however, based more on intuition than on knowledge derived from specific study of the possible impact of climate change on the food supply. No such study has yet been completed, although one is currently being conducted by the US Environmental Protection Agency which is due to report in 1992.

The information available at present is extremely limited. It has, for example, been estimated that increased costs of food production due to climate change could reduce *per*

capita global GNP by a few percentage points (Schelling, 1983). Others have argued that technological changes in agriculture will override any negative effects of climate changes and, at the global level, there is no compelling evidence that food supplies will be radically diminished (Crosson, 1989). Recent reviews have tended to conclude however that, at a regional level, food security could be seriously threatened by climate change, particularly in less developed countries in the semi-arid and humid tropics (Sinha *et al.*, 1988; Parry, 1990; Parry and Duinker, 1990).

Analyses conducted for the IPCC, designed to test the sensitivity of the world food system to changes of climate, indicate what magnitudes and rates of climatic change could possibly be absorbed without severe impact and, alternatively, what magnitudes and rates could seriously perturb the system (Parry, 1990; Parry and Duinker, 1990). These suggest that yield reductions of up to 20% in the major mid-latitude grain-exporting regions could be tolerated without a major interruption of global food supplies. However, the increase in food prices (7% under a 10% yield reduction) could seriously influence the ability of food-deficit countries to pay for food imports, eroding the amount of foreign currency available for promoting development of their non-agricultural sectors. It should be emphasized that these analyses are preliminary and more work is necessary before we have an adequate picture of the resilience of the world food system to climatic change.

6.   Conclusions

Our assessment of possible effects has, up to this point, assumed that technology and management in agriculture do not alter significantly in response to climatic change, and thus do not alter the magnitude and nature of the impacts that may stem from that change. It is certain, however, that agriculture will adjust and, although these adjustments will be constrained by economic and political factors, it is likely that they will have an important bearing on future impacts. Three broad types of adjustment may be anticipated: changes in land use (e.g. changes in farmed area, crop type and crop location), and changes in technology, (e.g. changes in irrigation, fertilizer use, control of pests/diseases, soil management, farm infrastructure, and changes in crop and livestock husbandry). For a discussion of these adjustments the reader is referred to a number of recent reviews (Parry, 1990; Parry and Duinker, 1989).

On balance, the evidence is that food production at the global level can, in the face of estimated changes of climate, be sustained at levels that would occur without a change of climate, but the cost of achieving this is unclear. It could be very large. Increases in productive potential at high mid-latitudes and high latitudes, while being of regional importance, are not likely to open up large new areas for production. The gains in productive potential here due to climatic warming would be unlikely to balance possible large-scale reductions in potential in some major grain-exporting regions at mid-latitude. Moreover, there may well occur severe negative impacts of climate change on food supply at the regional level, particularly in regions of high present-day vulnerability least able to adjust technically to such effects.

The average global increase in overall production costs could thus be small (perhaps a few per cent of world agricultural GDP). Much depends however, on how beneficial are the so-called 'direct' effects of increased CO_2 on crop yield. If plant productivity is substantially enhanced *and* more moisture is available in some major production areas, then world productive potential of staple cereals could increase relative to demand with food prices reduced as a result. If, on the contrary, there is little beneficial direct CO_2 effect *and* climate changes are negative for agricultural potential in all or most of the major food-exporting areas, then the average costs of world agricultural production could increase significantly, these increased costs amounting to perhaps over 10 per cent of world agricultural GDP.

Although we know little, at present, about how the frequency of extreme weather events may alter as a result of climatic change, the potential impact of concurrent drought or heat stress in the major food-exporting regions of the world could be severe. In addition, relatively small decreases in rainfall or increases in evapotranspiration could markedly increase both the risk and the intensity of drought in currently drought-prone (and often food-deficient) regions. Change in drought risk represents potentially the most serious impact of climatic change on agriculture at both the regional and the global level. The regions most at risk from impact from climatic change are probably those currently most vulnerable to climatic variability. Frequently these are low-income regions with a limited ability to adapt through technological change.

This paper has emphasized the inadequacy of our present knowledge. It is clear that more information on potential impacts would help us identify the full range of potentially useful responses and assist in determining which of these may be most valuable.

Some priorities for future research may be summarized as follows:

- Improved knowledge is needed of effects of changes in climate on crop yields and livestock productivity in different regions and under varying types of management; on soil-nutrient depletion; on hydrological conditions as they effect irrigation-water availability; on pests, diseases and soil microbes and their vectors, and on rates of soil erosion and salinization.

- Further information is needed on the range of potentially effective technical adjustments at the farm and village level (e.g. irrigation, crop selection, fertilizing, etc.), on the economic, environmental and political constraints on such adjustments, and on the range of potentially effective policy responses at regional, national and international levels (e.g. re-allocations of land use, plant breeding, improved agricultural extension schemes, large-scale water transfers).

To date, less than a dozen detailed regional studies have been completed that serve to assess the potential impact of climatic changes on agriculture. It should be a cause for concern that we do not, at present, know whether changes of climate are likely to increase the overall productive potential for global agriculture, or to decrease it. There is therefore no adequate basis for predicting likely effects on food production at the regional or world scale. All that is possible at present is informed speculation. The risks attached to such levels of ignorance are great. A comprehensive, international research effort is required, now, to redeem the situation.

References

Allen, L.H. Jr., Jones, P. and Jones, J.W.: 1985, Rising atmospheric CO_2 and evapotranspiration. Advances in Evapotranspiration. Proceedings of the National Conference on Advance in Evapotranspiration. American Society of Agricultural Engineers. St. Joseph, Michigan. pp. 13–27.

Beran, M.A., and Arnell, N.W.: 1989, *Effect of Climatic Change on Quantitative Aspects of United Kingdom Water Resources.* Report for Department of Environment Water Inspectorate. Institute of Hydrology, United Kingdom.

Beresford, R.M. and Fullerton, R.A.: 1989, Effects of climate change on plant diseases. Submission to Climate Impacts Working Group, May 1989, 6pp.

Bergthorsson, P., Bjornsson, H., Dyrmundsson, O., Gudmundsson, B., Helgadottir, A. and Jonmundsson, J.V.: 1988, The Effects of Climatic Variations on Agriculture in Iceland, In: *The Impact of Climatic Variations on Agriculture. Vol. 1, Assessments in Cool Temperate and Cold Regions*, M.L. Parry, T.R. Carter and N.T. Konijn (Eds), Kluwer, Dordrecht, The Netherlands.

Blasing, T.J. and Solomon, A.M.: 1983, *Response of North American Corn Belt to Climatic Warming*, prepared for the US Department of Energy, Office of Energy Research, Carbon Dioxide Research Division, Washington, DC, DOE/N88-004.

Crosson, P.: 1989, Greenhouse warming and climate change: why should we care? *Food Policy*, Vol. 14–2: pp. 107–118.

Cure, J.D.: 1985, Carbon Dioxide Doubling Responses: a crop survey. In: *Direct Effects of Increasing Carbon Dioxide on Vegetation*, pp. 100–116, B.R. Strain and J.D. Cure (Eds), US DOE/ER0238, Washington, USA.

Cure, J.D. and Acock, B.: 1986, Crop Responses to Carbon Dioxide Doubling: a Literature Survey, *Agricultural and Forest Meteorology*, 38, 127–145.

DOE, 1988: Possible impacts of climate change on the natural environment in the United Kingdom. Department of the Environment, London, UK.

Drummond, R.O.: 1987, Economic Aspects of Ectoparasites of Cattle in North America. In: *Symposium. The Economic Impact of Parasitism in Cattle*, XXIII World Veterinary Congress, Montreal, pp. 9–24.

Emanuel, K.A.: 1987, The Dependence of Hurricane Intensity on Climate: mathematical simulation of the effects of tropical sea surface temperatures, *Nature*, 326, 483-485.

Environmental Protection Agency: 1989, The Potential Effects of Global Climate Change on the United States. Report to Congress.

Gifford, R.M.: 1988, Direct effect of higher carbon dioxide levels concentrations on vegetation. In: *Greenhouse: Planning for Climate Change*, pp. 506–519, G.I. Pearman (Ed), CSIRO, Australia.

Hill, M.G. and Dimmock, J.J.: 1989, Impact of Climate Change: Agricultural/Horticultural Systems. DSIR Entomology Division, submission to New Zealand Climate Change Programme, Department of Scientific and Industrial Research, New Zealand, 16pp.

IPCC: 1990, Intergovernmental Panel on Climate Change, *Scientific Assessment of Climate Change: Policymakers Summary*, Geneva and Nairobi: WMO and UNEP.

Kaida, Y. and Surarerks, V.: 1984, Climate and agricultural land use in Thailand. In: Climate and Agricultural Land Use in Monsoon Asia, M.M. Yoshino (Ed), University of Tokyo Press, Tokyo, 231–253.

Kettunen, L., Mukula, J., Pohjonen, V., Rantanen, O. and Varjo, U.: 1988, The Effects of Climatic Variations on Agriculture in Finland. In: *The Impact of Climatic Variations on Agriculture: Volume 1, Assessments in Cool Temperate and Cold Regions*, M.L. Parry, T.R. Carter and N.T. Konijn (Eds), Kluwer, Dordrecht, The Netherlands, pp. 511–614.

Mearns, L.O., Katz, R.W. and Schneider, S.H.: 1984, Extreme high temperature events: changes in their probabilities with changes in mean temperatures, *Journal of Climatic and Applied Meteorology*, 23, 1601–13.

Morison, J.I.L.: 1987, Intercellular CO_2 concentration and stomatal response to CO_2. In: *Stomatal Function*, E. Zeiger, I.R. Cowan and G.D. Farquhar (Eds), Stanford University Press, pp. 229–251.

Morison, J.I.L.: 1990, Direct effects of elevated atmospheric CO_2 and other Greenhouse Gases. In: The *Potential Effects of Climatic Change on Agriculture and Forestry*, M.L. Parry and P.N. Duinker (Eds), Working Group II Report, IPCC.

Newman, J.E.: 1980, Climate change impacts on the growing season of the North American Corn Belt. *Biometeorology*, 7-2, 128–142.

Parry, M.L.: 1976, *Climatic change. agriculture and settlement*, Dawson, Folkestone, England.

Parry, M.L.:1990, *Climate Change and World Agriculture*, Earthscan, London, 165pp.

Parry, M.L. and Carter, T.R.: 1988, The Assessments of the Effects of Climatic Variations on Agriculture: Aims, Methods and Summary of Results. In: *The Impact of Climatic Variations on Agriculture. Volume 1, Assessments in Cool Temperate and Cold Regions*, M.L. Parry, T.R. Carter and N.T. Konijn (Eds), Kluwer, Dordrecht, The Netherlands.

Parry, M.L., and Sinha, S.K., 1988: Food security. Working Group Report in *Conference Proceedings, The Changing Atmosphere: Implications for Global Security*, Toronto, Canada, 27–30 June, 1988. WMO-No. 170, Geneva.

Parry, M.L., Carter, T.R. and Porter, J.H.: 1989, The Greenhouse Effect and the Future of UK Agriculture, *Journal of the Royal Agricultural Society of England*, pp. 120–131.

Parry, M.L. and Duinker, P.N.: 1990, *The Potential Effects of Climatic Change on Agriculture and Forestry*, Working Group II Report, IPCC.

Pearman, G.I. (Ed): 1988, *Greenhouse: Planning for Climate Change*, Melbourne, Australia: CSIRO.

Pitovranov, S.E., Iakimets, V., Kislev, V.E. and Sirotenko, O.D.: 1988a, The Effects of Climatic Variations on Agriculture in the Subarctic Zone of the USSR. In: *The Impact of Climatic Variations on Agriculture: Volume 1. Assessments in Cool, Temperate and Cold Regions*. M.L. Parry, T.R. Carter and N.T. Konijn (Eds), Kluwer, Dordrecht, The Netherlands.

Pitovranov, S.E., Iakimets, V., Kislev, V.E. and Sirotenko, O.D.: 1988b, The Effects of Climatic Variations on Agriculture in the Semi-Arid Zone of the USSR. In: *The Impact of Climatic Variations on Agriculture: Volume 2, Assessments in Semi-Arid Regions*, M.L. Parry, T.R. Carter, N.T. Konijn (Eds), Kluwer, Dordrecht, The Netherlands.

Pittock, A.B.: 1989, The Greenhouse Effect, regional climate change and Australian agriculture, Paper presented to *Australian Society of Agronomy. 5th Agronomy Conference*, Perth, Australia.

Rind, D., Goldberg, R. and Ruedy, R.: 1989, Change in climate variability in the 21st century. *Climatic Change*, 14: 5–37.

Rind, D., Goldberg, R., Hansen, J. Rosenzweig, C. and Ruedy, R.: 1990, Potential evapotranspiration and the likelihood of future drought. *Journal of Geophysical Research*, 95, D7:9983–10,004.

Rosenzweig, C.: 1985, Potential CO_2-induced climate effects on North American wheat-producing regions. *Climatic Change*. 7:367–389.

Salinger, M.J .: 1989a, The Effects of Greenhouse Gas Warming on Forestry and Agriculture. Draft report for WMO Commission for Agrometeorology, 20pp.

Salinger, M.J.: 1989b, CO_2 and Climate Change: Impacts on New Zealand Agriculture (personal communication) 5pp.

Salinger, M.J., Williams, W.M., Williams, J.M. and Martin, R.J.: 1990, *Carbon Dioxide and Climate Change: Impacts on Agriculture*, New Zealand: New Zealand Meteorological Service; DSIR Grasslands Division; MAFTech.

SCEGB: 1989, Study Commission of Eleventh German Bundestag, *Protecting the Earth's Atmosphere: An International Challenge*, Bonn: Bonn University.

Schelling, T.: 1983, Climate Change: implications for welfare and policy. In: *Changing Climate: Report of the Carbon Dioxide Assessment Committee*, Washington, DC: National Academy of Sciences.

Smit, B.: 1987, Implications on Climatic Change for Agriculture in Ontario, *Climate Change Digest*, CCD 87–02, Environment Canada.

Smit, B.: 1989, Climate warming and Canada's comparative position in agriculture, *Climate Change Digest*, CCD 89–01, Environment Canada.

UNEP: 1989, Criteria for assessing vulnerability to sea level rise: a global inventory of high risk areas, United Nations Environment Programme and the Government of the Netherlands, Draft Report.

Warrick, R.A. and Gifford, R. with Parry, M.L.: 1986, CO_2, climatic change and agriculture. In: *The Greenhouse Effect, Climatic Change and Ecosystems*, B. Bolin, B.R. Doos, J. Jager and R.A. Warrick (Eds), SCOPE 29, Chichester, John Wiley and Sons, pp.393–473.

Warrick, R.A. and Oerlemans, J.: 1990, Sea level rise. In: *IPCC Scientific Assessment of Climate Change*, Geneva and Nairobi: WMO and UNEP.

Williams, G.D.V., Fautley, R.A., Jones, K.H., Steward, R.B. and Wheaton, E.E.: 1988, Estimating Effects of Climatic Change on Agriculture in Saskatchewan, Canada. In: *The Impact of Climatic Variations on Agriculture: Volume 1, Assessments in Cool Temperate and Cold Regions*, pp. 219–379, M.L. Parry, T.R. Carter and N.T. Konijn (Eds), Kluwer, Dordrecht, The Netherlands.

Williams, G.D.V. and Oakes, W.T.: 1978, Climatic Resources for Maturing Barley and Wheat in Canada. In: *Essays on Meteorology and Climatology*: in Honour of Richard W. Longley, Studies in Geography Mono.3., 367–385, Haye, K.D. and Reinelt, E.R. (Eds), University of Alberta, Edmonton.

Yoshino, M.M.: 1984, Ecoclimatic systems and agricultural land use in Monsoon Asia. In: *Climate and Agricultural Land Use in Monsoon Asia*, M.M. Yoshino (Ed), University of Tokyo Press, Tokyo, 81–108.

Yoshino, M.M., Horie, T., Seino, H., Tsujii, H., Uchijima, T. and Uchijima, Z.: 1988, The Effect of Climatic Variations on Agriculture in Japan. In: *The Impact of Climatic Variations on Agriculture: Volume 1, Assessments in Cool Temperate and Cold Regions*, pp.723–868, M.L. Parry, T.R. Carter and N.T. Konijn, (Eds), Kluwer, Dordrecht, The Netherlands.

Effects of Global Climatic Change on Marine Ecosystems and Fisheries

by Tsuyoshi Kawasaki *(Japan)*

Abstract

The internal dynamics and community responses in the oceans tend to show longer term changes on the order of 10–10^2 years between alternative community structures. A mechanism exists whereby global warming could, by intensifying the alongshore wind stress on the ocean surface, lead to acceleration of coastal upwelling, which may in turn accelerate photosynthesis in the oceans, resulting in an increase in production of the pelagic fish.

In the English Channel changes in the ecosystem which had occurred between 1925 and 1935 were reversed between 1965 and 1979 (the Russell cycle), and this event was linked to the long-term climatic change. Around 1940 and in the 1980s when the global air temperatures were high, stocks of the Pacific and Atlantic sardines were at a very high level, possibly resulting from the worldwide intensification of ocean upwelling, but those of the Pacific and Atlantic herrings were very low seemingly influenced by an increase in SST. Global warming could cause a shift in prevailing wind patterns over the northeastern Pacific. Current systems in the coastal area may be reversed by an increase in southwesterly wind stress associated with a northward shift of the Aleutian low. The taxonomic compositions of plankton may change. Ranges of major stocks would be moved.

The fluctuations in catches of the Far Eastern sardine have had a great impact on the fisheries of the USSR, Korea and Japan. For example, the composition of the Kushiro Fish Market, which is a city located on the eastern Pacific side of Hokkaido and is a fishery base where total landings have been the highest in Japan recently, has become entirely large-quantity-fish-dependent. *Hamachi*, young yellowtail, culture in Japan has become primarily dependent on the sardine as its feed.

I. Climate System and Marine Ecosystems

Climate is determined by the interactions between the ocean, land, biomass and the atmosphere. The oceans account for more than 70% of the earth's surface and play an important role in determining climate.

Since the total heat capacity of the oceans is about 1000 times as large as that of the atmosphere, they work as a reservoir of heat and govern the motion of the atmosphere through an exchange of heat with the atmosphere. Moreover, because the density of sea water is about 1000 times as high as that of air above the earth's surface, the inertia of its motion is higher and its movement and state persist longer.

The kinetic energy of oceans and atmosphere comes from the solar radiation. Oceans are warmed by the solar radiation reaching the sea surface, which affects the ocean circulation as a result of heterogeneous distribution of water temperature, and coupled with the wind stress from the atmosphere, drives ocean currents. In addition, the atmosphere receives energy from the oceans in the form of latent heat mostly retained in the water vapour.

Steele (1985) reviewed the difference between temporal variability in terrestrial and marine environments and considered how this external forcing may affect population fluctuations in the two systems. The internal dynamics and community responses are expected to differ significantly with marine populations more likely to show longer term changes between alternative community structures. He pointed out that in the sea short-term variability is damped out by the very large heat capacity of the ocean, but in turn this large thermal capacity and the long period exchange rates between deep and near-surface waters (on the order of $10^2 - 10^3$ yr) leads to relatively large-amplitude changes on long time scales.

II. Changes of the Oceans Caused by Atmospheric Forcing

Further increases in the concentrations of the greenhouse gases (carbon dioxide, methane, nitrous oxide and chlorofluorocarbons) in the atmosphere are expected to result in substantial global-scale warming of the atmosphere and hence at the sea surface in future decades. Since the specific heat of sea water is large the response of oceans to the temperature rise in the atmosphere will be delayed and an increase in temperature of the oceans will not occur so quickly.

In response to the atmospheric warming, global mean sea level should change owing to thermal expansion of the oceans and the melting (or accumulation) of land ice. Prediction of these sea level changes is of importance, because many coastal regions could be adversely affected by even a small sea level rise. As a prerequisite to such predictions we need to be able to understand past sea level changes and to predict the future climatic conditions that will affect sea level.

Wigley and Raper (1987) investigated the relationship between greenhouse gas forcing, global mean temperature change and sea level rise due to thermal expansion of the oceans using upwelling-diffusion and pure diffusion models. The greenhouse-gas-induced thermal expansion contribution to sea level rise between 1880 and 1985 is estimated to be 2–5 cm. Projections are made to the year 2025 for different forcing scenarios. For the period 1985–2025 the estimate of greenhouse-induced warming is 0.6–1.0°C. The concomitant oceanic thermal expansion would raise sea level by 4–8 cm.

Peltier and Tushingham (1989) examined the relation between global sea level rise and the greenhouse effect. Secular sea level trends extracted from tide gauge records of appropriately long duration demonstrate that global sea level may be rising at a rate in excess of 1 mm per year. However, because global coverage of the oceans by the tide gauge network is highly nonuniform and the tide gauge data reveal considerable spatial variability, there has been a well-founded reluctance to interpret the observed secular sea level rise as representing a signal of global scale that might be related to the greenhouse effect. When the tide gauge data are filtered so as to remove the contribution of ongoing glacial isostatic adjustment to the local sea level trend at each location, then the individual tide gauge records reveal sharply reduced geographic scatter and suggest that there is a globally coherent signal of strength 2.4 ±0.90 mm per year for the past 100 years. This signal could constitute an indication of global climate warming. Mikolajewicz *et al.* (1990) considered both thermal effects using atmospheric general circulation models (AGCMs) and the full dynamic effects associated with changes in oceanic general circulation using an ocean general circulation model (OGCM), to test the sensitivity of the

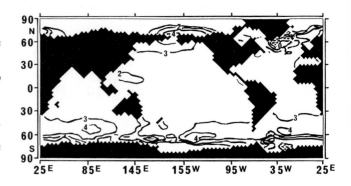

Figure 1: Geographical distribution of changes in SST in year 50 relative to the control experiment. Shading indicates negative values. (Mikolajewicz et al., 1990)

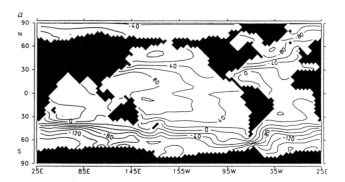

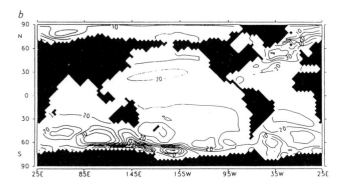

Figure 2: (a) Annual mean distribution of sea surface elevation (cm) of the control experiment; (b) Pattern of changes in sea surface elevation (cm) for year 50 relative to the control experiment. Note that changes in the Ross Sea area are negative. (Mikolajewicz et al., 1990)

ocean circulation to a specified forcing. Figure 1 displays the changes in sea surface temperature (SST) relative to the control for year 50, for doubled atmospheric CO_2. Negative SST anomalies appear in the vicinity of the Irminger Sea, the region in which most of North Atlantic deep water (NADW) is produced and sinks, and in the Weddell Sea, where arctic bottom water is formed and downwells.

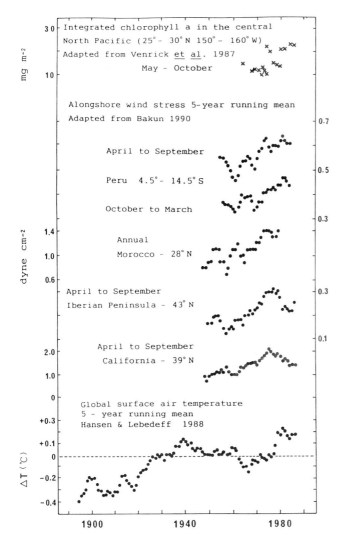

Figure 3: Bottom to top. Secular trends in global surface air temperature, alongshore wind stress and integrated chlorophyll a.

might, with global warming, become even more pronounced. Effects of enhanced upwelling on the marine ecosystems are uncertain but potentially dramatic. The intensification of coastal upwelling may accerelate photosynthesis in the oceans, leading to an increase in production of the pelagic fish.

According to Venrick *et al.* (1987), since 1968 a significant increase in total chlorophyll a in the water column during the summer in the central North Pacific has been observed. A concomitant increase in winter winds and a decrease in SST suggest that long-period fluctuations in atmospheric characteristics have changed the carrying capacity of the central Pacific epipelagic ecosystem (Figure 3).

III. Effects of the Climatic Change on Marine Ecosystems - Case Studies

The case studies presented here describe the effects of climatic change on marine ecosystems, i.e. change of the species composition, fluctuation in abundance of the pelagic fish stocks, and change of the range of fish stocks. I take up three cases which exemplify each.

Change of the Species Composition - the Russell Cycle
Cushing (1982) referred to the Russell Cycle in his famous book, "Climate and Fisheries". Sir Frederick Russell found that in the English Channel changes in the ecosystem which had occurred between 1925 and 1935 were reversed between 1965 and 1979. The profound events were named the Russell cycle. There are five prominent components in the cycle: the magnitude of the winter phosphorus maximum; the quantity of fish larvae ; the quantity of macroplankton, the presence or absense of the arrow worms, *Sagitta elegans* and *S. setosa*; and the appearance and disappearance of the pilchard population (Figure 4). There were a number of additional events. *Laminaria*

The annual mean sea surface elevation of the control run is displayed in Figure 2a, showing the well-known subtropical gyres and the Antarctic circumpolar current (ACC). The geographical distribution of the sea level change for the year 50 is shown in Figure 2b. The largest increases in sea level occur in the North Atlantic, as a result of reduced NADW formation and reduced surface density, and on the northern edge of the ACC, because of its southward displacement.

According to Bakun (1990), a mechanism exists whereby global warming could, by intensifying the alongshore wind stress on the ocean surface, lead to acceleration of coastal upwelling. Evidence from several different regions suggests that the major coastal upwelling of the world jumped up to higher levels in intensity in the late 1960s as a result of the global warming (Figure 3). Thus the cool foggy summer conditions that typify the coastlands of northern California and other similar upwelling regions

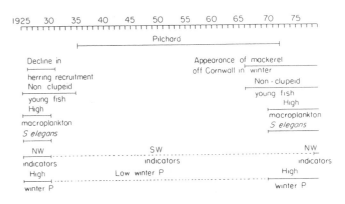

Figure 4: A summary of the main changes described as the Russell cycle. (Cushing, 1982).

ochroleuca, a seaweed recorded in the Plymouth area in the late 1940s, had spread all over the southwest by the early 1960s; it declined sharply between 1965 and 1975. Some species of shell gravel molluscs disappeared between the 1920s and 1950s.

The Russell cycle was linked to the general period of warming between the twenties and 1950 (Figure 3), which was probably accompanied by an intensification of atmospheric and oceanic circulation. The strength of the upper westerlies started to diminish in the mid 1920s. The Russell cycle was a part of the events occurring worldwide. Subtropical invaders were found off the Pacific northeast in 1926 and off California in 1931. Arctic species appeared in the Bay of Fundy between 1932 and 1936 and the southern animals were found during the same period in the Gulf of Maine. A more widespread and diffuse migration occurred of boreal animals to the Faroe Islands, Iceland, Greenland, Jan Mayen and the Barents Sea. The overwhelming impression is of changes both profound and widespread in that dramatic decade between 1925 and 1935.

Fluctuation in Abundance of the Pelagic Fish Stocks

Three stocks of the Pacific sardine, *Sardinops sagax*,--the Far Eastern sardine around Japan, the California sardine off North America and the Chilean sardine off South America - and also the European pilchard, *Sardina pilchardus*, which is a close relative of the Pacific sardine, have repeated coincidental, enormous amplitude and long-period fluctuations, while the Pacific herring, *Clupea pallasi*, and the Atlantic herring, *C. harengus*, show the fluctuation patterns similar to but completely out-of-phase with those of the sardines (Figure 5). Recently the sardines have been dominating, while the herrings have been at a low level on a worldwide scale. It is hard to consider that the fluctuations in abundance common to the sardine and herring stocks in the North Atlantic, as well as throughout the Pacific have been caused by the local environmental changes, and the global environmental changes could be responsible for the fluctuations.

In Figure 3, the global trend in surface air temperature (Hansen and Lebedeff, 1988) is indicated. If we compare Figure 3 with Figure 5, high positive and negative correlations are found between trends in temperature and catches of sardines and herrings: good sardine and poor herring catches occurred when the global temperature was high (around 1940 and in the 1980s) and conversely poor sardine and good herring catches were experienced when the global temperature was low (in the 1880s and 1960s). A mechanism for this was shown earlier.

Crawford *et al.* (1990) wrote, "The correlations between catches of sardines in the three regions of the Pacific Ocean provide strong supporting evidence for the hypothesis that they are being influenced by climate operating on an oceanic scale (Kawasaki, 1983; Kawasaki and Omori,

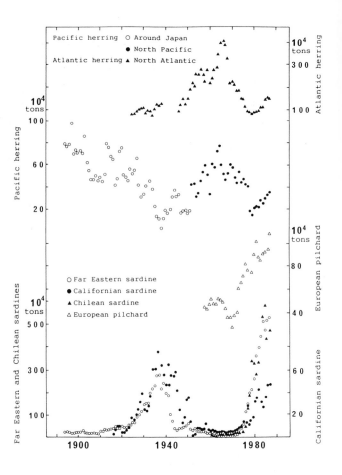

Figure 5: Secular trends in catches of the world sardines and herrings.

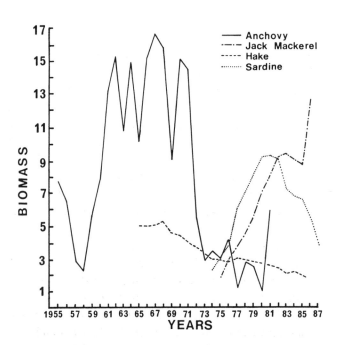

Figure 6: Time series of biomass estimates of the stocks of the Peruvian anchoveta, Chilean sardine off northern Chile, Chilean mackerel and hake. (Serra, 1990).

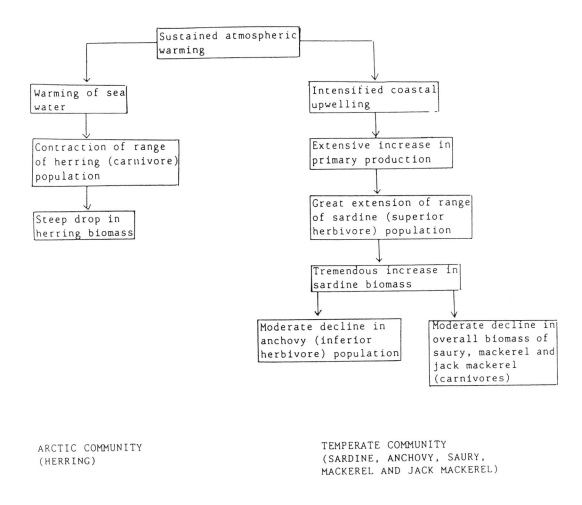

ARCTIC COMMUNITY
(HERRING)

TEMPERATE COMMUNITY
(SARDINE, ANCHOVY, SAURY,
MACKEREL AND JACK MACKEREL)

Figure 7: Flow diagram showing the influence of sustained atmospheric warming on the fluctuation of cycles with enormous amplitude and increased periods in pelagic communities.

1988). The fact that the regions are well separated, but that the influence probably acts simultaneously in each, suggests that the linkages are not mediated through the ocean but rather have their origin in the atmosphere. Possible linkage between the North Pacific and North Atlantic, with little opportunity for passage of water between them, reinforces this conclusion. Trans-Pacific linkages are not as evident for chub mackerel and anchovies as they are for sardines. Climate may therefore be expected to act upon an aspect of the sardine's biology that is not shared by the other two species."

Serra (1990) wrote, "The beginning of the pulse of the sardine off northern Chile started between 1969 and 1970 and the abundance increased steadily until 1981. No really weak recruitment is observed until 1986 and 1987, when the 1983 and 1984 year classes entered the fishery. The stock size started to diminish in 1982, and finally the landings also started to decrease in 1985 (Figure 6). The jack mackerel has undergone a large increase in abundance, which occurred at the same time as the sardine. The hake

declined since the end of the 1960s. The main recruitment failure of anchoveta occurred in 1971, before the 1972–73 El Niño event, which was once believed to be the main cause producing the collapse of the anchoveta.

Another crucial change occurred in the species composition of the ichthyoplankton pelagic community off northern Chile. In the period before 1969, a 'coastal' community led by the anchoveta predominated. On the other hand, in the period after 1970, an 'oceanic' community led by the sardine predominated. The first group is favored by coastal processes influenced by subarctic waters, while the second one shows a closer association with a greater influence of subtropical conditions near the coast. This is part of a structural change in the whole eastern Pacific Ocean current system."

A flow diagram showing the possible influence of the sustained atmospheric warming on the fluctuation of cycles with enormous amplitude and increased period in the sardines and the herrings and hence in the pelagic fish communities is illustrated in Figure 7.

Change of the Range of Fish Stocks Caused by Atmospheric Warming

Sibley and Strickland (1985) selected the coastal northeastern Pacific to better define the current understanding of the relationship between climate and marine fisheries production and to determine the consequences of CO_2 induced climate change on fisheries.

A global increase in wind velocities caused by heating could increase turbulent mixing and decrease stability. Furthermore, the prevailing North Pacific storm track, which overlies the Gulf of Alaska in winter and the western Aleutians in summer, could be displaced northward. Such a shift would cause increases in storminess and surface turbulence in the southeastern Bering Sea and decreases in the Gulf of Alaska. The removal of ice cover in the Bering Sea would enhance these effects in that area. The northward shift of the north Pacific storm track could produce increased winds and net southwesterly winds in the southeastern Bering Sea shelf domain. Increases in horizontal advection in the area could be reinforced by stronger horizontal density gradients and entrainment associated with increased freshwater input.

The advective effects of changes in winds and freshwater input appear to interfere with each other in the northeastern Pacific. Under present conditions cyclonic (counter-clockwise) flow is maintained in the Gulf of Alaska by cyclonic winter wind stress associated with the Aleutian low-pressure center and by the constraining influence of the shoreline. This current system may be reversed or substantially weakened by a net increase in southwesterly wind stress associated with a northward shift of the Aleutian low (Figure 8). This decrease in circulation would counteract coastal intensification of currents caused by increased runoff.

A shift in prevailing wind patterns would also change relative rates of coastal upwelling and downwelling and the associated offshore and onshore transport. Under present conditions, downwelling caused by the combined effects of wind and runoff is inferred to occur along the entire Gulf of Alaska and the coast of British Columbia in winter and along the eastern Gulf coast all year. In response to a northward shift of atmospheric pressure patterns, the relative frequency, magnitude, and seasonal and areal extent of coastal upwelling and offshore transport might increase all along the coast of the northeast Pacific. The taxonomic compositions of the phyto- and zooplankton may change, influenced by the change of ocean structure. The walleye pollock and Alaska pink shrimp would colonize north of the present range, possibly replaced by competitors to the south. The British Columbia herring would increase in northern populations, but decrease in southern ones. The Bering Sea yellow sole would colonize north of present range.

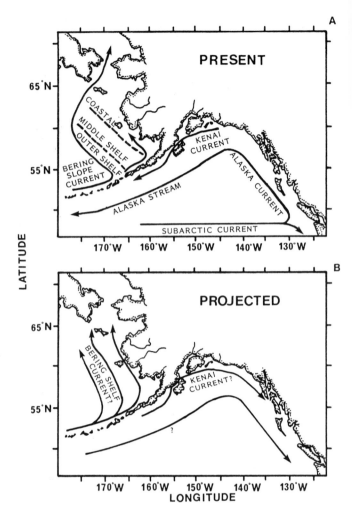

Figure 8: Present circulation in the northern North Pacific Ocean and eastern Bering Sea and possible future current patterns under a sustained atmospheric warming scenario involving a consistent northward shift of North Pacific atmospheric pressure centres. (Sibley and Strickland, 1985).

IV. Effect of the Climatic Change on Fisheries and Society - A Case Study

Fluctuations in catches of the sardine and the herring have had a great impact on the fisheries and society. Figure 9 shows trends in the sardine catch in the northwest Pacific during 1894–1988. The great increase in catch is a result of the increase in stock size, which is caused by an enormous extension of the range of sardine. When the sardine stock is abundant, its range is broad and it is distributed throughout the Sea of Japan on the west and even to the Western Hemisphere on the east, while, on the contrary, when it is at a low level, it is confined to a small, coastal area along southern Japan. During the most abundant years of the sardine stock, it was caught in large quantities along the eastern coast of the Korean Peninsula and the Coast Range of the USSR, as well as on the Japan Sea side of Japan.

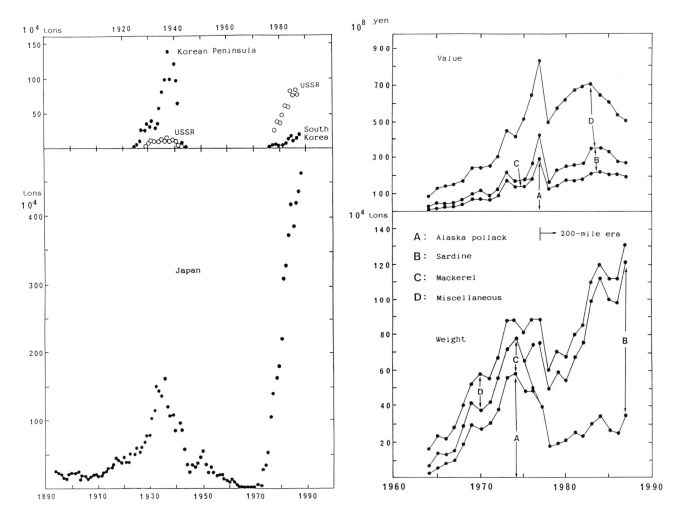

Figure 9: Trends in catch of the Far Eastern sardine, 1894-1988.

Figure 10: Year to year change in landings at Kushiro Fish Market in terms of value (top) and weight (bottom), 1964-1987.

As seen in Figure 9, fishing for the sardine on the Korean coast began in 1925 and a peak catch of about 1.2 million metric tons which is comparable to that around Japan, was obtained mainly on the northern Korean coast in 1937, but fishing was terminated in 1943 when sardine schools did not visit this coast. On the other hand, fishing by the USSR was carried out between 1930 and 1946. In the 1940s the sardine catch decreased steeply and reached a trough in 1965, when the catch was only 9000 tons. Since 1970 the sardine catch has increased again and it has been at the highest level recently. A substantial catch has been obtained on the coasts of South Korea since 1976 and off the USSR since 1978. Catch records of North Korea have not been available to date.

As described above, a great increase in sardine catch results from the extreme extension of their ranges and the impacts of the fluctuation in abundance on the society are found most markedly on the outer edges of their distribution. We therefore consider Kushiro, which is a city located on the eastern Pacific side of Hokkaido and is a fishery base where total landings have been the highest in Japan recently.

Figure 10 shows trends in landings at Kushiro since 1964 in terms of large-quantity species - the walleye pollock, chub mackerel and sardine. The walleye pollock has been caught by trawlers in the Bering Sea as well as around Hokkaido and its catch depends on the international regime of regulation over the sea. On May 1 1977 the regulations over the fisheries resources in the 200-mile EEZs of the USA and the USSR went into effect. Landings of the walleye pollock at Kushiro dropped to about 20 thousand tons in 1978, but have more or less leveled off thereafter.

Landings of the chub mackerel caught by purse seiners at Kushiro began increasing in the early 1960s and reached a peak during 1970–74. In 1976 sardine appeared suddenly in large quantities in the waters off southeastern Hokkaido and replaced chub mackerel. As seen in Figure 10, landing of the sardine at Kushiro Fish Market has increased rapidly thereafter and it accounts for 65 % of the total landing of all species in 1987. From Figure 10 we can see that the

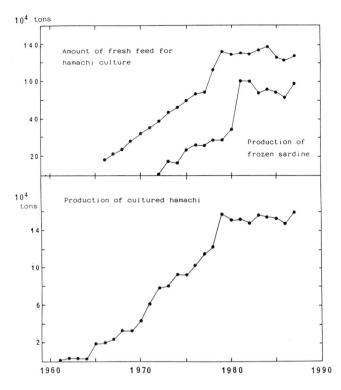

Figure 11: Amount of fresh feed for hamachi culture and production of frozen sardine, 1966-1987 (top), and production of cultured hamachi, 1961-1987 (bottom),

proportion of large-quantity fish such as the walleye pollock, chub mackerel and sardine to the total landings has risen from 42 % in weight and 28 % in value in 1964 to 91 % and 56 % in 1987, respectively. This means that the constitution of the Kushiro Fish Market has changed from 'less large-quantity-fish-dependent' to 'more large-quantity-fish-dependent'. Total value of landings has declined recently in spite of the increase in total catch, as a result of the increase in low-value landings (Figure 10). Most of the sardine landing has been reduced to 'oil and meal'.

Figure 11 shows trends in production of the cultured *hamachi*, young yellowtail. *Hamachi* culture in southern Japan started in 1962 and its production rose steeply until it amounted to about 160,000 tons in 1979, and it remains almost unchanged afterwards.

Figure 11 also shows a trend in amount of the fresh feed for *hamachi* culture and in production of the frozen sardine. In 1973 the frozen sardine accounted for 20 % of the fresh feed for *hamachi*, but this proportion rose to 77 % in 1987, indicating that *hamachi* culture is primarily dependent on the sardine.

V. Lessons

The problems of the global environment are becoming intensified as a result of human activities. The first step toward the solution of these problems, when viewed from the standpoint of natural science, will be to monitor selected environmental factors on a regular basis.

As stated earlier, oceans and hence ecosystems in them are fundamentally characterized by 'conservation' and 'regulation' as compared with land whose trait is 'change'. Oceans are less changeable and it takes a very long time for them to reach a steady equilibrium state in response to the atmospheric forcing. It is necessary, therefore, to continue for a long time with monitoring at selected, scattered stations in the world oceans on a regular basis.

As it is, however, continued hydrographical observations at fixed stations in the oceans have been scarcely carried out, which has made it difficult to better understand the effect of climatic change on the marine ecosystem as well as on the climate system itself.

At its General Meeting in 1986, the International Council of Scientific Unions (ICSU) resolved that it would begin the International Geosphere-Biosphere Programme (IGBP). This programme is aimed at clarifying global changes and the dynamics of the earth system. Among a number of domains of which the earth system is made up, there is the interaction between the biosphere in the oceans and atmosphere, and a monitoring network over the world oceans is proposed. It is most urgent, I think, to set up a network system to monitor the oceans from the scientific viewpoint.

References

Bakun, A.: 1990, Global Climate Change and Intensification of Coastal Ocean Upwelling, Science, 247, 198–201

Cushing, D. H.: 1982, Climate and Fisheries, Academic Press, 373pp

Crawford, R. J. M., Underhill, L. G., Shannon, L. V., Lluch-Belda, D., Siegfried, W. R. and Villagastin-Herrero, C. A.: Proceedings of the International Symposium on the Long-Term Variability of Pelagic Fish Populations and their Environment (in press).

Hansen, J. and Lebedeff, S.: 1988, Global Surface Air Temperatures: Update through 1987, Geophys. Res. Lett., 15, 323–326

Kawasaki, T.: 1983, Why Do Some Pelagic Fishes Have Wide Fluctuations in Their Numbers? Biological Basis of Fluctuation from the Viewpoint of Evolutionary Ecology, FAO Fish. Rep., 291, 1065–1080

Kawasaki, T. and Omori, M.: 1988, Fluctuations in the Three Major Sardine Stocks in the Pacific and the Global Trend in Temperature, Long-Term Changes in Marine Fish Populations, 37–53

Mikolajewicz, V., Benjamin, D. S. and Maier-Reimer, E.: 1990, Ocean Response to Greenhouse Warming, Nature, 345, 589–593

Peltier, W. R. and Tushingham, A. M.: 1989, Global Sea Level Rise and the Greenhouse Effect: Might They Be Connected?, Science, 244, 806–810

Serra, R.: Long-Term Variability of the Chilean Sardine, Proceedings of the International Symposium on the Long-Term Variability of Pelagic Fish Populations and their Environment (in press).

Sibley, T. H. and Strickland, R. M.: 1985, Fisheries: Some Relationships to Climate Change and Marine Environmental Factors, Characterization of Information Requirements for Studies of CO_2 Effects: Water Resources, Agriculture, Fisheries, Forests and Human Health, United States Department of Energy, 95–143

Steele, J. H.: 1985, A Comparison of Terrestrial and Marine Ecology Systems, Nature, 313, 355–358

Venrick, E. L., McGowan, J. A., Cayan, D. R. and Hayward, T. L.: 1987, Climate and Chlorophyll a: Long-Term Trends in the Central North Pacific, Science, 238, 70–72

Wigley, T. M. L. and Raper, S. C. B.: 1987, Thermal Expansion of Sea Water Associated with Global Warming, Nature, 330, 127–131

Sea Level Rise and Coastal Zone Management

by El-Mohamady Eid *(Egypt)* and Cornelis H. Hulsbergen *(The Netherlands)*

Abstract

Greenhouse-effect induced climate change repercussions will, via the catchment areas of big rivers, eventually arrive in amplified form at the far downstream coastal zones of the world. In the coastal zone these imported problems will add to the locally exerted climate change effects, while the climate change induced sea-level rise will attack the coastal zone from the other side, both by flooding and by saline groundwater intrusion.

These multiple problems will hit the coastal zone, which is already under increasing stress due to the population explosion, congestion, pollution, and over-exploitation of natural resources, which also in many instances leads to extensive subsidence of low-lying coastal areas.

The combined threats require fast and massive support for national Coastal Zone Management capabilities, which could help to limit the damage and to support planning for sustainable development. Recommendations are presented to facilitate implementation of Coastal Zone Management units in all coastal countries by the year 2000.

1. Introduction

Sea-level rise (SLR), one of the most accurately established consequences of global warming, will affect the coastal zones of some 180 nations and territories. Studies about the cause, the methods of limitation and the magnitude of SLR are conducted mostly by people and institutions who have no direct management responsibility. Therefore, their messages relating to SLR should be sent in two directions: to those who may contribute to curb global warming, and to those who are actually responsible for dealing with extra SLR. This paper is especially meant for the latter group. Decisions on how and when to respond to extra SLR have to be taken in circles of national, regional, or local Coastal Zone Management (CZM). This paper discusses some items pertaining to the nature and the importance of this task, thereby emphasizing the pressing need for boosting CZM-capacities, even if no SLR were to occur.

The Intergovernmental Panel on Climate Change (IPCC), Working Group III, Subgroup Coastal Zone Management has elaborated on strategies for adaption to SLR in their final report (Misdorp et al., 1990). Their executive summary and their proposal for future activities, including a suggested timeline 1991–2000 for the implementation of comprehensive CZM plans in all coastal countries, have been included as an Annex in the present paper.

2. Absolute and Relative Sea Level Rise

During the last hundred years the sea level has risen some 10 cm. The IPCC-Working Group I expects a further rise in global sea level in the range of 30 to 110 cm between 1990 and 2100, according to the so-called "business-as-usual" scenario (Warrick and Oerlemans, 1990). The "best estimate" in this range results in 66 cm in the year 2100. This rise in sea level is mainly due to thermal expansion of the upper ocean layers and to melting of glaciers and small ice caps. Depending on the success of various policies to reduce greenhouse gas emissions, other SLR scenarios are projected. Even if actions to limit greenhouse gas emissions are successful, SLR would continue to some extent, due to the warming commitment in the changed atmosphere at the time of emission reduction.

It must be emphasized that the above-mentioned figures on SLR only pertain to absolute values. To these figures local land subsidence must be added to find total relative sea-level rise. In the lower Nile delta subsidence rates range from 20 to 40 cm per century. Elsewhere much higher values occur, e.g. 2 to 6 cm per year due to groundwater withdrawal in Bangkok, Thailand (Chaiyudh and Niwat, 1988), and 2 m in 15 years due to natural gas exploitation in Niigata, Japan (Yamamoto, 1988). Subsidence is already of great concern in many coastal zones. SLR adds to this concern.

3. Coastal Zone Management

There is no unique, fixed definition of the Coastal Zone, let alone of CZM. Only a very few nations, if any, have an integrated CZM on an operational basis. In most countries some CZM-aspects are dealt with on an ad hoc basis, whereas many CZM-aspects are hardly dealt with at all. At present, CZM in many instances is still a concept rather than a reality. Yet the need for some sort of organized CZM is increasingly recognized. World-wide, coastal areas tend to be densely populated, economically productive, and environmentally sensitive. In many instances rapid population growth and economic expansion have led to resource use conflicts and environmental degradation. Consequently, these coastal areas are coming under increasing pressure. Therefore, appropriate coastal resource allocation and harmonizing of resource functions becomes increasingly desirable. This could be achieved in a CZM-framework.

The type, phase, and site-specific problems of coastal zone development show an extremely wide range all over the world, on local as well as on national scales. Consequently, the present development stage of CZM also shows extreme variations in terms of available qualified people, money, tasks, organization, and legal status.

Given this broad variety there are still some aspects which are similar all over the world. One such common factor, which typically appears in an early stage on the widely-varying CZM agendas world-wide, is the need to establish in sufficient detail the highly site-specific hydraulic boundary conditions which form the physical basis for the Coastal Zone: the statistical characteristics of tides, storm surges, waves, river discharge, groundwater levels, etc. These data bear little relevance, unless regarded with respect to the mean sea level including SLR, and to the land elevation, including local subsidence.

Regarded in this wider context, SLR at first sight would appear as merely one marginal aspect, viz a gradual, long-term variation in the local hydraulic boundary conditions, which should however constitute an essential part of the working knowledge in any national or local CZM framework. In this respect, SLR and its implications would "automatically" be put on the CZM agendas all over the world - if there were a sufficiently equipped CZM group in each coastal nation.

Without such a CZM group (the size, the task and the qualifications of which are not discussed here in detail), a proper accomodation of SLR (and all other above-mentioned aspects) in the respective local or national coastal resources development plans will not be warranted. Therefore, in this paper and in the Annex, much emphasis is placed on the need to implement CZM plans in every coastal country. Such CZM groups could be linked to or integrated with existing local or national centres, as long as their task is explicitly maintained. Proper education and training is a must. In the next Sections the importance of establishing CZM groups is illustrated.

4. Coastal Zone Management and Sustainable Development

The need for CZM as indicated in Section 3 in fact reflects the general need for investments in sustainable development. CZM costs money for data collection, studies, quantitative research, planning, and implementation of whatever measures are necessary to form a basis for sustainable development in the Coastal Zone. If this need is so obvious, why is it then that CZM is applied so little and so late? Perhaps because this investment is required at a time when money is least available.

A simple graph (Figure 1) illustrates a typical non-sustainable development process which could represent a nation, a group of nations, or a world average. Total Gross Production (GP) increases, partly due to population growth, and partly due to intense exploitation of resources (for development). However, total Damage (D) increases, too, due to pollution, congestion, over-exploitation of resources, etc. The remaining Net, useful (however defined) part of the Production, NP, shows less and less growth, then stalls, and finally collapses. How may this gloomy scenario be diverted from a catastrophe? Just boosting GP, e.g. by increasing population, seems not to be the solution but is rather a part of the problem. A better option might be the reduction of total Damage D. But when and how? The answer should start with an analysis of D, and this is precisely what the IPCC has done, thereby concentrating on one major aspect of the Damage, i.e. the greenhouse effect as a typical non-sustainable factor. However, even without the greenhouse effect many factors tend to let D grow out of bounds; some of them were always obvious, other factors were hidden or deliberately ignored. A common characteristic of many Damage factors is that their effect is lagging the cause, sometimes by many years or even decades.

Unfortunately, the same kind of time-lag applies to investments in sustainable development. Suppose we invest in CZM now, e.g. an annual amount E in data collection, research, studies, policy analysis and in preparing decision-support tools such as quantitative, predictive models, all in order to curb future Damage. Then it will take some time, ΔT, before useful results, i.e. a Damage reduction ΔD (Figure 2), come out. During this time span we need investments but have no benefits. This may take years or even decades. Two simple examples might illustrate this. In 1953 almost two thousand Dutchmen lost their lives in an extensive storm surge disaster. Plans were made to increase safety against flooding, but it took 35 years before all plans had been implemented, although a masterplan was already available and although safety against flooding by the sea is national priority number 1. As another example, fresh water quality in Holland largely depends on what the Swiss, German and French dump into the Rhine. It took decades

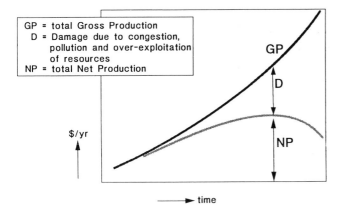

Figure 1: Typical non-sustainable development

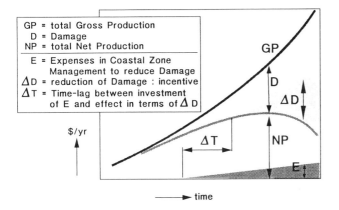

Figure 2: Coastal zone management for survival.

before an international Rhine water treaty came into operation.

Therefore, investments in CZM are typically long-term projects, which fit into long-term sustainable development policy. The sought profit, ΔD, can only be realised if three essential requirements are met:

- the nation must be aware of future trends in GP, D and NP;
- possibilities of Damage reduction must be identified; these are highly site-specific;
- money must be allocated in due time to actually start and improve Damage reduction (or, rather, prevention).

The medium-term threat of SLR is a useful trigger to actually start with CZM, thereby creating or boosting the necessary educational and institutional infra-structure for long-term sustainable development in the Coastal Zone.

5. Multiple Vulnerability of the Coastal Zone due to Climate Change

The Coastal Zone, when viewed from the sea, occupies the frontal position with all the benefits and drawbacks associated with this position. When viewed from the mountains, however, the Coastal Zone is by definition the last phase in the downward flow of the hydrological cycle: the Coastal Zone lies at the extreme downstream end of life-bringing freshwater resources. This position has serious implications for the vulnerability of the Coastal Zone with regard to the expected climate change.

First, direct climate change impacts affect the Coastal Zone, just as other terrestrial zones are affected, e.g. by changes in temperature, precipitation, evaporation, humidity, etc.

Second, the Coastal Zone is where all alterations in upstream hydrologic characteristics, natural or man-made, finally work out. If, e.g., upstream flooding problems are combatted through higher river dikes and drainage, downstream areas have to cope with the excess water. If on the other hand upstream water shortage problems are addressed, upstream measures result in downstream shortage of water. Such resulting changes in downstream river regime will have accompanying effects on downstream groundwater resources and on river sediment transport. This latter factor inevitably has consequences further downstream on river bed dynamics, on related river-bound infrastructure, and finally on coastal stability via the altered sediment supply. Next to these quantitative problems, water quality problems usually add up in downstream direction, too, culminating in a myriad of environmental problems which are caused upstream, but which work out in the Coastal Zone. These riverborne Coastal Zone problems will in general be more serious for longer rivers, crossing more international borders, and flowing through larger drought-prone basins: e.g. the Rhine, Nile, Mekong, Euphrates, Himalaya Rivers, etc.

Third, the indirect world-wide climate change effect of SLR, possibly accompanied by increased storminess, will attack the Coastal Zone from the salt water side. This issue, already very worrying as a stand-alone problem, is actually only a fraction of the combined climate-change induced problems which accumulate in the Coastal Zone.

The conclusion is that the Coastal Zone carries at least a triple burden due to climate change effects. Moreover it is world-wide especially the Coastal Zone where population growth and related economic developments with associated ecological devastation increasingly concentrate. These considerations are important signals to strengthen CZM efforts all over the world but especially in the more vulnerable nations.

6. Vulnerability Assessment: need for a more Quantitative Approach

It seems logical to direct the scarce money for CZM efforts especially to those areas which are most vulnerable. But where are they? The preceding Sections indicate that SLR is only one of many factors which render Coastal Zones vulnerable.

Up till now there is no unified, well-balanced, quantified yardstick available to rank coastal areas or nations in terms of vulnerability to SLR. First because there is no straightforward definition of SLR-vulnerability, which is a very complicated notion indeed; second, because there is a lack of data (DELFT HYDRAULICS, 1989). In this situation early attempts have used the "bath-tub" approach for a first indication of SLR-vulnerability, by estimating the national land area which lies below a certain elevated sea level, as a fraction of total national land area, but without addressing protection options. Refinements to this early approach include, e.g., the fraction of population or GNP which is involved, etc. Another approach estimated world-wide national costs for basic coastal protection against SLR (DELFT HYDRAULICS, 1990), but without estimating potential damage. Clearly, more and better site-specific quantitative assessments are needed to gain a better insight into local and national vulnerability to SLR and to other factors threatening the Coastal Zone.

A step by step improvement of the conceptual definition and subsequent quantitative development of Coastal Zone vulnerability, e.g., in the form of a Vulnerability Index, will serve both national and supranational goals. National, since it will help to allocate national efforts to those locations and problem areas where maximum gain in terms of "Damage reduction" (see Section 4) may be obtained. Supranational, since a world-wide applicable Vulnerability Index may assist in the allocation of the scarce funding money where it is needed most. National CZM teams could play a critical role in this respect.

7. Conclusions and Recommendations

1. Climate Change and Sea Level Rise cause an accumulation of additional problems to world-wide Coastal Zones already under severe stress as a result of congestion, pollution, excessive exploitation, and subsidence.

2. Investment in Coastal Zone Management is critically important for sustainable development.

3. For very vulnerable nations Coastal Zone Management-assistance should be made available by supra-national funding.

4. For proper allocation of national efforts and supra-national funding, an operational Coastal Vulnerability Index should be developed.

5. World-wide implementation of national Coastal Zone Management plans according to a suggested timeline 1991–2000 by IPCC-WGIII, Subgroup Coastal Zone Management (see Annex), needs full support.

REFERENCES

Chaiyudh Khantaprab and Niwat Boonnop: 1988, Urban geology of Bangkok metropolis: a preliminary assessment. In: Geology and urban development; UN/ESCAP, ST/ESCAP/570, Atlas of urban geology, Volume 1, pp. 107–135.

DELFT HYDRAULICS: 1989, Criteria for Assessing Vulnerability to Sea Level Rise: A global inventory to high risk areas. Report H838, for UNEP.

DELFT HYDRAULICS: 1990, A world-wide cost estimate of basic coastal protection measures. Included as Appendix D (pp. 64–119) in: Misdorp et al., 1990, op. cit.

Misdorp, R., Dronkers, J. and Spradley, J.R. (Eds.): 1990, Strategies for Adaption to Sea Level Rise. Intergovernmental Panel on Climate Change, Response Strategies Working Group, Report of the Coastal Zone Management Subgroup. Ministry of Transport and Public Works, Rijkswaterstaat, Tidal Waters Division, Library, P.O. Box 20907, 2500 EX The Hague, The Netherlands.

Warrick, R.A., and Oerlemans, J.: 1990, Sea Level Rise. Chapter 9 in: Climate Change; the IPCC Scientific Assessment. Report prepared for IPCC by Working Group I; J.T. Houghton, G.J. Jenkins and J.J. Ephraums, Eds. Press Syndicate of the University of Cambridge. The Pitt Building, Trumpington Street, Cambridge CB2 1RP, UK.

Yamamoto, S.: 1988, Investigation, assessment, prediction and counter-measures of land-subsidence in Japan with special reference to the case of Niigata. In: Urban geology in Asia and the Pacific; UN/ESCAP, ST/ESCAP/586, Atlas of urban geology, Volume 2, pp. 83–88.

ANNEX

Executive summary and CZM-Chairmen's proposal for future activities quoted from:

> Strategies for Adaption to Sea Level Rise (1990). Intergovernmental Panel on Climate Change, Response Strategies Working Group, Report of the Coastal Zone Management Subgroup. R. Misdorp, J. Dronkers and J.R. Spradley, Eds. Ministry of Transport and Public Works, Rijkswaterstaat, Tidal Waters Division, Library. P.O. Box 20907, 2500 EX The Hague, The Netherlands.

EXECUTIVE SUMMARY
Reasons for Concern

Global climate change may raise sea level as much as one metre over the next century and, in some areas, increase the frequency and severity of storms. Hundreds of thousands of square kilometres of coastal wetlands and other lowlands could be inundated. Beaches could retreat as much as a few hundred metres and protective structures may be breached. Flooding would threaten lives, agriculture, livestock, buildings and infrastructures. Salt water would advance landward into aquifers and up estuaries, threatening water supplies, ecosystems and agriculture in some areas.

Some nations are particularly vulnerable. Eight to ten million people live within one metre of high tide in each of the unprotected river deltas of Bangladesh, Egypt and Vietnam. Half a million people live in archipelagos and coral atoll nations that lie almost entirely within three metres of sea level, such as the Maldives, the Marshall Islands, Tuvalu, Kiribati and Tokelau. Other archipelagos and island nations in the Pacific and Indian Oceans and Caribbean could lose much of their beaches and arable lands, which would cause severe economic and social disruption.

Even in nations that are not, on the whole, particularly vulnerable to sea level rise, some areas could be seriously threatened. Examples include Sydney, Shanghai, coastal Louisiana and other areas economically dependent on fisheries or sensitive to changes in estuarine habitats.

As a result of present population growth and development, coastal areas worldwide are under increasing stress. In addition, increased exploitation of non-renewable resources is degrading the functions and values of coastal zones in many parts of the world. Consequently, populated coastal areas are becoming more and more vulnerable to sea level rise and other impacts of climate change. Even a small rise in sea level could have serious adverse effects.

The Coastal Zone Management Subgroup has examined the physical and institutional strategies for adapting to the potential consequences of global climate change.

Particular attention was focused on sea level rise, where most research on impacts has been conducted. The Subgroup has also reviewed the various responses and has recommended actions to reduce vulnerability to sea level rise and other impacts of climate change.

Responses

The responses required to protect human life and property fall broadly into three categories: retreat, accommodation and protection.

Retreat involves no effort to protect the land from the sea. The coastal zone is abandoned and ecosystems shift landward.

This choice can be motivated by excessive economic or environmental impacts of protection. In the extreme case, an entire area may be abandoned. Accomodation implies that people continue to use the land at risk but do not attempt to prevent the land from being flooded. This option includes erecting emergency flood shelters, elevating buildings on piles, converting agriculture to fish farming, or growing flood or salt tolerant crops. Protection involves hard structures such as sea walls and dikes, as well as soft solutions such as dunes and vegetation, to protect the land from the sea so that existing land uses can continue.

The appropriate mechanism for implementation depends on the particular response. Assuming that land for settlement is available, retreat can be implemented through anticipatory land use regulations, building codes, or economic incentives. Accommodation may evolve without governmental action, but could be assisted by strengthening flood preparation and flood insurance programmes. Protection can be implemented by the authorities currently responsible for water resources and coastal protection.

Improving scientific and public understanding of the problem is also a critical component of any response strategy. The highest priorities for basic research are better projections of changes in the rate of sea level rise, precipitation and the frequency and intensity of storms. Equally important, but more often overlooked, is the need for applied research to determine which options are warranted, given current information. Finally, the available information on coastal land elevation is poor. Maps for most nations only show contours of five metres or greater, making it difficult to determine the areas and resources vulnerable to impacts of a one metre rise in sea level. Except for a few countries, there are no reliable data from which to determine how many people and how much development are at risk. There are many uncertainties and they increase as we look further into the future.

Environmental Implications

Two thirds of the world's fish catch and many marine species, depend on coastal wetlands for their survival. Without human interference, (the retreat option), eco-systems could migrate landward as sea level rises and thus could remain largely intact, although the total area of wetlands would decline. Under the protection option, a much larger proportion of these ecosystems would be lost, especially if hard structures block their landward migration.

Along marine coasts hard structures can have a greater impact than soft solutions. Hard structures influence banks, channels, beach profiles, sediment deposits and morphology of the coastal zone.

Protective structures should be designed, as much as possible, to avoid adverse environmental impacts. Artificial reefs can create new habitats for marine species and dams can mitigate saltwater intrusion, though sometimes at the cost of adverse environmental impacts elsewhere. Soft structures such as beach nourishment retain natural shorelines, but the necessary sand mining can disrupt habitats.

Economic Implications

No response strategy can completely eliminate the economic impacts of climate change. In the retreat option, coastal landowners and communities would suffer from loss of property, resettlement costs and the costs of rebuilding infrastructure. Under accommodation, there would be changing property values, increasing damage from storms and costs for modifying infrastructure. Under the protection option, nations and communities would face the costs for the necessary structures. The structures would protect economic development, but could adversely affect economic interests that depend on recreation and fisheries.

Appendix D of this report (i.e. DELFT HYDRAULICS, 1990) shows that if sea level rises by one metre, about 360,000 kilometres of coastal defences would be required at a total cost of US$500 billion over the next 100 years (This sum only reflects the marginal or added costs and is not discounted). This value does not include costs necessary to meet present coastal defence needs. The estimate does not include the value of the unprotected dry land or ecosystems that would be lost, nor does it consider the costs of responding to saltwater intrusion or the impacts of increased storm frequency. **The overall cost will therefore be considerably higher**. Although some nations could bear all or part of these costs, other nations, including many small island states, could not.

To ensure that coastal development is sustainable, decisions on response strategies should be based on long term as well as short term costs and benefits.

Social Implications

Under the retreat option, resettlement could create major problems. Resettled people are not always well received. They often face language problems, racial and religious discrimination and difficulties in obtaining employment.

Even when they feel welcome, the disruption of families, friendships and traditions can be stressful.

Although the impacts of accommodation and protection would be less, they may still be important. The loss of traditional environments which normally sustain economies and cultures and provide for recreational needs could disrupt family life and create social instability. Regardless of the response eventually chosen, community participation in the decision making process is the best way to ensure that these implications are recognized.

Legal and Institutional Implications

Existing institutions and legal frameworks may be inadequate to implement a response. Issues such as compensation for use of private property and liability for failure of coastal protection structures require national adjudication. For some options, such as resettlement (retreat option) and structures that block sediments (protection option), there are transboundary implications that must be addressed on a regional basis. International action may be required through existing conventions if inundation of land results in disputes over national borders and maritime boundaries, such as exclusive economic zones or archipelagic waters. New authorities may be required, both to implement options and to manage them over long periods of time in the face of pressures for development. National coastal management plans and other new laws and institutions are needed to plan, implement and maintain the necessary adaptive options.

Conclusions

Scientists and officials from some 70 nations have expressed their views on the implications of sea level rise and other coastal impacts of global climate change at Coastal Zone Management Subgroup workshops in Miami and Perth. They indicated that in several noteworthy cases, the impacts could be disastrous; that in a few cases impacts would be trivial; but that for most coastal nations, at least for the foreseeable future, the impacts of sea level rise would be serious but manageable if appropriate actions are taken.

It is urgent for coastal nations to begin the process of adapting to sea level rise not because there is an impending catastrophe, but because **there are opportunities to avoid adverse impacts by acting now**, opportunities that may be lost if the process is delayed. This is also consistent with good coastal zone management practice irrespective of whether climate change occurs or not. Accordingly, the following actions are appropriate:

National Coastal Planning

1. **By the year 2000, coastal nations should implement comprehensive coastal zone**

management plans. These plans should deal with both sea level rise and other impacts of global climate change. They should ensure that risks to populations are minimized, while recognizing the need to protect and maintain important coastal ecosystems.

2. **Coastal areas at risk should be identified.** National efforts should be undertaken to (a) identify functions and resources at risk from a one metre rise in sea level and (b) assess the implications of adaptive response measures on them. Improved mapping will be vital for completing this task.

3. **Nations should ensure that coastal development does not increase vulnerability to sea level rise.** Structural measures to prepare for sea level rise may not yet be warranted. Nevertheless, the design and location of coastal infrastructure and coastal defences should include consideration of sea level rise and other impacts of climate change. It is sometimes less expensive to incorporate these factors into the initial design of a structure than to rebuild it later. Actions in particular need of review include river levees and dams, conversions of mangroves and other wetlands for agriculture and human habitation, harvesting of coral and increased settlement in low lying areas.

4. **Emergency preparedness and coastal zone response mechanisms need to be reviewed and strengthened.** Efforts should be undertaken to develop emergency preparedness plans for reducing vulnerability to coastal storms, through better evacuation planning and the development of coastal defense mechanisms that recognize the impact of sea level rise.

International Cooperation

5. **A continuing International focus on the impacts of sea level rise needs to be maintained.** Existing international organizations should be augmented with new mechanisms to focus awareness and attention on sea level change and to encourage nations of the world to develop appropriate responses.

6. **Technical assistance for developing nations should be provided and cooperation stimulated.** Institutions offering financial support should recognize the need for technical assistance in developing coastal management plans, assessing coastal resources at risk and increasing nation's ability, through education, training and technology transfer, to address sea level rise.

7. **International organizations should support national efforts to limit population growth in coastal areas.** In the final analysis, rapid population growth is the underlying problem with the greatest impact on both the efficiency of coastal zone management and the success of adaptive response options.

Research, Data and Information

8. **Research on the impacts of global climate change on sea level rise should be strengthened.** International and national climate research programmes need to be directed at understanding and predicting changes in sea level, extreme events, precipitation and other impacts of global climate change on coastal areas.

9. **A global ocean observing network should be developed and implemented.** Member nations are strongly encouraged to support the efforts of the IOC, WMO and UNEP to establish a co-ordinated international ocean observing network that will allow for accurate assessments and continuous monitoring of changes in the world's oceans and coastal areas, particularly sea level change.

10. **Data and information on sea level change and adaptive options should be made widely available.** An international mechanism should be identified with the participation of the parties concerned for collecting and exchanging data and information on climate change and its impact on sea level and the coastal zone and on various adaptive options. Sharing this information with developing countries is critically important for preparation of coastal management plans.

CZM-CHAIRMAN'S PROPOSAL FOR FUTURE ACTIVITIES

Introduction

Based on the views of the delegates and the recommendations of the Miami and Perth IPCC-CZMS workshops, the chairmen of the CZM Subgroup and their advisers have undertaken the task to facilitate the implementation of the CZM actions. They suggest that three parallel efforts be undertaken:

1. Data Collection - Efforts to build a current global data base on coastal resources at risk due to sea level rise need to be vigorously pursued. The IPCC-CZM Subgroup has developed a questionnaire which can serve as a first step in the collection of this information and in identifying the countries where additional work needs to be done. It is also suggested that a database or monitoring system be set up which would provide access to and information on adaptation techniques and which could be maintained in an international or regional "clearing house".

2. International Protocol - Efforts should commence immediately on the development of an international protocol to provide a framework for international and multinational cooperation in dealing with the full range of concerns related to impacts of sea level rise

and climate change impacts on the coastal zone. A protocol is needed to both establish the international frames of reference as well as to establish a clear set of goals and objectives. Possible elements contained in such a protocol are outlined in Table 1.

3. Organisational Requirements - A process should be set in motion to guide and assist countries, particularly developing countries in carrying out the IPCC-CZM actions. For this purpose IPCC could consider the formation of a small advisory group to assist in the development of more specific guidelines. Such an advisory group could be formalised at a later stage to support the secretariat for the parties to a future protocol on CZM and sea level rise.

The goals and actions presented in this report are based on problems common to all coastal nations; their achievement can benefit significantly from coordination at the international level.

Table 1: *Possible Elements to be included in a Protocol on Coastal Zone Management and Sea Level Rise*

Signatories endeavour to develop before the year 2000 a comprehensive coastal management programme. Giving priority to the most vulnerable areas, they agree to:

- **provide** support to institutions conducting research on sea level rise and other impacts of climate change on the coastal zone
- **cooperate** in international efforts to monitor sea level rise and other impacts of climate change on the coastal zone
- **contribute** to systematic mapping and resource assessment of coastal zones to identify functions and critical areas at risk
- **support** international initiatives to provide information and technical assistance to cooperating countries for the preparation of coastal management programmes
- **contribute** to the exchange of information, expertise and technology between countries pertaining to the response to sea level rise and other impacts of climate change on the coastal zone
- **promote** public and political awareness of the implications of sea level rise and other impacts of climate change on the coastal zone
- **manage** the coastal zone so that environmental values are preserved whenever possible
- **avoid** taking measures that are detrimental to the coastal zones of adjoining states
- **provide** emergency relief to coastal nations struck by storm surge disasters
- **establish** a secretariat supported by a small advisory group to facilitate the implementation of the protocol agreements.

The three activities described above are considered crucial steps in realizing the full potential of the IPCC process. The Miami and Perth Workshops demonstrated very clearly that many developing nations will not be able to respond effectively to the needs which have been identified, without some form of assistance.

Additionally and in accordance with the primary action for the development of comprehensive coastal zone management plans, a timeline (Table 2) of essential actions for the formulation of such plans is suggested. Countries which do not currently have coastal management plans could use this timeline as a basis for their own planning process over the next decade.

Composition and Functions of a Group of Advisors

Advisory Group: Composition and Functions

In order to facilitate the development of responses to the threat of sea level rise and other impacts of climate change on the world's coastal zones, a functional nucleus of experts is required. Its task should be limited to requests by coastal states for assistance in achieving the goal of having a comprehensive coastal zone management programme in place by the year 2000.

Upon receipt of a request for assistance the IPCC may send an investigative mission to the requesting country or encourage multilateral or bilateral aid organisations to do so. The mission should assess the country's institutional, technical and financial needs and means, i.e. its requirements in these three areas. The advisory group could prepare guidelines for such missions or provide other support, if asked for.

Countries should have the institutional capability to develop their own coastal management programmes and to establish a regulatory framework and the means for enforcement. The required technical capability should be brought to an adequate level by training programmes, expert advice and appropriate equipment. An estimate of the costs involved (excluding equipment) is presented in Table 3.

It should further be determined to what extent the necessary funding can be generated within the country itself and what part could be requested from outside financing institutions.

The mission report referred to above should then be considered against and in the light of worldwide data, synthesized from information supplied by countries with a marine coast. These data should initially be compiled on the basis of the responses to a comprehensive questionnaire sent to all coastal countries and augmented as required.

Finally the group of advisers would report to the IPCC panel on country assessments and priorities in terms of vulnerability to the coastal impacts of climate change and on related institutional needs.

Table 2: *Suggested 10 Year Timeline for the Implementation of Comprehensive Coastal Zone Management Plans*

1991:	Designate (a) national coastal coordinating bodies, (b) national coastal work teams and (c) an international coastal management advisory group to support the IPCC-CZM Sub-group and assist national work teams
1991–1993:	Develop preliminary national coastal man-agement plans; begin public education and involvement
1991–1993:	Begin data collection and survey studies of key physical, social and economic parameters assisted by an international advisory group. For example:

- Topographic information
- Tidal and wave range
- Land use
- Population statistics
- Natural resources at risk

1992:	Adoption of a "Coastal Zone Management and Sea Level Rise" protocol, with a secretariat of the parties, supported by the international coastal management advisory group
1992–1995:	Begin development of coastal management capabilities, including training programmes and strengthening of institutional mech-anisms
1995:	Completion of survey studies, including identification of problems requiring an immediate solution and identification of possible impacts of sea level rise and climate change impacts on the coastal zone
1996:	Assessment of the economic, social, cultural, environmental, legal and financial imp-lications of response options
1997:	Presentation to and reaction from public and policy makers on response options and response selection
1998:	Full preparation of coastal management plans and modifications of plans as required
1999:	Adoption of comprehensive coastal man-agement plans and development of legislation and regulations necessary for implementation
2000:	Staffing and funding of coastal management activities
2001:	Implementation of comprehensive coastal zone management plans

Table 3: *Operational Costs for Implementation of CZM-Actions 1, 6 & 10. Estimated funding to provide the necessary support to meet the year 2000 coastal zone management plan proposal.*

(1)	120 consultant months @ US$ 10,000 per month	= US$	1,200,000
	Expenses and travel	= US$	800,000
			+
		= US$	2,000,000
(2)	Training of 100 in-countries personnel to strengthen coastal zone technical and planning capabilities		
	100 people @ US$ 30,000 each	= US$	3,000,000
(3)	Expenses for secretariat and advisory group	= US$	3,000,000
(4)	Conferences and Workshops	= US$	1,000,000
(5)	Contingency	= US$	1,000,000
			+
Total for 5 years (1992-1997)		= US$	10,000,000

Beginning to Reduce Greenhouse Gas Emissions Need Not be Expensive: Examples from the Energy Sector

by Evan Mills, Deborah Wilson and Thomas Johansson *(Sweden)*

Abstract

Global greenhouse gas emissions can be reduced by 10% from current levels by implementing available cost-effective energy-efficiency improvements, even with continued economic growth and development. Further energy-sector emissions reductions can be achieved by increased use of renewable energy sources and continued research and development of energy-efficient technologies. Applying this approach to energy planning can produce substantial net benefits, in terms of costs and emissions, compared to official forecasts.

Our findings are illustrated with individual technology assessments and with global, national and regional studies, based on measures to increase end-use and supply efficiencies and to use substitute fuels. For the United States, national carbon emissions reductions of 11% are possible with increased energy efficiency and some substitution of natural gas for coal in the electricity sector. These reductions can be achieved with a net economic benefit of US$231/tonne by the year 2000. Increasing end-use efficiency and the use of renewable energy supply sources in Sweden's district heating and electricity sectors can achieve 35% emissions reductions and a net benefit of US$41/tonne by the year 2010, even with the planned phase-out of nuclear power. Applying the same approach to the electricity sector in the State of Karnataka India could avoid a more than two-fold increase in carbon emissions by the year 2000, with a net benefit of over US$3000/tonne. In each of these cases, investing to the point where cumulative investments equal economic savings (i.e. to $0/tonne) offers additional opportunities for emissions reductions.

Support for this work was provided by the Swedish National Energy Administration and the Stockholm Environment Institute.
For constructive review comments, the authors thank D.Abrahamson, K.Begley, R. Bowie, L.Brinck, T. Ekwall, Jo. Goldemberg, L. Gustavsson, P. Hofseth, B. Keepin, J. Koomey, M. Lundberg, H. Nilsson, L. Nilsson, A. Reddy, H. Rodhe, B. Wiman, and two anonymous reviews.

A "wait-and-see" policy on climate change will cost more money than beginning immediately to implement measures that would be economically justified even if there was no climate change problem. To achieve such "no-regrets" emissions reductions will require commitments to new policies aimed at enabling the market to function efficiently and providing supporting legislation where market forces do not suffice.

Introduction

Any strategy to reduce greenhouse gas emissions must involve the global energy system, because fossil-fuel-based energy consumption currently causes ≈50% of global greenhouse forcing.[1] To stabilize atmospheric concentrations of the major greenhouse gases at today's levels will require immediate reductions of over 60% of current anthropogenic carbon-dioxide (CO_2), nitrous oxide and CFC emissions and 15% to 20% reductions for methane.[2] Can such reductions be accomplished, and would they be compatible with developing countries' ambitions and continued economic growth in the industrialized countries? If so, what would they cost?

These questions became especially relevant when the World Energy Conference (WEC) published projections in 1989 suggesting that global primary energy demand would grow by 76% by the year 2020 as economic growth proceeded, and that the associated CO_2 emissions would grow by 69% despite a three-and-a-half-fold increase in nuclear power capacity.[3]

Recent econometric studies suggest that significantly reducing CO_2 emissions will be very expensive.[4] This conclusion tempts policy makers to postpone any response to the climate change problem, and to wait instead for undisputed scientific proof that climate change is occurring. However, while the marginal costs of major reductions may be high, measures with no net cost compared to the "non-action" alternative are available for beginning to reduce emissions.

An energy strategy aimed at reducing greenhouse gas emissions at the least cost can combine different approaches: fuel switching, improved energy-conversion efficiencies and improved energy end-use efficiencies. Many applicable technologies, especially end-use technologies, are presently available, but average efficiency levels do not come close to those of the best products on the market. Ongoing research and development work suggests that many more opportunities to reduce greenhouse gas emissions will become available. In this paper, we describe a variety of existing and emerging energy supply and end-use technologies, their costs and their associated emissions savings, and case studies of their possible application in the United States, Sweden and the State of Karnataka, India. Finally, we present examples of policies to consider for realizing such opportunities.

Specific Opportunities

Before making macro-level analyses covering entire sectors and geographical areas, it is necessary to review and compile data on specific measures for reducing greenhouse gas emissions. The net emissions associated with those measures, compared to a chosen base case, must then be estimated. These estimates can be combined with the incremental costs of choosing each measure, versus those for the base-case measures, in order to find the net cost (or benefits) of reducing emissions.

To provide a more complete assessment of greenhouse gas emissions, we have considered more than just combustion-related CO_2. As a first step, we have included fuel-cycle emissions associated with the mining, processing, transporting and burning of fossil fuels. We have also included greenhouse gases other than CO_2. In some cases more than half of the emissions occur before the point of fuel combustion. Moreover, including a broader range of emissions (expressed in carbon equivalents) can lead to different rankings of measures. We found, for example, that if only CO_2 is counted, compressed natural gas (CNG) automobiles appear "better" than gasoline-fueled cars. However, total greenhouse gas emissions are in fact greater for CNG automobiles after including the CO_2 releases from fuel production and related methane and N_2O emissions. The same reversal of rankings occurs for vehicles burning coal- and natural-gas-based methanol.

This section highlights some examples of the cost-effectiveness of measures to reduce greenhouse gas emissions, using economic evaluations reflecting a societal perspective. We compare fuels and technologies based on their costs without taxes and we use a societal discount rate of 6% (real) for calculating annual costs of supply and end-

use-efficiency investments. Our method of making these calculations is developed in Appendix A and a collection of examples is provided in Appendix B.

We find that all of the currently available end-use efficiency measures described in Appendix B (applicable to lighting, passenger cars, appliances, space heating, motors, etc.) reduce emissions while providing a net economic benefit (i.e. savings) in the $100 to $500/tonne range.[5] Switching to biomass-based fuels for heating can also reduce emissions and provide a net economic benefit. Examples of biomass fuels include forest residues, agricultural wastes, by-products from the paper and pulp and sugar industries, and materials grown on plantations specifically for energy purposes. Burning renewably-grown biomass feedstocks results in zero net carbon emissions because the carbon emitted from burning the fuel is just equal to the carbon reabsorbed from the atmosphere as replacement biomass is grown.

For a given fuel, advanced commercially available electric power plants can also achieve emissions reductions at no net cost. Switching from coal to high-efficiency natural-gas-fired power plants avoids emissions and produces a net benefit of ≈$100/tonne of carbon equivalent. Switching to renewables at today's prices tends to result in net costs, but is expected to provide net benefits of ≈$100/tonne within 10 to 20 years.[6] Avoiding emissions by building new nuclear power plants would cost about $40/tonne of avoided carbon-equivalent emissions under today's US conditions, versus a benefit of $11/tonne for the industry's target costs.[7]

Efficiency Opportunities

There are many cases in which a variety of technologies exist that serve the same function (i.e., provide the same service) but require very different amounts of energy to do so. Efficient light bulbs, refrigerators, cars, motors, pumps, fans, and compressors are just a few examples. There is a large global potential for reducing energy requirements by introducing the most efficient technologies as replacements for the current equipment stock in the natural turnover cycle and as the stock grows. Many studies have been carried out that estimate the energy-efficiency gains and emissions reductions achievable by applying efficient technologies. Physical and technical limits for these technologies are still far from being realized in most cases, thus continued research and development can continue to provide large returns.

Lighting

Lighting typically comprises 10 to 20% of electricity demand in developing and industrialized countries. Today, indoor lighting is often provided by incandescent lamps. As an example of achieving lighting services with reduced energy input, a 15-watt compact fluorescent lamp can be used to replace a 75-watt incandescent lamp. The compact fluorescent lamp provides the same energy service (lumen output), but requires five-times less electricity to do so.

Efforts to promote energy efficiency often focus on lighting. In Europe, 32 utility programs that used financial incentives to promote the use of compact fluorescent lamps achieved an average cost of conserved energy of 2.1 cents/kWh, including all direct and indirect costs for lamps, administration, advertising, postage, etc.[8]

Significant amounts of energy can also be saved in conventional fluorescent lighting systems. The Swedish utility Vattenfall recently completed a lighting retrofit demonstration at their headquarters. By modifying their existing fluorescent systems to use efficient lamps, ballasts, daylighting controls, occupancy sensors and other measures Vattenfall achieved a cost-effective 71% reduction in the electricity used for lighting. The quality of office lighting was also improved markedly.

Passenger Cars

The global fleet of gasoline-fueled vehicles emits larger quantities of greenhouse gases than any other category of end-use devices. In the United States, the production, distribution and combustion of transportation fuels account for approximately 27% of all fossil-fuel-related CO_2 emissions.[9] Efficiency improvements effectively avoided more than 500 MT of carbon-equivalent emissions between 1974 and 1985 (inclusive). Emissions from new gasoline-fueled cars can be cut by over three-fourths through higher efficiencies, and can be virtually eliminated with vehicles that utilize non-fossil fuel sources.

Significant efficiency improvements have already been achieved in passenger cars: the average nominal (laboratory test) fuel efficiency for new passenger cars sold in the United States improved by 50% (to 8.4 liters/100km) between 1973 and 1986.[10] Ninety percent of these efficiency gains can be attributed to fuel-efficiency improvements as opposed to a shift in vehicle mix.[11]

A wide range of technologies are available today for further increasing the fuel-efficiency of new automobiles,[12] some of which can bring about improvements with negative to small positive costs. From an efficiency standpoint, one of the most advanced passenger-car prototype is the Toyota AXV. This four- to five-passenger car has a fuel efficiency of only 2.4 1/100km.[13] A similar example is the Light Component Project prototype (LCP-2000) produced by Volvo. This four-passenger automobile passes the world's most stringent crash and emissions tests (i.e., California's), achieves a fuel efficiency of 3.6 1/100km, and was designed to cost about the same to manufacture as an average subcompact car at a production level of 20,000 cars/year.[14] The Elsbett engine used in this car can be operated with biomass-derived fuels as well as gasoline.[15]

Many of the technological advances that can reduce fuel consumption in gasoline-fueled cars will be applicable to methanol cars as well. Because of the better combustion properties of methanol, methanol vehicles have an intrinsic potential for achieving fuel efficiencies that are up to 25% better than those for gasoline and compressed natural gas (CNG) vehicles.[16]

Efficient electric vehicle technology is in a rapid stage of development. General Motors, for example, has announced its intention to market its prototype electric car: the Impact. This two-passenger car is expected to consume a mere 0.07 kWh/km (a gasoline equivalent of 1.7 1/100km): only 14% of the energy consumption of the 1987 U.S. stock-average gasoline-fueled vehicle.[17] As in the case of methanol, the amount of emissions from electric vehicles depends on the feedstock used to produce the electricity.

Supply Opportunities

Modernization of Power Generation

The average conversion efficiency from primary fuels to electricity in new power plants has slowly improved in recent years. Vigorous R&D efforts underway during the 1980s resulted in maximum efficiencies of ≈45% (on a higher heating value basis) for commercially-available combined-cycle power plants fueled with natural gas.

The choice of fuel and conversion technologies has an important effect on the amount of greenhouse gas emissions from electric power production (Figure 1). Replacing average conventional coal-fired steam plants with pressurized fluidized-bed combustion plants (PFBC, also coal based), reduces emissions by almost 20%. A further 48% reduction can be achieved by switching to natural gas and the best available combined-cycle power plants.

Power plant conversion efficiencies could be increased beyond currently available levels with continued development of aeroderivative gas turbines. Through a combination of measures, such turbine designs currently reach 54% in natural-gas-fired systems.[18] The capital costs of these power plants ($500/kW) are estimated to be lower than those of new central station, coal steam plants with flue gas desulphurization, nuclear or hydroelectric power plants.

The simultaneous production of electricity and useful heat (cogeneration) offers still higher emissions reduction possibilities. The total system efficiencies of some cogeneration applications are well above 80%. Total system emissions can be minimized by combining the best available technologies to meet a given demand for electricity and heat. Figure 1 compares cogeneration systems with central-station plants that provide no useful heat. Cogeneration based on natural gas offers an additional

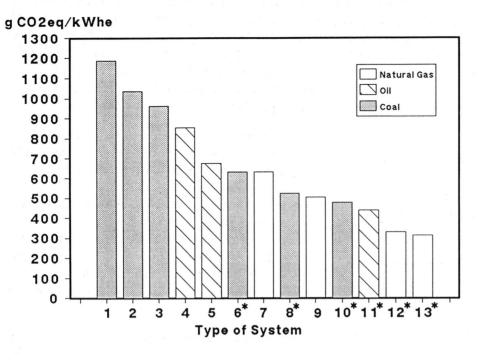

* = cogeneration

Legend

1. Average conventional steam turbine (coal, Eff. 34%); 2. Best steam turbine (coal, Eff. 39%); 3. Pressurized fluidized bed combustion (PFBC) (coal, Eff. 42%); 4. Average conventional steam turbine (oil, Eff. 38%); 5. Best available combined-cycle gas turbine (oil, Eff. 48%); 6. Cogeneration: average conventional steam turbine (coal, Eff. 78%, E/H = 0.50); 7. Average combined-cycle gas turbine (nat. gas, Eff. 36%); 8. Cogeneration: best available steam turbine (coal, Eff. 83%, E/H = 0.60); 9. Best available combined-cycle gas turbine (nat. gas, Eff. 45%); 10. Cogeneration: pressurized fluidized bed combustion (coal, Eff. 86%, E/H = 0.65); 11. Cogeneration: best available steam turbine (oil, Eff. 81%, E/H = 0.60); 12. Cogeneration: steam-injected gas turbine (nat. gas, Eff. 75%, E/H = 0.80); 13. Cogeneration: best available combined-cycle gas turbine (nat. gas, Eff. 77%, E/H = 1.0)

Figure 1: Greenhouse gas emissions vary substantially among commercially available technologies for producing heat and power. Central-station power plants are compared with cogeneration plants providing both useful heat and power. The energy requirement for electricity production using cogeneration technologies is taken as the total energy supplied minus that which would have been required to produce the heat independently (assuming a boiler efficiency corresponding to a lower-heating-value of 90%). All power plant efficiencies (Eff.) are based on higher heating values. For the cogeneration systems, the power-to-heat ratios are given as E/H. The greenhouse gases are expressed as an equivalent amount of carbon dioxide (CO_2eq/kWhe). Methane and methane-related fuel-cycle emissions from coal, oil, and natural-gas consumption are taken into account (see Appendix A).

38% emissions reduction beyond the best available combined-cycle plants fired with natural gas. The plants with the lowest emissions also have the lowest costs.

The emergence of highly efficient gas turbines has provided new opportunities for utilizing gasified solid fuels such as biomass and coal. Advanced gas turbines using gasified coal have been demonstrated in coal-integrated-gasifier/gas turbines (CIG/GT). Advanced gas turbines fired with gasified wood (BIG/GT) have not been demonstrated, but detailed design studies indicate that it may be possible to commercialize them more quickly than CIG/GT schemes. This is because biomass is inherently cleaner and easier to gasify than coal and because the scale of demonstration plants is also appropriate for commercial applications.[19]

Modernization of the Cane Sugar and Alcohol Industries
The sugar cane industries in developing countries offer a particularly promising area for the application of BIG/GT technologies. Sugar cane production yields two kinds of biomass fuel suitable for gasification: bagasse and barbojo. Bagasse is the residue from crushing the cane, and is thus available during the milling season; barbojo is the tops and leaves of the cane plant that could be harvested and stored for use after the milling season. Ogden et al. calculate that by the year 2027 the 80 sugar cane producing developing countries could generate 70% more than their total 1987 electricity production from all sources. Moreover, the costs would be competitive with conventional sources of electricity based on fossil fuels.

The ethanol simultaneously produced from cane would be equivalent to about 9% of total oil use in all developing countries in 1987. Ogden et al. point out that were the electricity produced instead from coal, and were the alcohol and methaneused instead of gasoline in the transport sector, the additional CO_2 emissions would equal nearly half of the total 1986 emissions from fossil fuels in developing countries.[20]

Solar Electricity

Currently, about 1500 MW of wind capacity and 300 MW of solar thermal capacity is installed in the United States, mostly in California. The electricity costs from these systems are in the 5 to 10 cents/kWh range, and ongoing development work is expected to lower costs to the point where they will be cost-competitive for bulk power generation.[21] Perhaps the most promising solar technology is the solid-state photovoltaic (PV) cell, which converts sunlight directly to electricity. The costs are presently 30 to 35 cents/kWh, and projected by five US National Laboratories to drop to 4 to 5 cents/kWh by 2020 to 2030.[22] About 40 MW of photovoltaic capacity is currently installed worldwide, with the greatest market in remote applications where the high costs are acceptable. Projections based on development work in the PV industry suggest that photovoltaics will become competitive with fossil-fuel power plants within a decade:

> "The development of large-scale, computer-integrated manufacturing lines should decrease the manufacturing costs of amorphous silicon solar cells to less than $1/W_p$ by the early 1990s, leading to the development of both residential and central station utility applications in the mid to late 1990s."[23]

Global, National and Regional Examples

What are the likely future economic and emissions impacts of individual measures such as those we have described, if combined and applied at the global, national or regional levels? Evaluations aimed at answering this question must begin by assuming a certain level of economic growth. The economic growth rates (and embedded assumptions regarding growth in material output and demographic factors) can then be used to project the future demand for energy services. Once a demand level for energy services is determined, specific technologies for providing them can be evaluated and compared. Before proceeding to scenarios, we make a few points about projecting the future demand for energy services.

Energy, Development and Economic Growth

History has shown that the link between economic growth and growth in energy demand has been broken, and there are good reasons to believe that there need not be a re-coupling. In contrast to the trends for total energy consumption, electricity use has tended to grow faster than

gross domestic product (GDP). However, growth in electricity/GDP in many countries including the United States, the United Kingdom, Japan, and West Germany slowed to zero in the mid-1970s and declined thereafter.[24]

The perhaps counter-intuitive decoupling of energy demand and GDP growth occurs as a result of three on-going types of change:

1. structural changes in consumption patterns, towards fewer energy- and materials-intensive products and activities, that take place as development in a country proceeds
2. technological changes that lead to less energy- and materials-intensive products
3. improvements in energy conversion and end-use efficiencies.[25] The combined effect of these changes has been to reduce national energy intensity, even where there are no formal policies to promote the changes.

Significant errors can be introduced into energy demand projections if these fundamental trends are ignored. For example, even if developing countries choose to follow the conventionally projected development path, structural and technological trends will enable them to be much less energy-intensive than their northern neighbors. With an effective set of national and international policies, the trend toward reduced energy intensity can be accelerated in both industrialized and developing countries.

A Global View

Goldemberg et al. found that by using efficient end-use technologies all of the energy services associated with a significant increase in the standard of living in the developing and industrialized countries alike can be attained by the year 2020 with only a 10% increase in global primary energy use.[26] This is much lower than the increases in global energy use projected by the WEC and others. The scenario results in a 10% decline in fossil-fuel CO_2 emissions compared to 1980, using a conventional mix of energy sources, but relatively less reliance on coal. To achieve greater reductions, end-use efficiency measures need to be combined with increased use of non-fossil energy sources.

Industrialized Countries

An Example for the United States

The United States reduced its energy intensity by approximately 25% between 1973 and 1986. Today's supply and end-use technologies offer the prospect for significant additional efficiency gains.[27]

Possibilities for avoiding carbon emissions in the near term are shown as a function of cost for the United States in Figure 2. Each step in the figure corresponds to an end-use efficiency measure or a supply strategy that saves energy

$/tonne C

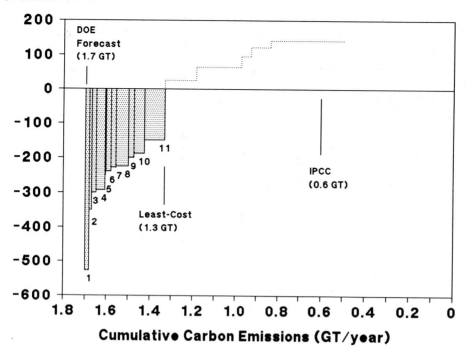

Legend

1. Raise the federal gasoline tax by 12 cents per liter within five years and spend part of the revenue on mass transit and energy-efficiency programs.; 2. Use white surfaces and plant urban trees to reduce air conditioning loads associated with the summer "heat island" effect in cities.; 3. Increase the efficiency of electricity supply through development, demonstration, and promotion of advanced generating technologies.; 4. Raise car and light truck fuel-efficiency standards, expand the gas-guzzler tax, and establish gas-sipper rebates: new cars average 5.2 l/100km and new light trucks average 6.7 l/100km by 2000.; 5. Reduce federal energy use through life-cycle cost-based purchasing.; 6. Strengthen existing federal appliance efficiency standards.; 7. Promote the adoption of building standards and retrofit programs to reduce energy use in residential and commercial buildings.; 8. Reduce industrial energy use through research and demonstration programs, promotion of cogeneration, and further data collection.; 9. Adopt new federal efficiency standards on lamps and plumbing fixtures.; 10. Adopt acid rain legislation that encourages energy efficiency as a means for lowering emissions and reducing emissions control costs.; 11. Reform federal utility regulation to foster investment in end-use energy efficiency and cogeneration.

Figure 2: The x-axis shows total national carbon emissions reductions achievable through the adoption of the 11 cost-ranked measures listed below. The upper limit (1.7 GT) represents the current US DOE forecast for the year 2000. The "IPCC" label indicates the level of reductions necessary to stabilize atmospheric concentrations of greenhouse gases, according to the Intergovernmental Panel on Climate Change. The y-axis indicates the net cost of implementing each measure. Negative costs reflect a net economic benefit compared to the DOE forecast. The average net economic benefit of avoided emissions for steps 1-11 is $231/tonne.[51]

and reduces emissions at a given net benefit or cost compared to the current US Department of Energy (DOE) forecast for the year 2000. The forecast assumes 2.5% real economic growth.

In the DOE forecast, US carbon emissions would increase by 200 million tonnes/year by the year 2000 (13% of 1988 emissions). In contrast, 11 policy strategies for the year 2000 would lead to an emissions reduction of 170 MT (11%) from 1988 levels - similar to the global opportunity just described--and a cost reduction of $85 billion/year compared to the DOE forecast. The emissions reductions are accompanied by an average economic benefit (savings) of $231/tonne.[28]

These results reflect only 11 specific policies that pertain mostly to improved end-use efficiency, and are achievable

over a period of 10 years. Renewable supply-side strategies are not included. Thus, the avoided carbon emissions represent only part of the long-run potential. The use of biomass in electric power and heat production, for example, has not been included and would lead to further emissions reductions. This option is illustrated by the following example of an integrated supply-demand approach.

An Integrated Example for Sweden

The situation in Sweden presents an interesting case of balancing climate change issues with other energy policy goals. In 1980, the citizens of Sweden conducted a public referendum calling for a full phase-out of nuclear power, which led to a subsequent parliamentary decision to phase out the country's twelve nuclear power plants by 2010.

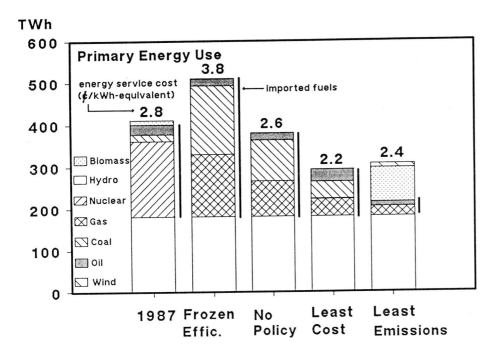

Key to x-axis

1. **Frozen-Efficiency Baseline.** End-use efficiencies do not improve beyond 1987 (base-year) levels. Electricity demand is met with existing non-nuclear power plants and by new efficient power plants half fueled by coal and half by natural gas. Cogeneration of heat and power is used extensively, both for industry and municipal district heating. 2. **No-Policy Scenario.** In this scenario, no new policies are implemented to increase the efficiency of electricity use or to increase the use of renewable supply sources. The scenario includes only those efficiency improvements that are expected to result from a cost-driven average 50% increase in real electricity prices. The energy supply mix for electricity production is the same as in the Frozen-Efficiency baseline. 3. **Least-Cost Scenario.** This scenario goes beyond the No-Policy scenario to show the impact on electricity demand if the most efficient technologies (for appliances, motors, lighting, etc.) available today or near commercialization were introduced at the natural rate of capital turnover up to the year 2010. Also included is some fuel switching in heating systems that can currently use electricity in combination with other fuels. Only those measures costing less than the electricity they save are employed. The energy supply mix from the preceding scenario is retained. 4. **Least-Emissions Scenario.** This scenario begins with the end-use measures included in the Least-Cost scenario and introduces gasified biomass fuel for electricity production (replacing fossil fuels). A small amount of wind-generated electricity is also included. Natural gas is used after available biomass resources are allocated.

Figure 3: Swedish primary energy, energy-service costs (efficiency investments and purchased energy) and energy import dependence: 1987 and scenarios for 2010. All services derived from electricity, district heat, and 15 TWh of industrial process heat are incorporated. Electricity from hydroelectric, nuclear, and wind power is converted to its fuel equivalent using a 36% thermal efficiency. Electricity demand increases 50% in the Frozen Efficiency baseline by the year 2010, whereas in the Least-Cost and the Least-Emissions scenarios demand is 25% lower than 1987 levels. In contrast, electricity use increases by 9% in the No-Policy scenario.

Today Sweden uses more nuclear-generated electricity per capita than any other country. Sweden has also committed itself to holding CO_2 emissions at or below 1988 levels and to abstain from constructing hydroelectric plants on the country's four remaining wild rivers. Hence, Sweden offers an acid test of energy planning in the face of potentially conflicting national policy goals.

Possibilities for meeting these goals were identified in a detailed assessment recently completed by Vattenfall (the Swedish State Power Board) and the University of Lund.[29] The boundary conditions encompass the energy services provided in the base year (1987) by electricity, cogeneration, district heating, and some industrial process heat: 63% of total primary energy supply in Sweden.[30] The scenarios incorporate the structural and demographic expansion associated with an anticipated 54% increase in real gross national product to the year 2010, as assumed by the Ministry of Finance.

Figure 3 shows the types of fuels used to supply the electricity and heat required in four scenarios, the portion of this energy that is imported, and the resulting costs per unit of energy services provided (for purchased energy plus efficiency investments). Energy import dependence drops from today's value of 54% to 38% in the Least-Cost scenario and to 11% in the Least-Emissions scenario.

The emissions avoided by each measure in the Least-

$/tonne Ceq

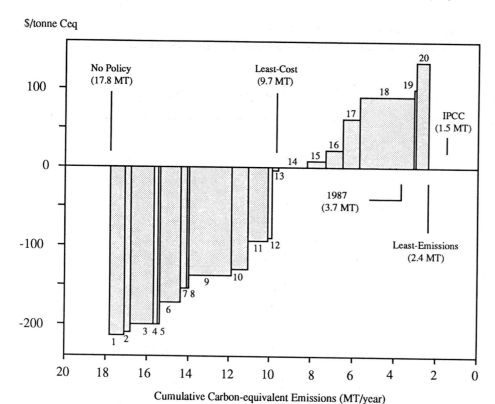

Cumulative Carbon-equivalent Emissions (MT/year)

Legend

1. Fuel switching to oil from electricity; 2. Fuel switching to gas from electricity; 3. Electronic office equipment; 4. Conversion of large heat pumps; 5. Construction of efficient new homes; 6. Efficient appliances (excluding lighting); 7. Efficient lighting; 8. Efficient lighting: residential incandescents; 9. Efficient motors, pumps, fans, compressors, etc.; 10. Efficient lighting--com'l & ind'l fluorescents; 11. Miscellaneous efficiency improvements; 12. Efficient commercial food preparation; 13. Heat pump--existing electric boiler; 14. Supplemental district heating (biomass); 15. Industrial cogeneration with biomass; 16. Cogeneration, biomass--(versus existing cogen.); 17. Cogeneration, biomass--(versus new central commercial and industrial incandescents station); 18. Cogeneration, biomass--(versus new cogen.); 19. Heat pump--existing direct electric; 20. Wind turbines.

Figure 4: The x-axis shows greenhouse gas emissions in Sweden from electric-power and heat production and some industrial process heat at various levels of energy efficiency and assuming various fuel sources. The scenarios indicated along the x-axis are defined in the Figure 3 caption. All scenarios include a complete phase-out of nuclear power by 2010. The upper limit (17.8 MT) represents the No-Policy scenario for the year 2010. The 20 numbered steps show possible emissions-reduction measures down to the Least-Emissions scenario level. The y-axis indicates the net cost of achieving the emissions reductions for each measure. Negative costs reflect a net economic benefit compared to the No-Policy case. The average net benefit of avoided emissions for steps 1-20 is $41/tonne. The "IPCC" label indicates the level of reductions necessary to stabilize atmospheric concentrations of greenhouse gases, according to the Intergovernmental Panel on Climate Change. The method for calculating carbon-equivalent emissions for various fuels is described in Appendix A.

Emissions scenario are depicted in Figure 4, where various end-use measures and supply-side measures are shown as steps ranked in order of increasing cost of avoided emissions. The No-Policy scenario results in annual emissions of 17.8 MT carbon-equivalent in the year 2010, a sixfold increase from 1987 levels. These emissions are reduced by 44% in the Least-Cost scenario but are 165% higher than base-year emissions. This situation is amended in the Least-Emissions scenario by bringing biomass (and a small amount of wind-generated electricity) into the supply mix. In this scenario, emissions decline to only 2.4 MT: 35% lower than actual 1987 emissions.

The projected (No-Policy scenario) emissions are avoided at an average net economic benefit to society of $41/tonne. This net benefit, however, is $102/tonne carbon-equivalent less than the benefit derived from the Least-Cost case. The $102/tonne difference between the two cases can be viewed as the net economic cost (value) to Sweden of achieving reductions beyond the point where the individual measures can be implemented with no net cost to society.[31] The carbon taxes to be introduced in Sweden in 1991 are $150/tonne, suggesting that the $102/tonne premium is well within the bounds seen as reasonable by the Swedish government.

Table 1: *Alternative commercial energy use scenarios for developing countries.* [49]

	1985	2025 [a]		
		Goldemberg et al.	RCWP [b]	RCW [b]
Final Energy (EJ)				
Fuel	55.0	149.4 (2.5)	150.0 (2.5)	170.1 (2.9)
Electricity	5.6	26.5 (4.0)	35.1 (4.7)	39.9 (5.0)
Total	**60.6**	**175.9 (2.7)**	**185.1 (2.8)**	**210.0 (3.2)**
Primary Energy (EJ)				
Fossil Fuel	71.0	124.5 (1.4)	132.6 (1.6)	242.8 (3.1)
Hydro	6.5	20.5 (3.7)	33.2 (4.3)	28.9 (3.9)
Nuclear	0.7	1.4 (1.9)	13.8 (7.7)	14.3 (7.8)
Biomass	0.0	76.5	76.5	4.8
Solar	0.0	6.7	17.0	5.2
Total	**78.2**	**229.6 (2.7)**	**273.1 (3.2)**	**296.0 (3.4)**
Carbon Emissions (GT)	1.5	2.2 (1.0)	2.9 (1.7)	5.4 (3.3)

[a]   Average growth rates (in %/year) for the period 1985-2025 are given in parentheses.

[b]   These are the projections made by the US Environmental Protection Agency (EPA) for its "Rapidly Changing World" (RCW) scenario and its "Rapidly Changing World with Policy" (RCWP) scenario.[50]

Developing Countries

Future greenhouse gas emissions in developing countries are of special concern because of the rapidly growing demand for energy services there. Energy, however, need not constrain development of the material standard of living in these countries.[32] Encouraging technological "leapfrogging" to reduce the cost of energy services (i.e., the adoption of technologies that are more efficient than the average in industrialized countries) must become a major objective in development policy.

Two scenarios for developing countries (for supply and demand respectively) indicate that fossil-fuel carbon emissions can be constrained to 2.2 GT/year in 2025.[33] In the supply scenario, natural gas and oil each contribute one-third of the commercial fuel supply, and most of the balance is met with biomass (Table 1). In comparison, the US Environmental Protection Agency (EPA) has estimated that emissions in developing countries will grow to 5.4 GT/year by 2025.[34] The EPA estimate does not include new policy initiatives aimed at reducing emissions.

An Integrated Example for Karnataka, India

The official Long Range Plan for Power Projects (LRPPP) for the electricity sector in Karnataka projects a more than threefold expansion of electricity generation and consumption between 1986 and 2000. In meeting this projection, the plan calls for annual expenditures of $3.3 billion per year and expansion of coal-fired electricity generation capacity, resulting in a 0.83 million-tonne (127%) increase in annual carbon emissions from current levels.

An alternative scenario developed for Karnataka by the Department of Management Studies at the Indian Institute of Science would provide a much higher level of electricity services than the LRPPP plan.[35] Electricity demand in this scenario, however, only doubles between 1986 and 2000, as a result of investments in efficiency-improvement measures and the introduction of solar water heating and LPG stoves. The additional demand is met with cogeneration in sugar factories, mini/micro hydroelectric systems, and decentralized rural power generation based initially on biomass and later on photovoltaics. Annual expenditures of $0.6 billion are called for, and annual carbon emissions increase by 0.004 million tonnes. The net economic benefit of this scenario compared to the LRPPP plan is over $3000 per tonne of avoided carbon emissions. The scenario illustrates that, even in developing countries, increasing electricity service levels and, thereby, living standards need not cause emissions increases.

Policies

Although some of the technologies incorporated in the scenarios described above would be adopted under current market conditions, others would not. The opportunities lost when inefficient new vehicles, equipment and buildings are added to the stock are of primary importance.

Existing markets can not be relied upon to lead to the sufficient use of technologies that cost-effectively reduce greenhouse gas emissions because:

a) biases towards conventional supply systems lead to underinvestment in energy efficiency

b) consumers lack sufficient information, financing and access to the full spectrum of available equipment

c) energy prices normally do not reflect marginal costs or the costs of externalities, such as climatic change.

Despite the fact that the global end-use scenario described earlier is consistent with plausible values of income and price elasticities, and with energy prices not much higher than those at present,[36] market failures are steering the energy sector towards a future with much higher energy demand and greenhouse gas emissions. New policies are required to remedy this situation.

The kinds of policies described below are within the sphere of government-level policy making. We indicate the general nature of such policies and offer a few examples of their application. Specific policies and their implementation must be chosen to fit the cultural, political and market conditions of each nation.[37]

Redefining the mission of energy suppliers and creating new markets for energy efficiency or emissions offsets can help build markets for energy services rather than energy *per se*.[38] Some US utilities have more than a decade of experience with operating informational and financial-incentive programs to promote energy efficiency.[39] As noted earlier, a number of European electric utilities have used financial incentives to promote energy-efficient lighting. In a broad-based initiative aimed at many sectors and end uses, the state-owned Swedish utility Vattenfall will soon begin investing as an energy service company. Vattenfall plans to spend $150 million (1 billion kronor) over the next five years to support the kinds of end-use technologies described earlier in the Least-Emissions scenario for Sweden. Some electric utilities in the US have gone a step farther by implementing innovative auction systems in which vendors of energy efficiency services are able to compete with power plant suppliers.

Creating markets for tradeable emissions offsets is another promising and economically efficient way to stimulate investments in energy efficiency. This is exemplified by recent amendments to the US Clean Air Act that set limits on total SO_2 emissions by individual utilities. Utilities emitting less than their SO_2 allowance - e.g. by increasing energy efficiency--can sell the excess emissions rights to other utilities.

Governments and other large buyers of energy-using equipment can use innovative procurement systems to promote the design and commercialization of more efficient products. The Swedish National Energy Administration, for example, has cooperated with the largest building-management companies in Sweden to submit efficiency-oriented procurement proposals to appliance manufacturers. As a first step, this group has invited the manufacturers to produce new refrigerator-freezer designs that feature improved energy efficiency. The result has been positive: new units will be commercialized in 1991 that achieve 55% lower electricity use than the most efficient models now on the market, while eliminating CFCs from the insulation.[40]

Emissions taxes can be used to signal the need to lower emissions, to fund implementation of measures that achieve emission reductions and to create funds for research, development and demonstration. To generate US$30 billion/year would only require a $1/bbl-equivalent charge on fossil fuels.[41] Such a tax would be important, but, for institutional and other reasons, the charge would not in itself create a significant impact on the market.[42]

Energy performance standards should be introduced when market failures block the attainment of optimal end-use efficiency goals. Automobile efficiency opportunities offer an instructive example. The total cost of owning and operating an automobile is essentially constant over a wide range of fuel economies,[43] resulting in the lack of an incentive for improved energy efficiency. As a remedy for this type of problem, standards have been used in many countries. For example, the United States has fuel-efficiency standards for automobiles, buildings and household appliances, and standards are pending at the state or national level for motors, lighting and other technologies.[44] Energy-performance standards are a logical extension of safety and other standards now in place to protect both consumers and the environment. Society benefits from energy-performance standards that maintain energy service levels while yielding a reasonable return on the investments associated with increased energy efficiency.

Research, development and demonstration (RD&D) priorities can be changed to reflect the most promising strategies for combating climate change. The current funding priorities of International Energy Agency member governments are indicated in Figure 5. Nuclear energy is receiving over half of the funding, despite its limited potential to reduce greenhouse gas emissions or the cost of

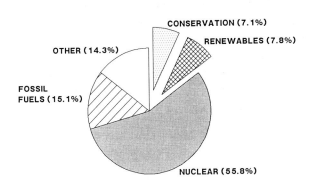

ENERGY RD&D BUDGETS: 1988
IEA Countries: (USD 6.88 billion)

CONSERVATION (7.1%)
RENEWABLES (7.8%)
OTHER (14.3%)
FOSSIL FUELS (15.1%)
NUCLEAR (55.8%)

Figure 5: Research, development and demonstration funding by International Energy Agency governments, 1988.[52] IEA Countries: USD 6.88 billion.

energy services,[45] and despite the fundamental security problem of creating weapons-proliferation-proof nuclear energy systems.[46] As of 1988, end-use efficiency received only 7% of the countries' RD&D budgets and renewable energy sources received 8%. The share of total RD&D allocated to energy efficiency and renewable energy declined between 1977 and 1988.[47] In a strategy to abate the greenhouse effect, renewable sources and energy efficiency should become the major focus for government-sponsored research and development because of their long-term potential to reduce emissions, their security-enhancing benefits, their present costs and the small size of the existing industrial base.[48]

Observations and Conclusions

The examples we have shown for the United States and Sweden suggest that industrialized countries can significantly reduce their energy-sector greenhouse gas emissions within 10 to 20 years by implementing measures with no net economic costs to society. Our example from a developing country (the State of Karnataka in India) suggests that emissions can be constrained to current levels at a net economic benefit, even with a doubling of electricity supply and even greater growth in the level of energy services delivered.

A commitment to implementing policies can be justified on its economic merits alone. A wait-and-see policy is economically inefficient, and also forgoes non-economic benefits to society, such as enhanced national/international security and the reduced environmental impacts beyond those associated with climate change.

Our results show much lower costs for reducing greenhouse gas emissions than those arrived at using econometric models. Our methods differ from econometric

methods in several important respects. We focus on emissions reductions that can be achieved now while ensuring a reliable supply of the energy services required for desired development and economic growth. Taking an end-use perspective enables us to incorporate new factors that cannot be accounted for by models based on observations of the past. It also enables us to identify existing market failures and to analyze policy options aimed at making markets work better.

Despite our conclusion that something, indeed quite a bit, can be done, the strategies that we have described are not sufficient to achieve the ~60% emissions reductions required to stabilize atmospheric concentrations of greenhouse gases in the long-term. Ongoing and accelerated technological development, reforestation and structural and behavioral changes not analyzed in this paper offer prospects of further emissions reductions. It is important to begin now with measures that are economically justified, rather than waiting until a detailed strategy for meeting long-run emissions-reductions targets can be developed.

Appendix A

Method for Calculating the Cost of Avoiding Greenhouse Gas Emissions

Quantifying and Comparing Greenhouse Gases Other Than Carbon Dioxide

In comparing the emissions from various fuels and technologies, we include the known effects of carbon dioxide (CO_2), methane (CH_4) and tropospheric ozone (O_3) from methane oxidation and, in the case of automobiles, nitrous oxide (N_2O).[53] We incorporate fuel-cycle emissions associated with mining, processing, transporting and burning the fuels. The greenhouse properties of each gas are expressed in terms of carbon-equivalents and added together. We use natural-gas leakage rates of 1% of production for new distribution systems. Emissions from on-site energy and those related to the energy embodied in materials are not included. Doing so would, in many cases, improve the attractiveness of efficiency and renewables in comparison to traditional energy systems.

Since the atmospheric residence times of greenhouse gases differ, it is necessary to choose a time period under which comparisons of the gases' relative greenhouse forcing will be made. In this paper, we use a 20-year period in order to incorporate the effect of the gases on the rate of climate change and to give a fair basis of evaluation to measures and policies aimed at buying time in a near-term perspective (i.e., 0 to 20 years). In particular, a long-term (100 years or more) perspective would discount the role of methane emissions in our comparisons.

The resulting carbon-equivalent emissions factors are (grams/MJ): Coal 30.6; oil 24.6; natural gas 17.2; and gasoline 25.7.[54] When shifting from a 20-year to a lifetime perspective (i.e., integrating from zero to infinity), carbon-equivalent emissions decline by 19.6% for coal, 16.7% for oil, and 19% for natural gas.

Estimating the Net Cost (Benefit) of Avoiding Greenhouse Gas Emissions

To compare various technical and fuel-choice measures for reducing emissions, to the corresponding base case, we use an indicator called the cost of conserved carbon-equivalent (CCCeq), which is calculated as follows:

$$\frac{\text{Cost of a given measure} - \text{Cost of base-case measure}}{\text{Emissions from base-case measure} - \text{Emissions from the given measure}}$$

$$= \frac{\text{Net cost (\$)}}{\text{Net emissions (tonnes)}}$$

The following is an illustration based on conserving coal-based electricity with adjustable-speed motor drives (7th item in Table B-1). The levelized cost of conserved electricity is \$0.0111/kWh versus an electricity supply cost of \$0.0436/kWh (busbar costs). The electricity production results in emissions of 318 grams of carbon-equivalent per kWhe. Choosing the efficiency strategy would result in a cost of conserved carbon-equivalent with a net economic benefit:

$$\frac{0.0111 \text{ \$/kWh} - 0.0436 \text{ \$/kWh}}{318 \text{ grams Ceq/kWh} - 0 \text{ grams Ceq/kWh}}$$

$$= -\$102/\text{tonne carbon-equivalent}$$

Emissions associated with the measure are zero for efficiency strategies, but are positive for substitution to fuels with net emissions of greenhouse gases.

Appendix B

Examples of the Cost of Avoiding Greenhouse Gas Emissions

In order to evaluate and rank the economic efficiency of technologies to reduce greenhouse gas emissions we have constructed a list of examples of specific measures and their costs (Tables B-1, B-2 and B-3). These measures are grouped into three categories: electricity, heating and transportation and include efficient end-use technologies, conversion technologies and fuel substitution options. In each category, a base case is chosen (e.g., power production with coal or natural gas) and its costs and emissions are compared with those of alternatives (e.g., efficient lighting equipment). The tables distinguish between available and emerging technologies.

The results incorporate CO_2 and other important greenhouse gases. The economic results are expressed in terms of the cost of conserved carbon-equivalent: CCCeq (i.e., the cost of a strategy to reduce emissions minus the cost of a base-case strategy, divided by the amount of emissions reduced (\$/tonne)). A negative CCCeq corresponds to a net economic benefit, i.e. that the cost of the emissions-reducing strategy is lower than the cost of the base-case strategy. Appendix A shows a sample calculation and describes our method of treating emissions in more detail.

The costs of avoided fuels are listed under the headings for each table. When comparing the results, care must be taken to focus both on the cost of avoided emissions and the amount of emissions avoided. A small quantity of avoided emissions and a relatively large cost differential will lead to a large CCCeq, as in the case of coal gasification for electric power production.

Table B-1: *Examples of avoided emissions and their costs: Electricity*

ELECTRICITY [a] (Cost of avoided resource: $0.044/kWh [coal])	Measure Resource Cost ($/kWh)	Avoided Emissions (g Carbon-eq per kWh)	 (%)	Cost of Avoided Carbon-equivalent (CACeq) ($/tonne)
End-Use Efficiency [b]				
– Available Technologies				
• Lighting (incand. --> compact fluorescent)	-0.011	318	100%	-171
• Lighting (efficient fluorescent tube)	-0.007	318	100%	-159
• Lighting (lamps, ballasts, reflectors)	0.013	318	100%	-96
• Refrigerator/freezer, no CFCs	0.018	318	100%	-79
• Freezer, automatic defrost, no CFCs	0.022	318	100%	-67
• Heat pump water heaters	0.034	318	100%	-30
• Variable-speed motor drive	0.011	318	100%	-102
• US field data, multifamily, htg. retrofits	0.038	318	100%	-19
• Retrofits in 450 US commerical buildings	0.026	318	100%	-54
• No-cost or behavioral measures	0	318	100%	-137
Electricity Production [c] [busbar costs]				
– Available Technologies				
• Biomass steam-electric (wood fuel)	0.041	318	100%	-9
• STIG (gasified coal)	0.041	9	3%	-313
• STIG (natural gas)	0.027	163	51%	-103
• Wind (1988)	0.054	318	100%	33
• Solar thermal electric (1988)	0.114	318	100%	221
• Solar photovoltaics (1988)	0.232	318	100%	588
• Nuclear--current US conditions	0.057	318	100%	41
– Emerging Technologies				
• ISTIG (gasified coal)	0.034	57	18%	-176
• ISTIG (natural gas)	0.024	187	59%	-106
• Chemically recuperated gas turbine (natural gas)	0.029	204	64%	-73
• Solar thermal				
(2000)	0.043	318	100%	-1
(2010)	0.036	318	100%	-24
(2020)	0.031	318	100%	-40
• Solar photovoltaics				
(2000)	0.072	318	100%	89
(2010)	0.050	318	100%	22
(2020)	0.036	318	100%	-24
• Wind				
(2000)	0.033	318	100%	-33
(2010)	0.027	318	100%	-51
• Nuclear--Industry target for US	0.040	318	100%	-11
– Fuel Choice [STIG base case]				
(Avoided resource costs: $0.071/kWh [coal gas]				
• STIG: Gasified coal --> natural gas (1990)	0.027	155	50%	-91
• STIG: Gasified coal --> biomass (sugar) (-2000)	0.033	309	100%	-25

Notes to Table B-1

[a] Unless noted, the annualized costs of efficiency and supply measures are calculated with a 6% real discount rate and no taxes. Costs for electric power plants are amortized over a 30-year period. Efficiencies based on higher heating value (HHV) are used throughout this analysis.

The reference (avoided) technology is a coal (2 x 500 MW) steam-electric plant, 34.6% efficiency, $1370/kW capital cost, and operations and maintenance (O&M) costs of 0.89 cents/kWh. For all power plant costs: 70% capacity factor, coal price $1.79/GJ, natural gas price $2.10/GJ.

[b] Three lighting examples: The first two measures result in a net benefit because reduced labor costs exceed incremental capital costs.[55] (7% real discount rate)

Refrigerator and freezer without CFCs: assumes a blend of hydrochlorofluorocarbons (HCFC-22, HFC-152a, and HCFC-124), with 97% lower ozone-depleting potential than the CFC-12 refrigerant currently used. For foam insulation, today based on CFC-11 as a blowing agent, alternatives are based on HCFC-141b and HCFC-123 which in fact have higher insulating values than current foams (7% real discount rate).[56]

Variable-speed motor drive: ABB Corporate Research (personal communication, Lars Gertmar, 1988). Capital cost $104/kW (750 kW motor). Assumes installation costs equal to 8% of capital cost; 15-year lifetime; 4000 hours/year; 35% reduction in electricity requirement.

Heat-pump water heater: adapted from Brown et al., assuming 2000 kWh/yr reduction in electricity required, 10-year life.[57]

US field data, multi-family space heating: includes measured data from 42 actual projects in the United States.[58]

[c] STIG (Steam-injected gas turbine) for gasified coal - 2 x 50 MW steam-injected gas turbine (STIG), air-blown gasifier, hot-gas clean-up, 35.6% efficiency, $1300/kW capital cost and 0.71 cents/kWh O&M cost.

STIG for natural gas—4 x 51 MW, 40% efficiency, $410/kW capital cost and 0.29 cents/kWh O&M costs.

Chemically Recuperated Gas Turbines: assumes 54% efficiency, $500/kW capital cost, 70% capacity factor.[59] O&M costs assumed same as for gas-fired STIG systems.

Biomass-steam: $1500/kW capital cost, $0.005/kWh O&M, $1.5/GJ fuel cost, Wind, Solar Thermal, and Solar Photovoltaics.[60] These prices are based on the "RD&D Intensification Scenarios" developed by five US National Laboratories.

Nuclear: [described in footnote 7].

ISTIG (intercooled STIG) coal—110 MW, 42.1% efficiency, $1030/kW capital cost and 0.60 cents/kWh O&M costs; gas—114 MW, 47% efficiency, $400/kW capital cost and 0.29 cents/kWh O&M costs.

Fuel Choice: The coal base-case is gasified coal versus natural gas:[61] Gasified coal versus biomass: 111 MW BIG/ISTIG system.[62]

Table B-2: *Examples of avoided emissions and their costs: Heating and fuel choice*

HEATING (Cost of avoided resource [oil]: $5.56/GJ)	Measure Resource Cost ($/GJ)	Avoided Emissions (g Carbon-eq per MJ)	(%)	Cost of Avoided Carbon-equivalent (CACeq) ($/tonne)
End-Use Efficiency [a]				
Available Technologies				
• Low-emissivity window glas				
• Flame retention head burners	1.75	25	100%	-155
• US field data, multi-family, htg. retrofits	0.93	25	100%	-188
• No-cost or behavioral measures	3.00	25	100%	-104
• Freezer, automatic defrost, no CFCs	0	25	100%	-226
Fuel choice [b]				
Menthanol (from sugar industry) (~2000)	11.53	31	100%	-95
Biomass (1990)				
High price (Sweden)	3.00	25	100%	-104
Low price (Brazil)	1.50	25	100%	-165
Biomass in district heating [incl. capital costs + O&M]				
1990 [coal base case]	9.94	35	100%	63
2010 [coal base case]	8.46	35	100%	-4

Notes to Table B-2

[a] Low-emissivity window glass: transmittance reduction of 0.23 W/m²-K in a climate with 3316 HDD(C).[63]

Flame retention head burners: assumes 18% fuel-requirement reduction, $300 incremental cost, 10-year life.[64]

US field data, multi-family space heating: includes measured results for 111 actual retrofit projects around the United States.[65]

[b] Methanol: Near-term technology assumed to be the pressurized, steam/oxygen-blown, fluidized bed biomass gasifier that has been developed by the Institute of Gas Technology, in plant sizes of 101.5 million gallons/year production capacity. Costs are significantly lower ($8.56/GJ) for a 555.6 million-gallon-per-year plant.[66] For comparison, the reference feedstock is coal. Biomass in district heating: calculations are for Swedish conditions, assuming 4000 hours/year operation, 20-year economic life, $370/kW capital cost, 88% thermal efficiency, $5.22/GJ fuel price (coal) in 1990 and $3.92/GJ in 2010.

Table B-3: *Examples of avoided emissions costs: Transport*

PASSENGER CARS [a]	Measure Resource Cost	Avoided Emissions		Cost of Avoided Carbon-equivalent (CACeq)	
End-Use Efficiency [b]	($/GJ)	(kg/GJ)	(%)		
(Cost of avoided resource: high=$12.63/GJ; low=$5.17/GJ)					
Available Technologies					
• US CAFE standards (16.8 to 8.7 l/100 km)	0.86	26	100%	-458	-168
• Fleet improvement (8.7 to 6.2 l/100 km)	5.17	26	100%	-290	0
• No-cost or behavioral measures	0	26	100%	-492	-201
Fuel choice [c]	($/100 km)	(kg/100km)	(%)		
(Avoided resource costs: $1.2/100 km [low], $2.9/100 km [high])					
Gasoline --> compressed natural gas	--	Higher emissions than from gasoline			
Gasoline --> methanol from natural gas	--	Higher emissions than from gasoline			
Gasoline --> methanol from coal	--	Higher emissions than from gasoline			
Gasoline --> methanol from biomass (present)	3.10	7.5	100%	22	253
Gasoline --> methanol from biomass (near-term)	1.70	7.5	100%	-164	67
Electric cars [comparison of operating costs]					
Gasoline --> electric car (natural gas)	0.19	6.4	100%	-427	-158
Gasoline --> electric car (biomass)	0.29	7.5	100%	-352	-122

Notes to Table B-3

[a] Reference-technology emissions for passenger cars: gasoline (7500 gCeq/100 km, at current US new-car average fuel economy of 8.4 liters/100 km).

[b] Gasoline prices: Gasoline prices (excluding taxes) vary amongst industrialized countries. We show cost calculations for the two extreme ends of the range. The "high" gasoline price (excluding taxes) is from Japan and the "low" price from the US. Fleet improvement: This reflects measures already used in the automobile industry, applied nationally (US) without changing the size distribution of the car stock.[67] The manufacturing costs of these measures should be comparable in the other major automobile-producing countries. The potential application of the measures to new cars sold outside of the U.S. depends on the extent to which they are already in use.[68]

[c] Fuel choice: Because the energy content of methanol and the energy efficiency of methanol cars differ from those of gasoline and gasoline-fueled cars, we show costs in terms of $/100km. For methanol cars, we use a vehicle efficiency that is 15% higher than that of gasoline-fueled cars. The assumed gasoline fuel efficiency is 8.4 liters/100km, the US average for new cars. The biomass feedstock price for methanol production is $2.33/GJ (both cases). The coal price is $1.48/GJ.

Through an extensive literature review, Sperling and DeLuchi concluded that the vehicle price and maintenance costs for single-fuel methanol cars will become comparable to those for similar gasoline cars (i.e., with same size, range, weight, vehicle life and power) in the near-term.[69] The only cost differential associated with owning and operating a methanol car, therefore, will be that of the fuel. Although methanol fuels (from natural gas or biomass) cost more per unit of energy than gasoline today,[70] part of this price differential will be balanced by the inherently better efficiency of methanol cars. Fuel-switching from gasoline to methanol manufactured from natural gas would increase the carbon-equivalent emissions per vehicle kilometer driven. The difference, however, is small and must be weighed against the potential role of natural gas as a transition fuel while biomass-based methanol supplies are made available.

Electric cars: The calculations for electric cars are based on GM's prototype Impact discussed in the text, assuming the electricity is made with the currently available power plants (the 1st and 3rd entries under the "Electricity Production" heading in Table B-1). The comparison is based strictly on operating costs.

Notes and References

1. Intergovernmental Panel on Climate Change (IPCC). Policymakers Summary of the Formulation of Response Strategies. Report Prepared for IPCC by WGIII.

2. Intergovernmental Panel on Climate Change (IPCC). 1990. Policymakers Summary of the Scientific Assessment of Climate Change. Report Prepared for IPCC by Working Group I.

3. World Energy Conference. 1989. World Energy Horizons: 2000-2020. Editions Technip, Paris.

4. Manne, A.S. and Richels, R.G. 1990. CO_2 Emission Limits: an Economic Cost Analysis for the USA. The Energy Journal 11(2) pp. 51–74. For a critique of this Manne and Richels article, see Williams, R.H. 1990. Low Cost Strategies for Coping with CO_2 Emission Limits. The Energy Journal 11(4) pp.35–59.

5. Monetary values are in U.S. dollars throughout this paper.

6. Solar Energy Research Institute. 1990. The Potential of Renewable Energy: An Interlaboratory White Paper. Solar Energy Research Institute Report SERI/TP-260-3674.

7. Based on a total busbar cost of 5.68 cents/kWhe for current light-water technology (6% real discount rate, 30-year amortization): plant size (1100 MW), capital costs ($3060/kW), and efficiency (33.4%). The capital cost drops to $1670/kW for the target light-water reactor. These estimates are from the Electric Power Research Institute (EPRI). The nuclear fuel-cycle cost $0.84/GJ (1.11 cents/kWhe: EPRI's projection for 1990 to 2000) and the operations and maintenance costs ($0.91 cents/kWhe) are the 1985 U.S. averages for nuclear plants. This busbar-cost derivation is shown in Williams, R.H. and Larson, E.D. 1989. Expanding Roles for Gas Turbines in Power Generation, p. 524, in T.B. Johansson, B. Bodlund, and R.H. Williams, Eds., 1989. Electricity: Efficient End-Use and New Generation Technologies, and Their Planning Implications, Lund University Press, pp. 503–554.

8. The indirect costs of implementing measures vary according to many factors. Performance standards, for example, can have a negligible cost per unit of energy saved while costs can be higher for conservation programs, especially during initial "learning" periods. European lighting efficiency programs have shown implementation costs of only 0.3 cents/kWh, i.e. ≈1/20 the cost of the energy they save. See Mills, E., Persson, A and Strahl J. 1990. The Inception and Proliferation of European Residential Lighting Efficiency Programs. Proceedings of the ACEEE 1990 Summer Study on Energy Efficiency in Buildings, American Council for an Energy-Efficient Economy, Washington D.C.

9. DeLuchi, M., Johnston, R.A., and Sperling, D. 1989. Transportation and The Greenhouse Effect, Transp. Res. Rec. 1175, pp. 33–44.

10. Davis, S.C., Shonka, D.B., Anderson-Batiste, G.J., and Hu, P.S. 1989. Transportation Energy Data Book: Edition 10, Oak Ridge National Laboratory Report ORNL-6565. Unless stated otherwise, single-figure fuel economies are weighted averages of highway (45%) and city (55%) mileage, as per the USEPA test procedure. Note: miles per gallon (mpg) = (235)/(liters per 100 km).

11. This analysis applies to the 1976 to 1988 period. See Ross, M. 1989, Energy and Transportation in the United States, Annu. Rev. Energy, 14, pp. 131–171.

12. Bleviss, D.L. 1988. The New Oil Crisis and Fuel Economy Technologies: Preparing The Light Transportation Industry for the 1990s, Quorum Books, 268 pp.

13. Toyota, press release, October 23, 1985.

14. Bleviss, 1988, op. cit., Ref. 12.

15. Mellde, R.W., Maasing, I.M., and Johansson, T.B. 1989. Advanced Automobile Engines for Fuel Economy, Low Emissions, and Multifuel Capability, Annu. Rev. Energy, 14, pp. 425–444.

16. See Unnasch, S., Moyer, C.B., Lowell, D.D., and Jackson, M. D. 1989. Comparing The Impacts of Different Transportation Fuels on the Greenhouse Effect, California Energy Commission Consultant Report P500- 89-001, 71 pp. The environmental implications of moving toward methanol vehicles are highly dependent on the feedstock used for methanol production. The prime attractiveness of methanol vehicles lies in the opportunity to use biomass feedstocks. However, creating a methanol vehicle fleet would also create a market for methanol produced from coal feedstocks. This eventuality should be avoided because producing, transporting, and burning coal-based methanol leads to higher greenhouse gas emissions than gasoline, compared on a per-vehicle-kilometer basis; see Difiglio, C., Duleep, K.G., and Greene, D.L. DeLuchi, M., Johnston, R.A., and Sperling, D. 1989. Cost Effectiveness of Future Fuel Economy Improvements. The Energy Journal, 11(1), pp. 65-86. It should also be noted that methanol vehicles are potential emitters of large quantities of formaldehyde: a highly toxic gas that can also contribute to the production of tropospheric ozone. Formaldehyde emissions can be kept very low, however, through the use of methanol engines with oxidation catalysts (Sperling, D. and DeLuchi, M.A. 1989, Transportation Energy Futures, Annu. Rev. Energy, 14, pp. 375–424).

17. This is the fuel economy in urban driving conditions. Due to the regenerative breaking system in the Impact, urban and highway mileage are about equal. Personal communication, Mr. Sloane, Public Relations, General Motors Detroit, August 2, 1990. The gasoline equivalent was computed using an electrical conversion efficiency of 45% and assuming 3.96 MJ/kWh which includes 10% transmission and distribution losses.

18. California Energy Commission (CEC). 1990. Chemically Recuperated Gas Turbine, Report number P500-90-001 (draft), 78 pages plus appendices.

19. Larson, E.D., Svenningson, P., and Bjerle, I. 1989. Biomass Gasification for Gas Turbine Power Generation, in Johansson et al. 1989. Op. cit. Ref. 7.

20. This discussion is based on Ogden, J.M., Williams, R.H., and Fulmer, M.E. 1990. Cogeneration Applications of Biomass Gasifier/Gas Turbine Technologies in the Cane Sugar and Alcohol Industries, Center for Energy and Environmental Studies, Princeton University. The electricity production potential is 2780 TWhe/year. The potential contribution of barbojo depends on how much of the resource can be recovered cost effectively. By utilizing only bagasse fuel in this estimate, 900 TWh could be produced, or 55% of 1987 electricity demand. (Based on note c to Table 13 of Ogden et al.) This scenario uses biomass-integrated gasifier technology with intercooled steam-injected gas turbines (BIG/ISTIG). The scenario is calculated assuming growth in sugar cane production of 3%/year (the historical rate of annual growth since 1960) over the forty-year period. Half of the growth is assumed to be committed to sugar production (equivalent to the World Bank projection of the sugar demand growth rate to 1995) and half to alcohol production. This much more efficient use of potential cane resources would considerably improve the economics of the cane industry, and in effect turn it into an electricity-production industry with sugar and/or alcohol as marketable by-products.

21. Solar Energy Research Institute. 1990. Op. cit., Ref. 6.

22. ibid.

23. Carlsson, D. 1989. Low-Cost Power from Thin-Film Photovoltaics, in Johansson et al. 1989. Op. cit., Ref. 7, p. 595.

24. Personal communication, Lars Nilsson, Department of Environmental and Energy Systems Studies, University of Lund, July 1990-based on Summers, R. and Heston, A. 1988. A New Set of International Comparisons of Real Product and Price Levels Estimates for 130 Countries, 1950-1985, Review of Income and Wealth, 1-25; and United Nations. 1988. Energy Statistics Yearbook, New York.

25. Williams, R.H. and Larson, E.D. 1987. Materials, Affluence, and Industrial Energy Use. Annu. Rev. Energy, 12, pp. 99-144.

25. Goldemberg, J., Johansson, T.B., Reddy, A.K.N., and Williams, R.H. 1988. Energy for a Sustainable World, Wiley-Eastern, New Delhi. Also summarized under the same title by the World Resources Institute, 1987, Washington, D.C. The standard of living in developing countries is assumed to increase to that of the WE/JANZ regions, and by 50 to 100% in the industrialized countries.

27. Mills, E., Harris, J.P, and Rosenfeld, A.H. 1988. Le Gisement d'Economies d'Energie aux Etats-Unis: Tendances, Perspectives et Propositions, Energy Internationale: 1988–1989, pp. 169–193 (in French). Also as Developing Demand-Side Energy Resources in the United States: Trends and Policies, Lawrence Berkeley Laboratory Report number 24920.

28. This cost is based on a comparison of methanol and today's US gasoline prices. Expected gasoline price increases will lead to a negative cost for this measure.

29. Bodlund, B., Mills, E., Karlsson, T., and Johansson, T.B. 1989. The Challenge of Choices: Technology Options for the Swedish Electricity Sector, in Johansson et al. 1989. Op. cit., Ref. 7.

30. Using OECD conventions for counting nuclear and hydroelectric power.

31. The marginal cost curve of avoided emissions is very steep. The marginal cost, in this case the cost of wind energy replacing fossil fuels, is $135/tonne. This reflects today's wind energy costs (7 cents/kWh). At projected future costs of 3.3 cents/kWh (see Table B-1), the marginal cost becomes a net benefit of $41/tonne.

32. Goldemberg, J., Johansson, T.B., Reddy, A.K.N., and Williams, R.H. 1985. Basic Needs and Much More with One Kilowatt Per Capita, Ambio, 14, pp. 190–200.

33. Goldemberg, J., Johansson, T.B., Reddy, A.K.N., and Williams, R.H. 1990. Energy for a Sustainable World: An Update, with Emphasis on Developing Countries, Bellagio Seminar on Energy Efficiency for a Sustainable World; Goldemberg, et al. 1985. Op. cit. Ref 32.

34. US Environmental Protection Agency. 1989. Policy Options for Stabilizing Global Climate. D.A. Lashof and D.A. Tirpak, eds. Office of Policy, Planning, and Evaluation, (draft).

35. A.K.N. Reddy, G.D. Sumithra, P. Balachandra, and A. d'Sa. 1990. Energy Conservation in India: A Development-Focused End-Use-Oriented Energy Scenario for Karnataka, Part 2 - Electricity, Department of Management Studies, Indian Institute of Science, Bangalore, India. Presented at the Bellagio Seminar on Energy Efficiency for a Sustainable World, Figures 24 and 32.

36. Goldemberg et al. 1988. Op. cit., Ref. 26, pp. 479–481.

37. See Goldemberg et al. 1988. Op. cit., Ref. 26; Geller, H. 1989. Implementing Electricity Conservation Programs: Progress Towards Least-cost Energy Services Among U.S. Utilities, in Johansson et al. 1989. Op. cit., Ref. 7, pp. 741-764; and Williams, R.H. 1989, Innovative Approaches to Marketing Electric Efficiency, in Johansson et al. 1989. Op. cit., Ref. 7, pp. 741–764.

38. Moskovitz, D. 1989. Profits & Progress Through Least-Cost Planning, National Association of Regulatory Utility Commissioners (NARUC), Washington, D.C. and de la Moriniere, O. Energy Service Companies: The French Experience, in Johansson et al. 1989. Op. cit., Ref. 7, pp. 811–830.

39. Geller. 1989. Op. cit., Ref 37.

40. As an incentive to participate, manufacturers submitting proposals were paid ~$16,000. As an incentive to exceed the target efficiency level, ~$80 per refrigerator will be paid to the manufacturer for each reduction of 15% in electricity use beyond the target value given. The procurement guidelines also invited manufactures to create an "energy label" to

compare the energy use of their design to that of models already on the market. Personal communication, Hans Nilsson, National Energy Administration, August 10, 1990.

41. This is the annual tax proposed to support three climate-change initiatives: (1) a CFC phase-out, (2) reforestation of 12 million hectares per year, and (3) fossil-fuel energy conservation in the developing world. See Goldemberg, J. 1990. Policy Responses to Global Warming, in Global Warming, J. Leggett (ed), pp. 166–184, Oxford University Press, London. p. 177.

42. For a more detailed discussion of carbon taxes, see Grubb, M. 1989. The Greenhouse Effect: Negotiating Targets. The Royal Institute of International Affairs, London.

43. von Hippel, F. and Levi, B.G. 1990. Automotive Fuel Efficiency: The Opportunity and Weaknesses of Existing Market Incentives. Resources and Conservation, 10, pp. 103–124.

44. For a description of the methodology used to determine appliance standard levels in the United States, see United States Department of Energy. 1989. Technical Support Document: Energy Conservation Standards for Consumer Products--Refrigerators and Furnaces, report number DOE/CE-0277. For proposed lighting and motor standards, see, for example, An Act Requiring Minimum Energy Efficiency Standards for Lighting Fixtures, Lightbulbs, Floor Lamps, Table Lamps and Electric Motors, House Bill number 5239 (1990), The Commonwealth of Massachusetts and An Act Reducing The Greenhouse Effect by Promoting Clean and Efficient Energy Sources, House Bill number 5277 (1990), The Commonwealth of Massachusetts, USA.

45. Keepin, B., and Kats, G. 1988. Greenhouse Warming: Comparative Analysis of Nuclear and Efficiency Abatement Strategies, Energy Policy, 16(6) pp. 537–632.

46. Williams, R.H. and Feiveson, H.A. 1989. Diversion-Resistance Criteria for Future Nuclear Power. Princeton Center for Energy and Environmental Studies, Princeton, New Jersey.

47. International Energy Agency. 1989. Energy Policies and Programmes of IEA Countries. IEA, Paris.

48. For examples of cost-effective government R&D in the buildings sector see Geller, H., Harris, J.P., Levine, M.D., and Rosenfeld, A.H. 1987. The Role of Federal Research and Development in Advancing Energy Efficiency: A $50 Billion Contribution to the U.S. Economy, Ann. Rev. Energy, 12, pp. 357-395; Brown, M.A., Berry, L.G., and Goel, R.K. 1989. Commercializing Government-Sponsored Innovations: Twelve Successful Buildings Case Studies. Oak Ridge National Laboratory Report number ORNL/CON-275, 132 pp.

49. See Goldemberg, et al. 1990. Op. cit., Ref. 33. In the low-carbon scenario, electricity demand grows by 4%/year, and would be met with the following generation mix: BIG/GT (32%), hydropower (31%), natural gas (17%), PV/wind/solar thermal (10%), coal (6%), and nuclear (2%).

50. United States Environmental Protection Agency. 1989. Policy Options for Stabilizing Global Climate. D.A. Lashof and D.A. Tirpak, eds. Office of Policy, Planning, and Evaluation, (draft).

51. Steps 1 and 3-11 are adapted from Geller, H. 1989. National Energy Efficiency Platform: Description and Potential Impacts. American Council for an Energy-Efficient Economy, Washington, D.C. Step 2 involves measures to reduce ambient temperatures, i.e., the "heat-island effect," in cities (personal communication, Arthur Rosenfeld, Lawrence Berkeley Laboratory, 1989).

52. International Energy Agency. 1989. Op. cit., Ref. 47.

53. Wilson, D. 1990. Quantifying and Comparing Fuel-cycle Greenhouse gas Emissions from Coal, Oil and Natural Gas Consumption, Energy Policy, 18(6), pp. 550–562.

54. Motor fuel emissions rates are converted from DeLuchi et al. 1989. Op. cit., Ref. 9 based on Wilson. 1990. Op. cit., Ref. 53.

55. Piette, M.A., Krause, F., and Verderber, R. 1989. Technology Assessment: Energy-efficient Commercial Lighting. Lawrence Berkeley Laboratory Report number 27032.

56. See United States Environmental Protection Agency. 1989. Op. cit., Ref. 50.

57. Brown et al. 1989. Op. cit., Ref 48.

58. Goldman, C.A., Greely, K.M., and Harris, J.P. 1988. Retrofit Experience in US Multifamily Buildings: Energy Saving, Costs, and Economics. Lawrence Berkeley Laboratory Report Number 25248 1/2, p. 24.

59. California Energy Commission. 1990. Op. cit., Ref. 18.

60. Solar Energy Research Institute. 1990. Op. cit., Ref 6. The costs shown in this report include taxes (i.e. a fixed charge rate is used). We recalculated the costs using a 6% real discount rate and no taxes.

61. Williams and Larson. 1989. Op. cit., Ref. 7.

62. Ogden et al. 1990. Op. cit., Ref. 20.

63. Geller et al. 1987. Op. cit., Ref. 48.

64. Brown et al. 1989. Op. cit., Ref. 48.

65. Goldman et al. 1988. Op. cit., Ref. 58.

66. Williams, R.H. 1990. Will Constraining Fossil Fuel Carbon Dioxide Emissions Really Cost So Much? Center for Energy and Environmental Studies, Princeton University.

67. Difiglio, et al. 1989. Op. cit., Ref 16.

68. Personal communication, David L. Greene, Oak Ridge National Laboratory, May 14, 1990.

69. Sperling, D. and DeLuchi, M.A. 1989. Is Methanol the Transportation Fuel of the Future?, Energy, 14(8), pp. 469–482.

70. Williams. 1990. Op. cit., Ref 66.

Climate and Land Use in North Africa

by M. Kassas *(Egypt)*

Abstract

Demographic patterns during the Quaternary reflected the geography of natural resources and climate. The redistribution of human populations in North Africa as climate changed from pluvial conditions to present-day climatic desiccation is an illustrative case.

The present distribution of population, their settlements and their agricultural activities in the Libyan Arab Jamahiriya (Libya) and in Egypt relate to rainfall in the former and the River Nile in the latter; in both instances they relate to water resources. Conflicts between agricultural and non-agricultural land uses are obvious in both Libya and Egypt.

Many aspects of agriculture in Egypt (geographical and seasonal distribution of crops, dates for sowing and for harvesting, etc.) are closely related to climatic conditions, particularly temperature regime.

Non-conventional land uses such as sites for recreation and settlements for tourism have become prominent in recent decades in Egypt. Sites suitable for recreation and tourism land use relate to prevalent climate and other environmental conditions. Seaside resorts along the Mediterranean and the Red Sea coasts are briefly discussed.

Climate change towards warmer temperatures will have impacts on various land uses and on management of natural resources.

1. Introduction

Land-use patterns relate to various activities of human communities; almost all such activities (settlements, farmlands, rangelands, etc.) require areas of land and associated space above and below ground. Environmental impacts of land use depend on density of population and intensity of their activities, and these, in turn, depend on available resources and technologies at hand.

Demographic history of the broader territories of North Africa seems closely linked with history of climate. The evolutionary history of the River Nile relates to climatic and geographic changes in North Africa and the Red Sea and Mediterranean basins and in the tropical and equatorial highlands further south. For details see, for instance, Williams and Faure (1980). Pluvial conditions with moist climate seemed to have prevailed in North Africa during 'the epoch 5000–2350 B.C. Man was able to settle over the greater part of the now-empty Sahara. Levalliose-Mousterian artefacts have been collected from between the dunes of the Libyan sand sea, from the Libyan Plateau many miles away from the river (Nile) as well as from the coast of the Red Sea' (Butzer, 1959). As the present arid climate evolved and as it prevailed, people moved towards territories where water resources were available: river valleys, oases, coastal lands and mountain oases.

The territories of the five countries of North Africa (Morocco, Algeria, Tunisia, Libya and Egypt) extend over the northern stretch of the African Sahara. Their present climate represents phases of the Mediterranean climate (UNESCO-FAO, 1963) that range from the meso-mediterranean in the coastal parts of Morocco, Algeria and Tunisia to the true desert of the inland Sahara. The three larger countries (Algeria, Libya and Egypt) have a total area of 500 million hectares, but the total area of farmlands is about 11.5 million hectares, that is, 2.3%. The situation in Morocco and Tunisia is better: arable lands represent 16% and 20% of total areas, respectively (see Table 1). With the exception of Egypt, agricultural land use (farmlands, woodlands and rangelands) in all four countries is controlled by the prevalent climatic features of rainfall. In Egypt, where the River Nile is the principal source of water, agriculture seems dependent on rainfall on the Ethiopian highlands and the African Equatorial plateau: two tropical climatic regimes that are different from the Mediterranean regime.

In all five countries groundwater represents the second source of water. Several formations provide water-bearing beds (aquifers) of considerable extent and richness. These bodies of water are mostly fossil, probably stored during earlier (Quaternary) periods of wetter climate (pluvials). Oases found in the Sahara depend on these groundwater

Table 1: *Basic data: land use in North African countries (after FAO and WRI 1988–89)*

Country	Total Area (1000 ha)	Arable Lands (1000 ha)	Arable Lands Per capt. (ha)	Pasture (1000 ha)	Forest (1000 ha)	Desert (1000 ha)	Desert % of total
Morocco	44 655	8 401	0.38	12 500	5 190	19 110	42
Algeria	238 174	7 610	0.35	38 452	2 424	190 188	80
Tunisia	16 361	4 923	0.70	3 250	530	7 346	45
Libya	175 954	2 127	0.59	6 780	543	166 096	94
Egypt	100 145	2 486	0.05	–	–	96 717	96

Table 2: *Rainfall belts, population and urban areas in Libya (from El-Massarati, 1986)*

Rainfall (mm/year)	Area (km²)	Population Number	Population /km²	Urban areas (ha) 1966	Urban areas (ha) 1978
500 or more	1 330	147 000	110.5	731	2 610
400 – 500	2 670	37 000	13.9	136	306
300 – 400	6 520	1 135 000	174.1	4 440	19 054
200 – 300	18 600	1 055 000	56.7	4 032	11 547
150 – 200	14 000	165 000	11.8	328	1 581
50 – 150 } 50 or less }	1 716 880	470 000	0.27		

resources that flow under artesian pressure or have to be pumped.

2. The Case of Libya

The geography of land use and demographic pattern in Libya show that rainfall is the determining factor. Table 2 shows that 84% (2 539 000) of the Libyan population lives in the areas (43 120 km²) with rainfall of 150 mm/year or more, that is, in 2.5% of the Libyan area (11.8–174.1 persons/km²). In the rest of the Libyan territories (1 716 880 km²) population (470 000) is thinly dispersed (0.27 person/km²), and even here half of this population (210 000) inhabits oases. Within the rainy 2.5% of Libya are all the urban centres and almost all the farmlands rainfed and irrigated, and hence the competition for land use.

The advent of oil development in Libya (first production in 1961) changed many aspects of life. Prior to this era agriculture (including nomadic grazing) was the principal occupation: in 1958 agriculture involved 60% of the manpower and contributed 26.1% of the GNP. Manpower involved in agriculture fell to 37% in 1961 and 19.1% in 1978; the contribution of agriculture to the GNP decreased to 4.3% in 1966 and to 1.7% in 1978.

The eclipse of agriculture was associated with the flow of oil wealth and the rise of urbanism. Urban settlements were also congregated in coastal territories where climate was milder. It is recorded (El-Massarati, 1986) that in 1911 there were four coastal towns with urban population of 60 000, representing 7% of the Libyan population; the rest were rural including the nomads. By 1954 the urban population increased to 270 000 (25% of the population). It increased to 674 300 (39.8% of the population) in 1966 and to 1 703 200 in 1978 (57.7% of the population). Urban centres increased in number and in size. In 1911 the four coastal towns covered an area of 250 hectares. In 1966 there were 14 towns covering an area of 9 667 hectares; this urban area increased in 1978 to 35 100 hectares (see Table 2). Between 1966 and 1978, urban population increased 129% and urban area increased 263%, mostly at the expense of arable land.

As an example: the population of the principal city, Tripoli, increased from 129 728 in 1954 to 300 000 in 1966, and to 697 000 in 1978. The area of the city was 30 hectares at the beginning of this century, extended to become 475 hectares by the end of the Second World War, 1 650 hectares in the early 1970s and 6 930 hectares in the early 1980s (El-Teer, 1986). Blake (1979) and El-Teer (1986) noted the active and large-scale urban development in Libya 'as financial wealth became available'. Urban planning and design were carried out by expatriate consultants. These introduced western conceptions and codes that bore little appreciation of the traditional architecture that had the attributes of being environmentally (related to climate) and socially sound. These new codes showed no sensitivity to the problems related to the limited arable land that needs to be conserved for agricultural use.

It should be noted (*a*) that the *per capita* area of arable land in Libya and Tunisia is high (0.59 hectares) compared to other countries (Table 1), (*b*) that before the oil era, Libya was—in most years—a net exporter of agricultural products, with more than 60% of manpower working in agriculture, and (*c*) data presented by Abu-Lughad (1986) showed that (in 1975) inhabitants of Libya comprised 2 223 000 nationals and 531 475 aliens (19%); aliens formed 63% of the labour force. The latter point means that the greater numbers of rural population who deserted pastoral and farming labour and moved to urban centres did not provide the main bulk of the labour force needed for the newly transformed economy.

As urban centres and other forms of development expanded and congregated mainly within the coastal territories, thus competing with agriculture for the same climatic zone, demand for water escalated. Groundwater resources were gradually exhausted: extraction exceeded replenishment. Water levels subsided. Aquifers were subject to saltwater intrusion. Facing this situation, the Government planned, and initiated implementation of, a large-scale water-transfer scheme: The Great River.

In its first phase of the scheme, to be completed in 1990, two million cubic metres of water/day will be extracted from aquifers in the Tazerbo area (120 wells, 450 m deep) and Serir area (150 wells, 500 m deep) and conveyed in a four-metre (diameter) pipe-line from the south-east desert of the country northward for 400–600 km. The total length of the pipelines will be 1600 km. In this first phase, the scheme will tap the water resources of Miocene and post-Miocene aquifers; it will later tap resources of the Nubia sandstone aquifers (Cretaceous), of the Kufra oasis further south and the Fezzan aquifers in the western part of the country.

The two million m^3/day of the first phase will be divided into 810 000 m^3/day to the Sert area and 1 180 000 m^3/day to the Benghazi area. The plan estimates the water use as follows: 100 million m^3/year for urban consumption, 200 million m^3/year for the use and rehabilitation of existing farmlands (some 90 000 ha) and 400 million m^3/year for establishment of new agricultural schemes (75 000 ha).

In its final phase the scheme will be pumping water volumes of some 5.5 million m^3/day and distributing it to various parts of the northern territories. The technical details of the various components of this grand enterprise are not the subject of this paper. But we may note that almost all the water resources of the southern territories that are to be tapped are fossil waters with little or no recharge. Fadl (1988) estimates that the groundwater level will drop 40–100 metres within the first 50 years. We may also note that water is piped for long distances from territories with harsh climate to be used in coastal territories where climate is milder.

3. The Case of Egypt

Agriculture and Land-use

Egypt has a total area of about 1,000,000 km^2, administratively divided into 26 governorates: 21 within the Nile valley and delta and its adjoining territories and five within the deserts and the Sinai peninsula. The former group cover an area of less than 4% of the country including almost all the farmlands, urban-industrial centres and a population of 48 205 049 (1986 census). The other group of five governorates cover 96% of the area of Egypt and are inhabited by 1% of the total population (565 389). This means that Egypt comprises a principal riverain oasis (the main inhabited area) and a few small oases in the western desert. 99% of the population inhabit a limited area of 42 000 km^2 (4% of the national territories), density is high: 1170/km^2. Population is increasing at the rate of 1.3 millions/year. Under these conditions, competition for land use, particularly between agricultural uses and non-agricultural uses, is intense.

During the last 30 years Egypt laboured to extend its cultivated area: it built the High Dam at Aswan to provide additional water (and power) resources for agriculture, and reclaimed extensive acreage of desert lands mostly within the peripheries of the Nile farmlands. According to a survey carried out by Pacific Consultants (1980), between 1953 and 1975 'a total of 912 000 feddans (acres) (was) reclaimed, representing an increase of approximately 15% of the cultivated land of Egypt'. Gardner and Parker (1985) give year-by-year data on land reclaimed in Egypt from 1953 to 1984: a total of 1,492,000 acres. At present the Government has a set target of 100,000 acres to be reclaimed every year.

But the cities, towns and villages of Egypt have all extended their areas and are encroaching on the prime farmlands. Data given by Gardner and Parker (1985) show that between 1953 and 1984 urban sprawl and conversion to non-agricultural land-use consumed 570 000 acres. Hefny (1982) gives a broad estimate of farmland area lost to urban encroachment as 'more than 40 000 feddans (acres) per year'. He gives LANDSAT-derived land-use statistics for the Qalyubia Governorate (south-east of the Delta) showing that during the six-year period 1972–1978 6 782 feddans (acres) of farmland were lost to urbanization. This represented 3.33% of the total farmland area of the Governorate.

Agriculture and climate

Egypt extends across some 9 degrees of latitude (22°–31°N), that is, from the Nubian desert (extremely arid), within the occasional reach of the northward migration of the Inter-Tropical Front, to the Mediterranean. Farmlands extend along a narrow strip from Aswan to Cairo and fan northward into the Delta. Annual rainfall is 100–180 mm at the Mediterranean coast and is practically nil south of Cairo; mean temperature range at Alexandria is 17.3–25.0°C and at Aswan is 19.2–34.2°C. The latitudinal differences in temperature set the broad limits of crop distribution: sugar cane, sorghum, millet, etc. in the warmer south; sugar beet, flax, several fruit crops, etc. in the cooler north (Delta).

Under the prevalent system of perennial irrigation, the seasonal pattern of crop rotation between winter crops (wheat, clover, beans, barley, flax, etc.) and summer crops (cotton, rice, groundnuts, etc.) is related to the seasonal pattern of temperature. One of the principal crop plants (cotton) is particularly sensitive to low temperature in the seedling stage. Determination of recommended dates of sowing and temperature forecast for subsequent days are principal issues of agrometeorology in Egypt. If sowing is delayed (early March) to ensure warmth, seedlings would be vulnerable to pests (including root pathogens); early sowing (early February) would subject seedlings to destructive chills of cold spells.

Late frost (spells of cold temperature) represents one of the principal weather hazards to several cash crops: tomatoes, apricots, peaches, etc. This is especially noted in inland areas and in the oases of the western desert. Flower buds often open in early spring (middle of February) and the incidence of a late February drop of temperature (night frost) may cause widespread damage to the crop.

Tourism, Recreation and Climate

The attraction for the flow of international tourism to Egypt has, for many generations, been the ancient history and its archeological sites. But in modern times nature (landscape, climate, wilderness, etc.) has become an added asset. The coastal lands of Egypt provide a variety of sites for recreation that are being developed. The deserts that cover the main area of Egypt combine attributes of climate (sunshine and warmth) and expansive landscape (serenity) that could be a suitable habitat for retirement or long winter vacations. Recreation is one of the alternative uses of arid lands (Drouhin, 1970); developments during recent decades in the State of Arizona (USA) provide an example.

The Mediterranean coastal lands extending for some 1000 km from Rafah at the eastern border to Sallum at the western border enjoy a mild climate during summer when inland territories are hot. Millions of inland residents move for their summer vacation to the coast of the Mediterranean where several summer-resort settlements have been established, and others are being built. Conflict between different land uses is evident; the coastal belt receives some rain and in many parts landform and topographic features (limestone ridges, coastal dunes) allow for the redistribution of runoff water and thus provide for some cultivation (fig orchards on the sand dunes, olive plantations at feet of ridges, etc.). Newly built summer resorts with their ancillary facilities often replace farmlands.

The coastal lands of the Red Sea and its north-eastern extension of the Gulf of Aqaba represent a corridor of tropical climate. The biogeography of the fauna (including corals) and flora (including mangroves) show the northward penetration of warm conditions. Here coastal lands combine the attributes of seaside environment and climate that is pleasantly warm during most of the year. Seaside resort sites take advantage of the sea (water sports, etc.) and the warm climate during winter and spring when cold weather prevails in Europe.

Development of recreation settlements in the Hurghada-Safaga district (on the Red Sea) and along the shores of the Gulf of Aqaba is one of the modern features of development in Egypt. The Egyptian coast of the Red Sea, that extends for more than 700 km, provides room for further development. In these almost rainless territories competition for land use does not involve agriculture but involves oil and mineral resources development and fisheries. But the natural environment, including coastal coral formations, embraces habitats that are ecologically fragile. 'The natural and man-made environment is the most essential resource for the development of tourism. Therefore, in planning that development, it is vital that ecological and environmental constraints should be recognized, and such recognition must be given priority over economic and technical consideration' (UN 1976, p9).

Domestic tourism is the principal user of the Mediterranean coastal resorts during summer months (June–September). The contrast between the inland hot summer and the coastal mild climate is a main factor on the influx of holiday-makers. International tourism (mostly from Europe) is the principal user of the Red Sea coastal resorts. The climate difference between the European cold winter and the warm climate of the Red Sea and the Gulf of Aqaba, together with the seaside attraction, seems to be a driving force in this tourism flow.

4.   General Discussion

Many aspects of land use are closely related to climatic conditions and to socio-economic parameters. Discrepancies that we have noted as regards land use (agricultural versus non-agricultural uses) in Libya and Egypt seem to be global in their spread. Wolman and

Fournier (1987) preface their book on land transformation by saying, 'The present volume was stimulated by this world-wide concern for the way in which land used in agriculture is being transformed throughout the world'. According to Buringh and Dudal (1987) the world land area is 14 900 million hectares of which 78% is not suitable for agriculture. Of the rest (22%) 13% is weakly productive, 6% is moderately productive and 3% is highly productive. This limited area is the food- (and other agricultural products) producing lands for the human race; its maintenance for agriculture use and its conservation against sprawl of non-agricultural use are an objective that transcends the limits of private ownership. World-wide land-speculation enterprises need to be constrained.

National priorities for land use should take into full consideration the ecological principle of 'limiting factor'. In countries of North Africa, as noted in the cases of Libya and Egypt, water is the limiting factor. All over the world arid and semi-arid territories (one-third of the world land area), water is the limiting factor. Agricultural economics in countries like Libya and Egypt still measure crop productivity per unit area of land. This needs to be changed to assess productivity per unit volume of water. This would have far-reaching consequences on crops and cropping patterns. At present water is not costed in the farm book-keeping. Water pricing could be an effective means for more efficient use of irrigation water, and may make such water-consuming crops as sugar-cane and rice uneconomic.

Climate remains a determining factor for many land uses, including agriculture and human settlement. Recreation resorts have become notable among land uses that will probably expand further as societies develop towards post-industrial phases (see Bell, 1974). As Meigs (1969) puts it, 'Of all economic industries of the desert seacoasts, none is likely to grow more than recreation'.

Prospects of future climatic changes, including increased warmth, pose special challenges to land-use systems, especially in those territories that lie at margins between different climatic systems. Warmer climate would mean direct ecological effects on plant and animal life, and indirect effects, as it would entail increased evapotranspiration and hence increased water consumption. The impacts of such changes on agricultural land use in Libya and on agricultural water use in Egypt may be far-reaching. Again the impacts of increased warmth on the geographical and seasonal distribution of crops in Egypt (and in countries where irrigation farmlands prevail) need to be explored and considered in planning for the future.

For a special aspect of land use such as tourism (and recreation at large) that is, in many ways, climate-dependent, changes in temperature may entail major modifications in the season of tourism. Changes in climate may also affect the ecology of natural fauna and flora.

These are often elements of attraction, e.g. for divers for corals, anglers, bird-watchers, etc.

References

Abu-Lughod, J.: 1986, Urbanization and social change in the Arab region, *Al-Fikr Al-Arabi*, Sept. 1986, pp. 154-173 (in Arabic).

Bell, D.: 1974, *The Coming of Post-Industrial Society*, Heinemann, London. XIII+507 pp.

Blake, G.H.: 1979, Urbanization and development planning in Libya; in *Development of Urban Systems in Africa*, ed. R.A. Abudho and S. El-Shakhs, Praeger Pub., N.Y., pp. 99–115.

Buringh, P. and Dudal, R.: 1987, Agricultural land use in space and time, in *Land Transformation in Agriculture*, SCOPE 32, ed. M.C. Wolman and F.C.A. Fournier, J. Wiley & Sons, pp. 9–43.

Butzer, K.W.: 1959, Environment and human ecology in Egypt during pre-dynastic and early dynastic times, *Bull. Soc. Geog. d'Egypt*, Vol. 32, pp. 43–87.

Drouhin, G.: 1970, Alternative uses of arid regions, in *Arid Lands in Transition*, ed. H.E. Dregne. AAAS, Washington D.C., Pub. No. 90, pp. 105–120.

El-Massarati, A.M.: 1986, Transformation of land to urban use: its nature and impact in Libya, *Al-Fikr Al-Arabi*, Sept. 1986, pp. 67–93 (in Arabic).

El-Teer, M.: 1986, Urban pattern in Libya, *Al-Fikr Al-Arabi*, Sept. 1986, pp. 14–25 (in Arabic).

Fadl, M.J.: 1988, Environmental impacts of the Great River Project, *Arab Journal of Science*, 12, December 1988, pp. 39–46 (in Arabic).

Gardner, G.R. and Parker, J.B.: 1985, Agricultural statistics of Egypt. *US Dept. of Agric., Statistical Bulletin*, 732, Aug.1985.

Hefny, K.: 1982, Land-use and management problems in the Nile Delta, *UNESCO, Nature and Resources*, Vol.18:1 2,pp. 22-27.

Meigs, P.: 1969, Future use of desert seacoasts, in *Arid Lands in Perspective*, ed. W.G. McGinnies and B.J. Goldman, AAAS, Washington D.C. and University of Arizona Press, pp. 103–118.

Pacific Consultants: 1980, New Lands Productivity in Egypt, US-AID Contract No. NE-C-1645, January 1980.

UNESCO-FAO: 1963, Bioclimatic map of the Mediterranean zone, UNESCO *Arid Zone Research*, Vol. 21, 58 pp.

UN: 1976, Planning and development of the tourist industry in the ECE region, *Proc. Symp. Dubrovnik 13-18 October 1975*, 198 pp.

Wolman, M.G. and Fournier, F.G.A. (eds.): 1987, *Land Transformation in Agriculture*, SCOPE, 32, J.Wiley & Sons, N.Y. XIX+531 pp.

Williams, M.A.J. and H. Faure (ed.): 1980, *The Sahara and the Nile*. A. A. Balkewa, Rotterdam, XVI + 607 pp.

Climate and Urban Planning

by Roger Taesler *(Sweden)*

Abstract

The development of urban climatology is briefly summarized from the perspective of urban growth and technological development with special emphasis on present rapid urbanization in low latitudes.

The role of climate in the development of the built environment is reviewed. Needs for climate data in planning and design are discussed, including problems of spatial and temporal representativity due to urbanization and climate change. The impact of climate on building energy use is reviewed together with comments on human thermal comfort. Lack of climate consciousness in urban planning and building design is discussed as a problem of professional specialization and global adoption of mid-latitude technology.

Certain urban climatic characteristics are summarized and discussed with reference to their impacts on human life. Deliberate planning as a possibility for improving outdoor and indoor climate is reviewed. Recent basic and applied research on atmosphere-urban interactions is briefly summarized, including activities of the World Meteorological Organization (WMO) and initiatives for the Tropical Urban Climate Experiment (TRUCE).

Conclusions and recommendations focus on the importance of the urban atmosphere as the living environment of half the global population. Needs are identified for improvements and developments in urban climatology and its applications.

1. Introduction

Climate affects urban development in two ways. The general climate of a region determines a range of needs and functional requirements on urban supply systems and on the design and stuctural strength of buildings. The city itself, on the other hand, creates a number of changes in the natural climatic conditions which may have important consequences for its further development.

Urban meteorology first developed in mid latitudes in response to industrial and urban growth (cf. Kratzer 1937, Anonym. 1978, Shitara 1979, Landsberg 1981, Oke 1974,1979,1981,1988, Pressman 1987). It is interesting to note that at about the time when Howard (1833) published his studies of the climate of London, Fourier and others raised the idea of the atmospheric greenhouse effect, subsequently developed by Arrhenius (1896), regarding in particular the role of carbon dioxide. Thus, it is clear that the perspectives of both local and global impacts on climate as a result of urban and industrial development were recognized at least 100 years ago. Air pollution was recognized already in ancient times as a severe problem in urban areas.

Research in urban meteorology aims at improving the understanding of the physical, radiative and chemical processes operating in the urban atmosphere system. This research forms the basis for predictions and control of future climatic conditions in urban areas and, thus, for climate concious urban planning.

Urban climatic changes depend on meteorological conditions as well as on the site and size of a city, land use patterns, structure and density of the built up areas and, further, on traffic, industry and other urban activities. Urban impacts on the atmosphere are not restricted to cities only but affect surrounding rural areas also. In many densely populated regions the natural landscape hardly exists anymore but is transformed into an urban-industrial landscape where urban climatic effects become regional features with significant changes in cloudiness, precipitation, solar radiation, air quality and visibility, see for example, Fortak (1980), Ayers et al. (1982), Yonetani (1982). Air pollution and acid rain are regional problems which limit permissible further emissions and may, eventually, restrict future urban development.

Furthermore, urban-industrial areas are major emittants of greenhouse gases, in particular of CO_2 and CFC's. There are direct relations and feedbacks between urban planning, building design and operation and indoor climate requirements on the one hand and local, regional and global climate change on the other.

A basic requirement of the built environment is to provide climatic conditions suitable for human needs and activities, indoors as well as outdoors. Climate conscious planning and building requires a good understanding of urban atmospheric processes and outdoor-indoor climate interaction. Today the architectural and enginering professions are facing a new challenge, as discussed by Page (1989), namely to respond to the threats of ozone layer destruction and the implications of global climate change. This requires development and utilization of passive and energy efficient techniques for heating, cooling and ventilation but also new approaches in urban development, eg. to reduce traffic emissions or to counteract increased urban temperatures in warm climates. At the same time it is necessary to assess the consequences of future climatic changes for urban life styles as well as for building construction and costs, building performance and safety against climatic loads, (Page, 1991).

Cities are often located in areas especially exposed to climatic hazards—i.e. on low lying land near a coastline , in valleys or near major rivers. In rapidly growing cities the least priviliged groups of the urban population usually are forced to live on the most exposed land, often in squatter type settlements with very low quality buildings, and, as a consequence, are especially endangered by flooding or extreme winds (Figure 1). It is also common to find

housing areas of low income groups concentrated around major industrial complexes. Inhabitants of such areas may be exposed to severe air pollution but also to accidental releases of dangerous gases.

2. Population Growth, Urbanization and Climate

Urban areas offer a range of socioeconomic and cultural opportunities that have a strong attraction to people. In many tropical and subtropical regions urban growth is taking place today at an uncontrolled rate. This frequently causes severe deterioration of the urban atmospheric environment as well as of the urban fabric with severe impacts on health, well being and safety of the urban population.

Around the year 1800 about 2 % of the total global population of 1000 million lived in urban areas—by the year 2000 the expected proportion is 50 % and 25 years later close to 60 % (Figure 2). Demographic trends, (Weihe 1986, Oke 1986) predict a global population of about 6000 million in the year 2000 and some 8000 million in 2025. This development is further accentuated by the even stronger growth of the very large cities. Estimates for the year 2000 show 25 cities with more than 10 million inhabitants, 20 of them in tropical/subtropical regions. Mexico City is estimated to reach over 30 million. According to Ojima (1989) the population of the greater Tokyo region was 31 million in 1985 and may reach 35 million in the year 2000. Decisions have already been taken on far reaching urban reconstruction to improve the urban environment in Tokyo.

Although urban growth today in industrialized countries is slow, large cities tend to grow at the expense of smaller ones. Relocation of residential housing from city centers to suburbs or satellite towns results in urban sprawl and

SECTORS OF URBAN ENVIRONMENT

CLIMATIC HAZARDS	Buildings and Settlements	Infra-Structure (roads, services, etc.)	The man-facturing economy, Industries, Factories	Loss of Life Potential
High winds	●	◑	●	○
High winds linked to coastal storm surge	●	●	●	●
Flooding: Heavy rainfall	◑	◑	○	○
Slow onset Flooding	◑	◑	◑	◑
Flash Floods	●	●	◑	●
Drought	◑[1]	○	●	●

Key: ● Major Impact

○◑ Partial Impact

○ No Impact

Note: [1]: In drought situations building needs can be created due to population movements. In addition, the lack of natural building materials (i.e., reeds/branches) can have adverse consequences on opportunities for building.

Figure 1: Impact of climatic hazards on various sectors of disaster-prone urban environments (after Davis, 1986).

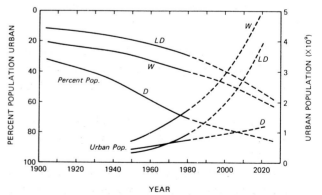

LD = Less Developed; W = World; D = Developed.

Figure 2: Urban population in absolute numbers and as percentage of total population in developed and less developed regions (after Oke, 1986).

increasing distances of travel. Building heights and building density increase in city centers in response to economic forces. As a result, atmospheric conditions change continously in urban areas.

3.   Climate as a Determinant for Urban Development

Prior to industrialization buildings and settlements had to be constructed under the constraints of locally available materials and energy resources and with access only to simple technology. Traditional designs are often influenced by climate and integrated to obtain suitable indoor and outdoor conditions. Still, the possibilities to create safe, healthy and comfortable conditions were much more restricted by climate than today. As a result of technological and economic development in Sweden, for example, indoor temperatures in residential buildings have increased from around +13°C in the year 1800 to around +22°C today. In 1860 the mortality rate in Stockholm was still higher than the birth rate. The urban population increased nevertheless due to migration to the city from rural areas.

Mortality rates show a seasonal variation in many countries. In cold climates maximum rates are found during winter, whereas summer maxima are found in some cases in warmer climates. Improved methods for indoor climatization help to reduce seasonal variations in mortality. In particular in cold climates, introduction of central heating systems has been accompanied by reduced winter mortality (Momiyama and Katayama, 1972).

Climate is a basic determinant not only for human health and comfort. It also determines wind forces, snow and ice loads on structures, corrosion and deterioration of building surfaces due to the combined action of wind, rain, solar radiation and air pollution. Avalanches, land slides, heavy rains and flooding of rivers are serious threats to the urban population in many parts of the world.

In the present context 'Climate' has to be taken as the full spectrum of meteorological conditions to be expected in the next 50–100 years. This is based on:

- the typical turn-over time for the built environment
- the importance of the range and variability of meteorological conditions rather than climatological averages
- the interaction of several meteorological processes and elements influencing the urban 'system' and its inhabitants.

The time scale 50–100 years refers in the first place to urban renewal and building reconstruction in the industrialized world and may be considerably less in the developing world.

Climatological Data for Urban Planning

Meteorological stations are usually located in populated areas. Many observing stations even outside urban areas, eg., at airports, become gradually influenced by building development. The representativity of climatic data is a problem in engineering applications in particular in design against extreme climatic forces and calculations of indoor climate and energy demands. There is a need on the one hand for data from background stations to obtain reliable long term statistics on the general climate and, on the other hand, for data representing 'typical' urban conditions.

Design values on extreme climatic impacts (air pollution, wind, rain, snow, cold, heat) are usually specified in national codes and standards. Such data are important for the safety, economy and energy efficiency of the built environment. Methods to correct for local or urban effects are essential for the calculation of wind loads (Cook 1985) but also for calculations of air pollution dispersion and building enegy demands. Data, on extreme hot or cold spells for the design of buildings and heating plants, need to be tailored according to specific building characteristics and calculated for different risk levels (Berg-Hallberg and Peterson 1986).

Present design data are based on historical records usually covering the last few decades. To some extent the lack of future representativity may be compensated for by a certain overdesign. A major change in climate may however considerably affect the safety of existing and new structures.

Energy Use

Indoor climate control accounts for 1/3–1/2 of the total energy use in most countries. Additional large amounts of energy are required for the manufacturing and transportation of building materials and in the construction of the built environment.

In order to cut down or even only to limit further increases in the burning of fossil fuels and associated emissions of greenhouse gases, it is of vital importance to reduce energy demands for indoor climate control. A key factor to reduce heating demands is the optimal utilization of passive solar heat and internal gains of heat from people and electric appliances. Heat losses can be greatly reduced by improved insulation and tightening of buildings. The result of energy conservation measures is, however, all too often counteracted by increased indoor temperatures and, in particular, overheating from passive solar gain even at temperatures far below the heating threshold. This then creates a need for increased ventilation or cooling. Thus, in a study of temperatures preferred by residents in Swedish dwellings (Widegren-Dafgård 1984) it was found that most people found temperatures to be too warm and, further, that the dwellers lowered temperatures by airing rather than by adjusting thermostats.

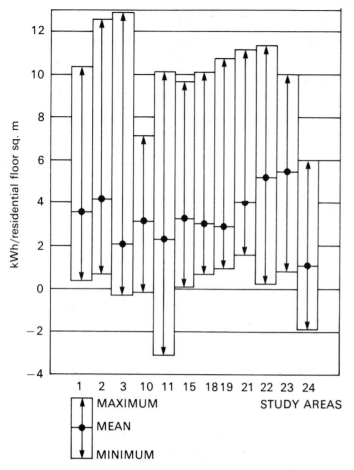

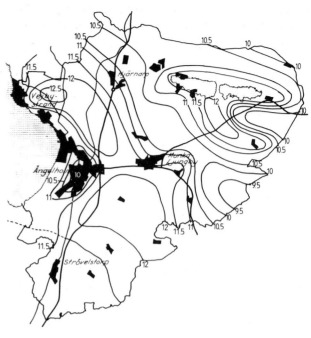

Figure 4: Calculated annual heat loss (20 years aver) in identical modern detached 1-family houses, in response to local wind and temperature differences within Ångelholm community, Southern Sweden (from Taesler, 1986).

Figure 3: Range in calculated annual heating demands due to micro-climatic differences in 12 residential housing areas in Finland (Kivistö and Rauhala, 1989).

Accurate calculations of building energy requirements need to take into account the interaction of simultaneous outdoor and indoor climate as well as the influence of the behaviour of occupants. Architects and engineers are usually unfamiliar with climatology and, in particular, urban climatic processes and, as a consequence, tend to greatly oversimplify the climatological input in planning and design. In practical design calculations usually the only input is heating or cooling degree days, which are easy to calculate but give only very crude estimates (Ayres 1977).

The 'PC-revolution' has created a completely new situation today both with respect to numerical simulations as well as access to meteorological data (cf. the WMO CLICOM system) and also the possibilities to model urban and microscale climatic conditions.

The building energy demand depends on outdoor temperature, solar irradiation on different surfaces, windspeed and exposure to wind—all of which are strongly dependent on urban land use and settlement design. A change of eg. +/- 1°C in mean outdoor temperature during

the heating season corresponds (in Sweden) approximately to a +/- 5 % change in heating demand. This relates to a mean outdoor temperature 0°C and indoor temperature of +20°C. In more temperate climates the relative influence of a change in outdoor (or indoor) temperatures will be even greater.

Recent work on building energy calculations in Finland (Figure 3) and Sweden (Figure 4) shows differences between individual buildings of, typically, 10–20 % (in extreme cases, up to 30%) due to local and microscale climate variations. Passive solar gain in particular is strongly influenced by detailed settlement layout (Wiren and O'Mara 1987). The net solar gain may compensate for approximately 20 % of the annual heat loss (Taesler 1988) but the total passive gain may be twice as much.

Cooling and air conditioning are of particular importance in low latitudes. Protection from heat is much more expensive and demanding in terms of building energy use than protection from cold. Town planning and settlement layout, including tree planting are potentially important tools for reducing heat loads in warm climates but can also greatly improve outdoor urban comfort and air quality.

Human Thermal Comfort

Power loads and total energy use for indoor climate control and illumination are strongly influenced by human comfort requirements. Many different indices have been proposed to express the influence of environmental factors on human thermal comfort (see reviews by Givoni 1976, Höppe 1984, Taesler 1987). The ET (Effective Temperature) and the PMV (Predicted Mean Vote) are used extensively as measures of indoor as well as outdoor comfort or discomfort. A major limitation with all such indices is their restricted applicability to certain situations or conditions and, further, the inability to account for dynamic impacts under non steady conditions (Wang, 1990). Even considering only steady state physiological thermo-regulation it is most unlikely, in view of behavioural and physiological adaptation and cultural or psychological factors (Weihe 1983, 1987), that all individuals will respond in the same way to any given set of environmental conditions.

All over the world modern buildings are strongly influenced by mid latitude designs and indoor climate is usually kept within very narrow limits of temperature and relative humidity. It may be questioned, however, if the comfort limits are universally valid. Studies by Auliciems (1983) indicate that thermal neutrality may be obtained at different temperatures in different climatic regions.

The 'Sick Building Syndrome' has become a major problem in many countries. A healthy indoor environment requires an adequate supply of fresh air and ventilation rates must be sufficient to prevent condensation and growth of rot, mold and fungus. High indoor concentrations of radon gas emitted from the ground are a dangerous threat in some areas. Modern building materials emit a wide variety of organic chemical substances that cause odor, asthma or other allergic reactions. Many of these problems are increasingly common in modern, air-tight buildings and some are further aggravated by elevated indoor temperatures or changes in humidity.

Urban planning and settlement layout may affect indoor air quality and pollution levels in several ways. Building heights and density strongly influence, for example, the exposure of buildings to wind and wind pressure distribution over the building envelope. This, in turn, influences the natural ventilation. In densely built up areas of low cost houses, not equipped with mechanical ventilation systems, ventilation may be totally insufficient under calm weather. In particular during heat waves, this creates very oppressive conditions with severe effects on indoor comfort and air quality.

Water

Water is a fundamental resource as well as a threat in urban areas. All over the world, but in particular in the developing world, cities are facing increasingly difficult problems due to shortage, abundance or pollution of water (Lindh 1983). Many cities in coastal areas are vulnerable to flooding by storm surges. Flash floods in rivers may result from rapid run off of rain water on impervious urban surfaces or

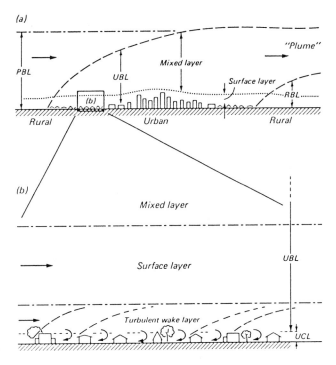

Figure 5: Conceptual structure of the urban atmosphere (after Oke, 1984).

increased run off in upstream river catchment areas due to forest cutting and may contaminate fresh water supplies and aggravate spreading of diseases.

4. Urban Planning as a Determinant for Climate

Characteristics of Urban Climate

The conceptual picture, proposed by Oke (1984) and generally adopted today, of the structure of the urban atmosphere (Figure 5) distinguishes between a layer below roof level (the urban canopy layer, UCL) and an internal urban boundary layer (UBL) above roof level. The two layers are connected via a more or less well defined 'wake layer' extending from just below roof level to a level corresponding to 2–3 times the average distance between buildings. In the UCL airflow, temperature, solar and long wave radiation show strong microscale variations in direct response to the dimensions and geometrical arrangement of buildings etc.

Virtually all climatic elements are modified by urbanization. Air temperatures are generally higher in built up areas than in surrounding rural areas (the well known 'urban heat island'). Wind speeds above cities are slowed down by increased surface friction. Below roof level and between buildings wind conditions are very erratic and may cause uncomfortable or even dangerous conditions for pedestrians. Under certain weather conditions the urban heat island generates an urban circulation with low level inflow and upper level outflow, analogous to the sea breeze

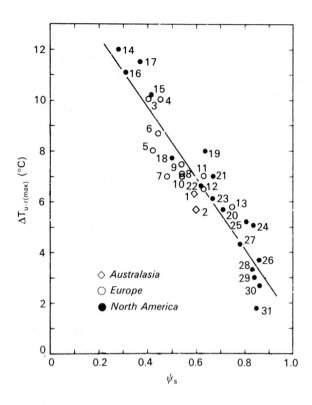

Figure 6: Maximum urban temperature excess ('heat island' intensity) versus urban canyon sky view factor (after Oke, 1982)

circulation. Solar radiation is reduced as a result of urban air pollution and increased cloudiness. Precipitation amounts and intensities usually increase whereas fog frequencies increase in some cases and decrease in others.

Several factors (Oke 1982) contribute to increase urban air temperature. The urban heat island effect has been found in cities of all sizes and in all climatic regions. Maximum heat island intensities, up to 12°C, usually found during clear, calm nights, are closely related to urban structure, expressed by the 'sky view factor' (Figure 6). Annual mean temperatures in central areas of medium sized or large cities are typically 1–3°C above rural background values.

Long term increases in temperatures are found as a result of urbanization (Landsberg 1981, Horie and Hirokawa 1979, Nord 1982). Heat stress increases in response to urban growth and changes in land use patterns (Balling and Brazel 1986). High nocturnal temperatures are a common cause for sleeping problems (Höppe 1988) and are a severe problem even in mid latitude cities during the warm season. Prolonged heat waves, air pollution and poor urban ventilation frequently create conditions leading to increased mortality in cities, especially among elderly people or those suffering from respiratory and cardiovascular diseases (Driscoll 1971, Nakamura, 1988).

Urban warming can create mesoscale convective windsystems. Triggering or enhancement,, in particular of convective precipitation, has been shown in several cities (Changnon 1978, Yonetani 1982). Figure 7 shows heavy

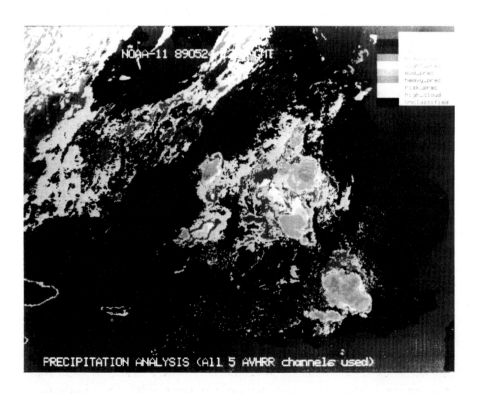

Figure 7: NOAA-11 Satellite image May 24, 1989 at 12.39 GMT. Colour-coded cloud classification and precipitation analysis by SMHI. Heavy thunderstorms are seen over London and urban-industrial Midlands.

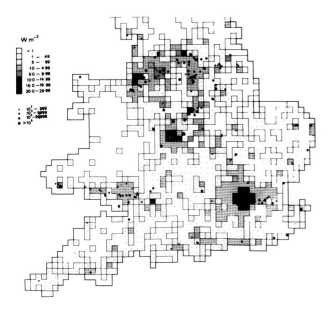

Figure 8: Distribution of artificial heat release from area and point sources in August 1975 (Harrison and McGoldrick, 1981).

thunderstorms over England in May 1989. The case was one of extremely strong rain and winds, causing severe damage and several casualties. By comparison with Figure 8 it is seen that the convective activity was closely related to the main urban-industrial areas, which indicates a combined effect of artificial heat release and emissions of air pollutants in addition to urban warming by absorption of solar radiation.

Urban Climate Planning

Urban atmospheric conditions are partly a result of decisions taken during the process of planning and design. However, with the exception of urban air pollution control, this has received only limited attention in urban planning and development until quite recently.

The urban climate may be improved by deliberate design of city structure, eg. open urban corridors and utilization of nocturnal downslope winds or variations in building heights to increase urban ventilation. Orientation and grouping of buildings can greatly contribute to a favourable outdoor microclimate. Urban climate characteristics as part of urban ecological systems have been established as a basis for urban planning in various cities, notably in the FRG (Noack et al. 1986, Horbert et al. 1985). Urban parks and greenbelts have many beneficial effects on urban climate, in particular on pollution levels but also on temperature and humidity (eg., Horbert et al. 1988, Oke 1989). Shelterbelts can greatly improve the microclimate in windy housing areas.

Protective measures, based on climatic considerations, against local air pollution include urban zoning, traffic regulation and building of tall stacks. An increasingly common approach is to use real time monitoring of meteorological dispersion parameters and pollution levels to control emissions. This is a questionable approach, however. It can serve to keep maximum concentrations at acceptable levels but it may easily lead to higher average concentrations and increasing long term total emissions.

Principles for design of the built environment in different climates are elaborated in great detail by many authors (eg. Schild et al. 1981, Elawa 1981, PLEA 1983, Givoni 1976, 1989, UNCHS 1984, de Schiller and Evans 1988) Certain main guidelines may be summarized as follows.

Hot dry climates
The dominating characteristics are the large diurnal swings in temperature and solar radiation, creating problems of overheating during daytime and cold at night. Reflection of solar radiation causes intensive glare. Wind is a daytime problem in general, causing dust storms and increasing thermal discomfort at temperatures above approximately 32°C. Buildings should be designed to delay the transfer of heat from the outside so that it will reach the interior of the building when it is needed. Ventilation should be maximized at night and minimized during daytime. The best locations are in valleys parallel to prevailing winds. Settlement layout should be dense with narrow streets, shaded sidewalks, orientation of buildings along NE-SW, NW-SE, courtyards with greenery and water, short walking distances and compact building geometry, minimal outer surface area.

Warm humid climates
Temperature exhibits little daily or seasonal variation— hence there is no need for structures with high thermal inertia. Instead, maximum ventilation is the primary requirement. This calls for open, widely separated buildings, preferably placed on stilts well above the ground. Building heights should be varied so as to improve street ventilation. Trees are needed for shade. Streets should be oriented at an oblique angle to the prevailing wind. Flooding is a serious hazard, requiring safe locations, rapid street drainage and rainwater-absorbing ground surfaces. The roof becomes the most important building component both for protection from heavy rains and from solar radiation.

Cold climates
Here the key problems are to obtain protection from cold and wind while maximizing solar irradiation. Buildings need to be designed to prevent heat transfer from the inside of the building to the outside. This leads, as in hot dry climates, to compact building form and dense settlements, use of heavy materials with high insulating capacity and, in addition, prevention of air leakage and draft. Shelterbelts are needed for wind protection. Solar gain is increased by darkening of outer surfaces and by having windows mainly on walls facing the sun.

5. Basic and Applied Urban Climate Research

Research on the urban atmospheric environment as related to urban development progresses along several different lines, in many different disciplines and is published through a great variety of channels. The following brief discussion does not attempt to cover the entire field but merely to indicate some main developments.

Beginning as a largely empirical and descriptive science, urban meteorology today is directed increasingly more towards physical and numerical modelling, combined with experimental studies, aiming at a deeper understanding of the dynamic and radiative processes in the UCL and UBL. The pronounced 3-dimensional structure and horizontal inhomogeneity of urban areas raise important problems in basic research and in practical applications, eg., for the transformation of standard climate ('air-field') data to a particular urban site.

Research in urban meteorology is considerably biased towards case studies under idealized synoptic weather conditions, whereas urban impacts on larger scales and the interaction between synoptic scale and urban atmospheric processes has received only limited attention. Present knowledge about urban climate still relies heavily on results from mid-latitude climates. Research in tropical/subtropical regions has a short history, beginning only about a decade ago, and is still at an early stage (Oke 1986, Jauregui 1986). Though available data indicate qualitative agreements with mid-latitude results in some respects, it is not possible to generalize these to low latitudes or to account for differences in urban morphology, synoptic controls or general climate.

Applied research is directed towards improving climate design data, mathematical and numerical microscale modelling of airflow and dispersion of pollutants around building complexes, wind pressure on buildings, energy demands and indoor climate, solar and long wave radiation fields in the UCL, utilization of solar collectors and passive solar energy. Human bio-climatic methods are being developed and applied for outdoor comfort assessement in urban areas (Burt et al. 1982, Jendritzky and Sievers 1989).

The World Meteorological Organization (WMO) and several other international organizations are paying strong attention to the interaction between the atmosphere and the built environment, in particular with regard to tropical/subtropical regions. A major concern is to initiate research on tropical urban climate modification and its impacts and, further, to improve collaboration between climatolgists and building and planning specialists (WMO 1986, WMO/UNEP/WHO 1987). Within the WMO/WCAP programme the CLICOM and CARS projects provide means to facilitate access to climatic data and guidance for planning and design of the built environment.

A major effort of the WMO will be the launching of a joint international research project, the Tropical Urban Experiment (TRUCE), recommended by the WCP Technical Conference on Urban Climatology in Mexico City 1984 (WMO 1986) and further supported by the IFHP/CIB/WMO/IGU International Conference on Urban Climate, Planning and Building in Kyoto 1989 (WMO 1990). The purpose of TRUCE (Oke et al. 1990) is to improve understanding of the mechanisms controlling the modification of the atmosphere of tropical cities and to provide a better basis for urban environmental planning and forecasting.

6. Conclusions and Recommendations

Urban areas and urban climate may be seen as local features in a global context. Yet, within the near future, they will be the living environment of half the global population. Furthermore, urban and industrial activities, in particular energy use for indoor climate control, traffic and industrial production, affect atmospheric conditions on much larger scales. Mutual impacts of climate and the built environment therefore need serious attention in local as well as in regional and, ultimately, global planning to reduce energy demands and greenhouse gases.

Cities offer a range of attractive opportunities and facilities. Especially for poor people in the developing world these attractions are very strong and may be the only available possibility for improved living conditions, health and medical care, education, job opportunities or, simply, to survive. The strength and momentum of this process is such that it may easily override any environmental concerns in a short time perspective. This happened in mid latitude countries during the stage of industrial and urban development and it is extremely difficult and expensive, and perhaps impossible to correct the damages later.

The interaction between the atmosphere and the built environment cuts across the borders of many disciplines, each of which have developed their specialized competence, conceptual frameworks and priorities. Increasing professional specialization, however, is also an obstacle to development in areas requiring interdisciplinary approaches (Wieringa 1987). This is a serious problem in a society where environmental protection and efficient use of resources become increasingly more important. Applica–tions of climatology in building are best achieved by close collaboration between different specialists. This requires a broadening of the professional competence of meteorologists as well as of builders and planners. There is also a need for interdisciplinary research on the urban atmospheric environment, involving economic, medical as well as atmospheric and building sciences.

The perspective of major global or regional climate changes adds an important new dimension to planning and design as well as management of the built environment. It may be insufficient to base decisions only on knowledge derived from historical climatic data. Reassessments of the need for protection against extreme climatic forces may have far reaching consequences for existing and future structures as well as for urban land use. Reduction of greenhouse gases will require great improvements in building energy performance, affecting both architectural design and heating, ventilation and air conditioning technology and may also necessitate reevaluations of criteria for indoor thermal comfort.

The challenge today is to integrate and make use of climate as a factor in the design and management of the built environment - rather than to overcome climate by indiscriminate use of energy and materials. The present rate of urban growth in tropical/subtropical countries makes this challenge a matter of urgency. To meet the challenge it is necessary to:

- improve the scientific understanding of the urban atmosphere
- assess impacts of climate change for building and planning
- improve the education of building professionals as well as climatologists on interactions between the atmosphere and the built environment
- develop and transform climatological data and knowledge into useable guidelines for urban planning with respect to protection from climatic hazards, energy conservation, human health and comfort
- develop, in particular, PC-based models and climate data bases for calculations of indoor climate and energy requirements and applications of passive heating and cooling of buildings.

References

Anonymous,: 1978, Bibliography on Urban Meteorological Studies in Australia, Royal Meteorological Society, Australian Branch, CSIRO, Melbourne.

Arrhenius, S.,: 1896, On the Influence of Carbonic Acid in the Air upon the Temperature of the Ground. The London, Edinburgh and Dublin Philosophical Magazine and Journal of Science, April 1896, 237–276, London. (Rep. 1988 Dep. of Heating and Ventilation, Royal Inst. of Techn., Stockholm).

Auliciems, A.,: 1983, Psycho-Physiological Criteria for Global Thermal Zones of Building Design, Proc. 9th Int. Biomet. Congress, Int. J. Biomet., Suppl. Vol.26 1982, 69–86.

Ayers,G.P., Bigg,E.K., Turvey,D.E., Manton,M.J.,: 1982, Urban Influences on Condensation Nuclei Over a Continent, Atm. Env. Vol.16, 951–954.

Ayres, J.M.,: 1977, Predicting Energy Requirements, Energy and Buildings, Vol.1, 11–18.

Balling, R.C., Brazel, S.W.,: 1986, Temporal Analysis of Summertime Weather Stress Levels in Phoenix, Arizona, Arch. Met. Geoph. Biocl., Serie B, Vol.36, 331–342.

Berg-Hallberg, E., Peterson, F.,: 1986, Dimensionerande Utetemperatur—Del 2, Climate and Buildings 2/1986 3–72, Dep. of Heating and Ventilation, Roy. Inst. of Technology, Stockholm.

Burt,J.E., O'Rourke,P.A., Terjung,W.H.,: 1982, The relative Influence of Urban Climates on Human Energy Budgets and Skin Temperature, I. Modelling Considerations, II. Man in an Urban Environment, Int. J. Biom., Vol.26, 3–35.

Changnon, S.,A.,: 1978, Urban Effects on Severe Local Storms at St. Louis, J. Appl. Meteor., Vol.17, 578–586.

Cook, N.J.,: 1985, The Designer's Guide to Wind Loading of Building Structures, Build. Res. Establ. (BRE) report, Butterworth's, London.

Davis, I.,R.,: 1986, The Planning and Maintenance of Urban Settlements to Resist Extreme Climatic Forces, In WMO (1986), 277–312, WMO, Geneva.

Driscoll, D.M.,: 1971, The Relationship Between Weather and Mortality in Ten Major Metropolitan Areas in the United States, 1962-1965, Int. J. Biomet., Vol.15, 23–39.

Elawa, S.,: 1981, Housing Design in Extreme hot Arid Zones with Special Reference to Thermal Performance, Dep. of Build. Sci., Univ. of Lund.

Fortak,H.G.,: 1980, Local and Regional Impacts of Heat Emission, Proc. Symp. Intermed. Range Atm. Transp. Processes and Technology Assessement, Gatlinburg, Tennessee.

Givoni, B.,: 1976, Man, Climate and Architecture, Appl.Sci.Publ., London.

Givoni, B.,: 1989, Urban Design in Different Climates,WMO/TD-No. 346, WMO, Geneva.

Harrison, R., McGoldrick, B.,: 1981, Mapping Artificial Heat Releases in Great Britain, Atmospheric Environment, Vol.15, 667–674.

Horbert, M., Kirchgeorg, A., v. Stulpnagel, A.,: 1985, Umweltatlas Berlin, Techn. Univ. Berlin, Inst. f. Ökol., Fachbereich Biokl., Berlin.

Horbert, M., Kirchgeorg, A., Chronopoulos-Sereli, A., Chronopoulos, J.,: 1988, Impact of Green on the Urban Atmosphere in Athens, Zentralbibliothek Kernforschungsanalage Julich GmbH, Julich.

Horie, G., Hirokawa, Y.,: 1979, Effect of Energy Consumption on Urban Air Temperature, 1978 Japanese Progress in Climatology 12–17, Japanese Climatology Seminar, Hosei University, Tokyo.

Howard, L.,: 1833, Climate of London Deduced from Meteorological Observations, 3rd ed., Harvey & Darton, London.

Höppe, P.,: 1984, Die Energiebalans des Menschen, Wissenschaftliche Mitteilungen Nr 49, Meteorologisches Institut, Universität Munchen.

Höppe, P.,: 1988, Bewertung des Einflusses des Stadtklima auf das Raumklima, Proc. Fachtagung 'Umweltmeteorologie' Munchen 1988, 88-98, Wissensch. Mitt. Nr. 61, Meteorol. Inst. Univ. Munchen.

Jauregui, E., 1986, Tropical Urban Climates: Review and Assessment, in WMO (1986), 26–45, WMO, Geneva.

Jendritzky,G. and Sievers,U.,: 1989, Human Biometeorological Approaches with Respect to Urban Planning, Proc. 11th ISB-Congress, 25–39, SPB Academic Publishing, The Hague.

Kivistö, T., Rauhala, K.: 1989, Town Planning and Operating Costs of Residential Areas (ASTA II), Res. Notes 935, Techn. Res. Center of Finland, Espoo.

Kratzer, A.,: 1937, Das Stadtklima, F. Vieweg & Sohn, Braunschweig.

Landsberg, H.,: 1981, The Urban Climate, Int. Geoph. Series Vol. 28, Academic Press, N.Y.

Lindh, G.,: 1983, Water and the City, Unesco, Paris.

Momiyama, M., Katayama, K.,: 1972, Deseasonalization of Mortality in the World, Int. J. Biomet., Vol. 16, 329–342.

Nakamura, Y.,: 1988, Trends of Human Deaths in Residences and Hospitals in Kyoto in Hot and Humid Summer Seasons, Proc. Conf. 'Healthy Buildings' Stockholm June 1988, 43–52, Swedish Council for Building Research, Stockholm.

Noack, E.-M., Mayer, H., Baumgartner, A.,: 1986, Quantifizierung der Einflusse von Bebauung und Bewuchs auf das Klima in der Urbanen Biosphäre, Schlussbericht (Final Rep.), Lehrstuhl fur Bioklimatologie und Angewandte Meteorologie der Universität Munchen.

Nord, M.,: 1982, Local Temperature Variations in an Urbanizing Area, Proc. CIB Symp. Building Climatology Moscow, 424–435.

Ojima, T.,: 1989, Changing Tokyo Metropolitan Area and its Heat Island Model, Proc. IFHP/CIB/WMO/IGU Int. Conf. Urban Climate, Planning and Building Kyoto Nov. 1989, to be published.

Oke, T.R.,: 1974, Review of Urban Climatology 1968-1973, Techn. Note No. 134, WMO, Geneva.

Oke, T.R.,: 1979, Review of Urban Climatology 1973-1976, Techn. Note No. 169, WMO, Geneva.

Oke, T.R.,: 1981, Bibliography of Urban Climate 1977-1980, WCP-45, WMO, Geneva.

Oke, T.R.,: 1982, The Energetic Basis of the Urban Heat Island, Quart. J. R. Met. Soc., Vol.108, 1–24.

Oke, T.R.,: 1984, Methods in Urban Climatology, Zurcher Geog. Schriften, Vol.14, 19–29, ETH Zurich.

Oke, T.R.,: 1986, Urban Climatology and the Tropical City, In WMO (1986), 1–25, WMO, Geneva.

Oke, T.R.,: 1988, Bibliography of Urban Climate 1981–1987,(Draft doc.), Dep. of Geography, Univ. of British Columbia, Vancouver.

Oke, T.R.,: 1989, The Micrometeorology of the Urban Forest, Phil. Trans. R. Soc. Lond. B 324, 335–349.

Oke,T.R., Taesler,R., Olsson,L.E., 1990, The Tropical Urban Experiment (TRUCE), Proc. IFHP/CIB/WMO/IGU Int. Conf. Urb. Climate, Planning and Building Kyoto Nov. 1989, to be published.

Page, J.K.,1989, Architecture, Renewable Energy and the Global Ecological Crisis, Paper presented to the 2nd European Conference on Architecture, Science and Technology at the Service of Architecture, UNESCO, Paris.

Page, J.K.,1991, The Potential Effects of Cimate Change in the United Kingdom, United Kingdom Climate Change Review Group, First Report, Section 11, Construction, UK Department of the Environment, to be published.

PLEA,: 1983, Passive and Low Energy Architecture, (S. Yannas, ed.),Proc. 2nd Int. PLEA conf., Pergamon Press, London.

Pressman, N.,: 1987, Reduction of Winter-induced Discomfort in Canadian Urban Residential Areas: Ann. ibl., Can. Mortgage and Housing Corp.,Ottawa.

Schild, E., Casselmann, H.-F-, Dahmen, G., Pohlenz, R.,: 1981, Environmental Physics in Construction Its applications in arch. design, Granada, London.

de Schiller, S., Evans, J.M.,: 1988, Healthy Buildings in a New City: Bioclimatic Studies for Argentina's New Capital, Proc. Conf. 'Healthy Buildings 'June 1988, 121–131, Swedish Council for Building Research, Stockholm.

Shitara, H.,: 1979, Fifty years of Climatology in Japan, 1978 Japanese Progress in Climatology 45–80, Jap. Clim. Seminar, Hosei Univ., Tokyo.

Taesler, R.,: 1986, Climate, Buildings and Energy Exchange—An integrated Approach, TM Nr. 297, Dep. of Heating and Ventilation, Royal Inst. of Technology, Stockholm.

Taesler, R.,: 1987, Climate Characteristics and Human Health— the Problem of Climate Classification, Proc. WMO/UNEP/WHO Symp. 'Climate and Human Health', Vol I, 81–119, WMO, Geneva.

Taesler, R.,: 1988, Passive Solar Gain in District Heating, Proc. Conf. 'North Sun '88—Solar Energy at High Latitudes', 227–232, Swedish Council for Building Research, Stockholm.

UNCHS (Habitat),: 1984, Energy Conservation in the Construction and Maintenance of Buildings, Vol. I, Use of solar energy and natural cooling in the design of buildings in developing countries, UNCHS, Nairobi.

Wang, X-L.,: 1990, A Dynamic Model for Estimating Thermal Comfort, Climate & Buildings 2/1990 46–68, Dep. of Heating and Ventilation, Royal Inst. of Technology, stockholm.

Weihe, W.H.,: 1983, Adaptive Modifications and the Responses of Man to Weather and Climate, Proc. 9th Int. Biomet. Congress, Int. J. Biomet., Suppl. Vol. 26 1982, 53–68.

Weihe, W.H.,: 1986, Life Expectancy in Tropical Climates and Urbanization, In WMO (1986), 313–353, WMO, Geneva.

Weihe, W.H.,: 1987, Heat Balance of Man in Relation to Health, Proc. WMO/ UNEP/WHO Symp. 'Climate and Human Health', Vol. I, 143–169, WMO, Geneva.

Widegren-Dafgård, K.,: 1984, Komfort och Inomhusklimat, Önskade och faktiska temperaturer i bostäder, Dep. of Heating and Ventilation, Royal Inst. of Technology, Stockholm.

Wieringa, J.,: 1987, On the Spectral Gap Between Wind Engineering and Meteorologists, Institut Royal Meteorologique de Belgique, Brussels.

Wiren, B., O'Mara, A.H.,: 1987, Ventilation Losses and Passive Solar Heat Gain in a Swedish housing Estate: The Influence of Local Climate, Orientation and Windbreaks, Research Report TN: 8, The National Swedish Institute for Building Research, Gävle.

WMO,: 1986, Urban Climatology and its Applications with Special Regard to Tropical Areas, (ed. T.R. Oke) Proc. Techn. Conf. Mexico D.F. Nov. 1984, WMO, Geneva.

WMO,: 1990, WMO Bulletin April 1990, Vol. 32, 122–124, Geneva.

WMO/UNEP/WHO, 1987, Climate and Human Health, Proc. Symp. Leningrad, Sept. 1986, Vols. I and II, WCAP-No. 2, WMO, Geneva.

Yonetani, T.,: 1982, Increase in Number of Days with Heavy Precipitation in Tokyo Urban Area, J. Appl. Meteor., Vol. 21, 1466–1471.

Human Well-being, Diseases and Climate

by Wolf H. Weihe *(Switzerland)* and Raf Mertens *(Belgium)*

Concept of Health

In its Constitution of 1946 the World Health Organization (WHO) distinguishes between two states of human health, a state of absence of disease and infirmity and a state of complete physical, mental and social well-being. As a principle "basic to the happiness, harmonious relations and security of all peoples" the aim is to reach "the highest attainable standard of health" (WHO, 1986).

The definition combines two diverse movements: to eliminate diseases and to reach complete human well-being. The first movement deals with something considered undesirable that should be done away with, the second is concerned with something judged desirable that should be fulfilled. A state of absence of disease can be judged with reference to the definition of diseases as entities defined with worldwide consensus in the International Classification of Diseases (ICD) (WHO, 1977). A comparable document based on a universally valid definition of mental and social human well-being is not available. Widely varying concepts of well-being of individuals and societies are based on differing states of socioeconomic, cultural, political, and religious developments within traditional, private and public institutions. They depend on the character of the individual and have deep roots in history. Concepts of well-being are changeable and dictated by conditions, resources, ideologies, and fashions. There is agreement on a superficial level over normative standards, such as levels of lifestyle, happiness and basic rights for everybody. However, the degree of happiness of the individual and societies can neither be diagnosed nor quantified. It occurs in different guises and can actively and passively be achieved over greatly different pathways. This has far reaching implications for the decisions within and between nations in response to environmental changes.

The Road to Health

The rapid development of biomedicine beginning in mid-latitude countries of Western Europe and North America during the middle of the nineteenth century has been highly successful in reducing and eliminating diseases, mainly communicable diseases. As the suffering from these diseases subsided up to the middle of this century, the promotion of well-being accelerated. Major developments and improvements in lifestyles are taking place. This development in mid-latitude countries - commonly classified industrialized - is increasingly spreading to and even systematically demanded by low-latitude developing and very high-latitude countries . In full agreement with its definition of health, the WHO is presently promoting a global programme of Health for All to be realized by the end of this century. This programme is based on the creditable performance of "Western medicine" and the trust that the western democratic industrial culture is the best form of social and political organization to promote health and provide human well-being with lifestyles offering the most of happiness, and security to the people. However, the procedure is invasive to Nature as the recent evaluations of the effects of Man's activities on climate indicate (IPCC, 1990). New risks for the future of human health are increasingly observed such as high concentrations of air pollution over urban-industrial areas, exploitation of natural resources, and food shortages from land degradation. With reference to the great variability of regional climates and particularly to the many regions with marginal climates for human subsistence, it is increasingly questioned whether excessive lifestyle norms can persist in their presently extending patterns. Their future value for maintaining human health is under debate, since signals are increasingly perceived of growing risks of food and energy shortages in the wake of a global Man-made climatic change.

Throughout history Man has searched, worked, and experimented with all his talents, strength and intelligence to find and arrange the most favorable conditions for nutrition and living, and to gain control over the adverse impacts from the environment such as heat and cold. Man was always aware of his dependence on nature. Climate determines the life of Man; Man cannot overrule climate. Nature is the independent variable, which was there first.

Man, arriving much later on the scene, is the dependent variable through genetic adaptation during evolution, but provided with an extraordinary efficient adaptive capacity during lifetime.

Western industrialization has been revolutionary for the promotion and realization of physical health. Up to and in the beginning of industrialization in Europe, poverty, desolate sanitary and living conditions, unbalanced nutrition and a large variety of infectious diseases were common. Morbidity and mortality from cholera, dysentery, typhoid and scarlet fever, pneumonia, tuberculosis, and other diseases were high. This changed after Koch and Pasteur discovered the role of the pathogenic microorganisms as causative agents of infectious diseases which proved to be the key to the introduction of hygiene. The elaboration of hygienic measures for sanitation and control of air, water, soil, food, and dwellings for pathogenic agents, toxins and their transmission proved highly efficient. By making these medical, technological, and economical achievements available in all parts of the world, the control over infectious diseases during recent decades was improved even in indigenously disease-affected tropical areas. Where these measures were consequently applied, populations began to live longer and to increase in number at an accelerating rate. This has led to major shifts in demographic patterns through the aging of populations.

In many mid-latitude, industrial countries where life expectancies of over 70 years for men and 75 for women are established, birth rates are declining, in some to the level of zero population growth. However, in most of the low-latitude, developing countries in Africa and Asia birth rates have remained high with rapid growth of their populations (World Resources, 1988). When for given geographical and climatic conditions, production and distribution of food cannot keep up with the pace of the population growth rate and rising demands for higher standards of well-being, the increasing shortages of food and other resources develop into health-adverse vicious circles. Where hunger already exists there will be more hunger, as is the case in many low latitude countries located in agriculturally unfavorable climates. Hunger may be a transitory state, following the unavailability of food after droughts and storms without lasting consequences for health. Permanent hunger leads to starvation and malnutrition. Temporary or permanent harm from malnutrition, will first affect the infants and young who can tolerate nutrient shortages the least and will suffer the longest time (Cravioto and Delicardie, 1975). The achievements in reduction of infant mortality and debility through improved nutrition and hygiene will inevitably discontinue and upset again whenever imbalances between the number of people and provision of food accelerate the already existing vicious circles in numerous countries.

Risks to health will remain high and new risks arise as long as societal control of fecundity does not effectively reduce increasing regional discrepancies between population growth rate and availability and utilization of resources. These risks mount up with infrequent events such as droughts in areas where populations already are vulnerable to climatic variations.

In countries where the primary target for physical health with effective control of diseases and mitigation of suffering has been attained, efforts are increasingly directed towards providing for human well-being. Economic prosperity in the industrial countries offers vast means and choices to the mass of the people for care, education, comfort, mobility, travel, and pleasure. Well-being as the second target of health programmes has made a great step forward from satisfying basic needs to fulfil individual desires beyond needs.

With the step from provision of the essentials to satisfying of desires, old risk factors have reemerged and new ones are introduced. Simultaneously the standards of permissible involuntary risk have been lowered worldwide. Individual desires are not necessarily health orientated. Indulgence turns out to be counterproductive to health. Examples are overnutrition, consumption of sedatives, stimulants, and physical inactivity, leading to a narrowing of the tolerance range and adaptive capacities of the people to cope with environmental and societal influences combined with unrest, irritablility and vulnerability of the individual. Imbalances are expressed in increasing mortality rates from accidents, homicide, suicide, drug dependence, overtaxing physical and mental capacities, cardiovascular diseases (CVD), and neoplasms from impact of pollutants, such as lung cancer from smoking. Should these developments proceed, the achievement of the presently high life expectancies and improvements in the health situation may subside. In the end only the causes of death will have changed from formerly communicable to now non-communicable diseases, but the extent of suffering and strain on well-being may remain.

Dependence on the Environment

The Western concept of health is orientated on the individual as the focal point of an anthropocentric ecosystem of unlimited boundaries. It tends to lose sight of the world around on which man's success for health depends. To live, to live well and live safely, mankind has inconsiderately drawn on the environment. Presently, rapidly increasing in number, mankind is forced to convert natural land and use up natural resources to such a colossal extent that the balance of nature as a whole begins to be disturbed. Floods, droughts, storms, frost and heat periods have always been episodic reminders that there are powers in the environment exceeding Man's potentials for control.

Man has learned that his living conditions and lifestyles are directed and determined by the ambient conditions. The recent risk of a global climate change due to human activities has aroused attention.

Climate-Man Inter-relationship

The human organism, the atmosphere, and the other environmental compartments are open systems with continuous fluxes of energy and components. The multiple interrelationships operating between Man and climate have many repercussions on health. Man is never a passive victim of climate but an active responder to climate. Provided with vast adaptive capacities he will adapt his body physiologically and modify the climate consciously by reflecting on the effects it has and deciding on measures to mitigate strain and unwanted risks. Every reaction and response involves complex pluri-level minor or major interlinked adaptation processes which are maintained for as long as the equilibria of the organism are disturbed. The responses are the expression of these adaptation processes. They are based on negative and positive feedback mechanisms, activated by external and internal stimuli, their perception, and internal assessment. They proceed as a continuum at the levels of autonomic-physiological regulation, instinctive behavioral, conscious emotional, conscious intellectual, and cultural and moral behaviors (Weihe, 1987).

Climate is a necessary challenge to mankind's potentials. Adaptation processes involve intracorporal mechanisms of the organism and extracorporal adjustments of the environment to satisfy needs with a minimum of strain. The efficiency of the coping measures have enabled mankind to settle in marginal climates of great cold, heat, and altitude. Of particular value are the inventions of technologies to protect life from a direct climatic impact by means of clothing, housing, heating, cooling, screening, and related technologies. The best proof for mankind's mastery over the environment is the worldwide improvement of health expressed in high fecundity and life expectancy. These improvements have their price. Wherever people settled down minor or major scale climatic changes have resulted.

Perception of Climate

Climate is defined as long-term spatial generalization of day-to-day meteorological conditions. Four types of climate change are of importance for humans (Hare, 1985):

1) Climatic noise which are the short-term variances of climate referred to as weather. People are familiar with them in their native climate and know by experience how to adapt and adjust

2) Climatic variability of climatic variables within typical time periods chosen for averaging, usually annual

3) Climatic fluctuations which are transitory climatic changes lasting over periods of decades

4) True climatic change with differences of means and variances between successive averaging periods longer lasting than climatic fluctuations.

Each of these types of change requires distinct patterns of human adaptation. Human perception of climate and sustained attention depends on the impacts and their interpretation. Without assessing the climatic perception for Man's purposes spatial generalizations of atmospheric conditions remain studies of climate-for-its-own sake (Terjung, 1976). Direct applications of climatic conditions to daily life are numerous. Best known applications are those to body temperature regulation resulting in thermal indices for indoor and outdoor climates based on physiological responses and feeling of thermal comfort or discomfort. The longer the overview period the more uncertain become individual statements on climate changes. Uncertainty of judgments begins over interannual changes and increases over interdecadal changes. A 30% increase of summer precipitation over a 30-year period in the St. Louis urban area was registered by slight changes in occupational and recreational behavior "but people remain unaware that they are making these changes in response to anything unusual" (Farham-Pilgrim, 1985). Perception will not be aroused as long as the people are free to adapt concurrently with mild or moderate climatic changes. With changes of weather and climate adaptation and adjustment processes are instantly initiated. They will persist as long as the changes continue. Through adaptation the vulnerability is reduced and the capacity for coping strengthened. While adaptation continues and gains efficiency, the sensitivity and vigilance to climatic changes relaxes. Soon they are less noticed, ignored or forgotten. It is characteristic of successful climatic adaptation that the attention to impacts of climatic changes fades and the motivation to store adverse effects in memory becomes unnecessary (Whyte, 1985).

Principles of Response to Climate

Climate acts through factors as stimuli that cause responses. Between impinging of the stimulus and response are multiple assessment and resolving processes. They operate through internal and external feedback mechanisms controlling the interior balances of the organism in exchange with the stimulus conditions. The self correcting negative and the accelerating positive feedback mechanisms act concomitantly to reestablish balances where disbalances in the organism are sensed and correcting processes activated. Climate changes with their ups and downs of conditions for the organism proceed linearly. Individuals and societies respond specifically in a circular fashion where deviations of equilibria lead to oscillation and looping of responses of the regulation systems within varying tolerance limits that are slowly moving back to stabilization of equilibria. The autonomic-physiological and instinctive behavioral adaptation processes are genetically controlled: the conscious emotional, intellectual and cultural-moral adaptation processes are learned from experience by trial and error, layed down and passed on as culture. They are guided by the vital forces to maintain the strength of the individual, avoid vulnerability and improve efficiency under changing conditions.

Bands of tolerance of the individual are the upper and lower limit of resistance to respond to climate noise and variability without lasting perturbation or harm. They can be widened actively by physical training and exposure to climatic noise, they will narrow under the protection from exposure with decreasing resistance to greater climatic variances. De Vries (1985) has described the band of tolerance as the upper and lower limits of climatic variability for which the institutional and technical practices of a society are designed. Margins of safety are formed in the individual as well as in the society. The safety margin of the society is dictated by the experiences and risk acceptance of the individual and community with climatic variabilities in the past as the key to the future.

The impact of climatic events is determined by their strength. The impact of strong events such as storms, heat and cold waves, will have more direct effect and involve more people, the larger the disaster, the more suddenly and unexpectedly they occur. Their strength of impact is larger the less adapted and prepared individual persons are. Consequently the least prepared and the most vulnerable persons will be affected first and most. Experience with climatic disasters shows that the economic effects on society from environmental damage can be very high but the effects on the individuals are comparatively little, though the psychological morbidity might be increased (Andrews and Tennant, 1978, Foster, 1976, 1980).

The most important vulnerability variables of the individual are age, gender, and health state. Adaptability and resistance of the individual are weak at the beginning and near the end of individual life, during disease, and when handicapped, they are strong during adult life. The effective coping of man by natural adaptation is assisted by his multiple technical, environmental and societal adjustments. Damage to health from climatic impacts is often pure chance because it depends on many additional individual and environmental factors with their protective or aggravating potentials. Apart from the climatic event the variety and multitude of concurrent stimuli and responses makes it extremely difficult to infer from an observed

effect to a direct climatic cause. This was not realized when a first review on climate and health was prepared (Weihe, 1979). In epidemiological studies each situation demands a specific analytical procedure. Because of the multicausality of effects in the evaluation of results from climate-orientated epidemiological studies, contemporaneity of events is not tantamount to causation or 'correlation is not causation'.

Anthropogenic Climate Change

Anthropogenic climate changes are forecast to occur within the next 40 to 100 years. Such time spans are too long for the people to perceive, or toarouse and sustain their attention. Attention will gradually fade in as much as adaptation of the people and adjustments of society and environment are approaching a better fit with climatic conditions. Baseline values will shift and old ones are soon forgotten. Irrespective of the patterns of change there is ample time during periods of one to two human lifespans to adjust tolerance bands and safety margins.

It is impossible to prognosticate for time periods extending over several generations of people what the direct and indirect impacts of the modelled climatic changes on health could be because of the simultaneous perpetual multiple human adaptation and adjustment processes to fit the climatic conditions. The implications of various scenarios of climatic change have been examined for specific diseases and disease groups such as CVD (Table 2) and skin diseases (Table 3). Analyses of possible consequences will greatly facilitate decisions on suitable measures for response strategies. Early analysis is the first step not only for coping with but also for prevention of potential effects of global climate change on human health. The lag period until a climate change will be evident is of value, because settling the many difficulties arising with societal adjustment take much time. Strategies to counteract or delay climatic changes can be introduced successively in due time as their effects become known. Measures for prevention have already been enforced on national and international levels, meaning that adaptation is operating though socioeconomical and political forces, even if this is hardly realized by the individual.

To date climate scenarios predict that the mean global temperature would increase between 1.4 and 2.8°C by 2030 and between 2.6 and 5.8°C by 2090, and that the sea-level would rise between 9 to 29 cm by 2030 and 38 to 96 cm by 2090 (IPCC, 1990). The mean annual global temperature increase will be associated with increases of humidity or dryness, changes in rainfall, and radiation intensities. Specifically, there will be an increase in UV-B radiation fluxes as a result of the destruction of the upper atmospheric ozone layer. The increases of mean temperatures are expected to be less at low latitudes and

Table 1: *Classification of effects of climate on health. Some examples.*

Direct	Comfort feeling
	Heat stroke, frostbite
	Food requirement
	Strain and strain relief
	Morbidity from systemic diseases
	Mortality
Indirect	Vector-, air-, water-, soil- and food-borne infectious diseases
	Food availability including draft animal diseases

Table 2: *Examples of diseases of the circulatory system (CVD) susceptible to cold and heat.*

Disease	ICD
Hypertensive HD	402
Myocardial infraction	410
Subacute ischaemic HD	411
Chronic ischaemic HD	414
Cardiomyopathy	425
Heart failure	428
Cerebrovascular diseases	436-438
Atherosclerosis	440

ICD = International Classification of Diseases (WHO)

higher at mid and high latitudes. These global and latitudinally and geographically modulated mean changes will be the outcome of unknown climatic perturbations with increases in amplitude, frequency, length, and phase rate at the time scale of (1) climatic noise, (2) climatic variability, and (3) climatic fluctuations. Only the first two require particular attention with respect to health. Forced migration of people living in low lying coastal areas in case of land inudation due to a progressive sea level rise has such a long time lag that adjustments will reduce the particular effects on well-being. But the main problem from rising sea-level in delta areas will be the increased frequency of severe flooding of coastal and upstream land.

Impacts of Climate on Health

A distinction of the impacts of climate between direct and indirect effects on individuals and population groups facilitates risk evaluation and risk management. Examples of effects of climate on humans are summarized in Table 1. In reality such a strict differentiation never exists and may be misleading. Even strong direct effects of climatic factors on the organism are modified through the impact of other environmental factors affected by climatic factors. A review of potential health effects of climatic change was prepared by WHO (1990a).

Direct Effects

Diseases of the vascular and other systems

The most important short lasting climatic events are heat and cold waves exposing individuals to marginal conditions to which they are commonly not adapted or which exceed their regulation capacities. Particularly vulnerable to heat are individuals suffering from cardiovascular, cerebrovascular, respiratory, endocrinal, renal and consumptive diseases such as chronic infections and terminal carcinoma. Without protection the diseased body

regulation systems will be overtaxed, the basic disease will be aggravated or lead to death. Most often investigated is the increase of mortality from cardio- and cerebrovascular diseases (CVD) of higher age groups during heat waves in the USA (Weihe, 1986). Examples of CVD's are given in Table 2. The protective measure of moving patients out of the heat to cool air-conditioned rooms has a long practice (Burch, 1969; Burch et al. 1959). Mortality peaks during heat waves can be avoided or even inverted to low mortality through care, education, and early warning for seeking cool shelter (Figure 2). These measures are to be seen as cultural achievements typical of societies in seasonal temperature climates with the means of applying technologies for protection against stressful changes of outdoor climate. In the diurnal temperature climates of the tropics with extreme and longer lasting heat waves intensified by high humidity, while technical facilities for retreat to comfortable indoor climates rarely exist, such sudden heat-related peak mortalities are not observed. Natural adaptation to heat and cultural habits have acted to reduce vulnerability of the people. Otherwise the large population growth in hot-humid tropical climates would never have occurred.

The phenomenon of increased urban mortality in the USA during heat waves does not necessarily mean that more frequent or extreme heat waves in the future would lead to higher rates of untimely death. To some extent the observed phenomenon of urban heat wave mortality could be the result of inefficient heat adaptation. With adaptation and habituation to living continuously under artificially sustained thermal comfort conditions, the individual and population heat tolerance band is narrowed down. This needs further investigation. Risk preparedness, of the population in general and of groups and individuals at risk in particular, to cope with heat would involve two alternatives, both have 'lessening' potentials (Bowden et al.

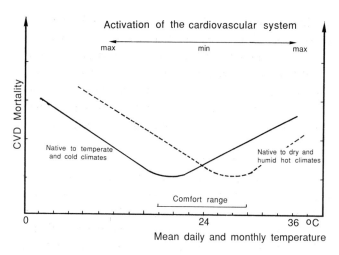

Figure 1: Relationship between CVD mortality rate and adaptation to native climates The cardiovascular system is activated and strained under both heat and cold conditions with a minimum of strain at comfort temperatures People native to temperate and cold climates prefer and tolerate lower, people native to dry and humid hot climates prefer and tolerate higher comfort temperatures On the global scale comfort temperatures range from about 17 to 30°C Cold native people show a higher vulnerability to heat, heat native people to cold.

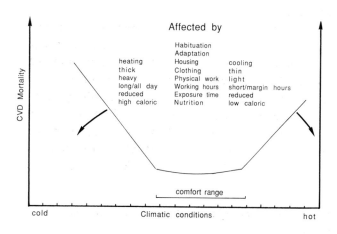

Figure 2: The physiological strain in cold and hot climates can be mitigated or prevented with elimination of temperature affected CVD mortality rate by a variety of single or combined behaviors which are different or opposite in cold and hot climates.

1981): one lessening effect would come from changing lifestyles with more direct exposure to climatic conditions to promote natural adaptation and activate regulation as in economically poor societies in tropical climates (Auliciems, 1982); the other is wider application of technologies such as air-conditioning for living permanently under accustomed comfort conditions. The former would require less energy for conditioning of buildings and

nutrition of the people but at the expense of a decline in their working capacity and efficiency. The second lessening response pattern would increase the energy demands for running the technology; the nutritional energy requirements would remain high, and the high level of working capacity and efficiency could be maintained. This version would conform with preserving present lifestyle of industrial countries of mid-latitude climates, but it would be expensive and further the process of climate change through the increasingly required utilization of CO_2-releasing energy resources. The relationships are illustrated in Figures 1 and 2.

The same applies to the effect of cold waves. The physiological capacities of the human body for cold adaptation are poor compared to those for heat adaptation. Mortality increases among the people at risk, particularly in countries at low socioeconomic levels, if they lack the means for cold protection and heating (Keatinge et al., 1989). If people cannot insulate their bodies with clothes and heat their dwellings, they will be unable to withstand cold for any length of time. The provision of thermal shielding will effectively break the impact of cold waves. It has been suggested that wide use of indoor heating will narrow the cold tolerance bands of the population (Keatinge et al., 1989). A favorable effect of global warming would be a decreasing requirement for heating in presently cool and cold climates beyond mid-latitudes.

Food requirement

The caloric requirements of individuals are much affected by the thermal climate through the amount of muscle work done and heat losses. For calculating caloric requirements the mean annual external temperature isotherms have been used as an index. The standard requirement for a Reference Man (25 years of age, 65 kg weight) is 3,200 Calories daily and Reference Woman (25 years of age, 55 kg weight) 2,300 Calories daily living in the 10°C mean annual temperature isotherm zone and consuming an adequate diet without gaining or losing weight. The caloric requirement increases as temperature decreases. The caloric requirements should be decreased by 5% for every 10°C above and increased by 3% for every 10°C mean annual external temperature below the reference temperature of 10°C (FAO, 1957). These normative considerations have to be modified, because people native to higher latitudes are heavier and larger and at low latitudes lighter than the global Reference Man. Caloric requirements are greatly modified by the living conditions, habits, availability of food, the physical activity and age of the people. There is adaptation to overnutrition as well as to undernutrition. The latter allows people a normal life of high productivity at much lower caloric intakes than estimated from FAO standards (Pratt and Boyden, 1985) without the suggested starvation (Edmundson, 1980).

Strain and strain relief

The perception of and confrontation with climatic changes is likely to provoke psychic stress in sensitive individuals. This is limited to a short-lasting arousal and an extended period of sustained attention in the case of insufficient coping (Baker and Chapman,1962). The psychoendo-crinological correlate is an increased release of neurohormones like endorphins (Heckenmueller, 1985) and noradrenaline, with subsequent stimulation of the adrenal medullary and cortical function (Frankenhaeuser, 1976; Mason et al. 1976; Van Praag, 1986). Such endocrine responses are thought to cause 'stress diseases'. Climatic factors are insignificant as causes or promotors of stress diseases. There is no evidence for stress disease entities in Man comparable to those produced experimentally in animals. However, evidence exists that psychic stress as a response to the perception of climatic events leads to transitory perturbations of body regulation mechanisms (Cassel, 1976). This may have some specific consequences for disease resistance, in particular for immune response functions (Kronfol and Schlechte, 1986). The individual evaluation of climatic events leading to psychic stress is largely determined by the event intensity, its perception, risk acceptance, and awareness of coping capacities (Foster, 1976, 1980; Melick, 1978). This is greatly influenced by the attitude to voluntary or involuntary exposure. Voluntary exposure activities are determined by the individual, involuntary exposure activities by the psychosocial environment. The public acceptance of voluntary risks was estimated to be 1000 times greater than that of involuntary risks (Starr, 1969). Public voluntary risk acceptance of the devastating environmental hazards in the Caribbean islands was far higher on the list than a good dozen involuntary risk conditions (Whyte, 1985). Psychoendocrinological perturbations in the event of voluntary risks are small but can be large in involuntary risks. Much depends on how the public will accept climatic changes. Even in severe cases of uprooting, people leaving land because of inundation or desertification will adapt psychosocially to new environments if they accept their chance for survival (WHO, 1979).

Longer lasting and wider ranging swings of climatic events would impose stronger but principally no different conditions for direct effects. On the contrary, longer lasting time periods would allow for physiological adaptation, provision and utilization of protective measures, and of aids to alleviate the strain. Indirect effects would involve the availability of food, quality foodstuffs and water supply.

It has been suggested that in hot, particularly in humid-hot climates, the life-long heat induced strain for the vascular system by wear and tear would reduce the high life expectancies presently observed among populations in warm- and cold temperate climates (Weihe, 1986). If this is true, the expected shift of higher mean temperature to mid-

and high latitudes would counteract prospects for high life expectancies. The likelihood of such a development could be increased by a lower heat resistance capacity of these populations not native to prolonged and strong heat impact. This is unproven and cannot be substantiated easily, because heat resistance capacity is a complex parameter that is affected by many interrelated factors (Figure 2).

Effects of UV-B radiation

The destruction of the stratospheric ozone layer by free radicals of CFC compounds released by human activities is expected to lead to an increase of UV radiation at the earth's surface. Of particular importance is the biologically effective UV-B wavelength band between 290 and 325 nm with the maximum at about 305 nm. The biological activity of UV-B is influenced by the thickness and cleanness of the skin, the quality and quantity of sweat, and the presence of chemical carcinogens in the air and on the skin (Cascinelli and Marchesini, 1989). Since correlations have been observed between UV-B irradiation induced non-melanoma skin cancer (NMSC) and malignant melanoma of the skin and latitude, an augmentation of incident biologically effective UV-B radiation could be a serious risk factor in the future (Urbach, 1989). UV-B radiation affected diseases are listed in Table 3.

Annual means of UVR increase from high towards low latitudes and from low towards high altitudes. Reiter et al. (1982) measured an increase of the annual mean of UV daily total and UV maximum of 12% from 0.7 to 1.8 km and 40% from 0.7 to 3.0 km in the Northern Alps (47.5°N. Lat.). The magnitude of impinging UVR is greatly modified by weather, month and hour of the day, and ozone in the human boundary layer, particularly over polluted areas. An increase of incident total UVR would result in a

Table 3: Examples of eye and skin diseases susceptible to UV-B radiation.

Disease	ICD
Eye	
Pterygium	372.4
Keratopathy	371.4
Snow blindness	370.2
Skin	
Non-melanoma skin neoplasm	M801-804
squamous cell	
basal cell	
Malignant melanoma	M872-879

ICD = International Classification of Diseases (WHO)

relatively higher increase of the biologically effective UV-B radiation. To estimate the importance of UV-B for skin damages, amplification factors have been defined. Based on epidemiological studies over a wide range of latitudes, the biological amplification factor for NMSC increases from 1.44 for >47°N Lat. to 3.0 at 35°N. Lat. (Albuquerque, USA). A similar study was carried out for malignant melanoma (Scotto and Fears, 1987). Minor increases below 2.5 to 4.0% such as the presently observed increases of 1 % UV-B fluxes at 47°N. Lat. (Blumthaler and Ambach, 1990) will not have significant consequences. The estimated percent increases of NMSC and malignant melanoma incidence due to ozone reduction data for period 1969–1986 in the USA are greatly below the recorded increases. While a 3 %/y increase of malignant melanoma was expected, the actual increase was 7%/y. Glass and Hoover (1989) found much higher increases for squamous cell skin cancer and for malignant melanoma of over several 100 % during three decades from 1960 to 1986 in the United States.

The divergence between results of scenarios and laboratory and epidemiological field studies indicate that beside the direct effects of UV-B other causative factors act indirectly, particularly in the case of malignant melanoma. A consistent UV-B dose-dependent association has been found between ocular UV-B irradiation and the two common types of cortical and posterior subcapsular cataract. Ocular exposure is also linked with several corneal changes, pterygium, keratopathy and acute snow blindness (Taylor, 1989). There is a highly significant correlation between time and length of exposure to solar irradiation and nonmalignant basal cell and squamous cell skin cancer. Both are primarily due to intense exposure to UV-B radiation (Urbach, 1989). Skin cancer incidence increases with distance from the poles in countries with white populations. A strong causative factor of their increase in Australia, Northern Europe, and the United States is seen in the extensive voluntary sunning of the uncovered body, particularly of white persons who burn quickly in the sun and do not tan well. This is a good example of high voluntary risk acceptance of sunbathers who are aware of the risk. Intense brief solar irradiation exposure is also a risk factor for malignant melanoma in exposed body areas and is greater the younger the individual. The bulk of malignant melanoma is due to other environmental factors than irradiation by the sun. Malignant melanoma was observed much more frequently in persons working in closed environments than those exposed on land and sea. Persons born in sunny climates are at greater risk than adult seasonal migrants (Lee, 1989). The increasing incidence of melanoma in recent times could be a result of the improved criteria of clinical diagnosis and the attention given to it by the public (Cascinelli and Marchesini, 1989).

In the case of increasing global UV-R irradiation the risk of contracting NMSC and malignant melanoma would be limited to the 20% of the people with poorly pigmented skin typical among the Caucasian race particularly among populations in northern areas such as Sweden and Scotland. It would not arise for the 80% of the present world population of ethnic groups genetically provided with dark skin or high potentials for melanin pigmentation on exposure to UV radiation. Protection against UV-B radiation damage is effectively achieved by covering the body surfaces with clothes and application of absorption lotions and creams on uncovered surfaces, wearing hats and dark UV absorbing eye glasses. In this case education in risk awareness is highly efficient for disease prevention.

The increase of UV-B irradiation has also been considered as a risk factor for suppression of the human skin immune system. On the basis of animal experiments the risk of immunosuppression is highly uncertain. Immunosuppression could further the development of NMSC and the course of infectious diseases, but protect against autoimmune reactions. To date an immuno-suppressive action of UV-B on human skin remains uncertain (Morison, 1989).

Indirect Effects

The indirect effects of climatic factors are of great importance as risk factors for the development and areal propagation of communicable diseases.

Air pollutants and respiratory diseases

Increasing air pollution emissions in urban and industrial centers impose an immediate danger to health, which will be aggravated by climatic change. Air pollutants act directly on the organism and indirectly through secondary compounds such as ozone and sulfuric acid after photochemical transformation of CO, SO_2, and NO_x. Increasing concentrations of CO_2, methane, and the various CFCs affect the transfer in the atmosphere of solar and terrestrial radiation, resulting in a warming of the surface and lower troposphere and increased UV radiation (Rodhe, 1990). Common concentrations of CFCs are not directly toxic to people. But indirectly the CFC-induced increase of incident UVR will intensify the photochemical formation of ozone in the human habitation layer. Ozone is an aggressive substance, in low concentrations it irritates, in high concentrations it destroys tissue membranes and cells. The irritation involves primarily the mucous membranes of the respiratory tract. The increase of ozone is correlated with NOx concentrations in urban areas and solar radiation. Recently much attention has been given to the biological significance of this phenomenon, particularly when heavy physical work is performed, and in the case of diseases of the respiratory system such as asthma, chronic pulmonary obstruction, and bronchitis in children. To date no reliable data exist to define toxic ozone levels; they are estimated to range between 120 and 450 µg/m^3 for a single 1-hr exposure. Local ozone is quickly destroyed in polluted urban air; ozone carried away by wind to the clean air of rural areas and up mountains will accumulate and remain stable longer. Industrialization of tropical cities will produce high NOx pollution levels in zones with particularly high incident radiation. To evaluate future risks monitoring of boundary layer ozone over the rapidly expanding cities of low latitudes is needed for comparison with data increasingly available at mid-latitude locations.

Communicable diseases

Climate affects those communicable or infectious diseases in which extracorporeal transmitters such as air, insects, animals, water, soil, and food are intermediated in the transmission route and development of parasites. The parasites include viruses, rickettsiae, bacteria, fungi, protozoa, helminths, and arthropods. While the various steps in the etiology and the nature of the transmitters of these diseases have been thoroughly investigated, the specific role of climatic factors is far from clear. As the majority of the diseases are indigenous and most frequently observed in hot-humid and dry tropical climates, they have been designated tropical diseases. However, they are not confined to the tropics for purely climatic reasons. Major reasons for their frequent epidemic appearance are the still widely existing low standards of hygiene in Man's physical and social environment. In many tropical countries where the standards of hygiene and parasite transmission control have been improved and are widely practiced, as is the case in most countries of temperate climates, many diseases have been eradicated and persisting endemic foci are kept under tight control (Matzke, 1979). An exceptional case of persistence is malaria. A spreading and reappearance of many of these diseases in the case of climatic shifts to higher latitudes would depend primarily on whether the presently high hygienic standards of health and environmental control could be maintained. Higher mean temperatures would increase the risk of a reappearance of the diseases if the institutionalized hygienic control were relaxed for reasons other than climate. Possible effects of climatic change can best be studied by means of epidemiological models integrating these factors (Rogers, 1988).

After an epidemic has passed there is resistance or herd immunity within the affected population which slowly declines in the absence of intermittent boosting. The import of a disease into a population which was not exposed to it for a long time will usually cause substantially more morbidity (and possibly mortality) than is seen in a population of the same size that has been exposed to the disease for a long time. The measles epidemic of 1846 in the Faroe Islands, where measles had not been prevalent since 1781, is a well known historical example of how devastating the intrusion of a "new" disease in an immunologically unprepared population can be: 77% of the entire population contracted the disease within a period of six months, and it was directly responsible for the death of at least 1.3% of the population in that period (Panum, 1948). These are distinct from the direct effects of climatic conditions on the disease resistance of individuals continuously harboring potential pathogens or their readiness to fall ill in case of contamination. These latter relationships are little understood.

Insect-borne diseases

Most important from the climatic point of view are the insect-borne communicable diseases depending on biological vectors. Crucial for their transmission is that the temperature conforming vector species - chiefly insects - and parasites find optimal thermal conditions of living, development, and survival for long weeks or months or most parts of the year. Depending on climatic and environmental conditions, the vector species spread from the low latitudes of the tropics and subtropics only during the warm seasons into the temperate climate zones at higher latitudes of both hemispheres. Vector-borne diseases in cold climates depend on temperature-regulating homeothermic mammals such as rodents. This makes them fairly independent of climatic conditions. The occurrence of tropical diseases is governed by an interplay of the host, the prevailing vector species, the parasite distribution, their mutual adaptations and resilience, densities and behavior of the people. Climate affects all of these factors to some extent.

Some important vector-borne diseases limited to tropical climates are listed in Table 1. The most important of them today is malaria. A large proportion of the world's population in the tropics is at risk. In the past endemic malaria seasonally spread far north in Europe and parts of North America as long as plasmodium and mosquito control were inapt (Bruce-Chwatt and De Zulueta, 1980). Epidemics of yellow fever occasionally occurred in Portugal, Spain and the USA (Brès, 1986). Stringent control measures, aided by improving environmental conditions such as drainage of marshland, and changes in lifestyle following economic progress, have led to the eradication of malaria and yellow fever in areas with temperate climate and many tropical areas. The possibility of geographical malarial occurrence based on temperature, rainfall and elevation is demonstrated in Figure 3 (Dutta and Dutt, 1978). The map does not represent the small scale geographic regional difference in climate and disease distribution. Moreover, during the past centuries, climate and, hence, disease distributions have been subject to shifts. The now semi-arid Indus valley was once a fertile plain,

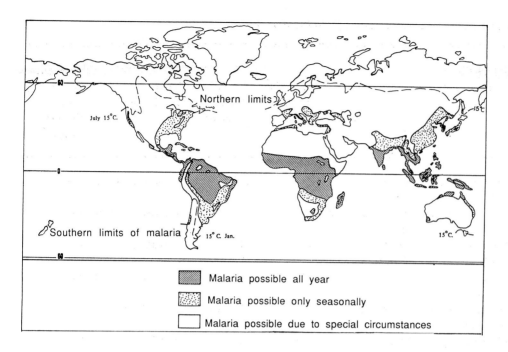

Figure 3: Possibility of malarial occurrence, based on temperature, rainfall and elevation (Dutta and Dutt, 1978) The limits in seasonal temperature climates are at about 50 degrees N and S With climatic shifts the conditions would favor spreading of Anopheline species to higher latitudes and to higher elevations in the tropical belt.

harbouring one of the early cultural development foci of mankind.

The importance of temperature is particularly impressive in the case of malaria. The mosquito-plasmodium-human body continuum operates not below 10°C and not above 32°C, is it optimal within the range of 15 to 32°C with 50–60 % RH. In the case of humans there is peripheral vasodilatation above 25°C with sweating. This enforces reduced clothing, may attract the mosquito and biting, and facilitate taking the bloodmeal. Temperature determines the spreading of the anopheline species, of which more than 300 subspecies are involved in the parasite transmission, their egg deposition, larval and pupal development, length of adult life and biting cycle. Due to the temperature lapse rate with elevation, Anopheline species do not breed at elevations above 2,000 m outside the tropical belt. Temperature guides the sporogony of the protozoa. Plasmodium will not survive below 10°C and above 32°C during sporogony in the mosquito. Sporogony of P. falcifarum, causing tropical malaria, depends on temperatures between 19 to 32°C, that of P. vivax, causing tertian malaria, begins already at 16°C. Sporogony accelerates from 25 days at the lower to 8 days at the upper temperature limit. It is fastest at 25–30°C and 60% RH. Other climatic factors are important such as still shallow waters, rainfall, high humidity, sunshine and shade. Relative humidities between 25 and 60% are required. The significance of rainfall for malarial incidence is demonstrated in Figure 4 (Dutta and Dutt, 1978). The

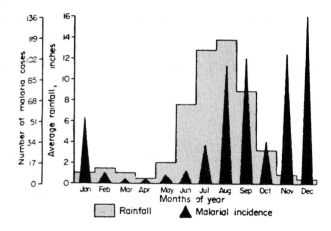

Figure 4: Mean monthly malarial incidence from 1970 to 1973 compared to 50-year mean rainfall at Bihar, India (Dutta and Dutt, 1978).

number of malaria cases increases with a 2-month time lag after the beginning of the monsoon rain and continues for so long after it has passed. There are also Anopheline species which prosper in dry zones resulting from drought. This high adaptability of the species demands untiring efforts in parasite transmisson control.

As air temperature and rainfall are the limiting factors for the malarial vector, global warming with a shift and extension of warm humid seasons towards higher latitudes and altitudes can be expected to provide more favorable

Table 4: *Examples of major vector-borne tropical diseases (WHO 1990b, c).*

Disease	Causative agent	Vector	Population at risk [†] (millions)	Prevalence of infection (millions)	Present distribution	Likelihood of spreading due to climatic change [††]
Malaria	*Plasmodium* spp. (protozoa)	*Anopheles* mosquitos	2,100	270	Tropics/subtropics	+++
Lymphatic filariasis	Nematode worms : family filaridae	Various mosquito species	900	90	Tropics/subtropics	++
Onchocerciasis (river blindness)	*Onchocerca volvulus* (filaridae)	*Simulium* blackflies	90	17.6	Africa/Latin America	+
Schistosomiasis (bilharziasis)	*Schistosoma* flatworms	Water snails	600	200	Tropics/subtropics	++
African Trypanosomiasis (sleeping sickness)	*Trypanosoma* spp. (protozoa)	Tsetse flies	50	(25,000 new cases/year)	Tropical Africa	+
Leishmaniases	*Leishmanis* spp. (protozoa)	Sandflies	350	12	Asia/S. Europe/ Africa/S. America	?
Arboviral Diseases Dengue fever	Dengue virus	*Aedes* mosquitos		(30-60 million infections per year)	Tropics/subtropics	++
Yellow fever	Y.F. virus	Various mosquito species		> 3,000 deaths in 1986-1988	Africa/ The Americas	+
Japanese encephalitis	J.E. virus	Various mosquito species			Asia/Pacific Islands	+
Other arboviral diseases	+90 other viruses	Anthropods, mainly mosquitos and ticks			Cosmopolitan	+

[†] Bases on a world population estimated at 4.8 million (1989)

[††] +++ = highly likely; ++ = very likely; + = likely; ? = unknown

conditions for the spread and longevity of mosquito populations. Malaria, specifically tertian malaria because of the better cold resistance of P. vivax, could possibly occur for longer periods where it is now only seasonal and it could occur seasonally where it presently does not exist. With reference to Figure 3 this shift in occurrence pattern would be beyond the approximate northern and southern limits at about 48 degrees latitude. The likelihood of spreading of malaria and other vector diseases is indicated in the last column of Table 4.

African trypanosomiasis, the other important protozoal disease of tropical climates, occurs in humans as sleeping disease and in cattle as naganan. Wild animals are an important reservoir of the parasite extending over most of tropical Africa. Presently the disease of cattle is of much greater importance than that of humans. Cattle are susceptible to numerous trypanosoma species which readily infect tsetse flies, while humans are susceptible only to a few species to which tsetse are usually resistant. Agriculturally uncontrolled land will soon be occupied by tsetse flies maintaining the parasite transmission route between wild animals which in the parasite-cattle continuum leads to high cattle mortality with the subsequent shortage of animal protein for human nutrition (Matzke, 1983).

Filarial infections such as lymphatic filariasis, causing swelling of the limbs with chronic disability and suffering, and onchocerciasis are endemic diseases chiefly of the humid tropics. They are transmitted by many species of mosquitos such as Culex, Aedes, Anopheline, Mansonia and Similium. Various species of Aedes favor wide climatic variations of habitat often in the vicinity of shack towns. Optimal ranges are 21–32°C and 70–100% RH. The helminth Onchocerca volvulus, causing debilitating blindness onchocerciasis known as river blindness, is transmitted by Similium species breeding on fast streaming rivers of warm water. The main transmission is during the rainy season when Similium breeding is maximal. The active flying capacities are limited to short distances away from the river beds but winds will carry the flies over long distances to remote areas leading to spreading of the disease.

The transmission of the helminth species causing the chronic intestinal or genitourinary schistosomiasis depends on the presence of aquatic snails. The eggs must reach water, penetrate a snail, within which they develop to cercaria; these emerge from the snail back into water. While floating in water they penetrate the intact human skin or are swallowed during wading and bathing. Snail species breed between 18–35°C, the optimum water temperature is between 22 and 26°C. Multiplication and cyclical development of the parasite in the snail takes about 40 days at 30°C water temperature, and longer if the water temperatures are lower. The disease is associated with low standards of personal and public hygiene. It is endemic in artificially-irrigated, agriculturally-used areas where children and workers have frequent contact in standing snail-infested waters.

The vectors of the arboviruses (arthropod borne viruses) which lead to outbreaks of diseases such as yellow fever, dengue fever, Japanese and other encephalitis diseases are various species of Aedes breeding at temperatures above 24°C and high humidity. Aedes aegypti is peri-domestic, breeds in puddles and every kind of water container around houses. The recent accidental introduction in the USA of Aedes albopictus, vector of the dengue virus has given much reason for concern especially as a climatic change towards higher ambient temperatures may enhance the vectorial capacity of this mosquito (Knudsen, 1986).

Plague, not listed in Table 4, is a disease which is not limited to tropical climates. It is essentially a disease of rats spreading from reservoirs of sylvatic plague and is transmitted to Man when contaminated rats invade human settlements. The causative agent is transmitted by fleas living and breeding on the rat's skin. Optimum conditions for the fleas are 20 to 25°C and high humidity.

Proximity diseases

Another category of climate-affected infectious diseases considers those in which the transmission of the causative agent depends on a mechanical vector via contaminated media such as air, aerosols, water, soil, faeces, foodstuff, and other articles with climatic factors, predominantly temperature and humidity. The viability of air-borne pathogens depends on temperature, humidity and solar radiation. A classical water-borne disease is cholera. Water

availability and water quality are of paramount importance in the prevention of all water-borne diseases of the intestinal tract. The burden of diarrhoeal diseases is mainly concentrated on populations living in tropical and subtropical regions, affecting particularly small children living in poor sanitary conditions. Salmonella, the causative agent of typhoid fever is a food-borne disease. Hookworm infection results from contact with soil contaminated with free living stages of the parasite from human faeces. Survival of infective hookworm require soil temperatures of 24–32°C and shade. The viability of larvae decreases at higher temperatures. Eggs and larvae of further species of worms pathogenic for man such as Ascaris, Strongyloides, tapeworm and others are directly dependent on soil temperature, rainfall, evaporation, humidity, and wind. Most causative agents of diseases of the respiratory system are transmitted by air such as the influenza virus or pulmonary bacteria causing acute respiratory diseases and pneumonia. Here the distances of the transmittance path for contaminated droplets from person to person are exceptionally short. While morbidity from the common cold is prominent during the winter season in temperate climates, most viral respiratory diseases have a cosmopolitan distribution. Like influenza they are not strictly limited to one season, outbreaks of epidemics may occur in summer as well as in winter.

All proximity diseases are associated with low standards of personal and public hygiene. Their etiology is well understood and effective methods to avoid spreading and to destroy the causative agents are well known. Outbreaks of these diseases are local and it has been possible to isolate epidemics to small areas. Rather little is known about the direct effects of climatic factors on these agents for the time of their extracorporeal phase. Perturbations in the disease control as a result of climatic change could come from societal disruptions through migrants, crowding, poverty, growth of urban slums, frequent storms, floods or flooding of town sewerage systems. Besides water quality, the quantity available to households is essential for personal hygiene.

Food availability

An indirect climatic effect of vital importance for health is the availability of food to fulfil the nutritional and caloric requirements of the people. Grain, food plant, and animal production directly depend on climate. Fungi, moulds and pests favored by climate destroy the quality of food and produce toxic admixtures. Climatic and other conditions favor rapid reproduction of rodent populations. Fields and storage spaces will be invaded by rats and mice transmitting these pests and taking vast shares of cereals for their consumption. Large climate-protected and rodent-proof storage facilities, possibly refrigerated, become necessary. Droughts—or exceptionally heavy rainfall—

such as droughts over periods of many years in the Sahel and the Great Plains in the USA, combined with crop pests will lead to harvest losses with subsequent reduction or absence of the availability of food. Conditions of overpopulation, crowding from rapid urbanization, disruption of social structures with poor access to health care would lead to undernutrition, even famine, and, in conjunction with failing hygienic control and societal perturbations, to the resurgence of infectious diseases with the outbreak of epidemics, endangered through infestation of ectoparasites as potential transmitters of bubonic plague in humid climates and of typhus fever in temperate climates. Relief would come from national or internatinal cooperation (Bowden et al. 1981). Endemic starvation will be more frequent with progressive land destruction in those low latitude countries where food production is already marginal and improper infrastructure impairs food distribution for relief. Climate-related destruction of agricultural land has forced migration of people. This will increase as these populations continue to grow to the point where they exceed the sustaining capacity of their native area without support from outside (Brown, 1989).

One aspect of particular relevance to health in the context of deforestation and land conversion is the inevitable loss of medicinal plant species. Many plants are potential producers of phytopharmaca and may disappear for ever, some of them before their potentials have been investigated (Soejarto and Farnsworth, 1989).

Estimation of Future Risks

The spread of human settlement from the equator to the highest northern latitudes and to great mountain plains at low latitudes provides a collossal natural experimental ground for comparative studies on the impact of climate on diseases. While the climatological data are standardized and are collected with increasing refinement, completeness and geographical density, disease and mortality statistics are incomplete, biased, and unbalanced. Most present knowledge on the direct and indirect effects of climate on diseases is derived from case studies with insufficient human health statistics in a limited number of geographical areas. The interpretation of results is controversial and often too widely generalized. This has given rise to much confusion over the impact of climate. Escudero (1985) has stressed this in the example of the two geographically neighboring countries Cuba and Haiti both with extremely stressful climates. The mortality profile in Cuba is nearly as low as in the United States and Canada; in Haiti the health situation appears "as a minor catastrophe"; in Cuba health statistics cover almost the entire population, in Haiti they hardly exist. Recently great advances in vital statistics at large have been achieved and are documented by and available from the series of World Health Statistics Annual

of the WHO. The highest standards of health statistics today are in the industrial countries with controlled health departments. Exhaustive reporting is not always necessary or cannot be obtained (Eylenbosch and Noah, 1988). High standard vital and health statistics of equal quality worldwide with total coverage of the population are essential for biometeorological correlation studies. They will facilitate the evaluation of representative demographic and morbidity profiles for cross-cultural and cross-climatic comparison. New methods such as the widest use of geographic maps as the basis for plotting climatic and health data, as traditionally used in medical geography (Howe, 1977), together with other pertinent information with varying resolution such as from remote sensing, will facilitate the design of epidemiological investigations and improve forecasting (Huggett, 1980). Cumulative geographic and climatic information plotting will provide for identifying risk climates, risk diseases, risk areas and defining risk zones for specific subjects of health. With high resolution this will make answers possible even to questions focussing on local details. Shifts of risk zones can then be forecast with higher authenticity.

Conclusion

The significance of climatic shifts in the future as a hazard for the health of humanity depends on the climatic variation patterns and their time scales: the wider, more sudden and infrequent the variances, the stronger the climatic stress and the higher the demands on adaptation and disease control will be. The risks are much affected and aggravated by the size of the population exposed, the size of the area under national control, and means available for effective responses. It is not to be expected that there will be something new in the interrelationship between Man and climate not experienced before by the people at large or by earlier generations though it might be novel for a settled population in a particular region. Risks for disease control and well-being could be alleviated, if more were known about how individuals and societies respond to climate.

Lessening or worsening of health states in the case of climatic change will be entirely a matter of the human behavior responses on a national and international level (Bowden et al. 1981). This is an extremely complex process which is not understood in its details because of the vast and changing number of participating factors in every country and area. Not even a baseline can be drawn at present because of insufficient health statistics and lack of epidemiological studies in which the state of climatic adaptation of the people is considered. Just as there is an increasing demand for meteorological data, comprehensive vital and health statistics are urgently required for carrying out biometeorological assessment studies. Distinct geographically and climatologically defined health statistics will make possible: (1) comparative translatitudinal and translongitudinal studies on the impact of climate on diseases, (2) an identification of those diseases which are affected by climate, (3) preparation for the control of such climate-affected diseases and their management, (4) timely forecasting of the likelihood of epidemics, (5) evaluation of preventive measures, and (6) strengthening of societal risk acceptance and public morals.

Most of the future development of disease control and well-being of living generations in the face of climatic change depends on the control of Man over himself (Riebsame, 1985). Never before in history has mankind been in the position of surveying his ecosystem on a global scale and able to study the effects of his activities on small and large scales in all existing climates. Being aware of their responsibilities, many societies have already stepped up measures to prevent further deterioriation of their environment. Here, these preventive measures have been explained as ongoing adaptation processes to balance out disturbed equilibria. Not knowing all the disbalancing forces mankind is bound to make premature decisions and new mistakes. But time will allow for the feedback mechanisms to operate effectively in favor of humanity and living species. Adhering to the existing knowledge in hygiene and sanitation, as long as there are no major perturbations in nature and of societal structure, people will be able to keep control over the two basic requirements for health: the provision of food and the control of major communicable diseases. Coping with the threats requires trust in Man's vast reservoir of adaptive and tolerance capacities. Mistrust and fear will weaken these capacities, will narrow the chances for future health and the correction and maintenance of the balances in nature.

References

Andrews, G and Tennant, C 1978 Being upset and becoming ill: An appraisal of the relations between life events and physical illness Med J Austr 1 : 324–327.

Auliciems, A 1982 Psychophysiological criteria for global thermal zones of building design Biometeorology 8, Int J Biometeorol vol 26, Suppl Part 2, 69–86.

Baker, G W and Chapman, D W (Ed.) 1962 Man and Society in Disaster Basic Books, New York, 442 pp.

Blumthaler, M and Ambach, W 1990 Indication of increasing solar ultraviolet-B radiation flux in Alpine regions Science, 248: 206–208.

Bowden, M J., Kates, R W., Kay, P A., Riebsame, W E., Warrick, R A., Johnson, D L., Gould, H A., and Weiner, D 1981 The effect of climate fluctuations on human populations: two hypotheses In: Climate and History T M L Wigley, M J Ingram, and G Farmer (ed.), Cambridge University Press, Cambridge, 479–513.

Brès, P L J 1986, A century of progress in combating yellow fever Bull Wrld Hlth Org 64 : 775–786.

Brown, L R (ed.) 1989, State of the world, 1989 : A Worldwatch Institute report on progress toward a sustainable society W.W Norton, New York, p 243.

Bruce-Chwatt, L J and De Zulurta, J 1980, The rise and fall of malaria in Europe A historico-epidemiological study Oxford University Press, Oxford.

Burch, G E 1969 On prescribing the climate Amer Heart J 77: 149–150.

Burch, G E., DePasquale, N., Hyman, A and DeGraff, A C 1959 Influence of tropical weather on cardiac output, work, and power of right and left ventricles of man resting in hospital Arch intern Med 104: 553–560.

Cascinelli, N and Marchesini, R 1989 Increasing incidence of cutaneous melanoma, ultraviolet radiation and the clinician Photochem Photobiol 50: 497–505.

Cassel, J 1976 The contribution of the social environment to host resistance Amer J Epidem 104: 107–123.

Cravioto, J and Delicardie, E 1975 Londitudinal study of language development in severely malnourished children In: Nutrition and Mental Functions G Serban (Ed.), Plenum Press, New York, 143–192 (TS 269/14)

De Vries, J 1985 Analysis of historical climate-society interaction In: Climate Impact Assessment R W Kates, J H Ausubel and M Berberian (Eds) Scope 27, Wiley & Sons, Chichester, 273–291.

Dutta, H M and Dutt, A K 1978, Malarial Ecology: A global perspective Soc Sc Med 12: 69–84.

Edmundson, W 1980 Adaptation to undernutrition: How much food does man need? Soc Sci Med 14D: 119–126.

Escudero, J C 1985 Health, nutrition and human development In: Climate Impact Assessment R W Kates, J H Ausubel and M Berberian (Eds) Scope 27 Wiley & Sons, Chichester, 251–272.

Eylenbosch, W J and Noah, N D (Eds.) 1988 Surveillance in Health and Disease Oxford University Press, Oxford.

FAO, 1957 Caloric Requirements FAO Nutritional Studies No 15 Foor and Agricultural Organization, Rome.

Farhar-Pilgrim, B 1985 Social analysis In: Climate Impact Assessment R W Kates, J H Ausubel and M Berberian (Eds) Scope 27 Wiley & Sons, Chichester, 323–350.

Foster, H D 1976 Assessing disaster magnitude: A social approach Professional Geographer, 28: 241–247.

Foster, H D 1980 Disaster Planning: The Preservation of Life and Property Springer-Verlag, New York.

Frankenhaeuser, M 1976 The role of peripheral catecholamines in adaptation to understimulation and overstimulation In Psychopathology of Human Adaptation G Serban (ed.), Plenum Press, New York and London, 173–192.

Glass, A G and Hoover, R N 1989 The merging epidemic of melanoma and squamous cell skin cancer J Amer med Ass 262: 2097–2100.

Hare, F K 1985 Climate variability and change In: Climate Impact Assessment R W Kates, J H Ausubel and M Berberian (Eds) Scope 27 Wiley & Sons, Chichester, 37–68.

Heckenmueller, J 1985 Cognitive control and endorphins as mechanisms of health In: Cognition, Stress and Aging J E Birren and J Livingston (Eds.), Prentice-Hall, New Jersey, 89–110 (GGN 3876)

Howe, G M (ed.),1977 A World Geography of Human Diseases Academic Press, London.

Huggett, R 1980 Systems Analysis in Geography Clarendon Press, Oxford.

IPCC, 1990 Intergovernmental Panel on Climate Change Executive Summary Working Group II, WMO/UNEP, Geneva.

Keatinge, W R , Coleshaw, S R K and Holmes, J 1989 Changes in seasonal mortalities with improvement in home heating in England and Wales from 1964 to 1984 Int., J Biometeorol 33: 71–76.

Knudsen, A B 1986, The significance of the introduction of Aedes albopictus into the southeastern United States with implications for the Carribean, and perspectives of the Pan American Health Organization J Am Mosq Control Ass 2 : 420-423.

Kronfol, Z and Schlechte, J 1986 Depression, hormones and immunity In: Plotnikoff, N P , Faith, R E., Murgo, A J and Good, R A 1986 Enkephalins and Endorphins Stress and the Immune System Plenum Press, New York and London, 69–80.

Lee, J A H 1989 The relationship between malignant melanoma of skin and exposure to sunlight Photochem Photobiol 50: 493–496.

Mason, J W , Maher, J T , Hartley, L H., Mougey, E., Perlow, M J., and Jones, L G 1976 Selectivity of corticosteroid and catecholamine responses to various natural stimuli In: Psychopathology of Human Adaptation G Serban (ed.) Plenum Press, New York and London, 147–172.

Matzke, G 1979, Settlement and sleeping sickness control - A dual threshold model of colonial and traditional methods in East Africa Soc Sci Med 13D: 209–214.

Matzke, G 1983 A reassessment of the expected development consequences in tsetse control efforts in Africa Soc Sci Med 17: 531–537.

Melick, M E 1978 Life change and illness: illness behavior of males in the recovery period of a natural disaster J Hlth soc Behav 19: 335–343.

Morison, W L 1989 Effects of ultraviolet radiation on the immune system in humans Photochem Photobiol 50: 515–524.

Panum, P L 1948, Observations made during the epidemic of measles on the Faroe Islands in the year 1846 Trans Ada Sommerville Hatcher N.p Delta Omega Society,

Pratt, B and Boyden, J., (eds.), 1985: The Field Director's Handbook Oxford University Press, Oxford.

Reiter, R Munzert, K and Sladkovic, R 1982 Results of 5-year concurrent recordings of global, diffuse, and UV-radiation at three levels (700, 1800, and 3000 m a.s.l.) in the Northern Alps Arch Met Geoph Biokl Ser B 30: 1–28.

Riebsame, W E 1985 Research in climate-society interaction In: Climate Impact Assessment R W Kates, J H Ausubel and M Berberian (Eds) Scope 27 Wiley & Sons, Chichester, 69–84.

Rodhe, H., 1990: A comparison of the contribution of various gases to the greenhouse effect Science, 248: 1217–1219.

Rogers, D J 1988, The dynamics of vector-transmitted diseases in human communities Phil Trans Roy Soc (Lond.) B 321 : 513–539.

Scotto, J and Fears, T R 1987, The association of solar ultraviolet and skin carcinoma incidence among Caucasians in the United States Cancer Invest 5: 275–283).

Soejarto, D D and Farnsworth, N R 1989 Tropical rain forests: potential source of new drugs? Persp Biol Med., 32: 244–256.

Starr, C 1969 Social benefit versus technological risk Science, 165: 1232–1238.

Taylor, H R 1989 The biological effects of UV-B on the eye Photochem Photobiol 50: 489–492.

Urbach, F 1989 Potential effects of altered solar ultraviolet radiation on human skin cancer Photochem Photobiol 50: 507–514.

Van Praag, H M 1986 Monomamines and depression: the present state of the art In: Emotion Vol 3 Biological Foundations of Emotions R Plutchik and H Kellerman (Eds.), Academic Press, London, 335–361.

Weihe, W H 1979 Climate health and disease In: Proceedings of the World Climate Conference World Meteororological Organization, Geneva, WMO No 537: 318–368.

Weihe, W H 1986 Life expectancy in tropical climates and urbanization In: Urban Climatology and its Applications With Special Regard to Tropical Areas T R Oke (Ed.), World Meteorological Organization, Geneva, WMO - No 652: 313–353.

Weihe, W H 1987 Social and economic values of applied human climatology In: Climate and Human Health World Climate Programme Applications, WMO, Geneva, WCAP - No.1, 233–250

Whyte, A V T 1985 Perception In: Climate Impact Assessment R W Kates, J H Ausubel and M Berberian (Eds) Scope 27, Wiley & Sons, Chichester, 403–436.

World Resources 1988–89, 1988 A Report by the World Resources Institute and the International Institute for Environment and Development Basic Books, New York.

WHO, 1977 International Classification of Diseases Manual of the International Statistical Classification of Diseases, Injuries, and Causes of Death Vols 1 and 2 World Health Organization, Geneva.

WHO, 1979 Psychosocial factors and health: New program directions In: Towards a new Definition of Health P I Ahmed and G V Coelho (ed.), Plenum Press, New York, 87–111.

WHO, 1986 Basic Documents Thirty-sixth Edition World Health Organization, Geneva.

WHO, 1990a, Potential health effects of climatic change Report of a WHO Task Group WHO/PEP/90/10, World Health Organization, Geneva.

WHO, 1990b, Tropical diseases 1990 TDR-CTD/HH90.1 World Health Organization, Geneva.

WHO 1990c, Global estimates for health situation assessment and projections 1990 WHO/HST/90.2 World Health Organization, Geneva.

Population and Global Climate Change

by David Norse *(Food and Agriculture Organization)*

Introduction

Almost four hundred years ago an Englishman called Francis Bacon wrote about a world government and about messengers of light and messengers of darkness (Bacon, 1625). The messengers of light were scientists who went out into the world to help sustain society. It is clear from the technical sessions of the Second World Climate Conference that at present scientists cannot accomplish all of this idealised role, particularly with regard to the population-climate nexus and how to address it. There are too many uncertainties, e.g. about climate change processes, about the possible socio-economic impacts of climate change, and regarding viable response options to such impacts, for scientists to recommend a complete range of actions, although some are quite clear, e.g. improvements in energy efficiency and in land-use planning. Nonetheless, there are two important messages relating to the population-climate nexus for those attending the Ministerial session and for the wider audience outside.

The first message is that it is probable that mankind is already locked into a 1–2°C rise in temperature (IPCC, 1990). Although this rise emanates largely from the actions of the developed countries, it could endanger the food security and lives of millions of people in the developing countries.

The second message is that as well as being locked into global warming, mankind is also locked into a doubling of population by about the middle of the next century. Climate change, therefore, will not be compounding the existing environmental and welfare problems arising from our present population of some 5 billion people. Instead, it will be compounding the problems of feeding, clothing and providing shelter for around 10 billion people (UN, 1989).

Population growth has a dual role regarding climate change. It is both a driving force for climate change, and a factor reducing the resilience of managed ecosystems to climate change. This contribution is largely about the latter, and focusses particularly on the role of population growth in increasing the vulnerability of agriculture and forestry to climate change.

Population Growth as a Driving Force for Climate Change

This aspect of the dual role is partly covered by other papers to this Conference, and by the IPCC report and its supporting documents, so it will not be treated in detail except in the case of the maintenance or enhancement of the soils' role as a sink for carbon dioxide. Population and income growth are the main driving forces for fossil and non-fossil fuel energy demand and for food production, but whereas energy demand accounts for most greenhouse gas emissions, food production is responsible for less than 10 percent. Moreover, it is important to recognise that there are substantial uncertainties regarding the contribution of population and income growth to some agricultural emissions, particularly methane and nitrous oxide, since there are serious gaps in our knowledge of the processes involved and the actual emission rates (EPA/IPCC, 1990).

Projections indicate that in the absence of draconian and currently socially unacceptable measures to slow down population growth, the world population will reach over 8 billion by 2025 and almost 10 billion by 2050. It is expected to stabilize at 11 to 12 billion some time in the

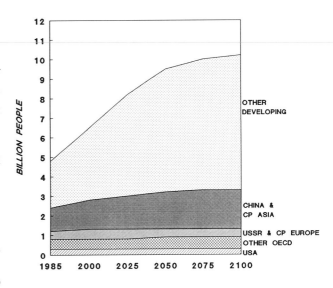

Figure 1: Global population levels by region (1985–2100).

22nd century (UN, 1989). The bulk of this growth will be in the least developed countries (Figure 1).

This growth poses a number of threats to the environment in general, particularly to the important role of soil as a sink for carbon dioxide that is considered briefly here, and to the resilience of man-made ecosystems to climate change which will be discussed in the next section.

Much of the attention to date has been focussed on the role of forests as a source and sink for carbon dioxide, and notably the former, in part because the history of man is the history of forest destruction. Population growth over the

centuries has almost universally involved the clearance of open and closed forest for arable or pastoral land. The problem now is that little land remains to be cleared in most countries (Alexandratos, 1988). Much of what remains is concentrated in a few countries and is fragile, infertile, and generally unsuitable for the production of staple food crops. Moreover, a substantial proportion of recent deforestation is the result of rural to rural migration, as population pressures force people away from densely populated areas or from marginal lands that are no longer able to support them.

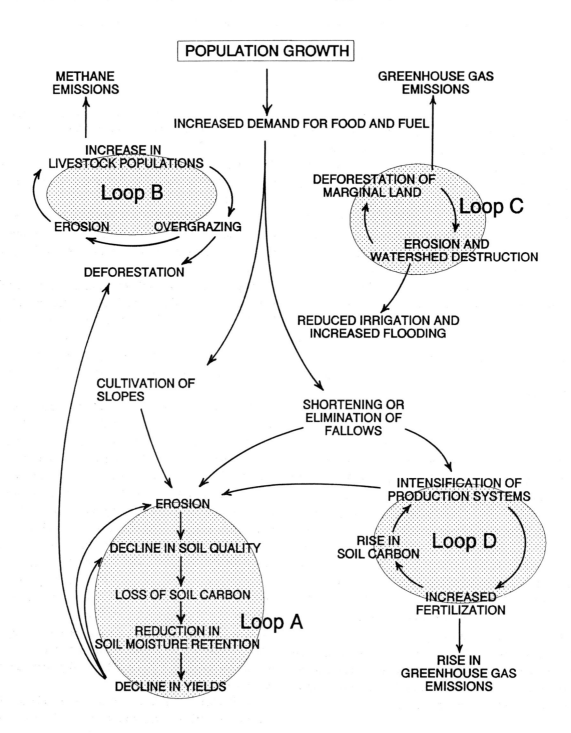

Figure 2: Impact of population growth on greenhouse gas emissions from agriculture and forestry.

Table 1: *Average Yields of Unfertilized Local Maize and Fertilizer Response Rates in Malawi*

Area	Yield (kg/ha) 1957-62	Yield (kg/ha) 1985/86-1986/87 [†]	Response Rate [††] 1957-62	Response Rate [†††] 1982/83-1984/85
Lilongwe	1,760	1,100	23	13
Kasuagu	1,867	1,120	24	18
Salina	1,693	1,060	25	17
Maize	1,535	775	32	18

† Mean of National Crop estimates for 1985/86 and 1986/87. The means include small areas of fertilixed local maize

†† Kg maize per kg N

††† Kg maize per kg N = P_2O_5. Quantity of P_2O_5 is very small.

Although there are good reasons for the focus on deforestation and forest management, this has tended to distract attention from the impact of population pressures on arable and pastoral lands in developing countries. Though these agricultural lands sequester less carbon per unit area than forests, they are much more extensive. Just as population pressures are degrading forest land, either temporarily or permanently, they are also having similar impacts on agricultural land. Some 6 million hectares of rainfed arable land are estimated to go out of production each year because of degradation, and one to two million hectares of irrigated land become so salinated that they are also lost, either temporarily or permanently (FAO, 1989). Such degradation can result in soil carbon levels falling about one percent per year, although these losses are small in per unit area terms compared with those resulting from deforestation (EPA/IPCC, 1990).

More serious, however, is the soil degradation and loss of productivity over vast areas of the remaining agricultural land. Population pressure in many developing countries in the context of state technological change in agriculture is causing soil carbon levels to decline over some 500 to 1000 million hectares. Part of the problem arises from shifting agriculture, which is also responsible for much of the deforestation. At low population densities shifting agriculture, as well as the more sedentary bush fallow systems, can be practised with a 5–15 year fallow between cultivation cycles, which is sufficient to maintain relatively stable soil carbon levels and avoid serious soil erosion (Lal, 1990). At today's high and increasing population densities such traditional agricultural practices are no longer sustainable, because most countries have little or no good land left to bring into cultivation and farmers consequently have to shorten or even eliminate fallows with negative consequences for soil carbon retention.

Population Growth as a Driving Force for Reduced Resilience to Climate Change and Climate Variability

This problem is of particular concern to developing countries, both in the context of climate change and in the context of sustainable development.

Figure 2 illustrates the major positive feedback loops associated with population growth and its impact on greenhouse gas emissions or on increased vulnerability to climate change.

Loop A is the dominant one, stemming from the failure to compensate for the shortening or elimination of fallows by changes in cropping systems or fertilizer practices because suitable technologies do not exist, or are uneconomic, or farmers are not given adequate incentives or support services to adopt them. It has three major implications for climate change.

First, it leads to soil erosion and declining yields which increases the pressures to develop marginal land. This largely takes two forms: clearance of marginal land under forest, which destroys carbon sinks and releases greenhouse gases; and cultivation of steep slopes that are very vulnerable to soil erosion through rainfall run-off, and could be adversely affected by any increase in rainfall frequency or intensity as a result of climate change.

Secondly, it leads to soil nutrient depletion which increases the need for additional fertilizers and can also reduce the efficiency of fertilizer use, thereby contributing unnecessarily to the release of nitrous oxides. Table 1 gives an example of this impact. Population pressure in Malawi since the late 1950s, together with inadequate support or incentives for fertilizer use, has reduced the effectiveness of mineral fertilizers by about 25 to 50 percent depending on the area (Twyford, 1988).

Thirdly, the soil nutrient depletion is associated with the loss of soil carbon, largely in the form of organic matter, and a decrease in the role of soil as a sink for carbon

Table 2: *Population experiencing a fuelwood deficit, 1980 and 2000 (millions)*

		Latin America	Africa	Near East & North Africa	Asia and the Pacific	Total	
1980	Total population	26	55		31	112	
	Rural population	18	49		29	96	Acute Scarcity
	Total population	201	146	104	832	1283	
	Rural population	143	131	69	710	1052	Deficit
2000	Total population	512	535	268	1671	2986	Acute scarcity
	Rural population	342	464	158	1434	2398	or deficit

Source: FAO, Fuelwood Supplies in Developing Countries, 1983

dioxide and in its ability to absorb rainfall. The latter could be particularly serious in those parts of the world where either rainfall is projected to decrease or evapo-transpiration rates are projected to rise.

Loop B is also closely linked with population growth, both through the increased demand for livestock products and through socio-cultural factors. The latter are particularly dominant in sub-Saharan Africa where farmers wish to maintain large numbers of cattle as an external sign of wealth, or as an insurance against severe drought (Jahnke, 1982), and in Asia where religious factors limit measures to control cattle numbers. Excessive numbers lead to over-grazing and land degradation, which in low rainfall areas increases the vulnerability to drought, and over much wider areas reduces the role of pastures as a sink for carbon dioxide. It also leads to serious shortages and poor quality of livestock feed. This in turn means that more livestock have to be held for a given level of meat or milk production and poor feed utilisation, which both cause proportionately higher methane emissions.

Loop C concerned with deforestation is driven by the demand for agricultural land and fuelwood. Since most of the deforested land is only marginally suitable for agriculture, there is almost invariably a net increase in greenhouse gas emissions. Unless actions are taken to counter the positive feedback of loop A and replace it by loop D described below, further population growth will continue to create relatively more greenhouse gas emissions than are necessary for a given level of output. Moreover, the problem is likely to get worse for two particular reasons. Increasing numbers of people are going to be experiencing a fuelwood deficit as illustrated in Table 2 (FAO, 1983). The deforestation is a major cause of the watershed destruction and soil erosion, and therefore of the rapid siltation of river beds and irrigation systems that may

increase significantly the incidence and intensity of flooding and reduce agricultural production.

Loop D is the only major negative feedback loop. Well managed intensification limits soil erosion, prevents serious land degradation and maintains or increases the role of agricultural soils as a sink for carbon dioxide. In Western Europe and parts of the USA, for example, such intensification has increased carbon dioxide sequestration over the past 40–50 years. Much of this sequestration, however, has been at the cost of some increase in emissions of nitrous oxide from fertilizers.

Urbanization

The population pressures on the rural environment described in the previous section are the main driving force for the well established trend for greater urbanization in the developing world (Figure 3) with both positive and negative implications. The movement of rural people to urban areas helps to take the pressure off marginal land and to limit carbon dioxide emissions from deforestation. However, there are also a number of negative consequences of urbanization, notably:

- urbanization tends to consume high quality arable land, since towns are commonly sited on coastal or flood plains and in river valleys; which in turn, increases the number of people vulnerable to sea level rise
- it raises the pressure for fuelwood collection in per-urban areas and greater land degradation, thereby increasing their vulnerability to increased rainfall run-off or greater aridity
- it changes people's consumption patterns in a direction that may put even greater pressure on the environment.

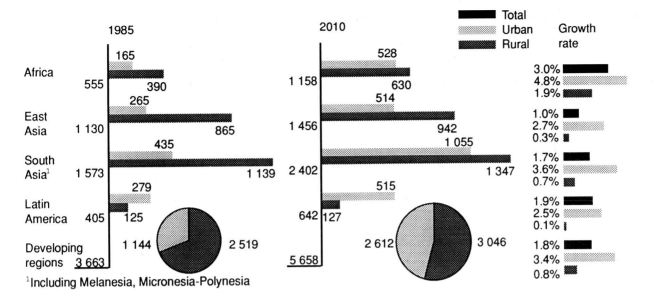

Figure 3: Urban and rural population projections by developing regions, 1985-2010. Source: UN, The Prospects of World Urbanization - Revised as of 1984-85, New York (1987).

Unfortunately the population most at risk from sea level rise or flooding from hill torrents are commonly low income groups who are least able to cope with such disasters. Squatter areas, for example, are commonly located on low lying ground unsuited for permanent housing.

Competition for natural resources

Population pressures will increasingly intensify the competition for natural resources and compound the difficulties in adapting to climate change. Within agriculture it will increase the competition between crop, livestock and forest production for a shrinking land resource base. In areas where farms are already too small to provide adequate family incomes, further population growth will add to the problem by greater land fragmentation, thereby increasing the risk of accelerated losses in soil carbon and greater vulnerability to drought.

In the same or other areas where farms are irrigated from groundwater, and there is increasing competition from urban and industrial development for the available water resources, overabstraction will run the risk of lowering the water table beyond an economic depth. Thus one could see many other areas following the fate of the Texas High Plain, which can no longer support irrigated agriculture and has caused the social and economic collapse of agricultural towns. Or alternatively, the combination of over-abstraction and sea-level rise could lead to serious salt-water intrusions into coastal water tables.

It follows from the foregoing that the links between population growth and climate change are complex and serious. There is a risk that unless there is decisive action to slow down population growth, the gains that could be realised through greater energy efficiency will mitigate climate change but not resolve the problem.

References

Alexandratos, N.(Ed.) 1988. "World Agriculture: Toward 2000", Food and Agriculture Organization, Rome, and Belhaven Press, London.

Bacon, F., 1625. "Of seditions and troubles" in 'Essays', London.

EPA/IPCC.1990. "Proceedings of the Workshop on Greenhouse Gas Emissions from Agricultural Systems". 12–14 December 1989, Washington DC,

FAO. 1983. "Fuelwood Supplies in Developing Countries": Food and Agriculture Organization of the UN, Rome.

FAO. 1989. "Sustainable Development and Natural Resource Management" in State of Food and Agriculture, 1989. Food and Agriculture Organization of the UN, Rome.

IPCC. 1990. "Climate Change: The IPCC Scientific Assessment". Eds. Houghton, J.T., G.J. Jenkins and J.J. Ephraums, Cambridge University Press, Cambridge.

Jahnke, H.E.1982. "Livestock Production Systems and Livestock, Development in Tropical Africa".Kieler Wissenschaftsverlag Vauk, Kiel.

Lal, R.1990. "Soil as a Potential Source or Sink of Carbon in Relation to the Greenhouse Effect", in EPA/IPCC, 1990. op.cit.

Twyford, I.T.1988. "Development of Smallholder Fertilizer Use in Malawi". Unpublished.

UN 1989 "World Population Prospects 1988". Population Statistics N. 106. New York

Public Information and Attitudes

by Robert Lamb *(United Kingdom)*

I have been involved fulltime with the environment movement for 12 years, with Earthscan, IUCN, UNEP and for the past seven years as Director of TVE. The other day, I read some newspaper articles I had written at the end of the 1970s. Though the issues I was writing about are much the same as today, the approach was very different: today I would start with the assumption that the reader is already aware of, and worried about, deforestation or ozone depletion; ten years ago I felt the need to be an advocate, to explain all the terms that are now so familiar.

In the intervening decade a significant shift has taken place in public attitudes at least in the western industrialized world. Today you no longer have to make the case for affirmative environmental action. It's the others, whoever they may be, who must put up the contrary case. It is a measure of the universal awareness of environmental issues that very few would do so publicly. Though, as I shall mention later, there are powerful behind-the-scenes forces who work to make sure action lags a long way behind all the talk.

What brought about this change in public attitudes?

The answer is quite simply public information. For most people for most of the time, their daily lives are untouched by the environmental 'crisis'. And yet, as numerous public opinion polls in the industrialized countries testify, the public is prepared to accept that there is a crisis. But it is a curious kind of crisis - the enemy is not at the gates, the economy is not collasping. There is nothing immediate, or life-threatening, and yet there is a groundswell demand for action: the mood is spreading that something must be 'done'.

But how have we reached this point? Mass media coverage of a succession of disasters like Bhopal or Chernobyl, of acid rain, of oiled and poisoned waste, of famine victims and tropical forests put to the torch have penetrated our mass psyche. But I would argue that the sum of this awareness did not bring home the idea of interdependence; that all nations, rich and poor, are facing a common threat.

Depletion of the ozone layer and global warming have done that. As recently as the mid-1980s, it would really have been an extraordinary man or woman in the street who knew about the 'greenhouse effect', let alone the multi-syllabled chlorofluorocarbons. But now they do. Ask a taxi driver, ask your children - people are really concerned.

Not to mince words, this is amazing. The received wisdom is that the public is most concerned by backyard or local issues - the quality of the water in the taps; pesticides residues on the food they buy and so forth. After all you can't see the ozone layer and to many of us living in northern latitudes, a warmer climate is by no means an unpleasant prospect.

So why the transformation?

In my judgement, it is because people with little or no scientific knowledge can grasp the idea of a "hole in the sky"; of chemical reactions zapping ozone molecules and leaving us exposed to cancer-causing UV radiation. And, well, everyone knows that it is much hotter in a greenhouse than outside.

For the most part, journalists in television, radio and newspapers have been adroit and responsible in getting the message across to their lay audiences. And successful to the point where global warming is now perceived as one of the great issues of our time.

I think it has come as a genuine shock to realize that human economic activity is now so all-pervading that we can actually change the climate. The public in the West has started to realize that this is a truly global problem that cannot be solved by a handful of rich nations acting in concert. No country owns its own patch of sky; no country can quarantine itself from what the poorer Third World and former Eastern bloc countries do.

This is why ozone layer depletion and global warming have finally registered the meaning of interdependence.

But the media and their consumers, the public, are notoriously fickle. Let us be in no doubt that the catalyst has been the recent sequence in temperate countries of the hottest years on record. Our man and woman in the street do not differentiate between climate and weather. A return to a few "normal" years will evoke the response - 'What was that all about? Environmentalists crying wolf as usual.'

We are still at the stage of 'worrying' about the problem. To date most media coverage has concentrated on the doomsday scenarios - island nations disappearing, massive disruption in food production and so on.

There is a danger that climate change appears so overwhelming that nothing can be done about it. While it is true that responsible media coverage should be preparing the public for what must be a radical, at times painful, overhaul of the way nations generate economic growth, no opportunity should be missed to convey the message that this lies within the power of the international community.

We all know that in cabinets and boardrooms a lot of trading goes on. And mostly the environment comes off worse. But climate change is not something that can be traded. Attitudes need to be changed to the extent that it is simply perceived to be immoral and wrong for a nation not to do all in its power to strictly adhere to a globally agreed strategy.

It is next to impossible to pinpoint a moment at which what was deemed to be acceptable becomes unacceptable. At the start of this century it was acceptable for women to be disenfranchised. At the end of the last world war, imperialism was acceptable. So were blatantly racist laws. None are acceptable now because public attitudes have undergone such a profound change. A ratchet has been applied, and there is no going back.

The advent of the communications revolution has radically speeded up the process of attitudinal change. In a majority of countries it is no longer acceptable for governments to control information. And the lightning spread of computer and telecommunications technology is making it possible for them to do so. The global village is already upon us.

What we still have to achieve is to apply the ratchet for action to deal with global warming in particular and the environmental crisis in general.

It is important here to differentiate between 'nation' and 'government' because if we are realistic we must accept the limitations on governmental action. Take Brazil for example. The federal government in Brazilia may be committed to conserving the tropical forest, but how about the provincial governments where some politicians are being elected on an anti-ecological platform? And how about the USSR and Central European countries, where miners have enormous political clout - how will they take to measures to cut back on coal production?

We have to beware of the backlash of what might be perceived to be a Western-inspired plot to keep these nations underdeveloped. For in the West climate change and ozone depletion are big concerns but elsewhere in the world, for a huge chunk of the world's population who aspire to Western standards of living, whatever the cost, these issues have scarcely registered on the barometer of public concern. At the London ozone meeting in June, India's former Minister of the Environment stated bluntly that if the rich countries are so concerned about depletion of the ozone layer they are going to have to pay countries like India not to repeat their mistakes.

Whatever way the problem is approached, new and additional resources will have to be generated through carbon taxes and so on. And know-how and funds will have to be transferred equitably in a spirit not of charity but of genuine partnership.

Attitudinal change has to reach the point where the oil, coal and other powerful lobbies, which work so effectively behind the scenes, cannot be allowed to frustrate what an overwhelming majority of the public wants.

And here we do start from a position of strength. For even if the doubting Thomases are right and the IPCC scientists have got it all wrong, there is not one single measure aimed at addressing global warming that will not do the biological productivity of the planet a power of good. Conserving energy through new and renewable technologies, safeguarding tropical forests, replanting trees, protecting genetic diversity, introducing stricter emission controls, making nuclear power safer are all measures that should be undertaken on a world scale, even if there were no such thing as global warming. All bring in their train enormous benefits for rich and poor nations alike.

This is the simple message that must be communicated in any which way we can.

And in this context who do I mean by we? Essentially I mean the six international sponsors of this Second World Climate Conference. Not since the days of Dag Hammersjkold has the United Nations stood in such high regard. The ozone protection agreement has been a tremendous boost, and there are great expectations for the proposed climate change convention.

We will be losing a truly historic opportunity if we let this momentum fritter away. With so much media attention centered on this conference the temptation is to believe that we have to do nothing. Certainly between now and ECO'92, I am confident the Western media will sustain their interest. But I am concerned that in the non-OECD countries the message is not getting through.

I would like to see the sponsors of this conference getting together to agree on a strategy for the remainder of the decade with the TV viewer, the radio listener and the newspaper reader in the Third, and what used to be the Second Worlds very much in mind. And the objective should be to make climate change as high a public priority as it is in the West.

A tall order indeed, but not one beyond our collective grasp. The audio visual communications revolution - satellite, cable, VCR's, cheap transistor radios - permit instant access to a global audience of billions. The greatest mistake that could be made is to underestimate how vital it is to keep the global public reliably informed of what is at stake.

This is what my own organization has been seeking to achieve through its Global Weather Watch project. Its main elements could form the basis of a unified strategy.[1]

---

[1] A copy of a video and proposal for the Global Weather Watch Project are available from the author

Climate, Environment and Ecology

by William H. Schlesinger *(USA)*

Abstract

I examine three levels of response of the terrestrial biosphere to climatic change:

1. Current vegetation and soils will show greater rates of respiration with increasing temperature, with lesser changes expected as a direct response to higher CO_2
2. The distribution of vegetation is likely to shift markedly in response to greenhouse warming of the Earth, and these shifts are likely to lead to an increasing loss of carbon to the atmosphere
3. Direct human impacts are likely to produce the large changes in future vegetation and biogeochemical cycling. These impacts will increase in direct response to rising human population. These levels of consideration represent sequentially increasing realism in scenarios for future levels of biospheric function.

Introduction

Most interest in global climate change focuses on changes that affect human social and economic systems. In our excitement and concern with this issue, we must remember that humans represent only one of the many species found on Earth. Beyond our office windows, the biosphere is composed of a rich diversity of plants and animals grouped in a wide variety of natural communities. Here, long-term evolution has led to elaborate, highly-structured ecological relationships that comprise "the balance of nature." What we call natural history is our attempt to study the social and ecological systems of the other species that share our planet. To predict the future condition of the planet, we must understand the effect of climate change on the persistence of biotic diversity on Earth and on the potential for the biosphere to sustain natural levels of ecosystem function from the tundra to the tropics.

Natural history is not only the concern of avid environmentalists and the "green" movement. A growing body of evidence suggests that the biosphere controls the surface chemistry on Earth, including the composition of the atmosphere. Changes in the biosphere will have direct, positive feedbacks that will accelerate the pace of global change and the disruption of human society. As we respond to rapid changes in climate, we may tend to concentrate on our own social and economic plight, failing to recognize that the plight of the natural biosphere is intimately linked to our own.

In this paper, I will review briefly what we know about potential responses of the biosphere to global change. My comments address three main questions:

1. What is the potential response of vegetation to increasing concentrations of CO_2 and increasing temperatures in the atmosphere?
2. What are the potential changes in the distribution of terrestrial vegetation in response to global climate change?
3. What is the effect of human destruction of natural ecosystems on biotic diversity, climate and global biogeochemical cycles?

Response of Terrestrial Vegetation to Global Change.

During the last 10 years, a large amount of scientific research has shown a direct effect of higher concentrations of atmospheric CO_2 on terrestrial vegetation (Strain and Cure 1985). Most plants show increases in photosynthesis and growth when exposed to high CO_2 in short-term experiments (Allen 1990). High CO_2 also decreases the loss of water from plant leaves, because CO_2 controls the size of the stomatal aperture in well-watered plants. As a result of greater photosynthesis and lower losses of water, the efficiency of water use during photosynthesis increases when plants are grown in high CO_2. Taken alone, these results might lead us to be rather sanguine about the future. We might expect that natural vegetation and crop plants would grow more rapidly and that a slower depletion of soil water would prolong the growing season in drought-prone areas.

The interpretation of long-term field experiments using high CO_2 is not always so straightforward. Tissue and

Table 1: *Effects of Soil Heating on **Picea mariana**. Data are tissue nutrient concentrations (%).*

Plot	N	P	Ca	Mg	K
Control	0.76	0.099	0.28	0.08	0.56
Heated	1.05	0.136	0.28	0.09	0.80

Source: Van Cleve et al. (1990)

Oechel (1987) reported on a field experiment in the Alaskan tundra, where *Eriophorum vaginatum* was exposed to 680 ppm CO_2 for 10 weeks. Within three weeks, plants exposed to high CO_2 showed downward physiological adjustments, so that their photosynthetic rate was similar to those grown at ambient (350 ppm) CO_2. Laboratory experiments imposing long-term exposure to high CO_2 show that the accumulation of carbohydrate in leaf tissues damages chloroplasts, reducing the rate of photosynthesis (DeLucia et al. 1985). Despite the downward adjustment of photosynthesis, tundra plants at high CO_2 produced a greater number of tillers (Tissue and Oechel 1987), so net carbon storage per unit of ground surface increased slightly (Grulke et al. 1990).

Indirect effects of increasing atmospheric CO_2, such as a warmer climate, are likely to be more significant than the direct effect of CO_2 on plant growth, particularly for long-lived woody plants. Ryan (1990) showed that maintenance respiration in *Pinus contorta* and *Picea engelmannii* rises in response to increasing environmental temperature. Based on the work of Amthor (1984), the IPCC concluded that plant respiration will double with a 10°C increase in environmental temperature. Photosynthesis is only slightly affected, so a greenhouse warming of climate is expected to reduce plant growth in most areas. The temperature effect on respiration is greatly in excess of direct effects of CO_2 on plant growth, but there are few laboratory or field experiments where interactions of these variables have been examined.

The effect of increasing temperatures on water-use efficiency also needs further investigation. A warmer climate and higher leaf temperatures may increase the rate at which plants lose water to the atmosphere, negating the possible benefits of improved water-use efficiency (Allen 1990). The growth of plants in response to high CO_2 often results in an increase in leaf area. Although the loss of water per unit of leaf area is lower, there are more leaves per unit of soil surface and a greater loss of soil moisture to the atmosphere. These observations, and the frequent limitation of plant growth by soil nutrients and water, suggest that there may be little effect of higher

concentrations of atmospheric CO_2 on plant yield per unit of land area.

Few studies have examined changes in the soil that accompany changes in photosynthesis, respiration, and plant growth in response to higher temperature and CO_2. Deficiencies of essential plant nutrients, especially nitrogen, limit the growth of plants in most natural ecosystems. Tissue and Oechel (1987) concluded that the response of *Eriophorum vaginatum* to high CO_2 was limited due to nutrient deficiencies in tundra soil. In contrast, the greater aboveground net primary production reported in a long-term study of salt marsh plants exposed to high CO_2 (Curtis *et al.* 1989) may be due to the relatively high nutrient availability in the estuarine waters flushing through this system. In a glasshouse experiment, white oak (*Quercus alba*) seedlings exposed to 690 ppm CO_2 for 40 weeks were 85% larger than those grown at 350 ppm, but nitrogen uptake showed little change (Norby *et al.* 1986). The plants suffered from an extreme nitrogen deficiency, and most of the enhanced growth was allocated to root tissue. We should expect that the response of terrestrial vegetation to high CO_2 will vary depending upon other environmental factors, including the length of the growing season and the availability of nutrients and water in the soil.

High CO_2 will have little direct effect on soil microbial transformations and decomposition, because CO_2 in the soil atmosphere is normally in the range of 2000 to 30000 ppm (Brook et al. 1983). However, the rate of decomposition is greater in regions of higher mean annual temperature (Schlesinger 1977), so global warming should result in an increase in the rate of decomposition in most ecosystems. Van Cleve et al. (1990) found that experimental soil heating in the boreal forest of Alaska increased the concentrations of available nitrogen and phosphorus for plant growth (Table 1). Further experiments of this type need to be conducted in a variety of ecosystems around the world.

High CO_2, high temperature, and increased nutrient supply may cause dramatic changes in the function of arctic ecosystems, where a large amount of carbon is stored in soil organic matter. Using intact cores in which these

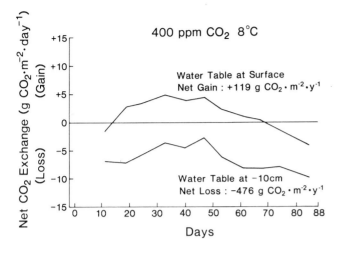

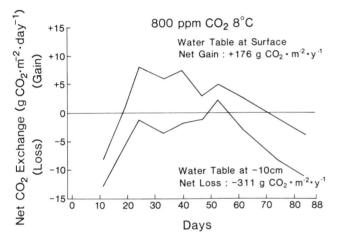

Figure l: Mean annual temperature at Barrow Alaska is 4°C. Billings et al. (1983) collected soil cores from the wet tundra near Barrow and measured daily mean carbon balance at 400 and 800 ppm CO_2 and two levels of water table at 8°C, representing a 4°C increase in temperature.

variables could be manipulated in glasshouse experiments, Billings *et al.* (1983) showed that high CO_2 (800 ppm) increased the storage of carbon in the soil-plant system, when the water table was maintained at the surface. Similar results are reported from a long-term field experiment in the tussock tundra of Alaska (Grulke et al. 1990). Added supplies of nitrogen also enhanced the storage of carbon in tundra soils (Billings *et al.* 1984). The rate of storage was lower when the temperature was raised by 4°C (Billings et al. 1982), and soil cores showed net carbon losses when the water table was allowed to drop below the surface (Figure l). These results suggest that the tundra could become a source of atmospheric CO_2 if climatic warming causes permafrost to melt.

Changes in the Distribution of Terrestrial Vegetation

At present, the boundaries between boreal forest and tundra and deciduous forest and boreal forest appear to be well correlated to lower extremes of annual temperature in North America (Black and Bliss 1980, Arris and Eagleson 1989). If we accept the predictions of future climate given by general circulation models, we can attempt to predict the future distribution of terrestrial vegetation, assuming that the correlations between current climate and vegetation will also be globally robust in the future. Using the temperature projections of Manabe and Stouffer (1980), Emanuel et al. (1985) show the changes in the areal extent of major biomes under doubled atmospheric CO_2 (Figure 2). There is a dramatic decrease in the area of boreal forest and tundra. Total forest area also declines, in favor of desert and grassland. Although most climate models predict an increase in global precipitation with CO_2-induced warming, the rainfall in mid-continental regions declines (Manabe and Wetherald 1986). Thus, projections for both temperature and moisture suggest that we can expect an increase in the area of deserts and a decrease in the area of semi-arid grasslands in the future (Schlesinger *et al.* 1990). The area of irrigated agriculture will expand over much of the United States (Adams *et al.* 1990).

Such projections are the best we now have, but they should not be accepted uncritically. Certainly there will be interactions between temperature and other environmental variables that will determine the future distribution of species. Using temperature alone, Davis and Zabinski (1990) show that the range of American beech (*Fagus grandifolia*) moves northward into Canada, and the species would be nearly absent from the eastern US. Experimental studies with beech and sugar maple (*Acer saccharum*), however, show a significant growth enhancement for beech seedlings in high CO_2, suggesting greater competitive ability in northern hardwood forests (Reid 1990, Bazzaz et al. 1990). Both species are likely to expand their range northward only on fine textured soils, in which greater soil moisture is stored (Pastor and Post 1988). It is clear that we need further studies in which various environmental factors, especially CO_2 and temperature, are varied concurrently.

Large-scale shifts in the distribution of terrestrial vegetation have important implications for ecosystem function and changes in global biogeochemical cycles. The invasion of a large area of boreal forest by deciduous species will result in the release of CO_2 to the atmosphere, because the carbon storage in boreal forest soils is much larger than in deciduous forests (Schlesinger 1977). Similarly, disappearance of soil organic matter from tundra ecosystems will release CO_2 to the atmosphere, regardless of whether tundra species increase or decrease their rate of photosynthesis under warm, high CO_2 environments. Ignoring additional losses due to the conversion of natural

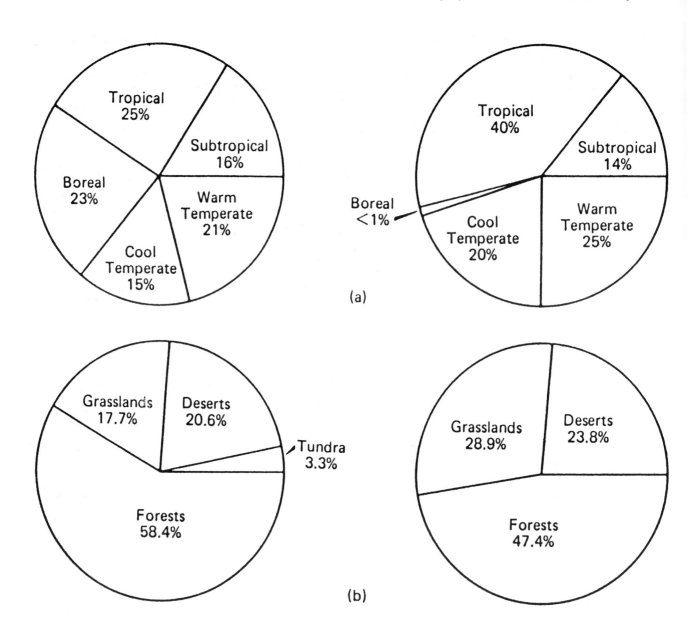

Figure 2: Changes in the areal extent of major forest types and world biomes under elevated CO_2 levels, calculated by Emanuel et al. (1985). Reproduced by permission of Kluwer Academic Publishers.

land to agriculture, a loss of 45.5×10^{15} gC from the global pool of soil organic matter occurs with the vegetation shifts predicted by Emanuel *et al.* (1985) and the mean soil carbon contents of Post *et al.* (1982). This loss of soil carbon is difficult to reconcile with recent model predictions that the terrestrial biosphere will *accumulate* 235×10^{15} gC in an atmosphere with double the present-day level of CO_2 (Prentice and Fung 1990).

Currently about 10% of the global flux of methane to the atmosphere is derived in boreal forest and tundra regions (Whalen and Reeburgh 1990), where it is directly related to soil moisture (Sebacher *et al.* 1986). Large changes in methane emission can be expected with soil warming, but whether methane emissions will increase or decrease will depend on concurrent changes in soil moisture levels.

Changes in the relative extent of semi-arid grasslands and deserts also affect global biogeochemistry (Figure 3). In the barren soils of deserts, abiotic processes control the movement of essential nutrient elements. A greater loss of soil to wind erosion will enhance the atmospheric burden of dust and the transport of soil-borne nutrients to other regions of the Earth. Martin and Gordon (1988) show that the primary productivity of the North Pacific ocean is limited by iron, which acts as a trace micronutrient for marine algae. An increase in the loss of soil dust from the deserts of China may enhance ocean productivity, as

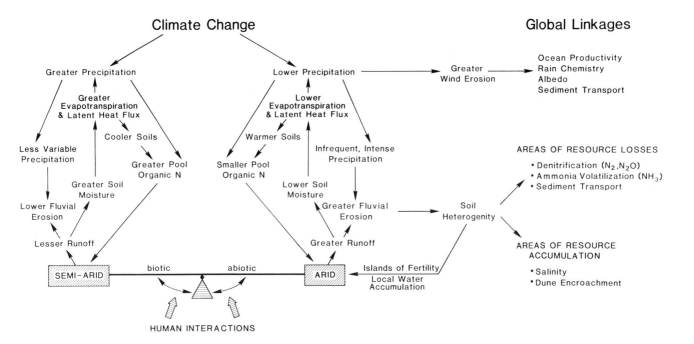

Figure 3: A model linking changes in ecosystem properties at the margin between semi-arid grasslands and desert leading to desertification and changes in global biogeochemistry. From Schlesinger et al. (1990).

greater quantities of iron-rich dust are deposited in the sea. Dust may also exert a warming or cooling influence on the atmosphere, but our present understanding of its effect is primitive (Schlesinger et al. 1990).

Human Impact on the Terrestrial Biosphere

It is easy to forget that the projections of future vegetation based on potential changes in climate do not include the impact of humans on land use. More realistic predictions are based on the anticipated vegetation in the face of expanding human populations and agricultural development. Esser (1987) showed that the changes in carbon storage in the terrestrial biosphere due to land clearing were much larger than any changes due to climate from 1860 to 1981. Now, carbon dioxide released from the destruction of native vegetation is a significant fraction of the amount released from the burning of fossil fuels (Figure 4). Given that the yields of most crop plants per unit of land have not increased in recent years (Brown and Young 1990), more land will be converted to agriculture in an attempt to feed the world's expanding population in the future. Thus, the best predictions of the future level of biospheric function will consider CO_2 fertilization, climate change, and the impact of human population growth.

Although there is wide variation in the estimates, currently it appears that about 7.5×10^6 ha of tropical forest are cleared and converted to agriculture each year (Melillo et al. 1985). In 1980, the net release of carbon from the biosphere was about 1.8×10^{15} gC/yr, contributing to the

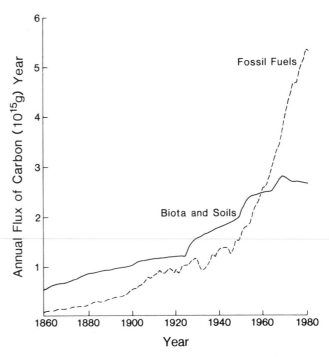

Figure 4: Net flux of carbon to the atmosphere as a result of land-use changes and harvest compared to that released from the combustion of fossil fuel. From Houghton et al. (1983). © 1983 by Ecological Society of America. Reprinted by permission.

pool of CO_2 in the atmosphere (Molofsky et al. 1984, Houghton et al. 1987). Almost all of the release occurred from land use change in the tropics; reforestation in some regions of the temperate zone acts as a small sink for atmospheric CO_2 (Delcourt and Harris 1980, Schiffman and Johnson 1989).

Changes in land use are likely to have direct impacts on global biogeochemistry and climate. For example, Lean and Warrilow (1989) show that clearing of forests in the Amazon basin will lead to a permanent reduction in regional rainfall, since much of the water vapor in the atmosphere is normally the result of transpiration by forest canopies. Much of this area could undergo permanent conversion to tropical grassland and dry woodlands, with lower primary productivity. Widespread burning of tropical forests appears to contribute to the global increase in tropospheric ozone (Crutzen *et al.* 1985), and the loss of N_2O to the atmosphere is greater from most agricultural and pasture soils than from the mature tropical forests that they replace (Robertson and Tiedje 1988).

Predictions of lower soil moisture in large areas of the central US and Asia imply lower crop yields, lower soil fertility and increased use of irrigation in the future (Adams *et al.* 1990), as farmers attempt to maintain production on increasingly marginal land. Seasonal migrations of nomadic populations in the Sahel now follow seasonal rainfall, which drives soil nutrient turnover and the productivity of vegetation (Breman and de Wit 1983). A change in the distribution of vegetation at the arid/semi-arid margin, or in the seasonal pattern of plant growth, will induce changes in the movement of nomadic peoples, with the potential for permanent impact on regions of low productivity. These changes will be exacerbated by rapidly increasing population (Mabbutt 1984). Climate change and human impact on the land may tip the balance of such marginal lands, leading to desertification (Figure 3).

Direct human impacts on the biosphere extend beyond changes in land use. It is likely that the widespread decline in forest growth in the northeastern United States and central Europe is due to exposure to air pollutants, acid rain, and excessive deposition of nitrogen (Pye 1988, Aber *et al.* 1989). Such perturbations of the natural biosphere are likely to increase markedly throughout the world as human populations and industry grow, especially in developing nations. To reassure us of the future, some workers have suggested that the human production of N and P fertilizer applied to the terrestrial biosphere to enhance plant growth may offer a significant sink for atmospheric CO_2 (Garrels *et al.* 1975). However, the potential net sink from this "eutrophication" of the biosphere is about 0.2×10^{15} gC/yr, a helpful, but insignificant, fraction of the annual release from fossil fuels (Peterson and Melillo 1985). Since the atmospheric deposition of P is seldom enhanced downwind of industrial emissions, excessive atmospheric deposition of N often leads to a phosphorus deficiency in vegetation (Mohren et al. 1986).

Conclusions

The conditions on our planet have changed dramatically through geologic time; it is only the present day rate of change that should cause our alarm. Changes in the concentration of atmospheric gases ranging up to 1.0% per year are unprecedented in our records of the geologic past. Just as canaries were used to detect unhealthful conditions in coal mines, the declining populations of migratory songbirds are perhaps our first indication of the extent of human destruction of natural habitat throughout the world (Robbins *et al.* 1989). During our deliberations this week we will focus on climatic change, but we must not forget that climatic change is only another indication of the global effects of an exponentaily increasing human population on Earth and our desire to achieve a higher standard of living for all peoples around the world. Twenty years ago, my colleague Dan Livingstone (1973) likened the oscillating CO_2 concentrations at Mauna Loa to a measure of the "pulse" of life on Earth. Now we believe that the body temperature of the biosphere is increasing, a symptom of an underlying infection.

All of the indications of global change--increasing CO_2, climatic change, species extinctions, and a decline in the productive capacity of the natural and managed biosphere-- should cause us alarm. Our population is growing exponentially in a closed ecological system - the Earth - the only environment we have. Unlike the early laboratory experiments of Gause (1934), no outside force will refresh the resources available for our persistence in a closed environment. Already through the sum of human activities, as much as 40% of the net primary production of the biosphere is reduced, foregone or harvested by humans (Vitousek *et al.* 1986). Perhaps at no time in the history of Earth has a single species usurped so much of its biotic productivity. Whether humans will persist with a desirable quality of life may well depend on our efforts to control human population growth and our destruction of the natural biosphere that supports all life on Earth.

Acknowledgements

I thank A.F. Cross, L.D. Schlesinger, B.R. Strain, and G.M. Woodwell for critical reviews of this manuscript.

References

Aber, J.D., Nadelhoffer, K.J., Steudler, P., and Melillo, J.M. 1989. Nitrogen saturation in northern forest ecosystems. Bioscience 39: 378–386.

Adams, R.M., Rosenzweig, C., Peart, R.M., Ritchie, J.T., McCarl, B.A., Glyer, J.D., Curry, R.B., Jones, J.W., Boote, K.J. and Allen, L.H. 1990. Global climate change and US agriculture. Nature 345: 219–223.

Allen, L.H. 1990. Plant responses to rising carbon dioxide and potential interactions with air pollutants. J. Environ. Qual. 19: 15–34.

Amthor, J.S. 1984. The role of maintenance respiration in plant growth: Mini-review. Plant Cell Environ. 7: 561–570.

Arris, L.L. and Eagleson, P.S. 1989. Evidence for a physiological basis for the boreal-deciduous forest ecotone in North America. Vegetatio 82: 55–58.

Bazzaz, F.A., J.S. Coleman, and S.R. Morse. 1990. Growth responses of seven major co-occurring tree species of the northeastern United States to elevated CO_2. Can. J. Forest Res. 20: 1479–1489.

Billings, W.D., Luken, J.O., Mortensen, D.A. and Peterson, K.M. 1982. Arctic tundra: A source or sink for atmospheric carbon dioxide in a changing environment? Oecologia 53: 7–11.

Billings, W.D., Luken, J.O., Mortensen, D.A. and Peterson, K.M. 1983. Increasing atmospheric carbon dioxide: Possible effects on arctic tundra. Oecologia 58: 286–289.

Billings, W.D., Peterson, K.M., Luken, J.O. and Mortensen, D.A. 1984. Interaction of increasing atmospheric carbon dioxide and soil nitrogen on the carbon balance of tundra microcosms. Oecologia 65: 26–29.

Black, R.A. and Bliss L.C. 1980. Reproductive ecology of Picea mariana (Mill.) BSP at tree line near Inuvik, Northwest Territories, Canada. Ecol. Monogr. 50: 331–354.

Breman, H. and de Wit, C.T. 1983. Rangeland productivity and exploitation in the Sahel. Science 221: 1341–1343.

Brook, G.A., Folkoff, M.E. and Box, E.O. 1983. A world model for soil carbon dioxide. Earth Surface Processes Landforms 8: 79–88.

Brown, L. and Young, J.E. 1990. Feeding the world in the nineties. pp. 59–79. In Starke, L. (ed.). State of the World 1990. W.W. Norton Co., New York.

Crutzen, P.J., Delany, A.C., Greenberg, J., Haagenson, P., Heidt, L., Lueb, R., Pollack, W., Seiler, W., Wartburg, A., and Zimmerman, P. 1985. Tropospheric chemical composition measurements in Brazil in the dry season. J. Atmos. Res. 2: 233–256.

Curtis, P.S., Drake, B.G., Leadley, P.W., Arp, W.J. and Whigham, D.F. 1989. Growth and senescence in plant communities exposed to elevated CO_2 concentrations on an estuarine marsh. Oecologia 78: 20–26.

Davis, M. B. and Zabinski, C. 1990. Changes in geographic range resulting from greenhouse warming effects on biodiversity in forests. In Peters, R.L. and Lovejoy, T.E., eds. Proceedings of the World Wildlife Fund's Conference on Consequences of Global Warming for Biological Diversity. Yale University Press, New Haven.

Delcourt, H.R. and Harris, W.F. 1980. Carbon budget of the southeastern U.S. biota: Analysis of historical change in trend from source to sink. Science 210: 321–323.

DeLucia, E.H., Sasek, T.W., and Strain, B.R. 1985. Photosynthetic inhibition after long-term exposure to elevated levels of atmospheric carbon dioxide. Photosynthesis Res. 7: 175–184.

Emanuel, W.R., Shugart, H.H. and Stevenson, M.P. 1985. Climatic change and the broad-scale distribution of terrestrial ecosystem complexes. Climatic Change 7: 29–43, 457–460.

Esser, G. 1987. Sensitivity of global carbon pools and fluxes to human and potential climatic impacts. Tellus 39B: 245–260.

Garrels, R.M., MacKenzie, F.T., and Hunt, C. 1975. Chemical Cycles and the Global Environment. W. Kaufmann, Los Altos, CA.

Gause, G.F. 1934. The Struggle for Existence. Hafner Publishing Co., New York.

Grulke, N.E., Riechers, G.H., Oechel, W.C., Hjelm, U. and Jaeger, C. 1990. Carbon balance in tussock tundra under ambient and elevated atmospheric CO_2. Oecologia 83: 485–494.

Houghton, R.A., Hobbie, J.E., Melillo, J.M., Moore, B., Peterson, B.J., Shaver, G.R. and Woodwell, G.M. 1983. Changes in the carbon content of terrestrial biota and soils between 1860 and 1980: A net release of CO2 to the atmosphere. Ecol. Monogr. 53: 235–262.

Houghton, R.A., Boone, R.D., Fruci, J.R., Hobbie, J.E., Melillo, J.M., Palm, C.A., Peterson, B.J., Shaver, G.R., Woodwell, G.M., Moore, B., Skole, D.L. and Myers, N. 1987. The flux of carbon from terrestrial ecosystems to the atmosphere in 1980 due to changes in land use: Geographic distribution of the global flux. Tellus 39B: 122–139.

Lean, J. and Warrilow, D.A. 1989. Simulation of the regional climatic impact of Amazon deforestation. Nature 342: 411–413.

Livingstone, D.A. 1973. Summary and Envoi. pp. 366–367. In G.M. Woodwell and E.V. Pecan, eds., Carbon and the Biosphere. CONF-720510. National Technical Information Service, Springfield, Va.

Mabbutt, J.A. 1984. A new global assessment of the status and trends of desertification. Environ. Conserv. 11: 103–113.

Manabe, S. and Stouffer, R.J. 1980. Sensitivity of a global climate model to an increase of CO2 concentration in the atmosphere. J. Geophys. Res. 85: 5529–5554.

Manabe, S. and Wetherald, R.T. 1986. Reduction in summer soil wetness induced by an increase in atmospheric carbon dioxide. Science 232: 626–628.

Martin, J.M. and Gordon, R.M. 1988. Northeast Pacific iron distributions in relation to phytoplankton productivity. Deep Sea Res. 35: 177–196.

Melillo, J.M., Palm, C.A., Houghton, R.A., Woodwell, G.M., and Myers, N. 1985. A comparison of two recent estimates of disturbance in tropical forests. Environ. Conserv. 12: 37–40.

Mohren, G.M.J., Van den Berg, J., and Burger, F.W. 1986. Phosphorus deficiency induced by nitrogen input in douglas fir in the Netherlands. Plant Soil 95: 191–200.

Molofsky, J., Menges, E.S., Hall, C.A.S., Armentano, T.V. and Ault, K.A. 1984. The effects of land use alteration on tropical carbon exchange. pp. 181–194. In T.N. Veziroglu (ed.). The Biosphere: Problems and Solutions. Elsevier Scientific Publishers, Amsterdam.

Norby, P.J., O'Neill, E.G. and Luxmoore, R.J. 1986. Effects of atmospheric CO_2 enrichment on the growth and mineral nutrition of Quercus alba seedlings in nutrient-poor soil. Plant Physiol. 82: 83–89.

Pastor, J. and Post, W.M. 1988. Response of northern forests to CO_2-induced climate change. Nature 334: 55–58.

Peterson, B.J. and Melillo, J.M. 1985. The potential storage of carbon caused by eutrophication of the biosphere. Tellus 37B: 117–127.

Post, W.M., Emanuel, W.R., Zinke, P.J., and Stangenberger, A.G. 1982. Soil carbon pools and world life zones. Nature 298: 156–159.

Prentice, K.C. and Fung, I.Y. 1990. The sensitivity of terrestrial carbon storage to climate change. Nature 346: 48–51.

Pye, J.M. 1988. Impact of ozone on the growth and yield of trees: A review. J. Environ. Qual. 17: 347–360.

Reid, C.D. 1990. The carbon balance of shade-tolerant seedlings of *Fagus grandifolia* and *Acer saccharum* under low irradiance and CO_2 enrichment. Ph.D. Dissertation, Duke University, Durham, N.C.

Robbins, C.S., Sauer, J.R., Greenberg, R.S. and Droege, S. 1989. Population declines in North American birds that migrate to the neotropics. Proc. Natl. Acad. Sci. USA 86: 7658–7662.

Robertson, G.P. and Tiedje, J.M. 1988. Deforestation alters denitrification in a lowland tropical rain forest. Nature 336: 756–759.

Ryan, M.G. 1990. Growth and maintenance respiration in stems of Pinus contorta and Picea engelmannii. Can. J. Forest Res. 20: 48–57.

Schiffman, P.M. and Johnson, W.C. 1989. Phytomass and detrital carbon storage during forest regrowth in the southeastern United States piedmont. Can. J. For. Res. 19: 69–78.

Schlesinger, W.H. 1977. Carbon balance in terrestrial detritus. Ann. Rev. Ecol. Syst. 8: 51–81.

Schlesinger, W.H., Reynolds, J.F., Cunningham, G.L., Huenneke, L.F., Jarrell, W.M., Virginia, R.A. and Whitford, W.G. 1990. Biological feedbacks to global desertification. Science 247: 1043–1048.

Sebacher, D.I., Harriss, R.C. Bartlett, K.B., Sebacher, S.M. and Grice, S.S. 1986. Atmospheric methane sources: Alaskan tundra bogs, an alpine fen, and a subarctic boreal marsh. Tellus 38B: 1–10.

Strain, B.R. and Cure, J.D. 1985. Direct Effects of Increasing Carbon Dioxide on Vegetation. US Department of Energy Report DOE/ER-0238, Washington, D.C.

Tissue, D.T. and Oechel, W.C. 1987. Response of Eriophorum vaginatum to elevated CO_2 and temperature in the Alaskan tussock tundra. Ecology 68: 401–410.

Van Cleve, K., Oechel, W.C. and Hom, J.L. 1990. Response of black spruce [Picea mariana (Mill) B.S.P.] ecosystems to soil temperature modification in interior Alaska. Can. J. Forest Res. 20: 1479–1489.

Vitousek, P.M., Ehrlich, P.R., Ehrlich, A.H. and Matson, P.A. 1986. Human appropriation of the products of photosynthesis. BioScience 36: 368–373.

Whalen, S.C. and Reeburgh, W.S. 1990. A methane flux transect along the trans-Alaska pipeline haul road. Tellus 42B: 237–249.

Climate, Tropical Ecosystems and the Survivability of Species

by Perez M. Olindo *(Kenya)*

Introduction

When most people think about climate and climate change, the picture that comes to their minds relates immediately to the climatic conditions they have lived under for most of their lives. The people who live in the temperate latitudes think about winters and summers, while those living in the tropics think about monsoon rains, deserts, hot, wet and humid months of the year, etc.

When we refer to the phenomenon of global warming, to a temperature increase of the magnitude of 0.3–0.6°C over the past century and a rate of global warming that could be 0.5–1.0°C per decade through the next few decades (IPCC Report), most people wonder why we should be so concerned about this subject when the differences between winter and the summer months of any given year exceed these figures without causing any catastrophic disruptions.

The public does not see why species of animals like elephants, whose range stretches from the Sahara into the tropical forests of Africa, cannot adapt to the temperature changes that are predicted. These questions represent lingering doubts in the minds of the public, from whom we expect to receive support for any plan of action which would counter or mitigate the climatic trends. A worldwide public awareness campaign is needed to present this problem in the simplest manner possible.

To most people living outside the tropics whose environment is not a subject of active international debate, and who are therefore not expected to make any sacrifices to save tropical forests, the call to prohibit all human activity in the massive tropical forests, which are essential in absorbing carbon dioxide, is as simple as it is logical. But for the communities resident in these forests, the issue constitutes a matter of life and death. It should, therefore, not come as a surprise that extensive dialogue is necessary, in order to identify sustainable ways of ensuring the welfare of the affected communities, if the survival of tropical ecosystems is to be assured. The elements to be included in such talks might cover such matters as compensation, which would have to be paid to those who must forego their right to a livelihood in selected forests of the globe.

Climate Change

In May 1990, an inter-disciplinary meeting was held in Nairobi to provide a forum for expressing an African perspective on the topic of climate change. The meeting was well attended with participants from meteorology, agriculture, a variety of Institutes dealing with public policy, global warming, technology studies, wildlife, environment and natural resources, population, law and environment-related disciplines of academia, etc. At the end of three days, the group formulated their thoughts in the Nairobi Declaration. In that statement, some of the concerns of Africa are published in summary form. The meeting was widely covered in the local media in Kenya and the general public was informed about the IPCC report dated May 1990, which reveals that the scientific community finds that if no measures are taken major climate changes could occur in the 21st century. The available evidence compiled by the IPCC also shows that the projected climate changes would be global in extent and affect all people. In other words, no nation, and no community, developed or developing, will be unaffected.

The IPCC report further indicates that over the last 100 years, the global mean surface air temperature has increased by 0.3–0.6°C and that the warmest years ever for the globe as a whole were recorded in the 1980s. It reveals that the global sea-level has increased by 10–20 cm over the same period. The Nairobi Declaration rightly points out the broad consensus among scientists, showing that substantial climatic change is going to occur and will have unpredictable and severe effects on tropical ecosystems. It will have the potential of devastating developing regions like Africa and other tropical zones, unless some 'anticipatory measures' to slow down its rate and the ultimate magnitude are taken without delay. The nations of Africa and other developing countries will need a lot of information before they are ready for the talks at which decisions for managing the anticipated climate change will be taken.

The Nairobi Declaration emphasized the contribution of developed countries through their high levels of fossil fuel consumption and the manufacture of inefficient combustion engines. The world is invited to take the view that what is facing us is truly a shared problem, because in one form or another, everyone has participated in causing it. The degree of responsibility differs between the developed and the developing countries. What must be determined is the apportionment of the burden and the timing of its resolution. This issue should not be tackled in an atmosphere of confrontation, but rather, through the process of consultation and sharing of ideas and technologies. Those countries which have more resources than others should assume the larger portion of the costs of mitigating the impending change.

Developing countries have limited financial and technological resources. Their institutional capabilities are low, and a majority of these countries are suffering from debt-related problems. However, if climate change affects the globe, it will not discriminate between development, wealth, poverty or technological sophistication. It will affect all. Therefore, the people of the developing world call for a far-reaching dialogue and the re-structuring of the systems that are responsible for bringing us so close to destroying the one earth we all share.

The Nairobi meeting called upon the UN system to support the participation of African countries in international negotiations, but in fact all the developing countries should be included. The NGOs were called upon to increase their public awareness efforts and to follow closely, and participate in the negotiations on climate change. The private sector was called upon to increase their research efforts, to develop sound environmental technologies and management practices and to disseminate these as widely as possible. The private sector should also be called upon to avoid involving themselves in activities that tend to increase environmental stress of any kind.

Food Production and Global Warming

A number of agricultural activities contribute to the increased concentration of greenhouse gases in the atmosphere. Farm wastes are used as manure. The production of this farm input releases methane. While some studies show that more carbon dioxide in the atmosphere would increase vegetative growth through the enhanced photosynthesis, other studies suggest that sufficient moisture and the other growth factors will not be available in the right quantities for this "fertilization" to work. Recent

findings indicated that when certain crop plants are exposed to ultraviolet light for long periods, their ability to photosynthesize is substantially reduced. The destruction of the ozone layer will allow a lot more ultraviolet radiation to reach the face of the earth and may have the effect of reducing the productive capacity of photosynthesis.

In the process of food production, nitrogen fertilizers are used as farm inputs, with the object of increasing yields. But in the process, nitrous oxide is released into the atmosphere. When animal and human wastes decompose they release methane and other gases. Unfortunately, in most parts of the world, biogas has not yet been harnessed to convert methane into a major energy resource. Instead, it is freely discharged into the atmosphere and increases the concentration of the greenhouse gases. Similarly, paddy rice production contributes to the concentration of methane in the atmosphere. Therefore, with the inevitable increase in human population and the corresponding need to feed that population, the quantity of both nitrous oxide and methane in the atmosphere will increase. There is a need to devise ways of extracting some of these gases from the atmosphere by conventional means for commercial purposes.

Although the volume of methane and nitrous oxide in the atmosphere is much smaller than that of carbon dioxide, the IPCC report shows that these gases are more efficient in enhancing global warming than carbon dioxide. The point that clearly comes out is that global warming is not a result of affluence alone. As a matter of fact, poverty also contributes to it.

Tropical Ecosystems

Current Issues of Biodiversity and Biotechnology

The work which has already been done by governments, NGO's and individuals towards the conservation of portions of tropical ecosystems must be highly commended. However, this amounts to only a tiny fraction of the work that must be done to assure the conservation of representative biodiversity of tropical ecosystems. Multinational corporations have made some contributions in this sector, but their attention has been directed largely to the species of plants that are of known commercial significance to them. Germplasm has been collected to safeguard species which are generally referred to as interesting, i.e. species that may in the future become of commercial interest. The hundreds of thousands of other species which are both unknown and of no immediate or known commercial importance remain completely unattended, unprotected, and therefore, highly vulnerable to extinction through deliberate human action or for lack of active conservation programmes. Plants do not have to be of commercial importance before they are given legal protection within the countries in which they grow.

Vegetation belts which cover significant water catchment areas of the countries within which they grow should be accorded functional protection. Other areas which are of outstanding biotic value should be protected as representative ecosystems of unique vegetation types or as national parks. Vegetation which provides food to specific animals or human communities must also be protected. The limitations in conserving tropical ecosystems are enormous. Leaving useful species of plants or animals at risk just because we do not know anything about them cannot be quantified in monetary terms nor should it be justified.

The theory of floral biodiversity generally relies on predictable plant succession under given climatic and biotic conditions, with each wave of changes witnessing a corresponding change in fauna composition until there is a climax of dominant plant communities with their dependent animal climax communities. Whenever localized changes have taken place in such climax communities, secondary vegetation has taken over and changes have continued to occur until the logical climax has been reached again. But given the uncertain climatic changes to come, it is anticipated that extensive changes to which most of the plants in the tropics may not be adapted, calls for human intervention in the form of establishing germplasm banks to save some of the fauna and flora we have today. It is my considered view that, while the multinational corporations will continue to save the plant species of commercial importance to them, the world community, through the United Nations machinery, should give very serious consideration to the establishment of a parallel germplasm bank to safeguard those plant species that no one else may be immediately interested in. This brings us to the issue of the interdependence between biological diversity and biotechnology, and the need for urgent international dialogue to determine how best to create a functional linkage between these related entities, which cannot function independently of each other.

There is a basic sensitivity which appears to be hindering the commencement of meaningful discussions on this matter. A big hurdle stands in the way of effective dialogue (in the sense that one party owns the resource while the other party owns the technology), but as the world marks time in resolving the problem, hundreds and maybe thousands of species of plants and animals are becoming extinct each year. The current position is this: the largest portion of the world's biological diversity which needs protection is in the Third World, located in the tropics, where the owners of this resource do not have ready access to the technologies required to manage that biodiversity in the changing world environment. However, the biotechnologies required to assure the long-term sustainability or for that matter, the survivability of this biodiversity, are the property of the developed countries, who have locked up this asset in patents and protocols of

intellectual property. That knowledge which will make a difference between life and death to a large portion of the one Earth we all share remains unavailable to those who need it most. Is humanity going to continue to watch the extinction of species after species of plants and animals just because those nations who own that biodiversity cannot afford, and therefore have no access to, the technology needed to assure the survivability of these species?

Since biotechnology has so much to offer to the conservation of biological diversity, the only way to ensure the long-term survival of tropical ecosystems would be to create the right climate for maximum co-operation between the two parties who own these properties. Patents and intellectual property rights have been mentioned. The protection of these institutions is vital and should be continued and indeed attractive incentives should attach to outstanding discoveries.

However, biotechnology provides new and improved ways of preserving specific genetic resources, perhaps those with traits which will enable them to survive in the changed climatic conditions of the future. This paper urges the early commencement of negotiations to remove the impasse that surrounds this crucial issue of survival. Take for example, tissue culture, which many researchers believe has a wide range of applications in the efforts of re-orienting tropical species of plants to survive in the changed climatic conditions of the future. To assure survivability of plant species, improved qualities to withstand higher temperatures, severe drought conditions, violent storms, etc are needed. If we succeed in training species in tropical ecosystems to withstand the predicted climate changes, we will have increased the survivability of species by several centuries. Biotechnology relating to embryo storage and transfer is another specialized science which holds good promise for the survivability of species. Research on breeding between genetic materials from domesticated varieties of crop plants, which have a wide range of qualities, including higher disease and pest resistance and the ability to grow in different soil types under more stressful climatic conditions, could be another contribution to the future survivability of species. It is through research and development that the important area of gene expression of a wide range of species will be better understood in order to be applied in the field to ensure a better chance of survival of species in tropical ecosystems. Survivability of species will, to a large extent, depend on whether or not the owners of the biotechnology and those who have the biotic resources agree to work together. If on the other hand, the parties fail to agree, then the survivability of species, even that of the human race, cannot be secure in the long-term.

At the present time, we see harsh competition accompanied by secrecy in the protection of biotechnology. We see the cheap acquisition of biodiversity and its storage by developed countries. Safety of storage of vital sets of germplasm may be achieved by duplicating the banks in different latitudes and altitudes, so that, whatever the pattern or severity of climate change, all the gene banks and other products of biotechnological innovation will not be instantaneously destroyed in the few countries where they have been stored.

Laboratory research and development are not enough for the task that lies before us. In situ ecosystems research will have to be undertaken and biological diversity examined in greater detail for better understanding of, for example, the climatic thresholds within which it would thrive and give a sustained high yield. The limits beyond which various portions of the biodiversity would perish must also be known if is to be effectively conserved.

Commercial Aspects of Biotechnology

The major break-throughs in the area of biotechnology have to a large extent been made by the private sector. The greatest motivation for the private sector has naturally been of a commercial nature. For the private sector to initiate major activities in research and development, the aspects that have no profitable outcomes would suffer and survivability of species would, in the long-term, be limited to those species that have clearly demonstrated commercial profitability. With the current limitations in knowledge, many useful species could be lost forever. It is for this reason that we propose that financial resources be identified for public biotechnology centres in strategic zones of the world, in order to guarantee the survivability of species in tropical ecosystems and globally. Species with broader genetic variability have greater attributes of survival in changing climatic conditions. Genetic similarity is a weakness that succumbs quickly to adversity. For this reason, every effort should be made to hunt for and save genetic variability for the future.

Economic Handicaps for Developing Countries

Tropical ecosystems are largely owned by developing countries which are in need of funds for many ongoing developmental activities. Additional financial resources are needed in these countries to enable them to actively participate in work on climate change. The approach of training and collaborative policy development between private business and the public in the developed world, and the export of the results to the third world countries to adapt, develop and apply the technologies to their ecosystems has begun, but is not sufficient.

The Survivability of Species

The survivability of species cannot be determined without precise empirical information on the survival thresholds of species of plants and animals over the long-term (i.e., 1000

years or more). To secure this information, extensive and continuing research is essential. The maximum and the minimum levels of tolerance of the various species of plants and animals to greenhouse gases must be determined. One would also have to find out how these same communities would respond to greatly changed climatic conditions.

In the discipline of conservation biology, researchers have attempted to address this issue in the temperate lands. They have also tried to define, model and predict the requirements for minimum viable populations but there are limitations and the prospect of adaptation differs from species to species of both plants and animals. To the extent that tropical ecosystems are the product of their evolving climatic conditions, these ecosystems have taken in their stride the small changes in temperature over the centuries, and the localized but sometimes violent tropical storms. With detailed investigation, it may be found that a substantial body of the species prevalent in the tropics must have developed a robustness which will accommodate change.

Considering the limited number of species of animals which have been studied in detail, and the array of unique characteristics that vary across the total range of any one species, it is clear that the integration of studies would help to ensure species survivability.

The direct attack on the tropical ecosystems through slash and burn, perpetrated by human beings in search of a livelihood, or for speculative purposes, is not helping the survival of species in the tropics. Tropical grasslands, for example, are subjected to regular burning by ranchers and herdsmen the world over. They believe that the practice results in improved grazing for their livestock and combats ticks and other parasites. On the contrary, and depending on the timing of the burning, many species of plants and animals are eliminated permanently in the process and the practice contributes its fair share to the processes leading to global climate change.

The Effect of Increased Temperature on Ecosystems

With global warming, the precipitation patterns in the tropics would change in a manner that cannot be predicted reliably. On the basis of current predictions, the middle of most continents will tend towards aridity or desert conditions. For south-east Asia, Africa and South America, such a change will not only be dramatic, it will have far reaching ecological implications. In the case of Africa, the process of desertification is already spreading at an alarming rate, southwards towards the equator. This fact tends to confirm the above prediction and points to a definite shift in the current vegetation belts of the tropics. With this shift will be the corresponding change in the resident composition of animal species, including human beings.

In the protected areas within this zone, a majority of the national parks and reserves are located in delicate marginal lands, e.g., in the dry savannahs. A few are in the forests of Africa, Latin America and South-east Asia. It is not known whether climatic change will result in the formation of new forests and grasslands of comparable functional capacity and extent elsewhere. If it does, it is uncertain how long it will take and what the species composition would be. Already protected areas have become islands with serious limitations, which have led to reduced genetic variability, and these limitations have therefore moved those species considered to be legally secure much closer to extinction long before the major climatic changes occur. Also, in this era of ecotourism, millions of well-intending visitors to the national parks are literally overrunning them and speeding up their demise as viable biological entities.

In the case of desertification, current observations show that most of the plant species in the way of advancing deserts are usually eliminated in the order of how much soil moisture they need to survive. The content of the new vegetational regimes to come remains largely unknown, but the biodiversity in the new climatic zones will be greatly reduced, compared to the prevailing rich biotic state in the same areas today. Considering also that climatic belts will not shift with their current floral or faunal compositions which have evolved over many centuries, the tiny present day protected areas, e.g., national parks and wildlife reserves, with their fixed boundaries, will not only suffer irreversible ecological changes, but they could disappear altogether.

The Sensitivity of Gene Banks

The commendable efforts which have been made by scientists to set up germplasm banks in many parts of the world, in order to ensure the continuity of species in the future, seem to have hit a number of serious snags. The environmental conditions in which these genetic resources are being collected are quite different from the changed environment in which they will be expected to regenerate. In the absence of extensive adaptation research and development, effective germination or propagation cannot be guaranteed. Traits suitable to the conditions that could occur in the future should be imparted to those materials before banking to assure a much higher confidence of success. Who is going to do serious adaptation research, where is it going to be done and how can the resources needed for its execution be secured? It has been observed that cold storage is successfully programming the preserved materials for qualities which favour growth in cold climates! How then would such genetic resources be suitable in a much warmer world? What is therefore

urgently required, is the allocation of sufficient resources to refine and diversify the technology of germplasm storage, to allow its reintroduction in all latitudes of the world including the tropics. In its own wisdom, nature uses the dry and the cold methods of storage as appropriate. The other area to be explored involves "in situ" breeding programmes, to produce plants which will be able to adapt and multiply naturally in the warmer and possibly drier climatic conditions of the future. Perhaps research should be directed towards dry and temperature-related storage for tropical germplasm.

While it is possible to predict the likelihood of survival for some of the gregarious and wide-ranging species of animals, human beings included, it is not equally easy to predict the fate of animal species that are highly adapted to specific climatic conditions. For example, in the low-lying coastal terrestrial microhabitats, which will be submerged by the rising sea-level. There is very little hope for the survival of these species of plants and animals that are dependent on such life support systems.

The response of a given species of plant or animal to most climatic variables is not constant across its range. This in effect means that, even if we had access to complete sets of data on a large number of species of plants and animals, uncertainty would still be inevitable due to the effects of specific adaptive traits and natural selection on genetic variation and mutation rates across the range inhabited by the group of plants and animals.

Although tropical ecosystems have a rich biological diversity, and the scientific communities of this region are keen to participate effectively in research focussed upon issues of survival, no financial resources are available to enable them to participate usefully in determining their destiny, and where possible, help to reshape that destiny. To predict accurately the survivability of any species in changed climatic conditions, unlimited access to biotechnology is vital. That biotechnology is not readily accessible when and where it is needed most, to prepare survival-oriented management programmes for a large number of plant and animal species.

The obvious trends of confrontation at the international level and the tendency to hide behind protocols and patents should be overcome in the forthcoming negotiations, so that funds are identified to purchase such biotechnologies and make them readily accessible to all those who need to use them and at the same time safeguard the important institution of patents and intellectual property.

Human Settlements and Survivability of Species

The survivability of species cannot be discussed as if the process would take place in a vacuum. If species which enjoy the largest biological diversity are to survive in a world being engulfed by human settlements, how many individuals of any given species of plant or animal would constitute a viable nucleus for the survival of that species? And for how long would such a population be viable to satisfy this open-ended concept of survivability? Would the ability of a species to manage marginal persistence over several centuries satisfy the criteria for survivability or would a 99% chance over 1000 years be adequate?

Since it was noted that major climatic changes are possible in the next 50 years, very few countries in the developing world have increased their national budgets for gathering and analyzing data, and incorporating their findings into their national planning process! How many scientific institutions use the available information in their management of species conservation programmes to ensure long-term survival? If the governments which initiate and subsequently budget and finance the implementation of such plans are not dealing with issues on the long time scale, this is an area of serious omission.

The concept of protecting sufficiently large habitats to sustain viable populations of plants and animals is a futile exercise, since nature is not static and the legal protection has no effective jurisdiction over natural changes. In the long-term, the vegetation and the animal species that may be found in these "island" parks may have very little similarity to the natural regimes that were accorded legal protection in the first place. The alternative is the prospect of "managed" natural ecosystems. This is what germplasm banks will lead to and justifies commencing management of current national Parks for their future survival. The question is, how much management? In the changed environment of the future, will the ethics of ecosystems manipulation be relevant or not? Alternatively, will ethics aim for survival and survival alone?

The rapid growth of the human population is another definite agent which militates against the survivability of species.

Conclusion

The question of survivability of species is a vexing issue which will require extensive international co-operation and co-ordination to resolve. Integrated research work to deal with various aspects of global warming and climate change will have to be undertaken. The findings of that research will have to be made public on a regular basis, and it should be agreed from the outset to implement the relevant findings. Naturally, governments will have the freedom to challenge any aspects of the work, and the findings should be widely discussed by the public. If humanity can deploy space satellites and retrieve them; if humanity can land on the moon and return safely to earth; then it is quite feasible that if humanity decides to tackle the issue of global warming and change in a joint and serious manner,

solutions will be found and the rate of change will be brought down to manageable levels.

With the question of survivability one is confronted with scientific, ethical, financial, political and logistical problems. This is not an issue that can be dealt with single-handedly or by one country or a handful of nations. The entire body of humanity must be mobilized. It will not be easy, but it can be done. Massive human, financial and material resources will be needed. As we continue to talk, species of plants and animals are rapidly becoming extinct. The present efforts of establishing national parks and biosphere reserves etc., are quite commendable but will not satisfy the long-term requirements of survivability of species.

Some Possible Impacts of Climatic Change on African Ecosystems

by C. H. D. Magadza *(Zambia)*

1. Introduction

Evidence on the phenomenon of global warming is sufficiently compelling to justify discussion of its possible impact on current ecosystems. For this we must first establish a dispassionate perspective of ecological change in nature. It is against such a perspective that we can evaluate the likely implications of global climate change.

In the billions of years that life has been on earth it has undergone significant qualitative change, in the range and complexity of its organization at the individual, community and ecosystem levels.

The projected rate of man-induced global warming is of the order of 0.5°C per decade, with estimates ranging between 0.1°C and 0.8°C. It is also estimated that the temperature difference between the last ice age and present temperature is of the order of 5°C. This temperature rise since the last ice age occurred over a period of about 20,000 years, giving a mean temperature rise of 0.0025°C per decade. Thus, the projected rate of climatic change due to the increasing greenhouse effect is about 100 times faster than the average rate since the last ice age. The projected global rise in temperature in the next century is estimated to be between two and six degrees Celsius. While it is thought that the frequency of extreme events could remain unchanged, they could be more severe in comparison to current climate. This accentuation of extreme events in a relatively rapidly changing climate regime is of considerable ecological significance.

2. Water Resources

Increased global temperature leads to increased ocean evaporation, and hence greater precipitation. Thus, it is expected that the impact of global warming would be a global increase in precipitation and runoff. However, the distribution pattern of evaporation, precipitation and runoff will continue to be influenced by the disposition and structure of the land masses, such that tropical continental arid lands are likely to benefit less from the increased precipitation than coastal areas. Yet, some of the simulation models used to construct possible future climate scenarios predict increases in precipitation in what are currently arid to semi-arid lands. For example, parts of the Namib desert and the Kalahari are likely to experience increases in mean rainfall of up to 1.5x current means. However, such precipitation increases also appear to be accompanied by increased mean temperatures, thus raising the evaporation.

Nemec and Schaake, cited in Riebsame 1989, have modelled the hydrology of Lake Victoria under various scenarios. Their results indicate that a 10% reduction in precipitation and a 12% increase in evapotranspiration, corresponding to a 3°C temperature rise, would lead to at least a 40% reduction in runoff, while a similar rise in precipitation for the same rise in temperature would result in no significant increase in runoff. It can be surmised, therefore, that in continental areas in the semi-arid to arid tropics, gains in precipitation are likely to be nullified by increased evaporation resulting in an overall deterioration in the soil moisture availability. It is therefore possible that the process of desert encroachment could be accelerated.

3. Southern Tropical Areas

There are only limited studies on the impacts of climatic change in the tropics, and whatever is available is not very clear. Williams and Faure (1980), in their preface to the Volume on The Sahara and The Nile recapture for us an image of a Sahara with lakes and rivers in which the hippopotamus wallowed and Nile perch were plentiful: a Sahara with settled hunter-gatherer cultures who fashioned domestic utensils and made pottery.

Gornitz (1985) has shown that in the last hundred years the tropical rain forest of West Africa has diminished in extent due to a combination of encroaching aridity and human impact. We also know that the fynboss vegetation of South Africa was more extensive in the Pleistocene and that tree ring data from South Africa indicate that the response of the woody plants to increased precipitation in wet episodes of the Holocene period was enhanced growth rates (Tyson 1989). During this period the Sahara and Namib were also moister than at present, indicating that the post-Pleistocene warming may have contributed to the aridity in these areas of Africa. Livingstone (1980) tells us of the existence in the Sahara of a similar xerophytic vegetation in the Pleistocene.

Hamilton, (1982), using various forms of evidence, such as plant microfossils, pollen deposits from a variety of African sites, and the geographical distribution of fauna, has also concluded that the principal response of the African forest flora to climate change in the upper Pleistocene was its spread during wet periods and its disruption into smaller refugia during more arid periods. The tropical forests of Eastern Zaire, Rwanda and Uganda, for example, coalesced with those of the slopes of Ruwenzori and the coastal areas of East Africa about 12 000 BP, while in the drier period that followed the refugium retreated to eastern Zaire, leaving the Uganda–Rwanda area with a savanna type of vegetation. The arid areas of Africa could experience temperature increases of up to 6°C by the middle of the next century. The mean rainfall expected for a doubled CO_2 atmosphere is about 90% of current mean. Thus a combination of elevated temperatures and reduced rainfall in this region is likely to lead to more severe aridity.

Several climate models suggest that with a $2 \times CO_2$ climate the tropical rain forest areas of Zaire and Uganda would not receive significantly higher precipitation, and in some cases may show deficit in comparison to current rainfall levels. The process of reduction of tropical forest, first noted in this area about 9 000 BP, (Livingstone 1975), with the arrival of Man will continued unabated and even be reinforced by global warming, resulting in an accelerated reduction of tropical forest and the expansion of the arid zones of Africa. Some of the scenarios indicate marginal increases in precipitation, but these precipitation increases could be offset by increased evaporation, leading to overall aridity.

Soil moisture and nutrients have emerged as the dominant factors (Cole, 1982; Tinley, 1982; Bell, 1982). In controlled environments increases of the CO_2 concentration have been shown to elevate photosynthetic activity of crop plants. In C_4 plants, which include most of the tropical grasses, it has been shown to lead to a higher water use efficiency. However, the field application of these results is not clear. In any case, the relationship between increased temperature and available soil moisture must be an integral part of any model that aims at predicting the impact of global warming in the sub-humid tropics.

4. Impact on Wildlife

Thackeray (1987) has examined remains of small mammals from the Pleistocene period in South Africa, dating back to about 50,000 years BP. Using factor analysis and by comparing to the present distribution of some of the extent types, he was able to characterise the mammals of the epoch into warm period and cold period types, with the assemblage of the fauna corresponding to changes in climate during the epoch. There was also some indication of association of mean body size with the precipitation regime. During this period temperature change is estimated to have been 1°C per 1000 years, ie, 0.01°C per decade, the fastest rate recorded in the post-Pleistocene period and 20 to 50 times slower than that predicted for the effects of our man-induced climatic change. The biological result of the climatic change discussed by Thackery has been a geographical re-distribution of the mammalian fauna of Southern Africa, as the ecologically preferred zones for the various members of the assemblage changed in their geographical ranges.

It is clear that changes in mean precipitation in these tropical lands will have a significant impact on their ecological productivity. The data lead us to speculate on the continued capability of the Savannas, particularly those of East and Southern Africa, to support large populations of large mammals, such as buffalo, elephant etc, as well as question the future of pastoralism in the rangelands of Africa.

Cumming (1982) investigated the relationship between the metabolic biomass of the three most dominant large herbivores and rainfall and noted the decrease in the dominance of large herbivores with a decrease of rainfall in the precipitation range 1200mm to 200mm. Furthermore, the dominance of the large herbivores decreased rapidly below the metabolic biomass density of about 5 kg/km[2], a figure corresponding to a mean annual rainfall of 500 mm. These data suggest that in the event of decreased rainfall or soil moisture due to global warming in the African savanna ecosystems, large bodied animals, such as elephant, hippopotamus and the large bovids, are liable to be affected more adversely than smaller species. This result also agrees with that of Thackery from his study of small mammal fossils in the South African Quaternary period.

Our concern over the rate of the anticipated climate changes must lead us to question the extent to which the present assemblage of animal and plant species is able to cope with rapid climate changes, particularly with episodes of extreme conditions, such as drought and heat waves.

We are aware of the capacity of desert organisms to cope with extreme conditions. The camel can travel for days without water intake. Lung-fishes, which have been in existence for about 50 million years, escape from desiccation when their temporary pools dry up by aestivation.

However, these are indeed exotic cases. Although the fossil record indicates that extant species have survived a number of climate changes in their evolutionary history, it also shows that some species and genera did not make it through the climate changes of the late Quaternary. Those that survive have an altered distribution compared with that indicated in the fossil record. The Persian fallow deer. *Cervus mesopotamica*, and the European fallow deer, *Cervus dama dama*, were much more extensively distributed in the Middle East and north Africa, but the subsequent desiccation has restricted their range. The African buffalo, *Synecerus caffra*, has a cousin, *Pelorovis antiqus*, which became extinct about 750,000 years ago. What genetic properties did the extinct forms lack, which enabled the present species to survive to the present period? Is there any genetic predisposition in extant animal species to tolerate extreme climatic episodes?

In this regard we consider the phenomenon described by, among others, Louw and Seely, (1972) called adaptive heterothermy. This is a situation in which an endothermal animal allows its body temperature to vary in response to changes in ambient temperature, without the impairment of its metabolic activities. This is achieved by a number of complex phenomena such as increased rate of replacement of damaged protein, an increase in the production of anti-denaturing substances and the ability of certain enzymes to exist in variants that have different temperature optima.

The biochemical and genetic basis of this capability are discussed by Hochachka and Somero (1973) in their book Strategies of Biochemical Adaptation. Such enzymes are called isozymes. These processes enable the animal to store or lose heat without having to expend energy in thermoregulation. Some African ungulates, the eland and the oryx, have been shown to allow their body temperature to fluctuate by as much as 7°C and rise to as much as 42.1°C in the oryx. We do not know how widespread these adaptive phenomena are, but at the other end of the vertebrate scale Hochachka and Somero discuss the ability of the tuna to produce various variants of enzymes, depending on the ambient temperature. Thus it appears that a bio-chemical basis for coping with changing climatic conditions may be widespread among the animal kingdom, and in it may lie, for some animals perhaps, some mechanism for coping with the anticipated climate change due to global warming.

The Australian Kangaroo shows an interesting adaptation which ensures prompt replacement of loss of young offspring. The female is receptive to fertilisation soon after parturition, but the blastocoele development is arrested by the suckling activity of the embryonic joey. If this embryo has an untimely death, the blastocoele immediately resumes development. Thus, during periods of stress where a high embryonic mortality might occur, the reproductive capacity of the population is also elevated to compensate for the high mortality (Louw and Seely 1972).

Very little similar work has been done on the impacts of global warming on the African continent, but the recent droughts in the Sahel compel us to pay particular attention to this region. The region is characterized by sparse water resources, and increasingly severe droughts. In this region the predicted mean annual temperature rises are as much as 5°C, but the corresponding changes in precipitation are marginal.

5. Wetlands

It is probable that dry-season runoff will be significantly reduced, but there will also be very little winter precipitation. The Senegal and Gambia basins will be faced with increased water demand. Lake Chad, an important water resource for a number of countries in the basin, now stands at about one third of its 1965 level, and could conceivably be further reduced by increased evaporation alone. Within the last two decades Lake Chirwa in Malawi almost dried up, causing considerable stress to communities that were dependent on the lake for water supplies and fishing. In southern Africa, south of the Zambezi, there are very few natural large water bodies, the Chobe swamp being one of them. Large herds of wildlife, as well as human agriculture, depend on the annual floods of the Chobe river that feeds this endorheic system. On Lake Bangweulu in Zambia it has been shown that under current climatic conditions only 10% of the hydrological income of this lakes appears as outflow on the Luapula river, the rest having been lost by evapotranspiration from the lake. With the predicted elevated temperatures of the region and reduced hydrological income it is pertinent to ponder on the future of this part of one of Africa's river basins and its impact on Lake Mweru. The spectacular speciation of ciclid fishes in the East African Great Lakes is testimony to the impact of the arid periods when these water bodies were dissected into smaller discontinuous water masses which then allowed the speciation we see now (Fryer and Iles 1972).

Further south in Southern Africa the shortage of natural lakes has lead to the construction of a very large number of medium-sized dams to large lakes. Many of the dams that should supply water for irrigation and potable water are designed for three years' runoff storage. Thus episodes of severe droughts, even of current duration, would have significant impacts on the availability of water for agriculture and industry.

6. Semi-arid Regions

Between the arid land of the Sahel in the north and the Kalahari and Namib desert in the south of Africa lie the

Figure 1: The extent of the over-wintering area of migrant birds in Africa. Source: International Council for Bird Preservation.

Savannas, the most intensely cultivated parts of tropical Africa, of which more than 50% is now under cultivation (Table 1). Population pressure and inappropriate land use are increasingly rendering the water resources of this zone also fragile. Forest clearing for cultivation and fuelwood gathering has produced a mosaic of more or less devegetated or deserted areas.

It is generally believed that forests influence precipitation in the surroundings. A weaker tropical forest moisture pump would have very long ranging ecological consequences by altering the precipitation levels of the savannas. Man-induced climatic change is therefore likely to lead to an encroachment of the Sahel syndrome into the

savannas. This region supports about 200 million people and any adverse effects of climatic warming could result in the worsening of already precarious production systems in the area.

The degradation of the Savanna ecosystems has a further implication for another important biological process, that of bird migration. According to the International Council for Bird Preservation (ICBP), there are two major migratory routes to Africa. One route originates in central Asia and eastern Europe, and ends in a broad swath in the eastern and southern savannas, utilizing the Bosporus straits crossing, while another, which uses the Gibraltar crossing, ends in the west African savannas and Sahel area. Figure 1

Table 1: Land use, population and livestock densities in the African savannas (Skoupy 1988)

Zone	Arid	Semi-arid	Sub-humid
Area $km^2 \times 10^6 n$	8.3	4.1	12.4
Population	24.8	65.7	90.5
Total TLU km^{-2}	50	91	64

shows the extent of the over-wintering area of these migrants.

Many species utilize savanna wetlands and shallow lakes, such as Lake Abiata in Ethiopia, Lake Turkana in northern Kenya and other rift valley lakes, the Kafue flats of Zambia and in west Africa the flood plains of the Niger and Senegal rivers, Lake Chad, and several other water bodies. Many more bird species utilize seasonal wetlands, locally known as dambos, vleis or bugas in eastern and southern Africa. Such seasonal wetlands, formed mainly by perched watersheds at the headwaters of streams, are rapidly disappearing due to drainage by cultivators or erosion. An increase in global temperature of the order anticipated by several of the GCM models will probably result in the reduction of many such habitats. Ringing experiments on the white stork, *Ciconia ciconia*, which migrates between Northern Europe and Southern Africa, indicate that as a probable result of habitat loss in southern Africa fewer birds are returning during the spring migration back to the northern hemisphere, indicating that already there is some ecological threat to some migratory birds.

We have mentioned above the probability that extreme events, such as droughts, will be more severe than at present. It is the frequency and duration of such events that will determine the future fate of such migrants. Our experience is that many of these water bodies and wetlands have precarious water budgets. Increased evaporation, combined with drought episodes could substantially reduce the number of and distribution of suitable habitats for these migrants.

SUMMARY

It is anticipated that climatically determined vegetation zones will be unable to follow the regional changes in climate and consequently will probably occupy less area than their current extent.

Fossil evidence and some historical data give us some insight on how past ecosystems have responded to climate change. There is some evidence that certain groups of animals have genetic mechanisms for coping with increased ambient temperatures up to about 40°C, through the physiological phenomenon of adaptive heterothermy, and the possession of isozymes that operate at different temperatures. In the semi-arid savannas of the African continent, where water resources are already precarious due to population pressure, and where production is determined by precipitation, paleontological records indicate that changes in precipitation patterns have in the past elicited corresponding changes in the distribution and composition of the animal communities. Our current knowledge of the function of the savanna systems is that precipitation and soil nutrients are the primary determinants in productivity, and that the composition and structure of the fauna particularly the herbivores, is determined by the effect of precipitation on vegetation performance. While future global warming will have little impact on the tropical rain-forest, it could lead to drier conditions in the subhumid tropical continental areas and a change in community structure. The effects will be particularly significant for aquatic life and migratory birds, whose over-wintering habitats will be affected by drought.

REFERENCES

Bell, R.H.V. 1982. The influence of soil nutrient availability on community structure in African savanna ecosystems. in Huntley, B.J.and B. H. Walker (ed) 1982. Ecology of Tropical Savannas. Springer-Verlag. Berlin.

Cole, M.M. 1982. The influence of soils, geomorphology and geology on the distribution of plant communities in Savanna ecosystems. in Huntley, B.J. and B. H. Walker (ed) 1982. Ecology of Tropical Savannas. Springer-Verlag. Berlin.

Cumming, D.H.M. 1982. The influence of large herbivores in savanna structure in Africa. in Huntley, B.J. and B. H. Walker (ed) 1982. Ecology of Tropical Savannas. Springer-Verlag. Berlin.

Fryer, G. and T.D. Iles. The Cichlid fishes of the Great Lakes of Africa. Oliver and Boyd. Edinburgh.

Gornitz, V. 1985. A Survey of anthropogenic vegetation changes in West Africa during the last century - climatic implications. Climatic Change, 3, 285-325.

Hamilton, A.C. 1982. Environmental history of East Africa. Academic Press, London, New York.

Hochachka, P.W. and G.N. Somero 1973. Strategies of Biochemical Adaptation. W.B. Saunders Company. 358pp.

International Council for Bird Preservation. 1985. Flying visitors: migratory birds of Europe and Africa; a map by ICBP. UNEP.

Livingstone, D.A. 1975. The late quaternary climatic changes in Africa. Annual review of ecology and systematics. 6: 249-2.

Livingstone, D.A. 1980. Environmental change in the Nile Headwaters. in Williams, M.A.J. and H. Faure. 1980 (ed). The Sahara and the Nile. A.A. Balkema, Rotterdam.

Louw, G.N. and M.K. Seely. 1972. Ecology of desert organisms. Longmans.

Riebsame, W. 1989. Assessing the implications of climate fluctuations: a guide to climate impact studies. UNEP, Nairobi.

Skoupy, J.J. 1988. Developing rangeland resources in African drylands. Desertification Control Bulletin, 17, 30-36.

Thackeray, J.F. 1987. Late Quaternary environmental changes infered from small mammalian fauna. Southern Africa Climatic Change, 10, 285-305.

Tinley, K.L. 1982. The influence of soil moisture balance on ecosystems patterns in Southern Africa. in Huntley, B.J. and B. H. Walker (ed) 1982. Ecology of Tropical Savannas. Springer-Verlag. Berlin.

Tyson, P.D. 1989. Modelling climate change in Southern Africa. Proceedings of Conference on Geosphere-Biosphere Change in Southern Africa, Cape Town; Dec 4–8 1989.

Williams, M.A.J. and H. Faure. 1980 (ed). The Sahara and the Nile. A.A. Balkema, Rotterdam.

Forests: Their Role in Global Change, with Special Reference to the Brazilian Amazon

by E. Salati, R. L. Victoria, L. A. Martinelli *(Brazil)* and J. E. Richey *(USA)*

Introduction

The discussion of global climate change generally centres on the increase in the global average temperature as a consequence of the increase in the atmospheric concentration of greenhouse gases. It is recognized that CO_2 plays a major role in the enhancement of the greenhouse effect, as it is also recognized that human use of fossil fuels is the major source of this gas for the atmosphere. But the role of deforestation in global change, and in particular deforestation of the humid tropics, has become a controversial issue both scientifically and politically. It has been estimated that the global area of tropical forests has decreased from the original 14×10^6 km^2 to approximately 8×10^6 km^2, a reduction of 45%. Predictions for the future range from very little primary forest left by the turn of the century, with the exception of western Amazonia and central Africa, to the other extreme that proposes deforestation rates of only 0.6%/yr (Myers, 1988). It was estimated that 25% of the global atmosphere CO_2 emissions in 1980 was derived from transformation of forests into annual cropland or grassland in the tropics (Houghton *et al.*, 1987). The polemical aspects of the question are due primarily to the lack of detailed information on the changes that have occurred and on the processes that would permit the establishment of a sustainable use of the renewable natural resources.

In this paper, we review deforestation in the Amazon in the global context. The Amazon is the largest contiguous extent of tropical forest remaining on Earth. Deforestation in this region might lead to changes in both climate and species diversity. We will address three basic questions:

1) What are the causes and extent of deforestation?
2) How would deforestation affect CO_2 release?
3) How would extensive changes in land use affect the hydrological cycle and the climate regime of the Amazon basin and neighbouring regions?

Causes and Extent of Deforestation

At the beginning of the colonization of Brazil by Europeans in 1530, and in the centuries thereafter, deforestation in Amazonia was the consequence of very few and small-scale human activities. Slash-and-burn agriculture and commercial timber resulted in the opening of settlements, some of which turned into important cities at the beginning of this century, including Manaus and Belém. Such activities were almost always developed along the Solimoes (as the Amazon is called above its confluence with the Rio Negro) and Amazon rivers and along the more important tributaries. The rivers were then the natural transport system in the region (Salati and Oliveira, 1987).

The trend of deforestation changed during the decade of the 1960s, with the onset of new human activities induced by the opening of highways which promoted easy and quick access to 'terra-firme' areas. With the expanded transport system combined with government incentives for development of the region, the population increase was rapid, reaching over 15 000 000 people in the last two decades (Sudam, 1987). The main economic activities associated with the occupation process include cattle ranching, timber extraction, perennial crops (cocoa, rubber, and homogeneous forests for pulp and paper), annual crops (sugar-cane, soybean, corn and rice), charcoal production for the cast iron industry, construction of dams for hydroelectric power stations, gold exploration and mining, and oil exploration.

A controversial issue has been the extent to which the Amazon region is being deforested. It is now well known that tropical forests are being deforested at increasing rates, but the critical question of knowing with a better accuracy the actual values of such rates is as yet unanswered. Estimates of deforestation in the Brazilian Amazon range from 5% (INPE, 1989) to 12% (Mahar, 1988). Annual rates of deforestation range from 17 000 km^2 (INPE, 1989) to 80,000 km^2 (Setzer and Pereira, 1989). In a more recent study Fearnside *et al.* (1990) estimated that the average deforestation rate for the last 11 years was $21\,000 \pm 2\,100$ km²/year, with a total deforested area of about $400\,000$ km² (this figure includes old deforestation), which comprises about 9% of the Brazilian Amazon primary forest area.

Though we will not dwell on it further in this paper, an additional aspect of deforestation that should also be considered is that extinction of flora and fauna may be expected whenever the diversity of a natural forest ecosystem is substituted by any other single species or less diverse ecosystem (Prance, 1986). Diversity is the basis for the functioning of the tropical forest ecosystem; the higher the number of species present in one environment, the higher are the probabilities of adaptation to changes and adversities of the environment.

Consequences for Carbon Dioxide Emissions

With respect to climate changes, one of the most important global aspects of deforestation is the transfer of carbon stored in the forest biomass to the atmosphere (Woodwell *et al.*, 1978; Houghton *et al.*, 1985, 1987). Since 1988, when the massive forest burnings in Rondonia generated enormous pressure by the international media, many scientists started worrying about obtaining better estimates of the contribution of forest burnings to CO_2 emissions to the atmosphere.

Here we review our calculations of the range of Amazonia's potential contribution to atmospheric CO_2 (Salati *et al.*, 1989b). Most such calculations in the literature are based on assuming particular deforestation and emission rates; given the variance in the data available, assuming a particular data set can lead to considerable under- or overestimates. We made several simplifying assumptions: (1) biomass will always be completely burned; (2) successional ecosystems will always have a small biomass when compared with the original forest (this is true taking into consideration that it takes about 100 years to completely regenerate a forest to its original status in the regions); and (3) soil organic matter will not increase significantly with time after burning. Variability in the calculations arises from the uncertainty in biomass estimation for the Amazon forest. There are very few biomass estimates for the Amazon, and they range from 250 to 350 tons/ha (Fearnside, 1987). We used the values found by Martinelli *et al.* (1988) of 300 ± 60 tons/ha of above ground biomass plus 40 tons/ha of litter and fallen trunks for the Environmental Protection Area of the Samuel power dam in Rondonia. Total biomass available for burning could therefore vary between 280 and 400 tons/ha, which would give from 140 to 200 tons of carbon/ha, if it is assumed that biomass contains 50% carbon. The final numbers to be used in our calculations are the total deforested area and the annual rate of deforestation.

Taking the above cited numbers and assumptions into account, we estimated that the Amazon region has already contributed emissions ranging from 3.5×10^{15} grams carbon (gC) to 12×10^{15} gC to the atmosphere (Table 1). This is in the order of 2 to 7% of the total atmospheric CO_2 emitted due to the deforestation and burning until 1980 (Woodwell, 1987). Table 1 also shows that annual emissions, considering the 1988 rates ranged from 0.24 to 1.6×10^{15} gC/yr, which is between 4% and 25% of the annual global CO_2 emissions, estimated to be around 7×10^{15} gC/yr (Woodwell, 1987). Even if, to be on the conservative side, we assumed that only 50% of the biomass is burned during forest fires, the numbers would still be significant, ranging from 2.0 to 12% of global emissions. Though these data may not be precise, they are useful as a warning of the potential dangers that uncontrolled deforestation in the Amazonia pose for our environment.

Table 1: *Estimates of potential CO_2 emission due to burning. The range of total emissions is calculated as the product of the low (Lo) and upper (Up) estimates of the carbon biomass available and the area deforested, and the annual emissions are calculated as the product of the ranges of biomass and the 1988 deforestation rates.*

Carbon biomass	Area deforested		1988 Rates	
(t/ha = 10^8 g/km²)	(10^3 km²)		(10^3 km²/year)	
	Lo (250)	Up (600)	Lo (17)	Up (80)
	Potential emissions		(x 10^{15} gC)	
Lo (140)	3.5	8.4	0.24	1.1
Up (200)	5.0	12.0	3.40	1.6

Note: estimate based on the assumption that 100% of the burned biomass is transformed into CO_2.

Unfortunately, little attention has been paid in the literature to the deforestation that occurred in Europe or North America during the expansion of their agricultural systems. The few data currently available indicate that 115 ± 35 x 10^{15} grams of carbon were emitted from 1850 to 1985 through changes in land use, particularly deforestation for agricultural purposes (Houghton and Skole, 1990). This value is of the same order of magnitude when compared with the total estimated CO_2 emissions (200 x 10^{15} grams of carbon) due to the combustion of fossil fuels and cement production during the period 1850 to 1987 (Marland, 1989).

In addition to being thought of as a source of CO_2, forests could also provide one of the possible ways of sequestering excess CO_2 from the atmosphere, thus mitigating the greenhouse effect (Melillo *et al.*, 1990). Since the tropical region has one of the highest rates of primary production on the globe, there has been considerable discussion on the subject of reforestation and afforestation programmes in that region. According to Jarvis (1989), the reafforestation of 320 x 10^6 ha would sequester 1 x 10^{15} grams of carbon per year in the biomass, which is equivalent to 17% of the current annual CO_2 emissions due to the combustion of fossil fuels. Although still controversial, it is believed that such programmes should improve economic and ecological conditions in the regions where implemented.

Consequences for the Hydrological Cycle and Climate

Depending on the extent of the cleared area, deforestation can affect the climate directly at micro- and mesoscales, with local and regional and possibly global consequences (Salati and Marques, 1984; Henderson-Sellers, 1987; Shukla *et al.*, 1990). The present atmospheric steady-state equilibrium in the region is dependent on the vegetation cover, i.e., forest. Figure 1 shows schematically the water balance for the Amazon Basin, including an area of approximately 6 000 000 km^2 from the river mouth at Marajó Island to the headwaters at 'Cordilheira dos Andes'. More detailed information can be found in Salati *et al.* (1986, 1989a). The most important fact to note is that the water vapour flux originating in the Atlantic Ocean is not of sufficient magnitude to explain the rainfall and the vapour outflux in the basin. As a direct consequence, it is necessary to assume the recirculation of evapotranspired water. The conclusion is strongly supported by the spatial distribution of the ^{18}O and deuterium isotopes in the region (Salati *et al.*, 1979).

The Amazon basin is a source of water vapour to other regions, in particular to the Brazilian central plateau, and eventually to the 'Pantanal' (large area of swamp and seasonally flooded land in central-western Brazil). Hence, changes in the water balance of the Amazon itself would be expected to have direct consequences for neighbouring regions.

Bearing such evidence in mind, it is clear that a major alteration in vegetation cover of the region would lead to changes in the climate at the micro- and meso-levels. Changes would be felt particularly through variations in the albedo, in the rainwater residence time, increase in runoff and a decrease in evapotranspiration (Salati *et al.*, 1979). Increases in maximum temperatures and daily thermal amplitudes, due to a decrease in precipitation, would also be expected. Simulation models, although containing uncertainties, also point to such a scenario (Henderson-Sellers, 1987; Shukla *et al.*, 1990). It is also known that the Amazon forest plays an important role in the transfer of energy (solar) from the Equator towards higher latitudes. The total amount of energy transferred is associated with the almost 3.5 x 10^{12} tons of water vapour leaving the Amazon per year. The influence of such water vapour and energy on the climatology of neighbouring or more remote regions is still unknown.

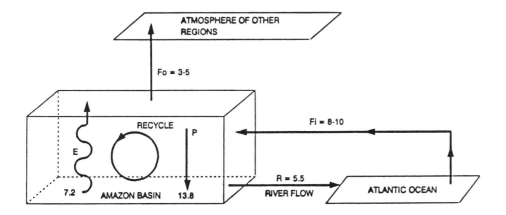

Figure 1: A schematic model of the water cycle in the Amazon basin; E is evaporation, P is precipitation (adapted from Salati *et al.*, 1989b)

Conclusions and Recommendations

Present knowledge reveals the key role of the forest in maintaining the dynamical equilibrium of the Amazonian ecosystem. Although the information presently available about Amazonia permits the above analyses, many doubts still persist. With rare exceptions, information is not continuous in time and/or space; it is still difficult to visualize the entire Amazonian ecosystem as a whole, which in turn impairs accurate predictions of the global or even regional changes that might be associated with its destruction. The issues of how increased atmospheric temperatures and CO_2 would affect the Amazon are still open questions. Many improvements will be necessary in Global Circulation Models to adapt them to the tropical conditions.

In addition, institutional problems in Brazil are serious. The lack of research funds is impairing the continuity of many research programmes and also the engagement of Brazilian research institutes in international global change programmes. Therefore, an expansion of international support to research groups and institutions dedicated to the study of the basic functioning mechanisms of the ecosystem and its role in global change, especially those with programmes already under way for a considerable time, is highly recommended.

References

Fearnside, P.M., 1987. Summary of Progress in Quantifying the Potential Contribution of Amazon Deforestation to the Global Carbon Problem. In: D. Athie, T.E. Lovejoy and P.M. Oyens (eds.) Workshop on Biogeochemistry of Tropical Rainforest: Problems for Research. Piracicaba, S.P., Brazil.

Fearnside, P.M., A.T. Tardin and L.G. Meira Filho, 1990. Deforestation Rate in Brazilian Amazon Instituto de Pesquisas Espaciais, Secretaria de Ciencia e Technologia, August, 1990, 8 pp.

Henderson-Sellers, A., 1987. Effects of change in land use on climate in the humid tropics. In: R.E. Dickinson (ed.). The Geophysiology of Amazonia. John Wiley & Sons, New York.

Houghton, R.A., Boone, R.D., Melillo, J.M., Palm, C.A., Woodwell G.M., Myers, N., Moore, B. and Skole, D.L., 1985. Net flux of carbon dioxide from tropical forests in 1980. Nature 316: 617–620.

Houghton, R.A., Boone, R.D., Frucci, J.R., Hobbie, J.M., Melillo, J.M., Palm, C.A., Peterson, B.J., Shaver, G.R., and Woodwell, G.M., 1987. The flux of carbon from terrestrial ecosystems to the atmosphere in 1980 due to changes in land use: geographic distribution of the global flux. Tellus 39: 122–139.

Houghton, R.A., and D.L. Skole, 1990. Changes in the global carbon cycle between 1700 and 1985. In: B.L. Turner (ed.). The Earth Transformed by Human Action. Cambridge University Press, in press.

Instituto de Pesquisas Espaciais, 1989. Avaliação da alteração da cobertura florestal na Amazonia utilizando sensoriamento remoto orbital. Primeira edição, São Jose dos Campos, 27 p.

Jarvis, P.G., 1989. Atmospheric carbon dioxide and forests. Phil. Trans. R. Soc. London, B 324: 369–392.

Marland, G., 1989. Fossil fuels CO_2 emissions: Three countries account for 50% in 1988. CDIAC Communications, Winter 1989, 1–4, Carbon Dioxide Information Analysis Center, Oak Ridge National Laboratory, USA.

Martinelli, L.A., Brown, I.F., Ferreira, C.A.S., Thomas, W.W., Victoria, R.L., and Moreira, M.Z., 1988. Implantação de Parcelas para Monitoramento de Dinamica Florestal na Area de Proteção Ambiental, UHE Samuel, Rondônia. Relatorio Preliminar.

Mahar, D.J., 1988. Government Policies and Deforestation in Brazil's Amazon Region. A World Bank Publication in cooperation with World Wildlife Fund and The Conservation Foundation. Washington, 56 p.

Melillo J.M., T.V. Callaghan, F.I. Woodward, E. Salati and S.K. Sinha, 1990. Effects on Ecosystems. In: J.T. Houghton, G.J. Jenkins and J.J. Ephraums (eds.) Climate Change. The IPCC Scientific Assessment, Cambridge University Press. Cambridge. UK. 365 pp.

Myers, N. 1988. Natural resources systems and human exploitation systems: physiobiotic and ecological linkages. The World Bank Policy Planning and Research Staff, Environment Department, Environment Department Paper No. 12, 61 p.

Prance, G.T., 1986. The Amazon: paradise lost? In: Kaufman, L. and Mallory, K. (eds.) The Last Extinction. MIT Press, Cambridge, MA.

Salati, E., Dall Ollio, A., Gat, J. and Matsui, E., 1979. Recycling of water in the Amazon basin: an isotope study. Wat. Resour. Res. 15: 1250–1258.

Salati, E., and Marques, J.,.1984. Climatology of the Amazon region. In: Sioll, H., ed., The Amazon Limnology and Landscape Ecology of a Mighty Tropical River and its Basin, Dordrecht, Dr. W. Junk Publishers, p. 87–126.

Salati, E., 1986. The Climatology and Hydrology of Amazonia. In: Prance, G.T. and Lovejoy, T.E. (eds.), Amazonia. Pergamon Press, Oxford, 442 p.

Salati, E. and A.E. de Oliveira, 1987. Os problemas decorrentes da ocupação do espaçã amazônico. Pensamiento Iberoamericano N12. Julio-Diciembre 1987, pp. 79–95.

Salati, E., R.L. Victoria, L.A. Martinelli and N.A. Vila Nova, 1989a. Soils, water and climate of the Amazonia. Proceedings of the Symposium Amazon: Facts, Problems and Solutions. Universidade de São Paulo, São Paulo, Brazil, July 31–August 3, 1989. p. 300–353.

Salati, E., R.L. Victoria, L.A. Martinelli and J.E. Richey, 1989b. Deforestation and its role in possible changes in the Brazilian Amazon. pp. 159–171, in R. DeFries and T.F. Malone (eds.) Global Change and our Common Future: Papers from a Forum. National Academy Press, Washington, DC, 227 p.

Setzer, A.W. and Pereira, M.C., 1989. Relatorio de Atividades do Projeto "Sensoriamento de Queimadas por Satelite-SEQUE", Ano 1988. Unpublished manuscript. 56 pp. INPE, São Jose dos Campos, Brazil.

Shukla, J., C. Nobre and P. Sellers. Amazon Deforestation and Climate Change, 1990. Science 247: 1322–1325.

Superintendencia do Desenvolvimento da Amazonia (Sudam). 1987. Censos demograficos das Unidades que compoem a Amazonia Legal. Relatorio da Divisão de Estatistica da Sudam, Brasilia, 42 p.

Woodwell, G.M., Wittaker, R.H., Reiners, W.A., Likens, G.E., Delwiche, C.C., and Botkin, D.B., 1978. The biota and the world carbon budget. Science 199: 141–146.

Woodwell, G.M., 1987. The warming of the industrialized middle latitudes 1985–2050: causes and consequences. Developing Policies for Responding to Future Climatic Changes. Villach, Austria, September, 1987.

Climate Change and Risk Management

by Kerry Turner, Tim O'Riordan, and Ray Kemp *(United Kingdom)* [*]

Introduction

At first glance, risk management seems an appropriate tool to utilise in the response to global environmental change (GEC) generally, of which climate change is one aspect. Risks combine uncertainty with hazard within a setting of social and personal choice when living apparently in danger. Risk management has evolved from the statistical engineering and psychological sciences to link probability estimation with assessment of consequences under uncertain conditions. In recent years, risk management has extended into the social sciences, notably in areas of economics, politics, and critical theory of social communication and institutional relationships. This excursion has produced a more interactive, or participatory approach to coping with hazards to environment and human health. A brief summary of key points in this evolutionary sequence follows.

In this paper global change, as outlined in Figure 1, provides the setting for the analysis. Climate change as one component of this, will be the focus for the discussion because it embodies many of the key aspects of global change, and, of course, is the theme of the Conference. The peculiar characteristics of global environmental change issues create serious difficulties for risk management, a challenge for which neither the available techniques nor the professional community is fully prepared.

We all know that enormous uncertainty surrounds GEC. Scientific indeterminacy is compounded by lack of knowledge about how societies may respond to the possible effects of climate change. For example, international trade, population movement and non-human species migration will all be affected in unforeseeable ways. It is indeed questionable whether existing theories of behaviour under uncertainty are suited to the analytical and policy-making tasks that now confront us.

The new circumstances combine a multi-dimensional and poorly understood physical system - global environmental change - under the influence of somewhat unpredictable socio-political and economic forces. There is now a need to operate the precautionary principle which can be related to some extent to economic theory and analysis, and the requirement for a more international and mutually cooperative diplomacy regarding possible response strategies over which the legal and diplomatic professions are still contemplating appropriate techniques. Both of these concepts present a fundamentally different set of problems for risk analysis. In addition, there is the question of vulnerability - the capacity of a society, or a community to adjust to threat, coupled with the forces that expose peoples and their resources involuntarily to danger not of their making. To reduce vulnerability raises important questions regarding civil liberties and equalisation of economic opportunity, as well as access to basic livelihood needs. These contentious issues are now very much part of contemporary theories of risk management, but will have to be given greater prominence in future work.

Risk management as theory and practice may be able to adjust, though its success cannot be guaranteed. It could be argued that a more participatory approach based on more open science and communication techniques, set in the context of precaution and adaptation in anticipation of change will be required. The paper will outline emerging theory and consider possible policy implications.

The Changing Character of Global Environmental Change

There are generally held to be five aspects of global environmental change which make the phenomenon distinctive, and particularly challenging from the perspective of risk analysis and management.[1]

i)   It is caused primarily by human activity, superimposed on physical or biogeochemical processes which are always evolving and adjusting.

This suggests that it is up to human agency to recognise the causes and consequences of activities which limit the survival chances of many peoples and

* We are grateful to G.Berz, J.Jager, M.Kelly & J.Parfitt for comments on an earlier draft. Remaining errors are of course our responsibility.

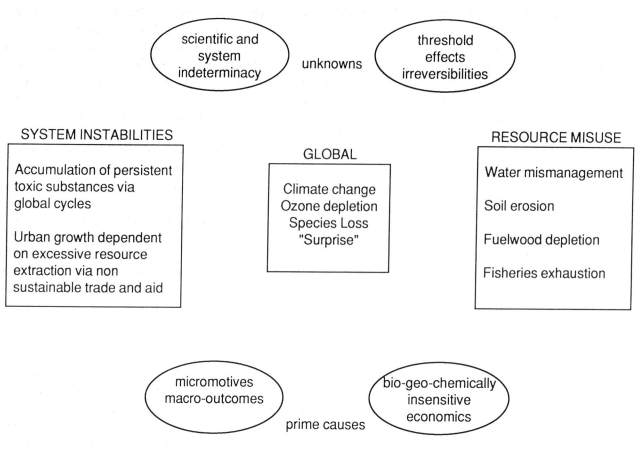

Figure 1: Global Environmental Change.

ecosystems, and to take the necessary preventative measures.

This in turn implies that global change reinforces the now well-articulated argument that economic and ecological systems are interrelated. It is still questionable, however, just how much senior policymakers actually realise that proposition. The ultimate danger is that economic growth and environmental quality cannot be decoupled to an extent sufficient to support a growing world population.

ii)   Its effects are global in extent, either because they literally span the earth (as in climate change and ozone depletion) or because they accumulate regionally to create potentially irreversible problems, damages, losses which are significant (as in habitat loss, soil erosion or toxification of biological systems). Additionally, either the consequences effect anyone or everyone on the earth; or the cost of trying to alter these changes is so great as to affect the viability of a regional, or even the global, economy.[2]

iii) It is characterised by rates of change in climatic conditions that exceed presumed human capabilities to adjust, especially in economics and natural environments that are already stressed, where vulnerability is greatest. This means that within 50–

100 years, when the consequences of today's political choices will be clearly evident, those responsible for political decisions, and their children could still be alive, confronted with the results of their actions and, presumably could be held to be culpable. Justification for political procrastination and the "not in my term of office" (the so-called NIMTOO syndrome) is therefore even less convincing in this case.

iv)   The scale of occurrence of its negative features are so persuasive and persistent as to be potentially irreversible, or at best so economically expensive to remedy that it could become a serious and permanent burden on future economies. This indicates that the phenomenon of global change, as outlined in Figure 1 and discussed below, will remain a permanent feature of national and international politics at the highest levels.

v)   Its remorselessness suggests that the rich nations cannot avoid its consequences by seeking to build barriers for their own protection. Global change unites us as potential victims[3].

This suggests that national sovereignty is not a sufficient basis for response. Nation states have a vital role to play. The factors that will influence their response, however, will have to be internationally determined, and, eventually, policed.

Truly Global Change Issues

Figure l seeks to illustrate the major components of global change. At its heart are the three truly global issues, namely climate alteration, stratospheric ozone depletion, and biodiversity loss. It is arguable that the three are potentially related, since, in theory at least, they are interconnected as to cause and consequence[4]. Habitat loss, especially in the tropics, will influence regional climates, and if sufficiently widespread, will be global in effect. Climate change can influence in turn habitat loss, because vulnerable species become more marginal under conditions of climate stress.

Also in this core is the unpredictable element of "surprise". This concept includes potential shifts in scientific and technological advance, use of novel techniques for forecasting and prediction, unprecedented shifts in policy or international relations, and metamorphoses in public perceptions and political demands. The scope for surprise may be reducible but it will never be dominated: this is one of the main challenges for risk management[5]. Another is that, whilst surprise may be imagined, it may not be given policy credibility. Guessing an outcome is not of much value if it is not believed or prepared for. Hence the permanence of surprise - at least until we devise institutional and policy response mechanisms that coherently and systematically incorporate surprise-adaptive strategies.

In risk management terms, surprise presents a further important challenge. Its unpredictability can be softened by the introduction of adaptive and experimental planning, coupled to extensive networks of reliable and compatible monitoring and improved, two-way, public communication.

We shall see that where the truly global change qualities of risk occur, the response strategies become based on two fundamental principles. The first is that of precaution, namely acting in advance of scientific certainty, but within the bounds of scientific uncertainty, and taking into account the consequences of possibly being wrong. The second is that of compensation, in terms of maintaining the resource base (interpreted as a set of economic opportunities) in place so as to bequeath to future generations the scope for manoeuvre.

The compensation principle ensures that future generations are not disadvantaged by the present generation's activities and resource utilisation. Actual compensation can be ensured by a transfer of capital assets (i.e. a stock of natural resources and human capabilities that yields a flow of goods and services) over time. The sustainable economic development requirement is met when no less than the current capital stock, in terms of its value, is passed on to future generations. The capital stock is a combination of man-made capital assets (for example

technology and economic infrastructure) and natural capital i.e., ecosystems, waste assimilative capacities etc.[6]

If it is the case that there is, and will continue to be, a high degree of substitutability between natural and man-made capital then the depletion of the natural capital stock can be offset by the accumulation of an equivalent amount of socially created capital. On the other hand, if substitution possibilities turn out to be much more limited, then both man-made and natural capital stocks should be maintained separately. In this argument man-made and natural capital are viewed much more as complements rather than substitutes. In cases where non-renewable resources (such as fossil fuels) are exploited, the revenue, or benefits generated should be divided into an income and a capital component. The latter should then be invested each year into complementary projects or longer run programmes utilising renewable resources. The aim is eventually to replace non-renewable resource use with a rising commitment to renewable resource utilisation by systematic planning.[7]

Advocates of the "strong sustainability" criterion also argue that it is inappropriate to calculate the net benefits of a project or policy alternative by comparing it with an unsustainable option and applying a discount rate which reflects an unsustainable rate of return. This is designed to ensure that renewable resource utilisation schemes get judged on the basis of their longer term gains, not their more short-term "losses". Thus if a sustainable resource utilisation project happens to generate a negative present value when a non-sustainable discount rate is used it should not be automatically rejected[8].

Analysts sympathetic to the concept of sustainable economic development nevertheless feel that discount rate manipulation is not on balance advisable. Instead they advocate the inclusion within any portfolio of investments one or more shadow projects or "offsets". The aim of the shadow project(s) is to compensate for the environmental damage from the other projects in the portfolio. A well known example would be the planting of forests to compensate for the increased CO_2 generation by a new fossil fuel power generating plant. For such compensating projects the normal cost benefit analysis decision rule does not apply, although it is rational to seek to minimise the cost of achieving the sustainability criterion[9]. The sustainability constraint on economic development becomes more binding the lower the degree and extent of substitution between man-made and natural capital.

Coalescing Regional Change Issues

Again an element of risk appears here, notably because of the uncertainties inherent in the scientific appraisal of effectiveness of offsets in overcoming known global change outcomes from conventional resource-depleting activities. Studies of offsets to examine how far such uncertainties can be reduced to lead to meaningfully sustainable outcomes are urgently required. Thus the scale and location of forest planting needs to be more accurately determined if the fossil fuel offset strategy is to be deemed workable.

Allied to the centre stage are two sets of global change activities that are regional in occurrence, but global in effect if coalesced or aggregated. These include stressing ecosystems beyond the limit of their carrying capacity, even assuming growth in both technology and management capability, at least for the foreseeable future. Examples include soil fertility loss due to erosion and "soil mining", depletion of fisheries, damage to coastal ecosystems that also provide physical protection from storms, and removal of tree cover with a consequent loss of carbon fixation, or siltation downstream, or change of earth reflectivity[10].

The related effect is the intervention in life support systems caused by toxification or by the excessive concentration of people in urban or rural areas whose resource demands not only exceed local provision, but which make excessive demands on resources elsewhere. These zones become the heartlands of regional-level non-sustainable development, which coalesce into global non-sustainability.

Risk Management Strategies

In practical risk management terms the strategies of precaution and compensation are not alone sufficient to combat global change issues. Three additional approaches have to be developed, namely resource-safeguarding pricing to ensure that resources are used more efficiently, resource safe-guarding regulation to set sustainability parameters to resource utilisation, and pre-emptive regulation where, for example, certain products or processes known to be toxic even in very small quantities are removed, simply because their presence, or possible presence, is deemed to be undesirable[11].

The concepts of precaution and compensation, together with pricing and regulation, constitute a radical version of conventional risk management - namely, insurance and enforcement. They form a tripartite strategy for protecting natural resources and the interests and well being of both present and future generations. This theme of managing the risks of global climatic change through an inter-connected network of insurance strategies, in the broadest sense of the term, coupled with economic and political regulation is the driving philosophy of our approach to global risk management.

The Challenge of Global Change for Risk Management

There is now a consensus that risk management is concerned with taking decisions about what are acceptable or 'tolerable' levels of risk to society termed societal risk.

By way of introduction it is worth outlining the available techniques for analysing societal risk levels and then move on to a description of what constitutes tolerability in this instance[12].

Decision Theory

This approach has major limitations in determining societal risk targets. This is due in part to the theoretical limitations on aggregating individual preferences to determine a societal level of acceptable risk[13]. Subjective values for probabilities and utilities would be necessary and although cost benefit approaches are helpful, there are strong reservations in arranging numerical values to the utilities of human life, particularly when the risks of harm through climate change affect different societies in a variety of ways[14].

Societal Risk Levels

The most common means of expressing levels of societal risk is through the use of complementary cumulative distribution functions (CCDF) or "fn curves" where f stands for frequency of deaths, or injury, and n stands for total number of injuries or fatalities. These graphical techniques allow historical data on major hazards, both industrial accidents and natural hazards, (and by extension "enhanced" natural hazards of global change), to be mapped in terms of their human consequences on the horizontal axis and their frequency of occurrence on the vertical axis. This traditional explanation of risk as frequency x consequences allows comparisons to be made between different hazardous events[15]. While great care has to be taken both in constructing the CCDFs and in extrapolating and drawing conclusions, it is becoming increasingly apparent that the distribution of natural disasters is skewed towards high consequence events such as the large numbers of immediate deaths which have been associated with earthquakes, tornadoes and flooding in recent years[16] (see Figures 2 and 3).

It has been estimated that each year some 20,000 people have been killed by major flood and cyclone events over the period 1964–84. This is probably an underestimate of the actual number of deaths related to such hazards given the likely second order impacts of these natural calamities such as epidemics. Refinement of CCDFs to provide more accurate information and analysis for enhanced natural hazards associated with climate change will be needed.

Risk Characterisation

Increasing efforts are being placed on the drive to replace the crude representation of societal risk as in human fatalities. Delayed deaths, injuries, disability, long-term health effects, evacuation, resettlement, clean-up, agricultural loss, sterilisation of land, property, aquifers, *inter alia*, all need to be represented within risk

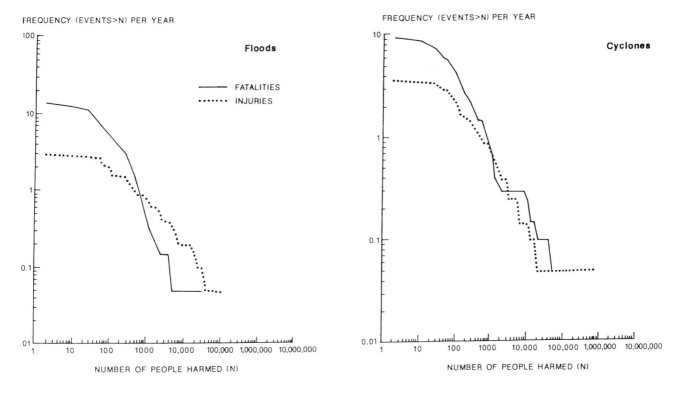

Figure 2: Frequency of floods and cyclones worldwide causing harm, 1964-1984. Source: D. Fernandes-Russel, Societal Risk Estimates from Historical Data for UK and Worldwide Events, Environmental Risk Assessment Unit, Research Report 2, Universtiy of East Anglia, UK (1988).

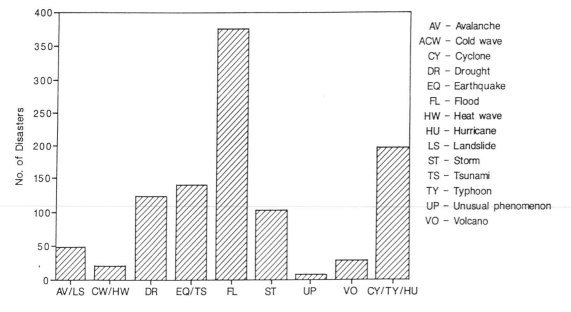

Figure 3: Worldwide natural disasters by type since 1900. Source: US Disaster Assistance, Agency for International Development, Department of State, Washington.

assessments yet even then the concern is narrowed on the risk to societies as a whole, and not to individuals, or even to individual societies. There needs to be a finer measure to reflect individual and communal willingness to accept, or tolerate, risk.

Tolerable Societal Risk

Tolerable risk may be defined as a level of risk which is never finally accepted or approved, but is continually under review. It implies that involuntary risks which affect society should be examined in terms of the range of risk/benefit trade-offs; that the processes through which safety is managed are reliable and continually raising its standards; that those risks should be communicated effectively both to decision-makers and to the public at risk; and finally that adequate "safety net" arrangements, such as insurance provision, emergency evacuation procedures, health and education programmes and relief agencies, are able to protect societies in the event of an actual hazardous occurrence[17].

The concept of risk therefore has evolved in three important respects over the last twenty years[18].

1. *From externalisation to internalisation.* To begin with, risk was judged to be an identifiable physical hazard, whether man-made or natural, that was external to the human condition, and for which some form of externalised blame could be attributed. This was the case for much of the early work on natural hazard research where coping strategies of adaptive response were sought in the face of an externalised physical event that impinged on individuals, societies and institutions. It was also implicit in the early work on risk perception which stressed the physical concreteness of danger (guns in violent hands, electro-magnetic radiation from power lines overhead, reformulated organisms from genetic engineering) to which people in different political and economic contexts might be expected to respond.

2. *Vulnerability and the abuse of power.* This stage was followed by the work of political theorists and anthropologists who sought to encapsulate the concept of risk in the political culture of history, social relationships, and political justice[19]. They visualised the distribution of risk as being disproportionately dependent on whether people were able to control the environmental conditions of their existence, or whether they were not in such a fortunate position.
   Vulnerability has two meanings in the context of global change[20].
   i)   An increase in the difficulty of surviving due to increasingly stressed physical and/or social conditions directly attributable to global change. This might be climate-induced deprivation of water, or increases in natural hazards (more violent tornadoes, hurricanes, more prolonged droughts, sharper and less predictable frosts, etc.), or it may be caused by social tension brought on by lack of access to basic resources such as fuelwood, soil fertility, grazing rights - tension that results in conflict and even more damaging misuse of resources.
   ii) A decrease in the capability to cope because traditional methods of adaptation are no longer available. For example, many non-industrial societies depend upon traditions of sharing in times of need, and of enjoying access to a variety of geographical areas so as to be able to obtain food from at least some locations. Overcrowding, war, new administrative boundaries or new landlord-tenant agreements can put a stop to this.

Both kinds of vulnerability are becoming more widespread in the modern world, irrespective of global change. The lesson of the vulnerability thesis is that those who are economically and/or politically exploited will suffer most and will become even more marginalised under conditions of social stress, political tension and the abuse of power.

3. *Risk communication and empowerment.* This additional stage of risk management has been dominated by the development of theory and some case examples of how best to communicate danger to the suspecting or to the unsuspecting so that the best management mix can be achieved[21]. Beginning with a relatively straightforward adaptation of signal theory and fairly unsophisticated communications idioms, more recent work has drawn from the experience of phase 1 and phase 2 of the evolution in risk management. This has introduced concepts of power, social relationships as influenced by regulatory and risk preventative strategies, and innovative experimental mechanisms of providing people with a sense of control over the risks they believe they face.

The point here is that "risk" is not merely an objective phenomenon, a hazard standing in free space. Risk is a hazard that is clothed both with social meaning and judgements about fairness and administrative competence. Drought may be triggered by a statistical deprivation of rainfall: but its effects are enormously influenced by a host of factors such as land tenure arrangements, indebtedness, social relations that control tree removal or grazing rights, and views about who holds rights to basic "resource entitlements" , food, water, shelter and fuel. A society that creates and perpetuates intolerance and oppression will create risks out of otherwise adjustable threats from the natural world.

Novel ways of addressing the issue of communication may include special devices for providing publicly accessible and credible risk mediating mechanisms, so that risk receivers, the vulnerable, can gain access to better information, and be assisted in developing or extending their own way of adjusting to danger. Examples include risk "ombudsmen" (offices of risk information, conciliation and compensation in the event of accident or peril), community risk management officers (to perform a similar task but more at "grass roots" levels), or community risk liaison panels. These and other devices should give a degree of delegated power to affected communities so as to allow them to present their view of appropriate safety measures in ways that they believe are culturally appropriate, that is acceptable socially and politically to local societies.

By the same token, communities without risk protection entitlement, with no effective control over safety and post accident compensation, are not enabled to respond to the threat of global changes of the kind outlined in the central boxes of Figure 1. Modern risk management should therefore combine these three traditions of choice making under uncertainty, for the management of global change.

Climate Change as Global Change

Clearly there are different levels of global change, where the concept of "level" suggests scale of alteration and character of social response. But for a number of analysts climate change *is* global change. Figure 1 tries to put this argument into perspective. Nevertheless climate change is sufficiently specific as to be regarded as a risk arena in its own right. This is partly because climate change embraces ocean-atmosphere-cryosphere energy and chemical coupling of truly global proportions, and because climate change has clearly major implications for society and economy as this Second World Climate Conference amply demonstrates. We concentrate on one aspect of climate change, sea-level rise and related risks, in order to bring out some of the principles discussed so far.

Case Study: Climate Change and Global Sea-Level Rise as a Risk Management Issue

The issue of global warming and sea-level rise has proved amenable to the probabilistic scenario approach. The consequences of sea-level rise (direct and indirect) are both extensive and significant. Some sixty per cent of the world's population live in coastal areas - three billion people - and more than forty of the world's largest cities are located within the coastal zone. Given the scale of social and economic investments found in urban areas, it is likely that every effort will be made to protect the city waterfront. But in the event that neighbouring coastal areas cannot be protected, the economy of the region and thus of the city - the focal point of economic activity—will suffer. Those living in undefended areas may be forced to migrate, thereby creating new economic and environmental pressures nearby[22].

Sea level rise also poses a series of more insidious threats. Water supplies may be affected by saltwater contamination. Infrastructure (roads, communications facilities, power stations, etc.) may have to be redesigned and rebuilt. Tourism revenue may be lost as coastal attractions such as beaches, dunes and wetlands disappear.

In assessing the impact of sea-level rise, it is important that urban focal points are seen as just one element in a coastal zone system which supports and surrounds the urban area. It should be further recognised that the supporting economic infrastructure is underpinned by natural resource systems. In the past, such an integrated view has not been taken by planners and decision-makers. There has been a tendency to undervalue or ignore coastal resources whose value as "non-market goods and services" can be difficult to assess.

Thus, coastal wetlands have been converted to agriculture or industrial uses, or have been degraded by pollution or reduced freshwater flow. The loss of wetlands, and their shoreline protection and storm buffer capacities, has contributed to increased shoreline retreat as coastal erosion is accentuated. In a number of low-lying areas, removal of groundwater and/or hydrocarbons has also accelerated the rate of local land subsidence. The result has been resource management conflicts, environmental damage and increased vulnerability.

The central question is: what characteristics of human societies at different levels of development make them very vulnerable or, on the other hand, quite resilient to environmental stress?[23] We have argued that vulnerability has both a biophysical and a socio-economic dimension. But it can also be viewed in terms of sustainability. The sustainability of an economic system is in part a measure of its ability to maintain productivity when subject to stresses that may make it less resilient over time.

Technical Fix and Increasing Vulnerability

The considerable accumulation of man-made capital and the high level of technological innovation in developed economies contribute to their resilience with respect to external shocks and stresses. Margins of flexibility are much greater than in developing economies where population growth, poor economic performance and direct dependence on natural resource exploitation to sustain livelihoods often produce very narrow risk margins in the face of external disturbance.

But the comparative resilience of advanced industrialized economies to environmental stress should not be overstated. Both man-made capital and the natural resource base - in effect, natural capital - contribute to the resilience of these economies, though as we argued earlier these two forms of capital are not perfectly substitutable. Overexploitation of the natural resource base can lead to ecosystem degradation and, beyond some threshold level, may lead to the breakdown of social order.

Technology is frequently seen as a means of reducing vulnerability to environmental change. Industrialised economies have, for example, invested large amounts of resources in dams and a variety of engineering structures in an effort to utilize water and mitigate storm and flood damage. This technological fix has not, however, eliminated biophysical vulnerability. Dams and related structures still retain a potential for catastrophic failure. A high probability of a relatively low-cost impact has been replaced by a low probability of an extremely high-cost impact.

Vulnerability can be heightened by false perceptions of development solutions. In developing countries, large sums of international aid have been directed into western-style water resource projects. Many of these schemes have turned out to be poor economic investments. The capital construction costs are high and output yields - the benefits to the society - tend to be low. But their status has been further compromised by the fact that they have also brought with them long-term environmental costs. This environmental damage has increased the vulnerability of poor farmers and pastoralists to drought, flooding and other natural hazards, and to sea-level rise.

Economic-environmental Assessments and Risk Management

While it is certainly true that all of the IPCC response strategies in the context of sea-level rise—retreat, accommodation and protection—will be expensive, it seems to us that the risks associated with a "do nothing" strategy are potentially even more significant. First of all, this strategy involves low current costs, but risks incurring large future costs as measures to reduce vulnerability and limit damage are adopted late in the day. There is also a "waiting" cost. The longer society waits and does nothing, the higher will be the committed warming and, in all probability, the higher will be the future cost of forced adaptation and residual damage. In addition, decision-making may be complicated by the unpredictable element of surprise. There is a high cost attached to an unanticipated event. Both the unpredictability surrounding the course of global warming and the likelihood that the risk of extreme events will rise suggests that "surprises may well occur".

But the most cogent argument in favour of prompt action stems from the analysis of vulnerability. It is clear that many coastal regions, including major cities, are already facing severe problems owing to sea-level rise.

Many of the world's cities are sinking because of the removal of groundwater and/or oil and gas lying beneath them. Many are located on unstable, unconsolidated sediments consisting of silts, clays, peats and sands associated with river flood plains or coastal marsh/delta complexes. In addition, many urban areas are located on coasts where long-term tectonic subsidence is between one and ten millimetres a year. Rivers have been diverted by barrages or have been dammed, with the result that sediment loads passing across deltas have diminished. The loss of sediment supply has accelerated shoreline erosion; reduced freshwater discharge has affected the productivity of coastal marshes and waters. The impact of these

problems in their own right is sufficient to justify remedial measures to future sea-level rise[24].

Decreasing the susceptibility of coastal regions to sea-level rise induced by global warming by addressing existing coastal zone management problems more effectively than has occurred in the past should be an immediate priority. As far as the world's coastal cities are concerned, this is the barest minimum that should be done as a precautionary response to global warming.

Determining Priorities for Response

If action is to be taken, priorities have to be determined. To date, most estimates of the potential impact of sea-level rise on urban areas and of mitigating strategies have been generalized in nature. Assets at risk have been identified. Potential response strategies have been reviewed. But there have been few attempts to place the analysis within a rigorous cost-benefit framework. In evaluating priorities for action, this approach has much to commend it. In this section, we look at how it might be achieved.

An economic approach to identification of appropriate policy responses requires a better understanding and acceptance of "cost-benefit thinking"[25]. In the first place, basic financial cost data will be required concerning the damage that can be expected in potential hazard zones associated with sea-level rise scenarios.

It is possible to compute current value estimates of the immediate financial vulnerability of a given hazard zone, ie. the value of the real sources of economic wealth that might be threatened by sea-level rise. Such financial cost estimates are, however, only a very rough approximation of the real economic cost to society if, for a given zone, a decision was made to forego any protection from rising sea-levels.

The financial vulnerability measure would, in principle, encompass the estimated value of lost land, property and infrastructure and a diverse range of environmental function and service values such as beach recreation and nature conservation values. In practice, it may well prove difficult to place meaningful monetary values on all the assets at risk. For some assets, there is no difficulty in establishing economic, or market values for their worth, but this is not the case for all assets. Environmental assets, in particular, fall into a category in which market values are not available, because they are often difficult or impossible to place in private ownership and possess open-access characteristics.

The translation of an approximate estimate of the immediate economic vulnerability into a measure of the "true" economic, or social, cost involves further difficulties. To take a simple example, land or property will lose value once it is clearly perceived to be at risk from inundation. But this situation is complicated by protective measures already in place or because property owners have an expectation that protection would be forthcoming. Thus, the market value observed may or may not reflect the risk involved. Which value should be used, the observed value or the value of some equivalent property outside the risk-prone area? Then, adjustment costs associated with redeploying productive resources that were once applied to the land now lost need to be considered.

The long-term nature of the global warming problem adds further complications. The most significant losers are likely to be future generations - especially in the developing countries. It is, therefore, necessary to factor into the calculations, and to implement, measures which would compensate future generations for the losses they may incur.

A mechanism for implementing intergenerational fairness would involve either direct or indirect compensation. Thus, the loss of some environmental assets because of sea-level rise, such as wetlands, could be compensated for by man-made capital substitutes or by wetland restoration and/or creation schemes away from the hazard zone. If substitution is not possible or practicable, then indirect compensation is required. This could take the form of cost-effective preventive and adaptive measures implemented in the present-day for the benefit of future generations, coupled with increased investment in research so that future decision-makers have an adequate knowledge base.

There is a further complication in the global warming cost-benefit model which has been underestimated in the past and deserves serious and immediate analysis. If equity between generations is a goal, then equity within this generation must also be taken into account. If, for example, the poorer nations of the world are to respond to the threat of sea-level rise then financial and technical assistance will be necessary. In the context of global warming, the principle that the polluter must pay would suggest an obligation on the part of the industrialized nations - the source of the bulk of greenhouse gas emissions to date - to provide appropriate compensation. A full cost-benefit analysis should take this obligation into account.

What is required for a full economic cost analysis is a general equilibrium approach which tracks the myriad direct and indirect damage costs through the socio-economic system until a balance is reached. Such a model would allow the estimation of the full effects of any changes - the physical, cost and price changes - due to inundation, flooding and erosion on a national economic basis. Models such as these have yet to be developed and applied to the sea-level rise issue. A more pragmatic and approximate approach to asset damage valuation will, therefore, have to be adopted until they become available.

Ideally, the cost-benefit approach would seek to balance the expected damage cost encompassed by the vulnerability measure against the costs of possible response measures. In

this way, benefits - the damage costs avoided - could be compared with mitigation and protection costs, both in a common monetary dimension. This approach would prove a useful aid to rational decision-making, even though it may never be possible to assess quantitatively all the costs and benefits.

Is society sufficiently risk-averse to find the resources necessary to cover the costs of responding to global warming? Assuming society is risk-averse and given the potential impact of global warming and sea-level rise, it can be demonstrated that society would be willing to pay a sum to avoid the future risks of climatic change in certain instances. Society will want the services of the environment in the future - wetlands to protect coastal cities, for example - but the prospect of changing climate and sea-level rise means that uncertainty, which society would always like to reduce, about the quality of the environment and its ability to provide needed services, has increased. The problem is that the risks associated with the impacts of climate change involve purely collective risk, a situation in which one individual's assumption of risk does not reduce the risk borne by others.

From the policy-making perspective, the message is that decisions taken in the current time period should attach greater weight to the future risks associated with the climate change. In other words, a precautionary approach to policy-making in the global warming context is recommended. This would suggest that, in the case of coastal cities, risk-aversion premiums would be well spent on investments in uncertainty- or risk- limiting activities. Reducing heightened vulnerability due to present-day problems of land subsidence and current sea-level rise is an obvious priority. On a wider scale, the precautionary approach suggests that it would be prudent to invest today in implementing emission reduction measures in order to reduce the overall impact of global warming.

Decision Making and Risk Management

Figure 4 describes three different kinds of uncertainty. On the vertical axis is a dimension representing the level of uncertainty. The horizontal axis gives some indication of the consequences of being wrong, or decision stakes. The point of the diagram is to show how uncertainty and scientific credibility interact via the themes of prediction, vulnerability and communication/empowerment outlined earlier.

At the lowest level of uncertainty risks can be analysed and managed by the conventions of positivistic science based on theory testing, observation, verification, falsification of error, and replication. The task is to improve the information level to ensure sufficient evidence to apply models and prediction. This is a familiar arena of science where priority is examined by established and

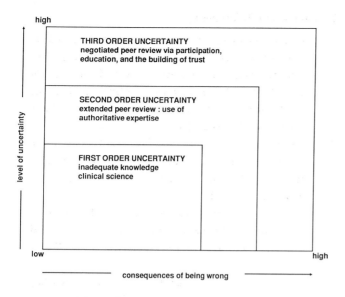

Figure 4: Interaction of uncertainty and scientific credibility.

knowledgeable peer groups. Examples for global environmental risk would include the provision of basic health care and sanitation at the community level.

At this level of uncertainty the science involved is generally incontestable. By and large, the public accept and endorse scientific prescriptions. Any shortcomings lie in insufficiency of data and incompleteness of models. Generally these deficiencies can be overcome by conventional peer review. In terms of our risk management perspectives, probability estimates become the focus of attention, and the communications issue is primarily one of explanation, clarification and comparison. The IPCC Working Group I Report is a prime example in this regard.

In the middle arena, the consequences of being wrong are greater, the wider environment is to some extent involved, and the level of uncertainty becomes a function of the experience and reputation of trained risk analysts. Here is the zone of second order uncertainty, where lack of understanding is compensated for by informed judgement subject to wider peer appraisal. That appraisal is reliant upon authoritative expertise accorded to individuals or institutions with reputations for independent, balanced judgement and adherence to scientific codes. The

precautionary principle is generally invoked.

The precautionary principle, however, is not always applied on such occasions. Precaution depends upon consensus regarding likely consequences of delay. In the narrow version of the extended peer review process, caution may prevail over precaution. Estimates of land degradation, or fisheries depletion tend to be conservative being based on official sources which do not always visualise the additive effects of resource misuse in regions beyond the scope of their monitoring information or administrative competence. Here, possibly, is where a greater level of communication between those who use resources and those who administer, or regulate their use, may be required.

The third level of uncertainty is most applicable to global change as a whole. This is the realm of more fundamental indeterminacy where unpredictable shifts in both physical and social system states are possible, but not forecastable with any precision. Here too conventional science needs time to develop the appropriate models to provide sufficient degrees of predictability that may be regarded as too long by a restive public, anxious to see precaution become enshrined in political pro-action.

This important category of uncertainty pervades global change and is especially apparent in the climate alteration studies. This particular quality of inherent indeterminacy and global anxiety goes beyond the application of precaution or even of pre-emptive regulation. It envisages a form of science that includes an element of discourse with the public and the politicians so that the context in which scientific research is conducted and evaluated is given full credit when determining the outcome.

The issue here is partly a matter of methodology, partly of peer review, and partly of the manner in which science is translated into policy. Different aspects of science have to be introduced. These involve ethical positions, consensus formation by public involvement, and trust in participatory decision procedures. It is this last aspect that lends to the issues of risk communication.

Table 1 outlines the various mechanisms through which the three realms of scientific discourse are validated. In the arena of first order uncertainty, the conditions for gathering

Table 1: *Types of uncertainty and styles of problem solving*

Type of uncertainty	Conditions for decision making	Prerequisite for authority	Source of authority	Problem solving style
Technical	Risk	Information	Instruction	Reductionist
Methodological	Uncertainty	Respect	Experience	pragmatic
Epistemological	Indeterminacy	Faith	Revelation	Holist

legitimacy, in public acquiescence, involve gathering more information by empirical means. The source of authority is hierarchical based upon structures that disseminate information via instruction.

In the second order uncertainty mode, the prerequisites for legitimacy are respect for scientific competence, experience in offering judgement, and pragmatism in attempting to solve problems. This mode of risk management is usually partially open to public participation, though the dominance of respectful authority is an important aspect of dispute resolution.

In the third order realm of uncertainty, risk management operates on a mixture of conventional science and faith. The method of analysis is holistic, or at least interdisciplinary, and the source of final authority is revelation through participatory discussion based on shared trust between all concerned. In this instance the precautionary principle is of particular importance as a guide to action. The task of the risk manager is to try to devise flexible and adaptive solutions to allow for appropriate adjustments as experience is gained and as knowledge increases. The demand for major investments in "good housekeeping" energy conservation measures even when climate change cannot be conclusively proved, is an example of this approach. Energy conservation measures form a pragmatic and adaptive response acting as an insurance against future uncertainty, so as to reduce the prospect of a major salvage operation if the forecasts of climate change induced land use changes or sea level rise turn out to be in the high domains of error bands.

Conclusions

This paper suggests that the principles of risk management, as they have currently evolved, are not sufficiently developed to provide a tool for response to global environmental change. For risk management to be helpful, it will have to combine principles of vulnerability and political fairness, to insurance and precaution, utilising innovative educational and participatory techniques such as interactive scenario game playing coupled to best guess estimates of costs and benefits of various courses of action. Such ideas will also have to be set in a framework of purposive regulation to ensure international compliance, and effective pricing of resources, most particularly fuels. To make progress a number of experimental approaches should be tried in different countries, reflecting the various dimensions of global environmental and economic change. These experiments should be independently monitored to provide a suitable basis for effective and acceptable policy implementation.

Footnotes and References

1. For a good summary see Roger E. Kasperson, Kirsten Dow, Dominic Golding and Jeanne X. Kasperson (eds.) 1990. Understanding Global Environmental Change : The Contributions of Risk Analysis and Management. Center for Technology, Environment and Development, Clark University,, Worcester MA.

2. Some preliminary calculations of these costs can be found in Lester R. Brown and Eric Wolf 1988. Reclaiming the future. In L.R.Brown (ed.) State of the World 1988 Norton, New York, pp176–188.

3. This is also a controversial point. It is covered in part in the September 1989 issue of Scientific American, entitled Managing the Planet especially the article by Dr.Jim McNeill.

4. This view is hinted at by Dr.Bill Clark in his introductory essay in the Scientific American issue mentioned in footnote 3 above. However the explicit scientific and economic interrelationships between these variables await the conclusions and recommendations of the IGBP Global Change Programme and subsequent economic analyses.

5. See John B. Robinson 1988. Unlearning and back casting: rethinking some of the questions we ask about the future. Technological Forecasting and Social change 33, pp.325–338.

6. D.W.Pearce, E.Barbier & A.Markandya 1990. Sustainable Development: Economics and Environment in the Third World. Edward Elgar, Aldershot, Hants.

7. K-G Maler 1989. Sustainable development. Mimeograph available from Economic Development Institute, World Bank, Washington DC.

8. H.Daly & J.Cobb 1990. For the Common Good. Green Print, Malden Rd. London.

9. E. Barbier, A. Markandya & D.W. Pearce 1990. Environmental sustainability and cost benefit analysis. Environment and Planning A, 22, pp.1259–1266.

10. These points and many others are covered in the biennial publication World Resources by the World Resources Institute and the International Institute for Environment and Development, and published by Basic Books, New York and Oxford University Press, Oxford.

11. These points are examined in a number of recent monographs on environmental economics. See in particular D.W.Pearce, A.Markandya and E.B. Barbier 1989. Blueprint for a Green Economy. Earthscan, London; and D.W.Pearce and R.K. Turner 1990. Economics of Natural Resources and the Environment. Harvester Press, London and New York.

12. See UK Health and Safety Executive 1988. The Tolerability of Risk from Nuclear Power Stations. HMSO London; and UK Atomic Energy Authority 1988. Social Risk: A Review of the Technical Basis for the Management of Risks to Society from Potential Major Accidents. United Kingdom Atomic Energy Authority, Warrington, UK.

13.   See D.V.Lindley, 1975. Making Decisions. Wiley, London
      and I.McLean, 1987. Public Choice. Blackwell, Oxford.

14.   See W. Edwards, and D.van Winterfeldt, 1988. Public
      values in risk debates. Risk Analysis 7, pp.141–158.

15.   For a comprehensive review see Delia Fernandez-Russel
      1987. Societal Risk Estimates from Historical Data for UK
      and Worldwide Events. Research Report No. 3,
      Environmental Risk Assessment Unit, University of East
      Anglia, Norwich, UK.

16.   See A. Coppola, and R.E.Hall, 1981. A Risk Comparison.
      NUREG/CR-1916, US Nuclear Regulatory Commission,
      Washington DC.

17.   See R. Kemp, 1990. Risk tolerance and safety management.
      Reliability Engineering and System Safety.

18.   The literature on risk management is now vast. A good
      survey of key themes can be found in the international
      journal Risk Analysis published by Plenum, New York. A
      review of the social science aspects of the subject can be
      found in Jennifer Brown (ed.) 1989. Environmental
      Threats: Perception Analysis and Management. Belhaven
      Press, London.

19.   In addition to the book edited by Brown, cited above, see
      also Mary Douglas 1980. Risk Acceptability According to
      the Social Sciences. Russell Sage Foundation, New York,
      and Thomas M.Dietz and Robert W.Ryasne 1987. The Risk
      Professionals, Russell Sage Foundation, New York.

20.   For a good review see Kenneth Hewitt (ed.) 1984.
      Interpretations of Calamity. Unwin Hyman, London.

21.   Two good texts on the general issues of risk
      communication have been published recently. See Helmut
      Jungermann, Roger E. Kasperson and Peter Wiedermann
      (eds.) 1990. Risk Communication KFA Julich; and
      National Research Council 1989. Improving Risk
      Communication NRC, Washington DC.

22.   See R.K.Turner, P.M. Kelly, R.C. Kay, 1990. Cities at
      Risk, BNA International, London.

23.   See D.M.Liverman, 1989. Vulnerability to global
      environmental change. In Understanding Global
      Environmental Change: A Report on an International
      Workshop. (see footnote 1)

24.   R.K.Turner, (ed) 1988. Sustainable Environmental
      Management: Principles and Practice. Bellhaven Press,
      London; D.W. Pearce, A. Markandya & B.Barbier, 1989.
      Blueprint for a Green Economy. Earthscan Publications,
      London.

25.   See J.D. Milliman, J.M. Broadus, and F. Gable, 1989.
      Environmental and economic implications of rising sea
      level and subsiding deltas: the Nile and Bengal examples.
      Ambio 18, 340-345; and J.Titus, ed. 1990. Changing
      Climate and the Coast. Washington DC, Environmental
      Protection Agency.

Climate, Climate Change and the Economy

by N. S. Jodha *(Nepal)* and W. J. Maunder *(New Zealand)*

Abstract

"Climate, Climate Change, and the Economy" is discussed in the context of the question "What does a climate change mean to people?". Topics discussed include: (1) people and climate change; (2) winners and losers where it is noted that there are climate-related winners as well as climate-related losers with today's climate; (3) the economic and social benefits of meteorological and climatological services; (4) long-term sustainability issues in developing countries and the linkages between climate variations and climate change issues; and, (5) climate change relative to economic competitors both now and in the future. The paper highlights four key but often forgotten factors: first, climate must be recognized as a resource, and not solely as a factor which imposes limitations on agricultural production; second, climate change should not necessarily be regarded as a threat but rather as a challenge; third, there is a need for a much improved understanding of the multitude of the inter-relationships between climate and society—including but not restricted to climate change—in both the developed and the developing countries; and fourth, many of the issues—particularly those involving climate change—have significant political and strategic implications.

Overview

In preparing this paper we had several choices, including whether to give a specific rigorous economic-based analysis of how climate and in particular climate change does affect and will affect political regions and individual nations. While this approach would be highly desirable, it was considered premature at this stage to present such an analysis. Instead, we have taken a more philosophical view and posed some of the questions we believe need to be asked before the more in-depth economic analysis can be made. Accordingly, attention is focussed in this paper on (a) what a climate change means to people, (b) the importance of both climate variations and climate change, (c) the winners and losers dilemma, (d) the response of human activities to climate variations and climate change, (e) the economic and social benefits of viable National Meteorological and Hydrological Services, (f) climate-related long-term sustainability issues in developing countries, (g) climate change relative to economic competitors, and (h) the need to consider climate change as a challenge and not necessarily as only a threat.

People and Climate Change

A climate change means different things to different people. With this in mind we would like first to take you to the small town of Pfunds in the Austrian Tyrol, which has a surrounding landscape of mountains, trees, swift flowing rivers, and in summer the cutting and harvesting of hay. Most of the town of Pfunds itself, like its agriculture around it, has probably not changed for decades, and the question one asks in looking out onto the very peaceful scene is: "What does a climate change mean to the PEOPLE who live and work in Pfunds?" At the same time, consider the people who live and work and live in our own countries—Nepal and New Zealand, and ask the same question "What does a climate change mean to the people of Nepal or New Zealand?" We consider that similar questions must be asked of the people of all nations, because whatever a climate change may mean to the Prime Minister of India, the Prime Minister of Australia, the President of Kenya, or the King of Norway, the more fundamental question is: "What does a climate change mean to the people?"

Clearly, the importance of any climate change is, and will be, looked at differently in the 160+ countries of the world (see Maunder 1986, Sewell 1966); nevertheless, we believe that to most people—as distinct from Prime Ministers and Presidents—climate VARIATIONS in terms of months, seasons, years and decades are still as important, if not more important, than the climate which is expected to exist in the year 2030, or 2060, or 2090. That is, irrespective of any climate change, climate VARIATIONS will remain. All countries will be subject to these "normal" climate variations (see Palutikof 1983, Phillips 1985) and, not withstanding the tremendous political importance being attached to climate change, we consider it would be very unwise for any country to place all their eggs in the 2030 or 2050 "climate basket" and not be equally or even more concerned with what will go into and out of the basket during the next 30 to 50 years. As always we have to live within our climatic income (see Maunder 1989).

Winners and Losers?

The unwritten official "oath of silence" on "winners and losers" in regard to climate change is finally breaking down as an increasing number of papers suggest or imply that some might gain and others might lose in the event of a global warming of the atmosphere. It has been frowned upon to discuss this issue—particularly winners—from the podium, but it has been acceptable—indeed almost mandatory—to discuss losers in a global warming situation. With a few extreme exceptions statements abound about how a climate change will produce losers, and although one can argue that OVERALL the world as a whole would be the loser as a result of a "significant" climate change, it is necessary to understand and appreciate that some areas and countries may well gain as a result of a warming. For example, do we know whether a warming of the winter climate would be beneficial or otherwise to the people of Alberta, Switzerland, Bolivia, Texas, Scotland, Botswana, or Japan? Moreover, if you asked the people in these areas whether they would prefer their winters to become warmer or colder what would they say?

There are, of course, climate-related winners as well as climate-related losers with today's climate. One might therefore ask why we should expect those climate-related losers to support attempts to "freeze" a changing global climate at today's regional distribution of climate resources. One might also ask what those with today's favourable regional climate regimes are doing to assist those with a less favourable climate.

We do not know whether there will be winners or, if so, who those winners might be. But the issue has been "swept under the carpet" at an important juncture in the process of assessing the implications of climate change for environment and society. We believe that the win/lose issue needs to receive more attention than it has in the past and that it emerge in public and scientific debate, not only from the viewpoint of possible economic winners and losers at the national level, but also possible winners and losers from the various economic sectors such as solar and wind energy, coastal engineering, winter tourism, air conditioning, biotechnology, reforestation etc.

Responses to Weather, Climate Variations and Climate Change

Although human activities are sensitive to weather, climate variations, and climate change on a wide range of scales, it is at times very difficult to assess these relationships (see CSIRO 1979, Maunder 1970). Difficulties arise first because relatively little definitive research has been done on assessing weather, climate, and climate change sensitivity; and second, what research has been done is often regarded by the critics as being either superficial or obvious. For example, how does one assess the importance—in dollar terms—of forest fires caused by lightning, or flooding caused as a result of tropical cyclones? Further, while the concept of sensitivity may be obvious, a scale to measure these sensitivities is not obvious. For example, how sensitive (to the weather, the climate, climate variations, and climate change) is the coal industry of Australia , coffee production in Brazil, the area of wheat sown in the Soviet Union, the inflation rate in Switzerland, or the tourist industry in Jamaica? Moreover, if we do not know, do we need to know? These and other questions are important if we are to better understand the real role of the atmospheric climate in the market place (see Johnson and Holt, 1986).

Economic and Social Benefits of Meteorological and Climatological Services

An important contribution to the concepts discussed in this paper are the findings of the WMO Technical Conference on the Economic and Social Benefits of Meteorological and Hydrological Services held in Geneva from 26 to 30 March 1990. The conference was attended by meteorologists, hydrologists, economists, and engineers from national Meteorological and Hydrological Services, universities, private meteorological consulting firms, and other interested institutions from all over the world.

Papers presented covered (a) methodologies for assessing the economic and social benefits of meteorological and hydrological services, (b) user requirements for specific weather and climate services and related economic studies, (c) user requirements for hydrological services and related economic studies, and (d) the role and status of national Meteorological and Hydrological Services in economic and social development.

Beginning from the 1960s there have been several occasions when WMO organized conferences, co-ordinated studies and published technical notes in connection with the economic and social value of meteorological and hydrological services. The close connection between these studies and the management aspects of national Meteorological and Hydrological Services had always been clearly recognized. It had been also clear that in different regions and in countries at different stages of development the emphasis would be placed on different aspects of the subject.

As global concern in connection with climate change is mounting, meteorology and hydrology are becoming politically more important. Consequently, many of the earlier relatively simple concepts of the social and economic value of the services provided by National Meteorological and Hydrological Services now need to be reconsidered. Similarly, new scientific results and new technologies have introduced very strong technology driven aspects in both the production of, and demand for, information. There are also new trends emerging in the development of several National Meteorological and Hydrological Services, related to the marketing of services, and user-pay policies, which are producing a changing vision of the role of meteorology and climatology and hydrology.

The main conclusions of the conference are contained in the Proceedings of the Conference (see World Meteorological Organization, 1990) and include the following:

- A concentrated combined effort is needed to further develop, evaluate, document and internationally exchange methodologies suitable for the assessment of economic and social benefits of meteorological and hydrological services
- User requirements for such services are often highly specific and should be identified, defined and supplied through close co-operation between meteorologists/ hydrologists, relevant users and intermediaries
- Efforts are required to inform potential users of the benefits they can realize through making use of available improved services. Marketing of these services is increasingly important
- Hydrological Services need to be extended (where it is not yet done) to water quality measurement and monitoring
- The principle of free exchange of basic meteorological information between national Meteorological Services should be preserved as the necessary basis for the provision of weather forecasts, warnings and other services all over the world
- In view of the trend in some countries to provide meteorological services on a commercial basis, there may be a need for a "code of conduct" on the use of data exchanged over the Global Telecommunication System of the World Weather Watch.

Developing Countries and Long Term Sustainability Issues

Issues regarding the long term sustainability of resources are of critical importance in many areas, and particularly so in several developing countries. In such areas, the importance of the climate in the future is very much overshadowed by the climate of the present and the immediate past. Of particular importance are the many fragile resource zones—examples of which are given in this section - which are faced with crisis situations with the present climate, let alone the impact of any climate change. There is an urgent need to strengthen the capabilities of these areas to meet these current climatic crises which, if correctly done, will also help them to withstand the impacts of climate change. An additional significant factor is that measures to strengthen the current capabilities of many developing countries will be more easily accepted, because the immediate and actual problems—many of which are climate related—are well recognised, and in many cases are much more relevant to society than potential problems relating to a climate change sometime in the future.

Important issues in this regard include:

- Climate, operating both as a productive resource and a constraining factor, forms an integral part of the overall production environment (especially for agricultural economies).
- For reasons such as population growth, the extractive role of the market, and increased public intervention, climate's role in many economies (with or without climate change) has changed. In particular its productive role has been eroded, and the constraining function has been accentuated. This is particularly so in bio-physically fragile resource zones such as mountains and arid and semi-arid tropical areas. From such zones—represented by selected hill areas of India, China, Nepal, and Pakistan, and arid and semi-arid tropical plains in India—evidence from field studies and knowledge review has been assembled which indicates that over the period of the last 50 years, several negative trends have emerged with respect to variables directly or indirectly associated with climatic factors and their interactions.
- The measurable or verifiable negative changes in the indicators of unsustainability, relate to the health of the natural resource base, production flows, and resource management practices. Some of them are directly visible, others concealed by people's response to such changes. Some changes have potentially negative consequences. These climate related factors and issues are now considered for (a) mountain areas, and (b) arid and semi-arid tropical plains.

A. Mountain Areas

1. Directly visible changes

a. Resource base

Increased land slides and other forms of land degradation; reduced water-flows for irrigation

b. Production flows

Prolonged negative trends in crop and livestock yields; increased input needed per unit production; increased time and distance involved in food, fodder, and fuel gathering

c. Resource use/management practices

Reduced extent of fallowing, crop rotation, and inter-cropping, and the extension of the plough to sub-marginal lands

2. Changes concealed by responses to (negative) changes

a. Resource base

Substitution of cattle by sheep/goats, deep rooted crops by shallow rooted ones, and a shift to non-local inputs; substitution of manure by chemical fertilizers

b. Production flows

Increased seasonal migration; introduction of externally supported public distribution systems, and intensive cash cropping on limited areas

c. Resource use/management practices

Reduced diversity; increased crop specialization

3. Development initiatives with potentially negative changes

a. Resource base

New systems without linkages to other diversified activities generating excessive dependence on outside resources

b. Production flows

Agricultural measures directed to short term quick results

c. Resource use/management practices

Excessive, crucial dependence on external advice and ignoring wisdom

B. Dry Tropical Areas

1. Directly visible changes

a. Resource base

Vanishing top soils due to water and wind erosion; deepening of water tables; ground water salinization; increased inferior annuals and thorny bushes

b. Production flows

Reduced average productivity of different crops; increased cropping on sub-marginal lands; higher dependence on inferior options (e.g. harvesting and lopping of premature trees); rising severity of successive drought-impacts; increased dependence on public relief; increased migration

c. Resource use/management practices

Decline of common property resources; reduced feasibility and effectiveness of traditional adaptation strategies (e.g. rotations, intercropping)

2. Changes concealed by responses in (negative) changes

a. Resource base

Increased emphasis on mechanisation of cultivation; large scale reclamation of wastelands

b. Production flows

Higher coverage of public distribution food systems and other anti-poverty programmes; greater dependence on external market sources; changes in land use pattern favouring grain production

c. Resource use/management practices

Replacement of self-help systems by public support systems

3. Development initiatives with potentially negative changes

a. Resource base

Research and development focus on crop rather than resource upgrading ignoring its limitations (e.g. irrigation in impeded drainage areas), and resource extractive measures (e.g. tractorization)

b. Production flows

Highly subsidised, narrowly focused production programmes: focus on crops ignoring other land based activities; grain yield ignoring biomass; relief operations focused on people and livestock while ignoring the resource base, thereby promoting high pressure on a poor resource base

c. Resource use/management practice

Disregard of folk knowledge in formal interventions; replacing local informal arrangements by rigid legal/administrative measures.

Climate Change Relative to Economic Competitors

A number of crucial questions need to be asked when discussing the choices you, as a decision-maker within your country, have in responding to a greenhouse gas induced climate change of either your country or countries which are your economic "competitors". In this context, THE decision-maker could be the Prime Minister or the President, or it could be someone who advises the Prime Minister or the President, or it could be an individual farmer, or manufacturer, or exporter who has to make a climate-change related decision.

The term "competitor" could be considered to be someone or some company or some farm within one's own country, but it is more appropriately defined as someone, or some company, or some farming organization in another country with which you (in your country) are in competition. Typical examples of "competitors" are exports (or exporters), imports (or importers), prices obtained for goods exported, prices of imported goods, foreign exchange rates, foreign interest rates, and tourism. In addition, the competition that needs to be considered in the context of climate change, is NOT only that of the present competition, but also the ways in which the competition is likely to change over time. In reality, we are dealing with a multi-dimensional "economic/climate change chess game" in which the pawns and the Kings are likely to change, as well as the "rules" of climate, economics, and politics. That is, we must deal with the real world of both today and tomorrow.

In considering the overall subject of greenhouse gas induced climate change (including both the direct climatic impacts such as an increased demand for air conditioning, and the indirect impacts caused by say a fuel tax), some countries—as noted earlier—believe that they will be "winners", while some believe they will be "losers" with a problem not of their own making. But, whether there be "winners" or "losers", a number of important issues need to be considered:

- Every country must be aware of how a climate change will affect their own country and also that of their competitors.
- There will be a strong demand for local, regional, and most importantly country by country studies, since most communities will want to know how a greenhouse gas/climate change will affect their locality, and that of their present and future competitors.
- A key factor in any economic consideration of the greenhouse gas/climate dilemma are the two questions: "What sort of economy/society can we realistically expect in the year 2030, 2060, or 2090?"; and "How should we react/adapt to any climate change in the light of the kind of economy/society we would like to have?".
- Irrespective of any climate change (through greenhouse gases or any other cause), climate variations will remain, and all countries—and their competitors—will continue to be subject to these climate variations (both large and small).
- All countries—developing and developed, north and south, west and east—have limited financial and people resources. In considering the greenhouse gas/climate change question, it is important that these resources be directed with appropriate vision to meet the real needs of EACH country. To do this an annual "balance sheet" will need to be made for EACH region and EACH country of (a) what is actually happening to the climate, (b) what are the current global, regional, national, and local predictions of the future climate, and (c) what are the actual and potential regional, national, and local impacts—specifically on a country-by-country basis—of these forecast climate changes.

The Challenge Ahead

The concerns and implications of the wide variety of items mentioned in this paper highlight three key factors: first, climate must be recognized as a resource, and not solely as a factor which imposes limitations on agriculture production, settlement patterns, and economic activities; second, there is a need for a much improved understanding of the multitude of inter-relationships between climate and society, and in particular in the manner in which changes in one may result in shifts in the other; third, many of the issues have political and strategic implications.

The Wizard and the Chess Game

Earlier, we mentioned that in discussing economic issues, climate variations, and climate change, we are dealing with a multi-dimensional "economic/climate change chess game" in which the pawns and the Kings are likely to change, as well as the "rules" of the game. In this regard, could we therefore relate a little story?

It seems that in a certain kingdom there was a school for the education of princes approaching manhood. Since the King and his court spent much of their time playing chess, it was decided that the subject called "games" should be added to the curriculum of the school. A wizard of the school was assigned to develop the course. One day the King summoned the wizard and asked him to describe the method used to teach chess in school. The king was amazed to hear that, in the classroom game of chess, all pieces moved in straight lines and the wizard used terms like "jumping men" and "double jumping", whereas the wizard never referred to things the King was familiar with such as rooks, bishops, pawns, and knights. Somewhat puzzled, the King asked the wizard if he had ever observed chess being played in the real world. The wizard replied: "No, but I do carry on correspondence with other wizards. This is better since everyone knows wizards are smarter than chess players."

The moral of this little tale for all of us is: "An education in checkers does not prepare one for a life of chess."

Conclusion

In conclusion, could we take you to a valley in Switzerland where we are looking out on to some vineyards, the passing of Swiss PTT Buses heading for Saas Fee, and the passing of Swiss trains heading for Zermatt. Again, we ask the kind of question which we posed at the beginning when we were considering the town of Pfunds in the Austrian Tyrol: "What does a climate change mean to all the Swiss farmers in the area, to all the restaurant owners, to all the hotel managers, to the operators of the Swiss PTT buses, to the manager of the Swiss Railways, and to the thousands of tourists heading for Zermatt and Saas Fee?". At the same

time we need to ask the people in all the other 160+
countries in the world, including the farmers, industrialists,
tourist operators, transport companies, alternative energy
engineers, and insurance salespersons, out there in the
market and consumer place: "What does—or more
correctly—will a climate change mean to you? "

There are no easy answers to these questions, even if
WHAT WE NEED TO DO once we find out the answers to
these questions is more easily answered. We must however
be mindful that climate change should not necessarily be
regarded as a threat—as it so often seems to be, but rather
as a challenge, the challenge being to all scientists,
economists, engineers, and politicians to produce a better
environment for the people of all nations.

References

CSIRO: 1979. The Impact of Climate on Australian Society and
Economy. Report of a Conference held at Philip Island,
Victoria, 27–30 November, 1978. Division of Atmospheric
Science, CSIRO, Mordialloc, Victoria, Australia.

Jodha, N.S.: 1989. Potential strategies for adapting to greenhouse
warming: perspectives from the developing world. In:
Rosenberg, N.J., et al.(Editors), Greenhouse Warming:
Abatement and Adaptation. Resources for the Future,
Washington, D.C.

Jodha, N.S.: 1990. Sustainable agriculture in fragile resource
zones: Technological imperatives. MFS Discussion Paper No.
3. International Centre for Integrated Mountain Development
(ICIMOD), Kathmandu, Nepal.

Johnson, S.R. and Holt, M.T.:1986. The value of climate
information. CARD(Center for Agricultural and Rural
Development) Staff Report 86-SR6. Iowa State University,
Ames, Iowa, USA.

Maunder, W.J.: 1970. The Value of the Weather. Methuen,
London, 388 pp.

Maunder, W.J.: 1986. The Uncertainty Business: Risks and
Opportunities in Weather and Climate. Methuen, London, 420
pp.

Maunder, W.J.: 1989. The Human Impact of Climate
Uncertainty. Routledge, London, 170 pp.

Palutikof, J.: 1983. The impact of weather and climate on
industrial production in Great Britain. Journal of Climatology,
3: 65–79.

Phillips, D.W.: 1985. Economic Benefits of Climate Applications.
A publication of the Canadian Climate Program. Atmosphere
Environment Service, Downsview, Canada.

Sewell, W.R.D. (Editor),: 1966. Human Dimensions of Weather
Modification. University of Chicago, Dept. of Geography,
Research Paper No. 105, 423 pp.

World Meteorological Organization: 1990. Economic and
Social Benefits of Meteorological and Hydrological Services.
Proceedings of the Technical Conference. Geneva, 26–30
March, 1990. WMO Publication No 733.

Assessing the Regional Implications of Climate Variability and Change

by William E. Riebsame *(USA)* and Antonio R. Magalhaes *(Brazil)*

Abstract

Climate fluctuations affect the quality and flow of natural resources in a region. Thus, an understanding of the regional effects of contemporary climate variability and hazards, as well as potential future climate change, is critical to resources management and sustainable development. This paper offers a broad evaluation of the approaches and tools of climate impacts assessment. It covers three areas: (1) techniques for assessing climate impacts on social and biophysical systems; (2) approaches for assessing regional sensitivity and adaptability, and (3) ways to improve regional capacity to conduct impact studies, especially in developing countries, and to make regional impact studies more useful to policy-makers seeking sustainable development.

Introduction

In the decade since the first World Climate Conference, a sizeable body of literature and methods, and a group of active researchers, have emerged in the field of climate impact assessment. Many new methods exist to measure effects of actual climate fluctuations as well as to assess the potential impacts of future climate change on natural resource and social systems. A wide array of studies have been conducted, ranging from historical analyses of climate extremes to projections of global warming effects. Yet, critical problems remain to be solved. The abiding difficulty of separating climate effects from other factors affecting social well-being weakens the conclusions and policy recommendations stemming from impact assessments. Moreover, analysts are now frequently asked to extrapolate impacts of climate change decades into the future, despite large uncertainties about regional details and the difficulty of predicting the behaviors and interactions of biophysical and social systems over the requisite time scales.

We offer here a broad evaluation of climate impacts research, covering three wide areas: (1) basic techniques and the strengths and weaknesses of available methods, including the problematic step of creating and applying realistic scenarios of regional climate change; (2) evaluating regional biophysical and social sensitivity to climate change, and how sensitivity and adaptability change through time; and (3) making regional impact studies useful to policy-makers facing critical development and resource allocation decisions. We conclude by discussing how governments can foster programs for assessing climate impacts as part of their efforts to achieve sustainable development.

Techniques of Climate Impact Assessment

The field of climate impact assessment has evolved rapidly over the last two decades as the need arose to evaluate the impacts of several disruptive climate fluctuations, and as governments began to press researchers for policy guidance on global warming. Results of these studies are dispersed through a wide array of multidisciplinary journals, books, and government reports, making it difficult to collect and review the relevant material.

At a conceptual level, climate impact assessment is an attempt to explicate an interactive model of climate and society. As developed by SCOPE (1978) and Kates, (1985), the climate/society interaction model (Figure 1) describes linkages and feedbacks between climate variation, social vulnerability, impacts, and adjustments that feed back to society or climate. Impact studies have tended to focus mostly on the direct impact links in the model, with less attention on adjustment (feedbacks) and underlying changes in social vulnerability or resiliency.

A Typology of Climate Impact Studies

The terminology of climate impact assessment is poorly codified. The phrase "climate impact assessment" has been used to encompass studies ranging from impacts of the historical effects of the Little Ice Age in Europe, to the projected effects of a doubling of the global greenhouse

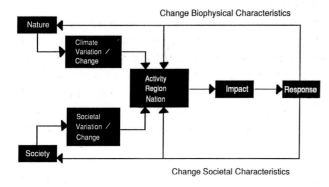

Change Biophysical Characteristics

Change Societal Characteristics

Figure 1: An interactive model of climate and society relationships illustrating the dynamic feedbacks between impacts and adjustment, as well as the potential for underlying natural and social changes that alter the likelihood of impacts (from SCOPE 1978, and Kates 1985).

effect--in a particular region--sometime in the middle of next century. A simple typology of approaches to climate impact assessment is offered here.

Diagnostic studies encompass the empirical measurement and assessment of contemporary and past climate impacts that best fits the customary title of the field. The most common diagnostic approach is the "case study," typically of marked climate anomalies such as widespread drought in 1972 (Garcia, 1981), the 1975 Brazilian freeze (Margolis, 1980), the Sahelian drought (Mortimore, 1989), and anomalies associated with recent El Niños (Caviedes 1975 and 1984; Glantz et al., 1987). This approach provides "reasoning from extremes," utilizing social and biophysical responses to short-term climate extremes to infer to longer-term climate change. Some diagnostic studies employ "longitudinal analysis," in which impacts of climate fluctuations in a region are followed over time, as Bowden et al., (1981) did in the Sahel, US Great Plains, and Tigris-Euphrates valley. "Natural experiments" are a more specialized form of diagnostic study conducted with laboratory-like controls, perhaps where adjacent regions experience different climate anomalies or a specific climate shift are described and retro-dictive hypotheses about their effects tested with independent data (e.g., Parry, 1981). This aids discrimination between climate signals and other factors affecting natural and social systems. "Comprehensive assessments" are another form of diagnostic study in which an attempt is made to evaluate the full implications of climate and climate change in a region or nation. This requires especially rich data bases on natural resources and economic activities, and few comprehensive studies have been conducted--simply because they are so demanding of data and analytical capability--though attempts have been made in a few countries and critical regions (e.g., Australia, New Zealand,

Brazil: see Anderson 1978; Maunder and Ausubel 1985; New Zealand Ministry for the Environment 1988; Magalhaes et al., 1989). The rich literature on historical climate impacts (Wigley et al., 1981) includes some comprehensive studies, such as Utterstrom's, (1955) study of population trends in northern Europe and de Vries', (1980) analysis of climate impacts on commercial activity in the pre-industrial Netherlands. Government demand for information on the potential effects of global warming has evoked a few attempts to project comprehensive national impacts, especially in agriculture (Adams et al., 1990), but also across a wide range of natural resources and social activities (Smith and Tirpak, 1989).

Integrated assessments are attempts to operationalize the full interactive model of climate and society (Kates 1985); this requires a systems approach to identifying the full range of climate-sensitive activities in a region, and sufficient understanding of their interactions (e.g., between rural and urban economies, or between habitats and species) to track impacts as they propagate through natural and social systems. Cohen, (1986) developed a flow chart for integrated analysis of climate impacts in the Great Lakes basin of North America, but such models are easier to create than to operationalize in empirical analysis.

Analog studies encompass a range of less well-developed techniques in which situations analogous to climate change, such as groundwater depletion (Glantz and Ausubel, 1984) or other cumulative environmental change (Morrisette, 1988), are analyzed to learn how societies might respond to climate change. Although the power of analogy tends to break down as one attempts to project into the future, analogues provide a valuable "reality check" on impact studies, and the point may soon be reached where sufficient studies exist to search for repetitive and enduring impact/response patterns in analogous cases, patterns that logically could be expected to hold in a future climate change (Glantz, 1988).

Projective studies employ scenarios of future climate and known links between climate and natural resources or social activities to extrapolate effects and to suggest appropriate responses. Projective methods have received great attention in the last few years as governments seek a better assessment of the risk of global warming, and are probably the most common form of impact assessment currently conducted, as illustrated by other papers in this volume.

Sensitivity analysis is used in diagnostic, integrated, and projective studies not to assess effects of a specific climate anomaly or future scenario, but to explore the sensitivity of a particular resource, economic sector, or social group, to a range of climate conditions. The goal is often to identify thresholds at which effects become significant by some biophysical or social criterion. Sensitivity analysis thus provides insight into regional vulnerability and the sustainability of natural resources under a range of climate variability and change, and may point to development strategies offering broad-spectrum impacts mitigation.

Climate impact assessments generally consist of a mixture of these approaches. For example, diagnostic and analog studies merge where the analog under investigation is itself a climate-induced problem, as in studies of rapid lake level changes in the USA (Cohen, 1988; Morrisette, 1988b). Projective and sensitivity methods are often knitted together, as in the IIASA/UNEP agriculture studies (Parry et al. 1988). Nevertheless, different approaches offer distinctive power to improve our understanding of climate/society interaction. Diagnostic studies in particular, perhaps conducted to explain specific resource management problems or to analyze and improve government responses (e.g., Magalhaes et al., 1988), can provide the basis for development of climate-society theory when examined under a common rubric (e.g. Kates, 1985; Riebsame, 1985). They also provide the basis for future extrapolations using models calibrated by past impacts. Integrated analyses link specific diagnostic studies in a systems matrix, thus providing more holistic insight on impacts and responses.

Impact Realms

Climate impact assessments focus on various impact realms, such as specific natural resources, economic sectors, or cultural groups (Table 1). The most commonly examined impact realms are resources such as water (Novaky et al., 1985; Waggoner 1990) and forests (Maini, 1988), or particular human activities such as food production (Warrick et al., 1986; Sinha et al., 1988; Easterling et al., 1989), energy production and consumption (Jager, 1985), manufacturing (Palutikof, 1983), transportation, and protection from natural hazards. Broader impact realms include ecosystems, defined by vegetative assemblage (e.g., tundra or prairie; Shugart et al., 1986), and broad social structures such as traditional agriculture (Jodha and Mascarenhas, 1985; Farhar-Pilgrim, 1985; Jodha, 1989) or urban systems.

Other topical impact areas include health, population, infrastructure, insects/pests, international trade, and migration (see IPCC Working Group II, 1990). Assessments have also been organized by zones or critical regions, as in studies of coastal areas affected by storms and potential sea level rise (Warrick and Farmer, 1990), by geographical areas such as drainage basins (Revelle and

---

Table 1: A Sample of Climate Impact Studies Organized by Impact Realm and Techniques Applied.

Impact Realm	Sample Studies
Socio-Technical Systems	
Commercial/Manufacturing	Parry & Read 1988[d]; Palutikof 1983[c]; Maunder 1986[b]
Settlement/Infrastructure	Kates 1981[b]; Walker *et al.* 1989[f]; Magalhaes & Neto 1989[b,c,d]; Titus 1990[b]
Transportation	Keith *et al.* 1989[f]; Maunder 1986[b]
Energy	Jager 1985[c,e]; Linder & Inglis 1989[f]
Intermediate Systems	
Water resources	Rhodes *et al.* 1984[b]; Schwartz 1977[d]; Nemec & Schaake 1982[c,d]; Revelle and Waggoner 1983[c,f]; Gleick 1987[e,f]; Cohen 1986[e]; Riebsame 1988b[d]; Schaake 1990[e]
Agriculture/food	Parry et al. 1988[b,c,d]; Jodha 1989[d]; Liverman 1990[d]; Bowden *et al.* 1981[a,b,d]; Jodha & Mascarenhas 1985[d]; Rosenweig 1985[c,f]; Adams et al. 1990[e,f]
Livestock/grassland	Le Hourerou 1985[c]; Schimel *et al.* 1990[e,f]
Forests	Shugart et al. 1986[e,f]; Woodman & Furiness 1989[e,f]; Maini 1988[f]; Smith et al. 1990[e,f]
Natural (Less Managed) Systems	
Fisheries	Kawasaki 1985[c]
Natural preserves	Peters & Darling 1985[c,f]

Techniques and Tools Applied:
- [a] historical reconstructions
- [b] extreme events analysis and other case studies
- [c] statistical correlations and transfer functions
- [d] sensitivity and longitudinal analysis
- [e] simulation modeling
- [f] greenhouse projections

Waggoner, 1983) or mountain ranges, by critical regions such as the Sahel, Baltic Sea, or Mediterranean, and, of course, by particular climate zones (e.g., arctic, semi-arid, etc.). One goal of zonation has been to examine places where the effects of even minor climate changes would be significant, what Parry, (1981) called the "risk frontier" where a particular activity--such as settlement, crop production, or water resources management--is inherently marginal and sensitive to disturbance. This approach is applied in the "Vulnerable Regions" segment of the international agriculture project being conducted by Rosenzweig and Parry, (1990). Logically, climate impact signals will be amplified in marginal areas, helping the assessor extricate them from the "noise" of other factors that affect crop yields, water supply, or economic well-being, as well as providing guidance on adjusting development plans to yield reduced vulnerability.

The Tools of Climate Impact Assessment

Most climate impact assessments employ more than one analytical tool, ranging from statistical analysis of archived data to intensive field observation. Table 1 notes some of the analytical methods employed in selected impact studies, and methodological reviews appear in Kates et al., (1985), Riebsame, (1988c), and Parry, (1986).

Historical reconstructions of past social impacts based on documentary evidence spearheaded the development of contemporary climate impacts assessment (e.g., Wigley et al., 1981). But, much of what we know about modern climate and society interaction comes from detailed case studies of more recent fluctuations. Events such as the Sahelian drought and El Niños of the 1970s and 1980s evoked a wide array of case studies diagnosing impacts and their relationship to characteristics of the climate fluctuation, and to biophysical or social factors. As a body of research, the accumulated case studies represent a rich knowledge base that has been reviewed by several researchers, but not yet evaluated under a common conceptual framework that might provide a whole greater than the sum of the parts.

A generic technique of climate impact assessment is the development of transfer functions between climate variables and environmental or social elements of concern, which allow impacts to be modeled and predicted. The function might be a simple statement of the sign ($\pm$) of effects from different climate conditions, but is more often expressed in mathematical form. The classic example is a regression equation relating climate conditions to crop yields, based on the statistical relationship between matched data on past climate and yields (Nix, 1985). Correlations between climate and runoff, and vegetation type are also common, and such statistical approaches have been applied to a lesser extent in energy use (Jager, 1985),

manufacturing (Palutikof, 1983), and other commercial activities (Maunder, 1986).

Regression or other statistical tools tend to lose power as assessment goals extend to include second- and third-order impacts, such as effects of reduced crop yields on farm income, and rural income effects on migration. Analytical skill weakens as more variables are included and as the focus moves along the "causal chain of impact" further from the first-order effect. Systems analysis which identifies multiple processes, system states, and linkages has been applied in some regions to track impacts as they ripple through the fabric of socio-economic development, exemplified by Cohen's, (1986) conceptual model for tracking climate impacts in the North American Great Lakes. The difficulty of actualizing such models in a quantitative fashion, however, has limited their use in climate impact assessment chiefly to conceptual aids.

Impact studies based on dynamic simulation modelling of physical and social processes are more rare, though mixed statistical and dynamic modelling approaches have been used in linking climate to runoff and water management (Schaake, 1990), grazing (Le Houerou, 1985), energy (Jager, 1985), and other sectors. Dynamic modelling is most developed in agriculture, where processes such as evapotranspiration, nutrient up-take, sugar production, and management interventions have been simulated (Warrick et al., 1986). Recent efforts to develop and use dynamic simulation models for climate impact assessment are encouraging: e.g., one group is modelling grassland, soil, and climate relationships on the USA Great Plains (Schimel et al., 1990), and Smith et al., (1990) are using both statistical models and simulations of forest dynamics (e.g., competition between species, seed dispersal, and aging) to model global warming effects on vegetation assemblages. However, detailed process models are available only for a few regions, and generally only for agricultural, hydrological, or forest applications.

Simulation models that emulate the dynamic interactions of regional natural and/or social systems are rare, but potentially quite useful, especially in socio-economic impact studies (Lovell and Smith, 1985). Johnson and Gould, (1984) developed a dynamic simulation model of irrigation in the Tigris-Euphrates River Valley to test the effects of climate fluctuations under changing social and technological conditions. Similar regional simulation models are used to aid planning and management of focused resource systems such as large, integrated irrigation systems in the Indus Basin in Pakistan (Ahmad et al., 1989) and on the Nile, and of complex water development plans such as the Mekong cascade (Mekong Secretariat, 1988). Though not constructed for climate impact analysis, some of these models are being used to assess potential climate change effects (Riebsame et al., 1990), and there is good promise for more integrated simulation modelling involving regional economic systems. For example, the Saskatchewan case in the IIASA/UNEP agriculture studies (Williams et al., 1988) illustrated the potential for tracking impacts as they propagate from immediate, biophysical effects into the regional economy using econometric, input-output algorithms. Such models are available for several regions, but most socio-economic simulations are too place-specific to be transferred to other areas.

Climate impacts simulation at the global scale has been especially unsatisfactory to date, but also holds promise for more integrated analysis (Robinson, 1985; Liverman, 1987). Rosenzweig and Parry, (1990) are using the IIASA Basic Linked System, a commodities trade model, to assess the international implications of climate warming, and some national assessments take account of global market conditions (Adams et al., 1990). Such work will become increasingly important as countries not only attempt to assess potential national effects, but consider their role in the international effort to limit and adapt to climate change.

However, good understanding of the linkages and assumptions built into simulation models is necessary to interpret results, especially when socio-economic processes are predicated on assumptions about policy, technology, and markets. Regional or sectoral simulation models can be so complex that large amounts of time are consumed to understand fully the results of just one climate-perturbed model run.

Assessment Skill

The power of theory, quality of data, and availability of methods varies across impact realms, affecting the skill with which climate effects can be measured (Figure 2). Reliability depends on how well the links between climate and biophysical or social systems are understood. Generally, methods are most robust for what Warrick and Riebsame, (1981) called "socio-technical systems" such as industrial, transportation, and energy systems, and for "intermediate" systems such as water resources and agriculture, where there is significant social control over natural processes and data are regularly collected for management purposes. In these cases, large amounts of empirical data can be analyzed and fashioned into transfer functions and simulation models. Assessment skill decreases dramatically for less managed resource systems (e.g., grasslands, forests, natural preserves, and fisheries) - where either data are lacking or theories linking important processes, such as plant or animal migration, to climate are weak.

Differential assessment skill is reflected in the nature of models and transfer functions linking climate to resources. Quantitative models are widely used for energy, water, and crops, while mostly qualitative and categorical statements about climate linkages are available for impact realms such as fisheries and natural preserves (Figure 2). Scale of analysis also cuts across impact realms and affects assessment skill. Generally, studies of local areas, individual sectors, or single elements of natural systems are more skillful, but less applicable to the inherently complex process of resource management. Ability to specify sensitivities and impacts lessens at more comprehensive systemic or spatial scales (e.g., ecosystems or national and global economies) simply because more variables and interactions are involved. We know little about how

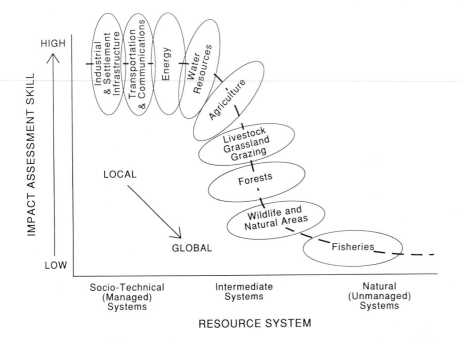

Figure 2: Schematic of assessment skill across climate impact realms.

oceanic fisheries, tropical vegetation, or national and international trade patterns are affected by climate. Understanding of climate/resource interaction is also especially fuzzy at environmental and social interfaces, where complex interactions occur, for example, between crops and soil, fisheries and wetlands, and different social institutions and the resources they manage.

The Special Case of Global Warming Assessments: Applying Regional Scenarios

While scientists might naturally prefer diagnostic analyses and the search for climate-society relationships in empirical data, concern over the impacts of global warming has led to greater emphasis on studies that project impacts of future climate conditions on critical natural and social realms. Thus, we give special attention here to projective techniques, noting also that several other papers in this volume address the topic in more detail. The crux comes, of course, when impact assessors are asked, as they are with increasing frequency, to project climate impacts several decades into the future. Typically this is associated with concerns over global warming, but long-term extrapolations of climate possibilities may also be needed to assess potential impacts on development projects with long time-horizons.

Impact projections are difficult simply because the physical and social future of a region is uncertain. In addition to some idea of evolving social sensitivity and adaptability (as discussed in the next section), analysts must develop at least a broad outline of the future climate to guide their impact projections. Climate scenarios are typically created by one of three approaches, each with strengths and weaknesses to be weighed in choosing scenarios for the impacts problem at hand (see Lamb, 1987; Cohen, 1990):

(1) instrumental or paleo-climatic analogues, where past conditions are assumed to foreshadow the future:

  Historical analogues based on the instrumental record offer great detail, and paleo-climatic analogues created from proxy-data on past climates offer physically realistic and consistent pictures of possible future climate. But neither may capture the full range of potential climate changes simply because future climatic controls may not emulate the past, especially if critical boundary conditions, such as the extent of glaciers or ocean surface, change significantly (Crowley 1990).

(2) global climate models (GCMs) simulating an enhanced greenhouse effect, typically equivalent to a doubling of CO_2 - the so-called "two-times-CO_2" case:

The most compelling method of creating scenarios for global warming studies is to extract regional climate statistics from global climate models (GCMs) simulating an enhanced (typically doubled) greenhouse effect. This approach has become the standard of impact projections, because GCMs produce internally-consistent scenarios based on greenhouse changes per se, a feature lacking in the other approaches. GCM simulations of current climate are relatively good at the hemispheric scale, and for most regions. Yet, some areas are poorly modelled. Current GCMs do not, for example, simulate the monsoons very well. In recent GCM-based studies covering every continent except Antarctica, the GCM baseline climates were found to be good estimates of actual climate in most regions (Smith, et al., 1991). But, large discrepancies were found in southern Asia: GCM baseline climates in the Indus Valley portrayed a precipitation regime seasonally out-of-phase with the actual pattern of monsoon rains. Moreover, some climate variables are modelled better than others in GCMs: while most global climate models produce similar temperature patterns under greenhouse forcing, they yield precipitation changes sufficiently different among models to cause large differences in impact projections (Schlesinger and Mitchell, 1985). The coarseness of GCM spatial resolution (roughly 8°latitude by 10°longitude) is perhaps their chief flaw in regional impact studies (Katz, 1988). The influence of some topographic features such as the Tibetan Plateau is difficult to inject into the models, and some features, for example New Zealand or the Panamanian Isthmus, are neglected at model grid scales. This can make GCM output data difficult to link to regional or local climate, and associated natural or social systems. As might be expected from the fact that the GCMs are reasonably accurate in simulating the broader global climate, the match between actual and modelled data improves as one aggregates climate data into larger areas.

(3) simple increments roughly consistent with evidence from either the past or from global models (e.g., $\pm$ 2 degrees C):

  Incremental scenarios consistent with past climate fluctuations or with global simulations (or both), while not necessarily realistic, have several useful characteristics. They are easy to work with and relatively insensitive to changes in global warming projections as models are refined. They can logically be applied either to individual climate station data or to regionally-smoothed data, but do not capture the potentially complex spatial patterns of regional or global climate change. Incremental scenarios (expressed in simple units of change such as 1 to 2°C. warming and 10–20% more or less precipitation) may facilitate intuitive impacts assessment because they are less intimidating when presented to resource managers; the greater detail and implied precision of analog or model scenarios may suggest more reliability than is warranted.

There are no widely accepted criteria for choosing climate scenarios, though practice favors use of GCMs because they address global warming explicitly, and offer internally consistent climate patterns. Once chosen, the key technical problem is translating scenarios into regional climate change. Instrumental analogues can be applied in a straightforward manner because a close match typically exists between the stations used to create the scenarios and sites under study. Incremental coefficients are merely applied to station records, a method which retains the richness of actual data, but neglects potential spatial gradients of climate change.

Impacts researchers using GCM output or paleo-climatic patterns continue to debate the merits of different ways to transform simulated large-area conditions into regional or even local scenarios. No definitive set of steps has emerged, and among the key questions faced by every assessor are: (1) how to translate data from the large GCM grid cells or general patterns of paleo-reconstructions to localities; (2) what climate statistics to use, and how to apply them, and; (3) how to treat obvious mis-matches between modelled or reconstructed data and actual climate. The first difficulty has been called the "climate inversion problem" because it is an inverted form of the climate modelers' predicament of scaling point meteorological observations up to the global level (Bach, 1988 p. 148). No definitive solution has emerged. Williams et al., (1988) aggregated station data to a grid similar to the GCM. Rosenzweig, (1985) produced isopleth maps of GCM output by interpolating between grid boxes, and compared them to isopleths of observed data. Others simply assign sites or areas (e.g., a small drainage basin) to the grid box in which they reside, or choose a representative grid cell for each region. Cohen et al., (1989) showed how this can be misleading since the grid cells are given unitary values for topography and other characteristics, and thus a plains station might be assigned to a montane grid box in areas where sharp topographic boundaries exist.

Choice of GCM statistics depends on assessment goals, but should also be made with reliability in mind. For example, besides the standard climatological parameters such as temperature and precipitation, GCMs produce data on runoff, soil moisture, ET, and other variables. These data are less reliable simply because the models focus on atmospheric variables, and treat most other physical systems and processes (e.g., soils and hydrology) more simplistically. Prudence suggests using only the primary climate variables and calculating the others as part of the impact assessment process.

Regional impacts researchers may also find discrepancies between GCM base runs (1 x CO_2) and the actual climate, and are forced to assume that the relative difference between the base run of the model and the enhanced CO_2 run is a valid projection of regional climate change, even if the base-run is unrealistic (see, for example, Williams et al. 1988). Temperature and precip-itation differences are then applied to actual data on the relevant time-scale (typically annual or monthly). More detailed transformations that take account of baseline variability, seasonality, and empirical reasoning of how a global warming might affect a particular region have been discussed in the literature (e.g., Bach, 1988; Cohen, 1990), but rarely applied.

Once the scenario has been developed, the tools discussed earlier can be applied to project impacts of the climate change, either instantly or gradually over time. Practice has favored assuming that the change occurs instantly at some point in the future, but more sophisticated studies are beginning to use transient scenarios and to incorporate rates and lags of ecological and social response.

Given these problems with GCM scenarios, it is prudent at this point to apply a variety of scenarios and to interpret projections as sensitivity analyses rather than impact predictions per se. Policy makers should be encouraged to use projections not as forecasts but as an aid to identifying a wide range of possible costs and benefits and a roster of policy options that are likely to help society cope with, or even benefit from, future climate. Of course, impacts are being projected to an uncertain future, and to be meaningful, global warming assessments require attention to changes in regional sensitivity, and the potential for adaptation as the climate changes.

Assessing Regional Sensitivity and Adaptability

The biophysical and social characteristics of regions change through time, regardless of climate, thus increasing or lessening sensitivity to climate impacts. It is equally important to understand changing regional sensitivity and adaptability as it is to understand patterns of climate change per se. Unfortunately, the processes of regional development and social change are no better known than those of climate change--though an extensive literature on development and sustainability has emerged, based on often conflicting theories and methods (see, for example, Corbridge, 1986; Redclift, 1987; World Commission on Environment and Development, 1987; Schoenberger,

1989). Nevertheless, global warming impact studies conducted without some plausible assessment of regional change, and of social responses that attenuate or enhance climate impacts, are as problematic as if they were based on illogical climate scenarios.

Measuring Regional Sensitivity

Social systems evolve quickly compared to biophysical systems, and can exhibit markedly differing vulnerabilities to shocks such as climate change. It has been argued, for instance, that Sahelian cultures were less vulnerable to drought in earlier periods than in the 1960s and 1970s (see Kates, 1981; Watts, 1983). The impact assessor attempting in 1915, when the region's previous major drought ended, to project the effects of the next severe drought would have required clairvoyance to anticipate the social and technical changes that occurred during the intervening fifty years (e.g., rapid population growth, veterinary innovations, and new water well technologies).

It has proven difficult even in hindsight to suggest how social changes, such as population increase or technological innovation, increase or decrease climate sensitivity and resource sustainability in a region. In a series of longitudinal analyses of regions affected by recurrent climate fluctuations, Bowden et al., (1981) found evidence for both worsening and lessening of impacts, though they and others argue that, in most cases, economic development tends to lessen sensitivity to climate fluctuation. Indeed, it is widely held that the less developed countries are the most vulnerable to climate hazards (see Jodha, 1989); this hypothesis underlies the goal of sustainable development. But, how can such characteristics be measured?

The broad sensitivity to climate fluctuation of most natural resource systems has not been analyzed. Assessment of, and ability to extrapolate, climate vulnerability requires an in-depth understanding of the cultural, economic, and environmental characteristics of the region under study. The body of research on climate impacts has now reached a point where we should be able to make more confident statements about differential sensitivity (see Liverman, 1990), but the literature remains weak on trends in vulnerability. At the broadest level of abstraction, it can be argued that resource managers, from small-holder farmers to basin-wide water resource planners, reduce climate sensitivity in two ways: 1) by maintaining flexibility and high diversity, while accepting relatively low output that nonetheless reliably remains above an absolute minimum, or; 2) by seeking large, reliable outputs through greater investment of inputs, storage, or other approaches that "harden" systems (e.g., drought resistant cultivars or oversized reservoirs).

Regional resource systems lie along a spectrum between the two poles of flexible/low-output, and hardened/high-output systems. The most flexible systems are generally "traditional" ones, proven to be quite resilient in the face of climate threats (Jodha, 1989). The high-input systems are thought of as more "developed," they survive shocks by virtue of technological protection and over-investment in buffering capacity. Regional development generally tends to move resource systems from the former to the latter. For example, water use in the Lower Mekong Basin is now mostly accomplished through small, traditional water management systems, while plans exist to create a sophisticated cascade of several main-stem dams that will regularize and expand the scale of water control and use (Mekong Secretariat, 1988). But, which is more resilient to climate fluctuation?

Parry, (1986) offered a diagrammatic approach to visualizing societies' vulnerability to climate impacts based on level of economic integration and development (Figure 3); he argued that greater economic integration and development reduce vulnerability to climate impacts. The analytical challenge here, of course, is to choose criteria for locating societies in the matrix, criteria that hold for a range of cultures. Development analysts have used measures ranging from per capita product to infant mortality, but each measure has weaknesses. However, climate impact assessors should monitor the emerging work on sustainable development for indicators that can be usefully incorporated into regional assessments (World Commission on Environment and Development, 1987; Redclift, 1987; Liverman et al., 1988).

The development process is certainly not uniform in effect on climate vulnerability. Evidence from natural

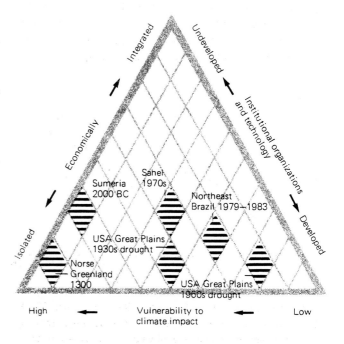

Figure 3: Schematic of climatic vulnerability across economic and institutional environments (from Parry 1986).

hazards research suggests that the greatest vulnerability to environmental shocks appears in transitional stages between traditional and industrial modes (Burton, Kates and White, 1978). At the first World Climate Conference, Kates, (1979) argued that developing countries experience greater impacts from climatic hazards because greater proportions of their modest GNPs can be affected by a single event. Moreover, in the transition phase from traditional to developed economies, they have lost the resiliency of traditional coping mechanisms (e.g., multiple crops, periodic migration), without having fully developed modern replacements (e.g., physical buffers and insurance; see also Kates, 1980b). A hazards approach would hold that regional vulnerability is related to the range and types of responses available to groups facing climate fluctuations. Jodha, (1989) showed that traditional farming systems encompass a wide range of adjustments to loss; they encompass what Brooks, (1986) would call very rich "response pools". However, researchers have noted significant gaps in this roster in certain regions, particularly in marginal areas, due to environmental and political-economic constraints (Heijnen and Kates, 1974; Jodha and Mascarenhas, 1985). Studies of the full range of responses to climate impacts, grounded in evaluation of social contexts, can point to areas of particular social vulnerability (Liverman, 1990). But, trends in vulnerability will remain difficult to characterize.

Climate sensitivity changes over time even in the most developed systems. For example, improvements in large water resource systems can only be made in occasional increments when favorable financial and political conditions prevail. Water systems thus evolve through periods of more or less capacity to absorb climate swings, as illustrated in Figure 4. Here, the relationship between water storage and use provides a rough measure of system vulnerability (similar measures might be applied to other

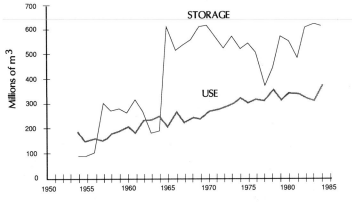

Figure 4: Thin upper line shows historical amounts of storage and demand in the Denver, Colorado, USA water system; Thick lower line shows ratio of storage to demand showing alternating periods of more or less vulnerability punctuated by new dams coming into service in 1956 and 1964 (from Riebsame 1988). Reproduced with permission from Kluwer Academic Publishers.

natural resource systems). Modern systems can be quite sensitive to climate fluctuations, especially where resource supply and demand are closely balanced, as in some heavily developed river basins (Brown, 1988; Riebsame, 1988b), or when hardened systems fail and few or no alternatives exist.

Assessing Regional Adaptability

The next generation of regional climate impact assessments must fully incorporate the potential for adaptations in resources management if they are to provide information on which policy decisions can be made. Unfortunately, the climate impacts and broader development literatures offer few guidelines for assessing regional change over time, and the process of long-term adaptation has not been described in detail. Kates (1985) included underlying natural and social change, and adjustment to climate impacts, in his model of climate-society interaction (Figure 1), but this simple diagram belies the substantial theoretical and methodological difficulties of describing adaptive processes. Biophysical adaptability is related to resiliency of overall ecosystems, and to the rate at which individual species can adapt or migrate (Shugart et al., 1986). Social adaptability to climate change depends on several factors: the resources and innovation that can be marshalled in response to climate fluctuations, peoples' perception of the threat, likely responses of social agents (e.g., individuals, families, economic units, and institutions), and social processes that put some groups at particular risk (Burton, Kates and White, 1978; Schelling, 1983; Sinha et al., 1988) or degrade the resource base (Blaikie, 1985).

A few impact studies explicitly address adaptation. Most economic models include some processes of substitution, supply and demand adjustment, trade and other market processes (e.g., Adams et al., 1990; Rosenberg et al., 1990), based on various assumptions. Generally, however, broader policy or cultural adjustment is neglected, partly because such processes cannot be quantified. The social costs of adjustment transactions, or political tolerance for impacts and adaptations, is difficult to assess, and current models may not be robust beyond certain limits of change. Yet, we may be able to conceive of societies as shifting among the cells in Figure 3, thus providing at least a rough idea of their sensitivity over time.

Social change and adaptation are "black box" processes in most climate impact assessments. The box must be illuminated by the impact assessor, perhaps with theories on the evolution of social vulnerability based, for example, on political-economy (Watts, 1983), markets, labor, institutions and technological innovation (Schoenberger, 1989; Yohe, 1990), social structures and constraints (Douglas, 1986), and perception and choice in natural resources management or hazards mitigation (Whyte, 1986; Emel and Peet, 1989). In contrast to this theoretical messiness, the tactical mechanisms of adaptation are well known to most planners (Titus, 1990). Research on response to past climate extremes, and to other natural hazards, offers a roster of potential adaptive responses, and some idea of the response pools available in different social systems or levels of development (Kates, 1980; Hewitt, 1983). In previous research on social response to natural hazards, adaptations were organized by a typology, developed through dozens of case studies (see White, 1974), which nicely encapsulates potential response to climate change (modified from Burton et al. 1978): (1) modify the hazard; (2) prevent or limit effects; (3) change location/avoid the loss; (4) share the loss; or (5) bear the loss. Modifications that can adapt resource systems to climate change include changes in storage (e.g., carry-over food stocks or water storage in reservoirs) and other types of insurance schemes, sharing between systems and regions (as in development or disaster aid, or inter-basin water transfers), in situ flexibility, and re-location. Additionally, provisions for monitoring and research are themselves a response to climate uncertainty, and can increase adaptability.

Simulation studies hold promise for allowing assessors to compare the climate sensitivity of alternative development paths. Such approaches may focus on one major element of regional development such as water resources. For example, Ti and Phien, (1990) are evaluating the robustness of alternative Mekong River development schemes under different climate scenarios derived from GCMs. Wescoat and colleagues in Pakistan (see Riebsame et al., 1990) are using the Indus Basin Model to assess impacts and adjustments in a large, integrated irrigation system; climate changes are impressed upon system development as planned between 1980 and 2000 by the government. On a broader scale, IIASA's Basic Linked System model of global trade is being used to assess global food supply adjustments to greenhouse warming (Rosenzweig and Parry, 1990). But, it can be difficult to evaluate the assumptions about development incorporated into such models, and how those assumptions affect the apparent adaptability of different development paths. Nevertheless, the resiliency of alternative regional developments must be incorporated into climate impact studies if they are to be useful in guiding decision-makers toward sustainable development in the face of an uncertain climate.

Finally, a more integrated approach to assessing change in regional sustainability is needed for future climate impact studies. This requires a synthesis of individual actions and institutional constraints and incentives that enhance or degrade natural resources. Here impacts analysts have choices and problems similar to those associated with projecting the future evolution of climate; extrapolation of past trends, analogues, or simulation

models might all be utilized. Given the regional differentiation of culture and development, however, it might be best to concentrate multidisciplinary expertise on selected regions to build historical trajectories of development that can be extrapolated as a basis for projective climate impact assessments. A new approach to regional studies is emerging that details the historical evolution of socio-natural landscapes, seeking explanation of regional trends in key, interacting social and ecological factors (see Turner et al., 1990b). Selected cases include land degradation in New Guinea and other parts of the tropical eastern Pacific (Blaikie and Brookfield, 1987); the Himalaya (Ives and Meserli, 1989), the Sahel (Franke and Chasin, 1980), and the world's mid-latitude grasslands such as the North American Great Plains (Clark, 1956; Worster, 1979; Riebsame, 1990b and in press). Identification of such critical zones, and analysis of their evolving sensitivity to environmental change (see Turner et al., 1990a), would provide a useful focus for a new generation of climate impact studies that fully incorporate changing sensitivity and adaptability. Such studies could also be applied in an analog framework: e.g., the USA Great Plains as analog for future development of the Asian steppe and other less-developed grasslands.

Regional Impact Assessment in Support of Sustainable Development

Because of the great uncertainty in estimates of future climate changes, it is premature at this time for decision-makers to make drastic changes in climate-sensitive development or resource management plans in anticipation of future impacts. However, recent experience clearly demonstrates the value of improved ability to cope with current climate variability, and associated hazards, and illustrates the usefulness of regional climate impact studies in sustainable natural resources management and policy-making.

Improving the Use of Impact Studies in Regional Planning

The growing body of regional climate impact studies represents a potentially valuable contribution to policy-making and planning for the sustainable development of climate-sensitive natural resources. Nevertheless, most climate-related policies around the world are based on decision-makers' informal perceptions of climate and its impacts, rather than on consideration of the impacts research. This gulf between research and practice is common in many fields, but may be especially pernicious in a changing climate, where perceptions and past experience will not lead to good decisions. Unfortunately, it is difficult for policy makers to comprehend the impacts of climate variation in their full complexity. People may misperceive the problem due to personal experience, cognitive biases, or lack of scientific knowledge. Faced with a complex climate threat, decision-makers may adopt simple and appealing, but perhaps misleading, interp-retations as a guide for policy (Fischof, 1981; Whyte, 1985). Natural resource planners tend to expect climate anomalies to pass and the climate to "return to normal", an unlikely prospect under global warming. Finally, policy-makers face multiple demands for their attention, and climate is generally a low-salience element--at least until its effects become extreme.

Integrating Impact Assessment and Policy

Though many studies do not go beyond identification of first-order impacts, and are not policy-oriented per se, recommendations for coping better with climate fluctuation are a natural out-growth of impact assessment. Some recent impact studies do advance policy suggestions, but only a few have been capable of effectively informing policy by creating links to, and involving, policy-makers. The IIASA/UNEP Project (Parry et al., 1988) provides a good example of liaison to policy. The Iceland case study, for instance, discusses at length the implications for agricultural policy and offers a rather comprehensive set of suggestions (Bergthorsson et al., 1988). The Brazilian Northeast case study proposes an integrated rural development strategy aimed at increasing resiliency and reducing vulnerability to droughts (Magalhaes et al., 1988). An effort was made in these studies to divulge the recommendations to the public and to policy makers, through policy workshops and written communication.

Through the World Climate Impacts Program, UNEP disseminates regional climate impact assessment methods via "roving seminars," and an impacts assessment network. UNEP is also sponsoring studies meant to help policy-makers assess options in the face of climate change by linking regional impact studies with policy exercises. A recent UNEP Project on Climate Impacts and Governmental Responses in Brazil (Magalhaes and Neto, 1989) employed policy-maker workshops for this purpose.

In some cases governments have sought impact assessments, as in the US Congress request for an analysis of national global warming effects (Smith and Tirpak, 1989), or recent studies in Brazil (e.g., analysis of the impact of the 1988 heavy rains in Rio de Janeiro) conducted as a part of a policy process linking research and planning (see Cunha et al., 1989). In other circumstances, impact assessments are prepared independently, but made available to policy agencies. Impact assessment and policy-making are, nevertheless, distinct activities, but what separates them is less content and more the power of choice, implementation and enforcement present within the domain of policy. For instance, problem identification and specification--ranging from future climate scenarios to measures of impacts--is fundamental to impact assessment, and must be based on theory and empirical information. In policy formulation, however, problem definition is a point of departure for detailed analysis of alternative responses. Moreover, while alternatives must be presented in impact studies, policy analysis requires choice of only one or perhaps a few lines of strategy, within the confines of economic and public welfare.

Impact studies should not necessarily detail policy, but should seek to offer policy-makers a framework within which responses to climate change can be formulated. In particular, regional impact studies should increase attention to measuring and specifying vulnerability (by area, population, activity, etc.), and to identifying alternative responses and enlarging the range of responses considered by policy makers. Impact assessments might also explore linkages between adaptation and greenhouse gas limitation policies, seeking responses that serve both functions. Regional impact studies can be made more useful to policy through these steps:

- Impact assessors, in cooperation with relevant policy-makers, identify "problems" such as: impacts of current climate fluctuations, potential future climate changes, vulnerable population groups, areas, or economic and social activities becoming more climate-sensitive

- Analysts demonstrate the link between climate and biophysical and social variables, and project them to the future if necessary. Then, a wide range of adjustments to reduce the problems - or to achieve general goals such as reduced vulnerability or increased resilience - and possible tools to be used, are proposed by both assessors and policy-makers. These responses are then analyzed to determine their ameliorative ability and feasibility.

- The results are communicated in a compelling manner (e.g., maps of forest change, graphs of altered water flows) so that policy makers can choose goals, select tools, and detail action plans.

The establishment of such interaction between impact assessment and policy deserves high priority in efforts to improve government effectiveness in responding to impacts of climate variation and potential change. Such linkage is not obtained easily: it has to be actively pursued and promoted by assessors and policy makers.

Improving Regional Assessment Capabilities

Climate fluctuations and hazards affect every region of the world, even in the absence of global warming. Climate impact assessments are useful guides to government response to climate hazards as well as to creating more resilient plans for regional development. Demand for impact studies will continue to increase as the call to action on global warming is amplified through motion toward an international greenhouse gas convention, and preparations for the 1992 World Conference on Environment and Development. But, many countries are not prepared to respond to climate change, or to calls for its mitigation, because they lack studies of regional sensitivity to climate fluctuation and change, the potential impacts of global warming, and the range of feasible responses.

UNEP and WMO have recognized the need to assess climate effects on society by initiating the World Climate Impacts Program (WCIP) as part of the broader World Climate Program (WCP). Several national climate programs prompted by the WCP have an impacts component, but most developing countries have no formal climate programs, and poorly developed capacity for impact assessment. Yet, they must also cope with climate impacts and the effects of future climate change and must develop indigenous climate impact assessment capabilities appropriate to their physical and cultural setting. This process could be hastened by better transfer of methods from experts in developed countries, so that at least a small team of experts able to conduct diagnostic studies, sensitivity analyses, and projective assessments along the lines described above can be created as developing countries play their role in social response to the threat of global warming

Key steps to improving national capacity to conduct regional impact studies include:

- Identify the experts and agencies that should be involved in a multidisciplinary assessment, create multidisciplinary assessment teams to take on specific responsibilities for developing climate impact assessments.

  This step may include forming an impacts assessment network that can be called on for special studies, advice, and objective review of methods and findings. It also might include building a roster of the agencies and personnel who should be called on to take part in ad hoc assessments.

- Assess the full range of data needs and availability. This is a critical step that is often neglected. The data necessary for climate impact studies (e.g., those regarding climate itself, crop yields, water use, energy production, income and human health) are often collected and archived by different agencies or research groups. The data is inevitably uneven in coverage, quality and accessibility. Impact assessors need either to catalog the data, noting its coverage and availability, or to create a central archive or clearing-house for key data sets. A wide-ranging impacts data catalog can be compiled relatively easily

and distributed to impact assessors as a guide to data sources.

- Develop links between assessors and policy makers. Decision makers in a position to reduce climate vulnerabilities must feel that the assessment process and its products are credible. Thus, assessors and decision makers must continually confer to assure that research addresses critical problems and that it produces realistic and feasible solutions. Additionally, researchers must insure that assessment results are succinct, understandable and conveyed quickly to decision makers.

In order for such recommendations to be followed, countries with little impact assessment experience may need a transfer of technology from others, and must be convinced for their own sake of the value of climate impacts research. As Glantz et al. (1985) showed, the organization of multidisciplinary studies is no mean feat, and the experiences of assessment teams should be shared between countries. It must be taken into account that the great differences between industrialized and developing countries cause different levels of awareness, vulner-abilities, priorities and instruments of action. The majority of the developing countries are still struggling to gather information to help them assess the threat of climate change and choose a response posture. International co-operation would speed the implementation of national climate programs in developing countries, but they will only succeed when climate is recognized as a national concern and if the benefits for the country can be clearly demonstrated. Careful regional impact assessments should provide this key insight.

REFERENCES

Ahmad, M., A. Brooke, and G. Kutcher, (1989) Water Sector Investment Planning Study: Guide to the Indus Basin Model Revised (IBMR). Environment, Operations, and Strategy Division, World Bank, Washington, DC. 113 pp.

Adams, R.M., C. Rosenzweig, R.M. Peart, J.T. Ritchie, B.M. McCarl, J.D. Glyer, R.B. Curry, J.W. Jones, K.J. Boote, and L.H. Allen (1990) Global climate change and US agriculture. Nature 345: 219–224.

Anderson, J.R. (1978) Impacts of Climate Variability on Australian Agriculture. In Commonwealth Scientific and Industrial Research Organization (ed.) The Impact of Climate on Australian Society and Economy, pp. 1–36. Mordialloc, Australia.

Bach, W. (1988) Development of Climatic Scenarios from General Circulation Models. In M.L. Parry, T.R. Carter, and N.T. Konijn (eds.) The Impact of Climate Variations on Agriculture, vol. 1, pp. 125–157. Dordrecht: Kluwer Academic Publishers.

Bergthorsson, P. et al., (1988) The Effects of Climate Variation on Agriculture in Iceland. In: Parry, M.L., Carter, T., Konijn,

N.T. (eds.) The Impact of Climate Variations on Agriculture, Vol. 1-Assessments in Cool Temperate and Cold Regions. Kluwer Academic Publishers, Dordrecht, Netherlands.

Bowden, M.L., R.W. Kates, P.A. Kay, W.E. Riebsame, R.A. Warrick, D.L. Johnson, H.A. Gould, and D. Weiner (1981) The effect of climate fluctuations on human populations: Two hypotheses. In Climate and History, ed. T.M.L. Wigley, M.J. Ingram and G. Farmer, 479–513. Cambridge UK: Cambridge University Press.

Blaikie, P.M. (1985) The Political Economy of Soil Erosion. Longman: London.

Blaikie, P.M. and H.C. Brookfield (1987) Land Degradation and Society. Methuen: London.

Brooks, H. (1986) The Typology of Surprise in Technology, Institutions and Development. In W.C. Clark and R.E. Munn (eds.) Sustainable Development of the Biosphere, pp. 325–350. Cambridge University Press.

Brown, B.G. (1988) Climate Variability and the Colorado River Compact: Implications for Responding to Climate Change. In M.H. Glantz (ed.) Societal Response to Regional Climate Change, 279–305. Boulder, CO: Westview Press.

Burton, I, R.W. Kates, and G.F. White (1978) The Environment as Hazard. Oxford: New York.

Caviedes, C.N. (1975) El Niño 1972: Its Climatic, Ecological, Human, and Economic Implications. Geographical Review 65: 493–509.

Caviedes, C.N. (1984) Geography and the lessons from El Niño. Professional Geographer 36: 428–436.

Clark, A.H. (1956) The Impact of Exotic Invasion of the Remaining New World Midlatitude Grasslands. In W.L. Thomas (ed.) Man's Role in Changing the Face of the Earth, pp. 721–736. University of Chicago Press.

Cohen, S.J. (1986) Climatic change, population growth, and their effects on Great Lakes water supplies. The Professional Geographer 38:317–23.

Cohen, S.J. (1988) Great Lakes levels and Climate Change: Impacts, Responses, and Futures. In M.H. Glantz (ed.) Societal Response to Regional Climate Change, 143–167. Boulder, CO: Westview Press.

Cohen, S.J. (1990) Bringing the global warming issue closer to home: the challenge of regional impact studies. Bulletin of the American Meteorological Society 71: 520–526.

Cohen, S.J., L.E. Walsh, and P.Y.T. Louie (1989) Possible Impacts of Climatic Warming Scenarios on Water Resources in the Saskatchewan River Sub-basin. Canadian Climate Center Report No. 89–9, Atmospheric Environment Service, National Hydrology Research Center, Saskatoon.

Corbridge, S. (1986) Capitalist World Development. Totowa, NJ: Rowman and Littlefield Publishers.

Crowley, T.J. (1990), "Are there any satisfactory geologic analogs for a future greenhouse warming?" Jour. Clim. 3, 1282–1292.

Cunha, L.R.A, Santos, M.M. and Castro Filho, J.F. (1989) An Integrated Program for Flood-Damage Reconstruction and Prevention. In: Magalhaes, A.R. and E. B. Neto, Socioeconomic Impacts of Climate Variations and Policy Responses in Brazil. UNEP/SEPLAN, Fortaeza, Brazil.

de Vries, J. (1980) Measuring the impact of climate on history: the search for appropriate methodologies. Jour. Interdisciplinary Hist. 10: 599–630.

Douglas, M. (1986) How Institutions Think. Syracuse, NY: Syracuse University Press.

Easterling, W.E., M.L. Parry, and P.R. Crosson (1989) Adapting Future Agriculture to Changes in Climate. In N.J. Rosenberg, W.E. Easterling, P.R. Crosson, and J. Darmstadter (eds.) Greenhouse Warming: Abatement and Adaptation, pp. 91–104. Resources for the Future, Washington, DC.

Emel, J. and R. Peet (1989) Resource Management and Natural Hazards. In R. Peet and N. Thrift (eds.) New Models in Geography, Vol. One, pp. 49–76. London: Unwin and Hyman.

Farhar-Pilgrim, B. (1985) Social Analysis. In R.W. Kates, J. Ausubel, and M. Berberian (eds.) Climate Impact Assessment: Studies of the Interaction of Climate and Society, pp. 323–350. New York: John Wiley and Sons.

Fischoff, B. (1981) Hot air: The psychology of CO_2-induced climatic change. In Cognition, Social Behavior and the Environment, ed. J.H. Harvey, 163–84. Hillsdale, NJ: Lawrence Erlbaum Associates.

Franke, R.W. and B.H. Chasin (1980) Seeds of Famine: Ecological Destruction and the Development Dilemma in the West African Sahel. Montclair, NJ: Allanheld and Osmun.

Garcia, R.V. (1981) Drought and Man: The 1972 Case History, Volume 1: Nature Pleads Not Guilty. Oxford: Pergamon Press.

Glantz, M.H. and J.H. Ausubel (1984) Impact assessment by analogy: comparing the impacts of the Ogallala Aquifer depletion and CO_2-induced climate change. Environmental Conservation 11: 123–131.

Glantz, M.H., J. Robinson, and M.E. Krenz (1985) Recent Assessments. In Kates, R.W., J. Ausubel, and M. Berberian (eds.) Climate Impact Assessment: Studies of the Interaction of Climate and Society, pp. 565–598. New York: John Wiley and Sons.

Glantz, M., R. Katz, and M. Krenz (1987) Societal Impacts Associated with the 1982–83 Worldwide Climate Anomalies. Boulder, CO: National Center for Atmospheric Research.

Glantz, M.H., ed (1988) Societal Responses to Regional Climate Change. Westview Press, Boulder, CO, USA.

Gleick, P.H. (1987) Regional hydrological consequences of increases of atmospheric CO_2 and other trace gases. Climatic Change 110: 137–61.

Heijnen, J. and R.W. Kates (1974) Northeast Tanzania: Comparative Observations Along a Moisture Gradient. In G.F. White (ed.) Natural Hazards: Local, National, Global, pp. 105–114. New York: Oxford University Press.

Hewitt, K. (1983) Interpretations of Calamity. London: Allen and Unwin.

Intergovernmental Panel on Climate Change: Working Group II (1990) Potential Impacts of Climate Change. WMO/UNEP. Geneva.

Ives, J.D. and B. Messerli (1989) The Himalayan Dilemma: Reconciling Development and Conservation. London: Routledge.

Jager, J. (1985) Energy Resources. In Kates, R.W., J. Ausubel, and M. Berberian (eds.) Climate Impact Assessment: Studies of the Interaction of Climate and Society, pp. 215–249. New York: John Wiley and Sons.

Jodha, N.S. and A.C. Mascarenhas (1985) Adjustment in Self-Provisioning Societies. In Kates, R.W., J. Ausubel, and M. Berberian (eds.) Climate Impact Assessment: Studies of the Interaction of Climate and Society, pp. 437–468. New York: John Wiley and Sons.

Jodha, N.S. (1989) Potential Strategies for Adapting to Greenhouse Warming: Perspectives from the Developing World. In N.J. Rosenberg, W.E. Easterling, P.R. Crosson, and J. Darmstadter (eds.) Greenhouse Warming: Abatement and Adaptation, pp. 147–158. Resources for the Future, Washington, DC.

Johnson, D.L. and H. Gould (1984) Effects of Climate Fluctuation on Human Populations: Study of Meso–potamian Society. In A.K. Biswas (ed.) Climate and Development. Tycooly International Publishing, Dublin.

Kates, R.W. (1979) Climate and Society: Lessons from Recent Events. Proceedings of the World Climate Conference. WMO, Geneva.

Kates, R.W. (1980a) Climate and society: Lessons from recent events. Weather 35(1): 17–25.

Kates, R.W. (1980b) Disaster Reduction: Links between Disaster and Development. In L. Berry and R.W. Kates (eds.) Making the Most with the Least: Alternative Ways to Develop, pp. 135–169. New York: Holmes and Meir.

Kates, R.W. (1981) Drought in the Sahel: competing views as to what really happened in 1910–14 and 1968–74. Mazingira 5: 72–83.

Kates, R.W. (1985) The Interaction of Climate and Society. In R.W. Kates, J. Ausubel, and M. Berberian (eds.) Climate Impact Assessment: Studies of the Interaction of Climate and Society, pp. 3–36. New York: John Wiley and Sons.

Kates, R.W. J. Ausubel, and M. Berberian, eds. (1985) Climate Impact Assessment: Studies of the Interaction of Climate and Society. New York: John Wiley and Sons.

Katz, R.W. (1988) Statistics of climate change: Implications for scenario development. In M.H. Glantz (ed.) Societal Response to Regional Climate Change, 95–112. Boulder, CO: Westview Press.

Kawasaki, T. (1985) Fisheries. In R.W. Kates, J. Ausubel, and M. Berberian (eds.) Climate Impact Assessment: Studies of the Interaction of Climate and Society, pp. 131–150. New York: John Wiley and Sons.

Keith, V.F., C. DeAvila, and R.M. Willis (1989) Effect of Climatic Change on Shipping within Lake Superior and Lake Erie. In J.B. Smith and D.A. Tirpak (eds.) The Potential Effects of Global Climate Change on the United States, EPA-230-05-89-058. Washington, DC: US Environmental Protection Agency.

Lamb, P.J. (1987) On the development of regional climatic scenarios for policy-oriented climate impact assessment. Bulletin of the American Meteorological Society 68: 1116–23.

Le Hourerou, H.H. (1985) Pastoralism. In R.W. Kates, J. Ausubel, and M. Berberian (eds.) Climate Impact Assessment: Studies of the Interaction of Climate and Society, pp. 155–185. New York: John Wiley and Sons.

Linder, K.P. and M.R. Inglis (1989) The Potential Impacts of Climate Change on Electric Utilities: Regional and National Estimates. In J.B. Smith and D.A. Tirpak (eds.) The Potential Effects of Global Climate Change on the United States, EPA-

230-05-89-058. Washington, DC: US Environmental Protection Agency.

Liverman, D.M. (1987) Forecasting the impact of climate in food systems: Model testing and model linkage. Climatic Change 11: 267–85.

Liverman, D.M. (1990) Drought impacts in Mexico: climate, agriculture, technology, and land tenure in Sonora and Puebla. Annals of the Association of American Geographers 80: 49–72.

Liverman, D.M., M.E. Hanson, B.J. Brown, and R.W. Merideth (1988) Global sustainability: toward measurement. Env. Management 12: 133–143.

Lovell, C.A.N. and V.K. Smith (1985) Microeconomic Analysis. In R.W. Kates, J. Ausubel, and M. Berberian (eds.) Climate Impact Assessment: Studies of the Interaction of Climate and Society, pp. 293–321. New York: John Wiley and Sons.

Magalhaes, A.E. et al. (1988) The Effects of Climate Variations on Agriculture in Northeast Brazil. In Parry, M.L., Carter, T., Konijn, N.T., eds. The Impact of Climatic Variations on Agriculture, Vol. II: Semi-Arid Regions. Kluwer Academic Publishers, Dordrecht.

Magalhaes, A.E. (1989) Government Strategies in Response to Climate Variations: Droughts in Northeast Brazil. In Magalhaes A.R. and E. B. Neto, eds. Socioeconomic Impacts of Climatic Variations and Policy Response in Brazil, pp. 2.01–2.36. UNEP/SEPLAN, Fortaleza, Brazil.

Maini, J.S. (1988) Forests and Atmospheric Change. In World Meteorological Organization (ed.) Proceedings of the Conference on the Changing Atmosphere: Implications for Global Security, pp. 193–208. WMO Publication No. WMO/OMM 710. Geneva.

Margolis, M. (1980) Natural disaster and socioeconomic change: Post-frost adjustments in Parana, Brazil. Disasters 4: 231–235.

Maunder, W.J. and J. Ausubel (1985) Identifying Climate Sensitivity. In R.W. Kates, J. Ausubel, and M. Berberian (eds.) Climate Impact Assessment: Studies of the Interaction of Climate and Society, pp. 85–101. New York: John Wiley and Sons.

Maunder, W.J. (1986) The Uncertainty Business: Risks and Opportunities in Weather and Climate. London: Methuen.

Mekong Secretariat (1988) Perspectives for Mekong Development: revised Indicative Plan. Interim Committee for the Coordination of Investigations of the Lower Mekong Basin, Bangkok.

Morrisette, P.M. (1988) The stability bias and adjustment to climatic variability: The case of the rising level of the Great Salt Lake. Applied Geography 8: 171–89.

Morrisette, P.M. (1988) The Rising Level of the Great Salt Lake: An Analogue of Societal Adjustment to Climate Change. In M.H. Glantz (ed.) Societal Response to Regional Climate Change, 169–195. Boulder, CO: Westview Press.

Mortimore, M. Adapting to Drought. Cambridge University Press.

New Zealand Ministry for the Environment (1988) The Effects of Climate Change on New Zealand. Wellington.

Nemec, J. and J.C. Schaake (1982) Sensitivity of water resource systems to climate variation. Hydro. Sci. 27: 327–343.

Nix, H.A. (1985) Agriculture. In R.W. Kates, J. Ausubel, and M. Berberian (eds.) Climate Impact Assessment: Studies of the

Interaction of Climate and Society, pp. 105–130. New York: John Wiley and Sons.

Novaky, B., C. Pachner, K. Szeztay, and D. Miller (1985) Water Resources. In R.W. Kates, J. Ausubel, and M. Berberian (eds.) Climate Impact Assessment: Studies of the Interaction of Climate and Society, pp. 187–214. New York: John Wiley and Sons.

Palutikof, J. (1983) The impact of weather and climate on industrial production in Great Britain. Journal of Climatology 3: 65–79.

Parry, M.L. (1981) Climatic change and the agricultural frontier: A research strategy. In Climate and History, ed. T.M.L. Wigley, M.J. Ingram, and G. Farmer, 319–36. Cambridge UK: Cambridge University Press.

Parry, M.L. (1986) Some Implications of Climatic Change for Human Development. In W.C. Clark and R.E. Munn (eds.) Sustainable Development of the Biosphere, pp. 378–407. Cambridge University Press.

Parry, M.L. , T.R. Carter, and N.T. Konijn, eds. (1988) The Impact of Climate Variations on Agriculture, Vols. 1 and 2. Dordrecht, The Netherlands: Kluwer Academic Publishers.

Parry, M.L. and N.J. Read (1988) The Impact of Climate Variability on UK Industry. AIR Report 1. Atmospheric Impacts Research Group, University of Birmingham, UK.

Peters, R.L. and T.D.S. Darling (1985) "The greenhouse effect and natural preserves." Bioscience 35: 707–717.

Redclift, M. (1987) Sustainable Development: Exploring the Contradictions. London: Methuen.

Revelle, R.R., and P.E. Waggoner. 1983. Effects of a carbon dioxide-induced climatic change on water supplies in the western United States. In Changing Climate, ed. Carbon Dioxide Assessment Committee, 419–31. Washington, DC: National Academy Press.

Rhodes, S.L., D. Ely, and J.A. Dracup. 1984. Climate and the Colorado River: The limits of management. Bulletin of the American Meteorological Society 65: 682–91.

Riebsame, W.E. (1985) Research in Climate-Society Interaction. In R.W. Kates, J. Ausubel, and M. Berberian (eds.) Climate Impact Assessment: Studies of the Interaction of Climate and Society, pp. 69–84. New York: John Wiley and Sons.

Riebsame, W.E. (1988a) The potential for adjusting natural resources management to climatic change: A national assessment. In The Potential Effects of Global Climate Change on the United States: Appendices, ed. J. Smith and D. Tirpak, Appendix J. Washington, DC: US Environmental Protection Agency.

Riebsame, W.E. (1988b) Adjusting water resources management to climate change. Climatic Change 12: 69–97.

Riebsame, W.E. (1988c) Assessing the Social Impacts of Climate Change: A Guide to Climate Impact Studies. Nairobi: United Nations Environment Program.

Riebsame, W.E. (1990a) Anthropogenic climate change and a new paradigm of natural resources management. Professional Geographer 42: 1–12.

Riebsame, W.E. (1990b) The Great Plains. In B.L. Turner and W.B. Myers (eds.) The Earth as Transformed by Human Action. Cambridge University Press.

Riebsame, W.E. (in press) If the climate should change: the future of the Great Plains in a deteriorating climate. Jour.of Great Plains Research.

Riebsame, W.E. for the International Rivers Project Team (1990) Complex River Basin Management in a Changing Climate. In Progress Reports to the US Environmental Protection Agency on the Potential International Impacts of Climate Change. Washington, DC.

Robinson, J. (1985) Global Modelling and Simulation. In R.W. Kates, J. Ausubel, and M. Berberian (eds.) Climate Impact Assessment: Studies of the Interaction of Climate and Society, pp. 469–492. New York: John Wiley and Sons.

Rosenberg, N.J. et al. (1990) Processes for Identifying Regional Influences of and Responses to Increasing Atmospheric CO_2 and Climate Change--The MINK Project. Working Paper No. 1: Background and Baselines. Resources for the Future, Washington, DC.

Rosenzweig, C. (1985) Potential CO_2-induced climate effects on North American wheat-producing regions. Climatic Change 7: 367–89.

Rosenzweig, C. and M.L. Parry (1990) International Agriculture, Global Food Trade, and Vulnerable Regions. In Progress Reports to the US Environmental Protection Agency on the Potential International Impacts of Climate Change. Washington, DC.

Schaake, J.C. (1990) From Climate to Flow. In P.E. Waggoner (ed.) Climate Change and US Water Resources, pp. 177–206. New York: John Wiley and Sons.

Schelling, T. (1983) Climatic change: Implications for welfare and policy. In Changing Climate, ed. Carbon Dioxide Assessment Committee, 449–82. Washington, DC: National Academy Press.

Schimel, D.S., W.J. Parton, T.G.F. Kittel, D.S. Ojima, and C.V. Cole (1990) Grassland biogeochemistry: links to atmospheric processes. Clim. Change 17: 13–25.

Schlesinger, M.E., and J.F.B. Mitchell. 1985. Model projection of the equilibrium climatic response to increased carbon dioxide. In Projecting the Climate Effects of Increasing Carbon Dioxide, ed. M.C. MacCracken and F.M. Luther. DOE/ER–0237. Washington, DC: US Department of Energy.

Schoenberger, E. (1989) New Models of Regional Change. In R. Peet and N. Thrift (eds.) New Models in Geography, Vol. One, pp. 115–141. London: Unwin and Hyman.

Schwartz, H.E. (1977) Climatic change and water supply: How sensitive is the Northeast? In Climate, Climatic Change, and Water Supply, ed. Panel on Water and Climate, 111–20. Washington, DC: National Academy of Sciences.

Scientific Committee on Problems of the Environment-- SCOPE (1978) Report of the Workshop on the Climate/Society Interface, held 10–14 December, 1978 in Toronto. SCOPE Secretariat, Paris.

Shugart, H.H., M. Ya. Antonosvsky, P.G. Jarvis, and A.P. Sandford. 1986. CO2, climatic change and forest ecosystems. In The Greenhouse Effect, Climatic Change, and Ecosystems, ed. B. Bolin, B.R. Doos, J. Jager, and R.A. Warrick, 475–521. New York: John Wiley and Sons.

Smith, J.B. and D.A. Tirpak, eds. (1989) The Potential Effects of Global Climate Change on the United States. EPA-230-05-89-01. Washington, DC: U.S. Environmental Protection Agency.

Smith, J.B., R. Benloff, J.G. Titus, T. Smith, C. rosenzweig, and W.E. Riebsame (1991), Potential International Impacts of Global Warming. Paper presented at the International Conference on Climatic Impacts on the Environment and Society. January 27–February 1, University of Tsuhuba, Japan.

Smith, T.M., H.H. Shugart, and P.N. Halpin (1990) Global Forests. In Progress Reports to the US Environmental Protection Agency on the Potential International Impacts of Climate Change, pp. 25–32. Washington D.C.

Sinha, S.K., N.H. Rao, and M.S. Swaminathan (1988)Food Security in the Changing Global Climate. Proceedings of the World Conference on the Changing Atmosphere, pp. 167–191. WMO/OMM No. 710. World Meteorological Organization, Geneva.

Ti, L.H. and H.N. Phien (1990) Study of the Effects of Climate Change on Development of the Lower Mekong Basin. Draft Report.

Titus, J.G. (1990) Strategies for adapting to the greenhouse effect. American Planning Assoc. Jour. (Summer): 311–323.

Turner, B.L., R.E. Kasperson, W.B. Myer, K.M. Dow, D. Golding, J.X. Kasperson, R.C. Mitchell, and S.J. Ratick (1990a) Two types of global environmental change: definitional and spatial-scale issues in their human dimensions. Global Env. Change 1: 14–22.

Turner, B.L., W.C. Clark, R.W. Kates, J.F. Richards, J.T. Mathews, and W.B. Meyer, eds. (1990b) The Earth as Transformed by Human Action. Cambridge University Press.

Utterstrom, G. (1955) Climate fluctuations and population problems in early modern history. Scandinavian Economic History review 3: 3–47.

Waggoner, P.E. (1990) Climate Change and U.S. Water Resources. New York: John Wiley and Sons.

Walker, J.C., T.R. Miller, G.T. Kingsley, and W.A. Hyman (1989) Impact of Global Climate Change on Urban Infrastructure. In J.B. Smith and D.A. Tirpak (eds.) The Potential Effects of Global Climate Change on the United States, Appendix H, pp. 2/1–2/37. EPA-230-05-89-058. Washington, DC: U.S. Environmental Protection Agency.

Warrick, R.A., and W.E. Riebsame (1981) Societal response to CO_2-induced climate change: Opportunities for research. Climatic Change 3: 387-428.

Warrick, R.A., R.M. Gifford, and M.L. Parry (1986) CO2, climatic change, and agriculture. In The Greenhouse Effect, Climatic Change, and Ecosystems, ed. B. Bolin, B.R. Doos, J. Jager, and R.A. Warrick, 393–473. New York: John Wiley and Sons.

Warrick, R.A., and G. Farmer (1990) The greenhouse effect, climate change and rising sea level: implications for development. Transactions of the Institute of British Geographers 15: 5–20.

Watts, M. (1983) Silent Violence: Food, Famine and Peasantry in Northern Nigeria. Berkeley: University of California Press.

White, G.F. (1974) Natural Hazards: Local, National, Global. New York: Oxford University Press.

Whyte, A.V.T. (1985) Perception. In Climate Impact Assessment: Studies in the Interaction of Climate and Society, ed. R.W. Kates, J.E. Ausubel, and M. Berberian, 403–36. New York: John Wiley and Sons.

Whyte, A.V.T. (1986) From Hazard Perception to Human Ecology. In R.W. Kates and I. Burton (eds.) Geography, Resources and Environment, vol II, pp. 240–271. University of Chicago Press.

Wigley, T.M.L., M.J. Ingram, and G. Farmer (1981) Climate and History. Cambridge University Press.

Williams, G.D.V., R.A. Fautley, K.H. Jones, R.B. Stewart, and E.E. Wheaton (1988) Estimating Effects of Climate Change on Agriculture in Saskatchewan, Canada. In M.L. Parry, T.R. Carter, and N.T. Konijn (eds.) The Impact of Climate Variations on Agriculture, vol. 1, pp. 221–279. Dordrecht: Kluwer Academic Publishers.

Woodman, J.N. and C.S. Furiness (1989) Potential Effects of Climate Change on US Forests: Case Studies of California and the Southeast. In J.B. Smith and D.A. Tirpak (eds.) The Potential Effects of Global Climate Change on the United States, Appendix D, pp. 6/1–6/42. EPA-230-05-89-054. Washington, DC: US Environmental Protection Agency.

World Commission on Environment and Development (1987) Our Common Future. Oxford University Press.

Worster, D. (1979) Dust Bowl: The Southern Great Plains in the 1930s. New York: Oxford University Press.

Yohe, G. (in press) Imbedding Dynamic Response with Impacct Information into Static Portraits of the Regional Impacts of Climate Change. In J. Reilly (ed.) Global Change: Environmental Issues in Agriculture, Forestry, and Natural Rsources. US Department of Agriculture, Washington D.C.

Climate Change and the United Nations Conference on Environment and Development

by Maurice F. Strong *(Secretary General - UNCED)*

It is a great pleasure for me to be able to address this conference as you near the end of this important first phase of your deliberations. Through your work, and that of the Inter-governmental Panel on Climate Change, we are now much closer to understanding the science of climate, the prospects of climate change and the risks to which it gives rise. This will be of immense help in guiding the development of policy measures to reduce risks and ameliorate consequences of climate change - the next major step which confronts the world community.

During the past week, speaker after speaker stressed that the evidence of climate change and its potential risk for the human community makes preventative and preparatory action on a broad front an imperative priority. The results of this important conference will provide the basis for a decision by governments to launch the process of negotiating a convention on climate change which is expected to be ready for signature at the United Nations Conference on Environment and Development to be held in June 1992. It will also make an important contribution to the agenda which the Preparatory Committee for the 1992 Conference is expected to develop for approval by the Conference. This agenda will incorporate a broad strategy and related action programmes for reduction of climate change and amelioration of its effects and will complement and support the negotiating process while enabling action to be initiated even prior to the completion of formal agreements.

The General Assembly resolution 44/228, creating the United Nations Conference on Environment and Development, was the result of a number of important converging events. The report of the World Commission on Environment and Development made the case for sustainable development and outlined the pathway to a sustainable future. The recommendations of its report have to be translated to detailed targets, strategies and programmes for individual countries, as well as for international organizations, in particular for the United

Nations System. This is one of the primary tasks of the 1992 Conference.

The basic concepts of the interrelatedness of the environment and development issues, central to both the 1972 Stockholm Conference on the Human Environment, and to the report of the World Commission on Environment and Development, have to be moved from conceptual to operational levels. Whereas the principal purpose of the 1972 Stockholm Conference was to put the environmental issue on the international agenda, the prime purpose of the 1992 Brazil Conference is to move the environment issue into the centre of the development agenda and of economic and sectoral policy and decision-making. It will focus on the need for fundamental changes in our economic behaviour and in international economic relations, particularly between North and South, to bring about a new, sustainable and equitable balance between the economic and environmental needs and aspirations of the world community.

While the list of individual issues which the United Nations General Assembly, in resolution 44/228, asked the Conference to address - including such issues as climate change, bio-diversity and toxic chemicals - may seem at first sight to be primarily environmental in nature, every environmental issue leads directly into development when you examine both causes and remedies. For the causes of virtually all environmental issues have their origins in the development process or in its failures and inadequacies, and it is only through better management of the development process that these issues can be addressed. Equally, environmental considerations can impose new constraints on traditional modes and patterns of development which call for changes that in turn can open up new development opportunities.

Preparations for the Conference will provide a basis for this kind of action process and, hopefully, a positive example of how it can work. The Preparatory Committee, which was established to oversee these preparations, is a

Committee of the General Assembly open to all member states. The Conference is expected to produce concrete results in the following areas:

a. an "Earth Charter" or declaration of basic principles, building on the Declaration of the 1972 Stockholm Conference, to govern our behaviour towards each other and towards nature so as to ensure global environmental security, and the possibility of development for all

b. legal instruments, particularly on such critically important issues as climate change and bio-diversity

c. an agreed agenda incorporating a series of specific actions to give effect to these principles and legal measures for the remainder of this century and leading into the 21st century, called *Agenda 21*, with overall and specific targets, costing for the different programmes, their priorities and the allocation of responsibilities to existing and required institutions for action

d. agreement on the means to ensure the implementation of this Agenda through provision of financial resources, access to technologies, particularly for developing countries, and strengthening of institutional mechanisms and processes.

In the area of climate change, some excellent and extensive preparatory work is already in place. The Intergovernmental Panel on Climate Change initiated by WMO and UNEP has concluded the first phase of a major process which has established a solid basis for the major action decisions that must now be taken by governments. These decisions will be very much influenced by the deliberations and recommendations of this Second World Climate Conference. This Conference itself will contribute considerably to the overall understanding of climate; the science of climate change, and on our perceptions of where we need to do further studies. While we have clearly reached a new stage, where we know enough about the problem to begin developing and implementing policy responses to it, parallel to this process, we need not just to continue, but extend substantially activities of research and monitoring, such as in the World Climate Programme, to reduce uncertainties, and guide future actions.

Negotiations on a legal instrument in the form of a convention are due to begin. The UNCED Preparatory Committee has stressed that these negotiating processes should ensure the full integration of the development dimensions in the arrangements that are contemplated. It has also stressed linkages to other issues. The Preparatory Committee will make specific proposals to the Negotiating Body once it starts its work on how this could be done. For example, the issue of climate change is closely connected to that of the oceans; the latter being a major carbon sink, and in so far as coastal area management is concerned for the future. Similarly, any climate convention will need to deal with questions of forestry, as it relates to forests being a biomass sink. Alternatively, any work on forestry or biodiversity which is undertaken by the international community will also need to consider the issue of climate change. The same is true for work on strategies related to sustainable agriculture and freshwater management. At the same time, any legal agreements which may be produced on climate change, as well as on other issues to be dealt with at the 1992 Conference, such as biodiversity and forestry, will have to include provision for financing and transfer of technologies to permit the implementation of required measures, particularly by developing countries. These agreements cannot be developed in isolation from each other and there will need to be close interaction between the respective negotiating processes. Each must also respect and address especially the development needs and constraints of developing countries.

Climate change is the primary truly global environmental problem we are facing. It is universal both in its causes and its impacts. All human beings contribute, in one way or another, to the emissions of greenhouse gases that are the principal source of the problem. But the industrialized countries are by far the largest contributors. And, while the impacts of climate change will also affect everyone and these impacts will be unevenly distributed, developing countries bear a disproportionate share of the risks and lack the resources required to deal with them. But climate change does not occur in a vacuum. It is happening at a time, when, in spite of advances in a number of areas, the world is facing growing env-ironmental deterioration in many areas. The atmosphere is still heavily polluted in many parts of the world; forests are disappearing at rates greater than ever, desertification and land degradation are removing vast tracts of land from productive use; mounting tides of pollutants flow into our waters and oceans, and the list could go on.

Poverty and underdevelopment are closely linked to environmental deterioration. Poverty compels people in the interests of immediate survival to overstress and destroy the resources on which their future development depends. It is a vicious circle in which human need and envir-onmental deterioration reinforce each other. The needs of the poor for food, for fuel and for income inevitably lead to destruction of forests and tree cover. And, although they use relatively small amounts of energy, they are compelled to use it inefficiently. Thus there is a direct link between poverty and the prospects of reducing climate change. It is simply not realistic to consider that people of the developing countries and particularly the poorest of them can participate effectively in the actions required to deal with climate change except as part of programmes in which their own needs for development and relief from poverty are also being met.

The industrialized countries will need to take the first steps in terms of stabilising and reducing their greenhouse gas emissions. In this regard, we welcome the initial step taken by the member countries of the European Community to stabilize carbon dioxide emissions by the year 2000, the recent decision of Japan to take similar action and the proposals of the Swiss government to introduce a carbon tax. These are all important steps in the right direction. The industrialized countries must lead the way. But they cannot, in the final analysis, reduce the risks of climate change to secure levels without the cooperation of the developing countries. And, while there is much that the developing countries can and should do on their own to reduce their impacts, particularly through increased energy efficiency, they will need access to substantially increased flows of financial resources and the best available technologies to become full partners in achieving climatic security.

As the statements made at this Conference make clear, there is a much higher degree of consensus on the diagnosis of the climate change issue than there is on the needs for action or the will to act. However disconcerting this may be, it is understandable. The action process could lead to some very fundamental changes in our economic life and behaviour, in energy, transport and industrial policies and in the competitiveness of nations and of corporations. Consequently, the reluctance to embark on them is a natural one.

Nonetheless, the evidence is compelling, if not yet definitive, that the risk of climate change poses the greatest threat ever to global security, a threat that can only be obviated through actions which must commence now before the scientific evidence is definitive. On an issue like this that affects the fate of the entire human community we simply cannot afford to take the chance of waiting too long. It is not feasible to wait for the post mortem on planet Earth to confirm our diagnosis. If there is ever an instance in which we must act in accordance with the precautionary principle, this surely is it.

But however difficult it may be to obtain agreement for actions that would involve real economic sacrifice, there can be no excuse for delaying those actions that can be justified on grounds that do not depend on their relationship to climate change. This is particularly true of energy efficiency in which some industrialized countries, notably Japan, have already demonstrated that substantial reductions to be made in energy use per unit of GDP are entirely compatible with high rates of economic performance and, indeed, provide an important dimension of competitive advantage.

I very much hope that the 1992 Conference will provide the basis for a major programme of energy efficiency which would provide at the same time an important means of reducing CO_2 emissions. This should be accompanied by strong new measures to reduce the fossil fuels component of the energy mix and develop alternative energy sources. Included in these must be measures to promote the use of alternatives to petroleum in motor vehicles. In this respect the requirements recently established in the city of Los Angeles may well be a harbinger.

Similarly, ways must be found to incorporate the environmental costs of fossil fuel use into their price. While this will be difficult, we need to make a start. At the same time, we need to undertake major studies to quantify to the extent possible all costs and benefits. Technical fixes to improve the efficiency of all vehicles; favouring public versus private transportation, as well as non-fossil versus fossil-fuel based modes need to be vigorously pursued. In the longer-term, however, the whole concept of human settlements needs to be rethought, in the sense of developing new ideas for the location of homes, work space, leisure, and for communications in general.

While it is essential to produce less greenhouse gases to start with, we need to explore possibilities for absorbing whatever is already in the atmosphere. From the point of view of climate change, the halting of deforestation (at the global level), and its reversal into net afforestation must be achieved as soon as possible. Other means of sequestering carbon from the atmosphere need to be explored also. For example, increasing the organic content of soils, by leaving more of the agricultural residues in the land can sequester significant quantities of carbon from the atmosphere, while also having, very positive side effects on the quality of the soil.

Land degradation is a major issue in a number of developing regions, but particularly in the semi-arid and arid areas. The solution of the various manifestations of the problem of land degradation will be a crucial part of the strategies both to adapt to climatic changes under way, and to limit their rate. Sustainable agricultural and general land-use strategies will need to be developed which take account of future climate change, and which result in practices which are more resilient to the new weather patterns that may develop in specific areas. Also, a close connection of the United Nations' Programme to Combat Desertification with the climate change and other related processes will be essential and will have to be undertaken by the Preparatory Committee of the UNCED.

Most of the actions outlined above will require a certain level of technology development, and adaptation, as well as their transfer on a fair and affordable basis to developing countries. Technologies will be needed for energy efficiency programmes, and for alternative, non- or low-greenhouse-gas-emitting supply options. One important element of this is the capacity of developing countries themselves to develop the technologies which are most appropriate to their needs. New and additional financial resources will need to be made available to ensure that the transition takes place. Funds will be required for international action, for the support of the legal instruments, and generally to support sustainable development activities in the developing countries, such as energy efficiency programmes. Presently no comprehensive international agreements are in place to allow this, and no appropriate institutional structures exist. I expect that the Preparatory Process leading up to the 1992 Conference in Brazil will formulate specific programmes to achieve these to be adopted by Governments as important components of Agenda 21.

A critical element in such strategies will be that of continuing scientific study and monitoring of the key variables, such as in the framework of the World Climate Programme or the International Geosphere-Biosphere Programme. All these will need to be supported by major programmes of information, education and training within all countries, and coordinated at the international level. While certain economic and geographical regions are well equipped with institutions to play such a role, most regions are not, and neither is there sufficient capability in the UN System at present. While there are noteworthy programmes on various sectors of the energy problem, such as on nuclear energy, no UN Agency has the mandate to deal with the whole spectrum of energy and related issues which will be required.

Climate change is perhaps the ultimate test on whether humanity will learn to live in harmony with itself and with nature. The issue is environmental; the responses are developmental. They affect and are affected by most sectors of human activity. The cause and effect system through which human activities impact on climate is complex in nature and global in scale. Causes and effects do not occur in the same places, nor happen at the same time.

As difficult as the problem looks, it seems that many of the solutions we need to reduce the problem of climate change, are also needed for other reasons: environmental as well as developmental. A strong programme for energy efficiency will reduce local as well as transboundary air pollution; a less congested city is not only less polluted, but is also a better place to be in; increased forest cover can act as a carbon dioxide sink, but also to safeguard biological diversity, and to provide economic wealth to future generations.

This is not to say that all the actions that we need to undertake belong to the above group. There will be actions which will require sacrifices from all, for the sake of the survivability of this planet. We must, however, proceed, at least with the "easy" ones as soon as possible. These, as well as other actions need to be brought together under a strategic action plan going well into the next century. It is my expectation that the UNCED process will produce this, and that the Governments present in Brazil in 1992 will

adopt it as well as signing a convention. Together these will provide the basis for a more secure and hopeful future for our planet as a hospitable home for our species and the other forms of life with which we share it.

Task Group 1

Climate, Hydrology and Water Resources

Chair:
Z. Kaczmarek *(Poland)*

Discussion Leader:
L. Oyebande *(Nigeria)*

Rapporteurs:
N. B. Ayibotele *(Ghana)*
M. Beran *(United Kingdom)*
C. Candanedo *(Panama)*

Introduction

A plentiful and wholesome supply of water is essential to life and is a key factor too in determining the impact of climate on natural ecosystems and on almost all other social and economic activities - food production, forestry, energy, industry, transport, human health and recreation. Moreover, the hydrological cycle is critically implicated in the climate system through the energy balance and biogeochemical cycling.

This central role for water was reflected in vigorous national and international programmes during the period since the First World Climate Conference in 1979. Their emphasis has been to understand and exploit the climate-water link for providing water to a fast growing world population and protecting it from flood and drought. In addition to these ongoing problems we now face the new challenge of coping with climatic change and sea level rise.

The main issues which flow from this new challenge are detailed in the paragraphs which follow together with research and policy implications. A progression is followed which mirrors the IPCC report by considering first the need to understand how the hydrological cycle will alter as climate changes. Consequences for water management and the vulnerability of supplies are next considered. Finally, response measures including adaptive and preventive policies for water are presented.

It is vitally important to society that these issues are followed through and the concluding recommendations define a broad structure for research and for studies led by international agencies. It is felt that the existing structure

instituted under the WCP should be pursued. A cross-cutting sub-programme, WCP-Water, contains the necessary elements of Data, Applications, Impacts and Research.

Climate and Hydrology

The issue: Climate change places new demands on hydrological science to improve techniques and models for evaluating the role of the hydrological cycle in processes implicated in the climate system, and to assess water availability, flood risk and water quality in a non-stationary world.

Comments: While water planners are accustomed to increasing (therefore non-stationary) water demands, the sciences of climatology and hydrology developed under the assumption that geophysical processes are stationary in the time-scale of decades to centuries. Consequently, most of the analytical and computational methods used by hydrologists (statistical analysis, simulation techniques, model calibration) are time-independent. To cope with climatic and other anthropogenic changes these methods should be appropriately modified. Evidence is mounting from hydrological data and models that not only the mean characteristics are shifting, but the frequency with which thresholds are surpassed is also increasing. In addition, hydrological fluxes are often oversimplified components of climatic models, and the importance of feedback mechanisms between hydrosphere and atmosphere is underestimated. Consequently, the present edition of

GCM's cannot yet give reliable predictions of regional precipitation and evapotranspiration patterns necessary for hydrological impact assessment.

Hydrological observations will form a vital component of the proposed Global Climate Observing System (GCOS), including both routine observations, like those assembled at the Global Runoff Data Centre (GRDC) and research observations such as those resulting from representative and experimental basins.

Research needs: Most of the existing methods for analyzing hydrological processes should be reevaluated in the light of anthropogenic changes. A new generation of coupled hydrologic/atmospheric models in support of GCMs has to be developed. Improvements in GCM parameterization and use of finer grids are of great importance for hydrological investigations. Furthermore, to get a better representation of hydrological fluxes, research on the role of the unsaturated zone should be pursued. Regional studies should be initiated to assess the changed risk of flooding and its consequences. The large-scale experiments of interdisciplinary character, like the Global Energy and Water Cycle Experiment/Continental Scale International Project (GCIP), should help to calibrate and evaluate such models.

Policy recommendations: To reduce climate/hydrology uncertainties, governments should provide necessary funds for supporting and enhancing national data collection activities as well as for national and international research and data programmes necessary for improved understanding of complex hydrological processes. Encouragement by governmental and international organizations of sharing of operational and field experimental data being collected at existing national observing networks and large-scale experimental research facilities should be continued.

Water Management Under Uncertainty

The issue: Through climatic change, water resource decision makers are faced with increased uncertainty in evaluating future water supply and water demand. Incorporating uncertainty into water resource decision making will have profound consequences both for methods and for conclusions.

Comments: Due to the high cost and long lifetimes of many water resource systems, any oversimplification or neglect of major natural and economic factors may lead to system failure and large financial and human losses. For this reason uncertainties including climatic changes should be incorporated into the planning of water resource development. At present, the response to higher uncertainty is to increase the margin of safety, but this is done informally. Techniques that can improve on this informal approach are very complex; solutions depend on hydrologic

processes, types of water resource structures and population plus economic trends. WMO (EC-XLI) has adopted a statement intended to advise those reponsible for water resource planning on an appropriate level of response. This can be the basis of policy in particular regions and may be modified as uncertainties about expected changes reduce. It may prove expensive to incorporate uncertainty, so reductions in uncertainty of climate predictions will have important economic benefits.

Research needs: Methodology should be developed for generating internally coherent regional scenarios of future water supply and water demand as a basis for water resources planning. Elaboration of procedures for separating climate from non-climatically forced changes should also be given high priority. In the most vulnerable regions improved methods for Climate Change Impact Assessment for important existing water systems should be elaborated.

Policy recommendations: Despite many uncertainties linked with the climate change, legal, technical and economic procedures used in water resources planning design and operation should be reevaluated in light of possible non-stationarity of hydrological processes. Sustainable water development strategies should be adopted. Water systems should be made more flexible and more robust to anthropogenic changes. Uncertainties linked with the global climatic changes should be incorporated in national and regional water resources decision making procedures.

Sensitivity and Vulnerability of Water Resources to Climate

The issue: From a global perspective, few quantitative studies have been undertaken that evaluate the sensitivity of water resources to climate variability and change. A knowledge of basin sensitivities is requisite to the identification and assessment of regions most vulnerable to climate-induced stress.

Comments: The current lack of skill in forecasting (a) future regional climatic and hydrologic conditions and (b) future regional social and economic conditions, make it virtually impossible to prepare a meaningful regional assessment of future water resource conditions. However, techniques exist to identify climatic and hydrological conditions that combine to place existing water resource systems under stress. It is imperative that studies be undertaken immediately on a regional basis worldwide, so that a global assessment of water resource sensitivities can be assembled quickly and consistently to identify vulnerable regions. Moreover, social and economic changes in some regions of the world may make those regions even more vulnerable to climatic change. A region's level of vulnerability is influenced by a number of factors including: (a) the present-day demand/supply ratio, (b) possible water supply changes in response to changes in climatic and hydrologic processes, (c) possible water demand changes induced by population growth and economic development, as well as by climatic changes, and (d) potential climate-induced degradation in water quality. It is important to remember, though, that climatic conditions can get better as well as worse and, in some of those areas of the world most sensitive to climatic variability, water resource conditions could actually improve.

Research needs: Studies are needed to develop numerical indices of vulnerability and that identify river basin responses to the combined effects of climatic and social and economic changes. Such studies should proceed using existing techniques, although improvements to such methods should also be developed. Methodologies for building internally consistent nonstationary water supply and demand scenarios should also be developed as a basis for evaluating basin sensitivities (see section on Water Management Under Uncertainty).

Policy recommendations: Coordinated national and international efforts are needed to ensure that responsible preventive and adaptive measures are available for implementation in highly vulnerable regions, particularly in developing countries.

Adaptive and Preventative Measures

The issue: Even under the assumption of a stationary climate, water resources authorities have constructed elaborate systems to cope with hydrological variability and to match water supply with increasing demands. Further, under present conditions developing countries are unable to cope with the devastating consequences of floods and droughts which afflict them from time to time. There is the need for the developing countries to increase their capacity to cope with the social and economic consequences of floods and droughts and also new adaptive measures will be necessary to deal with climate-induced water supply non-stationarity. Most of the technology and capacity for adjustment is in the hands of the developed countries.

Comments: Elaborate technical, institutional and legal measures have been provided to make water supply more reliable and to achieve the fundamental goal of matching supply to growing water demands. Most of the engineering structures and jurisdictional arrangements were however based on the assumption that future climatic and hydrological characteristics will be similar to the past. It is widely accepted that the set of adaptive means used today to manage water resources may also be useful under future climatic conditions. But in the vulnerable regions water resource systems should be made more flexible, more resilient and robust to the uncertain future conditions. This

may be expensive and in some parts of the world difficult to implement because of financial and other limitations. Under conditions of increased uncertainties, water conservation measures are particularly important, and often the cheapest way to cope with negative consequences of climate change. If such measures will not be effective enough, additional storage facilities or long-distance transfer of water may become necessary for responding to climate-induced water resources changes.

Water also plays a role in preventive measures primarily through the substitution of hydro-power for fossil fuel, but also through implication to the regional water balance of forests or other measures for carbon sequestration.

Research needs: There is a need for studies on means for developing more resilient, robust and flexible water resource systems, able to withstand or be cheaply upgraded to cope with future changes.

Policy recommendations: Governments should support and strengthen national agencies established to deal with natural hazards like floods and drought. Efficient means must be found for training and transfer of technology to assist developing countries to adapt. There is also a need to evaluate existing international water treaties, and international water law, to ensure that the provisions of those legal instruments are sensitive to the reality of climate change. Water power opportunities should be sought where possible and hydrologic services enabled to advise on possibilities.

Conclusions and Implications for International Agencies

The most important impact of climatic change will be its effects on the hydrologic cycle and water management systems. Water plays a central role for natural ecosystems, and for food production, forestry, energy, transport, human health and recreation. Moreover, the hydrological cycle is critically implicated in the climate system through the energy balance and biogeochemical cycles. Evidence of its importance is amply manifest in the consequences of floods and droughts. Climatic changes raise the possibility of environmental and socioeconomic dislocations and they have important implications for future water resources planning and management.

International agencies will continue to take the lead and there is a large number of non-governmental agencies and programmes in which the climate water link is an explicit area of study. A focus was provided under the first phase of WCP by the WCP-Water administered jointly by WMO and Unesco but with involvement and support from the other agencies. The cross-cutting position of water enhanced and strengthened the structure of WCP-Water activities through projects that related to Data, Applications, Impacts and Research. It is urged that WCP-

Water be continued into the next phase of WCP to provide the focus for the expanded interests that emerge from the current analysis of needs. International agencies also must play a lead role in the transfer of technologies and training skills to cope with the consequences of change.

Summary of Key Recommendations

1. Strengthen data systems in order to predict water resources impacts, detect hydrologic changes, and improve hydrologic parameterization in global process models

2. Integrate the work of individuals and agencies studying impacts by combining information on sensitivities of water availability and water demand to determine the global pattern of water stress

3. Transfer technology and financial resources to developing countries to provide a common basis for quantifying impacts

4. Increase international collaboration and experiments within international agencies directed toward an understanding of the role of water in the climate system and greenhouse gas production, and exploit this knowledge to prevent or adapt to climatic change

5. Reevaluate existing scientific methods of analyzing hydrological processes in the light of human-induced changes

6. Incorporate the possible effects of climatic changes in the design and management of water resources systems, considering their long construction times and subsequent lifetimes

7. Reevaluate international water law and treaties in light of future climatic changes.

Task Group 2

Agriculture and Food

Chair:
H. Nix *(Australia)*

Discussion Leader:
S. Schneider *(USA)*

Rapporteurs:
W. Degefu *(Ethiopia)*
N. Rosenberg *(USA)*
C. D. Sirotenko *(USSR)*

Context

Climate change will **add** to and **compound** already serious problems of mismatch between human populations, consumption and resources. Population growth and inequities among nations in access to resources will exacerbate these problems in part through increasing degradation of soil, water, genetic, and other natural resources. Maintaining global food security through 2030 will require at least a doubling of production to meet demands of increasing diet and health standards as well as the population increase.

Potential CO_2 fertilization confounds the problem of predicting agricultural response to climate change since major uncertainties remain in relation to the combined effects of changes in temperature, solar radiation, water and nutrient regimes. Estimates of the overall CO_2/climate change effect on net global food production potential range at present from +10% to -20%. In addition, it is reasonable to expect that the current +10% interannual variation in production would remain largely unaltered. However, very much larger regional deviations can be expected. These will likely have greatest impacts in those areas with greatest population growth rates.

The high productivity agriculture necessary to feed and clothe a rapidly urbanizing global population demands high inputs of materials (water, fertilizers, chemicals), energy and information. A major upgrade of resource inventories, research strategies and extension services is needed to improve agrosystem management that will increase crop yields and quality, cut costs, reduce consumption of resources, and protect the environment while ensuring long-term sustainability of land and water resources and minimizing greenhouse gas concentrations.

Actions taken **now** with respect to population, resource use, and research commitments will influence potential impacts and responses through 2030 and beyond! Population/consumption/technology trends between 1990-2030 will determine what the world must adapt to beyond 2030. We must avoid static comparisons and always consider the dynamics of human response to change in forecasting potential impacts and in devising responses.

Given this context we need to give urgent attention to resource inventory and data base development and to improve our understanding of complex agricultural production and food systems.

Knowledge Gaps

With respect to issues of climate and climate change, inadequacies exist in the following areas:

Basic data:

- climate and weather data are not available in edited, computer compatible form in many countries
- impact studies demand matching sets of site, weather, management and performance data; grave deficiencies exist in this area in most countries
- minimum data sets (climate; terrain; soil; land use; population) can be defined that provide necessary inputs for a whole range of impact studies, e.g., agriculture, forestry, hydrology, conservation

- because data acquisition and processing are expensive we first need to exploit existing data in order to design and develop optimal networks for monitoring climate and impacts
- the use of Geographic Information System (GIS) technology for integration of spatial and temporal data is a basic requirement, but scaling down hardware requirements will make this more accessible to a wider range of potential users.

Scientific understanding:

- although developing rapidly, basic understanding of system performance in relation to climate and weather remains **inadequate** for **most** production systems; we are especially ignorant about the climate sensitivity of many important crops and livestock systems upon which large populations in developing countries presently depend; integration of physical /biological/social/economic components of complex systems remains poorly developed
- much of our current understanding remains location, season, and crop specific; new research strategies are needed if truly generic understanding and operational models are to be developed
- understanding of key processes in biological responses to climate and weather ranges from adequate (photosynthesis, transpiration) through barely sufficient (nutrient cycling) to quite inadequate (phenology/morphogenesis)
- direct CO_2 response of whole crop systems (as distinct from single leaves, single plants or even populations of plants in growth chambers) needs clarification
- new combinations of weather variables pose additional problems for both rainfed and irrigated agroecosystems; it is not simply a matter of global warming shifting existing climate patterns polewards or uphill
- major deficiencies exist with respect to understanding of climate and weather on vector-driven fungal, viral and bacterial diseases (crop, livestock, human) and pests and pathogens generally
- there is insufficient understanding of (a) the sensitivity of patterns of agricultural trade and food aid to a changing climate and (b) the adequacy of food reserves under conditions of changing climatic variability.

Tools:

- Existing agricultural research strategies are inadequate; increased emphasis is needed on system modelling approaches to improve understanding of

the potential productivity of agricultural regions and to optimize the design and conduct of agronomic experiments; programs in retraining and redirecting agricultural research on a global basis (similar arguments apply to forestry, wildlife, hydrology) are urgently needed

- Development of truly generic models of production and food systems that are process-based, dynamic, robust and reliable is underway, but continued development and validation require a truly international effort
- GIS technology has a key role to play, but not alone; major challenges confront the integration of remote-sensing/GIS/modelling; at present, these developments tend to be captive of separate agencies at international and national levels.

Understanding of known and potential impacts:

- At a general level, there is a reasonable understanding of the likely impacts of certain climate and weather phenomena on agriculture and food production; thus, for example, water deficits or drought impacts on agricultural production systems are generally understood, but understanding of waterlogging is limited; soil processes such as nitrogen fixation, mineral element transformations and carbon cycling are also not well understood in relation to climate impacts
- A complex pattern of crop, pest and pathogen responses can be expected even with relatively modest changes in temperature, since, at any given location, photoperiod will **not** change while heatsum accumulation rates will; unless the phenology of crop species and cultivars, pests and pathogens can be predicted accurately impact assessment will be flawed
- The processes of soil erosion, physical degradation and salinization are reasonably well understood, but impact assessment must take into account the totality of changes (i.e. crop, cultivate, management) in the production system
- Understanding of crop losses due to harvest, post-harvest handling and storage is well developed, but, again, the dynamics of entirely new combinations of climate and weather conditions during and following the crop growing season needs urgent analysis
- Agriculture is not only a system consuming commercial energy, but it is a source of energy through creation of biomass; the latter aspect needs to be analyzed and understood in relation to climate variation; furthermore, in developing countries animal power is a major source of energy, whose relation to climate change needs to be understood

- Possible interactions with long-run changes in population, technology, and land use and with proposed emission control strategies are poorly understood.

Main Topics Requiring Further Study

Major uncertainties with respect to CO_2 'fertilization' must be resolved. Basic research under controlled environment conditions, where interactions between crop species, temperature, radiation, water and nutrient regimes at current and enhanced CO_2 levels are studied, should have high priority. Even more importantly, the extension of such experiments to samples of whole crop systems in the field using carefully chosen environmental gradients must have top priority.

Major uncertainties with respect to the real contribution of agricultural systems to greenhouse gases such as methane and nitrous oxides must be resolved. Comparative studies of ruminant livestock systems versus natural wetlands versus tundra and of rainfed paddy rice production systems versus irrigated versus wetlands are needed. There is also the need to improve knowledge of sequestration of carbon by natural vegetation and replacement systems such as field and tree crops, plantation forestry or sown pasture.

Major uncertainties must be resolved with respect to the potential threat posed by climatic changes to future global and regional food security. An overall assessment of this threat should be undertaken that includes examination of long-term opportunities and obstacles to growth in per capita food production, potential adaptive responses in both developed and developing countries (including technological, socio-economic and institutional options) and the vulnerability of poor population groups.

Continued advances in agrotechnology transfer requires the **balanced** development of generic models of all major agroecosystems and the specified input data needed to implement these models. While much of the model development is underway in national institutions there is a need for much wider testing and validation of such models if they are to have the necessary generality. Since the basic price of admission to these new PC based models and decision support systems is the provision of specified site, climate, weather, soil and management data at appropriate scales, there is an urgent need to develop more generally applicable interpolation techniques for baseline terrain climate and soils data and the simulation of daily weather data. A program of climate data interpolation could provide a global 5 x 5 km grid of elevation and monthly mean climatic data within 3 years at an estimated cost of US$ 3 million for example.

Comments on IPCC Report

The IPCC process assembled relevant information in a form that can be drawn on by policy formulators and decision-makers. While stressing the uncertainty of any projections relating to food security and agriculture the IPCC findings and recommendations convey an impression of the maintenance of, or an increase in, existing food production potential which may be unwarranted. The report does not really consider the dynamics of population growth, nutritional needs and food production through 2030 and beyond, nor does it consider problems of adjustment to transient changes. Nevertheless, the Task Group supports the review function of the IPCC process and recommends that assessments continue with co-ordinated meetings at say 5-year intervals.

Comments on WCP

The WCP has made a major contribution to the strengthening of national institutions through training and data base development programmes and this work should be amplified and continued. The WCIP has made significant contributions to our understanding of climate impacts (e.g. the IIASA/UNEP and SCOPE studies), but the overall progress of impacts research at both national and international levels - which has been slow to date - needs to be greatly accelerated. Given the global significance of climate change and the potentially severe impacts on developing countries, much greater emphasis must be given to impact studies and the development of impact assessment methodologies that permit consideration not only of future changes to physical climate but also of social, demographic, technological and other changes that will occur coincident with - but not necessarily because of - climate change. Linkages with research on human dimensions of global change should be encouraged.

The WCRP has focused the attention of scientists on the problems that still require resolution in order to reduce the uncertainties in our knowledge concerning climate change and variability. It has also served to mobilize part of the global community of scientists and engineers, particularly in developing countries, that are interested in resolving problems related to atmospheric and oceanic physics and chemistry, where good progress has been made. There is a need for greater inputs to resolve some of the uncertainties relating to the hydrological cycle and it is hoped that GEWEX will provide the knowledge necessary. The IGBP will also be addressing some of the problems relating to biogeochemical cycles and natural ecosystems, but only to a very limited extent those of managed ecosystems. This is a particularly critical gap that must be filled.

The WCDP has in general acted as a stimulus in the collection of climate data, but greater efforts are needed to ensure the provision of baseline climate data at the scales needed for inventory, evaluation, planning and management in order to increase still further agricultural productivity. In particular, data are required to develop global, regional and national estimates, at different scales, of climate variability and change. Every effort should be made to increase the involvement of the national agricultural and meteorological services in both the collection, application, and availability of data. Of particular importance is development of a clearinghouse mechanism for the dissemination and appropriate use of regional climate scenarios, including those based on GCM runs, paleoclimatic analogs and empirical climate data.

Comments on Planning and Policy Making

The Task Group considers that opportunities for scientific and technical solutions to many of the problems of adaptation of agricultural systems to climate change and climatic variability already exist or will be developed, but that technology transfer will be impeded by institutional and infrastructure limitations. Much greater emphasis must be given to training programmes, agricultural extension services and strengthening of institutions in-country if the gap in adaptive capacity of developed versus developing countries is to be reduced. Scientists in developing countries must become part of the analytical process and not just reviewing process.

Climate impact assessment extends far beyond the purview of any single international or national agency. Within the UN system it will demand integration of activities across relevant UN organizations, especially among WMO, FAO, Unesco and UNEP. Close collaboration with other major agencies such as the IMF, UNDP, the Regional Economic Commissions, WHO, IFAD, UNDRO, UNFPA, the World Bank and the CGIAR is essential.

Task Group 3

Oceans, Fisheries and Coastal Zones

Co-Chairs:
J. Gray *(Norway)*
G. Marchuk *(USSR)*

Rapporteurs:
G. Kullenberg *(UNESCO/IOC)*
F. U. Mahtab *(Bangladesh)*
D. A. Tsyban *(USSR)*

Discussion Leader:
N. Flemming *(United Kingdom)*

Current Adequacy of Knowledge Concerning Climate Change

The earth's climate is determined by the ocean-atmosphere system. The predicted global warming will have a significant effect on sea-level rise, and will modify ocean circulation, changing marine ecosystems with socio-economic consequences. Seventy percent of the human population lives within 60km of the sea, and the percentage is increasing. Human settlements, industry and agriculture in coastal lowlands are vulnerable to rising sea level or changes in storminess or storm surges. Alteration in currents and upwelling will affect fisheries, as exemplified by the El-Niño-Southern Oscillation (ENSO) phenomenon. Rainfall, and hence agriculture, forestry, and fresh water supply throughout the world, are determined by ocean-atmosphere interaction. Changes in ocean circulation thus impact on all human activities.

Ocean-atmosphere interactions have not yet been adequately included in global models of climate change. Through the Committee on Climatic Changes and the Ocean (CCCO - a joint SCOR-IOC committee) and the Joint Scientific Committee of the WCRP (JSC - a joint ICSU-WMO committee) initiatives have been taken to obtain relevant data on ocean-atmosphere interactions and ocean circulation (e.g. the TOGA and WOCE programmes) where joint planning and funding has led to the establishment of a much-needed global climate related ocean research programme. Other programmes such as the Integrated Global Ocean Services System (IGOSS) and the Global Sea-Level Observing System (GLOSS) are already in operation to obtain and facilitate the international exchange of relevant data on a global scale.

Although predictions concerning changes in wind patterns, rainfall, ocean currents and sea levels and other variables are being made these are produced with a low level of confidence and predictions differ considerably from model to model. Better predictions are needed if we are to predict impacts, and this requires an operational climate prediction system.

Basic Data

Despite the above initiatives the basic data are considered still totally inadequate for the climate models now being constructed, both in relation to prediction of global climate change and climate variability (IPCC WGI) and for prediction of impacts of change, especially in the coastal zones (IPCC WG2).

There is, therefore, a need for:

- a substantial increase in ocean data collection
- a concerted and coordinated programme of global coastal zone data collection
- improved global ocean models with much higher spatial resolution
- development of regional models nested within global models, especially for coastal zones (including estuaries) for assessment of likely local impacts

- improved methods of data assimilation into models of climate of variability in order to generate products useful for policy makers.

Scientific Understanding in Relation to Information Needs of Planners and Decision-Makers

Global climate models and tropical ocean models have had some success in providing global predictions of the consequences of greenhouse gases, as well as the large scale evolution and effects of the ENSO phenomenon. Models cannot yet produce the regional predictions of climate change needed by planners and decision makers, and the major ocean science programmes TOGA and WOCE have not yet been fully funded. Improvements in prediction will arise only when physical processes related to oceanic upper layer structures, oceanic circulation including formation of oceanic deep and intermediate waters, and to the hydrological cycle, can be understood well enough to be incorporated into global models. In addition we need to be able to include the biological and chemical processes in such models to determine how primary productivity and marine life and fisheries might interact with global and regional climate.

There is a logical and necessary linkage between understanding physical, chemical, and biological processes at each space scale. Global models are necessary to provide the boundary conditions for regional models, which in turn are needed to provide the boundary conditions for local models.

Planning and Management Applications of Techniques and Methodologies

There are major efforts being made within UN organizations (e.g., IOC, UNEP, WMO) and in other international organizations (IUCN, ICSU, SCOPE etc.) to address the scientific problems outlined above. The key problems lie in lack of appropriate data and programmes in particular for determining:

- how open ocean circulation and processes affect coastal and shelf sea processes
- how climate change affects primary productivity, biogeochemical cycles, recruitment of fish stocks and migration patterns
- in the coastal environment: how changes in rainfall and temperature will affect relative humidity and alter evapotranspiration rates hence affecting the hydrological cycle and local water balance. In turn how water balance will affect coastal vegetation distribution and abundance, animal distribution and the productivity of natural and cultivated systems and thereby human populations

- how to include boundary layers into existing weather forecast models and upper ocean models so as to obtain more accurate estimates of momentum, heat, and water flux between ocean and atmosphere at scales down to a few km
- how changes in coastal oceanographic conditions and river and groundwater runoff affect productivity in coastal areas and biological systems (including fish) and can lead to severe socio-economic consequences
- with respect to coastal and offshore structures: how changes in sea level, storminess, precipitation, floods, icing etc, affect design criteria for coastal defences and offshore structures. Attention is needed to develop flexible policies for coastal defences as sea level and climate change. The aim of such policies is to produce a flow of predictive information which enables politicians to invest in defence or other response in an incremental fashion
- how changes in sea level, storminess and precipitation affect the delicate balance of salinity in estuaries and salt intrusion into freshwater aquifers and surface waters, the flow of rivers, and the run-off, pollution load and sediment load into the sea
- how to interface the models of physics, chemistry and biology relating to ocean climate, with socio-economic models of factors such as transport, employment, tourism, sustainable agriculture and fisheries, and industrial development.

There is a need to:

- develop a coordinated coastal zone observation system
- develop a system for assessing coastal vulnerability to global change encompassing physical, biological and human population vulnerability
- conduct a number of intensive case studies of coastal zone regions
- formulate environmentally sustainable development strategies for the coastal zone.

Understanding of Known and Potential Impacts

Due to the uncertainties of how climate will change regionally, few countries have acted on the information at present available.

In general, any change, whether leading to a better or worse environment, demands that mankind adjust and invest. For example, even if total biological production does not change, the change in location of phenomena such as upwelling zones will create a need for re-investment. The expected changes are likely to lead to substantial increased costs.

Beneficial effects of climate change may include locally increased productivity of coastal ecosystems, including fisheries and extended seasons for navigation in coastal areas.

Harmful effects of climate change can include loss of productive low-lying coastal areas, altered coastal vegetation and agricultural productivity, loss of mangrove forests, intrusion of sea water into freshwater supplies, including drinking water, increased frequency of extreme events (e.g. a factor of ten reduction in the return period of storm surge levels of any given value), reduced inshore fisheries and increased pollution.

Since each coastal location represents a unique set of interrelated physical, biological and human components and processes the extent and nature of impacts in one location will differ from any other coastal site. Although some generic methodologies may be applicable to a range of coastal ecosystems of the same type, it is not clear that case studies which concentrate on site-specific characteristics could provide broad generalizations concerning future impacts.

High energy is contained in the ocean at spatial scales of only a few tens of kilometres. This leads to small scale patchiness in ocean physics, chemistry and biology and substantial changes over quite small distances along the coast. Such patchy structures will alter with climate change, and the changes on the coast will show small scale variability.

The time scale of feedback processes from the ocean to the atmosphere is long so that even if atmospheric emissions are controlled immediately it will be the end of the 21st century before climate is stabilized (IPCC WG 1). This provides a false sense of security, and a failure to act now will result in further, larger expenditures in mitigating impacts.

Main Topics Requiring Further Study

The Task Group proposes in this section two related data gathering, monitoring, modelling and prediction systems; 1) a global ocean system, including the upper ocean, to predict global climate and variability on seasonal to multi decadal timescales, 2) a global programme of coastal monitoring to predict local changes. These two systems can be planned in a logistically optimized manner. The detailed topics broadly support these two systems.

The Task Group suggests that the following topics require further study:

- a global ocean observing and data management system to accelerate progress toward an operational climate prediction system
- a global coastal zone monitoring system to predict local impact
- use of upper ocean data for operational monitoring and short- term prediction of climate variability
- regional coastal models nested within global models

- phased timing of satellite launches to achieve continued global coverage
- a system for assessing coastal vulnerability to global change encompassing physical and biological factors and their effect on human populations
- the interfacing of coastal models of high resolution with socio-economic models
- study of deltaic areas to discover ways of using natural river-supplied sediments to build up the land and provide protection against sea level rise
- development of regional and local coastal data management systems to provide information to planners
- development of new instruments to better measure climate change
- development of new technology and the two-way transfer of technology and data on coastal monitoring by training and support of local environmental (and other) agencies
- extension in scale and scope of existing programmes within the UN system concerned with climate issues to address the priorities listed above.

IPCC Report

The reports of IPCC have been fully taken into account during the work of Task Group 3. We consider that the IPCC should continue to provide impartial scientific advice to governments on matters relating to climate change through the United Nations organizations.

Future Evolution of the World Climate Programme

The Second World Climate Conference has confirmed the importance of interactions between the ocean and atmosphere in the world climate system, and has endorsed the present and planned collaboration between relevant UN organizations. The future evolution of the WCP should be planned so as to take full account of the importance of ocean-atmosphere coupling in climate change, and the impact of climate change upon human activities globally.

Inputs to Political Decision Making

The conclusions from the IPCC are that developing countries are especially vulnerable to impacts of climate change. There is, however, a great willingness on the part of UN agencies, international organizations and countries with advanced resources in oceanographic, fisheries and coastal seas management to help developing countries to tackle the major issues of climate change (data collection, access to equipment and technology, modelling techniques) and to both generate policies to cope with climate change and to implement those policies.

When considering the impact of climate change in the coastal zones of developing countries, special attention should be given to the control of natural disasters, food security systems, human population growth and migration, and the socio-economic impact on the populations of low-lying areas already experiencing unemployment, poverty, and over-urbanization. Every effort should be made to develop capabilities within developing countries, ensuring the participation of experts and decision-makers in the present research and analysis.

Developed countries have well established procedures for dealing with coastal defences (e.g., by storm surge forecasting) and water quality and there are a number of international conventions relating to them. Where relevant, transfer of knowledge on the technology and management of coastal areas needs two-way cooperation in the form of access to oceanographic data and deployment of research vessels to coastal areas of developing countries if realistic local climate change models are to be developed.

During the 1990's there will be a continuing need to convert existing or new scientific knowledge and systems into operational programmes. This process will require the involvement of operational agencies at national and international level who will need to progressively undertake the funding of operational activities. Scientific agencies will need to re-deploy their resources to tackle fresh scientific problems which confront the oceanographic community, whilst providing advice and guidance to the operational agencies.

Recommendations

The Task Group makes the following recommendations, not in order of priority. These actions should be undertaken in parallel:

- a permanent Global Ocean Observing System for the purposes of improving predictions of climate change should be established, as recommended by IPCC WG1
- IOC, with the cooperation of WMO, should be responsible for planning and management of the new Global Ocean Observing System, using the existing IGOSS as the basis
- a concerted and coordinated global programme of coastal zone monitoring should be established within the framework of ongoing UNEP/IOC/WMO activities designed to detect the long-term changes resulting from global climate change
- a system for assessing coastal vulnerability to global change should be developed through collaboration between IUCN and other relevant international organizations, encompassing the geosphere, the biosphere and socio-economic problems

- coastal regional models should be designed to reveal the local impacts of climate changes and which can serve as a basis for the development and implementation of management policies at local levels
- the IOC/WMO Joint Committee on IGOSS should be an advisory body on availability of oceanic data relevant to climate models
- international agencies and coastal and archipelagic states should develop short term coastal adaptation strategies and policies on the basis of existing data, which can be modified and extended as the results of further research become available.

It is important to act now with prudent measures of prevention and adaptation. The need for increased scientific research to improve certainty and accuracy of prediction should not be used as an excuse to delay precautionary action. The conduct of practical adaptation strategies must progress in parallel with increased scientific effort to improve predictions.

Task Group 4

Energy

Chair:
A. K. N. Reddy *(India)*

Discussion Leader:
I. Mintzer *(USA)*

Rapporteurs:
K. Heinloth *(Germany)*
H. Warren *(USA)*

Introduction

The energy sector is invariably presented as a "problem area" from the point of view of climate change. This is unfortunate. Such a view may follow naturally from the belief that the economic progress of countries must be judged by the magnitude of energy consumption. But the quality of life depends upon the **level of energy services** (light, process heat, motive power, etc.) rather than upon the amount of energy consumed. The distinction is crucial because increases in the level of energy services can be attained by increases in the supply and consumption of energy and/or by improvements in the efficiency of energy production, transport/transmission and utilization. The difference is also crucial because a stress on energy services forces an exploration of opportunities for achieving more services for the same or even less energy input. Thus, **the energy sector is actually an "opportunity area" in the challenge of combatting climate change**. This is the perspective of Task Group 4 .

General Principles for Dealing with Energy-Climate Interactions

In developing this perspective, the Task Group recognized the importance of the following principles for dealing with energy-climate interactions:

Action along with analysis: In the matter of energy-climate interactions, nations have to act even before analysis is complete, because if they wait for analysis to be completed, their subsequent actions will be too late to be effective.

Also, early action is more effective and less expensive. Hence, nations must act while pursuing further research and deeper analysis.

Tackling climate change in a manner that is compatible with the solution of other major global problems: The threat of climate change is not the only major global problem. Other problems such as development, population, food and peace are as important from a global perspective. Developing countries may feel that their development is a far more important problem. Therefore, nations must combat climate change in a manner that is compatible with, and contributes to, the solutions of the other global problems. In particular, the climate change threat must be tackled along with the advancement of development.

Combatting climate change as a bonus from the development strategy: This bonus principle, which is the other side of the coin of the **no-regrets** principle, requires that the combatting of climate change be achieved as a **bonus** from the measures that advance development in developing countries (and sustainable growth in industrialized countries).

Human Radiative Forcing Activities

It is established that there is a natural greenhouse effect that keeps the world warmer than it would otherwise be. Since the Industrial Revolution, human activities have enhanced this greenhouse effect, and continue to do so. The main anthropogenic radiative forcing activities consist of:

- increases in the concentrations of natural greenhouse gases (CO_2, CH_4, etc)
- deforestation and desertification, and
- increases in the concentrations of human-made greenhouse gases (for example, chlorofluorocarbons (CFCs)).

General Response Strategies for Combatting Climate Change

This listing of human radiative forcing activities suggests the basic response strategies for combatting climate change:

- reducing the emissions of natural greenhouse gases
- reducing deforestation
- afforestation, and
- eliminating the emissions of human-made greenhouse gases such as CFCs.

Energy-Related Response Strategies

Since the terms of reference of Task Group 4 (Energy) restricted it to those response strategies that fall within the purview of the energy sector, it was important to identify the sectoral origins of the various response strategies. The reduction of CO_2 is intimately connected with the utilization of fossil fuels which are a major source of human greenhouse gas emissions. The reduction of methane emissions is dependent on actions in the agricultural and animal husbandry sectors, but there is also a fossil fuel component connected with methane releases from coal-mining and with leakage of natural gas during production and transportation. The causes of deforestation are varied and depend upon the country. When, however, deforestation is due to the use of fuelwood for cooking and industry (as is the case in many parts of the world) its climatic impacts are energy-related and must be dealt with through the energy sector. If afforestation is directed solely towards the objective of sequestering CO_2, then it is not an energy-related response strategy. On the other hand, if the purpose of afforestation is to produce biomass fuel to displace fossil fuels, then it is very much an energy strategy that combats climate change.

What is to be Done in the Energy Sector

Identification of Technological Opportunities

The linkage between energy and climate change follows from the fact that the total emission of an energy-related greenhouse gas such as CO_2 is given by the product:

Population x (Gross Domestic Product/Capita) x (Energy/Gross Domestic Product) x (Greenhouse Gas Emissions/Energy).

This expression shows that there are two types of energy-related responses that can reduce emissions of greenhouse gases:

- reducing (Energy/Gross Domestic Product), i.e., the energy required per unit of Gross Domestic Product - this corresponds to **end-use efficiency improvements** in the consumption of energy in economic activities, and/or **conversion efficiency improvements** in the conversion of energy from one form into another; this applies, in particular, to the use of fossil fuels
- reducing (Greenhouse Gas Emissions/Energy), the greenhouse gases released per unit of energy consumed - this corresponds to **fuel-switching** to energy sources that do not emit or emit less greenhouse gases.

Many opportunities exist for both types of responses.

End-Use Efficiency

New technologies tend to use less energy for a given energy service than the average of the present stock. At the rate of capital turnover and expansion, new technologies with better performance can be introduced at little or no extra cost (over and above what would have been spent anyway). In some cases, the cost of the more energy-efficient alternative may even be lower than a more conventional alternative. There is a variation in energy efficiency among the devices that are sold at present. This implies that there is scope for policies that support the more efficient devices.

These opportunities also arise in the transportation sector, although in this case, there are also important structural issues (such as rail vs road transport and mass vs private transportation) that can be used to support the development of benign transportation systems.

Conversion Efficiency

There are also enormous opportunities, particularly with fossil fuel technologies, for the improvement of conversion efficiencies. These arise mainly through the cogeneration of electricity and steam and through advanced power production technologies such as combined cycle (gas plus steam) turbines, for instance.

As an immediate and short term strategy, a partial shift from coal to natural gas is advantageous in most countries, not only because natural gas produces less emissions per unit of energy produced, but also because new, low capital-cost conversion equipment with higher efficiencies is available for natural gas.

Non-Fossil Sources of Energy

There are some non-fossil sources of energy, such as nuclear energy, that do not emit carbon dioxide, but it is important to realize that no source of energy comes without environmental impacts and economic costs, and substituting such non-fossil sources for fossil fuels involves complex trade-offs.

Such trade-offs must be made even in the case of renewable sources of energy which do not emit carbon dioxide but which are at present relatively expensive. The cost effectiveness of renewable energy is improving through continued research and development and with rising conventional energy prices. Some applications are already attractive in certain circumstances and during the 1990's, increased use of organic wastes and biomass for cogeneration (especially in the sugarcane and paper and pulp industries) look attractive. There are good prospects for further cost-reduction of photovoltaics and lowered costs would bring them into major markets.

Behavioral and Structural Changes

Apart from technological opportunities for reducing energy needs, behavioral patterns such as preference for private transportation have a role to play. The structure of the economy and the orientation of development also influence the level of energy demand. In recent years, such structural changes have been a major reason for the overall decline in the energy intensity of the industrial nations, although most energy forecasts assume they will have a much more limited impact in the future. It should be pointed out that the projections of future energy-related carbon dioxide emissions are very sensitive to the assumptions that are made about the future composition of the economy, and this is an area of energy analysis requiring more critical analysis, especially with respect to global trends in production.

Estimates of Costs and Benefits

Economic growth in general, and development in particular, requires an increase in the level of energy services. In any specific energy situation, such an increase can, from a technical point of view, be achieved by an increase in the supply of energy to the energy utilizing devices and/or by an increase in the efficiency of these end-use devices. Whether the supply or demand-management option is chosen depends upon a number of criteria. Foremost among these criteria are the costs per unit of benefit. However, these costs can be counted in various ways, including:

- the investment and operating costs for the new equipment required to provide the service
- the **net** investment and operating costs required on the new equipment - this is the cost of providing the service by the new option **minus** that required to provide the service by conventional means- and
- the social costs that reflect direct economic costs as well as environmental and public health costs, tax

subsidies, national security expenditures and other indirect costs over the full life cycle of the energy source or technology.

The net cost of more energy efficient technologies is often negative in the sense that they are cheaper than the conventional technologies that would normally have been deployed. Despite this, the term **negative costs** (being used in the discussions on the subject) is not used here because it gives the erroneous impression to lay persons that energy efficiency can be obtained **free**. This is not the case because significant investments are required on new energy efficient technologies even when these investments lead to lower overall costs for the provision of energy services. Conversely, there is an opposite viewpoint that the bills for efficiency improvement are so high that they are beyond the reach of economies, but this view is frequently propagated without pointing out that in some situations the bills are actually lower than the investments that would have been made on the supply options.

It is also essential to avoid generalized arguments that suggest that efficiency improvements and renewable sources are either **always** cheaper or **always** costlier. New and more comprehensive methods are required to identify the least-cost mix of conventional fuel and electricity production, renewable sources, and efficiency improvements to meet the demand for energy services in each country or region. Such methods must reflect the full economic, environmental, and public health cost of energy supply and use in the price of fuel and efficiency improvements. Among the various methodologies useful for this task is **least-cost planning** involving the construction of cost-supply curves to identify the elements of the mix without any biases either in favour of or against conventional supply or renewables or efficiency improvements.

Social costs are relevant in a number of situations such as:

- setting priorities for research and development efforts according to the potential benefits of those efforts to society
- imposing energy or carbon taxes to reflect the potential and actual damage caused by the combustion of fossil fuels
- governments and associated agencies requiring that energy investment and supply-contract decisions be based on life cycle costs, and
- utilities considering the social costs, including environmental costs, as an integral part of investment decisions, including giving an appropriate credit to low/non-carbon supply sources such as renewable energy sources and energy conservation investments.

Much more attention must be given therefore to developing accounting techniques that encompass the full range of costs and benefits associated with measures for combatting climate change.

Overcoming the Barriers to the Dissemination of Climatically Benign Technologies

Even when efficiency improvements and other climatically benign technologies have been identified and shown to be technically feasible, economically viable and beneficial in many ways, their implementation is retarded and sometimes prevented by a number of barriers other than inadequacy of the physical infrastructure. The barriers exist at all levels - at the level of consumers, energy equipment manufacturers and suppliers, utilities, financial institutions, governments, international funding agencies and the aid agencies of developed countries. The understanding of these barriers and the identification of ways of overcoming them is a crucial challenge to the combatting of climate change. Tackling this challenge necessarily involves research into the reasons why people and institutions behave as they do in optimizing their welfare.

Already, there has been considerable progress in understanding these barriers, and many innovative ways of overcoming them have been designed and are being implemented (particularly by utilities in many developed countries). To overcome the barriers, responses are required from a number of actors - consumers, industry, governments and non-governmental organizations, and international agencies.

It appears that there are four main categories of measures to break down market barriers:

- information
- standards and regulations
- fiscal and other measures, and
- research and development.

With respect to information, there is widespread ignorance over the role of energy, the true costs of energy, and the range of cost effective technologies available to reduce demand and associated pollution. Sustained government information programmes are required for all consumers to overcome this barrier.

In the area of standards and regulations, policy measures that have proved successful in achieving sustained energy savings include the setting of minimum efficiency standards for vehicles, appliances, lighting, boilers, buildings and other equipment. Every region and country must determine for itself the appropriate standards and the future role for such standards. This flexibility is particularly important because various proposals for standards have been made. Some examples of technically achievable performance standards that have been proposed include automobile fuel efficiency targets of 45 miles per gallon by the year 2000 in North America, a reduction in average electricity consumption of electrical appliances by 40 per

cent in western Europe by 2000, and an upgrading of thermal insulation standards in western Europe and North America by more than 50 per cent over the next decade.

Regulations are critical in the area of gas and electrical utilities. Governments need to give clear guidelines on how environmental and social costs should be reflected in the planning process, and allow utilities to change business practices so that they can sustain profitability by selling end-use efficiency measures to their customers.

Fiscal and other measures are another important means of removing barriers to cost effective measures for combatting climate change. Phasing out subsidies to particular technologies or energy sources is a prerequisite towards achieving market conditions that reflect the environmental hazards of climatic change.

The policies of aid agencies should reflect the need for technological leapfrogging involving the implementation of advanced efficient technologies, for shifting towards program funding as opposed to project support, and for fostering self-reliance through the strengthening of indigenous human resources.

With respect to research and development, the capacity of the research and development community to develop alternative energy technologies needs to be strengthened. To provide society with the technological options it needs to combat climate change, funding for energy research, and especially research directed towards improving efficiency and making renewables cost-effective, needs to be increased and sustained over the long term.

The Roles of Developed and Developing Countries in Combatting Climate Change

Directions for the Evolution of National Energy Systems

An analysis of the historical patterns of change of the energy systems of industrialized countries shows a time variation in the energy intensity (energy utilized per unit of Gross Domestic Product, expressed for example in tonnes of oil equivalent per dollar). During the period of industrialization, the energy intensity increases up to a maximum which is then followed by a decline. What is interesting is that the more recently a country industrialized (for example, Japan or France compared to the USA or UK), the lower is the maximum that it reaches before going into a post-industrialization decline of energy intensity. The main reason for this lowering of the maximum with time is the advance in materials technology - lesser amounts of energy are required to produce a given quantity of material (e.g., steel, cement, paper) and smaller quantities of a material are required to perform the same function (e.g., in an automobile or a bridge).

These historical patterns of evolution of the energy systems of countries suggest two crucial guidelines for ensuring that the energy systems of countries contribute to the combatting of climate change:

(1)   Developed countries should continue their historical trend of lowering their post-industrialization energy intensities.

(2)   Developing countries should follow the historical trend of lowering the maximum intensity and achieving even lower maxima in the course of their industrialization than the countries that industrialized recently. This means that they should not follow the obsolete examples of the developed countries that industrialized long ago. In fact, they should seize all the opportunities for **technological leap-frogging** and use energy efficient technologies that perhaps even the developed countries have not yet used.

Recommendations for the Energy Systems of Developing Countries

Maintaining development as their first priority objective, developing countries have the option of choosing development patterns that have the additional benefit of reducing radiative forcing. Thus, developing countries can play an important role in combatting climate change provided they have access to climatically benign technologies and the ability to acquire them.

The Importance of Multiple Benefits

Investments in energy technologies on the supply or demand sides can often yield multiple benefits. This fact is frequently ignored. For example, it is often believed that energy investments either promote development or combat climate change. In other words, the investments in combatting climate change are considered to be over and above those required for development. This view, which is accepted by many developed and developing countries, gives rise to the development versus climate-protection conflict.

In point of fact, many energy technologies not only advance development but **as a bonus** also combat climate change. In such cases, extra investments are not required on the climate-protection account. (If the objective of economic growth is considered instead of development, the point being made here is equally valid in developed countries). Obviously, opportunities for such multiple benefits, particularly development plus climate protection, should be identified and pursued.

There are many situations where multiple benefits do not accrue and separate investments have to be made for combatting climate change. In such situations, it is pointless to recommend to developing countries climatically benign technologies without also suggesting sources of finance for these technologies.

Technology Transfer and Financial Aid from Developed to Developing Countries

In order to pursue climatically benign development paths, developing countries have to implement technologies appropriate to this objective. In selecting these appropriate technologies, developing countries must widen their choice and screen all possible candidates. They must not assume that all modern technologies are necessarily appropriate or inappropriate. In fact, there may be technologies that have been abandoned by the developed countries for historical reasons that may be appropriate by current criteria particularly for combatting climate change. From many points of view, it is desirable that the necessary technologies are developed indigenously in order to ensure compatibility with the culture, traditions and history of a country. In most cases, however, technologies are imported from private industry in the developed countries.

This raises the whole issue of technology transfer from the developed to the developing countries. This issue is neither new nor peculiar to the threat of climate change; it has been a recurrent theme in North-South negotiations. Among the many dimensions of the issue are the **costs** of technology transfer (royalties, patent and technical fees, the remuneration of experts, repatriation of profits, etc.) the **funds** for such transfers (aid, grants, loans, bilateral and multilateral assistance, etc.) the **content** of technology transfer (whether the know-how that is transferred is devoted to operation, maintenance, manufacturing or design) the **up-to-date character (modernity)** of the technology, and the **impact** of such imported technologies on the indigenous development of technologies.

The transfer of climatically benign technologies has all the above dimensions, and in addition, it has two extra dimensions:

- if the obligation of combatting climate change places upon developing countries the burden of implementing technologies **that would not otherwise have been acquired to satisfy development objectives**, then these technologies are unlikely be introduced unless special financial support is provided by the developed countries

- if the unrecovered development costs of the new climatically benign technologies make their transfer inherently costlier than the transfer of the more primitive environmentally harmful technologies, the extra financial burden will have to be alleviated by the developed countries.

For both these reasons, financial support for climatically benign development patterns must come from the developed countries to the developing countries if the latter are to successfully play their role in combatting climate change.

Recommendations for the Energy Systems of Developed Countries

According to the Intergovernmental Panel on Climate Change Working Group I (WGI) report, the energy sector accounts for about 50 percent of the enhanced radiative forcing resulting from human activities. The various scenarios investigated by Working Group I indicate that at some time between 2025 and 2100, the combined effects of all greenhouse gas emissions will have a radiative effect approximately equivalent to doubling the pre-industrial concentration of CO_2 alone. Furthermore, the Working Group has concluded that "immediate reductions of over 60% in the net emissions from human activities of long-lived greenhouse gases from human activities would (be needed to) achieve stabilization of concentrations at today's levels..."

A review of the prospects for achieving the large emissions reductions required to stabilize atmospheric concentrations of greenhouse gases shows that technically feasible and cost-effective opportunities exist to reduce CO_2 emissions in all countries. However, to stabilize atmospheric concentrations of greenhouse gases while allowing for some growth in emissions from developing countries, industrialized countries must implement reductions even greater than those required, on average, for the globe as a whole. Many studies have indicated that, in most industrialized countries, these opportunities (consisting of increases in the efficiency of energy use and the utilization of alternative fuels and energy sources) are sufficient to stabilize CO_2 emissions from the energy sector and subsequently to reduce those emissions by at least 20% by 2005. Responding to the results of existing analyses, several industrialized countries have already adopted targets that would stabilize and, in many cases, reduce national CO_2 emissions by 2005. As additional measures to achieve further cost-effective reductions are identified and implemented, even greater decreases in emissions can be achieved in the following decades.

Task Group 5

Land Use and Urban Planning

Chair:
Sir J. Burnett *(United Kingdom)*

Discussion Leader:
R. L. Kintanar *(Philippines)*

Rapporteurs:
J. S. Oguntoyinbo *(Nigeria)*
A. Viaro *(Switzerland)*

The objective of land use planning is to determine how best land can be used for a variety of purposes: rural, such as agriculture, forestry and fishing; urban, for towns and cities, offices, factories and the like; for transport by road or water; for services and their transmission; mining,or tourism, leisure and recreation of all kinds, or wildlife and conservation. Urban planning complements land use planning , dealing as it does with the design and layout of towns and cities, their communications and transport requirements and, of course, their buildings, both siting and construction. They have in common the objectives of so altering or adapting the environment as to best fit it for human activities and so promote the health, wealth and happiness of mankind.

The group was both surprised and disappointed to find that only a very few aspects of these issues had been addressed in the reports of the IPCC Working Groups and that they were scattered amongst other topics, so losing the holistic approach which is so essential to such activities. This impression was born out by the virtual absence of any reference to these planning issues during the conference save in the one paper by Professor Taesler, a member of the working group, who was only able to deal with a limited aspect of the topic in his paper. That the homes, workplaces and activities of man are subject to the vagaries of the weather and of the climate is a truism so obvious that the group concluded that IPCC had simply failed to notice it!

The Task Group therefore RECOMMENDS that:

The inter-relationship between climate and land use and urban planning be recognized as an area requiring expert study and assessment in relation to possible, enhanced, global change.

Whether or not global warming will occur as a result of the rapidly increasing atmospheric carbon dioxide, six factors will affect issues involving land use and urban planning, namely:

i. The rapid increase in world population, mostly in less well developed countries, resulting in increased demands for space, food, water, and energy.

Although this is particularly evident in the least developed countries (LDCs) it is certainly an important component in the encroachment of city boundaries into agricultural land in industrialised countries (ICs). The result is not infrequently described as 'planning blight'! In LDCs the reduction of forest through excessive cutting and destruction for fuelwood is yet another manifestation of both rural- and urban-generated population pressure, while the competing local, national, or even international demands for water e.g.by Egypt and its neighbours, is also, basically, a simple but appallingly serious consequence of population increase. These are all essentially planning problems and are unlikely to be resolved by categorizing them as simply agricultural, energy or water problems, and trying to deal separately with each of them as such.

ii.   The continuation of urbanization, which appears to be an unstoppable process, so that at least 50% of the world's population will be city dwellers by the 21st century. This poses the same problems as set out in (i), as well as leading to demands for work and social change.

Urbanization itself can lead to dramatic climatic change as exemplified by the development of 'hot spots'. Urbanization is an excessively difficult problem to deal with from the planning point of view, as the very large numbers of highly trained planners involved in trying to cope with it in mega-cities such as Mexico City, have found. More often, it has simply happened and little has been done to regulate it leading, for example, to the 'shanty towns' of the LDCs in S.America, Africa and Asia, developments which are often immediately adjacent to well planned areas, even though, architecturally, their buildings may not be well adapted to the climate. Depending upon the local regime they too may make more or less undesirable demands on the environment. The presence of large numbers of frequently ill-educated, urban poor inevitably results in heightened demands for work, leading to a surplus of cheap, exploited labour. Frequently it is employed in factories which produce environmentally undesirable emissions or wastes. All these conditions also foster social unrest.

iii.   The demand for water, whether for domestic use, agriculture or industry, is increasing throughout the world and no country is exempt either from some form of shortage, and/or depletion of irreplaceable groundwater.

This hardly needs to be documented save that too often it does not receive the public attention which its importance merits Indeed, at times it appears to be deliberately played down so far as the ordinary citizen is concerned. The abundance of major schemes, in the USSR, China, and potentially in Canada-USA-Mexico, or India, to rechannel complete river systems are examples of governmental recognition. Such schemes have major planning implications which, if adopted could provoke major feedbacks to climate in whole regions, quite apart from their poorly predicted effects on the lives of innumerable people. The possible consequences of such schemes need far more detailed study and assessment - hydrological, meteorological, economic and sociological - before they are initiated than they have received to date.

iv.   Pollution is increasing, both terrestrial and aerial, affecting living conditions, agriculture, water supplies, and the atmosphere, with comparatively well documented feedback effects.

A great deal can be done through wise planning, especially in conurbations, to reduce or ameliorate these effects, as is described later in this report.

v.   Factors i-iv will increasingly lead to increased social stress and, potentially, social disintegration.

vi.   Variation in climate and local weather conditions affect all these factors to a greater or lesser extent, just as there are feedback effects of greater or lesser magnitude on climatic factors. Intelligent land use or urban planning can reduce some of these interactions, or at least ensure that their consequences are greatly buffered.

Any appreciable increase in global temperature and associated climatic change will not only exacerbate the consequences of all these factors but will undoubtedly add new problems, such as an overall global decrease in water available for agriculture. Regionally distinct impacts such as a potential northward shift of climates suitable for wheat growth in Northern Canada, have also been suggested. At present such predictions are poorly quantified but adaptation to, or amelioration of, such effects will best be met if a sound basis exists for land use, and the potential and susceptibility of the land to change has been assessed.

The group is concerned that action can and should be taken forthwith to remedy or ameliorate the consequences of the six factors already affecting land use and urban planning and which will continue to affect these activities, no doubt in unexpected or different ways, if global warming takes place.

Action now could help both to reduce possible future effects as well as providing experience in managing environmentally favourable planning that could be adopted if global warming were to become a reality.

In any event it would have economically desirable consequences now.

In much of the world, baseline data neither exists for rational planning nor for resolving competing claims for land use. This situation should be rectified. It will be a demanding task but it is essential to have a better baseline not only for planning now but in preparation for change of any kind. It is becoming feasible to contemplate a vast improvement using modern technology and by transferring technology and training individuals in those countries without adequate staff. This will have to be supported by funding from IC's, who on the whole are well advanced in baseline land use data, although at various stages in their use in planning. Remote sensing data can now supply a basic framework, although that will need to be backed up by observations on the ground. LDCs would need to be able to obtain the former, which is very expensive, as well as training in the interpretation of such data, and methods of ground based data collection. Of course there is a framework of data collected by agricultural and/or forestry departments in most countries but the information is patchy and variable in quality and quantity. It is essential to act quickly, in view of the possibility of enhanced global warming.

The Task Group therefore RECOMMENDS that:

> An assessment and inventory of the (best) potential use for land be completed by every country by 2000 to internationally agreed, minimal standards appropriate for effective use by planners, and for the effective assessment of susceptibility to climatic change.

The group is deeply concerned by the numerous, worldwide incursions into, and loss of, good agricultural land, having regard to the fact that only about 11% of the land surface is used for cultivated crops and that most of the land best suited for such purposes is already under cultivation.

The Task Group further RECOMMENDS that:

> Land use data already available be used as soon as possible, if need be on an ad hoc basis, to protect good agricultural land from use for other purposes.

The group was equally concerned both by the known lack of baseline data on water resources, their depletion, quality, uses (measured quantitatively) and present and forseeable demands. They recognize that the predictions of the IPCC Working Groups of the possible effects of global warming on water availability and possible changes in availability in different regions are very tentative. But the present imperfectly known position, let alone any possible future situation, are sufficiently alarming to warrant a similar effort to that proposed for land use, over a similar timescale, in order to obtain an adequate baseline. The effort and costs involved are likely to be comparable with those enumerated for the land use survey. If anything, aid and training from ICs to LDCs will need to be even greater.

The Task Group therefore RECOMMENDS that:

> A contemporary survey and assessment to minimal, agreed international standards be made by all countries of all available water resources, their variability, quality, present and immediately forseeable uses and demands, together with an assessment of their environmental vulnerability, by 2000.

Building, both rural and urban, as well as urban planning needs to be pursued and developed in a more environmentally desirable manner. An appreciable proportion of rural building already utilizes traditional

materials and techniques which, in many parts of the world are well adapted to the local climatic regime. This is less so in some urban areas. Indeed, in many LDCs in particular, there has too often been a thoughtless miming, both in construction and design, of building fashion adopted in ICs, much of which is climatically inappropriate even in the countries of origin! One need only cite high rise, steel and concrete, highly energy-consuming buildings which occur in too many tropical cities. It is an urgent necessity that local, traditional techniques be combined with those which make the best use of local climatic conditions both for external and internal ventilation, for appropriate insulation, and for internal heating and cooling.

Such actions in themselves will promote energy saving, but that can be further promoted through further research into low-cost, low energy-consuming, easy-to-build systems for building, and devices for lighting/heating/cooling buildings. New techniques developed in ICs are already available but their direct transfer would probably not be the best way of achieving similar objectives in LDCs, mainly because of the high investment costs. An example which has been used in low-income suburbs in San Diego (California) and in India is the free issue of low-energy consuming light bulbs. Succesful financial returns followed rapidly through reduced costs to the utilities, because of the greater fuel economy achieved. In India, there was a reduced incidence of blackouts at peak periods.

This, in turn, reduced the costs of a number of industrial concerns which had invested in costly back-up generators to cover blackouts, and which were diesel driven. Thus, this simple and relatively cheap action resulted not only in reduced energy consumption but also in reduced costs to ordinary consumers, companies and the utilities together with reduced pollution effects. The degree, however, to which any device or technique can be adopted by any LDC will be limited by the extent to which it can be afforded.

In the urban environment there is a long list of requirements which need to be met if buildings are to be adapted adequately to the prevailing climatic regimes. These include:

- access to relevant climatic information and appropriate technologies
- an understanding of climatic impacts, both in and on urban areas, and an understanding of the means and possibilities of controlling or modifying such impacts
- costed assessments of the values, dangers and profits/losses of climatic impacts on urban buildings and areas.

Such an approach demands a well developed planning and information system, interdisciplinary co-operation and clearly established, objectively defined planning goals for urban development. This cannot be achieved without considerable cost, sophistication and training and much further research. However, some knowledge is already available and has been applied.

The general design and layout of urban areas should, in addition, be such as to promote a locally favourable climate. One particularly important consideration is to reduce the 'heat island' effect. One of the simplest, most attractive and cheapest techniques is to employ extensive tree planting coupled with lowrise, well spaced yet dense urban areas which are thereby rendered both more comfortable and attractive. In tropical regions this can be supplemented by roof gardens and a judicious interposition of higher and lower developments to ensure maximum light penetration and shading with maximum cooling. Space can also be made available for horticultural plantings, providing on-the-spot work and reducing the energy costs of distribution.

An important element in urban design is, of course, the transport system and this is equally a major source of pollution. This is a fairly intractable problem involving as it does behavioural problems such as commuting patterns and attitudes to car ownership and mass transport. Matters of energy saving and pollution reduction in relation to traffic are dealt with elsewhere. However, the technical solutions to these aspects can be enhanced by careful urban planning, including the right siting of industrial buildings, the notion of 'villages within a city' and so on. It also needs to be said that many attempts to promote effective mass transport have failed simply because the vehicles have been uncomfortable and unattractive, or wrongly routed in relation to work patterns. The 'park and ride' systems in which commuters leave their cars at city limits and enter the city by public mass transport are, however, ameliorating traffic congestion appreciably in some cities while, at the same time, promoting a more favourable attitude towards mass transport.

An inevitable concomitant of urbanization is the problem of waste. The traditional solution of transporting it to landfill sites with the consequential problems of methane production is not satisfactory. Local, controlled incineration is often more environmentally acceptable and can be used to generate energy on a local scale (this is discussed in both the IPCC reports and elsewhere in this volume). Equally possible and in many ways more attractive is to recycle much of it, although this may require more sophisticated collecting procedures and changes in behaviour patterns of the citizens. Organic waste, in particular, can be readily recycled by using it for horticultural purposes and this is an especially attractive solution in tropical countries where the turnover is more rapid. Indeed, in many places, Calcutta for instance, this use has grown up spontaneously as the result

of the activities of the homeless and jobless, to their great economic benefit, to the benefit of the city - which gets fresh vegetables thereby - and to the environment.

There is, of course, an important role for legislation in all these issues. Improved, climatically-friendly building and design can be promoted by updating existing building and planning regulations. The former are more likely to be motivated, at present, by considerations of safety than by environmental considerations. However, these objectives are not mutually incompatible.

It is not possible to detail here all the ways in which the climate can be used or ameliorated in the rural or urban environment through intelligent planning. It will equally be obvious that this is a field in which there is a great need for wider application and further research of an inter-disciplinary nature. Schools of Architecture and Planning need to consider local or regional climatic problems and their solution far more widely, and to devote more time to such considerations. There is need for international dissemination of best contemporary practice. These areas are, traditionally, not well supported in research. Moreover, the architectural and planning professions have, traditionally, been somewhat conservative about such matters. Governments and funding agencies need to do more, especially in the area of climatic adaptation and more energy efficient design and construction. But research needs to be holistic in approach as well as in application.

The Task Group therefore RECOMMENDS that:

a. The design and construction of buildings and urban planning should be better adapted to climate. This can be promoted by more training, updating of building and planning regulations, and further research. Greater attention should be paid to both density and layout of urban areas in relation to climatic considerations.

b. Urban systems should be designed and constructed to ensure maximum energy efficiency, domestically, industrially and in their transport systems, with minimum 'greenhouse gas' emissions, to which conurbations are major contributors.

c. Waste management should maximize recycling with minimal energy expenditure and maximum biological utilization.

d. More training should be given to architects and planners in the principles and practice of adapting construction and design to local or regional climatic circumstances, and to generally improving both internal and external environments by the use of low-energy demanding devices and techniques with particular attention to low-cost solutions for LDCs.

e. An international, interdisciplinary panel should be convened as part of the ongoing IPCC programme to further examine these issues, to promote standards, encourage wider dissemination of these ideas and to promote research involving architects, planners, behaviourists, climatologists and meteorologists.

Finally, the group considered briefly the possible scenarios of change which had been presented in the IPCC Workng Group reports with a view to determining if there were any areas within their remit in which further research was desirable. They were convinced that appreciably greater areas of land were likely to be exposed to aridity and desertification as a possible consequence of climatic change and human behaviour and fertility. They believed that more research was necessary into the prevention and containment of such areas and into the improvemnt of their productivity and, if possible, rehabilitation. Having regard also to the uncertain estimates of future food production and the limited amount of good agricultural land available, they believed that, coupled with oceanographic studies and the IGBP programme, greater effort should be put into fisheries research and development.

The Task Group therefore RECOMMENDS that:

a. The research effort into the prevention, containment, increased productivity and possibilities of rehabilitation of Arid and Semi-arid lands should be greatly increased, both nationally and internationally.
b. There should be an increase in fisheries research in connection with oceanographic research and the IGBP programme with the expectation that there is likely to be a greater dependence on fish as a renewable food resource.

All these recommendations will be expensive. They will involve technology transfer almost entirely from developed to developing countries, but their achievement will be beneficial, in the long run, to all. They will not be started, let alone achieved, without a massive continuing and informed marketing campaign directed to promoting positive action by Governments, coupled with an equally massive worldwide programme of environmental education.

Task Group 6

Human Dimensions of Climate Change

Chair:
R. E. Munn *(Canada)*

Discussion Leader:
A. Auliciems *(Australia)*

Rapporteurs:
S. Kane *(USA)*
J. McCulloch *(Canada)*
R. Mertens *(Belgium)*
T. Robertson *(UK/WWF)*

Introduction

The members of the Task Group came from the physical, biological, socioeconomic and health sciences, and thus feel that this report has much validity. Discussion was spirited, reflecting varied backgrounds from both developed and developing countries. The recommendations are unanimous.

Interactions Between Humans and Climate

The human dimensions of climate change have been, until now, seriously undervalued. Existing programmes dealing with the human dimensions of climate change have concentrated on health and have not had social demographic and economic aspects included in their terms of reference. The Task Group has no hesitation in recommending that these latter aspects be given high priority. Also, thorough economic analysis of the consequences (for example, on employment, trade, gross domestic product, species diversity, etc.) and feasibility of alternate response strategies is essential to provide an information base for credible policy making.

The three aspects of the human dimensions of climate change are:

- humans as contributors to climate change through their activities
- humans as those affected by climate change
- humans as agents to limit climate change and to adapt to its consequences.

There are major gaps in the knowledge applicable to each of the above-named aspects. The research required will, increasingly, require close collaboration among those in the physical, biological, social and health sciences; this will require a considerable adjustment by all members of the research community. Existing international organizations, both governmental (such as WMO, UNEP, FAO, UNESCO and WHO) and non-governmental (particularly ICSU, ISSC and their member scientific associations) must work together to foster national and international research.

Recommendation

The social, demographic and economic dimensions of climate change should receive priority.

Human Health

There is agreement that, in the interrelationships between society and a changing atmospheric environment, serious direct and indirect impacts will not be shared equally, but will exacerbate existing inequities; developing countries, which now lag seriously in both health and availability of high quality health care, will fall even further behind. Programs working toward improving equity in health should be supported, and need to take climate-induced impacts on health into account.

Direct impacts on human health include:

- heat disorder, including increased mortality from heat strokes in cities
- increased incidence of skin cancers and eye disorders, from increased UV-B radiation
- death or injury due to extreme weather events such as severe storms, floods and the like.

Indirect impacts of climate on human health include:

- impacts mainly due to climate effects on biological determinants of health
- increased incidence of vector-borne communicable disease
- increased incidence of air-, water-, soil- and food-borne communicable diseases
- impacts due to increased pollution potential
- increased morbidity and mortality from respiratory diseases due to air pollution
- spread of toxic or radioactive materials from waste dump sites and nuclear/industrial sites
- impacts associated with climate effects on socio-economic determinants of health
- increased famine and chronic undernutrition
- disease due to inadequate supply of safe water
- restricted access to health care.

The last three are potentially the most severe, and can be related to agricultural and socio-economic disruption, forced migration, overpopulation, and catastrophic weather events.

Recommendations

i) The health care sector should be represented in all bodies dealing with prevention of and response to climate change.

ii) Existing research programmes on the health effects of climate variability and on the sensitivity of health impacts to various scenarios of climate change should be expanded. As a first step, WMO/WHO should prepare technical reports synthesizing existing work on the direct and indirect health effects that would occur if the climate were to change. These should include reviews of the more speculative but potentially important health effects, such as direct and indirect effects on fertility.

iii) Coordinated regional efforts should be undertaken to strengthen and expand epidemiologic surveillance for timely assessment of incipient changes and of evaluation of preventive or corrective actions. Serious problems may arise in all parts of the world.

iv) One of the outcomes of climate change is an increase in the inequity in human health between developing and industrialized countries. Programmes working towards improving equity in health should be supported and should take climate-induced impacts on health into account.

Especially Vulnerable Populations

Some societies, often non-industrialized, live in geographical areas that are especially vulnerable to the impacts of climate change.

In some cases, an increase in the severity of natural hazards, such as tropical storms, floods and droughts, may make an area unfit for continued occupancy, creating environmental refugees. In other cases, local economic activities may no longer be viable, and people may be uprooted. In these two cases, both the refugees and those located in places to which the displaced emigrate are affected. Island states could be affected in both ways, as well as by rising sea level.

Many others whose lifestyle and culture are integrated with the natural environment are also especially vulnerable. An illustrative example is provided by the native peoples of the Arctic, for whom the survival of both person and culture depends on a delicate balance between the land and water on the one hand, and natural ecosystems on the other. The impacts on ice and permafrost suggested by current scenarios of future climate would be catastrophic for that balance.

Some research suggests that persons in large countries with a range of climates will be able to adapt better to climate change than those in small countries without climatic diversity. The greatest problems will be faced by those in small countries with large populations.

In some cases, a contributing factor is the lack of adaptive capacity, in part due to a low level of technological sophistication and economic resources. Particular attention should be paid to the development of adaptive strategies that are within the capacity of affected groups to implement. In many cases, the only possible adaptive strategy will be uprooting.

Recommendation

A Working Group on communities that are especially vulnerable to climate variability and climate change should be created by the WMO and collaborating bodies.

Socio-Economic and Demographic Data and Research

There are barriers to research of the kind necessary to address the problems posed by the human dimensions to climate change. Foremost among these is the trend to increasingly narrow disciplinary research in many institutions. Not only must researchers broaden their view within their particular discipline, but they must also take part in interdisciplinary activities. In this context, "Interdisciplinary" implies crossing the boundaries among the physical, biological, social and health sciences; all must contribute as full and equal partners to the work necessary to solve the problems.

Another major barrier has been the lack of availability of appropriate socio-economic and demographic data. Data from different disciplines (in some cases within disciplines) have not been compatible because of differing time or space scales, lack of intercalibration, inadequate quality control, unsuitable archiving formats, or similar shortcomings. On-going efforts such as UNEP/GRID should be supported and expanded. The programme for research on the human dimensions of global environmental change, under the aegis of the International Social Sciences Council, should be supported for its contribution to solving many of these problems with socio-economic data.

The Task Group recognizes that current climate models have their weaknesses. At the same time, there are few models available that can provide good approximations to the impact of changing climate for given economic trends. Models to predict the effects on society of changes of climate and consequent economic conditions are virtually nonexistent. Until these gaps are filled, reliable estimates of the details of social and economic impacts of climate variability and climate change, including both positive and negative effects, will not be possible.

The present generation of climate models cannot be used with much confidence for prediction on regional scales. This means that, in the development of national policies for coping with the socioeconomic impacts of climate change, strategies that are resilient (that is robust over a range of possible futures) should be sought.

It must be emphasized that, over the coming decades, social, economic and demographic variables are likely to undergo considerable change due to developments unrelated to climate. For example, biotechnology, transport engineering, population distribution and the like, are constantly changing. Assessments of the human dimensions of climate change must attempt to take these other changes into account. Ultimately, population-environment-resources-development models will prove extremely valuable, and their creation should be encouraged. High priority should be given to the role of technology in driving the extent of the human contribution to climate change, and responses to this change. As well, high priority must be given to the study of determinants of change in the economies of developing countries, and of the relationship between national economies and policies on natural resources.

Recommendations

i) Interdisciplinary research on human behaviour (including attitudes and perceptions) as a contribution to understanding the causes and effects of climate change should be promoted. To facilitate studies of the links between climate and man, existing scientific organizations, such as the International Society of Biometeorology (ISB) should be consulted.

ii) Mechanisms should be established within the World Climate Programme, including national components, to facilitate coordination amongst researchers on human behaviour contributing to climate change.

iii) A programme should be established to link economic, social, demographic, and climate studies to understand critical links between human societies and the environment.

iv) Studies within UNEP, and collaborating institutions, of the social, demographic, and economic dimensions of climate change should be strengthened.

v) A Working Group should be formed by UNEP and collaborating bodies to analyze the consequences, both positive and negative, of climate variability and climate change. The Working Group should also be tasked to analyze alternative strategies, including combinations of various economic incentives, to limit and adapt to climate change.

vi) Many socio-economic and demographic data essential to understanding climate change are not available or accessible. These should be collected, collated and made available through UNEP and collaborating bodies.

Cultural Heritage

Equally threatened is humanity's cultural heritage as represented by ancient monuments, buildings, works of art and other priceless relics of the past. Air pollutants, climate change and the inexorable rise of the world's oceans are in combination with other factors, such as development and wars, causing the destruction of antiquities around the globe. Unless research into the preservation of these manifestations of our cultural heritage is carried out, and the results implemented, they will be lost to us and to future generations.

Recommendation

A global programme for the detection and monitoring of climate-induced damage to humanity's cultural heritage and the assessment of the risks posed to the antiquities from climate variability and climate change should be created by UNESCO, in collaboration with WMO.

Public Information

If human attitudes and consumption patterns are to be changed, it is vital that information access, public education and awareness programmes receive high priority. Governments as well as non-governmental organizations have important responsibilities in these domains.

Recommendation

High priority should be given to information access, public education and awareness programmes, and other requirements necessary for changes in human attitudes and consumption patterns.

Economic Disparities

Both the level of development of a country and its economic wealth determine its ability to participate in the research mentioned above, and to implement the measures, identified by the research, to limit or adapt to climate change. We recognize that this issue is much broader than the remit of the Task Group and thus we make no specific recommendation.

Philosophy and Ethics

The Task Group regrets the apparent absence of philosophical or ethical analysis of either the basis of climate change or of the bases on which to build preventative or corrective actions.

Recommendation

There has been little philosophical/ethical input into the climate change debate. The Task Group recommends explicit analysis of these elements. This may in the long run be the most important of the Task Group's recommendations.

Task Group 7

Environment and Development

Chair:
M. Holdgate *(IUCN)*

Discussion Leader:
O. Mascarenhas *(Tanzania)*

Rapporteurs:
P. Bakken *(Norway)*
R. P. Karimanzira *(Zimbabwe)*
J. MacNeill *(Canada)*

Introduction

Task Group 7 considered the impact of climate, and projected climate change, on the environment and on the process of development to meet human needs. It took as a point of departure the necessity for development, itself based on the conservation of the resources of the environment, in order to lift millions of people above the basic survival level and to enhance their quality of life and that of their descendants. As with other Task Groups, the mandate was to provide recommendations which could be synthesized into the Conference Statement, and might influence the briefing to be provided for the Ministerial sessions.

The Group worked harmoniously and effectively, and thanks are due to the Discussion Leader, Rapporteurs and all participants.

Opening the meeting, the Chairman reminded the Group of the background briefing document he had prepared, at the request of the Conference Secretariat, and the generic questions which were asked in that connection. They covered:

- the adequacy of basic data in the Task Group's field
- the adequacy of scientific knowledge as a basis for decisions in that field
- the availability of methods for the application of that knowledge in practical action and policy
- the adequacy of understanding of known and potential impacts of climate change on environment and development
- the main topics requiring further study

- recommendations regarding the future evolution of the World Climate Programme and any successor body to the IPCC
- scientific and technical inputs that should be made to political decision-making and socio-economic planning in both developed and developing countries.

The Chairman also referred to a note that had been provided by Dr W. Riebsame, a member of the Group, and explained the outcome of an informal meeting which he had had on the previous day with the Discussion Leader and two of the three Rapporteurs. That discussion had led to the proposal that the Task Group should only take very brief note of the mechanisms of global change, as indicated by the IPCC but extended by various social and economic factors, and should then concentrate its discussion on two key issues:

- the environmental and social risks that might arise within the development process as a consequence of climatic change
- the priorities for action to reduce such risks, and increase the probability of environmentally sustainable development.

The Task Group's discussion followed this pattern.

Global Change and the Development Process

Climate change of the magnitude projected in the IPCC report would, as Working Group II in particular has indicated, have profound implications for the sustainability of the development process. Agricultural productivity would be diminished in many areas, thereby reducing the availability of food. Water - already a limiting resource in many regions - would become even more limiting. Loss of soil - again a critical problem already - might be exacerbated, and ecosystems would be disrupted as much by the extremely rapid rate of change as by the magnitude of the global warming indicated. The losses of biological diversity likely to follow would reduce the genetic resources available to meet human needs in an increasingly demanding future. The capacity of species to redistribute, and the enforced adjustment in agricultural and forestry practices and human settlement processes would obviously impose a major stress on the biosphere.

Serious though these changes might be, the Task Group emphasized that they must be viewed within a much wider environmental and socio-economic context. They would add to already acute pressures on the environment resulting from mounting human populations and rising human resource consumption. Even without climate change, the development process needs substantial adjustment in many regions if it is to be sustainable - that is, if it is to meet human needs today without destroying the environmental base essential to the satisfaction of tomorrow's needs. Since climate change is superimposed on demographic pressures, current over-consumption patterns, inequalities in the distribution of resources available for development, and significant regional variations in environmental resilience, the development processes in many parts of the world could well disintegrate. The danger of such disintegration is clearly greatest in countries with already high human pressures on the environment and limited social and environmental resilience.

The Task Group looked carefully at the nature of the environmental and social risks that climate change would bring. It concluded that the biophysical impacts, especially on agriculture and forestry, but also on the habitability of the coastal zones which are the home of the majority of humanity, would lead to a number of specific and severe risks including:

- the dislocation of human communities because of a reduction in the capacity of certain regions and ecosystems to support people (obvious examples are low-lying coastal and island states, and desert areas if aridity increases)
- the acceleration of current unsustainable practices such as over-use of marginal lands, leading to further soil degradation and deforestation
- a reduced social capacity for adaptation, because of the increased severity and frequency of extreme events
- the diversion of an increasing proportion of scarce resources from the sustainable development of economically weak areas to disaster relief.

Other socio-economic consequences would include risks to the already precarious quality of life in many places, increased poverty, and greater risks to health, especially among children and other vulnerable sectors of the community. Psychological stress is also likely to increase because of greater uncertainty and greater exposure to devastating hazards like cyclones and droughts. The consequence could well be major changes in the distribution of people, ranging from migration from rural areas to the cities and expansion of urban sprawl to migration across national frontiers by an ever-increasing number of "environmental refugees". It is clear that many communities in the world are heading for catastrophe unless there is a significant change in policy. Indeed today's policies are increasing the risk of disruption of development as a result of global change.

Priorities for Action

It is imperative to improve the prospects of sustainable growth. This can only come by conserving existing resources of the environment, and by ensuring that the development pathways adopted are viable in the face of likely climate change. Risk avoidance and risk reduction are also essential (and underlie the so-called "precautionary principle"). Communities need to improve their capacity to deal with change, and to adapt, wherever possible making the best of new conditions even if this means changes in lifestyle. Degraded environments will need rehabilitation, and given that climate change is being caused by countries other than those that will experience its worst impacts, questions of compensation and resource transfer will need to be addressed. The difficult issue is how costs should be distributed, bearing in mind the difficulty of ascertaining precisely who is responsible for what share of global change and who will be the greatest victims.

The Task Group concluded that it would be dangerous for any country to opt out of the measures needed for a concerted global response. Regional variation is inevitable, and some states (such as low-lying coastal countries) are especially vulnerable. But the concept of "winners" and "losers" is dangerous. Even countries that experience climatic amelioration are likely to face formidable challenges of adaptation. The identification of such apparent winners is, moveover, difficult and in an interdependent world short-term winners may prove to be long-term losers. Overall, we are not dealing with a zero-sum game but a significant burden on humanity as a whole. The stresses on the majority of states - "losers" in this sense - are likely to be transmitted through the system of global interaction to affect those on whom climate change bears less heavily. The social goal has to be a net win for humanity as a whole, and even if some countries may be less disadvantaged than others by climate change, this is no justification for an isolationist policy. Climate change will clearly increase vulnerability and stress the limited adaptive capability of many countries.

The priority is for actions that will reduce risk and increase the prospects for sustainable development. Such actions will be more likely if the resources of the environment are properly valued in economic terms. Improved resource accounting for the "natural capital" of assets like soil, water, biological diversity and other environmental components will in turn lead to sounder judgement between alternative uses of these resources. There is a linked need for monitoring of changes in national "natural capital", giving positive feedback to decision-makers.

Market distortions also need to be removed. They result in the wrong price signals, leading to the waste of energy, water and other resources. In many countries, groundwater and forests are being "mined" in a non-renewable way because financial policies favour such a course. The World Commission on Environment and Development showed that economic policy currently distorts the market against the environment and towards global warming and sea level rise. Subsidies for fossil fuel use, continuing externalization of costs by industrial technology, resource consumption patterns that aggravate waste, and world trading patterns that effectively lock developing country produce out of developed country markets are all working against the sustainable development and will aggravate vulnerability to climate change. At present, billions of dollars are being spent annually in subsidies as against millions in environmental protection, and a change in this balance is imperative.

The Task Group identified the following key principles and policies, as a guide to action:

1. Regional strategies and regional assessments will be needed. There is considerable regional variation in the environment and in socio-economic situations, and climate change itself will vary regionally. Global generalization can obscure the real needs, and blunt the effectiveness of response.

2. All countries must participate in this response; for reasons already stated, there are no grounds for apparent beneficiaries from climate change to opt out. Change poses a threat to the sustainability of socio-economic development throughout our interdependent world.

3. Action should go beyond so-called "no regrets policies" which will have a pay-back even if climate change does not materialize. To urge such policies implies that climate change is neither certain enough nor serious enough to deserve a response in its own right. The Task Group's view is that this is not the case. Climate change poses sufficiently severe potential problems to be addressed now. Policies that are not sustainable because they have an excessive impact on the environment or consume excessive resources and will hence prove "regrets policies", already pervade much of the industrialized world. The distorting impact of subsidization, promoting excessive use of energy and so favouring climate change, has already been noted.

4. It is imperative to adopt policies that enhance resilience in the face of scenarios of likely change, and avoid the immense long-term costs of reconstruction that would be required if the development process actually increased vulnerability to climate change - as is happening in some coastal zones whose natural defensive systems against storms and surges have been undermined by misuse.

5. Sustainable development requires as predictable a social context for its planning as possible. In a phrase of deliberate ambiguity, the Task Force noted that it required a "stable atmosphere". Climate change clearly reduces the confidence within which development can be planned, and can not only have a negative impact on development but consume a large degree of resources in order to provide for uncertainty.

6. If development is to be sustainable worldwide, the developed countries must adjust their technological and resource consumption patterns, and the developing countries must be aided to introduce the best available technology, providing for their development needs with the least practicable environmental impact. Unless the development needs of developing countries are met, the success of the global effort will be compromised.

7. Greenhouse gas emission limitation is essential and any measures that will assist in slowing down climate change will directly assist the sustainability of development patterns, giving communities some breathing space for adaptation and reducing the resources that will need to be devoted to what are essentially insurance policies against uncertainty. The Task Force did not discuss specific abatement targets, but it did affirm the need for the maximum attainable reductions in greenhouse gas emissions from the developed world, and the minimization of new emissions in the developing world through the transfer of new technology. It is appreciated that much of that technology has been developed by the

business sector of the developed world, which has vested property rights in it. Accordingly, mechanisms need to be found for funding the transfer of such technology without loss to its inventors. Patent and property rights are currently protected under an international agreement (the Paris Convention) and this may need review.

8. A new international Convention on Climate is welcomed if it is a practical, effective instrument that promotes action. It should not only provide a framework for the reduction of greenhouse gas emissions and concentrations, for the development of more effective new technology and for the transfer of both information and technology from developed to developing countries. It must also provide coherent policies and commit its States Parties to the actions needed for achieving these objectives.

9. Equity is important. It must be addressed in future measures, including the proposed Climate Convention. Unilateral actions in advance of the adoption of a Convention should be encouraged (as in Europe) and will need to be given credit in any targets set in the new Convention.

10. The Task Force is clear that the key to the future lies in dialogue between developed and developing countries, and within both between governments, scientific institutions, environmental groups and industry. This is essential now, in the context of preparation for the 1992 United Nations Conference on Environment and Development, and in the preparation of a World Climate Convention, but it is also essential thereafter. In the Task Force's view, the process of information and material exchange must be a continuing one, as far into the future as we can see.

11. That dialogue must promote the transfer of knowledge and information, and of methodology for monitoring, environmental resource accounting, technology development, evaluation and transfer, and institution and capacity building.

12. Research will be important to support this dialogue and policy reform, but Task Group 7 did not emphasize research needs in its debate. The Group was, however, concerned that agencies and programmes focussing on climate, like the World Climate Programme, IGBP and IPCC, should interact with agencies concerned with environment and development, and that the whole process should address issues of concern to developing as well as

developed countries. UNEP, UNDP, UNCTAD, WIPO and other major UN agencies have a major contribution to make, in partnership with climatological specialists.

Task Group 8

Forests

Chair:
J. Gilbert *(New Zealand)*

Discussion Leader:
H.H. Shugart *(USA)*

Rapporteurs:
H. El-Lakany *(Egypt)*
K. Ramakrishna *(India)*
E. Salati *(Brazil)*

Introduction

Forest issues are now high on the agenda of the world's political leaders because of their environmental, economic, social and cultural significance. The initial focus of concern for tropical forests has expanded to embrace all the world's forests.

The world's forests are a vital link in the exchange of carbon dioxide (CO_2) and other greenhouse gases (GHG) between the atmosphere and the biosphere. Changes in forest cover will therefore have an impact on this exchange and the concentration of these gases in the atmosphere. Increasing forest biomass through reforestation and afforestation, for example, enhances removal of CO_2 from the atmosphere, whereas decreasing forest cover reduces the amount of CO_2 stored in the forests and thereby contributes to increased concentration of atmospheric CO_2.

The relative contribution of forests to the greenhouse effect, however, must be kept in perspective. Deforestation and forest fires have only partially contributed to the observed increased concentration of GHG.

Storing additional carbon in forests by reforestation and afforestation can provide only a partial solution to reducing atmospheric carbon.

Forests could be placed in jeopardy by both the extent and speed of climate change. Predicted ranges of temperature increase and changes in precipitation will likely exceed those over which many species can survive, thereby causing local extinctions and range shifts. Furthermore, the anticipated rate of climate change exceeds the rate at which many species can migrate.

A number of approaches and initiatives are being proposed to address the complex issues involving the forest-climate interface. These include increasing the world's forest cover and biomass, resolving scientific uncertainties, developing effective strategies to cope with the consequences of anticipated climate change, and formulating internationally acceptable policies and institutional arrangements (e.g., conventions on climate change, biodiversity, and the world's forests). There is an urgent need to press forward with these initiatives in an integrated and coordinated manner.

Within the specific context of climate change, it is important to note that:

- total forest cover, distribution of forest types, productivity, species composition and vulnerability to fire, diseases and pests will certainly be affected
- air-borne pollutants and other factors may further increase the impact of climate change on individual species and forest types
- impacts on forests could lead to the dislocation of forest industries and of forest dependent communities, and will affect the availability of forest products and other forest derived benefits; also, there may be increased demands on forest land as a result of economic and social disruptions, and pressures to convert forest land to agricultural uses and to compensate for farmland affected by climate change or by rising sea levels
- forest-climate change issues have crucial cross connections with other global issues such as

population, poverty, energy, environmental and agricultural policies

- in view of the environmental, economic, social and cultural significance of the world's forests, their role as carbon sinks and as contributors to increasing concentrations of GHGs cannot be treated in an isolated and fractionated manner, and without consideration of their other functions, outputs and values.

Improving the Knowledge Base

Inventory and Monitoring

We recognize that policy actions depend upon and must be accompanied by effective and reliable inventory and monitoring systems. This should include the development and implementation of appropriate analysis methodologies that would provide an indication of the health and productivity of forests and forest ecosystems.

Recommendations

Inventory and monitoring systems and analysis methodologies should evaluate:

- amounts and trends of various types of forest and their condition (health)
- rates of deforestation, reforestation and afforestation
- changes in both unmanaged and managed forests with particular attention paid to important ecotones which may shift with changes of climate, i.e., ecotones determined by total rainfall or potential evapotranspiration, by seasonality of rainfall or potential evapotranspiration, by temperature such as the boreal forest/tundra interface
- key indicator species
- the effects of catastrophic events.

Co-ordinated Interdisciplinary Scientific Studies

The science aspects of the forest-climate interface raise three issues: (i) the status of present knowledge; (ii) the adequacy of available knowledge on which to base action; and (iii) the additional information needed.

Scientific issues and research considerations can fall into three categories:

1. *Evaluation* issues that involve collating and synthesizing existing knowledge to design future experiments, monitoring schemes and response strategies
2. *Interaction* issues that involve determining the important inputs to, and outputs from, the forest sector to other sectors - human systems, other biospheric systems and the geophysical earth systems
3. *Anticipation* issues that involve ascertaining what studies will be needed to better understand the future functioning and role of forests in an altered climate.

Scientific studies must be carried out at different spatial and temporal scales. The essential and important aspect of scale in global change is that a scientific characterization of a phenomenon at one level of organization does not connote an automatic characterization of the equivalent phenomena at another level. The critical challenge is to understand the relevant phenomena at all appropriate scales ranging from leaf to tree to stand to ecosystem, to global levels.

Recommendations

National and international bodies and development assistance agencies are urged to set up coordinated, interdisciplinary studies of the following issues on local, national, regional and global levels:

- monitoring and synthesizing factors pertaining to the forest/climate interface including the response and adaptation of keystone species and forest ecosystems under climate change as well as protection of forests from biotic and abiotic factors
- undertaking long term process studies of the dynamics of GHGs on all levels ranging from individual tree to forest ecosystems with emphasis on the roles of boreal, temperate, and tropical forests and of forest wetlands in the global carbon cycle
- evaluating the potential contributions of urban trees, protective tree plantations, rehabilitation of degraded land through reforestation and afforestation, and agroforestry systems for the amelioration of local climate
- undertaking research integrating the sustainable production of renewable biomass energy crops in short-rotation, intensively managed plantations with environmental protection of biodiversity and life support services in "natural" forests
- assessing socio-economic aspects and implications of implementing climate-related forestry projects
- designing and implementing monitoring systems to determine conditions and change in forest ecosystems in response to anticipated climate changes.

Education and Awareness - Raising

We recognize that education and the availability of appropriate information is essential for the implementation of response strategies. The general public, scientific and technical communities, and policy and decision makers need to be made aware of the necessity of protecting forests and reforestation and afforestation, at the local, national, regional and global levels.

Recommendation

We therefore recommend that the necessary material be developed and that it should be designed to show that sustainable forest development makes good environmental, social, political and economic sense and to demonstrate the benefits and risks associated with forest management options.

Management Policies and Practices

The Forests Task Group recommends two strategies based on existing knowledge: (i) optimize the role of forests, forest soils and wood as a carbon sink, and (ii) mitigate adverse effects of climate change on forests and take advantage of potentially favourable effects. This type of approach is consistent with that presented by the IPCC.

Effect of Forests on Greenhouse Gases

Forests can contribute to reducing atmospheric CO_2 while at the same time providing a wide range of economic and social benefits.

The size and nature of the CO_2 sink related benefits is tied directly to forest area and conditions. The contribution of forests as both a source of, and sink for, CO_2 depends on whether deforestation can be slowed, arrested and eventually turned into a net increase in global forest area.

A major source of greenhouse gases is the burning of fossil fuels. While increasing forest cover can contribute to the slowing of global climate change, this is not the major cause nor the major cure for the problem. In relation to the energy CO_2 problem, forests can only provide some relief to what must continue to be an energy problem which must rely largely on energy related solutions.

The Noordwijk forest remit calling for an annual net increase of 12 million hectares of forests is an ambitious target requiring further analyses. A framework for these analyses should be adopted in 1992 at the UN Conference on Environment and Development, while an aggregated global survey should be completed at the latest by 1995 in association with the IPCC science group review. An international framework for coordination and technical and financial support for this programme should be decided on as soon as possible and should include an assessment of the institutional capacity to respond.

Recommendations

The following actions are needed:

- implementing country policies which retain under forest cover the maximum possible extent of existing forest. In addition to carbon sequestration, this will have benefits for the maintenance of biodiversity, watershed protection, a sustainable supply of forest products, and provide associated social, cultural and economic benefits
- managing, where possible, forests to maintain the optimum standing biomass and soil organic matter
- bringing all production forests under sustained yield management by the year 2000

- implementing sustainable short-rotation forestry for the production of renewable biomass energy to replace nonrenewable fossil fuels as a key criterion in reforestation and afforestation activities.
- implementing tree planting or vegetation recovery programmes which are socially, economically and environmentally justifiable on their own merits
- carrying out policy analyses of targets for increased forest biomass to be considered by the UN Conference on Environment and Development, and to contribute to an aggregated global survey to be completed at the latest by 1995
- reducing wood waste in the manufacture of forest products, and implementing better wood utilization and the promotion of longlife wood products.

Adaption to Climate Change

The role of air pollutants in contributing to forest decline may be greatly influenced by the additional environmental stress brought on by climate change. Similarly, insects and disease formerly limited by climate constraints may extend into new areas and add to climatic stress. Any extensive reductions in soil moisture or rainfall frequency may contribute to an increase in fire and to the need for major expenditures on fire protection.

The adverse effects of climate change can be reduced and possible advantages of change optimized by preserving the full genetic dlversity and by anticipating the possible direction of climatic change.

The Task Group supports the conclusions of the IPCC on the need for a phased and progressive approach that maintains as much flexibility as possible. This includes the development of short-term responses and an examination of long-term options that can be refined as our understanding of the environmental, social and economic benefits and risks improves.

Recommendations

The Task Group recommends:

- the proper development of systems of protected areas planned for resllience to climate change; this would involve more and enlarged protected areas reconfigured to include a range of altitudes, soils and moisture regimes to enable their ecosystems to respond to climatic shifts
- the advance identification of tree species populations or ecosystems particularly at risk and special measures to provide for them
- scientific definition of the range of ecological tolerances of forest species (especially 'keystone' species)
- introduction of anticipatory management options including planting of adapted species or genotypes based on available knowledge

- control of emissions producing air pollution
- research and development into the efficient use of fuels derived from forests as a substitute for fossil fuels. This includes the use of biomass as a traditional fuel for cooking and warmth, as well as a fuel in the transportation and industrial sectors.

International Agreements, Institutions and Assistance

Agreements and Conventions

There is an urgent need for a comprehensive global forest agreement dealing with the conservation and sustainable management of all the world's forests on a fair and equitable basis. Such an agreement must complement initiatives under way on the conventions on climate change and on biodiversity and needs to take account of all forest values including the ability to contribute to the reduction of atmospheric CO_2. This issue will be a major focus in a framework convention on climate change and will also be considered under the proposed biodiversity convention.

It is imperative that such conventions are worked on and put in place as soon as possible. Such work, however, should be based on current scientific, technical and socio-economic data to ensure that agreements are soundly based, and programmes developed at national, regional and international levels can be achieved.

It is most important that the particular values and demands being placed on forests do not undermine the role of forests in providing food, shelter and sustenance for a large portion of the world's population. It is therefore important that linkages, including legal measures, exist between such agreements/conventions to ensure that the basic values of forests as a habitat and a provider are not lost.

The major threat to the world's forests is human activity. A rapidly increasing world population and its needs for more cropland, grazing land and fuelwood is placing severe pressures on many forested areas, particularly in developing countries. These activities are central to subsistence lifestyles and any changes in these lifestyles will require considerable time and resources.

There are several bilateral and multilateral agencies concerned with aspects of international initiatives that affect the world's forests. Coordination of these is imperative within the context of forest-climate interactions. There needs to be an immediate review of the coverage of climate change related forest research, investigations and planning being undertaken by these organizations.

Recommendations

Issues of prime importance are:

- halting the destruction of primary forests and habitats of biological importance in boreal, temperate and tropical forests; this involves forest conservation and management strategies to maintain the long-term viability of forest ecosystems
- addressing the needs of local and indigenous populations and including their involvement in the management and protection strategies
- realizing that an international forestry agreement should not serve as a substitute for the responsibility of industrialized countries to reduce greenhouse gas emissions, nor sidestep the need to reform existing international mechanisms
- understanding that it is essential that negotiators for such an agreement include the many sectors affecting forests
- recognizing that forests are not to be looked at only as gene banks and carbon sinks since they have multiple values and purposes
- exploring the possibility of establishing an international network on global forest monitoring.

Specific Needs of Developing Countries:

Forest research, transfer of technology and the necessary changes required in developing countries should be considered as candidates for funding from any climate fund established. Developed countries should be asked to consider such funding as additional to current bilateral and multilateral agreements and where possible make additional funding available to international organizations with appropriate forestry programmes.

Recommendations

The actions needed to assist developing countries are:

- to strengthen in-country scientific capacity to address global scale issues of forest-climate interface
- to establish an international task force to advise on the environmental, economic and technological implications of various forest policy options
- to transfer to developing countries, on a concessional basis, the technologies needed for forest protection, management and enhancement
- to increase agricultural production and energy efficiency and thus reduce the pressure on forested land
- to consider economic, financial and regulatory mechanisms to encourage increased planting by the private forestry sector and at the community and individual landowner levels; such work should be a central part of the on-going considerations of the application of economic and market mechanisms which were identified under the IPCC.

Task Group 9

WCP Overview

Chair:
K. Browning *(United Kingdom)*

Discussion Leader:
Ye Duzheng *(China)*

Rapporteurs:
A. L. Alusa *(UNEP)*
F. De Queiroz *(Brazil)*
G. Pearman *(Australia)*

Mainstream Climate Research

All nations have a common need to understand and predict the variability and trends of climate. The goal is to develop an operational climate forecasting capability for time-scales of less than a year to a century or more. Climate research activities exist in many countries. While much of that research has been and will continue to be motivated by national priorities, the nature of climate is such that there are significant needs for international collaboration through appropriate international programmes such as the World Climate Research Programme (WCRP) and the International Geosphere Biosphere Programme (IGBP).

Mainstream climate research rests on three main pillars: (1) global numerical **models**, (2) the understanding of small scale **processes** so that they may be properly represented (parameterized) in the global models, and (3) **observations**, on a global basis for the validation and development of global models and the detection of climate change, and also on a regional basis for the study of processes. The key importance of global observations is dealt with separately in the next section. In this section we restrict attention to climate models and process studies.

Models lie at the heart of modern climate research within the WCRP and, in combination with observational data, are the principal tool through which we can gain greater understanding and predictive capability of global climate. A principal cause of errors in atmospheric models is the parameterization of processes, especially those associated with clouds and with land and sea surface interactions. These parameterizations need to be substantially improved.

The corresponding need in ocean models is the parameterization of mixing. Coupled atmospheric and ocean models are needed in order to be able to simulate the observed long term variations of the climate system. Other areas of climate modelling needing emphasis are the development of methods to study local and regional climates in addition to the large scale climate that is portrayed by current climate models, and also the development of models of ice sheets that can be used to estimate changes of global sea level. Also important are studies of solar output variation and of historic observations on time scales of decades to centuries or longer. Knowledge of how climate has changed in the past will be of value for climate model validation, as well as shedding some light on the causes of climate change. IGBP has a core project for this called PAGES (Past Global Changes).

In order to improve the parameterization of clouds within climate models it is necessary to carry out studies of cloud dynamical, microphysical and radiative processes. Clouds are important not only because of the feedback effect of their interaction with radiation, but also because of the associated precipitation which is an essential component in the hydrological cycle. These studies, along with land surface and other studies, will form a part of the WCRP GEWEX programme which is concerned with understanding the earth's water and energy cycles. Studies of processes in the ocean depend on the implementation of WOCE and also on long term global observations which should be undertaken for all levels in the ocean. Also of great importance is the understanding and incorporation into coupled models of processes which control the major biochemical cycles and the changes in the concentration of trace gases. These are key processes in controlling long term climate changes. IGBP's core projects, JGOFS (Joint Global Ocean Flux Study), IGAC (International Global Atmospheric Chemistry Project) and others, address this problem.

The World Climate Research Programme and the International Geosphere Biosphere Programme are complementary. Together they constitute a comprehensive undertaking aimed at improving our understanding of the complete climate system. There is no scope for additional large programmes in this area, and new topics, such as the simulation of the biosphere in the climate system, can be pursued within the context of the existing programmes, where appropriate as joint sub-programmes between WCRP and IGBP. Co-ordination can be effected by strong links at top level (between the JSC of WCRP and the SAC of IGBP) as well as at sub-programme level. Such links are already being forged. Further bureaucracy in the shape of new structures or tiers in climate research should be avoided.

Most of the developing countries have no financial resources to participate effectively in the climate research programmes. Attempts should be made to support regional research centres in order to study the poorly understood regional aspects of climate in these countries.

Observing Systems for Climate Research

A prerequisite for the reliable prediction of climate variability and change is a comprehensive and continuing series of global observations of the critical components of the climate system. The observing system is severely deficient for the ocean, for certain chemical and biological indicators, and for atmospheric and terrestrial aspects of the hydrological cycle. An operational system exists for the physical atmospheric variables, but it needs to be safeguarded and enhanced in many respects to fulfil the needs of climate prediction. This observing system is especially deficient in many developing countries and over ocean areas.

International programmes such as GARP and WCRP/TOGA have shown that existing observational networks can be enhanced and data can be managed, quality controlled and disseminated internationally to meet specific climate research requirements, but such enhancements have been provided mainly through national research sponsorship of limited duration and require the strong involvement of participating scientists. In order to make reliable climate prediction possible, it is necessary to establish a secure, continuing and truly global observational system to provide data for the development, initialization and validation of climate prediction models. New

technology, particularly satellite observations, automated measuring devices and data management systems must be exploited and co-ordinated internationally, drawing heavily on organizations with experience in running operational systems.

The additional requirements of climate prediction over those of short range weather forecasting, in terms of observed quantities, their continuity and global coverage, need to be better understood by governments and their component agencies. The requirements for a climate observing system have been outlined in the IPCC WG1 report and will be further refined in the light of WCRP and IGBP experiments.

The Task Group considers that the active effort of all relevant international governmental and non-governmental agencies should be called upon to facilitate the creation of the system through:

- securing the general adoption and continuity of a scientifically defined set of requirements for observational variables needed for climate research and prediction
- the creation and implementation of international agreements for the acquisition and exchange of these data with sufficient frequency, density, quality, speed of transmission and availability to meet the above purpose.

Impacts Research

While there is a concern about global change, those changes manifest themselves at the regional and local levels. Comprehensive impact studies are required at these levels. However, these findings can be used to give generalizations about human behavior under conditions of changing environmental conditions. Impact studies should be internationalized whenever possible, that is, with scientists from different countries working together on common research problems.

There is a pressing need, as well as demand, for impact assessments of climate variability and change at the various levels of governments. Mitigative, adaptive and preventive strategic responses to climate change demand them. Scientific assessments will require higher levels of support than has been the case in the past.

WCIP should maintain as its focal point of concern the impacts of climate variability (such as extreme events and year-to-year variability) and change on managed and unmanaged ecosystems and on society, as well as on societal actions that affect climate. Impact assessment and the WCIP provide an important way to increase the level of scientific participation in the WCP by developing countries, most of which have few resources for such research. The establishment of Regional Climate Research Centres would be of great help in this regard. This involvement will also assist in raising the level of awareness and concern of these countries about environ-mental issues. There should be increased scientific interaction between the WCIP and the other components of the WCP. This would enhance the value of the output of these programmes.

The linkage between impact studies and policy-making and the impact of socio-economic development on the environment deserve more attention. There is a need for the development of policy alternatives that consider the issue of climate variability and change and their impacts as an input for policymaking and for sustainable development strategies.

Links Between Research and Policy

Policymakers have a requirement for timely assessments which are the considered consensus views of the international scientific community. It is obviously an advantage if policymakers worldwide have a single established source of information as the basis for their deliberations, and a continuation of IPCC for this purpose (at least in the science and impacts areas) is strongly recommended. To avoid the built-in inertia to which bodies are prone, the need for IPCC should be reviewed at regular intervals.

The pace of scientific developments makes it likely that a revision of the current IPCC science assessment will be necessary at about 5 year intervals, but short updates in areas of greatest scientific progress could be made in intermediate years. The different requirements of various countries should be borne in mind when the structure and content of the assessment is decided. In order to ensure that the assessments are of the highest quality, and to minimize the drain on working scientists' time, IPCC should make the maximum use of results from and activities of WCRP, WCIP and IGBP.

It should not be the task of IPCC to define or set up research programmes or monitoring networks, although its recommendations will be valuable to the responsible agencies.

Recommendations

1. The WCRP, emphasizing the physical aspects, and the IGBP, covering biogeochemical aspects, are complementary programmes that should provide a sufficient international framework for the required scientific research into the main components of the climate system. Joint co-ordinated sub-programmes are developing. There is a need to strengthen links being developed between them at all levels. Any new studies should be carried out within this strengthened framework.

2. More activity is needed in studies and parameterizations of processes in the atmosphere (especially clouds), oceans and at the land surface, because of their effects on climate sensitivity and rate of change. This will require studies, in which the WCRP and the IGBP have a strong role to play, involving field experiments and the use of high resolution and limited area models. Issues relating to biogeochemical cycles, to palaeoclimatology and to ecosystems need to be developed within IGBP.

3. The organizational framework for international scientific research is in place; what is now needed is proper recognition and allocation by government funding agencies of the financial and human resources that are required to make improved predictions of climate variability and change. The effectiveness of contributions by national funding agencies to these research programmes should be enhanced by ensuring effective interaction and co-ordination within the existing international scientific framework.

4. Current plans for the implementation of earth observing systems for the coming decades are insufficient for addressing the climate problem. A global climate observing system, providing data systematically and with long term continuity, is required for detecting climate change and for validating and developing climate models. Full implementation will take a long time to accomplish, and will need to build on the existing operational systems for collecting and processing earth system data. A start needs to be made now, giving due attention not only to the space and in-situ observing system itself, but also to the processing, storage and free exchange of climate data from all parts of the globe. Strong international co-ordination will be needed in the implementation of such a system.

5. The proposed research and observational systems will, if implemented, lead to a very much better understanding of the natural climate, its variability and change. The social and economic value of predictions based on this understanding will be considerable.

6. The climate impacts research community should increase its research efforts to understand the impact of climate on societies and societies on climate. Such research should remain focussed on climate variability as well as change. Increased emphasis on local and regional impacts will ensure that the involvement of scientists from developing countries will increase. Climate-related impact assessments must include a linkage to issues of sustainable

development. How to relay information about climate-society interactions should also be a main concern of WCIP. The existing arrangement between UNEP and WMO in which the lead for WCIP is taken by UNEP should continue.

7. The IPCC should be retained to serve as a bridge between the scientific research programmes and policymakers for the purposes of assessment and interpretation, independent of the organization of scientific climate research. It should provide careful, peer reviewed summaries and clear, direct presentation of conclusions.

8. Ways, such as establishing regional research centres and strengthening existing ones, need to be found to enable developing countries to participate actively in international research programmes such as WCRP and IGBP.

Task Group 10

The World Climate Programme: Overview and Future

Chair:
W. Böhme *(Germany)*

Discussion Leader:
I. Lang *(Hungary)*

Rapporteurs:
A. J. Apling *(United Kingdom)*
E. Bierly *(USA)*
S. E. Tandoh *(Ghana)*

Conclusions and Recommendations

Task Group 10 concluded that the World Climate Programme (WCP) definitely has promoted the application of climate information; the understanding of the global climate system, its state and its behavior; the assessment of the impact of climate variability and change; the development of tools to cope with climate issues; and support for a global framework convention on climate change. This is due to the efforts of scientists of more than 160 countries working within the WCP, a truly remarkable achievement in international, interdisciplinary, and inter-agency co-operation.

The original objectives of the WCP, as given in the second WMO Long-term Plan remain valid and important.

The scope of the WCP needs to be broadened due to several developments in which climate plays such a fundamental role. It must be closely coordinated with relevant programmes in light of new requirements in connection with:

- the proposed Global Framework Convention on Climate Change
- efforts necessary for sustainable development
- socio-economic impacts of climate change and climate variability, and
- the growing emphasis on climate prediction on all time scales.

The broadened programme will be referred to as the World Climate Program-Phase II (WCP-II). A more integrated approach within the WCP-II clearly is needed. It should include even more vigorous scientific programmes of sysematic measurements and research. More visibility for this programme is mandatory among the nations of the world if the WCP-II is to be successful.

These conclusions lead to two recommendations by the Task Group. The first concerns the structure of the WCP-II and the focus of its component programmes. The second deals with the coordination necessary for WCP-II.

Structure of the WCP-II

Task Group 10 recommends the following programmes as components of WCP II:

- World Climate System Monitoring (WCSM) Programme
- World Climate Applications and Services (WCAS) Programme
- World Climate Research (WCRE) Programme
- World Climate Impact Studies (WCIS) Programme
- World Climate Response Strategies (WCRS) Programme
- World Climate Education (WCED) Programme

The Task Group unanimously agreed that the recommended restructuring of the WCP-II means that all relevant UN agencies should be involved in the responsibility for all WCP components. Leadership of all present components of the WCP should be retained. Leadership responsibilities for the new components should be determined by the proposed Coordinating Council (see Section A.6).

Restructured Components of the WCP-II

World Climate System Monitoring (WCSM) Programme

It is recognized that to study climate it must be studied as a system. To understand that system, it is necessary to observe and monitor it as completely as possible. Monitoring should include observations of the atmosphere, hydrosphere, biosphere, and cryosphere. It also should include socio-economic aspects. Thus, it seems appropriate to restructure the World Climate Data Programme (WCDP) of Phase I to be a new World Climate System Monitoring (WCSM) Programme, as a component of WCP-II.

The WCDP was successful especially in the following activities:

- implementation of the climate data referral system (INFOCLIMA)
- introduction of small computers for accessing local climate data bases for national needs (CLICOM)
- production and distribution of the Climate System Monitoring reports (CSM)
- implementation of the data rescue project (DARE), and
- implementation of the climate change detection project.

The Global Climate Observing System (GCOS) that is included in the Second World Climate Conference (SWCC) Statement should become the basic element of the WCSM Programme. The Task Group felt that the WCSM Programme cannot be maintained nor new systems established without definite commitments by governments. This must include long term commitments to sustain and develop monitoring programmes, to introduce new observing systems as needed, to protect existing stations and to ensure against any changes in the surroundings that might endanger the homogeneity of any data base.

Governments are requested to remove obstacles to the free flow of data and information (including those that are connected with national Exclusive Economic Zones). It must be ensured that all countries and user groups that contribute data have access to and guidance on the use of regional and international data bases. Such guidance may include measures against misuse of the data. A full and open exchange of data would help to stimulate the participation of all countries in the WCSM Programme.

Diagnostic studies of the climate system, the importance of which was repeatedly mentioned during the SWCC, also would be facilitated by the free access to data. Diagnostic studies are very important for the derivation of consistent series of global and regional climate scenarios including those of extreme climate events. Climate scenarios are badly needed for climate impact studies. Another important objective of diagnostic studies would be to search for a "fingerprint" of the expected "greenhouse signal". Other studies of individual climate events, series of similar events, or studies of low frequency modes of climate variations may assist in analysing cause and effect relationships.

World Climate Applications and Services (WCAS) Programme

Although the Task Group recognized that much progress had been achieved by the World Climate Applications Programme (WCAP) of WCP-Phase I, it was felt that since there is a strong service aspect to this work, the name should be changed to signal a new emphasis to the programme. Thus, the new name World Climate Applications and Services (WCAS) Programme.

The WCAP during the first decade of the WCP-I accomplished many things, especially with respect to climate and water, agriculture and energy, urban and building climatology as well as climate and health. "Operational climatology", that is, using information from climatological data bases together with near-real time data, is able to provide current operational guidance for months, seasons and years. Thus, operational climatology techniques will be the basis for timely recommendations on how best to respond to future climate events and variations.

The implementation of CLICOM with its availability of applications software, will provide major advances in user-oriented climatological services for all, especially in developing countries. These promising aspects of climatological services to users justify the addition of the word services to the component's name.

The WCAS should promote, as a matter of urgency, the development of an international clearing-house mechanism to advance the efficient dissemination and appropriate use of regional climate scenarios based on diagnostic climate studies, general circulation model experiments, and other methods.

The application of scenarios or an array of scenarios to the investigation of the vulnerability to climate events in different fields is of great importance. The joint applications of these scenarios and assessments of vulnerability also support studies of climate impact and the determination of response strategies.

Continuing Components of WCP-II

The Task Group recommends that the World Climate Research (WCRE) Programme and the World Climate Impact Studies (WCIS) Programme be retained. These programmes are so fundamental and important that they were discussed separately by Task Group 9. Details are contained in the report of that task group in these proceedings.

New Components of the WCP-II

The recommendation for the formation of a World Climate Response Strategies Programme (WCRS) is based on the potential importance of findings accruing from assessments of global and regional impact studies. As such, the WCRS should be linked closely with the WCIS Programme. The WCRS should include the concepts of prevention, mitigation and adaptation. It should have as its goal studies to help governments determine what policy or policies should be followed as a consequence of natural variability (e.g., drought) as well as climate change (e.g., sea level rise). Apart from the WCP, there is no obvious home within the present UN system for a programme such as the proposed WCRS. It appears likely that, if such a programme has not already been established by 1992, the United Nations Conference on Environment and Development (UNCED) might well create one.

The second new component being recommended is the World Climate Education (WCED) Programme. The Task Group felt that such a programme was necessary either as a separate component of the WCP-II or, should that recommendation not be acceptable, as a viable part of all components of the WCP-II. The WCED should help to overcome the already tangible lack of human resources necessary to implement the WCP-II. Meaningful educational and training activities must be started in developing countries to allow them to participate in the full spectrum of programmes that comprise the WCP-II. The WCED should stress the importance of obtaining reliable observational data and utilization of adequate data management as a way to get developing countries involved. Such activities need to be emphasized and their need made known to governments as well as individuals. Educational activities will give greater visibility to the WCP-II and should make it more effective. Increasing public awareness and a better understanding of climate issues also should be a part of the WCED.

Co-ordination Activities

Task Group 10 underlined the need for more effective coordination between the individual components of WCP-II, the related parts of IGBP and other relevant programmes. It was felt necessary that in order to have an integrated and comprehensive approach to meet the different objectives of the WCP-II at both national and international levels, such action was needed.

The first part of this recommendation deals with the establishment of a Coordinating Council for the World Climate Program-Phase II (CC for WCP-II) to oversee, in general terms, activities that are germane to the WCP-II. It is envisioned that among its activities, the Coordinating Council would serve as an information centre so all components of WCP-II would be aware of the work of the other components. In this way meaningful interactions between components could ensue. The Co-ordinating Council should be composed of the chairmen and officers of the various steering committee and advisory bodies as well as the directors of the component programmes of the WCP-II. It in recognized that the composition of the CC for WCP-II would have to be determined once the component programmes were in place. Task Group 10 requests that the Executive Heads of the UN agencies meeting with ICSU consider the formation of such a CC for WCP-II.

The second part of the recommendation deals with the establishment of an "appropriate intergovernmental mechanism" by WMO, in collaboration with UNEP, UNESCO, IOC, FAO and other relevant UN agencies, and in consultation with ICSU, to fulfill the definite need that governments be more systematically involved in the funding procsss for the implementation of the WCP-II including relevant parts of the IGBP and other programmes. This point was made initially by developing nations, but developed nations felt this need as well.

The Task Group would encourage all nations to establish and maintain appropriate coordinating mechanisms to serve themselves both internally and internationally.

Finally, the Task Group suggests that the Executive Heads of the relevant UN agencies and ICSU continue to meet in order to facilitate integration and cooperation as was shown during the SWCC.

It is worth noting that the preparation and implementation of the UN Conference on Environment and Development (UNCED) will be a valuable opportunity to develop this coordination in the broad context of all other societal, environmental and developmental issues.

Task Group 11

Scientific Components of International Agreements

Chair:
R. M. White *(USA)*

Discussion Leader:
A. O. Adede *(UN)*

Rapporteurs:
D. Attard *(Malta)*
R. S. Rochon *(Canada)*
P. H. Sand *(ECE)*

The Task Group addressed the requirements of effective scientific input in the preparation and implementation of international agreements concerning climate change, both in terms of the **substantive** research needs and in terms of the necessary **process** of interaction between scientists and policy-makers. ("Science" in this context is understood to include the natural and social sciences.)

A.

The scale and complexity of the global scientific effort required will be of a new order of magnitude, calling for enhanced international co-operation and co-ordination. Early and extensive attention should be given to the following scientific and related problems, in particular, in order to facilitate negotiation, implementation and future operation of the agreements envisaged:

a) Reduction of uncertainties of global average climate projections

b) Greater specificity of projected regional climate changes, particularly as regards precipitation

c) Development and implementation of a considerably improved global climate observation and monitoring system, including, in addition to the usual climatic parameters, atmospheric composition, ocean properties, biological and other related parameters, and using surface-, subsea- and space-based techniques for:

- Detection of human-induced climate change
- Determination of the relative contribution of human factors
- Data for seasonal and interannual climate predictions

d) Improved research on and development of seasonal and interannual climate predictions

e) Increased specificity in identifying the economic and social impacts of climate change and their indicators (such as population migrations) and a better understanding of the economic, social and cultural forces contributing to climate change

f) Improved understanding and better projections of economic and social costs and benefits of mitigation and adaptation strategies, taking into account all technological and other options available or likely to become available

g) Improved understanding of what is needed to facilitate national implementation of international agreements in this field, especially in developing countries.

B.

International agreements should specifically provide for co-operation in research, systematic observation and information exchange relevant to the objective and purpose of such agreements. Moreover, governments should ensure maximum scientific input into the negotiation and follow-up of international agreements, and the scientific community should be prepared to respond to this need. It is therefore recommended that:

a) Besides participating in systematic global observation and research as outlined above, parties to international agreements should regularly provide, in accordance with agreed scientific criteria and guidelines, the national data and other information required to verify the effectiveness of agreements in achieving their objectives, such as emission limits or targets

b) In addition to continuous updating of the scientific information base, comprehensive assessments of the type so successfully completed by the Inter-governmental Panel on Climate Change should periodically be undertaken, as and when required, to assist in elaborating or revising agreements

c) International agreements should provide for timeiy review of their implementation, including expeditious adjustment or amendment of provisions in light of new information emanating from scientific advances and taking into account new knowledge of environmental, economic and social impacts

d) National scientific and environmental assessment capabilitles in deveioping countries should be strengthened to enable them to formulate and implement relevant national programmes and thus to carry out the provisions of international agreements.

Task Group 12

Synthesis

Chair:
T. F. Malone *(USA)*

Discussion Leader:
G. S. Golitsyn *(USSR)*

Rapporteurs:
Luo Jibin *(China)*
A. D. Moura *(Brazil)*
G. Yohe *(USA)*

A clear scientific consensus has emerged on mid-range estimates of the overall rate and level of climate change over the 21st century, and the conclusions of the Second World Climate Conference (SWCC) and the Reports of the Intergovernmental Panel on Climate Change (IPCC) have implications and call for measures which extend well beyond traditional bounds of World Climate Program (WCP). We note, in support of these conclusions:

- a growing awareness that climate change is the leading edge of a suite of global environmental changes which have their roots in profound technological development and social response to that development, and that dealing with the fundamental aspects of global change presents an unprecedented challenge to the survival of civilization
- an increased recognition of the importance of growing concentrations of multiple greenhouse gases
- a greatly increased knowledge of climate system behavior in both the past and the present
- an expanding capability to model mathematically the interacting physical, chemical and biological processes which determine climate, and a greater appreciation of fundamental feedback processes of that determination
- a heightened understanding of biosphere and oceanic sources and sinks of greenhouse gases

- an improving ability to offer regional and seasonal climate forecasts
- an exponentially growing computing capacity which outstrips our data collecting and analyzing capabilities in some cases, but which falls well short of our needs in others
- a rapidly developing but nonetheless increasingly neglected remote-sensing capability for observing climate variables and integrating these data with *in situ* measurement into more complete portraits of the total environmental system
- more urgent calls for economic growth from developing nations which require efficient and clean energy and production technologies even as many of them cope with large international debt service and continued high rates of population growth
- an increased application of economic and other social scientific disciplines to analysis of the sources and consequences of human-induced climate change and the decisions which must be made under enormous uncertainty to either mitigate against that change or adapt to its consequences, and
- an emerging political will worldwide to address development and environmental issues concurrently with workable, complementary strategies.

The insights derived from SWCC and IPCC have been bolstered by new and existing paradigms and programs designed to support holistic consideration of environmental problems. These have included:

- the evolution of the International Geosphere-Biosphere Program (IGBP) of the International Council of Scientific Unions (ICSU) into a complementary program to the World Climate Research Program (WCRP) exploring linkages of the biosphere with other earth components
- the creation of a complementary social science research agenda under the auspices of the International Social Science Council (ISSC)
- a renewed emphasis on the goal of achieving a sustainable and equitable society across the globe
- the first steps of a social scientific research program looking at linkages between human activity and changing climate
- the emergence of planning within international economic and financial communities to address debt relief for developing countries and to underwrite their improved scientific capability, and
- the spotty success of the international community to fund programs designed to ease the burdens of responding to environmental problems, collecting environmental data, and managing its analysis.

The experience of WCP over the past ten years, coupled with trends and outcomes in other areas, point to the need to change and adjust the way environmental science and policy analysis are conducted as we move toward the twenty first century.

The recommendations proposed here are derived from a short list of underlying themes which have surfaced frequently during SWCC; they are, perhaps, more important than the specific recommendations which follow. We submit, to that end:

1) that climate change is a global issue but that adaptation and response to both its impacts and any global mitigating strategy which might be adopted will occur at the local and regional level
2) that the global observing system and its associated data-management system are the backbone of climate research
3) that there is an acute need for **integration** within the scientific effort:
   a) across the domains of the climate system (e.g., ocean, land, atmosphere, water, etc.)
   b) among the relevant disciplines (e.g., physical sciences, biological sciences, economics, policy analysis, other social sciences, engineering, etc.)
   c) between **positive** (norm-free) scientific studies and impact analyses and **normative** decision-making processes

   d) between climate change and other environmental concerns, and
   e) between developed and developing countries
4) that uncertainty constrains our current understanding of what climate change might mean, but uncertainty must not restrain our resolve to propose, analyze and test various mitigating and/or adaptive response options
5) that expanding the realm of impacts analysis to include socio-economic effects intensifies the need for improved coordination among international and national research agencies and programs, on the one hand, and individual researchers from around the globe, on the other
6) that increasing the scope of climate research geographically and scientifically will challenge the financial and intellectual resource base from which we will continue to draw its support, and
7) that heightened awareness throughout society of environmental issues and their interaction with development goals is essential.

Several specific proposals designed to reflect one or more of these themes can now be articulated.

Regional Research Centers

In response to themes (1), (3), (4) and (5) listed above, we recommend the establishment of a network of **regional**, interdisciplinary research centers with a global orientation to address cross-cutting issues among the many domains of the climate system, the relevant scientific, engineering, and social scientific disciplines, and the complete array of nations at varying states of socio-economic development. Their focus should be on (i) research designed to develop the knowledge base required to inform policy decisions with respect to impacts, prevention and mitigation strategies, and alternative responses (ii) preparation of policy options with catalogs of their likely implications across the full range of possible futures for consideration by the public and private sectors of sovereign nations as they act independently and in concert through intergovernmental organizations, and (iii) the mutual interaction of regional and global policies which might be implemented in response to a changing climate.

Whenever possible, these centers should be established at existing institutions. They should represent a decentralized approach to the problems of climate change, but their governance and co-ordination should fall under the aegis of a world level Council comprised of representatives of the physical and social scientific communities as well as representatives from national governments and international organizations. it will be the task of the Council to ensure the efficient use of the large increase in resources which will be necessary. The centers should have strong educational components, including opportunities for research by young scholars and graduate degree candidates. They should be linked to universities, therefore, and they should otherwise complement existing institutions and the work of individual investigators. Special arrangements should be made by intergovernmental bodies to support the participation of individual investigators from developing nations.

Data Definition, Collection, and Management

We note that inadequacies in the data/information base for developing a predictive capacity, assessing impacts, and preparing policy options are the greatest impediment to an appropriate, holistic response to the potential ramifications of climate change. The range of variables observed over long periods of time needs to be extended to embrace socio-economic processes as well as land surface, oceanic and biological processes. Methods which correctly internalize the external costs of environmental damage need to be developed and applied in producing time series of economic statistics. The substantial potential of an internationalized system of remotely-sensed observations integrated with a network of surface observations through a modern information and archiving system demands urgent attention. Full and open access, both to the data (new and existing) and to the means of collection and processing, should be assured for the full suite of observations relevant to studies of climate change and impact assessment.

Improved Co-ordination

In response to theme (5), we submit that improved coordination of WCP with other existing agencies and programs, (a list including but not restricted to international organizations like IGBP, ISSC, and IPCC because official and semi-official national and regional organizations can also contribute to scientific exploration of climate change) should be a high priority of the next decade's work. Moreover, the scope of WCP needs to be widened not only to include socio-economic considerations, but also to improve its predictive capabilities on regional and local scales across all time frames. Significant progress has been made in developing a valuable capacity for short run climate forecasting, for example, and that progress should be furthered and its products offered widely. Task Groups 9 and 10 of the Second World Climate Conference have considered this coordination issue in some detail, and we hope to emphasize their recommendations.

Funding of Global Participation

Funding of pure and applied scientific and social scientific research will have to be increased significantly if the efforts of the next decade are to begin to meet the needs which are implicitly defined by these underlying themes. Substantially increased funding for research and policy development from national and international sources will be required from a base of five to ten thousand million dollars (1990 US) in annual support of core activity. National governments and international organizations should be expected to increase their financial contributions by 10 to 15 percent per year over the next decade. Total support amounting to nearly fifty thousand million dollars (1990 US) by the turn of the century would still amount to little more than one tenth of one percent of world economic output. To use these resources efficiently, as well as to develop the intellectual resources which must be brought to bear, global climate and related research programs need to be organized on a worldwide basis. While a number of organizations are already involved in a wide range of activities, much remains to be done. A network, well coordinated but not overly centralized, would be most appropriate to integrate effectively the diversity of the scientific disciplines and regional circumstances involved. WMO has established a trust fund to improve observational and monitoring systems in developing nations which must be renewed. Member nations should aggressively support this fund and inaugurate complementary funds to underwrite full participation of developing nations in the regional research centers.

Education

We acknowledge the importance of theme (7) and recommend that measures should be undertaken to increase understanding of environmental issues at all levels of society. An educated population is not only the best insurance that the global environment will survive, but also the best assurance that the political will to support the long term monitoring and research projects which are required will be increased and sustained. Curriculum development, distributed internationally by agencies and programs like Unesco, should be encouraged.

Conclusions

The above proposals and the underlying themes from which they were drawn, imply a substantial transformation of all relevant international agencies to enable them to deal with an issue which is inherently holistic and which requires a uniquely interdisciplinary approach.

In particular we conclude that:

- Climate change and social response to it are at the leading edge of global environmental change driven by technological developments and population dynamics. The resulting global societal change will have profound consequences.

- A clear scientific consensus has emerged on mid-range estimates of the overall rate and magnitude of climate change during the 21st century.

- Resolution of residual uncertainties is as likely to increase our estimates of the impacts of those changes as it is to diminish them. These uncertainties alone must not be the basis for defering societal response to these risks. Many of the actions that would reduce risk are also desirable on other grounds.

- It is timely that nations take steps toward a global climate convention and associated protocols.

- In parallel with any international convention, efforts should be made at regional and/or national levels to prepare for the social and economic consequences of climate change and measures to limit that change.

- A major observational and research effort organized at the world level is essential for understanding and interpreting climate change and for developing adaptive and limitation strategies. Provision for this effort should be part of any climate convention and associated protocols.

Recommendations

We recommend that:

- Present global observational systems, including the national weather services, be expanded to serve climate and climate-change needs more specifically, including socio-economic data. An internationalized system of remotely-sensed observations be developed, linking these measurements to *in situ* observations through a modernized information and archival network providing unrestricted access to current and past data.

- A network of regional, interdisciplinary research and information exchange centers be established to address cross-cutting issues of global change, including climate change, and to explore policy options. These centers should bring together individuals from both the developed and developing countries. They should include the relevant disciplines of the natural sciences, engineering, and the social sciences.

- Public understanding of global climate and environmental change and of decision making for public policy under conditions of uncertainty be fostered. Education and training programs be initiated

at all levels to meet the anticipated professional and technical personnel needs.

- Substantially increased funding be provided for research, observations, and policy development at national and international levels. From a base of five to ten thousand million dollars (1990 US) for support of core activities, a growth rate of 10 to 15 percent per year is realistic. Total research support should approach a few tenths of one percent of world economic output by the year 2000, and will require institutional innovation to ensure effective coordination. The WMO Trust Fund should receive additional support.

Report of the Consultation Group on the Special Needs of Developing Countries

Co-Chairs:
A. Al-Gain *(Saudi Arabia)*
J. Ripert *(France)*

Summary

The Consultation Group identified a number of issues that require attention by members of the international community. In the light of its discussions, the main areas of action that were elaborated are:

Developing Scientific and Technical Resources
- Intensive training programmes for young scientists from developing countries
- Prepare greenhouse gases emission inventories
- Identify impacts
- Establish environmental law

Technology Transfer
- Need for preferential and assured access
- Intellectual property rights

Climate Monitoring and Analysis
- Implementation of "CLICOM"
- Building national capacity to monitor climate and climate change
- Establishing climate reference stations
- Strengthening the data rescue programme (DARE)
- Participation of developing countries scientists in international research efforts

Action Programmes and Resource Requirements
- Efficient use of energy resources
- Better land use planning and management
- Rational forest management practices
- Creation of regional centres

Public Information and Education
- Access to detailed current environmental information and educational material
- Participation of NGO's to assist in public awareness programmes

The Involvement of Developing Countries

The Consultation Group on the Special Needs of Developing Countries was established at the request of the Coordinator of SWCC to provide recommendations in this vital area. Its members were selected in their personal capacities, but nevertheless, their views represent a broad spectrum of disciplines, sectors and regions.

The Consultation Group addressed issues which are becoming increasingly pivotal in determining the future environmental well-being of the planet.

It has become clear that climate issues reach far beyond atmospheric and oceanic sciences, affecting every aspect of life on Earth. Although traditional aspects of climatology continue to form a major element, there is a dramatic increase in the inclusion of environmental, economic and sociological topics.

Sea level rise, agricultural and fisheries production, energy use, forestry, drought, desertification, urban planning and a host of other socio-economic issues are dependent on the climate issue which is assuming great importance to the global political leadership. This is demonstrated by the ministerial component of this Second World Climate Conference, the opening of negotiations on a framework convention, and the importance attached to this subject on the agenda of the 1992 United Nations Conference on Environment and Development.

Until now, developing nations have been unequal partners in the climate and meteorological infrastructure, despite the large proportion of lands and population contained within their boundaries. Although they have collaborated in providing data, and participated to a degree in meetings and research, developing countries have benefited to a lesser extent from the results of the analyses and planning resources which were developed from their contributions, and even less so from the applications derived therefrom.

Concern over potential global climate change focuses an even greater importance upon active and meaningful developing country involvement because events within these nations will likely determine the success or failure of the international response to global warming and climate change. Just as the problem has largely resulted from prior industrialization, much needed economic growth within developing nations will play an important role in the rate of future climate change. This is because it is in developing countries where the need for development is greatest and where the development process lacks technological bases necessary for efficiency and minimization of environmental impact.

Developing countries, of which many must develop or face the likelihood of famine and disease for their people, are being called upon to develop in an environmentally responsible manner that is without precedent. They are also being called upon to participate in the alleviation of the legacy of environmental damage from prior industrialization. At the same time, developing countries are being asked to develop in a manner that pays environmental costs as up front costs rather than deferring them as has been done in the past. It is clear that developing countries must not go through the evolutionary process of previous industrialization but rather, must "leapfrog" ahead directly from a status of under development through to efficiencies currently found only in the most advanced industrial economies. In that context it was noted that there are certain investments that are both environmentally and financially sound, particularly in energy.

As stated by the IPCC Report, "industrialized and developing countries have a common but differentiated responsibility for dealing with the problem of climate change".

The Group has identified a number of issues that require attention by members of the international community. In the light of its discussions, main areas of action are elaborated below and it is hoped that these will be taken into account by governments and in international, regional and national programmes and further elaborated in future negotiations.

Developing Scientific and Technical Resources

- Climate issues demand concerted action by all countries based on a common understanding of the potential impacts of climate change and the measures to be taken to minimize adverse impacts. This common understanding will not emerge except under conditions of free exchange of information and know-how and full participation in assessment and analysis by all countries. It is recognized that the developing countries lack the scientific resources and indigenous expertise to fully participate in the scientific assessment and policy analysis. Therefore, a massive and sustained flow of scientific and technological expertise towards the development of the intellectual resources, technical and institutional capacity of the developing countries is a necessary complement to the efforts of those countries.

- Developing countries would also require the capability to carry out the following tasks:

  - to prepare greenhouse gases emission inventories and future emission projections
  - to identify impacts of potential global warming
  - to prepare cost estimates and priorities for response strategies to adapt and mitigate problems posed by climate change
  - to assist, where appropriate, in establishing environmental law, standards, monitoring and enforcement capabilities
  - to develop institutional and administrative mechanisms to link scientific information with policy and decision making by establishing specific arrangements for appropriate packaging and dissemination of information.

Technology Transfer

- A substantial resource and technology transfer from developed countries is a requirement, if developing countries are to be full partners in adapting and mitigating problems posed by climate change.

- The mechanisms of the transfer of technology and provision of technical assistance and co-operation to developing countries should take into account considerations such as the need for preferential and assured access, intellectual property rights, the environmental soundness of such technology and the financial implications.

- The transfer of technology should include all aspects of the developing countries' requirements (flow of full information, procurement, delivery of systems, manpower training, maintenance and technical assistance).

- The provision of environmentally safe technology to the world's consumers will undoubtedly remain in the realm of commercially oriented companies. UNEP/WMO should, therefore, maintain inventories of manufacturers who produce environmentally friendly goods and who provide environmentally oriented services and maintain data on impacts of these services. In addition to this, relevant United Nations Agencies could convey to the manufacturers specific requirements of individual communities and the results of their research.

- Taking note that industry plays a significant role in the development and transfer of science and technology, efforts by industry to promote further the development and transfer of environmentally sound technologies should be encouraged, and policies to encourage such efforts should be formulated.

Climate Monitoring and Analysis

- Implementation of "CLICOM" in each developing country. and enhancement of international climatic data exchange (in common format) must be encouraged

- Also needed is the updating of a climate operational manual and the establishment of a WMO regional climate operation experiment to improve capabilities to handle climate issues.

- National capacities to monitor climate and climate change need to be improved.

- WMO/WCP should provide assistance in establishing: firstly, at least one climate reference station per country; then secondly, at least one climate reference station per climatic region in each country. This would include implementing necessary rigorous quality control procedures.

- Increased support is needed for the data rescue programme to recover and preserve existing climatic records, particularly those old records which are scarce and, once lost, are irretrievable. The presentation should include converting the data into computer compatible form.

- Encouragement and assistance need to be provided to developing country scientists to work for periods of time (time frame of years) on climate change related problems facing their own country at the appropriate

leading research centres. For example, sea-level rise issues could include participation by scientists from such countries as Bangladesh, Kiribati, Egypt, Maldives, and Nicaragua.

- Arrangements need to be made for the participation of developing country scientists in international research efforts, such as GEWEX, TOGA, WOCE, etc., and enhancement of international climate monitoring networks (WWW, GAW, IGOSS). For example, leading atmospheric research institutes could help to support the analysis and data processing requirements of the developing country's BAPMoN stations. Developing countries' scientists could participate in attempts to model regional climate variations.

- Developing countries should be encouraged to assist in the provision of logistic support/maintenance/ protection, etc., for international monitoring programmes in oceanic and data-sparse land areas.

Action Programme and Resource Requirements

- Global action will require co-operation on an unprecedented scale. In order to achieve this goal, those nations which can provide assistance will have to assist those which lack resources necessary to deal with the environmental costs of their development. The negative environmental legacy of past development will have to be dealt with in addition to the impacts of future development.

- Future success or failure of response to the global environmental crisis will undoubtedly be determined by the environmental impacts that arise from economic expansion in the nations of the developing world. These nations are faced with rapidly increasing populations whose needs for food, safe drinking water, health care and a minimally acceptable standard of living must be met. Consequently, "While the global environment has assumed greater significance for the industrialized countries, the priority for alleviation of poverty continues to be the overriding concern of developing countries" (IPCC, 1990).

- Assistance by industrial nations to developing nations serves their own self-interest and should not be regarded as an altruisim, but rather as a pragmatic action whose end goal is that all nations develop the ability to pay the environmental costs of their own use of the environment in present value rather than deferring these costs to the future. By supporting these present costs, industrial nations can assure the future of their own environmental options, as well as those of developing nations. Front end environmental costs generally represent a very small percentage of total project capitalization and can provide real operating benefits in terms of efficiencies, increased plant life and increased health and safety of workers, as well as other measurable economic savings.

- Recognizing the need to develop national agendas and to organize the mobilization of additional resources to implement these agendas, developing countries will need to elaborate programmes and financial requirements for policy changes, investments, training, and research needed to address climate change; such programmes will need to be closely integrated and coordinated with the preparation of other environmental programmes and action plans.

- Additional financial resources channelled to developing countries would be most effective if focused on those activities which contribute both to limiting greenhouse gase emissions and/or adapting to any adverse effects of climate change, and promoting economic development. Areas for co-operation and assistance could include:

  - implementation of the Montreal protocol
  - efficient use of energy resources and development of national energy policy
  - better land use planning and management (coastal zone, rangeland, desertification and drought)
  - rational forest management practices and agricultural techniques which reduce greenhouse gas emisisons
  - facilitating technology transfer and technology development
  - enhancing observation networks to facilitate conducting research, monitoring, and assessment of climate change
  - encouragement of organization of international meetings on global climate change in developing countries
  - creation of regional centres to organize an information network on climate change, new technology developments, public awareness and education
  - requiring thorough environmental impact assessment of new development as a mandatory condition before investment capital can be released
  - soil and water conservation.

Public Information and Education

- A major problem in the developing countries, particularly those in which English is not widely known, is that the population as a whole does not have ready access to detailed current environmental information and educational material. An International Centre could produce and update information and educational material (regular publications, audiovisuals, posters, etc.) based upon their own and international developments. Regional Institutes could be given the task of converting this material into the languages and media most applicable to their region. In this way, current developments in environmental research and technology could be made available to the world. The fledgling UNEP/IEO publication, "cleaner production," is an example of the type of publication which could be expanded further and translated on a short time frame in the future.

- The Regional Institutes would also need to establish roving educational groups, particularly for countries with poorly educated populations.

- The increasing participation of NGOs to assist in public awareness programmes is to be welcomed and encouraged.

References

IPCC. 1989. A Proposal for Assisting Developing Countries to more Effectively Contribute to an Improved Understanding of the Physical Basis of Global Climate Change, Assessment of the Socio-Economic Impacts Resulting from Climate Change, the Formulation of Suitable Response Strategies. 11 pp.

IPCC. 1990. Report of the First Session of the IPCC Special Committee on the Participation of Developing Countries. 7 pp.

Panel Discussion

Global Climate Analogues and Global Climate Models

Chair:
J. Mahlman *(USA)*

Rapporteur:
J. Perry *(USA)*

Members:
G. Boer *(Canada)*
M. Budyko *(USSR)*
K. Hasselmann *(Germany)*
J. Mitchell *(United Kingdom)*
T. Wigley *(United Kingdom)*

J. Mahlman began by expressing hope for an "interactive scientific discussion" among Panel members noting that we are now at a "major milestone in a long journey". He argued that some have hailed the IPCC Report as the "definitive achievement" in climate change assessment and the problem is now passed on to the policymakers. Some others, however, believe that the uncertainties remain so great that no governmental action can be justified. Neither of these views can be correct. He asserted that we have learned much, but that it is time to press on to the "next generation of climate science". We must remember that the IPCC Report is not the achievement being celebrated here; the real accomplishment is the decades of scientific work that laid its foundation. Mahlman stressed that the goal of this Panel discussion is a "symbolic beginning" of the next stage in the quest. Scientists in the pursuit of new understanding are prone to argue much about the unknown and set aside that which is known. Because such arguments can be passionate and egocentric, policymakers and the public can get the impression that little is known about climate change. Bearing this in mind, the Panel focussed its discussion on (1) diagnosis of existing datasets, (2) use of paleoanalogues, and (3) self consistent models of climate. J. Mahlman cautioned that no one of these approaches, by itself, is "best" or sufficient. Indeed, a mature science must be characterized by a healthy interplay between theory and observation.

J. Mitchell then presented a comparative critique of modelling versus analogue approaches to estimating climate sensitivity. He outlined the strengths and weaknesses of models and of paleoanalogues most persuasively. He concluded that the Holocene and Eemian are not good direct analogues for increasing greenhouse forcing, because they were principally forced by Milankovitch orbital variations that produced a pattern of radiative forcing quite different from that of the contemporary greenhouse problem. He further noted that it is very difficult to differentiate between orbital forcing and other factors in the paleoclimatic record. He was equally frank, however, in acknowledging the deficiencies of current climate models. He speculated that the inferred similarities in all warm epochs in the paleoclimatic record may be due to strong feedbacks that may occur during such periods. Left unanswered is the question, On what time scales would such hypothesized feedbacks become dominant? If it is hundreds of years or longer, then paleoanalogues are likely of little value in the current global warming problem.

G. Boer discussed the ability of climate models to simulate regional changes from the continental scale down toward the mesoscale (100 km). He compared the designated IPCC regions with the grid scale of current models, concluding that "3rd generation" models would be required to get real regional detail. However, better resolution will not necessarily solve the problem. At these regional scales, new physical processes must be added. He assessed the current capability of climate models to predict regional changes as "modest". J. Mahlman noted a tendency within the scientific community to underestimate the amount and years of scientific work necessary for the models to realize substantial increases in regional scale predictive skill.

T. Wigley spoke of required technology for climate change detection. Variables should be selected on the basis of their length of record and their signal vs noise ratio. The latter can also be estimated from models. Extraneous factors (e.g., solar variability, volcanoes) can confuse the issue. He then considered the possible calibration of climate sensitivity from paleoanalogues. There are interesting indications of a number of events like the "little ice age" in the Holocene. There is also astronomical evidence that the population of stars like our sun falls into two classes: sunspot-cycle variables and quiescent. It may be possible to put these two insights together. J. Mahlman asked what climate monitoring system T. Wigley would like to have had in place 100 years ago in order to detect climate change to date. T. Wigley cited solar irradiance, aerosols, a global network of surface and near-surface temperature data, upper-atmosphere data to 20 mb, and data for evaluating and testing ocean models. K. Hasselman observed that the theory of data acquisition and processing is well worked out, provided that the covariances are known. Since we are concerned with periods at least as long as the period of record, we must estimate these covariances from models, as well as from available data.

J. Mahlman introduced M. Budyko as a "legend in climate dynamics". M. Budyko argued the presence of a distinct gap between what he called "normal" science in which discussion and patient inquiry are central tools for progress and that of the current politically charged situation in which we find ourselves. He claimed that we are on the brink of a precipice in which bad decisions based on inadequate knowledge or politics can cause enormous damage. He called for increased international cooperation in climate science between the Soviet Union and the western countries.

K. Hasselmann spoke of the role of oceans in climate. It is well known that the oceans act as the flywheel of the system, both thermally and chemically. He noted, however, that the oceans also play a more dynamic role by converting atmospheric white noise into climatic red noise, and by providing possibilities for instability (bifurcation). Model simulations show how salinity forcing by precipitation can produce major long-period changes in circulation and heat transport. The ocean is far from being just a simple integrator. Thus, coupled atmosphere-ocean models are absolutely essential for climate change prediction; they also probably need to include the carbon cycle in a predictive mode.

Questions from the floor dealt with: satellite temperature monitoring (T. Wigley stressed the need for ground truth, long-term calibration, and data continuity.); possibilities for improving climate models; the mystery of comparatively small tropical temperature variations in the paleoclimatic record; prospects for deterministic predictability of variations in the ocean on time scale of years to decades; and the climate role of aerosols and other greenhouse gases, including ozone.

J. Mahlman summarized by noting that we are in an era of explosive growth in scientific understanding and characterization of the climate system. He cautioned that the degree of difficulty is very high and that much patience will be required in the pursuit of further progress. He finished with a prediction that the societal need for predictive insights on climate change will grow faster than the scientific community will be able to supply them. This will produce a perception of slow progress just at the time when the growth in understanding and capability will be greater than at any time in the past.

Panel Discussion

Climate and Environmentally Sustainable Economic Development

Chair:
A. Khosla *(India)*

Members:
Y. K. Ahmad *(UNEP)*
F. de Oliveira *(Brazil)*
W. U. Chandler *(USA)*
A. C. A. Sugandhy *(Indonesia)*

The primary issues discussed by the Panel related to the two-way interaction between patterns of development and climate change. The initial presentations by the panelists attempted to define the pre-conditions for sustainable development, i.e., social equity, economic efficiency, environmental harmony and self reliance. To reorient development towards sustainability, the types of interventions available were identified to include changes in technology, institutional design, decision making system, knowledge structures and value systems.

Keeping in mind the special problems of developing countries, the panelists explored the kinds of impacts global climate change could have particularly on vulnerable populations, such as the poor, indigenous peoples, disabled, etc.

Panelists raised the issue of redirecting a society's scientific research effort to provide early warning of impending global changes and to the need for the scientists to study these changes within a broader context of societal needs. Other participants who felt that existing patterns of consumption and production were the root cause of climate change, suggested ways by which technology choice might now be made to minimize these impacts.

There was a strong consensus among the participants at the session that each society must develop the capacity, including the skills, the analytical tools and the databases to be able to understand and make endogenous decisions regarding the impact of climate change on it and vice versa. For this purpose, the global scientific community has a primary responsibility to establish much stronger programmes of education and research in climate related subjects, both in developed and, even more especially, developing countries.

Panel Discussion

Industry's Response

Chair:
B. Butler *(United Kingdom)*

Members:
U. Colombo *(Italy)*
H. Gassert *(Germany)*
J. Leggett *(United Kingdom)*
R. K. Pachauri *(India)*
W. R. Stevens *(USA)*

The Panel, representing a number of countries and businesses, considered the subject of global warming with a view to determining the principal concerns of industry and the priorities for industry's response.

While varying degrees of emphasis were placed on particular elements by individual panel members, there was a general consensus on the principal points and priorities.

Protection of the environment is a key issue facing industry in the 1990s. But the particular issue of global warming, or more precisely the climatic changes associated with global warming, is one of the least certain and most complex threats facing our planet in both timing and effect.

Industry supports the endeavours of government and the scientific community to reduce through further research and analysis the critical uncertainties which remain, for only then can the major implications for policy be considered in their proper perspective.

However, industry recognizes the degree of concern which now exists on this issue based on current knowledge. It concludes that there are a number of policy responses and actions which can be justified in their own right, can be initiated relatively quickly, and would make a significant contribution to energy efficiency and the reduction of CO_2 emissions. These would provide momentum while the scientific analysis continues and the ground can be prepared for more significant changes in the policy if these are eventually judged to be necessary.

While industry accepts the challenge to help protect and improve our planet, it is but one of the participants, and can make its contribution only with the help and consent of other parties. Insurance against global warming will have to be collective not individual. If the world decides to accept significant costs now to insure itself against potentially higher future costs, it will be primarily through a political process.

Industry's approach should be based on three principles: good science, sound economics and a proper dialogue.

On science, industry believes it is imperative that further research and analysis continue in an ongoing effort to reduce the areas of uncertainty.

Economics can help by providing criteria to give priority to those actions which have most effect on CO_2 emissions for the least cost. It would for example give priority to energy efficiency and conservation. Economics will enable costs and benefits to be balanced, not that a particular environmental result should be achieved at any cost.

There is currently much debate on the use of economic instruments as contrasted with command and control regulations. Industry believes there is scope for both, but:

- economic instruments have a critical role in correcting the decisions of producers and consumers by including, in their 'internal' costs, 'external' costs of their decisions

- regulations are appropriate where the economic 'penalty' to achieve the desired result would be disproportionate to the goal, for example to achieve greater efficiency with many household appliances
- government policy should be consistent and contributory; for example, taxes should be examined for their consistency with environmental goals; energy efficiency will be encouraged by the proper incentives for processes, and products; pricing policies should be rationalised.

A proper dialogue is critical, above all with the public as consumers of energy, to bring them into the debate, because many of the choices will effect them deeply and can only be made with their consent. And, in many cases, consumers can only make a choice if the 'correct' alternative is available; transport is a good example as are energy efficient appliances.

Other points emphasised by the Panel included:

- industry is not monolithic and proposed policies will have to be considered for their impact on individual groups or concerns
- technology will make a significant contribution and the wide dissemination of technology and information .should be a principal objective; however, advanced technology is not always appropriate and it should be considered in the context of the ability of the recipient to receive and apply
- payment for proprietary technology is a key issue
- research to discover and develop new technology will have a critical role
- industry does not have unlimited resources for either research or the implementation of new policies
- governments should.ensure that their own country industries are not competitively disadvantaged by internationally agreed policies
- global warming may be a very significant issue, but there are other major environmental issues currently demanding industry's attention and response.

In closing the discussion, the point was made that industry is accustomed to considering its objectives and action in terms of those appropriate to the short, medium and longer term. A similar approach should be adopted to determine the most effective responses to the issue of global warming and climate change.

Panel Discussion

Co-operation in International Research Programmes

Chair:
J. D. Woods *(UK/ICSU).*

Members:
B. Döös *(IIASA)*
G. Glaser *(Unesco)*
J. Labrousse *(France/IGFA)*
G. McBean *(Canada/WCRP)*
A. McEwan *(Australia/CCCO)*
A.D. Moura *(Brazil)*
T. Rosswall *(Sweden/IGBP)*

Rapporteur:
J. Marton-Lefèvre *(ICSU)*

The Panel considered the overall strategy and status of principal ongoing research projects and how these contribute to making climate prediction possible by improved understanding of how the climate system works and improved techniques of forecasting climate change.

On the *Atmosphere:* the WCRP's major thrust of research into the physical climate system whose principal elements are observation, understanding and modelling was stressed. In addition to TOGA (predictions of interannual climate fluctuations) two principal programmes of the WCRP try to respond to the questions of how much climate change to expect under greenhouse warming (GEWEX) and when such change might occur (WOCE).

On the *Ocean*: the need for accurate descriptions of the deep ocean,which is the major thrust of the WOCE programme was raised. The WOCE effort is seen as providing the foundation for a global ocean observing system.

On the *Biosphere:* the IGBP was introduced as attempting to understand the functioning of the global system as a whole with the central role of biospheric processes as these effect and are affected by climate. All panel members stressed the important complementarity and close cooperation between WCRP and IGBP.

On *Observations from Space:* the International Centre for Earth and Environmental Sciences in Trieste will be an important institution to help involve Third World scientists in global climate change studies.

The need to provide the right messages to decision makers, emphasizing not only what we don't know but also what we do know, was stressed; this will stimulate appropriate action on such issues as the reduction of emissions of greenhouse gases. On the implementation of *international projects,* the cooperation between funding agencies from several countries manifested in the recently formed International Group of Funding Agencies for Global Change Research (IGFA) was introduced. This group looks at coordination of resources for all of the global change programmes, encompassing both the natural and social sciences.

The importance of training more people to devote their talents to global change science was emphasized. Closer cooperation between natural and social sciences in providing training programmes on the use of climate forecasts; to encourage the fullest participation of Third World scientists in these endeavours; to increase human resources, information flow, and to improve both institutional capacities and financing were also stressed.

In summary, participants called for:

- increased cooperation and coordination among international (UN and scientific) and national bodies to ensure the success of future undertakings in climate change research
- a co-ordinated approach, such as the one being tried by IGFA, for identifying strategies and for funding
- identification of institutions and scientists from all parts of the world: developed and less developed
- continued integration of the sciences involved within existing programmes such as the IGBP and the WCRP, which, while requiring more scientists and more resources, also offers the only promise of major advances in understanding and in techniques
- continued work to facilitate operational forecasting of climate through improvement in global observation systems and forecasting models.

Second World Climate Conference
Geneva, Switzerland, 7 November 1990

Conference Statement

SUMMARY

1. Climate issues reach far beyond atmospheric and oceanic sciences, affecting every aspect of life on this planet. The issues are increasingly pivotal in determining future environmental and economic well-being. Variations of climate have profound effects on natural and managed systems, the economies of nations and the well-being of people everywhere. A clear scientific consensus has emerged on estimates of the range of global warming which can be expected during the 21st century. If the increase of greenhouse gas concentrations is not limited, the predicted climate change would place stresses on natural and social systems unprecedented in the past 10,000 years.

2. At the First World Climate Conference in 1979, nations were urged "to foresee and to prevent potential man-made changes in climate that might be adverse to the well-being of humanity". The Second World Climate Conference concludes that, notwithstanding scientific and economic uncertainties, nations should now take steps towards reducing sources and increasing sinks of greenhouse gases through national and regional actions, and negotiation of a global convention on climate change and related legal instruments. The long-term goal should be to halt the build-up of greenhouse gases at a level that minimizes risks to society and natural ecosystems. The remaining uncertainties must not be the basis for deferring societal responses to these risks. Many of the actions that would reduce risk are also desirable on other grounds.

3. A major international observational and research effort will be essential to strengthen the knowledge-base on climate processes and human interactions, and to provide the basis for operational climate monitoring and prediction.

Part I
MAIN CONCLUSIONS AND RECOMMENDATIONS

A. *Greenhouse Gases and Climate Change*

1. Emissions resulting from human activities are substantially increasing atmospheric concentrations of the greenhouse gases. These increases will enhance the natural greenhouse effect, resulting on average in an additional warming of the Earth's surface. The Conference agreed that this and other scientific conclusions set out by the IPCC reflect the international consensus of scientific understanding of climate change. Without actions to reduce emissions, global warming is predicted to reach 2 to 5 degrees C over the next century, a rate of change unprecedented in the past 10,000 years. The warming is expected to be accompanied by a sea level rise of 65 cm ± 35 cm by the end of the next century. There remain uncertainties in predictions, particularly in regard to the timing, magnitude and regional patterns of climate change.

2. Climate change and sea level rise would seriously threaten low-lying islands and coastal zones. Water resources, agriculture and agricultural trade, especially in arid and semi-arid regions, forests, and fisheries are especially vulnerable to climate change. Climate change may compound existing serious problems of the global mismatch between resources, population and consumption. In many cases the impacts will be felt most severely in regions already under stress, mainly in developing countries.

3. Global warming induced by increased greenhouse gas concentrations is delayed by the oceans; hence, much of the change is still to come. Inertia in the climate system due to the influence of the oceans, the biosphere and the long residence times of some

greenhouse gases means that climate changes that occur may persist for centuries.

4. Natural sources and sinks of greenhouse gases are sensitive to a change in climate. Although many of the response or feedback processes are poorly understood, it appears likely that, as climate warms, these feedbacks will lead to an overall increase rather than a decrease in greenhouse gas concentrations.

5. The historical growth in emissions has been a direct consequence of the increase of human population, rising incomes, the related exploitation of fossil fuels by industrialized societies and the expansion of agriculture. Under "Business-as-Usual" assumptions[†] it is projected that emissions will continue to grow in the future as a consequence of a projected doubling of energy consumption in the first half of the 21st century and an expected doubling of population by the latter half. As a result, the effect of human-induced greenhouse gas concentrations on the earth's radiation balance would by 2025 correspond to a doubling of carbon dioxide unless remedial actions are taken.

6. Over the last decade, emissions of carbon dioxide (CO_2) contributed 55% of the increased radiative forcing produced by greenhouse gases from human activities. The CFCs contributed about 24% of the past decade's changes, and methane 15%, with the balance due to other greenhouse gases. With controls on CFCs under the Montreal Protocol, the relative importance of CO_2 emissions will increase, provided the substitutes for CFCs have minimal greenhouse warming potential. Some 75% of total CO_2 emissions have come from the industrialized countries.

7. The above emissions can be expected to change the planet's atmosphere and climate, and a clear scientific consensus has been reached on the range of changes to be expected. Although this range is large, it is prudent to exercise, as a precautionary measure, actions to manage the risk of undesirable climate change. In order to stabilize atmospheric carbon dioxide concentrations by the middle of the 21st century at about 50% above pre-industrial concentrations, a continuous world-wide reduction of net carbon dioxide emissions by 1 to 2% per year starting now would be required. The

Intergovernmental Panel on Climate change (IPCC) also considered three other emissions scenarios, which would not lead to stabilization of CO_2 concentrations in the 21st century. A 15 to 20% reduction in methane emissions would stabilize atmospheric concentrations of that gas.

8. This Conference concludes that technically feasible and cost-effective opportunities exist to reduce CO_2 emissions in all countries. Such opportunities for emissions reductions are sufficient to allow many industrialized countries to stabilize CO_2 emissions from the energy sector and to reduce these emissions by at least 20 percent by 2005. The measures include increasing the efficiency of energy use and employing alternative fuels and energy sources. As additional measures to achieve further cost-effective reductions are identified and implemented, even greater decreases in emissions would be achieved in the following decades. In addition, reversing the current net losses in forests would increase storage of carbon. The economic and social costs and benefits of such measures should be urgently examined by all nations. An internationally coordinated assessment should be undertaken through the IPCC.

9. Countries are urged to take immediate actions to control the risks of climate change with initial emphasis on actions that would be economically and socially beneficial for other reasons as well. Nations should launch negotiations on a convention on climate change and related legal instruments without delay and with the aim of signing such a convention in 1992.

B. Use of Climate Information in Assisting Sustainable Social and Economic Development

Climate data, analyses, and eventually climate predictions, can contribute substantially to enhancing the efficiency and security of economic and developmental activities in environmentally sustainable ways. These benefits are particularly important in food and wood production, water management, transportation, energy planning and production (including assessment of potential resources of biomass, hydropower, solar and wind energy), urban planning and design, human health and safety, combatting of drought and land degradation, and tourism. This requires both data on the climate system, and its effective application. Data acquisition, collection, management and analysis must be more vigorously supported in all countries and special assistance provided to developing countries through international cooperation. Transfer of techniques for applying climate information should be accelerated through more widespread use of software (e.g. CLICOM)

---

† "Business-as-Usual" assumes that few or no steps are taken to limit greenhouse gas emissions. Energy use and clearing of tropical forests continue and fossil fuels, in particular coal, remain the world's primary energy source. The Montreal Protocol comes into effect but without strengthening and with less than 100 percent compliance.

for readily available personal computers and other means. Further development of methods for predicting short-term variations in climate and the environmental and social impacts should be vigorously pursued. These advances would provide enormous economic and other welfare benefits in coping with droughts, prolonged rain, and periods of severe hot and cold weather. Such predictions will require major steps forward in ocean-atmosphere-biosphere observing systems. Much greater efforts are also needed to increase involvement in these fields by developing countries, especially through increased education and training.

C. *Priorities for Enhanced Research and Observational Systems*

1. A consensus exists among scientists as summarized in the Report of Working Group I of the IPCC that climate change will occur due to increasing greenhouse gases. However, there is substantial scientific uncertainty in the details of projections of future climate change. Projections of future regional climate and climate impacts are much less certain than those on a global scale. These uncertainties can only be narrowed through research addressing the following priority areas:
   - clouds and the hydrological cycle
   - greenhouse gases and the global carbon and biogeochemical cycles
   - oceans: physical, chemical and biological aspects; and exchanges with the atmosphere
   - paleo-climatic studies
   - polar ice sheets and sea ice
   - terrestrial ecosystems.

2. These subjects are being addressed by national programmes, the World Climate Research Programme and the International Geosphere-Biosphere Programme and other related international programmes. Increased national support and substantially increased funding of these programmes is required if progress on the necessary time scale is to be made in reducing the uncertainties.

3. Present observational systems for monitoring the climate system are inadequate for operational and research purposes. They are deteriorating in both industrialized and developing regions. Of special concern is the inadequacy of observation systems in large parts of the southern hemisphere.

4. High priority must be placed on the provision and international exchange of high-quality, long-term data for climate-related studies. Data should be available at no more than the cost of reproduction and distribution. A full and open exchange of global and other data sets needed for climate-related studies is required.

5. There is an urgent need to create a *Global Climate Observing System* (GCOS) built upon the World Weather Watch Global Observing System and the Integrated Global Ocean Service System and including both space-based and surface-based observing components. GCOS should also include the data communications and other infrastructure necessary to support operational climate forecasting.

6. GCOS should be designed to meet the needs for:
   a) climate system monitoring, climate change detection and response monitoring, especially in terrestrial ecosystems
   b) data for application to national economic development
   c) research towards improved understanding, modelling and prediction of the climate system.

7. Such a GCOS would be based upon:
   1) an improved World Weather Watch Programme
   2) the establishment of a global ocean observing system (GOOS) of physical, chemical and biological measurements
   3) the maintenance and enhancement of monitoring programmes of other key components of the climate system, such as the distribution of important atmospheric constituents (including the Global Atmosphere Watch), changes in terrestrial ecosystems, clouds and the hydrological cycle, the earth's radiation budget, ice sheets, and precipitation over the oceans.

8. The further development and implementation of the GCOS concept should be pursued, with urgency, by scientists, governments and international organizations.

9. The impacts of climate variability on human socio-economic systems have provided major constraints to development. Climate change may compound these constraints. In semi-arid regions of Africa, drought episodes have been directly responsible for major human disasters. Research undertaken during the first decade of the WCP and through other international and national programmes has improved drought early warning systems, including FAO's Global Early Warning System, and increased the reliability of climate impact analyses. But much more remains to be done. Intensified efforts are required to refine further our ability to predict short-term climate

variability, anticipate climate impacts, and identify rational strategies to mitigate or prevent adverse effects. The threat of climate change brings new challenges to the future well-being of people. This requires greater efforts to understand impacts of climate change. Mitigation and adaptation strategies are also essential. Immediate steps to be taken include:

a) national and regional analyses of the impacts of climate variability and change on society, and study of the range of response and adaptation options available

b) closer co-operation and communication among natural and social scientists, to ensure that climate considerations are accounted for in development planning

c) significant increases in resources to carry out impact/adaptation studies.

10. Improvements in energy efficiency and non-fossil fuel energy technologies are of paramount importance, not only to reduce greenhouse gas emissions but to move to more sustainable development pathways. Such advances will require research and development, as well as technology transfer and co-development.

11. A specific initiative would create a network of regional, interdisciplinary research centres, located primarily in developing countries, and focussing on all of the natural science, engineering and social science disciplines required to support fully integrated studies of global change and its impacts and policy responses. The centres would conduct research and training on all aspects of global change and study the interaction of regional and global policies.

D. *Public Information*

People need better information on the crucial role climate plays in development and the additional risks posed by climate change. Governments, intergovernmental and non-governmental organizations should give more emphasis to providing accurate public information on climate issues. The public information and education and training component in the WCP and IGBP must also be expanded.

Part II
SPECIFIC ISSUES

1. *Water*

1.1   Among the most important impacts of climate change will be its effects on the hydrological cycle and water management systems, and through these, on socio-economic systems. Increases in incidence of extremes, such as floods and droughts, would cause increased frequency and severity of disasters.

1.2   The design of many costly structures to store and convey water, from large dams to small drainage facilities, is based on analyses of past records of climatic and hydrological parameters. Some of these structures are designed to last 50 to 100 years or even longer. Records of past climate and hydrological conditions may no longer be a reliable guide to the future. The design and management of both structural and non-structural water resource systems should allow for the possible effects of climate change.

1.3   Data systems and research must be strengthened to predict water resources impacts, detect hydrological changes, and improve hydrological parameterization in global climate models.

1.4   Existing and novel technologies, for more efficient use of water for irrigation, should be made available to developing countries in semi-arid zones.

2. *Agriculture and Food*

2.1   Important uncertainties remain regarding the prediction of the magnitude and nature of potential impacts of changing climate and higher CO_2 levels on global food security. The potential impact on food production in developing countries, with more than half the world's population, could be more uncertain than recent reviews suggest.

2.2   High priority should therefore be given to research on the direct effects of rising CO_2 concentrations on food and fibre crop productivity and equal priority should be given to research on agricultural emissions so as to determine agriculture's present and potential role as a source of and sink for greenhouse gases, and to clarify the costs and possible trade-offs arising from limitation measures.

2.3   New or strengthened institutional mechanisms are required to upgrade natural resource inventories, research strategies and extension services to raise agricultural productivity and minimize emissions. These mechanisms should include collaborative programmes between FAO and international and national agencies with stress on interdisciplinary activities on food security and related topics.

3. *Oceans, Fisheries and Coastal Zones*

3.1   The earth's climate including shorter-term variations is influenced by the coupled atmosphere - ocean system. Coastal zones and their associated high biological productivity, including fisheries, are especially affected. Thus, an improved data base of oceanic parameters is considered indispensable for operational climate forecasting. It is recommended that a global ocean observing and data management system be developed for improving predictions of climate change. Research on the oceans will provide quantification of important feedback loops in climate processes. Observation and research on the El Niño - Southern Oscillation phenomena, on upwelling areas and on biological productivity of the open sea are also important.

3.2 Coastal zones, which are the source of most of the global fish catch, are especially susceptible to effects of global warming and sea level rise. Predicting the impact of changes would be of enormous benefit to the increasing number of people living in coastal areas. Thus, it is also recommended that a programme of coastal zone research and monitoring be established to identify the effects of climate change on the coast and coastal ecosystems, and to assess the vulnerability of various natural and managed ecosystems such as coral reefs, mangroves and coastal aquaculture.

3.3 Action should be taken now to develop coastal zone adaptation strategies and policies.

4. *Energy*

4.1 In order to stabilize atmospheric concentrations of greenhouse gases while allowing for growth in emissions from developing countries, industrialized countries must implement reductions even greater than those required, on average, for the globe as a whole. However, even where very large technical and economic opportunities have been identified for reducing energy-related greenhouse gas emissions, and even where there are significant and multiple benefits associated with these measures, implementation is being slowed and sometimes prevented by a host of barriers. These barriers exist at all levels — at the level of consumers, energy equipment manufacturers and suppliers, industries, utilities, and governments. Overcoming the barriers obstructing least-cost approaches to meeting energy demands will require responses from all parts of society — individual consumers, industry, governments, and non-governmental organizations.

4.2 Developing countries also have an important role in limiting climate change. Maintaining development as a principal objective, energy and development paths can be chosen that have the additional benefit of minimizing radiative forcing.

5. *Land Use and Urban Planning*

Population growth, increasing urbanization, and competing demands for finite areas of arable land will produce increasingly severe problems of food supply, energy production, and water resources. Climate changes may exacerbate these problems in some regions. Prudent planning will require baseline analyses of land use, quality and quantity of water resources, and the assessment of vulnerability of urbanized societies to environmental change. In particular, improved adaptation of urban areas to local climatic regimes needs to be achieved by more appropriate layouts and building densities, and improved building construction through modifications to building and planning regulations. Because conurbations make a major contribution to energy-related greenhouse gas emissions, the design and efficiency of all aspects of urban systems should be enhanced.

6. *Health and Human Dimensions*

6.1 The direct impact of climate change on people, their health and cultural heritage, could be severe. There is likely to be increased health inequity between peoples of developing and developed countries. Climatic change could result in increasing numbers of environmental refugees with associated increases of ill-health, disease and death among them.

6.2 Global warming is likely to shift the range of favourable conditions for certain pests and diseases, causing additional stresses on people, particularly those of the semi-arid tropics. It must be appreciated however that serious problems may arise in all parts of the world.

6.3 Research into how human behaviour contributes to and responds to climate change must have increased emphasis. Public awareness and education programmes are particularly essential in this regard.

7. *Environment and Development*

7.1 Climate change, superimposed on population pressures, excessive consumption, and other stresses on the environment imperils the sustainability of socio-economic development throughout the world. In addition, slowing climate change will give countries more time to enhance their prospects for sustainable development. The developed countries need to reduce emissions and assist the developing countries to adopt new, clean technologies.

7.2 Climate change has such important implications for the sustainability of development that policy responses, including measures to reduce greenhouse gases, measures to reduce deforestation, and the commitment of financial and other resources, are justified for that reason alone. Economic policies, such as subsidies and trade restraints, can distort markets so they harm the environment and contribute to global warming and sea level rise. There is an imperative need for development policies that not only reduce global warming trends but also increase economic and social resilience.

8. *Forests*

While increasing forest cover can contribute to the slowing of global climate change, this is not the major cure for the problem.

Five priority actions are recommended:

1) Assessing national opportunities to increase forest carbon storage commensurate with national resource

development policies, developing an approach by 1992 and completing assessment by 1995

2) Managing the world's forests to optimize biomass and resultant carbon storage in addition to the maintenance of sustainable yields of forest products, biological diversity, water quality and the many other values that forests provide

3) Accelerating research to assess the added contribution that forests can make to atmospheric CO_2 reduction and the impacts of climate change on the world's forests

4) Designing and implementing international monitoring systems to determine conditions and changes in forest ecosystems in response to anticipated climate changes

5) Supporting the development of an international instrument on conservation and development of the world's forests linked with climate and biodiversity conventions.

Part III
ORGANIZATIONAL AND POLICY ISSUES FOR INTERNATIONAL ACTIVITIES

1. *The Future Structure of the WCP*

1.1 The WCP should be broadened and closely coordinated with related programmes of other agencies in response to increased emphasis on the prediction of climate and its impacts.

1.2 The World Climate Data Programme, renamed the World Climate System Monitoring Programme, should be redefined to take into account new objectives.

1.3 Greater emphasis in the strengthened WCP (WCP-2) should be given to adaptation, mitigation and education, with adaptation and mitigation activities closely linked to the Impact Studies Programme (WCIP).

1.4 The World Climate Applications Programme should be renamed the World Climate Applications and Services Programme (WCASP) to reflect the need for intensifying efforts to provide climatological services to a wide variety of users. There should be strong interaction between WCIP and WCASP.

1.5 The organizational framework for international scientific research is in place, constituted by the WCRP, emphasizing the physical aspects, and the IGBP, covering bio-geochemical aspects.

1.6 Governments should establish national committees for the WCP to mobilize support for national activities and to coordinate efforts. The UN agencies and ICSU should work towards ensuring regular contact and exchange of information with national committees.

1.7 The mechanism established for overall coordination of the WCP, involving meetings of the chairs of steering bodies for the various components, should be actively supported by WMO, the other UN bodies concerned and ICSU. Annual meetings of Executive Heads should consider their recommendations.

1.8 Restructuring and strengthening of the WCP will also be necessary to support new activities, such as the development of the proposed GCOS. The Conference recommended that a proposal for the new structure of WCP be formulated by the organizations involved, taking into account the above comments, and presented to the Eleventh World Meteorological Congress, May 1991, and at appropriate meetings of other participating organizations.

2. *Special Needs of the Developing Countries*

2.1 As stated in the IPCC report, industrialized and developing countries have a common but differentiated responsibility for dealing with the problems of climate change. The problem is largely the consequence of past patterns of economic growth in the industrial countries. However, in future the much needed economic growth in the developing countries could play an important role in determining the rate of climate change.

2.2 Developing countries are being asked to participate in the alleviation of the legacy of environmental damage from prior industrialization. If they are to avoid the potentially disastrous course followed by industrialized countries in the past, they need to adopt modern technologies early in the process of development, particularly in regard to energy efficiency. They also must be full partners in the global scientific and technical effort that will be required. It is clear that developing countries must not go through the evolutionary process of previous industrialization but rather, must "leapfrog" ahead directly from a status of under-development through to efficient, environmentally benign, technologies.

2.3 Although developing countries have collaborated in providing data, and participated to a degree in meetings and research, they have benefited to a lesser extent from the analyses developed from their contributions, and even less so from the applications derived therefrom.

2.4 Therefore, a massive and sustained flow of scientific and technological expertise towards the development of the intellectual resources, technical and institutional capacity of the developing countries is a necessary complement to the efforts of those countries.

2.5 Developing countries should be assisted to build up their capabilities

- to monitor, assess and apply climate information
- to prepare inventories of greenhouse gases emissions and future emissions projections
- to identify impacts of potential global warming
- to prepare cost estimates and priorities for response strategies to adapt and mitigate problems posed by climate change

- to participate in the World Climate Programme.

2.6   The mechanisms of the transfer of technology and provision of technical assistance and co-operation to developing countries should take into account considerations such as the need for preferential and assured access, intellectual property rights, the environmental soundness of such technology and the financial implications.

2.7   Taking note that industry plays a significant role in the development and transfer of science and technology, efforts by industry to promote further the development and transfer of environmentally sound technologies should be encouraged, and policies to encourage such efforts should be formulated.

2.8   Additional financial resources will have to be channelled to developing countries for those activities which contribute both to limiting greenhouse gas emissions and/or adapting to any adverse effects of climate change, and promoting economic development. Areas for co-operation and assistance could include the efficient use of energy, land use planning, forest management, soil and water conservations, strengthening of observational systems and scientific and technological capabilities.

3.   *Co-operation in International Research*

3.1   The existing and planned research projects of the WCRP and the IGBP address the highest priority scientific issues related to the understanding and prediction of climate variability and change.

3.2   These programmes should be implemented completely and rigorously. It is particularly important that adequate funding, including long-term funding commitments, be provided.

3.3   In view of the progress made in climate research, it is now timely to proceed to the detailed design of an operational global climate observing system (Section C, paras. 5 - 8), together with the data communications and other infrastructure needed to support operational climate forecasting. Governments should enter into early discussions aimed at international cooperation in operational climate forecasting.

4.   *Co-ordinated International Activities and Policy Development*

4.1   The Conference endorsed the three streams of international activity:

a.   Global measurement and research efforts through the WCP, IGBP, and other related international programmes

b.   Assessment functions of a continuing IPCC to support negotiation of and provide technical input to a Convention

c.   Development of a Convention on Climate Change.

It is essential that all parties to a Convention and related legal instruments should, as part of their obligations, be required to participate fully in the free exchange and flow of information necessary for technical input to the convention. Such a convention should include a technical annex to provide for:

- International co-operation in research, systematic observation and exchange of related information
- Adjustments based on up-dates of scientific knowledge
- Strengthening national scientific and environmental capabilities of developing countries.

4.2   The development of policy regarding climate change requires on the part of policy makers an understanding of the underlying science and a weighing of the scientific uncertainties associated with the prediction of climate change and its likely impacts. An important aspect of future work is therefore a continued dialogue between scientists and policy makers.

4.3   The UN Conference on Environment and Development (Brazil 1992) provides a valuable opportunity to relate the above three themes to the other environment/development issues and objectives being examined by the Conference. It is therefore essential that the three streams should interact effectively with UNCED.

4.4   It is proposed that the sponsoring agencies for the SWCC consider the possibility of holding a Third World Climate Conference at an appropriate time about the year 2000.

Ministerial Sessions

Second World Climate Conference
6 November, 1990

Introduction by Professor G. O. P. Obasi
Secretary General, World Meteorological Organization (WMO)

It is my great honour to welcome you to this historic meeting, on behalf of the sponsoring organizations. This morning, we begin the portion of the Second World Climate Congress devoted to governmental statements and discussions at the highest political level. With representatives coming from more than 120 countries, I believe we have an indication of the significance being attributed to this Conference.

Last week, about seven hundred leading specialists on climate and climate change from the natural sciences, social sciences as well as legal and other fields met to review and discuss existing knowledge on this topic, and its implications for public policy. In particular, they discussed and endorsed the first assessment report of the Intergovernmental Panel on Climate Change, and reviewed the outcome of the first decade of work under the World Climate Programme.

The scientific community and specialists at last week's Conference prepared a statement which incorporates three clear messages. These are that:

1. Climate is fundamental factor in sustaining earth's natural systems, human health and the economies of nations. There is a need to take immediate effective measures to deal with adverse consequences of its natural variations, as well as with changes induced by human actions.

2. There is an adequate consensus of scientific knowledge on greenhouse gases and global warming that should enable early action by governments.

3. It is also urgently important to support the measurement and research programmes essential to improvement of our knowledge of change and variability of the global climate system.

Our understanding of the issues of increasing greenhouse gas concentrations in the atmosphere and resulting climate change is a product of an unprecedented degree of co-operation. This involved co-operation between scientists from countries around the world, within the framework of international programmes, and co-operation between international agencies to provide this framework. The considerable scientific achievements of the World Climate Programme, established as a result of the First World Climate Conference in 1979, provides one example. While co-ordinated by the World Meteorological Organization (WMO), from the very beginning this programme has had active participation by the United Nations Environment Programme (UNEP), the United Nations Educational, Scientific and Cultural Organization (UNESCO) and its Intergovernmental Oceanographic Commission (IOC), the non-governmental International Council of Scientific Unions (ICSU), and the Food and Agriculture Organization (FAO).

WMO and UNEP have enjoyed particularly close collaboration through the establishment of the Intergovernmental Panel on Climate Change in 1988, and in preparing for negotiations by nations of a Framework Convention on Climate Change, as requested by the United Nations General Assembly and our respective Governing Bodies. This Second World Climate Conference involving five sponsoring international agencies is just the most recent example of the effective collaborations between Members of the UN family of agencies, along with the scientific community through the International Council of Scientific Unions.

Much of the discussions at the earlier scientific part of this Second World Climate Conference focussed on social and economic analyses of the impacts of climate variations and change; touching on the droughts of sub-Saharan Africa that destroy economies and devastate populations, terrible floods and wind from tropical storms, and gradually rising sea level. Much more attention must be devoted to means of adapting social and economic activities to better cope with both the natural variations and man-induced climate change. The biological scientists were also active in the discussions along with colleagues from social, physical, chemical and engineering disciplines especially in considering the critically important International Geosphere Biosphere Programme of ICSU. These experts from many fields have given their best intellectual efforts to bring together greater understanding of the climate issue, an understanding that is important for the establishment of a sound basis for future actions by governments.

The UN agencies and ICSU, together with the world's scientific community from all disciplines, have laboured for many years to help bring us to this moment. We now believe that the scientific knowledge about greenhouse gases and climate change is sufficient for governments to begin actions to reduce the burden of greenhouse gases in the threatened global atmosphere. Many of the initial steps, such as improved energy efficiency and afforestation, have been shown to be beneficial in their own right as well as contributing to greenhouse gas emission reductions. The world leaders and Ministers of governments bear an enormous responsibility for the safeguarding of the future of the planet we all share. We look forward to the Ministerial Declaration that will emerge from this Conference. The watchful eyes of the whole human community are upon us, today. We must respond positively to this unprecedented global challenge, which also serves as a rare opportunity to further global understanding and co-operation, now that humankind is faced with a common environmental concern.

We have an appointment with history to keep. Let us make sure that we do so, that we may be proud to say that we left a lasting legacy to the future generations from whom we borrow the environment we now live in.

Second World Climate Conference
6 November, 1990

Address by H.E. Mr. Arnold Koller
President of the Swiss Confederation

It is a privilege for Switzerland to welcome you all here today in Geneva for the Second World Climate Conference. And it is a privilege for me to have the opportunity to address this distinguished assembly on a most important problem which affects humanity as a whole.

Since we have come to realize that the impact of human activities on our planet are reaching unprecedented levels, we have also become more and more aware that ours is a finite and unique world, whose capacity to sustain life depends on subtle natural equilibria.

As the international scientific community convened here last week aptly reminded us, we have also realized our profound ignorance of the complex processes and mechanisms which regulate our climate and all living systems.

First of all, therefore, we must endeavour to improve our understanding of these mechanisms, through continued study and analysis of the processes at work in our climate system and of the factors affecting it. This will require unprecedented interdisciplinary co-operation; national, regional and international scientific programmes must direct their future efforts towards this end.

Next, at the decisional level, we must act aggressively and within a global perspective to curb the multiple pollutions which are upsetting natural equilibria, to arrest the pillage of natural resources and to preserve them for future generations.

Finally, in spite of the sometimes diverging interests and particular situations, we must lay the foundations of a new international solidarity. Only in this way can we succeed in finding lasting and effective solutions to global environmental problems in the long term.

It is undoubtedly true that we are not tackling a global environmental problem for the first time. The Conference on the Protection of the Ozone Layer, held in London last June, serves to illustrate our firm commitment to taking concrete measures in response to a global threat. However, for all its severity, the ozone problem is comparatively easy to address.

In contrast, climate change, which scientists—to whom I wish to pay hommage—have been predicting, is a much broader and more complex problem. Undoubtedly, modern technology must and can provide solutions. However, the root cause of the problem lies in our very attitudes, in our domineering view of the world and nature, which is why true and lasting solutions can only be found through radical changes in our societies. We shall have to learn to bow to the reality of a finite world and to the immutable laws of nature. We shall have to redefine our development. This Conference and the resulting Declaration are a first decisive step in this direction.

Our action must be dictated by three principles: the precautionary principle, the principle of equity and that of solidarity.

I am fully aware of the magnitude of the task to be accomplished and of its far-reaching implications.

We have already stressed on numerous occasions the need to obtain the most exhaustive and accurate scientific data so that we can take sensible and rational decisions in the area of climate change. We have entrusted the Intergovernmental Panel on Climate Change with the task of producing these data.

The scientists who participated in this effort have now convinced us that, in spite of the complexity of the issues and the remaining uncertainties, what we know now is more than enough to warrant no further delay in making specific and concrete commitments in what we regard as priority areas.

Thus, the industrialized countries, which bear the chief responsibility for the problem, not to mention other forms of pollution, must reduce their emissions of greenhouse gases, in particular carbon dioxide.

In this respect, it seems to us essential that all industrialized countries participate fully in this effort in a spirit of solidarity, and that the largest emitters commit to reductions in proportion to their emissions.

Switzerland is determined to assume its share of the responsibility and take the measures necessary to reduce its emissions of carbon dioxide. Our minimal objective is to stabilize these emissions at their 1990 level by the year 2000. Priority will be given to energy saving measures and improvements in energy efficiency.

Moreover, the Swiss Federal Council intends to introduce a tax on carbon dioxide emissions, a measure which should contribute significantly to stabilizing fossil fuel consumption.

Our experts are presently studying the feasibility and costs of 20% and 50% cuts in carbon dioxide emissions by the year 2005 and 2025 respectively. On the basis of the results, the Federal Council will decide on a future course of action to reduce these emissions.

We consider measures to reduce emissions of carbon dioxide and other greenhouse gases as a vital element of the global strategy to curb global warming. However, it is also essential to realize that the global warming will not be solved by these measures alone.

Industrialized countries must also strengthen and widen their co-operation with developing countries. It is imperative that we support their efforts to combat poverty, malnutrition and all the other factors which contribute to the multiple aggressions against their ecological base. In this respect we must clearly provide developing countries with additional financial resources so that they too are in a position to participate in the global effort to combat climate change.

This effort should include in particular the efficient use of natural and energy resources and the sustainable management of forests. To this end, we industrialized countries must promote the transfer of the most efficient technologies to developing countries and countries of Central and Eastern Europe, so that these countries can secure their economic development while controlling their greenhouse gas emissions.

The convention on climate change, which the international community is ready to negotiate and will adopt in conjunction with the 1992 UN Conference on Environment and Development, will provide a novel and essential impetus for this co-operation.

Switzerland regards the widest possible participation by developing countries in this negotiation as essential. We intend to contribute significantly to the trust fund that UNEP and WMO will establish to support such participation, and will also make contributions to the secretariat of the Negotiation Committee.

We shall have to show responsibility and solidarity at all levels—global, regional, local and also individual—in order to rise to the formidable challenge of global warming.

To this end, all actors must be able to benefit from adequate, accurate and accessible information. Establishing the mechanisms which will be able to disseminate this much needed information is imperative and urgent.

Switzerland intends to contribute to such an endeavour and is presently examining the possibility of creating, in collaboration with UNEP, an international climate information centre in Geneva. The vocation and purpose of this centre will be to collect and disseminate pertinent information on climate change issues, in particular for the benefit of developing countries.

In addition, the Swiss Federal Government is supporting an initiative by the University of Geneva to create an International Academy of the Environment. The Academy will provide continued education in the field of environment to decision makers worldwide.

Finally, at the national level, my country will see to it that courses on environmental matters, and in particular climate change, be incorporated in school and university curricula.

I wish to conclude by saying how honoured and pleased I am that Switzerland was able to contribute to the organization of this Conference, and to the elaboration of the Ministerial Declaration. I can assure you of our firm intention to continue to participate actively in this process, in particular within the framework of the upcoming negotiations.

May I also congratulate, in the name of my country, the United Nations Environment Programme and the World Meteorological Organization for their outstanding achievements. I sincerely hope that their efforts will result in the adoption of an effective convention on climate change and protocols, in conjunction with the United Nations Conference on Environment and Development, which will be hosted by Brazil in June 1992.

We are less than twenty months away from that historic moment when, for the first time ever, Heads of State from around the world will come together to draft an action plan for an intelligent stewardship of our planet. It is my sincere hope that this Second World Climate Conference will be a significant step towards the success of that most important event.

Second World Climate Conference
6 November, 1990

Address by His Majesty King Hussein Bin Talal of the Royal Hashemite Kingdom of Jordan

Mr. President,
Ladies and Gentlemen,

I would like first of all to extend my sincere congratulations to you, Mr President, for your election to preside over this Second World Climate Conference, and to assure you of my full personal support and that of the government and people of Jordan in the important task which you have accepted.

I would also like to express my deep appreciation for the efforts of the several respected bodies which have worked together to organize this meeting - the World Meteorological Organization; the United Nations Environment Programme; the United Nations Educational, Scientific and Cultural Organization; and the International Council of Scientific Unions.

Such cooperation reflects the reality that many regional and even domestic problems derive from transnational causes and dynamics, which in turn can only be properly resolved through a multi-national approach. This is very clear in fields such as environmental degradation and climatic changes, human rights, economic adjustment and growth, social equity, emigration and refugees, or political stability.

Technical as it might be perceived by some, the climate issue provides an opportune occasion to appreciate the full, broader global and human context in which we are called upon to face up to present challenges, learn from past experience, and anticipate future consequences and likelihoods.

The world first seriously appreciated the threat of environmental degradation in the 1960s and 1970s, which was followed in the 1980s by a realization of the potentially momentous changes that were taking place in the Earth's climate. This concern moved in a politically linear manner. It started with expressions of environmental concern by small special interest groups in the industrialized democracies of the north, which subsequently spawned organized pressure groups, green-oriented political parties, official governmental bodies, and international organizations. It reached a first milestone in the Hague Declaration in March 1989, which brought together Heads of State and Heads of Government from many countries around the world.

We are still in the early stages of identifying the scope and consequences of environmental damage and long-term climatic changes. Given the rising political popularity of green politics in some parts of the world, we should beware of the danger of allowing our participation in efforts such as this to be driven by short-term political or electoral gain.

Rather, we must be motivated solely by our genuine fears of what could happen if we did not gather, did not care, and did not act collectively. We were in the Hague, and we are here today, because of our awesome responsibility for the interests and perhaps even the very survival of future generations.

In all honesty, I had to think hard about whether the circumstances in the Middle East allowed me to make this trip here to join you in this important meeting. But I came here in the end because I realised that this is an appropriate occasion to share with you some of my lessons and thoughts about what I referred to earlier as the broader human context of the global challenge we face. I also came here because throughout my life I have always sought to use every opportunity of dialogue among friends to discuss issues of common concern in a constructive spirit of honesty and hope.

Inviting as it may be, we cannot take the easy road of isolating technical issues such as the natural climate from the broader human environment and political climate in which we live. You who are from the industrialized democracies of the north have already experienced this fact in the emergence of politically successful green parties in your countries.

The physical quality of the earth which we pass on to future generations will be a major determinant of their chances for productive, balanced and dignified lives.

Equally important will be the moral legacy and the socio-political criteria which our children and their children inherit from us.

While I do not wish to address political problems in the narrow sense, I feel that we live in a time and a place in which we can no longer draw a clear line between political concerns, our environmental life base, and the prospects for future generations around the world. Nowhere is this more evident than in the Middle East. And at no other time in recent memory - indeed, in my lifetime and yours - has the Middle East held the seeds of a potentially global catastrophe as it does today.

Let us review the picture in our region today: nearly a million soldiers from Iraq, the United States, Great Britain, France, Saudi Arabia, Egypt, Syria and a score of other countries confront each other in the Gulf - bristling with sophisticated armaments, supported by thousands of advanced tanks and aircraft, and driven by a deadly combination of human fear and political ferocity. Our region is also rich in weapons of mass destruction, in the hands of emotionally charged Arab, Israeli and international parties who in a war situation may not hesitate to use them. The military build up and the political tension both continue to escalate, day-by-day, hour-by-hour.

This confrontation is taking place literally on top of the single richest natural petroleum reservoir in the world, which accounts for over half the world's mineral energy resources. A war in the Gulf would not only result in devastating human death and injury, tremendous economic loss, and prolonged political confrontation between Orient and Occident. It could also lead to an environmental catastrophe the likes of which the world has not experienced since the accident at the Chernobyl nuclear power plant, which sent shock waves around the world, and awoke us all to the true meaning of *global* threats and challenges. A war in the Gulf could result in the use of chemical and biological weapons, and widespread destruction to oil fields and oil storage depots.

The preliminary calculations of our scientists indicate that if half Kuwait's oil reserves (or about 50 billion barrels) were to go up in flames during a war, the environmental impact would be swift, severe and devastating. Emissions of carbon monoxide, carbon dioxide and sulphur dioxide would surpass internationally accepted safety standards by factors of hundreds, and, without factoring in wind effects, would blacken the skies over a radius of at least 750 kilometres from Kuwait - that is, all of Kuwait, Iraq, Bahrain, Qatar, the Emirates and the waters of the Gulf, and most of Saudi Arabia, Jordan, Syria and Iran.

Other than poisonous emissions of carbon monoxide and sulphur dioxide, the emission of carbon dioxide will increase by at least a factor of 100 over the current total global emission of 5.5 billion tons per year. Lingering in the atmosphere for around 100 years, this massive CO_2 emission would promote the greenhouse effect, and contribute to global warming, climatic changes, lower global food production, and human and animal health deterioration.

The environmental and human toll of such a scenario would be beyond our wildest fears. If oil production facilities suffered long term damage, which is likely in the event of war, the catastrophe, we warn of, would be compounded by the devastating impact of the loss of oil imports for economies and peoples around the world.

Yet, these consequences do not have to materialize. The catastrophe scenario can be averted. We owe it to ourselves, and to the future generations whose life prospects may be determined by the policies we pursue today, to give vitality and vision to our quest for a better world. We owe it to those who place their trust in us to address and deal with the root causes of the conflicts and the potential catastrophes which confront us today.

As we meet here today, consider, if you would the following snapshot of our region:

- In Sudan, millions of Arabs face yet another round of deathly famine and starvation
- In Yemen, over half a million nationals are returning to live in temporary camps, as refugees in their own country
- In Jordan, we have seen nearly one million people pass through our country in the past three months, fleeing the crisis in Kuwait and Iraq.
- Nearly half a million Kuwaiti nationals find themselves refugees outside their occupied country.
- In the Israeli occupied West Bank and Gaza, about one thousand Palestinians have been killed in the *intifada* during the past three years, 100,000 have been injured, 10,000 have been held in administrative detention, and over 90,000 olive and fruit trees have been destroyed by the Israeli occupation forces.
- Throughout the region, living standards, as measured by *per capita* gross domestic product, have declined steadily throughout the past decade. We grow less and less of the food we need. Our water balance grows more precarious every day.
- Millions of Palestinians remain disenfranchised, and denied their legitimate national expression and the protection of their own sovereign government.
- Portions of Lebanon and Syria remain under Israeli occupation, while millions of Lebanese are denied the right of a normal and secure life.

This is the human, political and social climate in which we find ourselves. In such a context, the primary concern of most people is sheer survival. Environmental and climatic concerns, no matter how valid, are not only difficult to address rationally, but also suffer gravely from the negative underlying factors which I have just mentioned. In the scramble to stay alive today and to make it to tomorrow, natural environments are further degraded. Long term concerns are buried beneath the desperate rush for water, food, arable land, forage and fuel. We are thrown back into the debilitating cycle of environmental degradation, economic regression, human frustration and anger, and, ultimately, political extremism. With the Middle East once again threatening the security of the entire world, we must all continue exerting all possible efforts to resolve the Gulf crisis through peaceful negotiations. Then we must quickly move on to resolve the Arab-Israeli conflict, which, if allowed to fester, will continue to be a source of regional and global instability. In the longer term, having secured the essential dictates of justice, we need to address the idea of a Middle East zone of peace, free of nuclear, biological, chemical and other weapons of mass destruction. In such a context, with justice providing the foundation for peace which in turn assures stability and security, the people of the region could allocate their substantial human and material resources to the quest for a better future.

We in Jordan are not a military or an economic power in the region, neither a demographic nor an ideological force. Yet, we feel a deep responsibility today to our own people, to the rest of the world, and to future generations - a responsibility which impels us to do everything within our power to try to avert a disaster in the Gulf. We cannot compromise on our right and our duty to speak our minds honestly, to stand up for principled policies, to demand a single international measure of morality and justice, and to toil for the rights of our children and their children. The cardinal right we hope to pass on to them is the right to live in a world that has a chance to survive and to sustain itself, both physically and morally.

Some of you may feel that I have strayed far from the concerns of climate. I do not think so. If I could distill into a single observation the lessons of my many years of public life and responsibility in a turbulent region, spanning parts of five decades from the 1950s to the 1990s, it would be this: the genuine security and stability of nations and regions emanates not from military power or economic assets; rather, it stems from the conviction of indigenous peoples that their essential basic needs are met, that they enjoy dignity and respect, that they have recourse to justice, and that their children will survive and perhaps even have an opportunity for a better life.

Mr President,

When we signed the Hague Declaration in March 1989, I noted that our actions then reflected not only our duty, but also the right of future generations to inherit a viable planet and to anticipate a reasonable quality of life. Our gathering here today takes place within a more alarming global context of confrontation, militarism and a potential global ecological and economic catastrophe.

This vividly highlights the full spectrum of dangers facing our future generations - dangers which emanate from a combination of political, economic, social, moral and physical degradation. How far can we really separate the climate of the earth, the political environment, and the ecosystem of the human mind and heart?

All our commitment and activism on behalf of our earth will remain inadequate if hundreds of millions of people on earth feel that we pay more attention to the degradation of their physical environment than we do to the desecration of their human spirit. In the long run, the fastest and most efficient route to safeguarding our planet through truly global action would be to assure the spread of democratic principles, human rights and social equity throughout the world - to offer all people the vision of justice, and the promise of dignity.

Why is it that during the past three decades, organized concern for the environment emanated first and foremost from societies in North America and Western Europe? Is it perhaps because those peoples enjoy basic standards of genuine national sovereignty, material welfare, social justice, human rights, and political accountability?

And why is it that urban and rural environmental degradation moves apace most quickly throughout the Third World? Is it partly because throughout the Third World the political and social climate remains erratic? Isn't the linkage direct, obvious and compelling? I believe it is.

In view of the imbalances in global political stability and socio-economic justice, and the enormous threats to the environment from crises such as that in the Gulf today, if our global quest to safeguard our earth is to be credible and effective, it must reflect an appreciation for the broader human, political and psychological climate of our earth and its people.

Allow me to emphasize once again my personal commitment, and that of my government and people of Jordan, to the goals of this meeting. We recognise our modest but real role within the global order as working both to preserve the physical aspects of God's world as well as its moral, spiritual and ethical attributes. We shall always do all that is within our means to promote these noble goals, and I am certain that our collective efforts shall triumph in the end.

Thank you again, and may God give you strength in all your endeavours.

Second World Climate Conference
6 November, 1990

Address by The Rt. Hon. Margaret Thatcher Prime Minister of the United Kingdom of Great Britain and Northern Ireland

Introduction

Mr. Chairman, Your Majesty, President Koller, Distinguished Colleagues, Your Excellencies, Ladies and Gentlemen,

May I begin by thanking Heads of Agencies and Organisations for sponsoring this Second World Climate Conference, and indeed all those connected with it. It is a most important event for all our countries and I wish you success in your endeavours.

Mr. Chairman, since the last World War, our world has faced many challenges, none more vital than that of defending our liberty and keeping the peace. Gradually and painstakingly we have built up the habit of international cooperation, above all through the United Nations. The extent of our success can be seen in the Gulf, where the nations of the world have shown unprecedented unity in condemning Iraq's invasion and taking the measures necessary to reverse it.

But the threat to our world comes not only from tyrants and their tanks. It can be more insidious though less visible. The danger of global warming is as yet unseen, but real enough for us to make changes and sacrifices, so that we do not live at the expense of future generations.

Our ability to come together to stop or limit damage to the world's environment will be perhaps the greatest test of how far we can act as a world community. No-one should under-estimate the imagination that will be required, nor the scientific effort, nor the unprecedented cooperation. We shall have to show statesmanship of a rare order. It's because we know that, we are here today.

Man and Nature: Out of Balance

For two centuries, since the Age of Enlightenment, we assumed that whatever the advance of science, whatever the economic development, whatever the increase in human numbers, the world would go on much the same. It was progress. And that was what we wanted.

Now we know that this is no longer true. We have become more and more aware of the growing imbalance between **our** species and **other** species, between population and resources, between humankind and the natural order of which we are part.

In recent years, we have been **playing** with the conditions of the life we **know** on the surface of our planet. We have cared too little for our seas, our forests and our land. We have treated the air and the oceans like a dustbin. We have come to realise that man's activities and numbers threaten to upset the biological balance which we have taken for granted and on which human life depends.

We must remember our duty to Nature before it is too late. That duty is constant. It is never completed. It lives on as we breathe. It endures as we eat and sleep, work and rest, as we are born and as we pass away. The duty to Nature will remain long after our own endeavours have brought peace to the Middle East. It will weigh on our shoulders for as long as we wish to dwell on a living and thriving planet, and hand it on to our children and theirs.

The Importance of Research

I want to pay tribute to the important work which the United Nations has done to advance our understanding of climate change, and in particular the risks of global warming. Dr. Tolba and Professor Obasi deserve our particular thanks for their farsighted initiative in establishing the Intergovernmental Panel on Climate Change.

The IPCC report is a remarkable achievement. It is almost as difficult to get a large number of distinguished scientists to agree, as it is to get agreement from a group of politicians. As a scientist who became a politician, I am perhaps particularly qualified to make that observation!

Of course, much more research is needed. We do not yet know all the answers. Some major uncertainties and doubts remain. No-one can yet say with **certainty** that it is **human** activities which have caused the apparent increase in global average temperatures. The IPCC report is very careful on this point. For instance, the total amount of carbon dioxide reaching the atmosphere each year from **natural** sources is some 600 billion tonnes, while the figure resulting from human activities is only 26 billion tonnes. In relative terms that is not very significant. Equally we know that the increases of carbon dioxide in the atmosphere date from the start of the industrial revolution. And we know that those concentrations will continue to rise if we fail to act.

Nor do we know with any precision the extent of the likely warming in the next century, nor what the regional effects will be. We cannot be sure of the role of clouds.

There is a continuing mystery about how atmospheric carbon, including the small extra contribution from human sources, is being absorbed: is most of it going into the ocean, as used to be thought? Or is it being increasingly absorbed by trees or plants, or soils, especially in the northern hemisphere? These are questions that need answers, sooner rather than later.

Global climate change within limits need not by **itself** pose serious problems - our globe has after all seen a great deal of climate change over the centuries. It is notable that the blue-green algae which dominated the Precambrian period at the dawn of life are still major components of the marine phytoplankton today. Despite the climate changes of many millions of years, these microbes have persisted on earth virtually unchanged, pumping out life-giving oxygen into the atmosphere and mopping up carbon dioxide.

The real dangers arise because climate change is combined with **other** problems of our age; for instance:

- the population explosion
- the deterioration of soil fertility
- increasing pollution of the sea
- intensive use of fossil fuel, and
- destruction of the world's forests, particularly those in the tropics.

Britain will continue to play a leading role in trying to answer the remaining questions, and to advance our state of knowledge of climate change. This year, we have established in Britain the Hadley Centre for Climate Prediction and Research for this purpose. We need to improve in particular our understanding of the effect of the oceans on our weather, improve too our capability to model climate change. I have seen for myself the outstanding work being done on both these subjects at the National Center for Atmospheric Research in Boulder, Colorado.

We must also make sure that research is carefully targeted. many people can do the **same** thing, and at the same time **vital** problems can be neglected. The task of global observation is immense. It will require a coordinated effort more ambitious than any attempted before, as the meeting of scientists and experts last week recognised.

The Need for Precautionary Action

But the need for more research should not be an excuse for delaying much needed action. There is **already** a clear case for precautionary action at an international level. The IPCC tells us that we cannot repair the effects of past behaviour on our atmosphere as quickly and as easily as we might cleanse a stream. It will take, for example, until the second half of the next century, until the old age of my grandson, to repair the damage to the ozone layer above the Antarctic. And some of the gases we are adding to the global heat trap will endure in the Earth's atmosphere for just as long.

The IPCC tells us that, on present trends, the earth will warm up faster than at any time since the last ice age. Weather patterns could change so that what is now wet would become dry, and what is now dry would become wet. Rising seas could threaten the livelihood of that substantial part of the world's population which lives on or near coasts. The character and behaviour of plants would change, some for the better, some for worse. Some species of animals and plants would migrate to different zones or disappear for ever. Forests would die or move. And deserts would advance as green fields retreated.

Many of the precautionary actions that we need to take would be sensible in any event. It is sensible to improve energy efficiency and use energy prudently; sensible to develop alternative and sustainable sources of supply; sensible to replant the forests which we consume; sensible to re-examine industrial processes; sensible to tackle the problem of waste. I understand that the latest vogue is to call them 'no regrets' policies. Certainly we should have none in putting them into effect.

And our uncertainties about climate change are not all in one direction. The IPCC report is very honest about the margins of error. Climate change may be **less** than predicted. But equally it may occur **more** quickly than the present computer models suggest. Should this happen it would be doubly disastrous were we to shirk the challenge now. I see the adoption of these policies as a sort of premium on insurance against fire, flood or other disaster. It may be cheaper or more cost-effective to take action now than to wait and find we have to pay much more later.

The Need for Environmental Diplomacy

We are all aware of the immense challenge. The enormity of the task is not a matter for pessimism. The problems which science has created science can solve, provided we heed its lessons. Moreover, we have already established a structure of international co-operation on the environment to deal with ozone depletion. For the first time ever, rich and poor nations alike set out together to save our planet from a serious danger. This painstaking work culminated in the historic agreement reached in London this year. That agreement is a real beacon of hope for the future.

The main focus in London was on protecting the ozone layer. But the agreement will have other consequences. We should not forget that CFCs are 10,000 times more powerful, molecule for molecule, than carbon dioxide as agents of global warming. But of the other greenhouse gases, carbon dioxide is by far the most extensive and contributes around half the man-made greenhouse warming. **All** our countries produce it. The latest figures which I have seen show that 26 per cent comes from North America, 22 per cent from the rest of the OECD, 26 per cent from the Soviet Union and Eastern Europe and 26 per cent from the less developed countries.

These figures underline why a **joint** international effort to curb greenhouse gases in general and carbon dioxide in particular is so important. There is little point in action to reduce the amounts being put into the atmosphere in one part of the world, if they are promptly increased in another. Within this framework the United Kingdom is prepared, as part of an international effort including other leading countries, to set itself the demanding target of bringing carbon dioxide emissions back to this year's level by the year 2005. That will mean reversing a rising trend before that date.

The European Community has also reached a very good agreement to stabilise emissions. I hope that Europe's example will help the task of securing world-wide agreement.

Targets on their own are not enough. They have to be achievable. Promises are easy. Action is more difficult. For our part, we have worked out a strategy which sets us on the road to achieving the target. We propose ambitious programmes both to promote energy efficiency and to encourage the use of cleaner fuels.

We now require, by law, that a substantial proportion of our electricity comes from sources which emit little or no carbon dioxide. That includes a continuing important contribution from nuclear energy.

Such measures as these - which increasing numbers of countries are adopting - should be seen as part of the premium on that insurance policy which I mentioned. They buy us protection against the hazards of the future: but they also pay dividends even though the gloomier predictions about global warming are not fulfilled - dividends such as less air pollution, lowered acid rain, reduced energy costs.

Mr. Chairman, people may disagree about the effects of increased man-made carbon dioxide in the atmosphere. But everyone agrees that we should keep in healthy condition the forests and seas which absorb and utilise a large part of it here on earth. We would be wise to do that for other reasons too: for the beauty of the forests and the infinite variety of species which inhabit them, and to preserve the food chain and balance of nature in the sea.

That's why we want to contribute to conserving the world's forests, and to planting new ones. Trees help to reduce global warming. We intend to plant more at home: we have just announced our plans to replant one of the ancient forests of England - destroyed in an earlier phase of our history.

We shall offer our expertise and aid funds to help plant and manage forests elsewhere in the world, particularly in tropical countries. A year ago I told the United Nations General Assembly that the United Kingdom would aim to increase its funds for tropical forestry by £100m. We now have 150 projects underway in more than 30 countries.

Our aim is to give the people in those countries a better standard of living by conserving and using the forests than by cutting them down.

The Need for a Global Convention

But our immediate task this week is to carry as many countries as possible with us, so that we can negotiate a successful framework convention on climate change in 1992. We must also begin work on the binding commitments that will be necessary to make the convention work.

To accomplish these tasks, we must not waste time and energy disputing the IPCC's report or debating the right machinery for making progress. The International Panel's work should be taken as our sign post: and the United Nations Environment Programme and the World Meteorological Organisation as the principal vehicles for reaching our destination.

We will not succeed if we are too inflexible. We will not succeed if we indulge in self-righteous point-scoring for the benefit of audiences and voters at home. We have to work sympathetically together. We have to recognise the importance of economic growth of a kind that benefits future as well as present generations everywhere. We need it not only to raise living standards but to generate the **wealth** required to **pay** for protection of the environment.

It would be absurd to adopt polices which would bankrupt the industrial nations, or doom the poorer countries to increasing poverty. We have to recognise the widely different circumstances facing individual countries, with the better-off assisting the poorer ones as we agreed to do under the Montreal Protocol.

The **differences** can't be drafted away in that famous phrase so beloved of diplomats "a form of words". They need to be resolved by tolerant and sympathetic understanding of our various positions. Some of us use energy more efficiently than others. Some of us are less dependent on fossil fuels. And we each have our own economic characteristics, resources, plans and hopes for the future. These are the realities that we must face if we are to move forward towards a successful conclusion to our negotiations in 1992.

Just as philosophies, religions and ideals know no boundaries, so the protection of our planet itself involves rich and poor, North and South, East and West. All of us have to play our part if we are to succeed. And succeed we must for the sake of this and future generations.

One of our great poets, George Herbert, in his poem on "Man" wrote this:

> "Man is all symmetry,
> Full of proportions, one limb to another, And all to all the world besides;
> Each part may call the farthest, brother; For head with foot hath private amity,
> And both with moons and tides."

We **are**, in symmetry with nature. To **keep** that precious balance, we need to work together for our environment. The United Kingdom will work with all of you in this cause - to save our common inheritance for generations yet to come.

Second World Climate Conference
6 November, 1990

Address by H.E. Mr. Michel Rocard
Prime Minister of France

Our meeting here today is a very special occasion, thanks first of all to the warm hospitality the Confederation of Switzerland invariably extends to its guests, for which I thank you, Mr. President. Thanks also to various international organizations: the United Nations Environment Programme, the World Meteorological Organization, UNESCO, the United Nations Food and Agricultural Organization and the International Council of Scientific Unions. I should like to take this opportunity to assure their leaders once again of our support and to express our deep appreciation of their enthusiastic work. I should also like to thank the scientists with us here today for their diligence and their efficiency, for they play an essential part, particularly in alerting world opinion to climate problems.

We have covered much ground in recent years. Just eighteen months ago, acting on a joint initiative of the Netherlands, Norway and France, twenty-four Heads of State and of Government, including Mr. François Mitterrand, launched an appeal at the Hague. We have come several stages further since then, with the Declaration of Noordwijk and Bergen, for example, and two summit meetings of the major industrialized countries.

Yet nothing decisive has yet been accomplished. Our Conference must provide that decisive turning point by launching the negotiating process for a convention on climate change. The time for words is over; what we need now is action. The race against time is on. The very survival of our planet is at stake.

The strengthening of the "Montreal Protocol" in London last June is proof that it can be done. All that is needed is determination. And I am here today to confirm that France has that determination.

I am fully aware of the magnitude of the challenge facing us. I know that numerous special cases and even conflicting interests must be taken into account in the short term. And that long-term strategies must remain flexible enough to accommodate scientific progress and the fruit of experience. This will only be possible if there is a common resolve to act, based on certain major principles:

- the price of taking no action is far too high to be worth the risk
- environmental protection and preventing climate change must henceforth be an integral component in the economic development process
- a new dimension must be brought to international co-operation to make it fully effective.

Any remaining doubts must not be allowed to dissuade us from acting. It is obvious that we still have much to learn about the greenhouse effect; scientific knowledge of the phenomenon and its eventual consequences need further improvement. I am thinking, for example, of the major question of the role played by the oceans. The research effort must therefore be intensified as a part of our action, just as clean, energy-efficient technologies need to be developed. This is precisely what the French Government had in mind in 1989 when it spent 285 million francs on research into climate change. And the Academy of Science has just submitted an important report to the Government analyzing the greenhouse effect.

There is no doubt whatsoever that more resources need to be devoted to monitoring and gathering data worldwide, particularly in the countries of the South. This was the reason behind the dual initiative taken by President François Mitterrand in launching the Sahara and Sahel Observatory, with the help of different countries and international organizations, and in proposing to establish a permanent environmental monitoring observatory in the South Pacific. This second Observatory will be a correspondent of the European Environment Agency and the future French Environment Institute, collecting and disseminating data on the natural environment. It will of course be open to the countries of the region. Finally, my country wishes to make satellite data more readily accessible to all, which is why it has launched its scheme to set up an environmental satellite data collection and distribution centre.

Faithful to a consistent policy—as illustrated by these few examples—France wants to help the developing countries to take a fully active hand in the international research effort on the world environment. It is a major aspect of French policy, and I solemnly confirm this here today should there be any doubt in anyone's mind.

Although research must be stepped up, and its results constantly fuel our action, this must by no means be considered as a prerequisite or allowed to delay action. The Intergovernmental Panel on Climate Change (IPCC)—which I should like to congratulate here on the quality of its work—made it very clear in its report that the price to pay in the fight to control climate change will be all the higher the longer we wait before taking the necessary action. This conclusion was amply illustrated in this very place in the first part of this Conference, which resulted in the adoption of a highly remarkable Conference Statement rallying the scientific community.

It is in our best interest that the IPCC should continue its scientific work and fully play its part in the ongoing assessment of research in this field around the world. We shall need its periodic reports, even to help us fulfil our commitments under the international agreement, France is eager for.

It is a matter for the international community of adopting a "strategy of caution", failing which irreparable damage is likely to be wrought on the environment by the changing climate. The responsibility of the world's governments is in the balance; they must, we must, prove ourselves worthy of our peoples' trust.

The international action in question must be undertaken in a universal spirit of fairness and solidarity. Universal because the answer to a worldwide problem must also be worldwide. France certainly intends to incorporate its own national effort into the broader framework of larger groups such as the European Community and the OECD. On 29 October last the twelve EEC member states reached an initial agreement. This is an encouraging sign. The concerted reduction of carbon dioxide emissions—for which the target set by the IPCC is 60% of current world emissions—can only be achieved through solidarity.

The industrialized nations, responsible for most of the world's greenhouse gas emissions, are also those which possess the technological know-how and the financial wherewithal to put a stop to them. They must hearken to the needs of their partners. Amongst other things, through their own efforts they must give the developing countries the time they need before they are in a position to participate fully in the combined effort, and provide them with financial, technical and human means to help them

turn to development policies which are less costly in terms of natural resources and more energy-efficient.

Finally, the action must be equitable, which means that the effort we each need to make is proportional to the amount of progress yet to be made, i.e. that it is up to those with the most wasteful lifestyles in terms of fossil fuel consumption to make the greatest cuts. Furthermore, the effort yet to come is a function of that already made and the results obtained by certain countries in saving energy and developing forms of energy that contribute to reducing the greenhouse effect.

Achieving substantial reductions in fossil fuel consumption means reflecting the cost of the harm they do to the environment in their price, so as to modify user behaviour. At the same time, direct or indirect incentives to use CO_2-producing fossil fuels must be phased out. On our march to curb the greenhouse effect we must be realistic in the targets we set ourselves. But above all, each and every one of us must assume his responsibilities. The industrialized countries should set emission reduction targets on a country-by-country basis. France announced its own target in September, when we undertook to level out our per capita CO_2 emissions at the equivalent of less than two tonnes of coal per annum by the year 2000.

But above and beyond these specific measures and guidelines, protecting the world environment and limiting the greenhouse effect must become an integral feature of development policy in every country. It is imperative first of all that the industrialized countries review and if necessary change their energy production and consumption methods, and all the more drastically the higher their per capita greenhouse gas emissions. It is a lengthy process and it must be set in motion without delay. The first step is no doubt to reinforce energy-saving policies and develop forms of energy that do not emit greenhouse gases, like nuclear energy and renewable fuels. Such policies—which must be based on careful research and economic analysis—require single-minded determination, and can prove costly, in social terms for example. But the net result is positive and the long-term benefits are substantial.

Our country has proved as much in the last ten years or so by separating economic growth and energy consumption. Between 1980 and 1988 CO_2 emissions were reduced by 25%, while economic growth reach 17%. We have little leeway, but we are determined to exploit it to the full by counting on the development of new technologies, particularly in the transport sector, where our results have not been so good. The French railway programme, in which the high-speed train is a major factor, is a partial answer. But much remains to be done, such as improving urban public transport networks to make them more attractive to the user, or encouraging the use of electric vehicles in cities, and the marketing of cleaner engines.

There are other greenhouse gases, of course, which must not be forgotten. France will give priority to reducing CFC emissions, which are particularly instrumental in the greenhouse effect. The quantities of CFCs stored in refrigerators and aerosol sprays are a major source of concern. Although CFC production has decreased considerably in France—down 50% since 1986—the Government intends to do more, particularly in connection with the recovery and destruction of these gases after use. The Government will also be striving to make significant cuts in methane emissions as soon as possible by limiting the dumping of putrescible waste in favour of incineration or, whenever possible, sorting and recycling. These two measures will of course have other favourable impacts on the environment or the quality of life. This is why I have taken them without regret, as they say in international circles. I can only hope that other countries in a position to do so will follow suit without delay.

The developing countries, for their part, have the opportunity of adopting growth policies that keep greenhouse gas emissions to a minimum. The industrialized countries will of course have to help them obtain the extra financial resources they will need to enjoy the benefits of suitable technologies, or better still, help them develop their own technologies using their own know-how. For there are no miracle solutions or "turn-key" strategies. Every State must show determination and ingenuity in seeking the road to sustainable, environment-friendly growth. The countries of the South are faced with special constraints which we must try to understand; the effort to curb the growth of the greenhouse effect must not further hinder a growth process which is already hampered by the structural adjustment programmes imposed by the crisis that has affected the countries concerned in recent years.

In facing up to these realities, however, there are two basic truths of which we must not lose sight:

- while it is true that the industrialized countries are largely responsible for the build-up of greenhouse gases, the proportion of emissions generated by the developing countries is sure to increase considerably in the space of a few years
- in many cases the effects of climate change, be it in the form of atmospheric pollution, desertification, rising sea levels or cyclones, will be more serious and more difficult in the developing countries than elsewhere.

In many conntries, particularly the smaller ones (such as the islands of the South Pacific), regional and sub-regional co-operation is vital if the greenhouse effect is to be tackled effectively. It is essential in my opinion that such co-operation be intensified in the future.

The developing world is not alien to the debate on the greenhouse effect. Quite the contrary. With the support of the industrialized countries and the international organizations concerned it must contribute in full to the combined effort, by bringing population growth under control, for example.

If they are to bear fruit, efforts to limit the greenhouse effect must include an acceleration of international co-operation, which in turn means greater North-South solidarity and the elaboration of more suitable international instruments.

In the first place we must take into account the special situation of the developing countries to which I have just referred. Generally speaking, the connection between development and the environment must be recognized and borne in mind at every level. Bilateral and multilateral aid programmes in particular should incorporate this concern. In this respect France supports the innovative concepts launched by the "Brundtland Commission", and makes a conscious effort to take them into account in its own development co-operation policy. It is in this same spirit that my country is approaching the decisive stage marked by the United Nations Conference on the Environment and Development scheduled to be held in Rio de Janeiro in 1992. In order to contribute more effectively to international debate, the President of the French Republic has invited 1,000 representatives of NGOs to meet in December 1991. These people who are active in the field have some essential things to say. We must listen to them.

More specifically, France considers that world environment questions merit additional financial resources, allocated through new channels. For this reason my Government has suggested setting up a new mechanism within the World Bank, in the form of a "global environment facility", which through direct co-operation between industrial operators in the North and the South, could help, for example, to adapt and disseminate technologies in keeping with their common goals. We are confident that this project—for which we have earmarked 900 million FF—will be a success, and I take this opportunity to appeal to all the donor countries to make it operational as soon as possible. France is also willing to finance research in certain developing countries to help them assess the strategies they need to implement and their cost.

It is our belief that where the greenhouse effect and climate change are concerned the onus for immediate action lies mainly with the industrialized countries. As in the case of the Montreal Protocol, the developing countries should be granted a "waiver" or a period of "grace" in respect of the commitments different countries are expected to make. In any event the international machinery we set in motion should be sufficiently flexible to combine a degree of variety in the nature of the objectives to be achieved and deadlines in keeping with the real possibilities of the countries concerned. International debate on this point must be accelerated.

We must be ambitious in the way we address the institutional aspects of the international approach to dealing with the greenhouse effect. Over the last year a consensus has emerged on the need to conclude an international agreement on the climate as soon as possible. That is a step in the right direction, but it is not enough. The content of the agreement must be tangible, with targets, deadlines and follow-up facilities. I hope the final declaration of our Conference will reflect this need. Further delay before action is taken is quite unacceptable. It would be deceiving our contemporaries and taking on a heavy responsibility with regard to future generations. And if we look beyond the single problem we are gathered here to address today, we are clearly faced with the fundamental question of the enforcement of international law on the environment. What point is there in holding meetings and conducting research if there is no certainty that the standards we adopt and the commitments we enter into will actually be respected?

As you know, together with other countries, and in particular Australia, my country has taken the initiative of proposing negotiations to draw up a convention to protect the Antarctic environment and transform the continent into a natural and scientific reserve. We did so in the conviction that the time has come for firm commitments and if need be for binding mechanisms to enforce them and follow them up. In my opinion the same applies to the matter under discussion here today. The problem of the climate and the negotiations that lie ahead are an ideal opportunity for us to mark our firm resolve to work together to change international co-operation and make it altogether more effective. Let us not waste this opportunity. Let us not hesitate to include principles and mechanisms in our future convention that will guarantee the effective application and monitoring of the commitments undertaken.

The task before us is a huge one, and we are pressed for time. The task is nevertheless within the bounds of human capacity. Our very survival could be at stake. I urge us all to overcome our reservations, to listen to our partners, for whom the task is even more arduous, and to rise to the hopes vested in us. I do not underestimate the difficulties we shall encounter in rising to this challenge. We simply have to choose between these difficulties and the alternative: irreparable damage to the future of our planet.

It is for us to choose. France has done so, clearly and irrevocably.

Second World Climate Conference
6 November, 1990

Address by The Rt. Hon. Edward Fenech-Adami Prime Minister of Malta

Mr Chairman,
Distinguished Colleagues,
Ladies and Gentlemen,

We are gathered here because of our shared sense of the great need for some more of that light of the mind which alone can enable us to meet the threat of an overpowering physical heat destroying our planet's life.

Therefore, I would like in the first place to express my appreciation to the organisers of this conference, - the World Meteorological Organization, the United Nations Environment Programmme, the United Nations Educational Scientific and Cultural Organisation, the Food and Agricultural Organisation and the International Council of Scientific Unions - for their setting it up in such good order: a real victory over entropy!

Facing the problems of climate change, in the framework of both sustainable development and environmental conservation, requires collaborative inputs at national, regional and universal levels. The large number of states represented here today augurs well for our task.

Since the mid-seventies, scientists had been sounding various alarms about the weather. In fact, a great deal of research work in connection with this problem was undertaken by a number of scientific organisations. The lack of co-ordination in their work, however, led to a lot of overlapping, with the consequent waste of precious limited resources. Moreover. the scenarios produced were never quite consistent, sometimes they were diametrically conflicting, and always perturbing for both citizens and policy-makers.

In 1988 my Government decided that, in view of this confusing situation and the serious threat which climate change presented, urgent action needed to be taken to have this problem discussed at the highest political level. In September of that year, a request was made to put the issue of climate change on the agenda of the 43rd Session of the United Nations General Assembly. Malta's initiative on climate was aimed at creating a co-ordinated global response to climate change. It was felt that all States had to participate in the task, if the aim of climate preservation was to be achieved.

There were three main elements in our proposal. First, any serious attempt to address the problem of climate change had to be based on a thorough understanding of its causes and effects. Consequently, the first element contained in the resolution Malta presented to the United Nations General Assembly was specifically that a comprehensive review of the state of knowledge attained and the promotion of scientific research on climate be undertaken, together with a study of the economic and social impacts of climate change, including global warming.

Secondly it was requested that recommendations be made with regard to possible response strategies to delay or mitigate the impacts of adverse climate change.

Thirdly, we asked for the drawing up of a list of elements for inclusion in a possible future convention on climate.

This draft was unanimously accepted by the General Assembly in Resolution 43/53 entitled "Protection of global climate forpresent and future generations of mankind". Today, just over two years later, we are gathered to review the remarkable progress made since then and to consider the next steps we should take.

With regard to the first point, the Intergovemmental Panel on Climate Change under the able leadership of Prof Bert Bolin and with continuous support from Dr Mostafa Tolba and Prof G O P Obasi, undertook to carry out the comprehensive review requested by the General Assembly's Resolution.

I am sure we all agree that the IPCC has successfully tackled its formidable task, especially when one considers the volume of work which had to be undertaken, and the limited time available. I feel confident that you will join me in expressing my sincere thanks to the hundreds of men and women, experts in many fields, who have contributed to the work of the IPCC.

The IPCC's first assessment report will in itself stand out not only as a masterly analysis of the state of the art, but also as a milestone in the new emerging field of environmental diplomacy. The report has, on the one hand, put the problem of climate change in its proper scientific perspective, and shown, more reliably than ever before, the complexity and delicacy of the multi-facetted problem. It has also, on the other hand, indicated possible response strategies which may be adopted, having taken full account of the uncertainties, in the attempt to delay, limit or mitigate the adverse effects of climate change, the overall effects of which policies can only be beneficial to mankind. It has catered for the second as well as the first point of the Maltese Resolution.

I hope that the United Nations General Assembly will agree that this Panel, which has been so flexible and effective, should continue to function. This would enable us to benefit from the very good work which it has already shown itself to be capable of doing. I would like to stress that there is still a very great need for scientific research and analysis to continue.

The time is nevertheless ripe, Mr Chairman, for decisive action for the preservation of climate to be taken now. Decisions as to such action are political in nature and must be taken by us, political leaders. The scientists and other experts who have been doing their essential part, must, no doubt, continue to do so in the future. It is however up to us, no longer to delay taking certain appropriate decisions, the indisputable rationality of which has by now been clearly established in terms of any plausible scenario.

A good start has already been made. Through the Montreal Protocol on Substances that Deplete the Ozone Layer a number of nations have agreed to stop the production of ozone-depleting chlorofluorocarbons, and a fund has been established with the purpose of facilitating the implementation of the commitments made under this Protocol. The achievements already secured through this protocol are to be applauded and its further effective implementation encouraged.

The Vienna Convention on the Protection of the Ozone Layer, and the Montreal Protocol on Substances that deplete the Ozone Layer, have shown that such international conventions and protocols can be very effective instruments in addressing global environmental problems. But, in the present conjuncture, we must move from partial and limited ones to more general and holistic undertakings.

The third element, Mr Chairman, contained in Malta's climate resolution to the General Assembly in October 1988 was preparation for a comprehensive framework convention on climate. At that time, the word 'possible' had to be included to qualify the word 'convention', as a considerable number of countries doubted whether such a convention was actually needed. As time went by,

however, the support for a climate convention grew to the point, that now there is no significant opposition to such a convention. On the contrary the general opinion now is, that there is an urgent, if not to say desperate, need for such a convention.

Resolution 43/53 of the General Assembly went even further than opening the way to a global convention on climate. It recognised that climate change is a common concern of mankind since climate is an essential condition that sustains life on earth. Thus, it declared the philosophical principle which provides the theoretical underpinning for the practical measures that the Convention would include.

This new legal concept introduced by Malta can serve as the basis for fresh developments, now urgently needed, in international law. For instance, it entitles the international community to concern itself with activities which produce adverse climate change, even when such activities occur within a State's boundaries. The principle of domestic jurisdiction cannot be allowed to be used as an absolute defence when global environmental well-being is at stake.

A recent comment in the authoritative "American Journal Of International Law" notes that recognition of climate as a "common concern of mankind" implies that any state should have a standing to make representations to any other, concerning the latter's climate-affecting policies and activities, without having to allege that it is uniquely affected.

The concept of the common concern of mankind, because of its innovative nature, needs to be further examined. In an effort to make a contribution in this field, my Government, in conjunction with the United Nations Environment Program, will soon hold in Malta a meeting of leading international lawyers to further study this legal concept and its possible role in the solution of global environmental problems.

I am informed, Mr Chairman, that during the meeting of the IPCC, and the preparatory meetings for this conference there was a general consensus on the need for widespread international co-operation to combat effectively climate change and its adverse effects. This is an indication that an increasing number of countries are gradually accepting the responsibility to take measures to protect climate, as a life sustaining system, for present and future generations.

This enthusiasm for co-operation cannot, however, be interpreted to imply that all nations are responsible to the same extent for the protection of climate. It would certainly be unreasonable to expect developing countries, which produce less than a quarter of the greenhouse gasses, to carry the same responsibility in combating climate change, as the developed countries which have been the main producers of greenhouse gases. One must also take into consideration the fact that developing countries certainly do not possess the financial resources, and the necessary technology, to address this complex problem. Besides, the administrators of most of these countries carry the very heavy responsibilities of alleviating poverty, and providing a decent standard of living to their growing and poor populations. To do this, they will certainly need to consume more and more energy, and they have no option, under present circumstances, but to burn more and more fossil fuels to produce this energy.

It is the firm belief of many governments, including mine, that if the option of a decent survival is to be left open for future generations of mankind, the present generation must make a total commitment to the preservation of climate in a condition that can sustain life. This commitment would constitute, what is very probably, the greatest challenge ever to have faced humanity. This commitment will need to be characterised by the willingness on the part of all countries to undertake co-operation so widespread and deep as has been hitherto unknown.

I have already mentioned, Mr Chairman, that no problem can be addressed effectively unless it is thoroughly understood. The problem of climate change is a highly complicated problem in which humanity has had hardly any experience. I have already stressed that not only much more research has to be undertaken, but that it also needs to be undertaken with the co-operation, and participation, of as many countries as possible, from all regions of the world. The work done by the IPCC's Working Group I, and other research work undertaken under the auspices of the World Meteorological Organisation have shown that such co-operation is possible.

It is unfortunate, however, that most developing countries are not in a position to participate in such research to the extent desirable, because of limited financial resources, and a shortage of personnel with the required skills in this particular area.

The training of such personnel from developing countries is certainly one of the areas where co-operation is needed. It is essential that assistance is provided to the developing ones in this field, thereby making it possible for them to participate in climate related research programmes, and in the drawing up of proposals for response strategies that may be adopted in the combat against climate change.

Co-operation between developed countries is also highly desirable in the development and transfer of environmentally clean and safe technologies. Such technologies should be made available for use in developing countries on conditions which do not cause any financial or technical difficulties. It must be emphasised here that the use of environmentally clean and safe technologies in developing countries is in the interest of all countries, both developed and developing.

Apart from the areas of research, and in the development and transfer of clean and safe technologies which I have just mentioned, co-operation between developed and developing countries will also need to cover such sectors as the building, agricultural, educational, industrial and forestry sectors. I will make no attempt to define the co-operation needed in these areas here, as such co-operation is the subject of widespread discussion elsewhere.

The threat of climate change, and the commitment for the protection of climate for present and future generations presents our generation with as big a challenge as an opportunity. Negotiations for a climate convention will provide all nations with an opportunity to understand each other's problems the like of which has never existed before.

The universally accepted principle that climate change is a common concern of mankind, implies that all nations have a responsibility to co-operate for the preservation of climate and opens the door for co-operation among all nations on an unprecedented scale.

Let us not fail future generations on this issue. Let us make every effort possible to pass on to our children a climate which can really sustain their life, as we have inherited. But let us do this without asking for any sacrifice from those who cannot make any sacrifices - the poor and deprived no matter where they are.

Let us, Mr Chairman, in the tackling of this problem co-operate in such a manner as to turn this common concern of ours into our common success.

Second World Climate Conference
6 November, 1990

Address by The Rt. Hon. Bikenibeu Paeniu
Prime Minister of Tuvalu

I am most honoured and privileged to be amongst the main key speakers this morning.

I congratulate the World Meteorological Organisation (WMO) and the United Nations Environment Programme (UNEP) and all those who have worked hard to ensure the holding of this most important Conference.

I must say it is quite a daunting experience to address a collection, or should I say a cluster of most eminent scientists, experts, researchers, diplomats and policy makers today on the subject of climate.

When I was asked to take part in this Conference, I was comforted by the very fact that I could speak with my heart on the matter. I can assure each and every one of you that I speak to you today from real experience because I live on one of the most smallest island groups in the Pacific. We are therefore, along with others, extremely vulnerable to environmental hazards and the dangers of the Greenhouse Effect and sea level rise. These are problems which we have done the least to create but now threaten the very heart of our existence.

We in the Pacific, the Carribean and elsewhere had done the least to create these hazards but now stand the most to lose.

My colleagues who have eloquently spoken before me are prominent leaders from some of the leading nations of our world. From the onset, I want to express my gratitude for their important remarks and their concern for climate change. Despite the smallness of my country, and of others, we are all here gathered to address a paramount global challenge. Climate change will affect all of us.

However, because of our vulnerability and high susceptibility to climate changes, we would be affected proportionally more than many of the countries represented here today.

I am very confident that not only do I speak for my country, but also I speak for my neighbouring countries in the Pacific and as well as others, especially those from the Southern Hemisphere.

I sincerely hope I will be able to share with you this morning a somewhat different but interesting perspective on the problem of climatic change.

Why Tuvalu is Particularly Susceptible to Climatic Changes

As some of you may not know, my country has a total land area of only 24.4 square kilometres - only a speck on the maps of the world. This 24.4 square kilometres is further divided among nine islands, ranging in size from Vaitupu which is 4.9 sq.km to Niulakita of only 0.4 sq.km. All our nine islands are low lying coral atolls dispersed over an area of 1.2 million square kilometres of the South Pacific Ocean, none rising more than 2 metres above sea level.

In sharing a number of thoughts with you today, I want to tell you the reality of life in Tuvalu in terms of the impact on our lives of the global climatic changes so far. My country is amongst the most vulnerable to climate and sea level rise in the world. The survival of Tuvalu and the survival of other neighbouring countries in the South Pacific - Kiribati, Tokelau, Tonga, Fiji, Marshall Islands and so forth hang in the balance. This eminent gathering therefore could make the difference between Tuvalu's imminent demise and its continued existence.

The scientists have confirmed that global warming is now a certainty. The findings of the international scientific community finalised in Sweden this year, have now been confirmed by the experts gathered here last week. We have learned of the best Scientific and Technical estimates assuming "business-as-usual" for greenhouse gas emissions which are:

- An average rate of increase of global mean temperature during the next century of about 0.3 deg C per decade
- An average rate of global mean sea level rise of about 6 cm per decade

Now it is up to the decision makers of the world, particularly those of the leading industrialised nations, to respond with decisiveness and conviction. We know from all the scientific evidence collected to date that the fragile and delicate atmospheric balance on which life and our existence depends has been upset. The threat to the atmosphere comes predominantly from the breath of industrialised civilisation.

We in the island nations cannot afford to wait for more studies and research. We would like to have firm and pragmatic solutions to curb the negative impact of climatic changes on our lives.

We live on these small fragile islands and we know that the problems are **now** and not tomorrow. The time to act is now or never.

Tuvalu needs help to face up to this challenge. In particular, we in the Pacific and others in the Southern Hemisphere feel isolated and vulnerable.

As we in the Pacific Island region enter the 1990's, we increasingly face a range of environmental problems far more extensive and more serious than ever before. Problems of ozone depletion, climate change, sea level rise, hazardous waste dumping, driftnet fishing - to name some which fall into this category. These impinge severely on the development process itself of our countries in the Pacific and the overall impact has been, and will continue to be, unfavourable.

Already we in Tuvalu have been experiencing the adverse effects of global warming and the greenhouse effects. I am happy to share with you some of the most salient features.

Frequency of Tropical Cyclones

In 1940 there were 2 cyclones, but over the last fifty years the number has risen dramatically - a ten-fold rise to 8 in 1960, 18 in 1980 and 21 so far in 1990. These storms have seriously caused erosion to the already extremely limited amount of land we possess. This in turn bears an impact on population growth and pressure on other resources.

Cyclones have also devastated many of our crops and trees. The loss of trees can cause the interruption of the nutrient cycle above and below the soil.

The Water Table

We have not yet installed measuring instruments to test the rise in sea level but we have noted increased salinity in ground water supplies, attesting to intrusion of sea water into the water table. This increased salinity impacts unfavourably on both the supply of drinking water as well as efforts to grow things - agriculture being already difficult due to harsh soils.

Temperature Changes and Rainfall Levels

Despite the rise in the number of cyclones, Tuvalu has been experiencing an increasing number of dry spells, with average annual rainfali declining fro 420 millimetres per annum at December 1980 to 300 millimetres at December 1988 - a drop of 28.6% in the eight year period. Simultaneously, average daily temperatures have risen by 0.5 degree celsius over the last thirty years.

In light of all the above we entreat this Conference to propose measures which promote urgent and pragmatic actions, rather than having this gathering end up as another academic endeavour.

Action Being Taken by Tuvalu to Prevent/Mitigate Adverse Climatic Changes

Energy

Less than 50% of our total population of 8,309 people have access to electricity but 1.8% of household energy consumption is solar generated. We intend to increase this proportion to 3% by 1993. All electricity in the outer-islands is solar generated.

Agriculture

We have embarked on a modest re-afforestation programme whereby, between 1970 and 1987 some 115.3 hectares of land was replanted with coconut trees. Early in 1991 we intend to begin a new tree planting exercise, including coconuts, mangoes and citrus. As a further measure, farmers are actively encouraged to inter-crop coconuts with root crops. Again we consciously utilize natural manure instead of chemical fertilizers and promote biological pest control rather than dangerous pesticides in our agricultural practices.

Assistance Tuvalu Needs to Promote a Healthy Environment

We recognise that much more can and needs to be done. For instance we currently use borrow pits dug out by the U.S. Army during the Second World War to help in the construction of an airfield - to dump our garbage, because of limited land space.

We know there are more environmentally friendly methods of solid waste management, but due to financial and other constraints we cannot access them.

Already the EEC is providing funding in the areas of coastal reclamation and solar energy, while the Australian Government has promised to supply us with equipment to test the rise in sea level. But there has been a rather slow response from donors to our request for rehabilitation assistance for damage caused by the most recent Cyclone Ofa - which occurred in February of this year.

Therefore we urge the developed countries to come to

our assistance in designing and implementing a comprehensive environmental strategy. We need assistance in the training of a number of national experts in environment and climatic changes as well as the strengthening of our national capabilities to monitor onsite climatic changes and sea level rise.

Regional and Global Perspective

On a regional perspective, the Pacific Island governments have demonstrated unity on environmental issues where we have perceived outside activiites likely to cause damage to the environment and threaten our small islands and fragile ecosystems.

Our opposition to these abuses by others of our environment has led to our undertaking collaborative efforts in a number of areas.

In particular, I refer to the South Pacific Nuclear Free Zone Treaty, the Convention for the Protection of the Environment and Natural Resources of the South Pacific Region which recently came into force, our actions on Driftnet Fishing and our collaboration through the South Pacific Regional Environmental Programme.

We welcome and encourage the participation of other governments in these important areas where our small island countries have already joined together with some of our larger neighbours to protect our common heritage.

Some of our developed partners present here today may be asking themselves what they can do to help countries like Tuvalu withstand this problem. I would like to share with you a very recent experience. At a Summit between leaders of the South Pacific sovereign nations with President Bush, held at Hawaii on 27 October, the environment and climatic changes were amongst the contentious issues discussed. At one point, President Bush asked - "What can we, the United States do in order to help countries in the South Pacific?" I replied immediately - "Mr President, can I answer that question?" He looked at me and said - "Go ahead". Then I said - "Well, the United States could help countries like Tuvalu in building coastal protection schemes such as those the EEC are already providing us with though we would need more assistance in this area; we would need to build more water catchments. I went on saying: "I am pretty sure I shall be the only Head of a South Pacific Government that will be present at the forthcoming Second World Climatic Conference in Geneva early next month. At that Conference, Ministers will deliberate on a Ministerial Declaration which will, *inter alia*, include provisions for developed countries to cut down their emission rates of greenhouse gases; for developed countries to provide the necessary resources to developing countries to help them establish the required measures and infrastructures which will lessen the impact of the Greenhouse Effect." Then I said, "Mr President, I

look forward to the support of the United States on this Ministerial Declaration".

On the same token, and in anticipation of a similar question from our developed partners present here today, I will add the following:

- Provide assistance at the regional level to help us strengthen the South Pacific Regional Environmental Programme
- Help us train our nationals in the areas of environment and climatic changes
- Help us to strengthen both at the national and regional levels our capabilities of monitoring and detecting cyclones and other natural disasters in order to allow us to prepare in advance
- Help countries in the South Pacific establish a pragmatic and reliable Disaster Preparedness and Rehabilitation Programme.

Ministerial Declaration

I support the Draft Ministerial Declaration presented to us by our experts merely because I consider it a start of an imperative process towards a Framework Convention on Climate. I must admit however that I am not happy with the commitments made by a number of the industrialised countries which has resulted in what I regard as a weak Declaration. It appears that science has been ignored by a number of decision makers. I would be content if such stance is done on goodwill. But if science is being deliberately ignored, then I will regard such an attitude as one of total selfishness, and inhuman.

The problem we are addressing right now is of a global nature. We need the cooperation and support of all nations in this world. We in Tuvalu, and this goes as well for our sister countries in the Pacific, the Indian Ocean and the Carribean, contribute little or nothing to the problem and yet we will be the first to suffer. Our survival is at stake.

I may come from a poor and small nation but the fact of life is that my country and its people in particular are a distinct race of our earth planet. Like yourselves, we are citizens of the world - our world. We have our own language and culture. We have a God given right to live on earth. We may be small but we have a spot there to maintain in our biosphere. We contribute to the diversity and balance of the Earth's ecosystem. It will therefore be an injustice should we in Tuvalu and the island nations, be denied our right to live in our homeland.

The nations of the developed world carry an enormous burden of responsibility. The problem is clear and the options for action are clear. This Conference therefore presents a clear opportunity for effective and speedy action. These countries which have been contributing to the accumulation of greenhouse gases for so long continue to pollute our atmosphere. I say this with no sense of criticism for past actions - indeed the world has only in the past few years realised the problem. At times I feel that this is God's way of telling us, human beings, to be more responsible in the maintenance of the Earth's balanced ecosystem. After all the Earth was not created by a human being. And if we do not respond positively to the challenge, then I feel the catastrophe will take place as predicted bv science. Such is not at all fair to Tuvalu.

My bigger neighbours in the Pacific - Australia and New Zealand - have been quick to act with decisive national policies. However, the effects of their efforts will not be enough on the global scene. It is up to the large industrialised economies of the world to respond to the clear challenge. Let us all bind ourselves together and pragmatically solve the problem as responsible citizens - brothers and sisters - of the world. We will surely bring about peace and prosperity to all if we join together in addressing the problem.

Second World Climate Conference
6 November, 1990

Statement by Dr. E. Saouma
Director General, Food and Agriculture Organization (FAO)

For the third consecutive year, the world has produced fewer cereals than it has consumed. Buffer stocks have fallen by one third, representing at the moment less than 65 days of world consumption. Among the multiple causes of this situation, climatic scourges are in the forefront. Is this to be viewed as the effect of a lasting change? I leave it to the meteorologists and environmentalists to go more deeply into this question and give us an answer. All I know is that there are already 500 million men, women and children who do not have enough to eat, and that it only takes a drought here or a flood there to trigger off famines in poor countries or to disturb seriously the agricultural sectors and the economies of rich countries.

Is it the meteorological climate that is changing, or is it the economic and social climate of our planet that is not changing enough? Have our societies today become more vulnerable to the vagaries of the climate? Before speculating on the possible effects of global warming in 10 or 20 years or beyond we should take a look at our society and the state of the world today, as well as the reasons behind the rising importance of meteorology.

Agriculture, more than any other human activity, has always been dependent on climatic factors. And whether it be through the empirical wisdom accumulated by generations of farmers or thanks to modern research and technology, agronomy aims to exploit natural resources to the utmost, particularly those affected by climate: solar energy, water and the gases that make up the atmosphere— such as carbon dioxide, oxygen and nitrogen.

But it must be recognized that, except in the case of irrigation and sheltered crops, technology is far from having solved the problem of agriculture's inherent vulnerability to the climate. On the contrary, this has been aggravated and risks have increased with the expansion of single-crop cultivation and intensive stock-raising, at the expense of diversified crops and animal husbandry; market forces and economic competition have encouraged increased specialization in and intensification of agricultural production, thus making it vulnerable to both climatic and market hazards. Elsewhere, poverty and over-population have led to the clearing or grazing of marginal land, whose soils are precisely the most sensitive to climatic assault.

Already rendered fragile by the economic and technological sophistication of the North and by the rapid degradation of land in the South, agriculture is certainly not ready to absorb tomorrow the shock of a change in climate, even a gradual one, any more than it is ready today to absorb a new oil shock. We had better not wait for some new crisis to open our eyes, for it will then be too late to make the necessary changes. Forecasts of meteorological trends and their consequences include, for sure, many elements of uncertainty, but this is no excuse for inertia. On the contrary, a frank diagnosis of the vulnerability of agriculture to climate today and tomorrow should lead us to make some major changes in agricultural policies without delay.

What is to be done? First of all, we must identify the right policies to reduce the present vulnerability of agriculture to climatic hazards. Then we must formulate the measures to be adopted to address the prospect of climate change, in order to protect the environment and thus to promote sustainable development.

Being prepared for climatic hazards and changes does not only mean reducing the vulnerability of agricultural production; it also means equipping society to withstand the shocks of agricultural disasters, inevitable despite all our efforts, and to reduce their impact.

I shall not dwell on measures already taken and those to be taken regarding food security: monitoring and early warning systems, larger buffer stocks, the efficiency of distribution networks, appropriate use of food aid, and so on. Even if we were not confronted with climate change, the growing number of crises demands a reinforcement of these activities. The prospect of climate change makes this reinforcement even more necessary.

We must also pursue research and impact assessment, and pinpoint the fields of activity and regions which are the most vulnerable. In consultation with other institutions, FAO has undertaken studies on this subject which will be developed still further in co-operation with the Inter-governmental Panel on Climate Change and with its effective support. As of now, we can perceive the main lines to be followed if the present fragility of world agriculture is to be remedied; climatic risk makes action in this direction even more urgent.

In the first place, agriculture must be diversified at every level, from the farm to national, regional and global levels. Diversification should help to reduce the risks of either under- or oversupply as well as the negative effects of drought and bad weather, disease, resource degradation—all risks which could be exacerbated by climate change. This is just one more reason for continuing our efforts to preserve and to exploit biological diversity, itself threatened by climate change. Diversification should also be enhanced by increased recourse to biotechnologies to make available species adapted to different climate, ecological and socio-economic conditions.

The second major requirement for heightening the resistance of agriculture today and tomorrow to climatic hazards and changes is to increase the technical efficiency of agricultural production. Most inputs—fertilizer, pesticides, irrigation water, livestock feed and machinery—are often inefficiently used, with too much here and not enough elsewhere, or with large-scale losses. This wastage currently threatens the ecological and socio-economic viability of our production systems and, in the long term, sustainable agricultural development. It contributes, in some respects, to the production of greenhouse gases.

The search for improved efficiency should cover not only inputs and the use of wastes and by-products, but also the entire resource base for agriculture. It should enable us to reduce pressure on marginal and forest lands. More efficient use of land already under cultivation would make an appreciable contribution to limiting deforestation, and I should like to dwell for a moment on this particular set of problems which is typical of those we have to deal with here.

Most of the threats hanging over the forests of our planet today stem from a crisis in agriculture. Today, agriculture is responsible for more than three quarters of deforestation; tomorrow, this phenomenon could contribute to climate changes which, in turn, may well threaten the forests. But in the immediate future, it is the conflict between agriculture and forestry that we must resolve.

The preparation of this Conference appears to me to have been marked by increasing awareness of the problems of deforestation and the role of forests in major natural processes, particularly those governing the production and fixation of greenhouse gases; I can only be pleased at this.

If governments reach agreement on a climate convention, forests cannot be excluded. However, international action to save forests should not be diverted or delayed if the negotiation of an agreement on climate proves difficult.

We have in fact many reasons apart from climate to engage in concerted action with a view to the rational management of forest resources. I am not speaking only of deforestation or reforestation, or of ensuring the recycling of industrial wastes present in the atmosphere. Forests have an ecological role but also an economic and social one. Therefore they must be treated globally, including the aspect of climate but without being limited to this, however important it may be.

FAO is working on this, together with other institutions, within the framework of the Tropical Forestry Action Plan. But it is worldwide action that we must promote, and I have therefore taken the initiative of launching the technical preparations for the drafting of an international legal instrument on forests. it will be up to governments to decide on its form and the content. FAO for its part will tirelessly pursue ways of ensuring co-ordination and complementarity between its own action for forests and initiatives taken to draw up a climate convention. It is already adopting a similar approach in its efforts with regard to a convention on biological diversity.

Interdependence between the problems of today and those of the next millennium should lead us to seek common and concerted solutions on several fronts at the same time. The priorities of the agriculture, forestry and fisheries sectors are among the most important. They are closely linked, not only to the future evolution of our climate but also, even now, to many constraints imposed by climate on the development of our societies. I trust that any decisions made by governments will take due account of the interests of these sectors, which are of crucial importance for the economies of the great majority of countries represented here.

Second World Climate Conference
7 November, 1990

Closing Remarks by Dr. F. Mayor
Director General, United Nations Educational, Scientific and Cultural Organization (UNESCO)

On behalf of the Executive heads of the organizations sponsoring the Second World Climate Conference, I have the pleasure and privilege of saying a few concluding remarks to you during this closing session.

Let me first of all express the great satisfaction which we as sponsors of this Conference feel at this moment of the adoption of the Ministerial Declaration. I wish to express also our deep gratitude to all delegations as you have all contributed to make this Conference a successful one. The success is shared between the scientists and other experts who have worked so hard during the scientific and technical sessions of the Conference last week and the successful negotiations and discussions of the ministerial delegations during the last days. We know that some of you would have liked the Ministerial Delegation to go beyond the consensus views and recommendations now included in the Declaration as regards the need to stabilize emissions of greenhouse gases not controlled by the Montreal Protocol on substances that deplete the ozone layer; and as regards the needed second step of achieving reductions of CO_2 and other greenhouse gas emissions. However, on such a global issue, solutions needs to be found through globally concerted action. Globally, consensus can only be achieved through a process of debating the issues without fear or favour—with realistic prognosis and remedy as our only concerns. I believe this Ministerial Meeting and the Conference as a whole has been a very important milestone in this process. There are three streams of international activity which deserve your full support and the continuing support of the sponsoring organizations in different capacities:

1) Increased efforts of global measurement and research in order to reduce scientific uncertainties;

2) A continuation of IPCC to support negotiation of and provide input into a Convention;

3) The development of a Convention on Climate Change.

We in UNESCO and IOC are most concerned by the priority areas for scientific activities as they have been identified in the scientific Conference Statement. We are committed to contributing to a revamped World Climate Programme and to the International Geosphere Biosphere Programme. We will also strengthen related international scientific activities within the programmes of UNESCO and IOC. The IOC's role will be crucial in the establishment of a global ocean observing system as part of the Global Climate Observing system.

There was agreement in the scientific and technical sessions in recommending to you and to all governments that the relevant international observations and research programmes be implemented in full. Increased national support for, and substantially increased funding of, these programmes is required if progress on the necessary time scale is to be made in reducing the uncertainties.

A second area of investment of needed refers to the technically feasible and cost-effective opportunities which exist already to reduce CO_2 emissions in all countries. Notwithstanding remaining scientific and economic uncertainties, nations should take immediate actions to control the risks of climate change with initial emphasis on actions that would be economically and socially beneficial for other reasons as well.

Developing countries are being asked to participate in the alleviation of the legacy of environmental damage from prior industrialization. In order to enable them to participate in this global action, developing countries must be assisted in various aspects. First, developing countries should be helped to build up their scientific and technical capacities. A major training programme will be needed as well as financial support for the creation of relevant institutional capacities. Second, there is a need for increased transfer of technology. Developing countries must be helped to "leapfrog" ahead directly from a status of under-development through to efficient, environmentally benign technologies.

People in industrialized and developing countries alike need better information and education on the crucial role climate plays in development and the additional risks posed by climate change. UNESCO ought to certainly be an active partner in major public awareness and education endeavours.

Climate change is one of a number of global problems confronting humanity—others include poverty, illiteracy and the risk of war. Moreover, these major problems are—as we know—interlinked. The poverty of the farmers in the tropical forest areas contributes through deforestation to CO_2 increase in the atmosphere. Such interlinkages will be the focus of the more comprehensive United Nations Conference on Environment and Development to be held in 1992. We must work—all together, synergically—to ensure that the time between now and the 1992 Conference is used to achieve a consensus among the nations of the world on the need for the international community to make a solemn commitment in Rio de Janeiro to redressing the present situation and trends in environment and development.

Let me also express here at the end of the Second World Climate Conference that the UN bodies involved take pride in having offered another example of the capacity of the organizations of the UN system to work together on this occasion in partnership with the international scientific community represented by the International Council of Scientific Unions.

In fact, this partnership is indispensable in dealing with an issue which has such a unique scientific dimension. Moreover, as a scientist, I cannot pass up the opportunity to say—and I know my colleagues share my view—that the Intergovernmental Panel on Climate Change and this Conference, with its built-in links between science and policy, represent only the first steps on a long road along which scientists and world leaders will have to move together in ever closer consultation and dialogue.

On behalf of the Executive Heads of the Organizations sponsoring the Second World Climate Conference I would like to thank you all again for making this Conference a success. What we now need jointly is a long-sighted and courageous attitude so as to ensure that the legacy we are duty bound to pass on to future generations is not irreversibly damaged. As I said in my speech at the opening session of the Conference, "Our descendants will not judge us on our hopes and recommendations, but on our actions". Here and now.

Second World Climate Conference
Geneva, Switzerland, 7 November 1990

Ministerial Declaration

PREAMBLE

1. We, the Ministers and other representatives from 137 countries and from the European Communities, meeting in Geneva from 6 to 7 November 1990 at the Second World Climate Conference, declare as follows:

2. We *note* that while climate has varied in the past and there is still a large degree of scientific uncertainty, the rate of climate change predicted by the Intergovernmental Panel on Climate Change (IPCC) to occur over the next century is unprecedented. This is due mainly to the continuing accumulation of greenhouse gases, resulting from a host of human activities since the industrial revolution, hitherto particularly in developed countries. The potential impact of such climate change could pose an environmental threat of an up to now unknown magnitude; and could jeopardize the social and economic development of some areas. It could even threaten survival in some small island States and in low-lying coastal, arid and semi-arid areas.

3. We *appreciate* the work of the World Climate Programme (WCP) during the past decade which has improved understanding of the causes, processes and effects of climate and climate change. We also *congratulate* the IPCC, established by the United Nations Environment Programme (UNEP) and the World Meteorological Organization (WMO) on its First Assessment Report on Climate Change. It has identified causes and possible effects and strategies to limit and adapt to climate change, and in the light of the United Nations General Assembly resolutions, has identified possible elements for inclusion in a framework convention on climate change.

4. Recognizing climate change as a common concern of mankind, we commit ourselves and intend to take active and constructive steps in a global response, without prejudice to sovereignty of States.

I. GLOBAL STRATEGY

5. Recognizing that climate change is a global problem of unique character and taking into account the remaining uncertainties in the field of science, economics and response options, we *consider* that a global response, while ensuring sustainable development [1] of all countries, must be decided and implemented without further delay based on the best available knowledge such as that resulting from the IPCC assessment. Recognizing further that the principle of equity and the common but differentiated responsibility of countries should be the basis of any global response to climate change, developed countries must take the lead. They must all commit themselves to actions to reduce their major contribution to the global net emissions and enter into and strengthen co-operation with developing countries to enable them to adequately address climate change without hindering their national development goals and objectives. Developing countries must, within the limits feasible, taking into account the problems regarding the burden of external debt and their economic circumstances, commit themselves to appropriate action in this regard. To this end, there is a need to meet the requirements of developing countries, that adequate and additional financial resources be mobilized and the best available environmentally-sound technologies be transferred expeditiously on a fair and most favourable basis.

II. POLICY CONSIDERATIONS FOR ACTION

6. We *reaffirm* that, in order to reduce uncertainties, to increase our ability to predict climate and climate change on a global and regional basis, including early identification of as yet unknown climate-related issues, and to design sound response strategies, there is a need to strengthen national, regional and international research activities in climate, climate

change and sea level rise. We *recognize* that commitments by governments are essential to sustain and strengthen the necessary research and monitoring programmes and the exchange of relevant data and information, with due respect to national sovereignty. We *stress* that special efforts must be directed to the areas of uncertainty as identified by the IPCC. We *maintain* that there is a need to intensify research on the social and economic implications of climate change and response strategies. We *commit* ourselves to promoting the full participation of developing countries in these efforts. We *recognize* the importance of supporting the needs of the World Climate Programme, including contributions to the WMO Special Fund for Climate and Atmospheric Environmental Studies. The magnitude of the problem being addressed is such that no nation can tackle it alone and we stress the need to strengthen international cooperation. In particular, we *invite* the 11th Congress of the World Meteorological Organization, in the formulation of plans for the future development of the World Climate Programme, to ensure that the necessary arrangements are established in consultation with UNEP, UNESCO (and its IOC), FAO, ICSU and other relevant international organizations for effective coordination of climate and climate change related research and monitoring programmes. We *urge* that special attention be given to the economic and social dimensions of climate and climate change research.

7. In order to achieve sustainable development in all countries and to meet the needs of present and future generations, precautionary measures to meet the climate challenge must anticipate, prevent, attack, or minimize the causes of, and mitigate the adverse consequences of, environmental degradation that might result from climate change. Where there are threats of serious or irreversible damage, lack of full scientific certainty should not be used as a reason for postponing cost-effective measures to prevent such environmental degradation. The measures adopted should take into account different socio-economic contexts.

8. The potentially serious consequences of climate change, including the risk for survival in low-lying and other small island States and in some low-lying coastal, and arid and semi-arid areas of the world, give sufficient reasons to begin by adopting response strategies even in the face of significant uncertainties. Such response strategies include phasing out the production and use of CFC's, efficiency improvements and conservation in energy supply and

use, appropriate measures in the transport sector, sustainable forest management, afforestation schemes, developing contingency plans for dealing with climate related emergencies, proper land use planning, adequate coastal zone management, review of intensive agricultural practices and the use of safe and cleaner energy sources with lower or no emissions of carbon dioxide, methane, nitrous oxide and other greenhouse gases and ozone precursors, paying special attention to new and renewable sources. Further actions should be pursued in a phased and flexible manner on the basis of medium and long-term goals and strategies and at the national, regional or global level, taking advantage of scientific advances and technological developments to meet both environmental and economic objectives.

9. We *note* that per capita consumption patterns in certain parts of the world along with a projected increase in world population are contributing factors in the projected increase in greenhouse gases.

10. We *agree* that the ultimate global objective should be to stabilize greenhouse gas concentrations at a level that would prevent dangerous anthropogenic interference with climate.

11. We *stress*, as a first step, the need to stabilize, while ensuring sustainable development of the world economy, emissions of greenhouse gases not controlled by the Montreal Protocol on Substances that Deplete the Ozone Layer. Contributions should be equitably differentiated according to countries' responsibilities and their level of development. In this context, we acknowledge efforts already undertaken by a number of countries to meet this goal.

12. Taking into account that the developed world is responsible for about 3/4 of all emissions of greenhouse gases, we *welcome* the decisions and commitments undertaken by the European Community with its Member States, Australia, Austria, Canada, Finland, Iceland, Japan, New Zealand, Norway, Sweden, Switzerland, and other developed countries to take actions aimed at stabilizing their emissions of CO_2, or CO_2 and other greenhouse gases not controlled by the Montreal Protocol, by the year 2000 in general at 1990 level, yet recognizing the differences in approach and in starting point in the formulation of the above targets. We also acknowledge the initiatives of some other developed countries which will have positive effects on limiting emissions of greenhouse gases. We *urge* all developed countries to establish targets and/or feasible national programmes or strategies which will

have significant effects on limiting emissions of greenhouse gases not controlled by the Montreal Protocol. We *acknowledge*, however, that those developed countries with as yet relatively low energy consumption (measured on a per capita or other appropriate basis) which can be reasonably expected to grow, and some countries with economies in transition, may establish targets, programmes and/or strategies that accommodate socio-economic growth, while improving the energy efficiency of their economic activities.

13. We *urge* developed countries, before the 1992 UN Conference on Environment and Development, to analyze the feasibility of and options for, and, as appropriate in light of these analyses, to develop programmes, strategies and/or targets for a staged approach for achieving reductions of all greenhouse gas emissions not controlled by the Montreal Protocol, including carbon dioxide, methane and nitrous oxide, over the next two decades and beyond.

14. We *recommend* that in the elaboration of response strategies, over time, all greenhouse gases, sources and sinks be considered in the most comprehensive manner possible and also that limitation and adaptation measures be addressed.

15. We *recognize* that developing countries have as their main priority alleviating poverty and achieving social and economic development and that their net emissions must grow from their, as yet, relatively low energy consumption to accommodate their development needs. Narrowing the gap between the developed and the developing world would provide a basis for a full partnership of all nations and would assist the developing countries in dealing with the climate change issue. To enable developing countries to meet incremental costs required to take the necessary measures to address climate change and sea-level rise, consistent with their development needs, we *recommend* that adequate and additional financial resources should be mobilized and best available environmentally sound technologies transferred expeditiously on a fair and most favourable basis. Developing countries also should, within the limits feasible, take action in this regard.

16. The specific difficulties of those countries, particularly developing countries, whose economies are highly dependent on fossil fuel production and exportation, as a consequence of action taken on limiting greenhouse gas emissions, should be taken into account.

17. We *recommend* that consideration should be given to the need for funding facilities, including the proposed World Bank/UNEP/UNDP Global Environmental Facility, a clearing house mechanism and a new possible international fund composed of adequate additional and timely financial resources and institutional arrangements for developing countries; taking into account existing multilateral and bilateral mechanisms and approaches. Such funding should be related to the implementation of the framework convention on climate change and any other related instruments that might be agreed upon. In the meantime, developed countries are urged to co-operate with developing countries to support immediate action in addressing climate change including sea-level rise without imposing any new conditionality on developing countries.

18. We *recommend* further that resources be assessed. Such assessments, to be conducted as soon as possible, should include country studies and mechanisms to meet the financing needs identified, taking note of the approaches developed under the Montreal Protocol.

19. Financial resources channelled to developing countries should, *inter alia*, be directed to:

    (i) Promoting efficient use of energy, development of lower and non-greenhouse gas emitting energy technologies and paying special attention to safe and clean new and renewable sources of energy;
    (ii) Arranging expeditious transfer of the best available environmentally sound technology on a fair and most favourable basis to developing countries and promoting rapid development of such technology in these countries;
    (iii) Co-operating with developing countries to enable their full participation in international meetings on climate change;
    (iv) Enhancing atmospheric, oceanic and terrestrial observational networks, particularly in developing countries, to facilitate conducting research, monitoring and assessment of climate change and the impact on those countries;
    (v) Rational forest management practices and agricultural techniques which reduce greenhouse gas emissions;
    (vi) Enhancing the capacity of developing countries to develop programmes to address climate change, including research and development activities and public awareness and education.

Funding should also be directed to the creation of regional centres to organize information networks on climate change in developing countries.

20. Appropriate economic instruments may offer the potential for achieving environmental improvements in a cost-effective manner. The adoption of any form of economic or regulatory measures would require careful and substantive analyses. We *recommend* that relevant policies make use of economic instruments appropriate to each country's socio-economic conditions in conjunction with a balanced mix of regulatory approaches.

21. We *note* that energy production and use account for nearly half of the enhanced radiative forcing resulting from human activities and is projected to increase substantially in the absence of appropriate response actions. We *recognize* the promotion of energy efficiency as the most cost-effective immediate measure, in many countries, for reducing energy-related emissions of carbon dioxide, methane, nitrous oxide and other greenhouse gases and ozone precursors, while other safe options such as no or lower greenhouse gas emitting energy sources should also be pursued. These principles apply to all energy sectors. Transport energy use attracts special attention of many of us in the light of its role in many developed countries and of its expected importance in many developing countries.

22. We *recognize* that there is no single quick-fix technological option for limiting greenhouse gas emissions. However, we are *convinced* that technological innovation as well as individual and social behaviour and institutional adaptations is a key element of any long-term strategy that deals with climate change in a way that meets the goal of sustainable development. Therefore, we *urge* all countries, the developed countries in particular, to intensify their efforts and international cooperation in technological research, development and dissemination of appropriate and environmentally sound technologies, including the reassessment and improvement of existing technologies and the introduction of new technologies.

23. We *urge* that environmentally sound and safe technologies be utilized by all sectors in all countries to the fullest extent possible and further *urge* all countries, developed and developing, to identify and take effective measures to remove barriers to the dissemination of such technologies. To this end, the best available environmentally sound and safe technologies should be transferred to developing

countries expeditiously on a fair and most favourable basis.

24. We *note* that the conservation of the world's forests in their role as reservoirs of carbon along with other measures are of considerable importance for global climatic stability, keeping in mind the important role of forests in the conservation of biological diversity and the protection of soil stability and of the hydrological system. We *recognize* the need to reduce the rate of deforestation in consonance with the objective of sustained yield development and to enhance the potential of the world's forests through improved management of existing forests and through vigorous programmes of reforestation and afforestation, and to support finarcially the developing countries in this regard through enhanced and well-coordinated international cooperation including strengthening Tropical Forest Action Plan (TFAP) and International Tropical Timber Organization (ITTO). We *recommend* that the protection and management of boreal, temperate, sub-tropical and tropical forest ecosystems must be well-coordinated and preferably compatible with other possible types of action related to reduction of emission of greenhouse gases, rational utilization of biological resources, provision of financial resources, and the need for more favourable market conditions for timber and timber products. The developing countries should be able to realize increased revenue from these forests and forest products.

25. We also *recognize* that forests and forest products play a key social and economic role in many nations and communities. We *recognize* that States have, in accordance with the Charter of the United Nations and the principles of international law, the sovereign right to exploit their own resources pursuant to their own environmental policies, and the responsibility to ensure that activities within their jurisdiction or control do not cause damage to the environment of other States or of areas beyond the limits of national jurisdiction.

26. We *recommend* that appropriate precautionary and control measures be developed and implemented at regional, sub-regional and country levels as appropriate to counter the increasing degradation of land, water, genetic and other productive resource bases by drought, desertification and land degradation. Observatories on climate and climate change and observatories on ecosystems should be encouraged to work together on drought risks consequences. Studies must be undertaken on drought and desertification. We *stress* that stepped-

up financial and scientific contributions be provided to facilitate these efforts.

27. We *recommend* that similar measures be adopted to address the particular problems and needs, including funding, of low-lying coastal and small vulnerable island countries, some of whose very existence is placed at risk by the consequences of climate change.

III. GLOBAL FRAMEWORK CONVENTION ON CLIMATE CHANGE

28. We *call* for negotiations on a framework convention on climate change to begin without delay after a decision is taken by the 45th Session of the General Assembly of the United Nations recommending ways, means and modalities for further pursuing these negotiations. Taking note of all the preparatory work, particularly the recommendations adopted 26 September 1990 by the *ad hoc* working group of government representatives and regional economic integration organizations to prepare for negotiations on a framework convention on climate change, we *urge* all countries and regional economic integration organizations to join in these negotiations and *recognize* that it is highly desirable that an effective framework convention on climate change, containing appropriate commitments, and any related instruments as might be agreed upon on the basis of consensus, be signed in Rio de Janeiro during the United Nations Conference on Environment and Development. We *welcome* the offer of the Government of the United States of America to host the first negotiating meeting.

29. We *recommend* that such negotiations take account of the possible elements compiled by the IPCC, and that the framework convention on climate change be framed in such a way as to gain the support of the largest possible number of countries while allowing timely action to be taken. We *reaffirm* our wish that this convention contain real commitments by the international community. We *stress*, given the complex and multi-faceted nature of the problem of climate change, the need for new and innovative solutions including the need to meet the special needs of developing countries.

30. We also *welcome* the invitations of Thailand and Italy to host workshops, respectively on the feasibility of forestry options, and on all technologies for energy production and use and their transfer to developing countries.

31. We *believe* that a well-informed public is essential for addressing and coping with as complex an issue as climate change, and the resultant sea-level rise, and *urge* countries, in particular, to promote the active participation at the national and when appropriate, regional levels of all sectors of the population in addressing climate change issues and developing appropriate responses. We also *urge* relevant United Nations organizations and programmes to disseminate relevant information with a view to encouraging as wide a participation as possible.

Appendices

Appendix 1

SECOND WORLD CLIMATE CONFERENCE: INTER-AGENCY COMMITTEES

INTERNATIONAL ORGANIZING COMMITTEE

J.C.I. Dooge	Chair
H.L. Ferguson	Co-ordinator
E.W. Bierly	(USA)
V.G. Boldirev	(WMO)
W. Degefu	(Ethiopia)
G. Glaser	(UNESCO)
R. Herrera	(Venezuela)
J.T. Houghton	(United Kingdom)
Yu. Izrael	(USSR)
G. Kullenberg	(UNESCO/IOC)
J.W.M. la Rivière	(ICSU)
W.J. Maunder	(WMO)
D. Norse	(FAO)
P. Usher	(UNEP)
P. Vellinga	(Netherlands)
M. Yoshino	(Japan)

OPERATIONS ADVISORY COMMITTEE

H.L. Ferguson	(Co/SWCC) Chair
V.G. Boldirev	(WMO)
S. Broere-Moore	(WMO)
E. Dar-Ziv	(WMO)
J.C.I. Dooge	(Org. Cmte)
G. Glaser	(UNESCO)
G. Goodman	(SEI)
J. Jäger	(SEI)
J. Joyce	(SWCC)
A.W. Kabakibo	(WMO)
G. Kullenberg	(UNESCO/IOC)
J.W.M. la Rivière	(ICSU)
J. Marton-Lefèvre	(ICSU)
D. Norse	(FAO)
J. Perry	(USA)
P. Usher	(UNEP)

CONFERENCE STATEMENT WRITING TEAM

J.P. Bruce	(WMO)
N. Desai	(UNCED)
J.C.I. Dooge	(Org. Cmte)
H.L. Ferguson	(Co/SWCC)
G. Glaser	(UNESCO)
G. Goodman	(SEI)
J. Jäger	(SEI)
G. Kullenberg	(UNESCO/IOC)
J.W.M. la Rivière	(ICSU)
J. Marton-Lefèvre	(ICSU)
D. Norse	(FAO)
M. Oppenheimer	(USA/EDF)
J. Perry	(USA)
P. Usher	(UNEP)

MEDIA CORE GROUP

S. Broere-Moore	(WMO) Chair
F. Barron	(UNEP)
E. Guilbaud-Cox	(UNEP)
E. Krippl	(IIASA)
M. Millward	(ICSU)
S. Oberthuer	(Germany)
F. Quinn	(WMO)
P. Ress	(UNEP)
J. Stickings	(WMO)
H. Von Loesch	(FAO)
J. Winterson	(WMO)
G. Wright	(UNESCO)

LOCAL ARRANGEMENTS GROUP

H.L. Ferguson	(Co/SWCC) Chair
A. Alusa	(UNEP)
S. Broere-Moore	(WMO)
A. Clerc	(Switzerland)
E. Dar-Ziv	(WMO)
R. Gambazzi	(Switzerland)
J. Joyce	(SWCC)
A.W. Kabakibo	(WMO)
F. Martin	(SWCC)
W.J. Maunder	(SWCC/SEI)
Z. Milosevic	(UN)
J. Murithi	(WMO)
M. Peeters	(WMO)
E.Perrin	(Switzerland)
E. Ponce	(WMO)
J. Sauteur	(WMO)
A. Secretan	(Switzerland)
D. Stuby	(Switzerland)

PROCEEDINGS EDITORIAL TEAM

H. L. Ferguson	(Co/SWCC) Co-Editor
J. Jäger	(SEI) Co-Editor
J. Joyce	(SWCC)
S. K. Varney	(United Kingdom)
C. Reynolds	(WMO)
C. Jameson	(WMO)
C. Cudjou	(WMO)

Appendix 2

POSTER PAPERS PRESENTED

Rearrangement of the Distribution of Precipitation in Hungary
P. Ambrozy, Hungarian Meteorological Service, Budapest

Secular Variations of Global Radiation and Sunshine Duration in Europe
Atsumu Ohmura et al., Federal Institute of Technology (ETH), Switzerland

GAIA - Its Implications for Industrialized Society
P. Bonyard, The Ecologist Magazine, UK

World Hunger Belt and Global Climate Change
J. J. Burgos, Centro de Investigaciones Biometeorologicas & University of Buenos Aires, Argentina

Insulation Means Environmental Protection
H. David, European Insulation Manufacturers Association (EURIMA), Belgium

Climate Change and Carrying Ability of Agro-Resources
L. Erda, Agrometeorological Institute, Academy of Agricultural Sciences, China

Meteorological Extremes as an Indicator of Climatic Change
F-W. Gerstengarbe, Meteorological Service, Germany

Climate Variations in Argentina and the Adjacent Sub-Antarctic Zone
J. A. J. Hoffman, Meteorological Service, Argentina

The Analysis of the Temporal and Spatial Characteristics for Temperature and Precipitation in Summer Over China
Huang Jiayou, Peking University, China

Regionalization and Inter-annual Variation of the Atmospheric Dryness in China Mainland
Jiang Jianmin, Beijing Institute of Meteorology, China

The Finnish Research Programme on Climate Change
M. Kanninen, Academy of Finland

Greenhouse Gas Emissions and their Control Potential in Finland
R. Korhonen et al., Technical Research Centre, Finland

Global Change: African Vegetation, El Niño-Southern Oscillation and Carbon Dioxide
J. Lancaster, Californian Space Institute, Scripps Institute of Oceanography & the Environmental Science and Policy Institute, USA

Global Runoff Data Centre
  H. Liebscher, Global Runoff Data Centre, Germany

Global Warming - The Greenpeace Report
  J. Leggett, Greenpeace, UK

Long-Hydrological Series in The Baltic Sea Area
  R. Lemmela, National Board of Waters and the Environment, Finland

The Characteristics of the Hydrological Cycle in North China
  Liu Chiunzhen, Hydrological Forecasting and Water Control Centre, China

Recent Trends of the Primary Greenhouse Gases
  J. Peterson et al., Climate Monitoring and Diagnostic Laboratory, NOAA, USA

The Renewable Energy Paradox in Paradise: A Case Study of Hawaii
  V. Phillips, Hawaii Natural Energy Institute, USA

The World Warming and Climate Change in China
  Tu Qipu, Nanjing Institute of Meteorology, China

Climate Change Impact on Norwegian Water Resources
  N. R. Saelthun et al., Water Resources and Energy Administration, Norway

Impacts of Climate on Agriculture: A Southern Perspective
  M. J. Salinger et al., Meteorological Service, New Zealand

Trends in Radiation Climate Components 1937/1989 at Potsdam
  W. Schone et al., Meteorological Service, Germany

Trends in the Climate Over India
  H. N. Srivastava et al., Meteorological Department, India

Climate Change and Environmental Goals
  R. J. Swart et al, National Institute of Public Health and Environemntal Protection, The Netherlands

Teleconnections Between the Thermal Forcing of the Australian Continent and Circulation Systems
in the Northern Hemisphere
  Xu Xiangde et al., Nanjing Institute of Meteorology, China

Global Warming: Towards a Strategy for Ontario
  B. Yang, Cabinet Office, Government of Ontario, Canada

Effect of Meridional Heating Difference and Macro-Topography on Asian Summer Monsoon
  Zhu Qiangen et al., Nanjing Institute of Meteorology, China

Sensitivity Experiments and Assessment of Climatic Changes in China Induced by Greenhouse Effect as
Simulated by the GCMs
  Zong-Ci Zhao, Academy of Meteorological Science, China

Appendix 3

STATEMENT OF ENVIRONMENTAL NON-GOVERNMENTAL ORGANIZATIONS
Co-ordinated by A. Bonnin Roncerel *(Belgium)*

Introduction

The Second World Climate Conference marks the historical divide between global assessment of climatic change and global negotiation over its solution. On both sides of this divide, our knowledge of the global warming process has been and will remain imperfect. Research will proceed uninterrupted; indeed we hope it will be accelerated. Yet, it is a goal, it is an obligation, of this conference to pass judgement on more than research. One of the stated objectives of the Second World Climate Conference is to "assess ... the implications for public policy" of our scientific knowledge. On the eve of negotiations on a global climate convention, the world awaits your judgement.

If you doubt the power of your words, consider recent history. Before the 1985 WMO/UNEP/ICSU conference at Villach, global warming drew little attention anywhere, and most people had never heard of global warming at all. In just five years global warming has become a household term around the world, and an issue that politicians cannot afford to ignore. The warnings first issued in 1957 by Revelle and Seuss, that the world had embarked on an unprecedented global experiment with the climate system, are finally understood. The world is now waiting for your recommendations on how to cancel the experiment.

Scientific Assessments

In addition to warning the world of the dangers of the experiment being conducted, the scientific community has produced a dazzling array of new and remarkable findings that demonstrate its significance and allow improved predictions of its outcome. We need only point to the IPCC scientific assessment and Dr. Oeschger's paper at this Conference. Consider what the ice cores and ambient data tell us:

- that greenhouse gas concentrations are higher than they have ever been in more than 160,000 years
- that these concentrations are increasing each year as a result of human activities

- that changes in greenhouse gas concentrations, global temperature, and the structure of the biosphere have been tightly coupled, one amplifying the other, over millennia
- that human-induced global warming projected for the next century is of the same magnitude as the world-altering changes during the glacial cycles
- that CO_2 from fossil fuel combustion is the single most important factor enhancing the greenhouse effect
- that emissions of carbon dioxide must be reduced by more than 60% from current levels if we are to stabilize its atmospheric concentration.

There are, of course, many uncertainties. These should neither be ignored, nor should they be allowed to serve as an excuse for inaction. Indeed, the nature of these uncertainties increases the urgency of halting the uncontrolled experiment we are currently conducting.

A major obstacle to resolving the uncertainties we face is presented by the lags inherent in the climate system. the consequences of today's emissions will remain obscure for decades. By the time they are manifest, more emissions will have occurred, and yet more will be inevitable. As Hans Oeschger noted, the stronger the feedbacks in the climate system, the longer will be this lag, and the more severe the ultimate consequences. With unavoidable and irreversible consequences destined to forever outpace our understanding, the only prudent course, the only defensible course, is to slow this experiment by reducing emissions of greenhouse gases.

Beginning with the Ministerial part of this Conference, world leaders will be looking to the scientists assembled here for guidance on how to proceed with these reductions. This is no easy task, and some may feel that, given the uncertainties, it is better not to propose targets that we might later regret. This is not a viable option. If we fail to provide guidance, then politicians will choose targets anyway. We have an obligation, based on the best information available now, to participate in this process.

Accordingly, we urge this conference to recommend limits on the rate of global warming, and corresponding targets for greenhouse gas emissions.

Targets

The targets that this conference proposes will be adjusted in response to evolving scientific knowledge just as the CFC reduction targets of the original Montreal Protocol have been replaced with a complete phase-out of the controlled chemicals. But if it were not for the early action of the United States, Canada, Sweden, and Norway to ban CFCs in aerosols by 1978 based on very preliminary scientific understanding, followed by the initial Montreal agreement in 1987, which was based on a more sophisticated assessment, the world would be committed to far more ozone depletion and global warming than we are today.

As its happens, these early assessments severely underestimated potential ozone depletion, as the polar and global ozone losses discovered since 1985 have shown. If scientists has not called for early action, we would indeed have major regrets today. There is always the possibility of overestimating consequences in preparing targets. But we should be more concerned, as Martin Parry's discussion of global food security illustrated, with the regret that would result if the risk is underestimated.

If we are to avoid such regret in the future, we must set preliminary limits now. As mentioned in Dr. Vellinga's paper, while we may already be committed to a level of climate change involving significant risk, investigation of the response of ecosystems and society to warming indicates that rates of warming greater than 0.1°C per decade, and total warming of more than 1–2°C over pre-industrial levels would present severe dfficulties in adjustment, adaptation and accommodation. Similarly, sea-level rise should be limited to a maximum rate of 2–5 cm per decade and a total of 30–50 cm above current levels. These limits were first proposed at the WMO/UNEP workshops at Villach and Bellagio in 1987 and confirmed by the findings of the report to the WMO/UNEP/ICSU Advisory Group on Greenhouse Gases just published by the Stockholm Environment Institute. We urge their adoption as targets by this section of the conference.

Understanding the implications of these targets for emissions is complicated by uncertainty regarding climate sensitivity. For the 1°C temperature limit, CO_2-equivalent concentrations would have to be held to 330–400 ppm—perhaps substantially less than current levels. We would be called irresponsible if we advocated the immediate reduction of CO_2 emissions by more than 60% noted by the IPCC as necessary to reach this target. It would also be irresponsible if we failed to point out that the world is already entering the danger zone and therefore any delay in cutting emissions poses ever increasing risks. For a 2°C

temperature limit the CO_2-equivalent concentrations must be held to 400–560 ppm. The only prudent course in the face of current uncertainties is to aim toward the lower end of this range. This implies that global CO_2 emissions must be reduced about 20% from current levels by 2005 as proposed by the Toronto Conference in 1988, and more than 60% by 2030.

In addition, the radiative effect of the other greenhouse gases must be stabilized. Although the sources of methane and particularly nitrous oxide are not well quantified, this can probably be accomplished through technologies that reduce specific sources of these gases, and by limiting the global warming potential of CFC substitutes. While actions must be taken to reduce emissions of all greenhouse gases, insistence on treating all gases in a single basket within a climate change convention should not be used as an excuse to delay reducing emissions from those sources for which control options are available.

Uncertainty about how humans will respond to the challenge before us may be as great as uncertainties about how climate will respond to human forcing. Under the IPCC 'Business as Usual Scenario', CO_2-equivalent concentrations reach 1200 ppm by the end of the next century (radiative forcing reaches 10 Wm^{-2})—more than twice the maximum tolerable level. Fortunately, the Business as Usual Scenario is not destiny. As illustrated by Thomas Johansson at this Conference, energy futures are possible that dramatically reduce CO_2 emissions while promoting economic development. The recent Stockholm Environment Institute report confirms that a reduction of about 25% in CO_2 emissions is feasible by 2005 with 40% feasible by 2020. Fortunately, these feasible emission reductions correspond to those required to stay within the maximum warming limits of 1–2 degrees. We urge this conference to reaffirm the goal adopted by the Toronto Conference by recommending a cut of at least 20% in global CO_2 emissions by 2005 as a first step toward the at least 60% decrease eventually needed. In order to achieve this target while allowing for the differentiated responsibilities of developing countries, industrialized countries must reduce their CO_2 emissions at least 20% by the year 2000.

Developing Countries

Industrialized and developing countries have shared, but differentiated, responsibilities for responding to the threat of global warming. Industrialized countries have been responsible for the vast majority of the greenhouse gas increases that have occurred so far. It must be recognized that developing countries will require increased financial, human, and information resources, both for development of appropriate national strategies and for deployment of environmentally beneficial technologies. It is also clear that developing countries will—and should—evaluate climate response strategies in the broader context of concern over poverty, hunger, population growth, environmental degradation and resource depletion and other pressing concerns. A new global 'social contract' is needed that recognizes the legitimate aspirations of developing countries for an enhanced quality of life for their peoples, and makes fundamental reforms in international finance, growth and development policies.

We must also realize that the consequences of global warming, including sea-level rise, shifts of climatic zones, changes in water availability, and an increase of weather-induced disasters will be particularly pronounced in developing countries. For example, Bangladesh has a little over 2% of the world's population, but it generates less than 0.1% of the world's greenhouse gas emissions. Nonetheless, it is likely to suffer from increasingly frequent hurricanes and flooding and will be severely affected by sea-level rise.

Furthermore we note that the industrialized nations have often promoted energy intensive growth in developing countries, extracting economic gains while in the process breaking down traditional conservation practices which were more environmentally benign. Therefore we urge that all action taken at the national and international level be undertaken with the full participation of the public, involving experts from different backgrounds. This would ensure that such efforts promote socio-economic development in the poor countries to enable the poorest people to achieve a sustainable existence and quality of life without assuming the environmentally destructive patterns of the industrialized countries.

These considerations demand:

- that priorities for developing countries should be in the local and regional development of energy efficient technologies as alternatives to inefficient and polluting technologies and projects
- that priorities include the development of sustainable natural resource management
- that strategies for addressing climatic change in developing countries not inhibit, but facilitate, necessary socio-economic growth through support to local initiatives
- that further research on region-specific and country-specific impacts and response strategies for each developing region and country must be done
- that there is a need to support research and development institutions, both governmental and non-governmental, in developing countries to elevate their understanding and ability to cope with their local problems
- that such assistance will help developing countries develop adaptive strategies and thereby be in a better

position to cope with climatic change and contribute to limiting climatic change
- that specific national targets on greenhouse gas reductions can *only* be supported if international agreements permit continued socio-economic development in developing countries.

Some of the rich and powerful countries are now reluctant to set specific national targets on greenhouse gas reductions. Their stance can be interpreted in the light of the resources they have available to enable them to respond quickly to the adverse effects of climatic change. Meanwhile developing countries, while facing the brunt of global climatic change, are left wondering why the nations that caused the problem do not want to do something about it.

Biodiversity

Too little of the climate change debate has focused on the possible impacts to biological diversity. The world's 5 to 30 million species face threats to their continued existence as never before. Environmental degradation due to causes such as habitat destruction, pollution, and urban development is causing species extinction to accelerate. It has been estimated that 15–20% of all species may have been lost by the year 2000, a rate approximately 1000 times the natural rate. Climate change may significantly speed up this already frightening loss of species.

Genetic, species and ecosystem diversity provides the natural base on which human civilization depends. Each species may be the storehouse for between 1,000 and 400,000 genes. And these genetic treasures, apart from their intrinsic individual value as unique species, are also crucial for the medicines, crop development and industrial products that fuel human development.

Climate change may have a devastating effect on biological diversity. All species will be affected in one way or another, but some groups of species will be especially vulnerable.

Geographically localized species, such as those endemic to islands, may be at risk because populations are not spread widely enough to be maintained under new climatic conditions.

Genetically impoverished species may lack the genetic diversity to adapt to new conditions, migratory species which depend on appropriate environmental circumstances along their migratory routes may lose out.

Highly specialized species, such as those dependent on a single food source, are likely to be especially sensitive to climate change, as are species that are already rare or threatened, and the peripheral populations of species whose range is already shrinking.

Species dependent on timing of ice-melt in polar regions may suffer, and alpine species may have their ranges fragmented and drastically reduced in area as they are forced to migrate upslope.

Coastal ecosystems such as mangroves and saltmarshes could be severely disrupted as sea level rises, and coral reefs could die for the same reason or because of warming water or changed nutrient flows.

The catalogue of potential catastrophes for biological diversity could go on. Some people have suggested that many species may thrive as a result of climate change. In fact, the most likely winners from a warmer world will be parasites and disease vectors such as trypanosomes and malarial mosquitoes. Their ranges will expand into areas where host resistance is low, and their ability to evolve quickly will put them at an advantage over other species. There is no doubt, then, that climate change will have a major negative impact on biological diversity.

Climate change will cause us to question traditional conservation techniques because protected areas too will be under threat. Ecological assemblages will break up as their component species migrate at different rates or adapt differently to climate change, and many species, already at the limit of their geographical ranges and within isolated protected areas, may be threatened with extinction if new habitat is not available for them to move into.

Forests

The greatest storehouse of biological diversity on earth are the tropical forests. Holding at least half of all species, forests are being destroyed at an unprecented rate. Conservation of biological diversity, soil stabilization, protection of watersheds, the maintenance of hydrological cycles and climate itself, and the continued existence of indigenous cultures depend on the conservation of tropical forests. Climate change represents an additional, and potentially disastrous new stress for these ecosystems.

It is obvious that because of the extremely complex ecological relationships between species and the importance of seasonality in the functioning of these ecosystems that depend on rainfed agriculture, small changes in timing or severity of either wet or dry seasons can have major impacts.

There has been much talk amongst governments in the lead-up to the negotiation of an international convention on climate change of the need to adopt forestry policies aimed at storing carbon in forests as biomass. There have been calls for a global net afforestation and reforestation target of 12 million hectares per year. At the current rates of tropical deforestation, this would mean the need to plant and successfully maintain, over their entire planned lifespan, 29 million hectares of plantation.

Although some of this forest could be in the temperate or boreal zones, there is little doubt that the vast majority would need to be in the tropics, where productivity is higher, growth rates faster and establishment and maintenance costs lower. There is also little doubt that the task is well nigh impossible. The current rate of reforestation and afforestation in the tropics is only around 1 million hectares a year.

There is no guarantee that if the current root causes of deforestation are not dealt with, new reforestation and afforestation projects will not suffer the same fate as remaining natural tropical forests.

There is absolutely no doubt that the most concrete forestry-related action that can be taken to slow the rate of climate change is to conserve the world's remaining tropical forest. This should be achieved through a combination of protected area management and environmentally sustainable forest management on a long-term basis. Adaptation to climatic changes will be facilitated by measures to conserve biodiversity and to use genetic enhancement techniques.

Consistent with this, the entire international tropical timber trade should be based on environmentally sustainable utilization of tropical forests.

In order to achieve these targets there should be new and additional financial and technical resources made available to the developing, tropical countries by industrialized nations for forestry and biodiversity conservation activities.

NGOs are convinced that there is a need for a comprehensive global forestry agreement dealing with the conservation and sustainable management of boreal, temperate and tropical forests on a fair and equitable basis. The multidisciplinary nature of the upcoming climate convention negotiation process will present an ideal opportunity to integrate other key issues such as energy, agriculture, industrial development and trade policies into the forest debate.

Forests can contribute to the solution of the climate change problem, rather than being the helpless victims of a changing climate. Above all, forestry conservation and maintenance of biodiversity must be recognized as being necessary for a whole host of environmental, social and economic reasons.

Agriculture

The agriculture sector, like the forest sector, is fundamentally related to climate change. Agriculture emissions of greenhouse gases will become an increasingly important contributor to climate change. Improved agricultural practices that can be implemented now would result in reductions in emissions of nitrous oxide and methane. Some of the immediate steps that can be taken include improvements in fertilizer and livestock management practices, increased use of biogas digestors and changes in water management practices in rice cultivation. Parallel efforts must also be made to intensify research on the sources of such gases and development of longer term control strategies. At the same time, impacts on agriculture from global warming present important implications for food production and distribution that must be carefully assessed.

Energy

Energy production and use are responsible for more than one-half of the world's greenhouse gas emissions. Therefore, strategies to reduce the threat of global warming must focus heavily on the energy sector. In the long run, the world must make a major shift away from reliance on fossil fuels as our major energy source; in the near term, aggressive measures to increase the efficiency of energy use must be implemented in both developed and developing nations. It should be noted that in the developed nations, efficiency gains will allow continued increases in *energy services* while yielding substantial reductions in *energy consumption*. The economic restructuring currently occurring in Eastern Europe presents a critical opportunity to introduce energy efficient technologies that will benefit both the environment and the economies of these countries.

In developing nations, the need for more rapid growth in energy services means that efficiency gains are essential to prevent capital shortages and allow sustainable development, as well as to slow the rate of increase in primary energy demand and CO_2 emissions. Furthermore, most of the population depends on biomass energy. In some countries, such as Kenya, biomass supplies up to 85% of the domestic energy needs. Yet, current proposals on technology development for these countries are focused almost exclusively on centralized power generation. While it is true that these countries require efficient technology to increase levels of energy production as a prerequisite for sustained social and economic development, and electrification may reduce the pressure on vegetation, no consideration has been given to the long term strain such new technology will place on the local economies. These economies cannot afford any additional investments in the types of "development" for which most of the current problems can be blamed.

The benefits of increased energy efficiency and a shift to renewable energy resources are not limited to reducing greenhouse gas emissions. Such measures will also yield reductions in conventional pollutants that cause acid rain, urban smog and other environmental problems; reduce competition and possible conflict over access to fossil fuel supplies, particularly oil; and also produce economic benefits by reducing (or slowing the rate of increase of) consumer energy costs, generating increased domestic employment, and, for many countries, increasing the energy services provided per unit of investment while reducing the financial burden of imported energy supplies.

The transition to a much more efficient, increasingly renewable-based energy system will take decades to fully realize, and the investment of substantial financial and human resources. It will not be cheap—just less costly to our environment and our economies than any of the alternatives, as demonstrated by the paper by Mills, Wilson and Johansson. Most important, it will take consistent, sustained leadership from government policymakers and the private sector, and the integrated use of the full range of policy tools at our disposal—market mechanisms, performance standards, regulation, restructuring electric and gas utilities, information and public education efforts, technology development, and commericialization strategies.

The objectives of these policies should be to:

- improve the efficiency of energy use in all sectors (e.g. more efficient vehicles, appliances, lighting and motors)
- improve the efficiency of energy supply and, where possible, shift to lower-carbon fossil fuels
- shift to renewable energy sources where they are economically competitive on the basis of full social costs (and promote technology development and large-scale production to further reduce the cost of renewables)
- reduce the demand for energy services in industrial countries, particularly in the transport sector, through use of improved development and land-use strategies, increased availability of transit, and energy pricing and tax policies that more fully reflect the true social and environmental costs of energy production and use.

The choice of technological measures and policy tools to encourage these goals should be based on the need to minimize all environmental risks, reduce social risks, and maximize the cost-effectiveness of the programme. Under these criteria it is clear that nuclear power is not a viable response to the threat of global warming. Indeed investments in nuclear power divert resources from more productive uses. Energy research and development programmes should be re-oriented accordingly, to focus on energy efficiency and renewable energy sources. Similarly, development assistance programmes of the World Bank and bilateral programmes of the industrialized nations must be re-oriented to the kinds of development-focused end-use-oriented (DEFENDUS) strategies discussed by Thomas Johansson based on work by Amulya Reddy.

Conclusion

We conclude by re-emphasizing that stabilizing CO_2 emissions is not a sufficient response to the challenge before us. While the commitments to action made to date are encouraging, it is the responsibility of the scientists at this meeting to measure these commitments against targets based on environmental necessity rather than political expediency. Politicians can be made to face these global threats, but they will only do so with the help of the scientific community. The world is looking to this Conference to provide clear guidelines to the politicians who will begin negotiating a convention next year. If we fulfill our task, this Conference will indeed mark the turning point from assessing the dangers of global warming to implementing the solution. Do not shirk that responsibility.

Appendix 4

PARTICIPANTS AT THE SCIENTIFIC/TECHNICAL SESSIONS

A. I. ABANDAH	Meteorological Department, Jordan
E. ACHEAMPONG	Friends of the Earth, Ghana
A. ADAMS	Department of Community Services & Health, Australia
O. C. ADEBISI	Federal Environmental Protection Agency, Nigeria
A. O. ADEDE	Legal Affairs, UN Headquarters USA
J. A. ADEJOKUN	Meteorological Department, Nigeria
J. ADEM	Centro de Ciencias de la Atmosfera, Mexico
S. P. ADHIKARY	Department of Hydrology & Meteorology, Nepal
P. AGATONOV	Oceanographic Committee, USSR
G. P. AGUIRRE	Direccion Meteorologica de Chile
E. D. AHLONSOU	Division of Climate and Meteorological Services, Benin
Y. J. AHMAD	UNEP, Kenya
E. O. A. AINA	Federal Environment Protection Agency, Nigeria
M. AKBAS	Universite Paris IV, Switzerland
L. E. AKEH	Meteorological Services, Nigeria
N. N. AKSARIN	Central Asian Hydrology Research Institute, USSR
V. V. AKSENOV	State Research Centre for Study of Natural Resources & Environment, USSR
I. K. AL-ATWI	Meteorological Department, Jordan
M. AL-ERYANI	Civil Aviation & Meteorological Authority, Yemen
I. H. AL-MAJED	Department of Meteorology, Qatar
S. M. AL-YAGOOT	Al Dhafra Air Base, United Arab Emirates
M. ALAIBAN	MEPA, Jeddah, Saudi Arabia
M. ALAIBAN	MEPA, Riyadh, Saudi Arabia
S. ALAIMO	Servicio Meteorologico Nacional, Argentina
A. E. ALEYA	Mission Permanente d'Egypte, Switzerland
A. A. ALGAIN	MEPA, Saudi Arabia
Y. ALHAJ	Civil Aviation & Meteorology Authority, Yemen
A. ALI	Ecole Hassania des Travaux Publique, Morocco
S. ALOUFI	Council of Ministers Bureau of Experts, Saudi Arabia
M. S. ALSABBAN	Ministry of Petroleum & Mineral Resources, Saudi Arabia
A. L. ALUSA	UNEP/GEMS (Atmosphere), Kenya
P. AMBROZY	Meteorological Service, Hungary
J. C. ANDRE	Etab. d'Etudes et de Recherches Meteorologiques, France
A. J. APLING	Department of the Environment, United Kingdom
S. APUNTE FRANCO	Permanent Mission of Ecuador, Switzerland
M. ARDON	The Hebrew University at Jerusalem, Israel
O-A. AREZKI	Servicio de Climatologie, Algeria
J. ARIAS	Energy Consultant, Mexico
E. ARRHENIUS	World Bank, USA
G. ARUM	Energy and Environment Organization, Kenya
L. ASSUNCAO	UNCED, Switzerland

A. AULICIEMS	Applied Climate Research Unit, University of Queensland, Australia
J. AUSUBEL	Carnegie Commission, The Rockefeller University, USA
D. N. AXFORD	World Meteorological Organization, Switzerland
N. B. AYIBOTELE	Water Resources Research Institute, Ghana
N. AZIMI	UNITAR, Switzerland
L. BAEHR	Centre for Science & Technology for Development, USA
F. W. G. BAKER	Consultant, France
D. J. BAKER	Joint Oceanographic Institutions, USA
P. M. BAKKEN	Ministry of Environment, Norway
A. BALBONI	Permanent Mission of Italy, Switzerland
V. BANZER	Mission Permanente de Bolivia, Switzerland
A. BARCENA IBARRA	UNCED, Switzerland
J-P. BARI	Universite de Geneve, Switzerland
E. T. A. BARGAN	Meteorological Department, Libyan Arab Jamahiriya
G. BARNES	Planning Environment & Lands, Hong Kong
A. BARRE	Permanent Mission of Qatar, Switzerland
T. BARTSCH	University of Heidelberg, Germany
B. BAUDOT	Saint Anselm College, USA
M. BAUTISTA PEREZ	Instituto Nacional de Meteorologia, Spain
A. BECKER	Water Resources Division, UNESCO, France
D. F. BECKER	Sierra Club, USA
E. A. BELL-GRAM	Permanent Mission of Nigeria, Switzerland
M. A. BELSHIET	Meteorological Authority, Egypt
R. E. BENEDICK	World Wildlife Fund/Conservation Foundation, USA
M. BENISTON	ProClim (National Climate Programme), Switzerland
A. BENSARI	Meteorologie Nationale, Morocco
M. BERAN	Institute of Hydrology, United Kingdom
A. BERGER	Institute of Astronomy & Geophysics, Belgium
M. A. BERLATO	Universidad Federal Do Rio Grande Do Sul, Brazil
A. BERNAERTS	Consultant, Saudi Arabia
F. BERNTHAL	National Science Foundation, USA
C. E. BERRIDGE	Caribbean Meteorological Organization, Trinidad, W.I.
D. F. BEST	Meteorological Department, Barbados
A. BICHSEL	Communauté de Travail Swissaid, Switzerland
E. BIERLY	National Science Foundation, USA
J. T. BLAKE	Meteorological Service, Jamaica
M. A. B. BODDINGTON	Foundation for International Security, United Kingdom
G. J. BOER	Atmospheric Environment Service, Canada
W. BÖHME	Meteorologischer Dienst, Germany
R. BOJKOV	Environment Division, WMO, Switzerland
V. G. BOLDIREV	World Climate Programme Department, WMO, Switzerland
B. BOLIN	Department of Meteorology, Sweden
A. BONNIN-RONCEREL	Climate Network Europe, Belgium
Y. BOODHOO	Meteorological Service, Mauritius
M. BOOTSMA	Stichting Natuur en Milieu, Netherlands
D. BORIONE	Ministere des Affaire Etrangeres, France
F. BOTTCHER	IFIAS, Netherlands
A. BOUANANE	College du Leman, Switzerland
M. BOULAMA	Meteorologie Nationale, Niger
R. BOULHAROUF	Ministry of Foreign Affairs, Switzerland
S. BOYLE	Association for the Conservation of Energy, United Kingdom
P. BRANNER	United Nations Sudano-Sahelian Office (UNSO), USA
H. BREITMEIER	UNITAR, Germany

R. BRESLAU	International Chamber of Commerce, France
M. BkEUCH-MORITZ	Federal Ministry of Transport, Germany
R. BRINKMAN	Soil Resources Management & Conservation Services/FAO, Italy
F. BRO-RASMUSSEN	Society for Nature Conservation, Denmark
E. BROWN WEISS	Georgetown University Law Center, USA
K. BROWNING	Meteorological Office, United Kingdom
J. P. BRUCE	Consultant, Canada
B. BRUMME-BOTHE	Bundesministerium fur Forschung und Technologie, Germany
M. BRUSTLEIN	Alusuisse - Lonza, Switzerland
S. BUDAIR	Mission d'Arabie Saoudite, Switzerland
M. BUDYKO	State Hydrological Institute, USSR
M. BUITENKAMP	Vereniging Milieudefensie FOE, Netherlands
F. BULTOT	Institut Royal Meteorologique, Belgium
J. J. O. BURGOS	Centro de Investigaciones Biometeorologicas, Argentina
W. BURHENNE	Environmental Policy Law & Adm. (IUCN), Germany
J. BURNETT	Oxford University, United Kingdom
B. R. R. BUTLER	British Petroleum Co., PLC, United Kingdom
K. BYLL	Ingenieur Meteorologie, Togo
J. BYRNE	University of Lethbridge, Canada
C. CACCIA	Member of Parliament, House of Commons, Canada
F. L. CACEROS	National Meteorological Office, Chile
R. CAMACHO	National Oceanographic Committee, Chile
C. CANDANEDO	Instituto de Recursos Hidraulics y Electrificacion, Panama
D. CARRILLO	EEC, Belgium
D. CARSON	Meteorological Office, Hadley Climate Centre, United Kingdom
G. CARTWRIGHT	Consultant, USA
J. C. CENTENO	Latin American Forestry Institute, Venezuela
S. CHACOWRY	World Meteorological Organization, Burundi
P. CHALI	Permanent Mission of Zimbabwe, Switzerland
A. H. CHAMPEAN	Permanent Mission of Canada, Switzerland
K. H. CHANG	Ministry of Energy and Resources, Republic of Korea
M. L. CHAVARRI	Permanent Mission of Peru, Switzerland
R. CHEN	World Hunger Program - Brown University, USA
G. B. CHIPETA	Meteorological Department, Zambia
H-M. CHO	Meteorological Service, Republic of Korea
M. H. K. CHOWDHURY	Meteorological Department, Bangladesh
R. CHRIST	Ministry of Environment, Austria
N. CHRISTOFORIDIS	EEC, Belgium
W. CIESLA	Forest Protection Department/FAO, Italy
R. A. CLARKE	Bedford Institute of Oceanography, Canada
J. CLARKE	Battelle, USA
B. C. COHEN	UNECE, Switzerland
M. COLACINO	Istituto di Fisica del l'Atmosfera, Italy
U. COLOMBO	ENEA, Italy
M. CONTE	Servizio Meteorologico, Italy
R. COPPOCK	National Academy of Science & Engineering, USA
R. CORDOVA	Direccion General Aeronautica Civil, CA Honduras
S. G. CORNFORD	World Meteorological Organization, Switzerland
M. J. COUGHLAN	NOAA/OCAR R/CAR, USA
M. CROWE	World Meteorological Organization, Switzerland
R. CZELNAI	World Meteorological Organization, Switzerland
R. V. DA CUNHA	Scientific Affairs Division, NATO, Belgium

M. A. DADDISH	Civil Aviation & Meteorological Office, Libyan Arab Jamahiriya
A. W. DAKKAK	Natural Resources Director, Saudi Arabia
O. DANK	Ministry of Foreign Affairs, Libyan Arab Jamahiriya
W. DANSGAARD	Geophysical Institute, Denmark
H. DAVID	European Insulation Manufacturers Association, Belgium
M. DAVIDSON	UNCED, Switzerland
K. DAVIDSON	National Climate Data Center, USA
A. DAVIS	Florida State University, USA
C. DAWSON	Department of State, USA
K. DAWSON	Canadian Climate Centre, Canada
M. DE BOTERO	Corporacion Colegio, Colombia
C. DE BOULLOCHE	International Council of Women, Switzerland
M. DE CASTRO	Permanent Mission of the Philippines, Switzerland
R. A. DE GUZMAN	World Meteorological Organization, Switzerland
H. DE HAEN	Food & Agriculture Organization, Italy
Y. DE HALLER	SA L'Energie De L'Ouesi Suisse (EOS), Switzerland
F. DE OLIVIERA	Instituto Nacional de Pesquisas Espacia (INPE), Brazil
J. DE PERETTI	Direction Generale de l'Energie et des Matieres Premieres, France
E. F. DE QUEIROZ	Instituto Nacional de Meteorologia, Brazil
W. DEGEFU	National Meteorological Services Agency, Ethiopia
P. DELACROIX	Mission Permanente de la France, Switzerland
T. DELBREUVE	UNDP European Office, Switzerland
F. DELSOL	World Meteorological Organization, Switzerland
A. M. DEMMAK	Agence Nationale de Resources Hydrolique, Algeria
N. DESAI	UNCED, Switzerland
P. E. DEXTER	World Meteorological Organization, Switzerland
N. T. DIALLO	Meteorologie Nationale, Papua New Guinea
H. DIALLO	World Meteorological Organization, Switzerland
A. DIAMANTIDIS	ROE/UNEP, Switzerland
A. DIN	University of Geneva, Switzerland
A. B. DIOP	Ministere de l'Equipment, Senegal
L. DISSING	EEC, Belgium
M. DJEDAINI	Division de Climatologie, Algeria
M. C. DLAMINI	UNEP/Focal Point, Swaziland
R. DOLZER	Universitat Mannheim, Germany
J. C. I. DOOGE	University College Dublin, Ireland
B. R. DÖÖS	IIASA, Austria
H. DOVLAND	NILU, Norway
E. DOWDESWELL	Atmospheric Environment Service, Canada
I. DRAGHICI	Ministry of Environment, Romania
A. DRUCKER	United Nations Volunteers, Switzerland
N. K. DUBASH	Environmental Defense Fund, USA
M. N. DUC NGU	Hydrometeorological Service, Viet Nam
P. EDAFE	Permanent Mission of Madagascar, Switzerland
L. EDGERTON	National Resources Defense Council (NRDC), USA
A. A. EGBARE	Meteorologie Nationale, Togo
R. EGLOFF	Office de Cooperation pour les Energies Renouvelables, Switzerland
W. E. EICHBAUM	Conservation Foundation/World Wildlife Fund, USA
EL-M. EID	Environmental Affairs Agency, Egypt
H. EL-LAKANY	American University in Cairo, Desert Development Center, Egypt
D. ELDER	Marine and Coastal Programme, IUCN, Switzerland
C. EMMENEGGER	Service Hydrologique et Geologique National, Switzerland
C. D. ESCOBAR	Servicio Nacional de Meteorologia e Hidrologia, Bolivia

F. ESTELITA	Ministerio das Relagoes Exteriores, Brazil
N. EVTIKHOV	State Committee for Hydrometeorology, USSR
W. EYAMBE	Permanent Mission of Cameroon, Switzerland
F. FANTAUZZO	Servizio Meteorologico del l' Aeronautic, Italy
R. FANTECHI	EC Climate & Natural Hazards Research Programme, Belgium
H. L. FERGUSON	SWCC, World Meteorological Organization, Switzerland
V. FERRARA	ENEA, Italy
A. FISA	MEPA, Saudi Arabia
D. J. FISK	Department of the Environment, United Kingdom
D. FITZGERALD	Meteorological Service, Ireland
B. P. FLANNERY	Exxon Research & Engineering Co., USA
N. C. FLEMMING	Institute of Oceanographic Sciences, United Kingdom
H. FLOHN	Meteorologisches Institut der Universitat Bonn, Germany
E. W. FRIDAY	NOAA/National Weather Service, USA
D. FROMMING	Foreign Affairs Section Deutscher Wetterdienst, Germany
A. FRY	International Environmental Bureau, Switzerland
A. FUJIMORI	Global Industrial & Social Progress Research Institute, Japan
N. GABR	Permanente Mission d'Egypte, Switzerland
H. GASSERT	Bundesverband der Deutschen Industrie, Germany
L. B. GATAN	Permanent Mission of the Philippines, Switzerland
W. L. GATES	Lawrence Livermore National Laboratory, USA
D. J. GAUNTLETT	Bureau of Meteorology, Australia
R. P. GEHRI	Southern Company Services, USA
A. N. A. GHANI	Permanent Mission of Sudan, Switzerland
J. GILBERT	Ministry of Environment, New Zealand
N. A. GINTING	Forest Research & Development Centre, Indonesia
M. GLANTZ	National Center for Atmospheric Research, USA
G. GLASER	UNESCO, France
T. J. GLAUTHIER	World Wildlife Fund/Conservation Foundation, USA
P. GLEICK	Pacific Institute for Studies in Development of Environment & Security, USA
R. GODIN	UNESCO/IOC CCCO, France
G. S. GOLITSYN	Institute of Atmospheric Physics, USSR
G. T. GOODMAN	Stockholm Environment Institute (SEI), Sweden
D. GRABOWSKI	Permanent Mission of Poland, Switzerland
I. GRAHAM-BRYCE	Shell International Petroleum, Netherlands
A. GRAMMELTVEDT	Meteorologiske Institutt, Norway
H. GRASSL	Max Plank Institut fur Meteorologie, Germany
J. GRAY	University of Oslo, Norway
D. GREENE	Conservation Foundation, Australia
C. A. GREZZI	Nacional de Meteorologia, Uruguay
G. R. GROB	CMDC Delthi House, Switzerland
M. GRUBB	Royal Institute of International Affairs, United Kingdom
L. M. GUERRA	Instituto Autonomo de Investigaciones Ecologicas, Mexico
D. GUTHRIE, JR.	IAEA, Austria
M. D. GWYNNE	UNEP/GEMS, Kenya
A-B. HABIB	Switzerland
S. A. HAJOST	Environmental Defense Fund, USA
C. HAKKARINEN	Electric Power Research Institute, USA
I. HALAMA	SZOPK, Czechoslovakia
R. HALLO	Stichting Natuur en Milieu, Netherlands
L. S. HAMILTON	Environment and Policy Institute, USA

B. HARE   Australian Conservation Fondation, Australia
J. B. HARRINGTON   Petawawa National Forestry Institute, Canada
J. S. HARRISON   British Coal Corporation, United Kingdom
D. HARVEY   University of Toronto, Canada
M. M. HASHIMOTO   Environment Agency, Japan
M. H. A. HASSAN   Third World Academy of Sciences, Italy
F. HASSAN   American University - Cairo, Egypt
K. HASSELMANN   Max Planck Institute for Meteorology, Germany
T. HAWANDO   National Committee on Climate Change, Ethiopia
A. D. HECHT   Environmental Protection Agency, USA
L. HEILEMAN   Institute of Marine Affairs, Trinidad, W.I.
K. HEINLOTH   University of Bonn Physikalisches Institut, Germany
R. HEINO   Meteorological Institute, Finland
G. P. HEKSTRA   Ministry of Environment, Netherlands
A. HENDERSON SELLERS   Macquarie University, Australia
J. HENNINGSEN   EEC, Belgium
E. B. HENRY   Meteorological Office, Trinidad, W.I.
J. U. HIELHEMA   AGR Division/FAO, Italy
H. N. HIZAM   Permanent Mission of the United Arab Emirates, Switzerland
T. Y. HO   World Meteorological Organization, Switzerland
L. HOFFMANN   Deutscher Wetterdienst, Germany
A. J. HOFFMANN   Servicio Meteorologico Nacional, Argentina
P. HOHNEN   Greenpeace International, Netherlands
M. HOLDGATE   IUCN, Switzerland
E. HOLOPAINEN   University of Helsinki Department of Meteorology, Finland
P. F. HOLTHUS   South Pacific Regional Envi.ronment Programme, New Caledonia
M. HOOTSMANS   Ministry of Environment, Netherlands
J. T. HOUGHTON   Meteorological Office, United Kingdom
T. HPAY   International Tropical Timber Organization, Switzerland
I. J. HRBEK   Hydrometeorological Institute, Czechoslovakia
N. HTUN   UNCED, Switzerland
J. HUANG   Beijing University, China
C. H. HULSBERGEN   Delft Hydraulics, Netherlands
S. HUQ   Bangladesh Centre for Advance Study, Bangladesh
L. HYTTINEN   UNCED, Switzerland

L. G. IL   State Meteorological Services, Democratic People's Republic of Korea
M. B. IMADI   Permanent Mission of Syria, Switzerland
D. IMBODEN   Federal Institute for Water Resources, Switzerland
B. IMEREKOV   State Committee for Science & Technology, USSR
A. M. A. IMEVBORE   Institute of Ecology, Nigeria
H. INDRAJAYA   Permanent Mission of Indonesia, Switzerland
I. ISAKSEN   University of Oslo Institute of Geophysics, Norway
S. ISASHIKI   Ministry of Foreign Affairs, Japan
K. ISHIKAWA   Friends of the Earth, Japan
Yu. A. IZRAEL   State Committee for Hydrometeorology, USSR

H. K. JACOBSON   Institute for Social Research, University of Michigan, USA
J. JÄGER   Stockholm Environment Institute, Germany
J. J. JANSEN   Southern Company Services, USA
B. JARVIS   Department of Energy, Mines & Resources, Canada
E. J. JATILA   World Meteorological Organization, Switzerland
E. JAUREGUI   Centro de Ciencias de la Atmosfera, Ciudad Universitaria, Mexico
L. JEFTIC   UNEP Oceans & Coastal Area Project, Greece

G. J. JENKINS	Meteorological Research Flight, United Kingdom
P. G. JEREZ	Instituto de Recursos Naturales y del Ambiente, Nicaragua
P. JESSUP	Friends of the Earth, Canada
N. S. JODHA	ICIMOD Mountain Farming Stystem Division, Nepal
T. B. JOHANSSON	Lund University, Sweden
S. JOHNSON	FAO, Italy
A. M. K. JORGENSEN	Meteorological Institute, Denmark
D. JOST	Umweltbundesamt, Germany
J. JOUZEL	Laboratoire de Geochimie Isotopique, France
J. JOYCE	SWCC, World Meteorological Organization, Switzerland
A. JUNOD	Institut de Meteorologie, Switzerland
Z. KACZMAREK	IIASA Water Resources Project, Austria
M. KADI	Centre du Climatologique National, Algeria
W. KAKEBEEKE	Ministry of Housing, Physical Planning & Environment, Netherlands
J. KALVOVA	Meteorology & Environment Protection Dept., Charles University, Czechoslovakia
D. KAMANA	Ministry of Conservation, Cook Islands
S. KANE	NOAA/Office of Climatic & Atmospheric Research., USA
M. KANNINEN	Academy of Finland, Finland
R. P. KARIMANZIRA	Department of Meteorological Services, Zimbabwe
M. KASSAS	University of Cairo, Egypt
S. KATAYAMA	Office of Earth Science & Technology, Japan
S. KATO	Global Environment Department, Japan
T. KAWASAKI	Tohoku University, Japan
W. KELLOGG	National Center for Atmospheric Research, USA
W. KEMPEL	Permanent Mission of Austria, Switzerland
A. KHALID	MEPA, Saudi Arabia
M. A. KHAN	Permanent Mission of Pakistan, Switzerland
M. E. KHIDIR ABDALLA	Meteorological Department, Sudan
S. KHMELNITSKI	UN External Relations & Interagency Affairs Office, Switzerland
S. S. KHODKIN	State Committee for Hydrology, USSR
M. KHORAMIAN KERMANCH	Permanent Mission of Iran, Switzerland
A. KHOSLA	Development Alternatives, India
B. R. KININMONTH	Bureau of Meteorology, Australia
R. J. KINLEY	International Policy Environment, Canada
R. L. KINTANAR	PAGASA, Philippines
M. R. KIPGEN	L'Administration des Services Techniques de l'Agriculture, Luxembourg
H. KOLB	Ministry of Science & Research, Austria
K. P. KOLTERMANN	WOCE Institute of Oceanographic Sciences, United Kingdom
K. J. KONDRATIEV	Academy of Sciences, USSR
H. R. KONOW	Electrical Association, Canada
F. KRAUSE	Lawrence Berkeley Laboratory, USA
S. KRIEF	Bulle Bleu, France
J. KRUUS	Atmospheric Environment Service, Canada
M. KUHN	IAMAP, University of Innsbruck, Austria
E. KULICK	UNIDO, Switzerland
G. KULLENBERG	Intergovernmental Oceanographic Commission, France
S. M. KULSHRESTHA	Meteorological Department, India
T. KVICK	Meteorological and Hydrological Institute, Sweden
J. W. M. LA RIVIERE	International Council of Scientific Unions, France
A. LABAJO	Instituto Nacional de Meteorologia, Spain
J. LABROUSSE	M.R.T. TOEE, France
G. LALIOTIS	National Meteorological Service, Greece

R. LAMB	Television Trust for the Environment, United Kingdom
J. LANCASTER	Scripps Institution of Oceanography, USA
I. LANG	Academy of Science, Hungary
M. LANGE	Alfred Wegener Institute, Germany
M. LAPIN	Hydrometeorological Insititute, Czechoslovakia
D. A. LASHOF	Natural Resources Defence Council, USA
D. LAUGHTON	ORGCOMM, University of Geneva, Switzerland
A. LEBEAU	Meteorologie Nationale, France
I-C. LEE	Institute of Energy & Resources, Republic of Korea
P. LEFALE	Greenpeace, New Zealand
J. LEGGETT	Greenpeace, United Kingdom
R. LEMMELA	Hydrological Office, Finland
K. LEWIS	ILO, United Kingdom
C. S. LI	Permanent Mission of the Democratic People's Republic of Korea, Switzerland
W. H. LINDNER	Centre for our Common Future, Switzerland
H. F. LINS	Geological Survey, USA
C. LIU	Ministry of Water Resources, China
D-Q. LIU	Ministry of Foreign Affairs, China
D. LLUCH-BELDA	IOC, Mexico
P. LO SU SIEW	Meteorological Service, Singapore
G. LOESER	BUND Baden-Wuerttemberg, Germany
M. A. LOMBARDO	Forum of Brazilian NGO's to UNCED '92, Brazil
B. LONG	OECD, France
N. T. LONG	Pemanent Mission of Viet Nam, Switzerland
W. M. LONGWORTH	Meteorological Service, Vanuatu
C. LOPEZ-POLO	ECE Energy & Human Settlements Division, Switzerland
I. LORENZO	UNIDO, Switzerland
R. LOSKE	Institit fur Okologische Wirtschaftsforschung, Germany
J. LU	Permanent Mission of China, Switzerland
J. LUDERITZ SALDANNA	Federaçao dos Engenero Agronomos, Brazil
L. P. LUDVIGSEN	UNHCS (Habitat), Switzerland
J. B. LUNGU	Permanent Mission of Zambia in Rome, Italy
J. LUO	State Meteorological Administration, China
S. LYAGOUBI-OUAHCHI	Mission Permanente de Tunisie, Switzerland
M. MacCRACKEN	Lawrence Livermore National Laboratory, USA
D. MacDONALD-McGEE	Atmospheric Environment Science, Canada
J. MacNEILL	Institute for Research on Public Policy, Canada
C. H. D. MAGADZA	University of Zimbabwe, Lake Kariba Research Station, Zimbabwe
A. R. MAGALHAES	Esquel Brasil Foundation, Brazil
A. MAGAT	Office de Cooperation pour les Energies Renouvelables, Switzerland
T. MAGINNIS	National Climate Program Office, USA
F. A. MAGUINA LOPEZ	Servicio Nacional de Meteorologia e Hidrologia, Peru
J. MAHLMAN	Geophysical Fluid Dynamic Laboratory, USA
F. U. MAHTAB	Consultant, Bangladesh
J. S. MAINI	Forestry Canada, Canada
A. MAJEED	Department of Meteorology, Maldives
F. M. Q. MALIK	Meteorological Services, Pakistan
T. MALONE	Saint Joseph College, USA
H. MAMADOU	Pollution Prevention Department, Burkina Faso
C. MANAN MANGAN	Agency For Assessment & Application of Technology, Indonesia
M. MANTON	Bureau of Meteorology, Australia
G. MARACCHI	National Research Council, Italy
P. MARKANDAN	Meteorological Service, Malaysia

A. MARKHAM	World Wildlife Fund International, Switzerland
S. MARTCHOUK	Ministry of Foreign Affairs, USSR
J. MARTINEC	Federal Institute for Snow & Avalanche Research, Switzerland
J. MARTON-LEFÈVRE	ICSU, France
O. MASCARENHAS	University of Dar-es-Salaam, United Republic of Tanzania
W. M. K. MASILINGI	National Environment Management Council, United Republic of Tanzania
B. S. MATHUR	Council of Scientific & Industrial. Research, India
T. MATSUNO	University of Tokyo, Geophysical Institute, Japan
K. MATSUSHITA	UNCED, Switzerland
M. A. MATUTE	Ministerio Communicaciones y Transporte, Honduras
W. J. MAUNDER	Bureau of Meteorology, Australia
N. MAYNARD	Office of Science and Technology Policy, Executive Office of the President, USA
F. MAYOR	United Nations Educational, Scientific and Cultural Organization, France
G. A. McBEAN	University of British Columbia, Atmospheric Science Program, Canada
J. A. W. McCULLOCH	Consultant, Canada
A. D. McEWAN	CSIRO Division of Oceanography, Australia
M. McFARLAND	International Chamber of Commerce, France
S. McKAY	Friends of the Earth, Japan
D. McROBERT	Global Warming Program, Canada
L. G. MEIRA-FILHO	Earth Observation Space Research Institute, Brazil
J. E. MEJIA UCLES	Permanent Mission of Honduras, Switzerland
G. MELANDRI	Lega per l'Ambiente, Italy
P. I. MELNIKOV	Academy of Sciences, Research Institute of Permafrost, USSR
J. C. MENDES	Instituto Nacional de Meteorologie, Portugal
M. G. K. MENON	International Council of Scientific Unions, France
R. MERTENS	Ministry of Health, Institute of Hygiene & Epidemiology, Belgium
A. P. METALNIKOV	State Committee for Hydrometeorology, USSR
A. MEYER	Union of Concerned Scientists, USA
A. MEYER	Ecologist Magazine, United Kingdom
R. MILLER	National Science Foundation, USA
M. MILLWARD	International Council of Scientific Unions, France
I. MINTZER	University of Maryland, USA
J. F. B. MITCHELL	Meteorological Office, Hadley Climate Centre, United Kingdom
N. MIYOSHI	Ministry of Foreign Affairs, Japan
P. D. MKHATSHWA	Ministry of Agriculture, Swaziland
B. K. MLENGA	Meteorological Department, Malawi
A. MOHAND	Agence Nationale Resources Hydrolique, Algeria
N. J. MOHARRAM	Civil Aviation & Meteorological Authority, Yemen
A. W. MOHOTTALA	Department of Meteorology, Sri Lanka
M. MOKHTARI-AMIN	Permanent Mission of Iran, Switzerland
A. MONGADANDI	Institut National de Meteorologie, Republic of Zaire
C. O. MOORE	House of Representatives, USA
P. MORAND-FRANCIS	Federal Office of Environment, Forests & Landscape, Switzerland
P. MOREL	World Meteorological Organization, Switzerland
U. MORGENTHALER	World Wildlife Fund, Switzerland
F. F. MOROS	Instituto de Meteorologia, Cuba
K. E. MOSTEFA KARA	Ministere des Transports, Algeria
A. D. MOURA	Instituto de Pesquisas Espaciais, Brazil
P. A. MSAFIRA	Diretorate of Meteorology, United Republic of Tanzania
E. A. MUKOLWE	Meteorological Department, Kenya
M. MULLER	Ministere de l'Environnement/SRETIE, France
R. E. MUNN	University of Toronto, Institute of Environmental Studies, Canada
H. R. MUNOZ	Direccion Meteorologica, Chile
J. K. MURITHI	World Meteorological Organization, Switzerland

P. A. MWINGIRA	Directorate of Meteorology, United Republic of Tanzania
B. MYAGMARJHAV	Hydrometeorological Service, Mongolia
L. MYERS	Ministry of Environment, Trinidad, W.I.
N. NACIRI	Bulle Bleu, France
A. NANIA	TECHNAGRO Weather Modification Experiment, Italy
L. NATSAGDORS	Hydrometeorological Institute, Mongolia
G. NAUMOV	Department of International Economic Cooperation, USSR
I. NAZAROV	State Committee for Hydrometeorology, USSR
L. NDORIMANA	Institut Geographique, Burundi
G. NEEDLER	WOCE Institute for Oceanogarphic Sciences, United Kingdom
J. NEMEC	Consultant, Switzerland
F. NEUWIRTH	Zentralanstalt Meteorologie/Geodynamik, Austria
R. L. NEWSON	World Meteorological Organization, Switzerland
J. NG'ANG'A	University of Nairobi, Department of Meteorology, Kenya
C. NICHOLSON	British Petroleum Company, United Kingdom
D. N. NGU YEN	Permanent Mission of Viet Nam, Switzerland
Y. NISHIMURA	Ministry of International Trade and Industry, Japan
Y. NISHIMURA	Permanent Mission of Japan, Switzerland
S. NISHIOKA	National Institute for Environmental Studies, Japan
T. NITTA	Meteorological Agency, Japan
W. A. NITZE	Alliance to Save Energy, USA
H. NIX	Center for Resource & Environmental Studies, Australia
J. K. NJIHIA	Meteorological Department, Kenya
S. M. NKAMBULE	Swaziland
D. NORSE	FAO Agricultural Department, Italy
M. W. NOWICKI	Ministry of Environmental Protection, Natural Resources & Forestry, Poland
M. C. NUNES	Direccion Generale de la Qualite de l'Environment, Portugal
B. S. NYENZI	Meterological Service, United Republic of Tanzania
C. NYSTEDT	Sweden
J. O'BRIEN	Florida State University, USA
G. O. P. OBASI	World Meteorological Organization, Switzerland
H. OESCHGER	Physikalisches Institut, Universitat Bern, Switzerland
L. OGALLO	University of Nairobi, Kenya
H. OGATA	Permanent Mission of Japan, Switzerland
S. OGBUAGU	University of Calabar, Nigeria
J. S. OGUNTOYINBO	University of Ibadan, Nigeria
A. OHMURA	Geographisches Institut ETH, Switzerland
S. OKAMATSU	Ministry of International Trade and Industry, Japan
P. C. OKOT	Department of Meteorology, Uganda
G. OKRAINETZ	House of Commons, Canada
J. OKUMURA	Ministry of International Trade & Industry, Japan
P. M. OLINDO	Africa Wildlife Foundation, Kenya
Y. OLIOUNINE	UNESCO/IOC, France
L. E. OLSSON	World Meteorological Organization, Switzerland
I. A-A. OMAR	Permanent Mission of Libya, Switzerland
M. M. OMAR	Mission d'Egypte, Switzerland
S. O. E. OMENE	Permanent Mission of Nigeria, Switzerland
M. S. OPELZ	IAEA Office, Switzerland
M. OPPENHEIMER	Environmental Defense Fund, USA
I. ORNATSKI	UN Economic Commission for Europe, Switzerland
L. OYEBANDE	University of Lagos, Nigeria
G. OZOLINS	World Health Organization, Switzerland

R. K. PACHAURI	TATA Energy Research Institute, India
S. PANCHEV	Academy of Science, Bulgaria
R. PAPE	Secretariat on Acid Rain, Sweden
Y-D. PARK	Meteorological Service, Republic of Korea
M. L. PARRY	University of Birmingham, AIR Group, United Kingdom
E. PÂSZTÔ	Ministry of Trade & Industry, Hungary
J. PASZTOR	UNCED, Switzerland
P. PATVIVATSIRI	Ministry of Commerce, Meteorology Department, Thailand
C. PAULETTO	Bureau des Affaires Economique Exterieures, Switzerland
G. PEARMAN	CSIRO Division of Atmospheric Research, Australia
C. PEDERSEN	UN Office for Development & International Economic Cooperation, USA
M. PERDOMO	Direccion de Hidrologia y Meteorologia, Venezuela
J. PERNETTA	UNEP/IOC/WMO Coastal Observatory System, United Kingdom
J. S. PERRY	National Academy of Science, USA
N. PETIT MAIRE	CNRS Laboratoire de Geologie du Quaternaire, France
D. PHILLIPS	Atmospheric Environment Service, Canada
O. A. PHILLIPS	Ministry of Agriculture, Nigeria
V. PHILLIPS	University of Hawaii at Manoa, Natural Energy Institute, USA
G. PINCHERA	Comitato Nazionale Energia Nucleare ed Energia Alternativa (ENEA), Italy
A. PISSARENKO	State Forest Committee, USSR
N. POLUNIN	World Council for the Biosphere, Switzerland
R. POMERANCE	Consultant, World Resources Institute, USA
M. E. D. POORE	United Kingdom
D. POWER	Climate Institute, USA
J. PRATT	World Bank, USA
O. PREINING	Ministry of Science & Research, Austria
G. PRICE	St. Andrews Management Institute, United Kingdom
A. PURCELL	FAO, Switzerland
G. PALACIUS	Direccion Meteorologica, Chile
A. QURESHI	Climate Institute, USA
A. A. RAMA SASTRY	Ministry of Communications, United Arab Emirates
K. RAMAKRISHNA	Woods Hole Research Center, USA
G. K. RAMOTHWA	Meteorological Services, Botswana
C. A. RASAHAN	Ministry of Agriculture, Indonesia
J. RASMUSSEN	World Meteorological Organization, Switzerland
J. S. RASON	Permanent Mission of Madagascar, Switzerland
J. C. RAY	Bulle Bleue, France
A. K. N. REDDY	Institute of Science, India
D. A. REIFSNYDER	Department of State Office of Global Change, USA
H. REISER	Deutscher Wetterdienst, Germany
C. F. REUDINK	Meteorological Service, Netherlands Antilles
I. REVAH	CNES, France
W. E. RIEBSAME	University of Colorado, Natural Hazards Research Center, USA
J. T. RIISSANEN	Meteorological Institute, Finland
D. RIJKS	World Meteorological Organization, Switzerland
J. RIPERT	Ministere des Affaires Etrangeres, France
B. RIZZO	Environment Canada Policy Group, Canada
L. ROBB	Ecole Internationale de Geneve, Switzerland
S. ROBERTS	Friends of the Earth Ltd., United Kingdom
T. ROBERTSON	World Wildlife Fund, United Kingdom
R. J. ROCHON	External Affairs, Canada
J. C. RODDA	World Meteorological Organization, Switzerland

D. RODRIGUES	Forum of Brazilian NGO's to UNCED '92, Brazil
Y. S. ROMANOV	Astronomical Observatory, Odessa State University, USSR
H. ROSCHEISEN	Deutscher Naturschutzring (DNR), Germany
N. ROSENBERG	Resources for the Future, USA
T. ROSSWALL	Academy of Sciences (IGBP), Sweden
N. RUANGPANIT	Kasetsart University, Thailand
J. RUGIRANGOGA	Department of Meteorology, Rwanda
I. RUMMEL-BULSKA	UNEP, Kenya
J. T. RUSHWORTH	Sociological-Ecological Institute, Poland
O. RYBAK	Ministry of Foreign Affairs, Ukrainian SSR
H. SADEK	Consultant to the Climate Institute, Egypt
R. SADOURNY	Laboratoire de Meteorologie Dynamique, France
M. SADOWSKI	Institute of Meteorology & Water Management, Poland
O. SAETERSDHAL	Ministry of the Environment, Norway
S. SAFRONON	UNCTAD Inter-Sectoral Issues Unit, Switzerland
M. SAHNOUN	Hydra-Alger, Algeria
M .M. SAHOR	Water Resources & Hydrometeorological Service, Gambia
E. SALATI	University of Sao Paulo, Brazil
J. SALINGER	Meteorological Services, New Zealand
N. R. SALTHUN	Water Resources/Energy Administration, Norway
F. SAMAJ	Hydrometeorological Institute, Czechoslovakia
H-F. SAMUELSSON	UNECE/UNCTC, Switzerland
P. H. SAND	UNECE Environment & Human Settlement Division, Switzerland
E. SANHUEZA	Instituto Venezolano de Investigaciones Cientificas, Venezuela
M. K. SARMA	Ministry of Environment and Forests, India
I. SAVOLAINEN	Nuclear Engineering Laboratory, Finland
H. SCHAAP	State Electricity Committee, Australia
W. SCHLESINGER	Duke University, USA
C. SCHLOSSER	UNEP/GEMS (Atmosphere), Kenya
J. SCHMANDT	Houston Advanced Research Center, USA
B. SCHMIDBAUER	Deutscher Bundestag - Enquete Kommission, Germany
S. H. SCHNEIDER	National Center for Atmospheric Research, USA
U. SCHOTTERER	ProClim, Switzerland
C. J. E. SCHUURMANS	Royal Netherlands Meteorological Institute (KNMI), Netherlands
M. SECK	Ministere de l'Equipment de Transport et du Logement, Senegal
J. SEGOVIA	Instituto Nacional de Meteorologia, Spain
B. T. SEKOLI	Meteorological Services, Lesotho
H. A. SEMICHI	Permanent Mission of Algeria, Switzerland
A. SENE	Mission Permanente du Senegal, Switzerland
P. SERPI	Permanent Mission of Italy, Switzerland
B. SEVRUK	Abteilung Hydrologie, Geographisches Institut ETH, Switzerland
A. M. SEYED	Permanent Mission of Iran, Switzerland
M. SHAKOURI	Permanent Mission of Iran, Switzerland
P. SHAM	Royal Observatory, Hong Kong
S. SHAM	University of Kebangsaan, Malaysia
R. J. SHEARMAN	Meteorological Office, United Kingdom
H. SHIHAB	Ministry of Planning & Environment, Maldives
J. A. SHIKLOMANOV	State Hydrological Institute, USSR
H. H. SHUGART	University of Virginia, USA
J. SHUKLA	University of Maryland, USA
U. SIEGENTHALER	Physics Institute, University of Bern, Switzerland
D. R. SIKKA	Institute of Tropical Meteorology, India
G. SILVESTRINI	Lega per l'Ambiente, Italy

J. SINCLAIR	OED/UNEP, Kenya
N. C. SINGH	Caribbean Environment Health Institute, St Lucia
S. K. SINHA	Agricultural Research Institute, India
J. SIRCOULON	ORSTOM, France
C. D. SIROTENKO	All Union Research Institute of Agricultural Meteorology, USSR
R. SKINNER	OECD International Energy Agency, France
M. SKODA	Hydrometeorological Institute, Czechoslovakia
N. P. SKRYPNYK	Board on Hydrometeorology, Ukrainian SSR
M. S. SMEDA	Meteorological Organization, Libyan Arab Jamahiriya
B. SMIRNOV	Permanent Mission of USSR, Switzerland
C. D. R. SMITH	Conservation Commission of the Northern Territory, Australia
C. SMYSER	OECD International Energy Agency, France
N. SOLOMATIME	UNDRO, Switzerland
K. SONTOKUSUMO	Meteorological & Geophysical Institute, Indonesia
D. SPÄNKUCH	Weather Service, Germany
D. SPASOVA	Federal Hydrometeorological Institute, Yugoslavia
A. SPERANZA	National Research Council, Italy
J. R. SPRADLEY	Department of Commerce (Ocean & Atmosphere), USA
M. R. SRINIVASAN	Department of Nuclear Energy (IAEA), Austria
O. STAROSOLSZKY	Research Centre for Water Resources Development, Hungary
M. STEIN	Bundesforschungsanstalt fur Fischerei, Germany
W. R. STEVENS	Du Pont de Mours & Co., USA
R. W. STEWART	University of Victoria, Canada
J. M. R. STONE	Canadian Climate Centre
R. B. STREET	Canadian Climate Centre
N. STROMMEN	Department of Agriculture, USA
M. STRONG	UNCED, Switzerland
M. A. STUBBS	Ministry of Transport, Meteorology Department, Bahamas
A. SUGHANDY	Ministry of Population and Environment, Indonesia
N. SUNDARARAMAN	World Meteorological Organization (IPCC Secretariat), Switzerland
C. SUTRISNO	Meteorological and Geophysical Institute, Indonesia
K. SUZUKI	Global Environment Agency, Japan
M. S. SWAMINATHAN	IUCN, India
G. SWANIDZE	Hydrometeorological Institute of Regional Research, USSR
R. SWART	National Institute of Public Health & Environmental Protection, Netherlands
A. SZOLLOSI-NAGY	UNESCO Division of Water Sciences, France
H. TABA	World Meteorological Organization, Switzerland
M. S. TABUCANON	National Environment Board, Thailand
R. TAESLER	Meteorological & Hydrological Institute, Sweden
H. TAKAHASHI	Ministry of Agriculture, Forestry & Fisheries, Japan
R. G. TALLBOYS	World Coal Institute, United Kingdom
S. E. TANDOH	Meteorological Service Department, Ghana
H. TARZI	Ministry of Civil Aviation & Tourism, Afghanistan
R. TATEHIRA	Meteorological Agency, Japan
J. A. TAYLOR	University College of Wales, United Kingdom
H. TEBOURBI	Mission Permanente de Tunisie, Switzerland
B. TENCER	Roy F. Weston Inc., USA
S. TEWUNGWA	UNEP (IPCC Secretariat), Switzerland
R. THABEDE	Water Resource Branch, Swaziland
M. D. THOMAS	Directorate of Meteorology, United Republic of Tanzania
J. THOMPSON	Ministry of Environment, Norway
C. TICKELL	Radcliffe Observatory, United Kingdom
P. T. TIMEON	Ministry of Foreign Affairs, Republic of Kiribati

D. TIRPAK   Environment Protection Agency, USA
M. K. TOLBA   UNEP, Kenya
I. N. TOPKOV   Ministry of Foreign Affairs (Governing Council of UNEP), Bulgaria
J. TOPPING   Climate Institute, USA
R. TORRIE   Torrie Smith Associates, Canada
H. TRABELSI   Institut National de la Meteorologie, Tunisia
V. A. TRAN   Meteorological Service, Viet Nam
G. TSUZUKI   Departamento Nacional de Aguas e Energia Eletrica, Brazil
D. A. TSYBAN   State Committee for Hydrometeorology, USSR
Q. TU   Nanjing Institute of Meteorology, China
B. G. TUCKER   CSIRO Division of Atmospheric Sciences, Australia
K. TURNER   University of East Anglia, United Kingdom
E. TUTUWAN   Ministere du Plant et de L'Amenagement du Territoire, Cameroon

B. UCCELLETTI   National Oceanographic Committee, Chile
S. UNNINAYAR   UNITAR, Switzerland
M. UPPENBRINK   UNEP Regional Office Europe, Switzerland
P. E. O. USHER   UNEP/GEMS, Kenya

R. R. VAGHJEE   Meteorological Service, Mauritius
Y. VAKALYUK   State Committee for Hydrology, USSR
J-I. VALENCIA-FRANCO   Hidrologia y Meteorologia, Colombia
N. VAN LOOKEREN CAMPAGNE   Shell, Netherlands
D. VANWINKLE   CASA, Japan
S. K. VARNEY   Meteorological Office/IPCC WGI Secretariat, United Kingdom
J. VASARHELYI   Independent Ecological Centre, Hungary
A. A. VASILIEV   Hydrometeorological Centre, USSR
E. V. VELASQUEZ   Instituto de Sismologia, Vulcanologia, Meteorologia e Hidrologia, Guatamela
P. VELLINGA   Climate Change Programme, Netherlands
A. VIARO   City Planning Office of Geneva, Switzerland
B. VOITURIEZ   IFRENER, France
C. VON MEIJENFELDT   Ministry of Agriculture, Netherlands

C. C. WALLEN   UNEP, c/o WMO, Switzerland
N. WANGCHUK   Royal Government Planning Commission, Bhutan
H. E. WARREN   NOAA/NESDIS, USA
R. WARRICK   Climatic Research Unit, University of East Anglia, United Kingdom
R. T. WATSON   NASA Upper Atmospheric Program, USA
W. H. WEIHE   Consultant, Switzerland
F. WEIR   Friends of the Earth Ltd, United Kingdom
W. K. WENGER   International Environment Bureau, Switzerland
D. M. WHELPDALE   Atmospheric Environment Service, Canada
R. M. WHITE   National Academy of Engineering, USA
T. M. L. WIGLEY   Climatic Research Unit, University of East Anglia, United Kingdom
P. DE WILDT   Consultant, Netherlands
J. O. WILLUMS   International Chamber of Commerce (Environment & Energy), Norway
J. D. WOODS   Natural Environment Research Council, United Kingdom
J. WOOLRIDGE   Earthwatch FOE, Ireland
D. S. WRATT   Meteorological Service, New Zealand
W. C. WREDE   Servicio Nacional de Meteorologia e Hidrologia, Paraguay
S. XIE   Ministry of Energy, China

M. YAGISHITA   Environment Agency, Japan
T. YAMAMURA   CASA, Japan

B. YANG	Cabinet Office, Government of Ontario, Canada
D. YE	Institute of Atmospheric Physics, China
B. B. YEAGER	National Audubon Society, USA
M. YERG	NOAA, National Weather Service USA
G. W. YOHE	Wesleyan University, USA
K. YOKOBORI	Ministry of International Trade & Industry, Japan
H. YONG	Ministry of Foreign Affairs, Democratic People's Republic of Korea
M. YOSHINO	The University of Tsukuba, Institute of Geoscience, Japan
D. A. YOUSSOUF	Inter Equipment, Mauritius
S. B. ZABSONRE	Direccion de la Meteorologie Nationale, Burkina Faso
M. C. ZALIKHANOV	State Committee for Hydrometeorology, USSR
G. ZAMBRANO GARCIA	Servicio de Meteorologia, Venezuela
G. ZANELLA	Laboratory of Climatology, Parma University, Italy
H. ZAOUCH	Mission Permanente de Tunisie, Switzerland
E. ZARATE HERNANDEZ	Instituto Meteorologico Nacional, Costa Rica
Q-C. ZENG	Institute of Atmospheric Sciences, Academy of Sciences, China
J. ZHANG	State Meteorological Administration, China
Z. ZHAO	State Meteorological Administration, China
S. ZHONG	Ministry of Foreign Affairs, China
M. ZICH	Permanent Mission of Czechoslovakia, Switzerland
J. ZIELINSKI	Institute of Meteorology & Water Management, Poland
J. W. ZILLMAN	Bureau of Meteorology, Australia
G. ZIMMERMEYER	Gesamtverband des Deutschen Steinkohlenbergbaus, Germany
M. C. ZINYOWERA	Department of Meteorological Services, Zimbabwe
J. ZOU	State Meteorological Administration, China

Appendix 5

LIST OF COUNTRIES AND ORGANIZATIONS REPRESENTED AT MINISTERIAL SESSIONS

COUNTRIES:

Afghanistan
Algeria
Argentina
Australia
Austria

Bahamas
Bahrain
Bangladesh
Barbados
Belgium
Benin
Bhutan
Bolivia
Botswana
Brazil
Bulgaria
Burkina Faso
Burundi

Cameroon
Canada
Chile
China
Colombia
Cook Islands
Costa Rica
Côte d'Ivoire
Cuba
Cyprus
Czechoslovakia

Democratic People's Republic of Korea
Denmark
Dominica
Dominican Republic

Ecuador
Egypt
El Salvador
Ethiopia

Fiji
Finland
France

Gambia
Germany
Ghana
Greece
Guatemala
Guinea
Guinea-Bissau

Honduras
Hungary

Iceland
India
Indonesia
Iran
Ireland
Israel
Italy

Jamaica
Japan
Jordan

Kenya
Kiribati
Kuwait

Lebanon
Lesotho
Libyan Arab Jamahiriya
Liechtenstein
Luxembourg

Malawi
Malaysia
Maldives
Malta
Mauritius
Mexico
Mongolia
Morocco
Mozambique

Nauru
Nepal
Netherlands
Netherlands Antilles

New Zealand
Nicaragua
Niger
Nigeria
Norway

Oman

Pakistan
Panama
Paraguay
Peru
Philippines
Poland
Portugal

Qatar

Republic of Korea
Romania
Rwanda

Saint Lucia
Saint Christopher and Nevis
Saint Vincent and the Grenadines
Sao Tomé and Principe
Saudi Arabia
Senegal
Singapore
Solomon Islands
Somalia
Spain
Sri Lanka
Sudan

Swaziland
Sweden
Switzerland
Syrian Arab Republic

Thailand
Togo
Tonga
Trinidad and Tobago
Tunisia
Turkey
Tuvalu

Uganda
Ukrainian Soviet Socialist Republic
Union of Soviet Socialist Republics
United Arab Emirates
United Kingdom of Great Britain & Northern Ireland
United Republic of Tanzania
United States of America
Uruguay

Vanuatu
Venezuela
Viet Nam

Western Samoa

Yemen
Yugoslavia

Zaire
Zambia
Zimbabwe

SPONSORING AGENCIES:

World Meteorological Organization (WMO)
United Nations Environment Programme (UNEP)
Food and Agriculture Organization (FAO)
United Nations Educational, Scientific and Cultural
  Organization (UNESCO), and its
Intergovernmental Oceanographic Commission (IOC)
International Council of Scientific Unions (ICSU)

UNITED NATIONS ORGANIZATIONS:

United Nations (UN)
United Nations Centre on Transnational Corporations
  (UNCTC)
United Nations Conference on Environment and
  Development (UNCED)
United Nations Conference on Trade and Development
  (UNCTAD)
United Nations Development Programme (UNDP)
United Nations Institute for Training and Research
  (UNITAR)
United Nations Economic Commission for Europe
  (UNECE)
United Nations Disaster Relief Co-ordinator (UNDRO)

SPECIALIZED AGENCIES:

International Labour Organisation (ILO)
World Health Organization (WHO)
World Bank
International Telecommunication Union (ITU)
United Nations Industrial Development Organization
  (UNIDO)
International Atomic Energy Agency (IAEA)
International Organization for Standardization (ISO)

INTERGOVERNMENTAL ORGANIZATIONS:

Caribbean Meteorological Organization
Commission of the European Communities (CEC)
European Free Trade Association (EFTA)
International Commission on Irrigation and Drainage
  (ICID)
League of Arab States (LAS)
Organization of African Unity (OAU)
Organization for Economic Co-operation and Development
  (OECD)
South Pacific Regional Environment Programme (SPREP)

† *For Non-Governmental Organizations see Appendix 6*

Appendix 6

PARTICIPATING NON-GOVERNMENTAL ORGANIZATIONS

Alliance to Save Energy
Alternative Fluorocarbons Environmental Acceptability
  Study
Audubon Society
Australian Conservation Foundation
Baha'i International Community
Bangladesh Centre for Advanced Studies
Bangladesh Network on Global Climate Change
Bulle Bleue - France, Morocco
BUND, Germany
Centre for Our Common Future
Centro Ecumenico de Documents e Informacao, Brazil
Citizen Alliance for Saving the Atmosphere & Earth, Japan
Climate Institute, USA
Climate Action Network, United Kingdom
Climate Network, Europe
CSFR, Czechoslovakia
Danish Nature Concern
Deutscher Naturschutzring, Germany
Earthwatch
Environmental Defense Fund
Environmental Law Institute
Environmental Science and Policy Institute
Eurelectric
European Fluorocarbon Technical Committee
Forum of Brazilian NGO's to the UNCED-92
Foundation for Environmental Conservation
Friends of The Earth - International, Canada, Ghana, Japan,
  Netherlands, United Kingdom
Greenpeace - International, Japan, South Pacific,
  Switzerland, United Kingdom

Independent Ecological Centre, Hungary
International Chamber of Commerce
International Institute for Environment and Development
International Peace Research Association
International Social Science Council
International Union for Conservation of Nature
Kenya Energy and Environmental Organization
Leadership Council for Effective Climate Change Policies
Lega per L'Ambiente, Italy
Natural Resources Defence Council
Pacific Institute for Studies in Development, Environment,
  and Security
Paeto de Grupos Ecologistos, Mexico
Royal Institute of International Affairs
Securite Environnement
Sierra Club
Social-Ecological Institute, Poland
Stichting Natuur en Milieu
Stockholm Environment Institute
Swiss Aid Agencies Coalition
Union of Concerned Scientists
University of the Andes, Venezuela
Uranium Institute
Woods Hole Research Centre
World Council of Churches
World Coal Institute
World Federation of United Nations Associations
World Resources Institute
World Wildlife Fund & The Conservation Foundation
World Wildlife Fund for Nature - International,
  Switzerland, United Kingdom

Appendix 7

ACRONYMS

ACMAD	African Centre of Meteorological Applications for Development
AGGG	Advisory Group on Greenhouse Gases
AWS	Automatic Weather Station
BAPMoN	Background Air Pollution Monitoring Network
CCC	Canadian Climate Centre, Downsview, Ontario, Canada
CCCO	Committee on Climate Changes and the Ocean
CCDP	Climate Change Detection Project
CCl	Commission for Climatology of WMO
CEC	Commission of the European Communities
CFC's	Chlorofluorocarbons
CGIAR	Consultative Group on International Agricultural Research
CLICOM	Climat-Computer System
CLIMAP	Climatic Applications Project (WMO)
CNRS	Centre National de Recherches Meteorologiques, France
COADS	Comprehensive Ocean Air Data Set
CRU	Climatic Research Unit, University of East Anglia, UK
CSIRO	Commonwealth Scientific & Industrial Research Organisation, Australia
CSM	Climate System Monitoring
DARE	Data Rescue Programme
EDF	Environmental Defense Fund, USA
EFTA	European Free Trade Association
ENSO	El Niño-Southern Oscillation
EOS	Earth Observing System
EPA	Environmental Protection Agency, Washington, USA
ERBE	Earth Radiation Budget Experiment
ERS	Earth Resources Satellite
FAO	Food and Agriculture Organization
FCCC	Framework Convention on Climate Change
GARP	Global Atmospheric Research Programme
GAW	Global Atmospheric Watch
GCIP	GEWEX Continental-Scale International Project
GCOS	Global Climate Observing System
GCM	General Circulation Model
GCTE	Global Change and Terrestrial Ecosystems
GDP	Gross Domestic Product
GDPS	Global Data Processing System
GEMS	Global Environmental Monitoring System
GEWEX	Global Energy and Water Cycle Experiment
GFDL	Geophysical Fluid Dynamics Laboratory, Princeton, USA
GIS	Geographic Information System
GISS	Goddard Institute of Space Sciences, New York, USA

GLOSS	Global Sea-Level Observing System
GMCC	Geophysical Monitoring of Climatic Change
GOOS	Global Ocean Observing System
GO_3OS	Global Ozone Observing System
GRDC	Global Runoff Data Centre
GRID	Global Resource Information Database
GTCP	Global Tropospheric Chemistry Programme
GTS	Global Telecommunication System
HRGC	Human Response to Global Change
IAEA	International Atomic Energy Agency
IBP	International Biological Programme
ICID	International Commission on Irrigation and Drainage
ICRCCM	Intercomparison of Radiation Codes in Climate Models
ICSU	International Council of Scientific Unions
IEA	International Energy Agency
IFAD	International Fund for Agricultural Development
IGAC	International Global Atmospheric Chemistry Programme
IGBP	International Geosphere-Biosphere Programme
IGFA	International Group of Funding Agencies for Global Change Research
IGOSS	Integrated Global Ocean Services System
IGY	International Geophysical Year
IHD	International Hydrological Decade
IIASA	International Institute for Applied Systems Analysis
ILO	International Labour Organization
IMCO	Intergovernmental Maritime Consultative Organization
IMF	International Monetary Fund
INC/FCCC	International Negotiating Committee, FCCC
INFOCLIMA	Climate Data Information Referral System
IOC	Intergovernmental Oceanographic Commission
IOC	International Ozone Commission
IPCC	Intergovernmental Panel on Climate Change
ISCCP	International Satellite Cloud Climatology Project
ISLSCP	International Satellite Land Surface Climatology Project
ITCZ	Intertropical Convergence Zone
IUCN	International Union for Conservation of Nature and Natural Resources
JGOFS	Joint Global Ocean Flux Study
JSC	Joint Steering Committee for the WCRP
MAB	Man and the Biosphere Programme
MGO	Main Geophysical Laboratory, Leningrad,USSR
MPI	Max Planck Institut, FRG
MRI	Meteorological Research Institute, Japan
NASA	National Aeronautics and Space Administration, USA
NATO	North Atlantic Treaty Organisation
NCAR	National Center for Atmospheric Research, Boulder, USA
NGO	Non-Governmental Organization
NOAA	National Oceanic and Atmospheric Administration, USA
OAU	Organization of African Unity
OECD	Organization for Economic Co-operation and Development
OSU	Oregon State University, USA
PAGES	Past Global Changes
SAC	Scientific Advisory Committee for the WCIP
SAGE	Stratospheric Aerosol and Gas Experiment
SCOPE	Scientific Committee on Problems of the Environment
SCOR	Scientific Committee on Ocean Research

SEI	Stockholm Environment Institute
SPREP	South Pacific Regional Environment Programme
SST	Sea Surface Temperature
TCP	Tropical Cyclone Programme
TOGA	Tropical Ocean and Global Atmosphere Programme
TRUCE	Tropical Urban Climate Experiment
UKMO	Meteorological Office, Bracknell, UK
UNCTAD	United Nations Conference on Trade and Development
UNCTC	United Nations Centre on Transnational Corporations
UNDP	United Nations Development Programme
UNDRO	United Nations Disaster Relief Coordinator
UNECE	United Nations Economic Commission for Europe
UNEP	United Nations Environment Programme
UNESCO	United Nations Educational, Scientific and Cultural Organization
UNFPA	United Nations Population Fund
UNIDO	United Nations Industrial Development Organization
UNITAR	United Nations Institute for Training and Research
USDOE	Department of Energy, USA
VCP	Voluntary Co-operation Programme (WMO)
WCAP	World Climate Applications Programme
WCDP	World Climate Data Programme
WCIP	World Climate Impact Studies Programme
WCP	World Climate Programme
WCRP	World Climate Research Programme
WHO	World Health Organization
WIPO	World Intellectual Properties Organization
WMO	World Meteorological Organization
WOCE	World Ocean Circulation Experiment
WWW	World Weather Watch

Appendix 8 †

UNITS

SI (Systeme Internationale) Units:

Physical Quantity	Name of Unit	Symbol
length	meter	m
mass	kilogram	kg
time	second	s
thermodynamic temperature	kelvin	K
amount of substance	mole	mol

Fraction	Prefix	Symbol	Multiple	Prefix	Symbol
10^{-1}	deci	d	10	deka	da
10^{-2}	centi	c	10^2	hecto	h
10^{-3}	milli	m	10^3	kilo	k
10^{-6}	micro	μ	10^6	mega	M
10^{-9}	nano	n	10^9	giga	G
10^{-12}	pico	p	10^{12}	tera	T
10^{-15}	femto	f	10^{15}	peta	P
10^{-18}	atto	a			

Special Names and Symbols for Certain SI-Derived Units:

Physical Quantity	Name of SI Unit	Symbol for SI Unit	Definition of Unit
force	newton	N	$kg\ m\ s^{-2}$
pressure	pascal	Pa	$kg\ m^{-1}s^{-2}(=Nm^{-2})$
energy	joule	J	$kg\ m^2\ s^{-2}$
power	watt	W	$kg\ m^2s^{-3}(=Js^{-1})$
frequency	hertz	Hz	s^{-1}(cycle per second)

Decimal Fractions and Multiples of SI Units Having Special Names:

Physical Quantity	Name of Unit	Symbol for Unit	Definition of Unit
length	ångstrom	Å	10^{-10} m = 10^{-8} cm
length	micrometer	μm	10^{-6} m = μm
area	hectare	ha	10^4 m^2
force	dyne	dyn	10^{-5} N
pressure	bar	bar	10^5 N m^{-2}
pressure	millibar	mb	1hPa
weight	ton	t	10^3 Kg

Non- SI Units:

°C	degrees Celsius (0°C = 273K approximately)
	Temperature differences are also given in °C (=K) rather than the more correct form of "Celsius degrees".
ppmv	parts per million (10^6)by volume
ppbv	parts per billion (10^9) by volume
pptv	parts per trillion (10^{12}) by volume
BP	(years) before present
ky BP	thousands of years before present

The units of mass adopted in this report are generally those which have come into common usage, and have deliberately not been harmonised, e.g.,

GtC	gigatonnes of carbon (1 GtC = 3.7 Gt carbon dioxide)
MtN	megatonnes of nitrogen
TgS	teragrams of sulphur

† - Reproduced from Scientific Assessment of Climate Change, IPCC Working Group I Report, 1990